CAMBRIDGE LIBRARY COLLECTION

Books of enduring scholarly value

Physical Sciences

From ancient times, humans have tried to understand the workings of the world around them. The roots of modern physical science go back to the very earliest mechanical devices such as levers and rollers, the mixing of paints and dyes, and the importance of the heavenly bodies in early religious observance and navigation. The physical sciences as we know them today began to emerge as independent academic subjects during the early modern period, in the work of Newton and other 'natural philosophers', and numerous sub-disciplines developed during the centuries that followed. This part of the Cambridge Library Collection is devoted to landmark publications in this area which will be of interest to historians of science concerned with individual scientists, particular discoveries, and advances in scientific method, or with the establishment and development of scientific institutions around the world.

A History of the Theory of Elasticity and of the Strength of Materials

A distinguished mathematician and notable university teacher, Isaac Todhunter (1820–84) became known for the successful textbooks he produced as well as for a work ethic that was extraordinary, even by Victorian standards. A scholar who read all the major European languages, Todhunter was an open-minded man who admired George Boole and helped introduce the moral science examination at Cambridge. His many gifts enabled him to produce the histories of mathematical subjects which form his lasting memorial. First published between 1886 and 1893, the present work was the last of these. Edited and completed after Todhunter's death by Karl Pearson (1857–1936), another extraordinary man who pioneered modern statistics, these volumes trace the mathematical understanding of elasticity from the seventeenth to the late nineteenth century. Volume 2 (1893) was split into two parts. Part 1 includes the work of Saint-Venant from 1850 to 1886.

A History of the Theory of Elasticity and of the Strength of Materials

Volume 2: Part 1

Saint-Venant to Lord Kelvin (1)

Isaac Todhunter

Edited by Karl Pearson

CAMBRIDGE
UNIVERSITY PRESS

University Printing House, Cambridge, CB2 8BS, United Kingdom

Published in the United States of America by Cambridge University Press, New York

Cambridge University Press is part of the University of Cambridge.
It furthers the University's mission by disseminating knowledge in the pursuit of
education, learning and research at the highest international levels of excellence.

www.cambridge.org
Information on this title: www.cambridge.org/9781108070430

This edition first published 1893
This digitally printed version 2014

ISBN 978-1-108-07043-0 Paperback

A HISTORY

OF

THE THEORY OF ELASTICITY.

London: C. J. CLAY AND SONS,
CAMBRIDGE UNIVERSITY PRESS WAREHOUSE,
AVE MARIA LANE.

Cambridge: DEIGHTON, BELL AND CO.
Leipzig: F. A. BROCKHAUS.
New York: MACMILLAN AND CO.

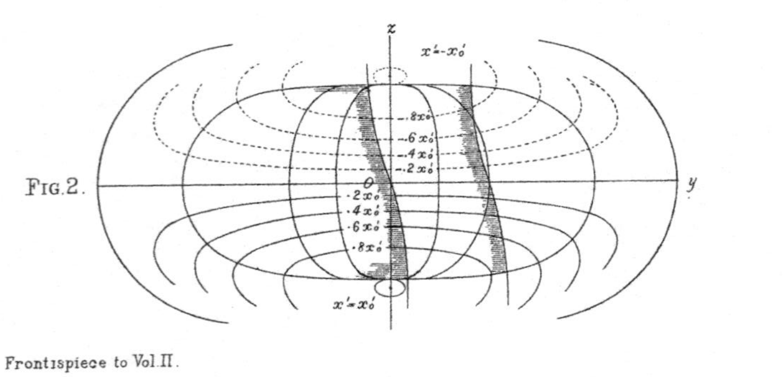

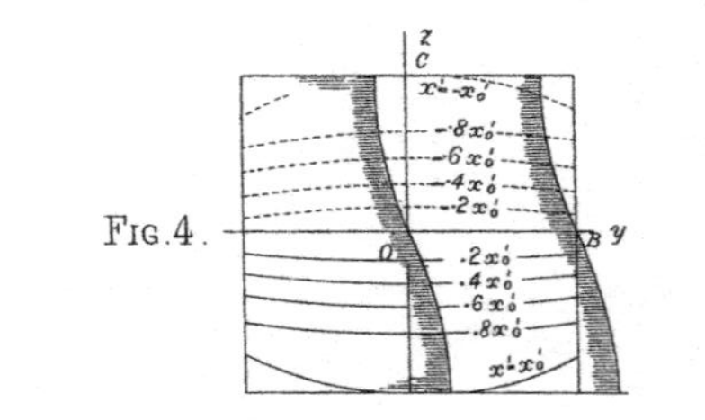

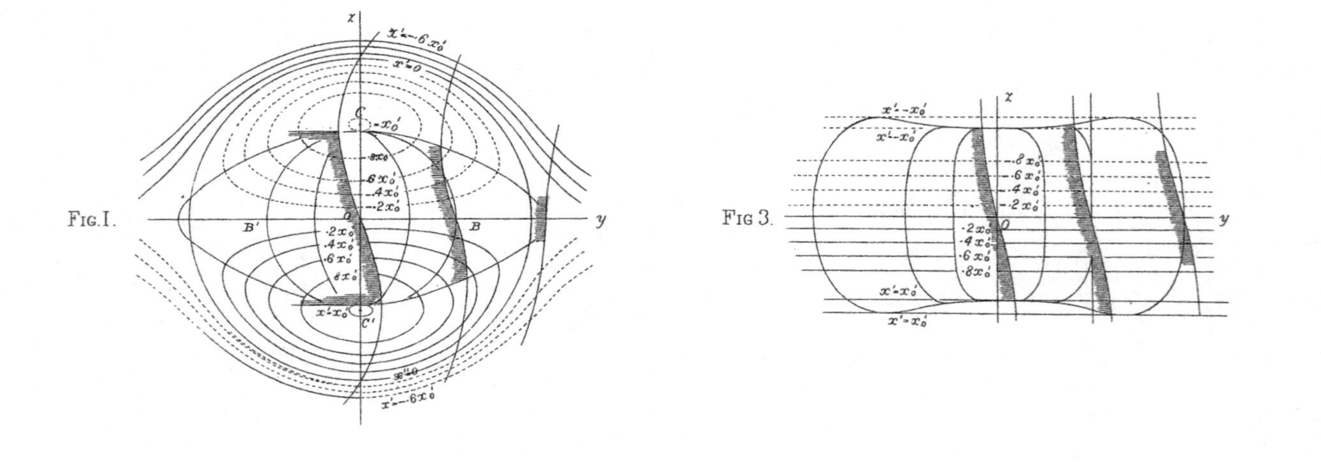

Frontispiece to Vol. II.

Drawn by G. Baker after Saint-Venant.

Distorted Cross-Sections of Beams under Flexure

A HISTORY OF
THE THEORY OF ELASTICITY
AND OF
THE STRENGTH OF MATERIALS

FROM GALILEI TO THE PRESENT TIME.

BY THE LATE

ISAAC TODHUNTER, D.Sc., F.R.S.

EDITED AND COMPLETED
FOR THE SYNDICS OF THE UNIVERSITY PRESS

BY

KARL PEARSON, M.A.

PROFESSOR OF APPLIED MATHEMATICS, UNIVERSITY COLLEGE, LONDON,
FORMERLY FELLOW OF KING'S COLLEGE, CAMBRIDGE.

VOL. II. SAINT-VENANT TO LORD KELVIN.

PART I.

CAMBRIDGE:
AT THE UNIVERSITY PRESS.
1893.

Cambridge:
PRINTED BY C. J. CLAY, M.A. AND SONS,
AT THE UNIVERSITY PRESS.

TO

LORD KELVIN P.R.S.

WHOSE RESEARCHES HAVE SO LARGELY

CONTRIBUTED TO THE RECENT PROGRESS

OF THE SCIENCE OF ELASTICITY

THE EDITOR DEDICATES HIS LABOUR

ON THE PRESENT VOLUME.

Man hat aber erst angefangen die Gesetze der Elasticität in ihrem ganzen Umfange zu studiren; bei jedem Schritte stösst man in diesen Untersuchungen auf neue Eigenschaften der elastischen Körper; je weiter man vorgeht desto mehr Verwickelung. Bei solchen Umständen ist wohl in diesem Augenblick keine völlig abgeschlossene Arbeit über irgend eine Eigenschaft der elastischen Körper möglich.

Kupffer.

I cannot doubt but that these things, which now seem to us so mysterious, will be no mysteries at all; that the scales will fall from our eyes; that we shall learn to look on things in a different way—when that which is now a difficulty will be the only common-sense and intelligible way of looking at the subject.

Lord Kelvin.

Works of this nature form, as it were, the principal fund of the science property of mankind, the *interest* of which we may turn to further profit. We might compare them to a capital invested in land. Like the soil, of which landed property consists, the knowledge stored up in these catalogues, lexicons, etc., may have but slender attractions for the vulgar, the man unacquainted with the subject can have no idea of the labour and cost at which the soil has been prepared; the work of the husbandman appears to him terribly toilsome, tedious and clumsy. But although the work of the lexicographer and physical science cataloguer calls for the same painful and persevering industry as the labour of the husbandman, we must not therefore hastily assume that the work itself is of an inferior character, or that it is as dry and mechanical as it at first appears when we have the catalogue or lexicon ready printed before us. For it is necessary in such compilations that all the isolated facts should be selected by careful observation, and afterwards tested and compared with one another, the essential sifted from the unessential,—and all this it is plain, he only can efficiently accomplish who has clearly conceived the end and aim of his work, and the scope and method of the branch of science which it concerns; but for such an one each minute detail will have its own peculiar interest from its position in relation to the whole science of which it is a part. Were it not so, such work would indeed be the worst kind of mental drudgery it were possible to conceive.

von Helmholtz.

PREFACE.

NINE years have elapsed since the manuscript of the earlier part of this History was placed in my hands; seven years since the first volume was published[1]. Some words of apology are needful for this delay. Interest in my subject and a desire to complete without breach of continuity a work which I had commenced led me to persist in the task of editing even after I had recognised how little prompt execution of that task was compatible with the large demands which the work of a London teacher makes upon limited physical strength. Rapid and efficient fulfilment needed the single-hearted devotion of one to whom this History would have been the first and not a secondary duty. To complete the work, as I could have wished it completed, would have needed the undivided energies, the fresh and undisturbed intellectual power of several years' labour. As it is the Editor has failed to fulfil the promise made on the title-page and bring the History down "to the present time." The Second Volume carries the analysis of individual memoirs completely to the year 1860, but beyond that year the work of certain elasticians only has been dealt with up to the present date. These elasticians, however, —Saint-Venant and Boussinesq, Rankine and Lord Kelvin, F. Neumann, Kirchhoff and Clebsch—are those upon whose researches the modern science of elasticity rests. It may be safely

[1] Chapter X. of the present volume appeared in 1889 as an extract entitled: *The Elastical Researches of Barré de Saint-Venant.*

said that without a thorough study of their writings, it is impossible to be an accomplished elastician, or to follow without great difficulty the drift of modern elastical research. Their memoirs and treatises form the frame, which the Editor had hoped he might be able to fill up by briefer accounts of the discoveries due to, perhaps, less distinguished but none the less useful workers in the same field. This process of filling up is only completed for the years 1850–60, but the Editor ventures to think that the reader of his Chapter XI. will be surprised at the wealth of material, theoretical, technical and physical, which was brought to light in that decade. Many facts have been discovered, more, perhaps, rediscovered since 1860, but till the last few years it may be doubted whether any period has been more fruitful of genuine progress in the science of elasticity than these ten years.

The number of the memoirs included in this volume by no means measures the work of preparation it has involved. The study and analysis of many memoirs not included in its contents had to be undertaken. But the chief task has been the verification of the analysis of all the more important mathematical memoirs. In some cases the whole of this analysis has been undertaken *de novo*, occasionally with different results. As examples of this I may cite Resal's researches on the figure of the earth, the whole of Winkler's work on the strained form of the links of chains, and Lord Kelvin's analysis of the strains produced by the tides in an elastic earth. In all the work of verification, not only of others' analysis but of my own, I have had the most self-sacrificing and devoted assistance from Mr. C. Chree of King's College, Cambridge. Without his aid not only would this volume have been much longer delayed, but I veritably shudder to think of the blunders which would certainly have escaped my unaided revision. My thanks are due to him, not as to a mere friendly proof-reader, but as to one whose cooperation in the task of editing has given the volume the major portion of any freedom from error it may possess. I trust that many serious errors may not still remain to be found, but in a work of reference like the present errors and misinterpretations of a writer's meaning are sure to occur. I can only hope that my criticisms, especially when they deal with the work of living men of science, will be received in the spirit in which they were written; namely, in that spirit the sole motive of which is the impersonal one of attaining truth and eliminating error. A somewhat lengthy list of additions and corrections to the first volume is issued with this, and I should be glad of any suggestions or emendations of the present volume which my readers may care to send me, and which might be issued with later copies[1].

Of others besides Mr Chree who have helped me in the work of revision, I must refer in the first place to M. Flamant, Professeur à l'École des Ponts et Chaussées, whose help especially in the chapters devoted to Saint-Venant and Boussinesq has been very considerable. To my colleagues Professors G. Carey Foster and T. G. Bonney, and to Mr W. H. Macaulay of King's College, Cambridge, I am indebted for assistance in special points. To Lord Kelvin I owe a number of corrections in Chapter XIV. In several instances I had misunderstood or misinterpreted passages in his papers. He has enabled me to express something of the gratitude which I among other elasticians feel to him for his contributions to our science, by accepting the dedication of the present volume.

The editorial preponderance in this volume—the articles due to Dr Todhunter[2] are practically confined to a few dealing with

[1] Mr A. E. H. Love in his *Treatise on the Theory of Elasticity*, Vol. I. § 107, refers to certain terms in Saint-Venant's theory of flexure which are discussed in Art. 96 of the present volume as expressing only a "rigid-body rotation" and states that they "need not therefore be considered." It seems to have escaped Mr Love that Saint-Venant's theory allows for what experimentally is easily demonstrated to exist, namely, a small but finite change of direction in the central line of a bar under flexure either at a section where a load is applied or at a built-in end. The terms referred to do not therefore correspond to a "rigid-body rotation," and the deflections as given by Mr Love are really measured from a line, *i.e.* the tangent at a load or at the built-in end, the position of which he has not determined.

[2] Articles due to the Editor have their numbers enclosed in square brackets.

Clebsch's *Treatise*—arises chiefly from two causes. In the first place Dr Todhunter omitted all memoirs dealing with the physical or technical branches of our subject, and more than a third of the present volume will be found to deal with physical or technical problems. In the second place a still larger portion of the work falls beyond the period to which Dr Todhunter had carried his researches. On this point I may, perhaps, be permitted to refer to the remarks I have made in the preface to *The Elastical Researches of Barré de Saint-Venant*, and content myself here with citing from them the following words:

...it has seemed to me that the best memorial to the first Cambridge historian of mathematics would be that the last history bearing his name should have the widest possible sphere of usefulness. That usefulness will, I am firmly convinced, be best obtained by its comprehensive character, by its attempt to be a *Repertorium* of elasticity rather than an *Historique Abrégé* of its purely mathematical side.

For the Index to the present volume I alone am responsible. In a work of this comprehensive character a complete and systematic index is a first necessity. To prepare it is a duty which experience has taught me no one can fulfil so efficiently as the writer of a book.

Lastly, I have to express the great sense of the indebtedness I feel to the Syndics of the University Press for the patience with which they have submitted to the delay in the publication of this History, and the kindness with which they have permitted these volumes to grow so much beyond my original estimate. Should the reader complain that the work after all remains a fragment, then the blame must fall on the shoulders of the Editor, who much underestimated the extent of his material and overestimated his own powers, when he reported to the Syndics nine years ago on the original manuscript.

KARL PEARSON.

UNIVERSITY COLLEGE, LONDON.
June 7, 1893.

CONTENTS.

PART I.

CHAPTER X.

SAINT-VENANT, 1850—1886

CHAPTER XI.

MISCELLANEOUS RESEARCHES, 1850—1860.

PART II.

CHAPTER XII.

THE OLDER GERMAN ELASTICIANS.

CHAPTER XIII.

BOUSSINESQ.

CHAPTER XIV.

SIR WILLIAM THOMSON (LORD KELVIN).

References throughout this volume to the articles of the *first* volume have an asterisk affixed, *e.g.* Art. 128* Numbers without an asterisk refer to the articles of the present volume.

PART I. contains Articles 1—1191.

PART II. „ „ 1192—1818.

CHIEF ELASTICIANS TREATED OF IN THIS VOLUME.

	Birth. Death.
Saint-Venant	1797—1886
Rankine	1820—1872
E. Phillips	1821—1889
Bresse	1822—1883
H. Resal	1828— *
Clapeyron	1799—1864
E. Winkler	1835—1888
C. Neumann	1832— *
Ångström	1814—1874
Joule	1818—1889
Matteucci	1811—1868
G. Wiedemann	1826— *
Kupffer	1799—1865
Wertheim	1815—1861
Morin	1795—1880
G. H. Love	1818— *
Sir W. Fairbairn	1789—1874
A. Wöhler	1819— *
Hodgkinson	1789—1861
Grashof	1826— *
Wade	1789—1875
Mallet	1810—1881
Cavalli	? — ?
Tresca	1814—1885
Kirkaldy	1820— *
Franz Neumann	1798— *
Kirchhoff	1824—1887
Clebsch	1833—1872
Boussinesq	1842— *
Lord Kelvin (Sir William Thomson)	1824— *

* Living Scientists.

ERRATA.

PART I.

p. 3, l. 5, from bottom *dele* reference to Hopkins.

p. 26, l. 7, from top *for* $M = \cdot 843462 \mu\tau\omega^2 b^2/3$ *read* $M = \cdot 843462 \mu\tau\omega 2 b^2/3$.

p. 68, l. 2, from top *for* ω on left-hand side of equation *read* w.

p. 79, l. 19, *for* $a_r = du/dr$ *read* $u_r = du/dr$.

,, footnote *for* co-latitude *read* latitude.

,, ,, ,, in first body-stress equation of sphere *read* $2\widehat{rr}$ *for* 2^{rr}.

p. 113, l. 13, *for* neutral line *read* neutral axis.

p. 114, l. 4 of footnote, *for* central axis *read* central line.

p. 125, l. 2, *for* S_0/G *read* S_0/μ.

p. 244, *add* to footnote: see, however, our Art. 410.

pp. 379–81. Phillips's analysis for the case of a doubly built-in girder has been shown by Bresse and Saint-Venant to be in error: see our Arts. 382 and 540. ll. 3 and 4, p. 380, and the footnote p. 381, must be modified in this sense. Arts. 552–4 were written at a very different date to Arts. 381 and 540, and the facts stated in the latter had escaped me.

CHAPTER X.

SAINT-VENANT, 1850—1886.

SECTION I. *Torsion.*

[1.] WE commence our second volume with some account of the later work of the great French elastician whom we are justified in placing beside Poisson and Cauchy. From the last memoir referred to in our first volume till June 13, 1853 we have nothing to report. A slight note, however, entitled: *Divers résultats relatifs à la torsion*, which was read to the *Société philomathique* (*Bulletin*, February 26, 1853, or *L'Institut*, no. 1002, March 16, 1853), sufficiently indicates that our author had been diligently at work during these years on his new theory of torsion. On the 13th of June, 1853, his epoch-making memoir was read to the Academy (*Comptes rendus*, T. XXXVI. p. 1028). The memoir was inserted in T. XIV. of the *Mémoires des Savants étrangers*, 1855, pp. 233—560, under the title:

Mémoire sur la Torsion des Prismes, avec des considérations sur leur flexion, ainsi que sur l'équilibre intérieur des solides élastiques en général, et des formules pratiques pour le calcul de leur résistance à divers efforts s'exerçant simultanément.

We have referred to it in our first volume as the memoir on *Torsion*, and shall continue to do so.

The memoir was referred by the Academy to a committee consisting of Cauchy, Poncelet, Piobert and Lamé. Their report drawn up by Lamé (*Comptes rendus*, T. XXXVII., December 26, 1853, pp. 984—8) speaks very highly of the memoir. We cite the concluding words:

Le travail dont nous venons de rendre compte, mérite des éloges à plus d'un titre: par les nombres et les résultats nouveaux qu'il offre aux arts industriels, il constate, une fois de plus, l'importance de la théorie de l'équilibre d'élasticité; par l'emploi de la méthode mixte, il indique comment les ingénieurs, qui veulent s'appuyer sur cette théorie, peuvent utiliser tous les procédés actuellement connus de l'analyse mathématique; par ses tables, ses épures, et ses modèles en relief[1], il donne la marche qu'il faut nécessairement suivre, dans ce genre de recherches, pour arriver à des résultats immédiatement applicables à la pratique; enfin, par la variété de ses points de vue, il offre un nouvel exemple de ce que peut faire la science du géomètre, unie à celle de l'ingénieur. (p. 988.)

The report gives a succinct account of the memoir. A second account by Saint-Venant himself will be found in: *Notice sur les travaux et titres scientifiques de M. de Saint-Venant*, Paris, 1858, pp. 19—31, and 71—80. This work together with one of the same title published in 1864, when Saint-Venant was again a candidate for the *Institut*, gives an excellent *résumé* of our author's researches previous to 1864. We shall refer to them briefly as *Notice* I. and *Notice* II.

[2.] The memoir itself is principally occupied with the torsion of *prisms*, a great variety of cross-sections being dealt with. This particular problem in torsion has been termed by Clebsch: *Das de Saint-Venantsche Problem* (*Theorie der Elasticität*, S. 74), and following him we shall term it *Saint-Venant's Problem*. The memoir consists of thirteen chapters.

3. The first chapter occupies pp. 233—236; and gives an introductory sketch of the contents of the memoir. If the values of the shifts of the several points of an elastic body are given the stresses can be easily found by simple differentiation. But the inverse problem—to find the shifts when the stresses are given—has not been generally solved, because we do not yet know how to integrate the differential equations which present themselves. Saint-Venant accordingly proposes the adoption of a *mixed method* (*méthode mixte ou semi-inverse*), which consists in assuming a part of the shifts and a part of the stresses, and then determining by an exact analysis what the remaining shifts and the remaining

[1] Copies of these numerous models are at present deposited in the mathematical model cases at University College. They represent much better than the poor woodcuts of the original memoir the distortion of the various cross-sections.

stresses must be. Before proceeding to the torsion of prisms Saint-Venant illustrates this *mixed method* in the third and fourth chapters of his memoir by applying it to simple problems.

[4.] The second chapter occupies pp. 236—288; it analyses strain and stress and investigates the general formulae for the equilibrium of elastic bodies. In 1868 Saint-Venant contributed to Moigno's *Statique* another elementary discussion of the fundamental formulae of elasticity; the later work is somewhat fuller and contains the more matured views of the author; the earlier is, however, very good. I will note the leading features of the treatment adopted:

(α) On p. 236 Saint-Venant defines the shifts as the *déplacements moyens* or as the *déplacements des centres de gravité de groupes d'un certain nombre de molécules.* He thus starts from the molecular standpoint, but this definition does not appear to be absolutely necessary to the course of his reasoning.

(β) On pp. 237—248 we have the analysis of strain. Here the slides first defined by Navier and Vicat (see our Vol. I. p. 877), and then theoretically considered by Saint-Venant in the *Cours lithographié* (see our Art. 1564*), are for the first time introduced by name and directly from their physical meaning into a general theory of elasticity. The slide of two lines primitively rectangular is defined as the *cosine of the angle between them* after strain (p. 238).

(γ) On p. 239 Saint-Venant carefully limits his researches to very small strains within the elastic limit, so that what he says later (pp. 281—288) on the conditions of *rupture*, must when applied to his torsion problems be interpreted only of the elastic limit. Indeed, as for certain materials, set is produced by any initial loading below the yield-point and is not practically dangerous (i.e. the material is not 'enervated,' to use Saint-Venant's language), we can only look upon the conditions of torsional rupture given in the memoir as of value when either (1) the material is elastic and *follows Hooke's Law* nearly up to rupture (cf. the steel bar H of the plate p. 893 of our Vol. I.), or, (2) the material has a state of ease extending almost up to the yield-point.

(δ) On pp. 242—5 we have the general expressions for s_r and $\sigma_{rr'}$. The first is due to Navier in his memoir of 1821, the second is attributed by Saint-Venant to Lamé (*Leçons...l'élasticité*, 1852, p. 46) but as we have seen it had been previously given by Hopkins in 1847 (see our Art. 1368*). From the second flows naturally a discussion of principal and maximum slide, together with a proof of Saint-Venant's theorem that a slide is equal to a stretch and a squeeze of half the magnitude of the slide in the bisectors of the slide angles (see our Art. 1570*).

Finally the strain is expressed for *small* shifts in terms of the shift-fluxions (pp. 246—8). There is reference in a footnote to the strain-values for *large* shifts (see our Art. 1618*).

(ϵ) We next pass to an analysis of stress on pp. 248—254. Stress is defined from the molecular standpoint as follows:

Nous appellerons donc en général *Pression, sur un des deux côtes d'une petite face plane imaginée à l'intérieur d'un corps ou à la limite de séparation de deux corps, la résultante de toutes les actions des molécules situées de ce côté sur les molécules du côté opposé, et dont les directions traversent cette face;* toutes ces forces étant supposées transportées parallèlement à elles-mêmes sur un même point pour les composer ensemble. (p. 248.)

The reader will find it interesting to follow the evolution of the stress-definition by comparing this with Arts. 426*, 440*, 546*, 616*, 678—9* and 1563*.

From this definition Saint-Venant deduces Cauchy's theorems (see our Arts. 606* and 610*) and an expression for $\widehat{rr'}$. On p. 253 p_{rr} is erroneously printed for $p_{rr'}$.

In a footnote to p. 254 a generalisation of the expression for $\widehat{rr'}$ is obtained. Suppose x, y, z to be any three concurrent but non-rectangular lines, and let x', y', z' be lines normal respectively to the planes yz, zx, xy. Then in our notation:

$$\begin{aligned}\widehat{rr'} = {} & \frac{\cos rx'}{\cos xx'}\left(\widehat{xx}\,\frac{\cos r'x'}{\cos xx'} + \widehat{xy}\,\frac{\cos r'y'}{\cos yy'} + \widehat{xz}\,\frac{\cos r'z'}{\cos zz'}\right)\\ & + \frac{\cos ry'}{\cos yy'}\left(\widehat{yx}\,\frac{\cos r'x'}{\cos xx'} + \widehat{yy}\,\frac{\cos r'y'}{\cos yy'} + \widehat{yz}\,\frac{\cos r'z'}{\cos zz'}\right)\\ & + \frac{\cos rz'}{\cos zz'}\left(\widehat{zx}\,\frac{\cos r'x'}{\cos xx'} + \widehat{zy}\,\frac{\cos r'y'}{\cos yy'} + \widehat{zz}\,\frac{\cos r'z'}{\cos zz'}\right).\end{aligned}$$

The proof is easily obtained by the orthogonal projection of areas.

(ζ) Saint-Venant next proceeds to express the relations between stress and strain (pp. 255—262). It cannot be said that this portion of his work is so satisfactory as the later treatment in Moigno's *Statique* (see p. 268 *et seq.*) or the full discussion of the generalised Hooke's Law in his edition of *Clebsch* (pp. 39—41). In fact the linearity of the stress-strain relations is obtained in the text by assumption: *Admettons donc avec tout le monde que les pressions sont fonctions linéaires des dilatations et des glissements tant qu'ils sont très-petits* (p. 257). A long footnote (pp. 257—261) treats the matter from the standpoint of central intermolecular action. Appeal is made to Cauchy (*Exercices de mathématiques* t. IV. p. 2: see our Art. 656*) for the reduction of the 36 coefficients to 15. Saint-Venant, however,—consistent rari-constant elastician as he has always been—retains the multi-constant formulae, remarking:

Mais des doutes ont été élevés sur le principe de cette réductibilité des 36 coefficients à 15 inégaux. Bien que ce doute ait pour motif principal une autre manière de l'établir, et qu'il ne paraisse atteindre, tout au plus, que les corps régulièrement cristallisés dont nous n'aurons pas à nous occuper dans la suite de ce mémoire, et, même, ceux seulement de ces corps où des groupes atomiques éprouveraient des rotations ou des déformations particulières lorsque l'on déforme l'ensemble, nous conserverons en général, à l'exemple de M. Lamé, l'indépendance des coefficients, ce qui, comme il l'a remarqué, ne rend pas plus compliquées les solutions analytiques des problèmes.

The reference to atomic rotations was suggested by Cauchy's paper of 1851: see our Art. 681*.

(η) We have next to deal with the reduction in the number of coefficients which arises in certain symmetrical distributions of homogeneity or in cases of isotropy. Saint-Venant adopts Cauchy's definitions of homogeneity and isotropy, which should have found a place in our first volume under Art. 606* (see the *Exercices* t. IV. p. 2):

On dit alors que le corps est *homogène*, ou que *l'élasticité y est la même dans les mêmes directions en tous ses points* (p. 263).

On the other hand a body is *isotrope* when it has *une élasticité constante ou égale en tous sens autour du point* (p. 272).

Saint-Venant refers to a *semi-polaire* distribution of elastic homogeneity as an example of elastic distribution. He has, as we shall see later, thoroughly treated the entire subject in a memoir of May 21, 1860.

The various cases in which one or more planes of symmetry exist are worked out, but I think brevity as well as uniformity of method are gained by adopting Green's expression for the internal work due to the strains.

(θ) As an example of Saint-Venant's method in this section we may take the following problem. He has shewn that in the case of one plane of symmetry, that of yz, the shears perpendicular to this plane reduce to:

$$\widehat{xy} = f\sigma_{xy} + h\sigma_{zx}, \quad \widehat{zx} = e\sigma_{zx} + h\sigma_{xy} \quad \text{..................(i)},$$

where

$$f = |xyxy| \quad h = |xyzx| = |zxxy|$$
$$e = |zxzx|,$$

in the umbral coefficient notation: see Vol. I. p. 885.

Now by a suitable change of axes these shears can be expressed each in terms of a single slide. This problem is not reproduced in Moigno's *Statique*.

Turn the axes of yz round x through an angle β, then we easily find:

$$\left.\begin{aligned} \widehat{z'x} &= -\widehat{xy}\sin\beta + \widehat{zx}\cos\beta \\ \widehat{xy'} &= \widehat{xy}\cos\beta + \widehat{xz}\sin\beta \end{aligned}\right\} \text{.....................(ii)}.$$

$$\left.\begin{aligned} \sigma_{xy} &= \sigma_{xy'}\cos\beta - \sigma_{xz'}\sin\beta \\ \sigma_{xz} &= \sigma_{xy'}\sin\beta + \sigma_{xz'}\cos\beta \end{aligned}\right\} \text{.....................(iii)}.$$

Substitute from (iii) in (i) and then the values so deduced in (ii). We obtain

$$\left.\begin{aligned}\widehat{xy'} &= \left(\frac{f+e}{2}+\frac{f-e}{2}\cos 2\beta + h\sin 2\beta\right)\sigma_{xy'} \\ &\quad + \left(-\frac{f-e}{2}\sin 2\beta + h\cos 2\beta\right)\sigma_{xz'} \\ \widehat{xz'} &= \left(\frac{f+e}{2}-\frac{f-e}{2}\cos 2\beta - h\sin 2\beta\right)\sigma_{xz'} \\ &\quad + \left(-\frac{f-e}{2}\sin 2\beta + h\cos 2\beta\right)\sigma_{xy'}\end{aligned}\right\}\dots\dots\dots(\text{iv}).$$

Obviously, if we take $\tan 2\beta = \dfrac{2h}{f-e}$ we reduce this last pair of equations to

$$\left.\begin{aligned}\widehat{xy'} &= f_1\,\sigma_{xy'} \\ \widehat{xz'} &= e_1\,\sigma_{xz'}\end{aligned}\right\}\dots\dots\dots\dots\dots\dots\dots\dots\dots(\text{v}),$$

where f_1 and e_1 are roots of the quadratic $\mu^2-(f+e)\,\mu+fe-h^2=0$.

Such is substantially Saint-Venant's reduction. It is obvious, however, that this result follows at once when a known problem as to the invariants of a conic is applied to the work-function.

(ι) A remark as to isotropy on p. 272 may be reproduced as bearing on the uni-constant controversy:

Mais l'isotropie paraît rare. Non-seulement les corps fibreux, tels que bois, les fers étirés ou forgés, mais même les corps grenus ou vitreux, refroidis de la surface au centre après leur fusion, peuvent présenter des élasticités différentes en divers sens.

Saint-Venant refers to the experiments and remarks of Regnault, Savart and Poncelet already noted in our first volume: see Arts. 332*, 978* and 1227*

(κ) On pp. 272—8 we have deductions of the body-stress equations, the body-shift equations and the surface-stress equations.

On p. 276 Saint-Venant deduces the body-shift equation for a planar distribution of elasticity such as he requires for his torsion problem.

He takes for the shears the expressions found in Equation (v) above, and for the traction $\widehat{xx}$ perpendicular to the planar system the expression

$$\widehat{xx} = as_x + bs_y + cs_z + d\sigma_{yz} + e\sigma_{zx} + f\sigma_{xy},$$

with *six independent* constants. Substituting in the body-stress equation $\dfrac{d\widehat{xx}}{dx}+\dfrac{d\widehat{xy}}{dy}+\dfrac{d\widehat{xz}}{dz}=X$, and expressing the strain in terms of the shift-fluxions, he finds:

$$a\frac{d^2u}{dx^2}+f_1\frac{d^2u}{dy^2}+e_1\frac{d^2u}{dz^2}+f\frac{d^2u}{dxdy}+e\frac{d^2u}{dxdz}$$
$$+(f_1+b)\frac{d^2v}{dxdy}+(e_1+c)\frac{d^2w}{dxdz}+d\left(\frac{d^2v}{dxdz}+\frac{d^2w}{dxdy}\right)+f\frac{d^2v}{dx^2}+e\frac{d^2w}{dx^2}=X.$$

C'est la seule équation dont nous aurons besoin pour les problèmes sur la torsion, comme on verra.

It will be noted that it contains *eight* independent constants, and that X is a body-*force*, not a body-*acceleration*, and acts *towards* the origin. It is needless to say that Saint-Venant much reduces the number of his constants before he applies this equation to his problem. In Moigno's *Statique* (p. 637) he adopts in place of X the more usual notation of $-\rho X$ where ρ is the density.

[5.] The concluding pages of this chapter (pp. 278—288) contain matter which appears here for the first time, and which, as it is of considerable interest, deserves an article to itself. The section is entitled: *Conditions de résistance à la rupture éloignée ou à une altération progressive et dangereuse de la contexture des corps.*

(*a*) We have already noted the misleading character of this title: see Art. 4. (γ). In the first place initial loads frequently produce set which although neither progressive nor dangerous may alter the shape or elastic homogeneity of the body; and in the second place, if the body be in a state of ease, still in many cases the generalised Hooke's law will be far from holding even approximately up to the elastic limit. Saint-Venant recognises the first point by distinguishing between small sets, "qui ne font qu'*écrouir* le corps ou rendre plus stable l'arrangement de ses parties" (p. 278) and large sets, which he holds either augment progressively so that "la matière *s'énervera* bientôt" (p. 239), or else by change of form destroy the value of a structure. But he hardly seems to have taken note of the second point, for he does not hesitate on pp. 280 and 286 to use stretch- and slide- moduli which connote a proportionality of stress and strain. The same point recurs in almost each torsion problem, where a *condition de non-rupture ou de stabilité de la cohésion* is given (e.g. pp. 351, 396 etc.). It is essentially a limit to the proportionality of stress and strain which is in each case given, but this limit in many materials has no sensible existence or may in the case of a material which does not possess an extended state of ease be safely passed.

(*b*) One further remark before we proceed to Saint-Venant's process. He starts from the formula (p. 280)

$$s_r = s_x\cos^2\alpha + s_y\cos^2\beta + s_z\cos^2\gamma + \sigma_{yz}\cos\beta\cos\gamma + \sigma_{zx}\cos\gamma\cos\alpha + \sigma_{xy}\cos\alpha\cos\beta \quad\text{.............. (i)},$$

but on p. 242 he has obtained this by supposing the stretches and slides to be so small that their squares may be neglected. It is conceivable that in some materials before rupture and, possibly, before a dangerous set is reached, this might not be allowable.

(c) Our author begins by noticing that the proper limit to be taken for the stability of a material is a *stretch* and not a *traction* limit. He attributes to Mariotte[1] the first recognition of this fact "que c'est le degré d'extension qui fait rompre les corps" and remarks that although it is legitimate, and occasionally convenient, to take a traction limit given by $T = E\bar{s}$ where $\bar{s}$ is the stretch-limit and E the stretch-modulus, T need not be the stress across any plane, whatever, at the point in question.

Et cette sorte de notation est sans inconvénient si l'on n'oublie pas que T *représente simplement le produit* $E\bar{s}$, ou la force capable de donner (aussi par unité superficielle) à ce même petit prisme supposé isolé, la dilatation limite $\bar{s}$ relative à sa situation dans le corps, mais qu'il ne représente que *quelquefois* et *non toujours* l'effort intérieur ou la pression supportée normalement par sa section transversale pendant qu'il fait partie du corps. (p. 280.)

This remark is all the more important as the distinction has been neglected by Lamé, Clebsch and more recent elasticians: see our Arts. 1013*, 1016* footnotes and 1567*.

(d) The stretch in any direction being given by the equation (i) above, we have next to ask what in an aeolotropic body is the distribution of limiting stretch? Saint-Venant having regard to equation (i) *assumes* it to be *ellipsoidal* in character; in other words he takes

$$\bar{s} = \bar{s}_x \cos^2\alpha + \bar{s}_y \cos^2\beta + \bar{s}_z \cos^2\gamma,$$

where $\bar{s}_x$, $\bar{s}_y$, $\bar{s}_z$ are three constants to be determined by experiment, and the axes of ellipsoidal distribution are chosen as those of co-ordinates. The condition of safety now reduces to the maximum value of $s/\bar{s}$ being = or < 1. By the ordinary max.-min. processes of the Differential Calculus we obtain for $s/\bar{s}$ the equation:

$$4\bar{s}_x\bar{s}_y\bar{s}_z\left(\frac{s}{\bar{s}} - \frac{s_x}{\bar{s}_x}\right)\left(\frac{s}{\bar{s}} - \frac{s_y}{\bar{s}_y}\right)\left(\frac{s}{\bar{s}} - \frac{s_z}{\bar{s}_z}\right) - \sigma^2_{yz}\bar{s}_x\left(\frac{s}{\bar{s}} - \frac{s_x}{\bar{s}_x}\right)$$
$$- \sigma^2_{zx}\bar{s}_y\left(\frac{s}{\bar{s}} - \frac{s_y}{\bar{s}_y}\right) - \sigma^2_{xy}\bar{s}_z\left(\frac{s}{\bar{s}} - \frac{s_z}{\bar{s}_z}\right) - \sigma_{yz}\sigma_{zx}\sigma_{xy} = 0 \ldots\ldots\text{(ii)}.$$

The roots of this equation are known to be real and we must have the greatest of them = or < 1.

Suppose the material is subject only to a sliding strain, then $s_x = s_y = s_z = \sigma_{zx} = \sigma_{xy} = 0$. Hence it follows that

$$\frac{s}{\bar{s}} = \frac{\sigma_{yz}}{2\sqrt{\bar{s}_y\bar{s}_z}}.$$

In other words if $\bar{s}$ is the limit of s, then $2\sqrt{\bar{s}_y\bar{s}_z}$ is the limit of σ_{yz} or gives the slide-limit. Let us represent it by $\bar{\sigma}_{yz}$.

[1] *Traité du mouvement des eaux*, sixième et troisième alinéa du *second discours*.

Similarly we have $\bar{\sigma}_{zx} = 2\sqrt{\bar{s}_z \bar{s}_x}$ and $\bar{\sigma}_{xy} = 2\sqrt{\bar{s}_x \bar{s}_y}$.

Saint-Venant then rewrites his equation (ii) as:

$$\left(\frac{s}{\bar{s}} - \frac{s_x}{\bar{s}_x}\right)\left(\frac{s}{\bar{s}} - \frac{s_y}{\bar{s}_y}\right)\left(\frac{s}{\bar{s}} - \frac{s_z}{\bar{s}_z}\right) - \left(\frac{\sigma_{yz}}{\bar{\sigma}_{yz}}\right)^2\left(\frac{s}{\bar{s}} - \frac{s_x}{\bar{s}_x}\right) - \left(\frac{\sigma_{zx}}{\bar{\sigma}_{zx}}\right)^2\left(\frac{s}{\bar{s}} - \frac{s_y}{\bar{s}_y}\right)$$

$$- \left(\frac{\sigma_{xy}}{\bar{\sigma}_{xy}}\right)^2\left(\frac{s}{\bar{s}} - \frac{s_z}{\bar{s}_z}\right) - 2\frac{\sigma_{yz}\sigma_{zx}\sigma_{xy}}{\bar{\sigma}_{yz}\bar{\sigma}_{zx}\bar{\sigma}_{xy}} = 0 \quad \text{.........(iii)}.$$

He remarks that this equation may be adopted as if the six limiting strains $\bar{s}_x$, $\bar{s}_y$, $\bar{s}_z$, $\bar{\sigma}_{yz}$, $\bar{\sigma}_{zx}$, $\bar{\sigma}_{xy}$, were all independent, and the values of the slide-limits $\bar{\sigma}$ had to be found by experiment. At any rate equations of the form $\bar{\sigma}_{yz} = 2\sqrt{\bar{s}_y \bar{s}_z}$ need only be used when there is an absence of experimental data. (p. 284.)

(*e*) In the following paragraph (25) Saint-Venant explains how we are to find $s/\bar{s}$ for every point in the body and then take its maximum value for all these points,

> l'on obtiendra, en l'égalant à l'unité, la condition nécessaire et justement suffisante de la résistance du corps à la rupture (p. 284).

We have noted that this language is hardly exact. The point where this maximum takes place is called after Poncelet *point dangereux*, a name which it is convenient to render by *fail-point*. This term will not necessarily connote rupture, but merely a point at which 'linear elasticity[1]' first *fails*. The consideration of this point leads Saint-Venant to a concise definition of the solid of equal resistance:

> Souvent il y a plusieurs *points dangereux*, ou plusieurs points pour lesquels la plus grande valeur de $s/\bar{s}$ est la même, d'après la manière dont les forces sont appliquées. Lorsque, dans un corps de forme allongée, il y a un pareil point à chacune de ses sections transversales, ce corps est dit d'*égale résistance:* tels sont les prismes lorsqu'ils sont simplement étendus ou tordus par des forces appliquées aux extrémités.

(*f*) We have next the application of (iii) to the case of torsion about x as axis. Here

$$s_x = s_y = s_z = \sigma_{yz} = 0,$$

whence it follows

$$s/\bar{s} = \sqrt{(\sigma_{xy}/\bar{\sigma}_{xy})^2 + (\sigma_{xz}/\bar{\sigma}_{xz})^2}.$$

We have thus the limiting condition

$$1 = \text{or} > \left(\frac{\sigma_{xy}}{\bar{\sigma}_{xy}}\right)^2 + \left(\frac{\sigma_{xz}}{\bar{\sigma}_{xz}}\right)^2.$$

It is obvious that the principal slide in any direction $\sqrt{\sigma_{xy}^2 + \sigma_{xz}^2}$ is given by the ray of an ellipse of which $\bar{\sigma}_{xy}$ and $\bar{\sigma}_{xz}$ are the

[1] I use the words 'linear elasticity' in the sense in which 'perfect elasticity' has been used by the writers of mathematical text-books, i.e. to connote the elasticity which obeys the generalised Hooke's Law or the linearity of the stress-strain relation.

semi-axes. Saint-Venant uses throughout his memoir a slightly different form. Let μ_1, μ_2 be slide-coefficients and S_1, S_2 the shears capable of producing the slides $\bar{\sigma}_{xy}$ and $\bar{\sigma}_{xz}$; then the condition of *non-rupture par glissement* (i.e. of no *failure* of linear elasticity) is expressed by

$$1 = \text{or} > \left(\frac{\mu_1 \sigma_{xy}}{S_1}\right)^2 + \left(\frac{\mu_2 \sigma_{xz}}{S_2}\right)^2$$

The chapter concludes with a few general remarks on the physical characteristics of rupture by torsion.

[6.] The third chapter occupies pp. 288—99: it relates to the simple case of a prism on any base, whose terminal faces and sides are subjected to any uniform tractive loads. Lamé and Clapeyron in their memoir of 1828 (see our Art. 1011*) had treated the simple case of isotropy. Saint-Venant as an example of the *mixed* or *semi-inverse* method gives the solution for the case when there are three planes of elastic symmetry, the intersection x of one pair being parallel to the axis of the prism. He assumes that the tractions are constant and the shears zero throughout. This satisfies the body stress-equations; the constant values of the tractions are in this case given by the surface stress-equations. The stress-strain relations then give in terms of the elastic constants and the loads the values of the shift-fluxions. We thus arrive at a system of simple linear partial differential equations, whose solution is extremely easy. The complete solution gives for each shift a part proportional to the corresponding coordinate and a general integral which is only the resolved part of the most general displacement of the prism treated as a rigid body. On p. 292 Saint-Venant determines the value of the stretch-modulus when the tractive load on the sides of the prism is zero, and on p. 293 he considers the simple cases of (1) the axis of the prism being an axis of elastic symmetry, and (2) the material being isotropic: see our Art. 1066* On p. 293 we have a remark that some writers have doubted the exactness of the above results, considering them only as plausible but not necessarily unique. Saint-Venant asserts that they are unique, which is undoubtedly true in this case, but I am not quite satisfied with the nature of his proof, for it would at first sight apply to any elastic body. It depends essentially on the following line of reasoning: Take any particular integrals of the equations of elasticity u_0, v_0, w_0, put the shifts equal to u_0+u', v_0+v', w_0+w'; we now obtain equations of elasticity without body-force or surface-load. "On verra que u', v', w' seront les déplacements des points d'un prisme qui ne serait sollicité que par des forces nulles. *Ces déplacements seraient nuls eux-mêmes.* Nos expressions offrent donc la solution complète et unique." (p. 294.) This is true for the prism, but it does not always follow that where there are no surface- or body-forces, the body is without strain, or has only rigid displacement. For example, take a cylindrical shell, a spherical membrane of small thickness, or an anchor ring of small cross section, and turn them inside out, we have a state of strain with no applied force.

On p. 295 Saint-Venant shows that his results for the prism still hold if the shifts are large, but their fluxions remain small.

[7.] A method of solving a still more general problem is indicated on p. 296. Suppose a homogeneous aeolotropic body of any shape to be subjected to a surface-load L which is the resultant of $\widehat{xx}$, $\widehat{yy}$, $\widehat{zz}$, $\widehat{yz}$, $\widehat{zx}$, $\widehat{xy}$; these stresses being given constant values throughout the body and at the surface. Then we have six equations from which to find in terms of the 21 elastic constants the six strains. These are six simple partial differential equations which give at once the shifts. Saint-Venant suggests how the stretch-modulus for any direction may thus be obtained as a function of the 36 (21 or 15) elastic coefficients: see our Arts. 135—7, 198 (*c*), 306—8 and 796*.

[8.] The final section of this chapter (§ 33, pp. 297—9) relates to a point which Saint-Venant has frequently taken occasion to refer to. The principle involved is the following:

C'est que *le mode d'application et de répartition des forces vers les extrémités des prismes est indifférent* aux effets sensibles produits sur le reste de leur longueur, en sorte qu'on peut toujours, d'une manière suffisamment approchée, remplacer les forces qui sont appliquées, par des forces statiques *équivalentes*, ou ayant mêmes moments totaux et mêmes résultantes avec une répartition justement telle que l'exigent les formules d'extension, de flexion, de torsion, pour être parfaitement exactes. (*Notice* I. p. 22.)

Saint-Venant does not clearly state the portion of the prism over which he holds the influence of distribution to extend, the term *sur le reste de leur longueur* is somewhat vague. In the memoir itself he uses the words *en excluant seulement les points très-proches de ceux où agissent les forces* (p. 299). We can perhaps, however, reach some conception of the field to which he supposes the influence to extend by paying attention to a footnote on p. 22 of *Notice* I.

Suppose the terminal of a prism subjected to any system of load statically equivalent to that distribution which produces the system of strains theoretically calculated. Impose upon the terminal two equal and opposite loads having the theoretical distribution. One of these will produce the theoretical strains, the other will be in statical equilibrium with the actual load distribution. The terminal is thus acted upon by two equivalent and opposite systems of force. These systems will produce certain small shifts in the end of the prism, and these shifts measure the extent to which the prism is influenced by the difference between the theoretical and practical distributions. Saint-Venant tells us in his footnote that the influence of forces in equilibrium acting on a small portion of a body extend very little beyond the parts upon which they act.

L'auteur a fait deux expériences de ce genre sous les yeux de l'Académie en lisant un de ses mémoires. Elles ont consisté simplement à pincer avec des tenailles un prisme de caoutchouc, et à dilater transversalement une lanière mince de même matière, en tirant ses bords en deux sens opposés. Tout le monde peut les répéter et voir que l'impression ou l'élargissement *ne se fait point sentir à des distances excédant la profondeur dans le premier cas et l'amplitude dans le second.*

The reader will find this matter still further treated of in the *Navier*, pp. 40—41 and the *Clebsch*, pp. 174—7. The principle is of first-class importance, as it is scarcely possible in a practical structure to ensure any given theoretical distribution of load. The terminals will generally take a form which lies beyond theoretical investigation and only the statical equivalent of the load system will be really ascertainable, e.g. the tractive load on a bar may be applied by means of a nut carrying a weight, the nut itself being supported by the thread of a screw cut on the bar.

[9.] Saint-Venant's fourth chapter deals with the problem of flexure by the *semi-inverse* process. The important results here first published were afterwards considered at greater length in the well-known memoir on flexure: see our Art. 69 *et seq.*

Throughout the chapter the writer supposes three principal planes of elasticity, one of which coincides with the cross-section, and the two others intersect in the line of sectional centroids, i.e. in the axis of the prism. He thus makes use of formulae which in his notation apparently involve twelve independent coefficients, but these he at once reduces to three independent moduli (E, ϵ, ϵ') and two coefficients (f, e): see pp. 303, 311—313.

As Saint-Venant justly remarks:

La détermination exacte et générale des déplacements des points d'un prisme sous l'action de forces qui tendent à le *fléchir*, a échappé jusqu'à présent aux recherches les plus laborieuses des géomètres. (p. 299.)

But although his solution does not solve the problem for *all* terminal distributions of load, it is yet as close an approximation in practice as, say, Coulomb's solution of the torsion of a circular cylinder. It cannot be too often repeated that the distributions of tractive and shearing loads, such as occur in theory, are not attainable in practice, and that we must be content with their statical equivalent over small areas (see our Art. 8). But let us hear Saint-Venant himself:

Aussi les résultats ci-dessus ne sont pas applicables d'une manière tout à fait rigoureuse.

Mais l'analyse précédente nous prouve toujours que si sur deux sections quelconques, extrêmes ou non, les forces sont appliquées et distribuées de cette manière, il en sera absolument de même sur toutes les sections intermédiaires, et que les déplacements, dans toute l'étendue du prisme, seront représentés par les autres expressions trouvées ci-dessus. Les formules donnent donc l'état de choses vers lequel converge l'état intérieur réel du prisme à mesure que l'on considère des parties plus éloignées de ses extrémités ou des points d'application des forces qui font fléchir.

Il s'établit ici, dans l'espace, une sorte d'état permanent semblable à celui qui est produit, dans le temps, par l'action continue de causes constantes qui finissent par effacer l'effet des causes initiales d'un grand nombre de phénomènes. (p. 314.)

Saint-Venant's solution of the problem of flexure is thus the real solution of the problem, for were any other solution obtained it could differ from his only by terms which would be really insignificant as compared with the differences in terminal loading which must occur, not only between theory and practice, but between any two practical cases of flexure. It is just as reasonable or unreasonable to quarrel with Coulomb's torsion solution as with Saint-Venant's flexure results.

[10.] With regard to the uniqueness of the solution obtained by the semi-inverse method—supposing the theoretical shearing and tractive loads were applied to the terminals—Saint-Venant has some remarks on p. 307 which it is well to consider. After remarking that the shifts satisfy all the conditions and equations of the problem, he continues:

Et ils sont les seuls qui y satisfassent, car le problème des déplacements est complètement déterminé si, en donnant les pressions et tractions sur tous les points de la surface, on suppose fixes l'un des points du prisme (le point O), et les directions d'un élément linéaire et d'un élément plan qui y passent (un élément sur l'axe des z et un élément sur le plan yz) en sorte qu'il ne puisse y avoir ni translation ni rotation générale à ajouter aux déplacements provenant de la flexion. (p. 307.)

He then proceeds as on p. 294 to put the shifts equal to the particular solutions found plus additional unknown parts (u', v', w'), these latter he argues must be zero as they are shifts due to a zero system of loading as appears by the vanishing of the load terms from the equations on substitution. This sketch of a proof of the uniqueness of solution of the equations of elasticity has been adopted and expanded by Clebsch: see Kap. I. § 21 of his *Theorie der Elasticität.* I have suggested above that there is need of applying the proof with some caution: see Art. 6.

[11.] In treating the problem of flexure Saint-Venant assumes the longitudinal shifts and the lateral loading, hence he deduces the transverse shifts and the terminal loading. The values of the longitudinal shifts were doubtless suggested by the Bernoulli-Eulerian solution of the problem, but in this chapter they appear to arise very naturally from the consideration of the simpler case of uniform flexure, or the bendings of each longitudinal 'fibre' into a circular arc; see pp. 292—304.

Saint-Venant makes two generalisations of his problem. The first (p. 306) to the case when besides terminal shearing load, there is also terminal tractive load. It is necessary, however, to remark that when such load is negative, and the prism of considerable length as compared with the dimensions of cross-section, the question of the *buckling* action of such load arises. This is a point to which we have referred in our first volume: see Art. 911* Saint-Venant does not allude to it. The second generalisation is to the case of large shifts, or as it is here termed: *Extension de cette solution à une flexion aussi grande qu'on veut.* I cite the following remarks as suggestions which have been adopted by later writers (e.g. Kirchhoff):

Les formules donnant u, v, w ne s'appliquent, comme les équations différentielles dont on les a tirées, qu'à des déplacements très-petits ne produisant qu'une petite flexion. On peut cependant en tirer des déplacements d'une grandeur aussi considérable qu'on veut, tels que ceux d'une verge élastique longue et mince qu'on ploie au point de faire presque toucher les deux bouts, ce qui est très-possible sans altérer aucunement la contexture de sa matière, car les déplacements relatifs et les déformations peuvent rester petits dans chacune des portions d'une longueur bien moindre que le rayon ρ de la courbure, dans lesquelles on peut diviser par la pensée un pareil corps; et c'est leur accumulation qui produit, à l'extrémité, des déplacements considérables (p. 308).

12. The section of the chapter pp. 308—313 which deals with the general problem of flexure is reproduced in the memoir in Liouville's *Journal* and will be considered later: see our Arts. 69 *et seq.*

Two results are given on p. 312 without demonstration. The first of these relates to the case of an elliptic section; it coincides with equation (56) of the memoir in Liouville's *Journal* (see our Art. 86, Eqn. 25) when we put C the constant of that equation zero. The second of these relates to the case of a rectangular section; it is an approximation: the memoir in Liouville's *Journal* gives the exact solution, but not this approximation. It is however easy to supply the steps which lead to the approximation. In equation (91) of the memoir in Liouville's *Journal* the exact value of $F(y, z)$ is given depending on $F_1(y, z)$ which is determined by (102). If we were to expand $F_1(y, z)$ in powers of y and z, the term which involves z only would disappear by (103); then the next two terms would involve $y^2 z$ and z^3 respectively. This suggests our taking a form like that of (85) in the memoir on *Torsion* as an approximation; take this and calculate $\widehat{xz}$, that is $G'\left(\dfrac{dw}{dx} + \dfrac{du}{dz}\right)$. We find this to be

$$G'g_0 + \frac{G'P}{2EI}\left\{(K - 2\epsilon)y^2 + (E - fK)\frac{z^2}{e}\right\};$$

then in order that this may vanish when $y = 0$ and $z = c$ we must have

$$g_0 = -\frac{Pc^2(E - fK)}{2EeI}.$$

Then Saint-Venant assumes that $\int \widehat{xz}\, d\omega = -P$; and this leads to the value of K which he uses in this case: see p. 312 line 3 from the foot.

13. On p. 311 Saint-Venant says that $F = 0$, and $dF/dz = 0$, when $y = 0$ and $z = 0$. Suppose that h and k denote very small quantities; then the value of u at the origin being denoted by u_0 the value at a point very near the origin would be

$$u_0 + \left(\frac{du}{dy}\right)_0 h + \left(\frac{du}{dz}\right)_0 k.$$

Now $\left(\frac{du}{dy}\right)$ is zero since u is an even function of y, so that if we have $\left(\frac{du}{dz}\right)_0$ zero as well as u_0 then the value of u vanishes all over *an element* near the origin.

[14.] Pp. 316—318 are deserving of close attention; they give results which were partially published in the memoir of 1843 (see our Art. 1581*) and which followed up the suggestion of Persy: see our Art. 811*. Saint-Venant namely finds the plane of flexure when the load-plane does not coincide with the plane of one set of principal axes of the cross-sections.

Let Oz, Oy be the principal axes at O the centroid of any cross-section of area ω; let κ_z, κ_y be the swing-radii about these axes, and ϕ, ψ the angles which the load and flexure planes make respectively with the plane through Oz and the axis of the prism. Then Saint-Venant easily shews that:

$$\tan \psi = \frac{\kappa_z^2}{\kappa_y^2} \tan \phi; \quad \frac{1}{\rho} = \frac{M}{E\omega} \sqrt{\frac{\cos^2 \phi}{\kappa_y^4} + \frac{\sin^2 \phi}{\kappa_z^4}},$$

where $1/\rho$ is the curvature, M the bending moment and E the longitudinal stretch-modulus[1].

Assuming that only longitudinal stretch produces danger, Saint-Venant deduces that if $s_0 = T_0/E$ be the limit of safe stretch then

$$M = \text{or} < \text{the minimum of } \frac{T_0 \omega}{z \dfrac{\cos \phi}{\kappa_y^2} + y \dfrac{\sin \phi}{\kappa_z^2}}.$$

For the rectangle $(2b \times 2c)$ we have

$$M = \text{or} < \frac{4T_0 b^2 c^2}{3(b \cos \phi + c \sin \phi)},$$

[1] The first equation expresses geometrically that the plane of flexure is perpendicular to the diameter of the momental ellipse (neutral axis) conjugate to the plane of loading: see our Art. 171.

for the ellipse ($2b \times 2c$)

$$M = \text{or} < \frac{T_0 \pi b^2 c^2}{4\sqrt{b^2 \cos^2\phi + c^2 \sin^2\phi}}.$$

Such results as these he has reproduced and considerably added to in his edition of *Navier*, pp. 52—60, pp. 122—126 and 128—136. Indeed, we may affirm that Saint-Venant was the first to insist on the practical importance of investigating the relation between the planes of flexure and of loading, when the latter plane is not one of inertial symmetry.

[15.] The chapter concludes with the deduction of Saint-Venant's all-important discovery that the cross-sections of a beam under flexure do not remain plane even within the limit of elasticity. There is also an investigation of the change in the cross-sectional contour (pp. 318—323). We shall return to these points later, but meanwhile may quote the concluding words of the chapter as some evidence of the satisfaction which Saint-Venant legitimately felt at the results of his new process:

On voit, par ce chapitre IV, que la méthode mixte de solution des problèmes de l'équilibre des corps élastiques peut, non-seulement confirmer des résultats connus, en apprenant à quelles conditions ils sont exacts, mais encore les compléter, et donner sur les circonstances de la flexion des résultats nouveaux.

[16.] Saint-Venant's fifth chapter defines torsion and deduces the general equations by the semi-inverse method; it occupies pp. 323—333.

The definition of torsion which does not involve the maintenance of the primitive planeness of the cross-sections is contained in the following paragraph:

Et nous nous donnerons *une partie des déplacements* ou de leurs rapports, en ce que nous supposerons que ces déplacements ont produit une *torsion* autour d'un axe parallèle à ses arêtes, torsion qui consiste *en ce que les déplacements transversaux des divers points appartenant primitivement à une même section quelconque perpendiculaire à l'axe ne diffèrent de ceux des points homologues d'une autre section, que par une rotation d'un même angle pour tous, autour du même axe;* en sorte que les points qui se correspondaient primitivement sur les droites parallèles à l'axe puissent être ramenés à se correspondre encore, en les faisant tourner d'un angle qui est le même pour les points des deux mêmes sections (p. 324).

We will now sketch the method by which our author reaches the general equations of torsion.

[17.] The axis of torsion will be taken as axis of x; the direction of torsion will be from the axis of y towards that of z. The area of a cross-section will be denoted by ω, and we shall write $\omega\kappa_z^2 = \int y^2 d\omega$, $\omega\kappa_y^2 = \int z^2 d\omega$, these being the sectional moments of inertia. The torsion referred to unit of length will be τ; that is, if we draw the radius-vector of a displaced point in one section, and also that of the homologous point in a section at distance ξ from the first, then the second radius-vector makes with a parallel to the first an angle of which the circular measure is $\tau\xi$; this angle is measured from the axis of y to that of z. This language implies that the torsion is constant, but the meaning of τ, when it is not constant, will be assigned in the same manner as before at any point, provided we consider ξ as infinitesimally small.

The above definition of torsion leads us at once to the results:

$$dv/dx = -\tau z, \quad dw/dx = \tau y \text{ .... (i).}$$

The consideration that there is no lateral load gives for every point of a sectional contour the equation

$$\widehat{xz}\,dy - \widehat{xy}\,dz = 0 \text{ .... (ii).}$$

On p. 329 Saint-Venant fixes a point, line and elementary plane as in our Art. 10, and remarks that the total torsion between the terminal sections may be considerable provided each short element into which we may divide the prism by two cross-sections receives only a small distortion relative to itself, the length of the prism being great as compared with the linear dimensions of the section. The total shifts can then be obtained by summation from the solutions of the above equations for each short element.

Referring to the equations in our Art. 4 (θ) we easily obtain

$$\widehat{xy} = f_1\,(du/dy - \tau z), \quad \widehat{xz} = e_1\,(du/dz + \tau y) \text{ ....(iii).}$$

Whence if M be the moment of all the stresses on a cross-section about the axis of x,

$$M = \int_0^\omega d\omega\,[e_1\,(du/dz + \tau y)\,y - f_1\,(du/dy - \tau z)\,z] \text{ ....(iv).}$$

It will be seen that this agrees with the old theory—which gave $M = e_1\tau \int_0^\omega d\omega\,(y^2 + z^2)$,—only when $e_1 = f_1$ and $du/dz = du/dy$. This, since du/dx is assumed constant, amounts to $u = 0$, or the old theory that the cross-sections remain plane and perpendicular to the axis. Substituting in the equation of our Art. 4 (κ), and in (ii) above, we find for body and surface shift-equations:

$$\left.\begin{array}{r} ad^2u/dx^2 + f_1\,d^2u/dy^2 + e_1\,d^2u/dz^2 + fd^2u/dxdy + ed^2u/dxdz \\ + (ey - fz)\,d\tau/dx = 0 \\ e_1\,(du/dz + \tau y)\,dy - f_1\,(du/dy - \tau z)\,dz = 0 \end{array}\right\} \text{...(v).}$$

Saint-Venant (p. 331) at once simplifies these equations by taking $d^2u/dx^2 = f\, d^2u/dxdy = e\, d^2u/dxdz = 0$; these follow at once from the supposition that du/dx, or the longitudinal stretch, is constant or zero, or again from the second supposition that it is constant only along lines parallel to the axis of torsion and that a principal plane of elasticity is perpendicular to this axis (i.e. $e = f = 0$).

In general we shall adopt the notation $e_1 = \mu_2$, $f_1 = \mu_1$, so that our equations become

$$\left.\begin{array}{r} \mu_1\, d^2u/dy^2 + \mu_2\, d^2u/dz^2 = 0 \\ \mu_2\,(du/dz + \tau y)\, dy - \mu_1\,(du/dy - \tau z)\, dz = 0 \end{array}\right\} \text{.........(vi).}$$

Saint-Venant for the purpose of simplifying the form of his results takes $\mu_1 = \mu_2 = \mu$ in the following four chapters. Further to avoid the complexity which would be initially introduced by treating at the same time the problem of flexure Saint-Venant takes

$$s_x = s_y = s_z = \sigma_{xy} = 0.$$

We shall see in the sequel that Clebsch has combined the two problems of torsion and of flexure by preserving the general form of the equations.

The next four chapters of the memoir VI.—IX. are occupied with the torsion of prisms of various cross-sections. I shall briefly give the results here for the purpose of reference; the reader will find little difficulty in deducing the proofs for himself, if the original memoir be not accessible. At the same time I shall draw attention to one or two important points involved in Saint-Venant's discussion.

[18.] The sixth chapter occupies pp. 333—352, and is entitled: *Torsion d'un prisme ou cylindre à base elliptique.*

The following results are obtained, the axes of the cross-section being $2b$ and $2c$, and the notation being otherwise as before:

$$u = -\frac{b^2 - c^2}{b^2 + c^2}\tau yz, \quad \left.\begin{array}{l} v = -\tau xz \\ w = \tau xy \end{array}\right\} \text{..................(i).}$$

$$M = \mu\tau \frac{\pi b^3 c^3}{b^2 + c^2} = \frac{4\mu\tau\omega}{1/\kappa_y^2 + 1/\kappa_z^2} \text{..................(ii).}$$

$$\left.\begin{array}{c} \widehat{xx} = \widehat{yy} = \widehat{zz} = \widehat{yz} = 0. \\ \widehat{xy} = -\mu\tau \dfrac{2b^2 z}{b^2 + c^2}, \quad \widehat{xz} = \mu\tau \dfrac{2c^2 y}{b^2 + c^2} \end{array}\right\} \text{......................(iii).}$$

We see at once from (i) that the primitively plane sections suffer distortion (*gauchissement*), and become hyperbolic paraboloids. In the accompanying figure the contour lines of these surfaces of distortion are marked; broken lines denoting depressions.

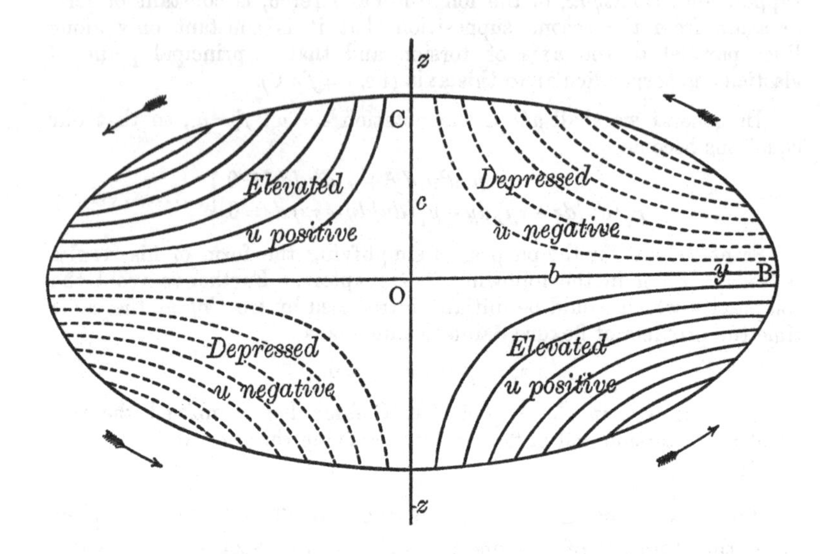

The principal slide σ is given by

$$\sigma^2 = \sigma_{xy}^2 + \sigma_{xz}^2 = \left(\frac{2\tau}{b^2+c^2}\right)^2 (b^4z^2 + c^4y^2) \text{ .... (iv).}$$

The *point dangereux* or *fail-point* is obtained by making $b^4z^2 + c^4y^2$ a maximum, thus it is at the extremity of the minor axis, i.e. is the point *nearest to the axis of torsion.*

From (iv) we obtain by means of our Art. 5 (f), if $S_1 = S_2 = S_0$:

$$S_0 = \text{or} > \mu\tau\, 2b^2c/(b^2+c^2) \text{ .... (v),}$$

whence it follows that $M = \text{or} < \dfrac{\pi bc^2\, S_0}{2} \quad \left(= \text{or} < \dfrac{2\omega\kappa_y^2\, S_0}{c}\right)$........ (vi).

The general appearance of the prism under torsion is given in the figures on the next page, the torsion being diagrammatically exaggerated.

[19.] There are one or two important points to be noticed in this chapter. In the first place Saint-Venant solves equation (vi) of Art. 17 by a series ascending in powers of y and z; one term (a'_2yz) suffices for the elliptic cross-section, he makes use of others later. Secondly he points out pp. 339—341 that his results agree with the theory of Coulomb only in the case of a circular section, for every other elliptic cross-section the value of the torsional moment is *smaller* than that given by the old theory and there is

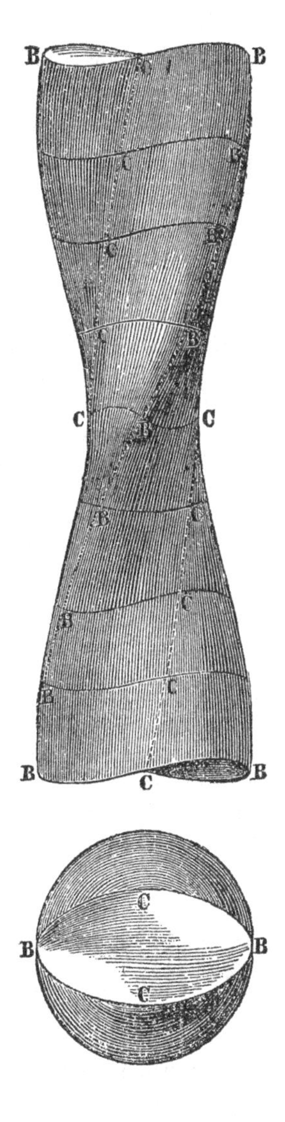

distortion. He shews by numerical examples on p. 352 how much sooner the safe limit is reached in the true than in the old theory.

[20.] On pp. 341—343 we have, thirdly, a footnote on Cauchy's suggestions that the torsion τ should be made to vary transversally: see our Art. 684*. Saint-Venant shews that this would require,—at least in the case of a circular cross-section and an axis of elasticity coinciding with the axis of figure—a shearing load at each element of lateral surface. This is a supposition which could hardly be attained in any practical case.

[21.] Fourthly we have on pp. 342—345 a very concise and admirable consideration of the point referred to in our Art. 9; namely, the *practical* equivalence of statically equipollent systems of terminal loading at very short distances from the terminals.

Nos résultats relatifs à la torsion d'un prisme elliptique par des couples quelconques peuvent être adoptés au même titre et avec la même confiance qu'on adopte les formules, soit de l'extension simple, soit de la flexion par des forces latérales, et la formule plus analogue du cas de torsion des cylindres circulaires (p. 345).

In all these cases there is the same assumption as to the equivalence of the shifts produced by the theoretical and by the actual equipollent load systems.

[22.] Fifthly §§ 59 and 60 (pp. 346—7) may be noted. The first deduces from the equations $\int_0^\omega \widehat{xy}\, d\omega = \int_0^\omega \widehat{xz}\, d\omega = 0$ that the axis of torsion for the shifts assumed must coincide with the line of sectional centroids[1]: see our Art. 181 (*d*). The second treats of the case of large torsional shifts, see our Art. 17, p. 18. Saint-Venant remarks that the values $v = -\tau xz$ and $w = \tau xy$ of our Art. 18, equation (i), no longer hold, but by an easy process of summation (p. 347) we find the new values:

$$\begin{aligned} v &= -z \sin \tau x - y(1 - \cos \tau x) \\ w &= \quad y \sin \tau x - z(1 - \cos \tau x) \end{aligned}.$$

[23.] Lastly we may note on p. 349 the general argument by which Saint-Venant would explain why the *fail-points* are those nearest and not farthest from the axis of torsion as in the old theory (*la théorie ordinaire*, St-V.). He points out that at the extremity of the major axis the slide produced by the distortion of the plane section is zero and so we have only the slide produced by the 'fibres becoming helical,' while at the extremity of the minor axis the two components of the slide both exist and compound, *operating* together. Hence generally we see how it is possible for the slide to be greater at the latter than the former point.

[1] This paragraph was cancelled in the copies of the memoir remaining in Saint-Venant's possession.

[24.] Chapter VII. of the memoir (pp. 352—360) is occupied with the analytical solution of the equation $u_{zz} + u_{yy} = 0$. The first form obtained is that in a series of exponentials and sines or cosines of multiples of y and z.

The second is in terms of cylindrical coordinates. Let $y = r\cos\phi$, $z = r\sin\phi$, then:

$$u = \Sigma A_n r^n \cos n\phi + \Sigma B_m r^m \sin m\phi,$$
$$M = \mu\tau \int r^2 d\omega - \mu\Sigma n A_n \int r^n \sin n\phi d\omega$$
$$+ \mu\Sigma m B_m \int r^m \cos m\phi d\omega.$$

These results are obvious. Special cases of uni-axial and bi-axial symmetry lead to the vanishing of certain coefficients.

[25.] Chapter VIII. (pp. 360—413) deals with the important case of the torsion of prisms of rectangular cross-section ($2b \times 2c$).

The chapter opens with some account of Cauchy's memoir of 1829—30 (see our Art. 661*) which had led Saint-Venant to recognise the general distortion of the cross-sections in the torsion problem. Cauchy had found as an approximation $u = -\frac{b^2 - c^2}{b^2 + c^2}\tau yz$, Saint-Venant's expression for the shift parallel to the axis in the case of an ellipse. This really is only an approximation when b and c are very unequal. It makes the greatest slides take place at the corners, but when we note that $\widehat{xy} = \widehat{yx}$ and $\widehat{xz} = \widehat{zx}$, then since $\widehat{yx}$ and $\widehat{zx}$ are zero on the lateral surfaces, it follows that at the angles the nullity of $\widehat{xy}$ and $\widehat{xz}$ connotes that the stress can only be tractive to the cross-section, or that:

il n'y a, en ces points, aucun glissement, et la section a dû se ployer de manière à rester normale aux quatre arêtes saillantes devenues courbes (p. 362).

This perpendicularity of the cross-section to the sides, at projecting points or angles, holds for all prisms. The recognition of it led Saint-Venant to the investigation of a more exact expression for the torsion of rectangular prisms than that discovered by Cauchy.

[26.] The equations to be solved are

$$\begin{cases} d^2u/dy^2 + d^2u/dz^2 = 0, \\ du/dy = \tau z \text{ for all values of } z \text{ between } c \text{ and } -c \text{ when } y = \pm b, \\ du/dz = -\tau y \text{ for all values of } y \text{ between } b \text{ and } -b \text{ when } z = \pm c. \end{cases}$$

At the suggestion of Wantzel, Saint-Venant reduced these equations to a known form by the substitution of $u = -\tau yz + u'$, when they become

$$\begin{cases} d^2u'/dy^2 + d^2u'/dz^2 = 0, \\ du'/dy = 2\tau z \text{ for all values of } z \text{ between } c \text{ and } -c \text{ when } y = \pm b, \\ du'/dz = 0 \text{ for all values of } y \text{ between } b \text{ and } -b \text{ when } z = \pm c. \end{cases}$$

These equations can be solved by the assumption

$$u' = \Sigma A_m (e^{my} - e^{-my}) \sin mz$$

and the usual determination of the constants by Fourier's Theorem.

[27.] Saint-Venant obtains the following general results:

(i)
$$u = \tau bc \left[-\frac{yz}{bc} + \frac{1}{2}\left(\frac{4}{\pi}\right)^3 \frac{c}{b} \Sigma_{n=1}^{n=\infty} \frac{(-1)^{n-1}}{(2n-1)^3} \frac{\sinh \frac{(2n-1)\pi y}{2c}}{\cosh \frac{(2n-1)\pi b}{2c}} \sin \frac{(2n-1)\pi z}{2c} \right]$$

$$= \tau bc \left[\frac{yz}{bc} - \frac{1}{2}\left(\frac{4}{\pi}\right)^3 \frac{b}{c} \Sigma_{n=1}^{n=\infty} \frac{(-1)^{n-1}}{(2n-1)^3} \frac{\sinh \frac{(2n-1)\pi z}{2b}}{\cosh \frac{(2n-1)\pi c}{2b}} \sin \frac{(2n-1)\pi y}{2b} \right]$$

(ii)
$$\widehat{xx} = \widehat{yy} = \widehat{zz} = \widehat{yz} = 0,$$

$$\widehat{xy} = -\mu\tau c \left\{ 2\frac{z}{c} - \left(\frac{4}{\pi}\right)^2 \Sigma_{n=1}^{n=\infty} \frac{(-1)^{n-1}}{(2n-1)^2} \frac{\cosh \frac{(2n-1)\pi y}{2c}}{\cosh \frac{(2n-1)\pi b}{2c}} \sin \frac{(2n-1)\pi z}{2c} \right\}.$$

$$= -\mu\tau c \left(\frac{4}{\pi}\right)^2 \frac{b}{c} \Sigma_{n=1}^{n=\infty} \frac{(-1)^{n-1}}{(2n-1)^2} \frac{\sinh \frac{(2n-1)\pi z}{2b}}{\cosh \frac{(2n-1)\pi c}{2b}} \cos \frac{(2n-1)\pi y}{2b}.$$

$$\widehat{xz}^{1} = \mu\tau b \frac{c}{b} \left(\frac{4}{\pi}\right)^2 \Sigma_{n=1}^{n=\infty} \frac{(-1)^{n-1}}{(2n-1)^2} \frac{\sinh \frac{(2n-1)\pi y}{2c}}{\cosh \frac{(2n-1)\pi b}{2c}} \cos \frac{(2n-1)\pi z}{2c}.$$

$$= \mu\tau b \left\{ 2\frac{y}{b} - \left(\frac{4}{\pi}\right)^2 \Sigma_{n=1}^{n=\infty} \frac{(-1)^{n-1}}{(2n-1)^2} \frac{\cosh \frac{(2n-1)\pi z}{2b}}{\cosh \frac{(2n-1)\pi c}{2b}} \sin \frac{(2n-1)\pi y}{2b} \right\}.$$

(iii)
$$M = \mu\tau bc^3 \left\{ \frac{16}{3} - \left(\frac{4}{\pi}\right)^5 \frac{c}{b} \Sigma_{n=1}^{n=\infty} \frac{1}{(2n-1)^5} \tanh \frac{(2n-1)\pi b}{2c} \right\}$$

$$= \mu\tau b^3 c \left\{ \frac{16}{3} - \left(\frac{4}{\pi}\right)^5 \frac{b}{c} \Sigma_{n=1}^{n=\infty} \frac{1}{(2n-1)^5} \tanh \frac{(2n-1)\pi c}{2b} \right\}.$$

[28.] It will be noted that Saint-Venant obtains in each case double values for his quantities which are unsymmetrical in b

[1] Saint-Venant puts sinh for cosh in the denominator here by a misprint (p. 368, equation 159).

and c. Symmetrical values may be at once obtained by adding and halving his solutions. Or, symmetrical values may be obtained directly by the assumption of the particular integral

$$u = A_p \sinh \frac{\pi p z}{b} \cos \frac{\pi p y}{c} + B_p \sinh \frac{\pi p y}{c} \cos \frac{\pi p z}{c},$$

where p is a positive integer.

It will be found that the surface conditions are then very easily satisfied, and the symmetrical forms of the results thus deduced possess for some cases practical advantages.

Saint-Venant next proceeds to consider special cases of rectangular cross-section which will occupy us in the following seven articles.

[29.] *Cas où l'un des côtés du rectangle est très-grand par rapport à l'autre.* (pp. 372—375.)

From the first of the expressions for M, we obtain

$$M = \frac{16}{3} \mu\tau bc^3 (1 - 0{\cdot}630249\, c/b),$$

and for a first approximation to u

$$u = -\tau zy.$$

These results agree with Cauchy's $M = \frac{16}{3}\mu\tau \frac{b^3c^3}{b^2+c^2}$ and $u = -\frac{b^2-c^2}{b^2+c^2}\tau yz$ when c/b is very small.

Saint-Venant in a footnote deduces Cauchy's results, but at the same time brings out the insufficiency of his method, for Cauchy neglects the fourth powers of the dimensions of the prism, but it is not at all clear *what the quantity is in comparison with which he neglects them,* for the term omitted $\frac{2\tau(y^3z - yz^3)}{3(b^2+c^2)}$ seems really of the same order as that retained $-\frac{b^2-c^2}{b^2+c^2}\tau yz$ (p. 375).

[30.] On pp. 376—98 we have the full discussion of the prism of square cross-section. The numerical results are calculated from the tables for the hyperbolic functions given by Gudermann[1]. They are calculated from both expressions obtained in Art. 27. Saint-Venant seems to have taken from three to eight terms of his series, but he has not entered upon any investigation as to whether those series satisfy Seydel's condition of *equal convergence.*

[1] *Theorie der Potenzial- oder cyklisch-hyperbolischen Functionen*, S. 263.

The values of u are calculated and given in a table on p. 377. The accompanying figures give the contour lines of the distorted cross-section and the boundaries of the cross-section as cutting the lateral faces of the distorted prism in elevation (diagrammatically exaggerated)

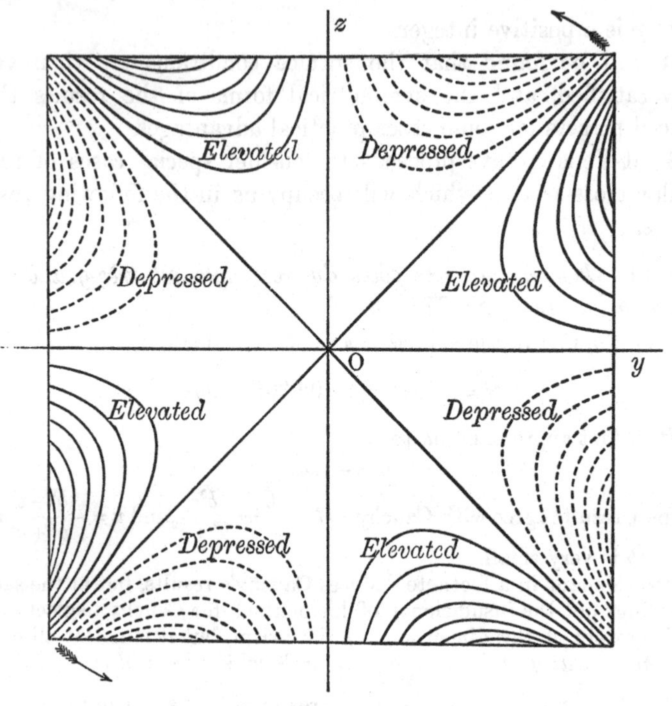

For numerical values we have,

$$M = \cdot 843462\ \mu\tau\omega^2 b^2/3$$
$$= \text{or} < 1\cdot 66532\ S_0 b^3$$

$\sigma = 1\cdot 350630\, b\tau$ is the maximum slide and occurs at the middle points of the sides of the cross-section, which are thus the *fail-points*. These values are all less than those obtained from the old theory.

[31.] On pp. 382—387 Saint-Venant refers to the experiments of Duleau[1] and Savart[2] as confirming his results. From Duleau's experiments on circular bars the mean value of μ obtained was 6,659,230,000 kilogs. but from his experiments on

[1] See our Art. 229*. [2] See our Art. 334*.

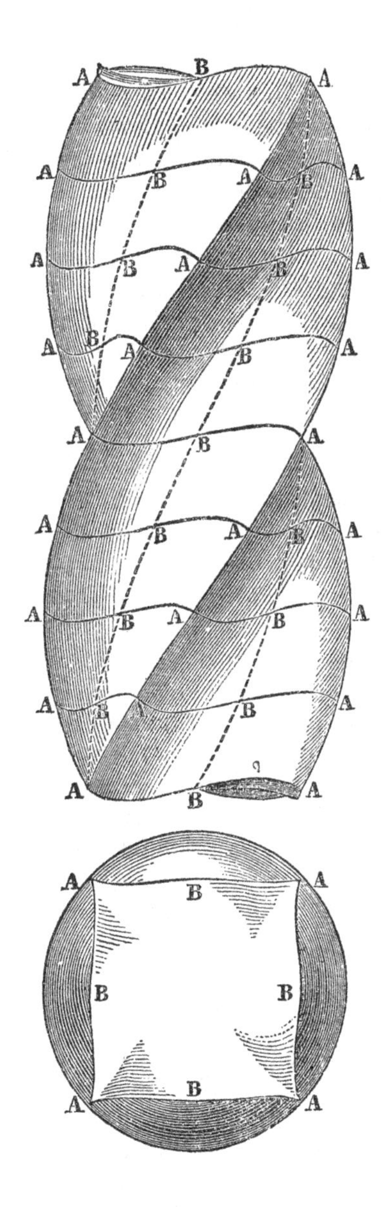
A
B
A
A
B
A
B
A
A
B
A
B
A
A
B
A
B
A
A
B
A
A
B
A
B
A
A
B
A
B
A
A
B
A
B
A
A
B
A
A
B
A
B
B
B
A
A

square sectioned bars it was only 5,636,625,000 on the old theory. Saint-Venant's however brings it up to 6,682,750,000, which may be considered in fair agreement with the result obtained from bars of circular section; especially when we remember the non-isotropic character which was inevitable in the iron bars of Duleau's experiments (see table p. 383). At any rate Saint-Venant's theory accounts for the greater part of the inferior resistance to torsion of square as compared with circular bars of equal sectional moment of inertia.

Some experiments on copper wires of square and circular cross-sections are tabulated on p. 386. Here the mean for the circular cross-section is $\mu = 4{,}174{,}825{,}000$; the old and the new theory give for μ the values 3,384,121,000 and 4,012,180,000; again to the advantage of the latter. The isotropy of these wires is however very questionable.

[32.] Saint-Venant deduces on pp. 387—391 the value of the numerical factor which occurs in M (see our Art. 30) by an algebraic expansion for u and a calculation after the manner of Fourier (*Théorie de la chaleur*, chap. III. art. 208, Eng. Trans. p. 137) of the indeterminate coefficients. It does not seem a very advantageous process. A remark on p. 397 as to the difference between *résistance à la rupture éloignée* and *rupture immédiate* is to the point. Saint-Venant remarks namely that experiments on the latter can throw little light on the mathematical theory of elasticity. At the same time it is regrettable that he should have retained the word *rupture* in reference even to the first limit. Some support, however, for his theory may even be derived, he thinks, from Vicat's experiments on rupture; see our Art. 731* and p. 398 of the memoir. For Vicat found that for *pierre calcaire, brique crue* and *plâtre* the moment of the forces required to break a prism of square cross-section and length at least twice the diameter was *less* than in the case of an infinitely short prism, i.e. a case where the plane section cannot be distorted. This result of Vicat is of great interest and would be well worth further experimental investigation.

[33.] We now come to the general case: *Cas d'un rapport quelconque des deux dimensions de la base* (pp. 398—413). Saint-Venant has calculated numerically all the particulars of the special case when $b/c = 2$. We reproduce the contour lines for the distorted cross-section as given by Saint-Venant on p. 400 according to the table on p. 399.

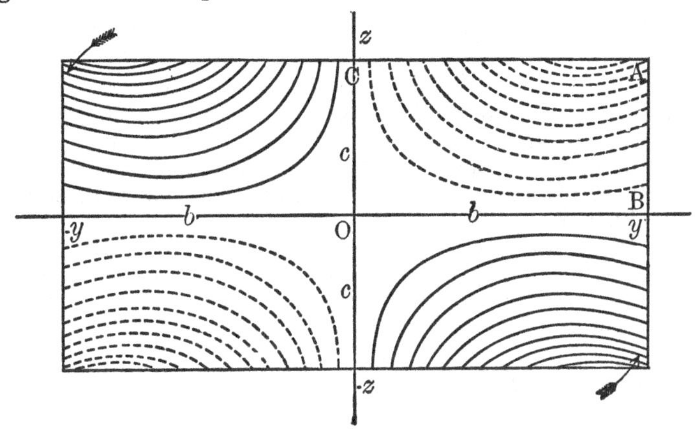

The reader will at once note the change that these lines present, and Saint-Venant on pp. 400—1 determines the value of b/c for which the change from tetra-axial to bi-axial congruency takes place.

In order to ascertain this we must find when $du/dz = 0$ at the point $y = b$, $z = 0$. For, with the tetra-axial congruency of the contour lines u is positive as we pass from $z = 0$, $y = b$ along the line $y = b$ into the first quadrant, but in the case of biaxial symmetry du/dz is negative, for u decreases or becomes negative as we pass along the same line. Our author thus obtains the equation

$$\sum_1^\infty \frac{1}{(2n-1)^2} \operatorname{sech} \frac{(2n-1)\pi c}{2b} = \left(\frac{\pi}{4}\right)^2,$$

the numerical solution of which gives $b/c = 1{\cdot}4513$.

[34.] The following general results are obtained $(b > c)$:

(i) $\begin{cases} M = \mu\tau b c^3 \beta, \\ \text{where } \beta = \left\{ \dfrac{16}{3} - 3{\cdot}361327 \dfrac{c}{b} + \dfrac{c}{b}\left(\dfrac{4}{\pi}\right)^5 \displaystyle\sum_1^\infty \dfrac{1 - \tanh \dfrac{(2n-1)\pi b}{2c}}{(2n-1)^5} \right\} \end{cases}$

(p. 401),

(ii) $\begin{cases} \text{maximum slide } \sigma = c\tau\gamma, \\ \text{where } \gamma = 2 - \left(\dfrac{4}{\pi}\right)^2 \displaystyle\sum_1^\infty \dfrac{1}{(2n-1)^2 \cosh \dfrac{(2n-1)\pi b}{2c}} \end{cases}$ (p. 412), and this maximum slide takes place at the centre of the longer side of the rectangular cross-section. (p. 410.)

(iii) $S_0 = \text{or} > \mu\gamma\tau c$, hence

$$M = \text{or} < \frac{\beta}{\gamma} bc^2 S_0 = M_0.$$

These complex analytical results are rendered practically of service by a table on pp. 559—60 of the memoir, the most serviceable portion of which we shall reproduce later. This table gives the values of β and of β/γ for magnitudes of the parameter b/c varying from 1 to 100, after which they become sensibly constant. We are thus able to determine M and its limit M_0.

Saint-Venant, however, gives in footnotes empirical formulae which agree with less than 4 per cent. error with the above theoretical values. He appears to have reached them by purely tentative methods, but he holds that they satisfy all practical needs. They are

$$\text{(iv)} \quad \beta = \frac{16}{3} - 3{\cdot}36\frac{c}{b}\left(1 - \frac{1}{12}\frac{c^4}{b^4}\right).$$

$$\text{(v)} \quad \frac{\beta}{\gamma} = \frac{8}{3}\Big/\left(1 + {\cdot}6\frac{c}{b}\right) \qquad \text{or, } M_0 = \frac{40b^2c^2}{15b + 9c} \cdot S_0.$$

{It should be noted that our $\sigma = g_x$, our $\beta = \mu$, our $\tau = \theta$, our $\mu = G$, our $S_0 = T_0$ of the memoir.}

[35.] On pp. 403—6 we have a further discussion of experiments of Duleau and Savart on the torsion of rectangular bars of iron, oak, *plâtre*, and *verre à vitre*, the paucity of the experiments, and the large variation in the values of the slide-moduli as obtained from Saint-Venant's formula do not seem to me very satisfactory. A series of experiments directly intended to test the torsion of rectangular bars for variations of the parameter c/b would undoubtedly be of considerable value.

[36.] We now reach Saint-Venant's ninth chapter which is entitled: *Torsion de prismes ayant d'autres bases que l'ellipse ou le rectangle.* It occupies pp. 414—454.

The chapter opens with an enumeration of the various forms of contour for which it is easy to integrate the equations of Art. 17. We will tabulate them on the next page.

	Contour given by : Constant =	Longitudinal shift given by : u =
1	$\frac{\tau}{2}(y^2+z^2) + \Sigma e^{my}(A_m \cos mz - B_m \sin mz)$	$\Sigma e^{my}(B_m \cos mz + A_m \sin mz)$
2	$\frac{\tau}{2}(y^2+z^2) + \Sigma A_m \cosh my \cos mz$	$\Sigma A_m \sinh my \sin mz$
3	$\frac{\tau}{2}(y^2+z^2) - a_1 z + b_1 y - a_2 2yz + b_2(y^2-z^2) - a_3(3y^2z - z^3)$ $+ b_3(y^3 - 3yz^2) - a_4(4y^3z - 4yz^3) + b_4(y^4 - 6y^2z^2 + z^4) - \text{etc.}$	$a_0 + a_1 y + b_1 z + a_2(y^2 - z^2) + b_2 2yz + a_3(y^3 - 3yz^2)$ $+ b_3(3y^2z - z^3) + a_4(y^4 - 6y^2z^2 + z^4)$ $+ b_4(4y^3z - 4yz^3) + \text{etc.}$
4	$\frac{\tau}{2}(y^2+z^2) + \sqrt{-1}\,\phi(y + z\sqrt{-1}) - \sqrt{-1}\,\psi(y - z\sqrt{-1})$	$\phi(y + z\sqrt{-1}) + \psi(y - z\sqrt{-1})$
5	$\frac{\tau}{2}r^2 + \Sigma(-a_n r^n \sin n\phi + b_m r^m \cos m\phi)$	$\Sigma(a_n r^n \cos n\phi + b_m r^m \sin m\phi)$

Solutions (3) and (5) are really identical. No. 4 has given rise to the solutions in terms of *conjugate functions:* see Thomson and Tait's *Natural Philosophy*, 2nd Ed. Part II. pp. 250—3.

[37.] In the present chapter Saint-Venant dismisses Nos. 1 and 2 on the ground that the resulting curves are very difficult to trace. He contents himself with two closed curves of the fourth degree and one of the eighth as given by No. 5. On pp. 421—434 he calculates and traces these curves at considerable length. The most practically valuable results are those obtained on p. 439.

We have there the following characteristic sections treated:

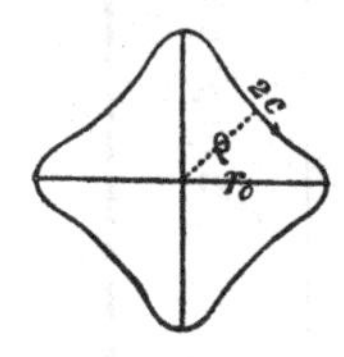

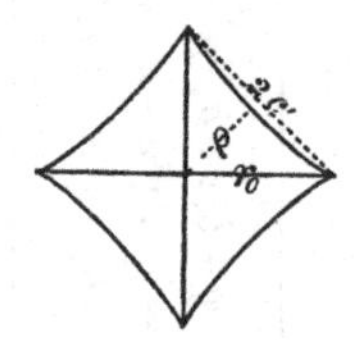

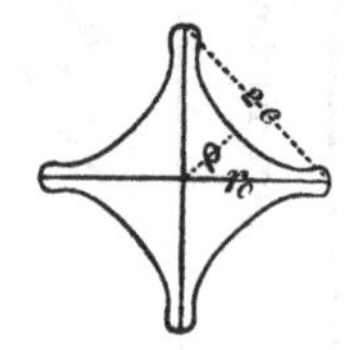

(*a*) The equation of the first curve is:

$$\frac{y^2+z^2}{r_0^{\ 2}} - \cdot 4\,\frac{y^4-6y^2z^2+z^4}{r_0^{\ 4}} = \cdot 6 \quad \text{(Square with rounded angles).}$$

$$\omega = 2\cdot 0636 r_0^{\ 2};\ \omega\kappa^2 = \cdot 7174 r_0^{\ 4} = 1\cdot 0586\,\omega^2/2\pi;$$

$$M = \cdot 5873\,\mu\tau r_0^{\ 4} = \cdot 8186\,\mu\tau\omega\kappa^2 = \cdot 8666\,\mu\tau\omega^2/2\pi.$$

(*b*) The equation to the second curve is:

$$\frac{y^2+z^2}{r_0^{\ 2}} - \cdot 5\,\frac{y^4-6y^2z^2+z^4}{r_0^{\ 4}} = \cdot 5 \quad \text{(Square with acute angles).}$$

$$\omega = 1\cdot 7628 r_0^{\ 2};\ \omega\kappa^2 = \cdot 5259 r_0^{\ 4} = 1\cdot 0634\,\omega^2/2\pi;$$

$$M = \cdot 4088\,\mu\tau r_0^{\ 4} = \cdot 7783\,\mu\tau\omega\kappa^2 = \cdot 8276\,\mu\tau\omega^2/2\pi.$$

(*c*) The equation to the third curve is, if $y = r\cos\phi$, $z = r\sin\phi$,

$$\frac{r^2}{r_0^{\ 2}} - \frac{48}{49}\cdot\frac{16}{17}\,\frac{r^4\cos 4\phi}{r_0^{\ 4}} + \frac{12}{49}\cdot\frac{16}{17}\,\frac{r^8\cos 8\phi}{r_0^{\ 8}} = 1 - \frac{36}{49}\cdot\frac{16}{17}$$

(Star with four rounded points).

$$\omega = 1\cdot 2202 r_0^{\ 2};\ \omega\kappa^2 = \cdot 2974 r_0^{\ 4} = 1\cdot 2551\,\omega^2/2\pi;$$

$$M = \cdot 15983\,\mu\tau r_0^{\ 4} = \cdot 5374\,\mu\tau\omega\kappa^2 = \cdot 6745\,\mu\tau\omega^2/2\pi.$$

We add to these the results for the circle and square.

(*d*) Circle: $M = \mu\tau\omega\kappa^2 = \mu\tau\omega^2/2\pi.$

(*e*) Square: $M = \cdot 84346\,\mu\tau\omega\kappa^2 = \cdot 88327\,\mu\tau\omega^2/2\pi.$

From the above numbers we can deduce some important practical inferences, which we will do in Saint-Venant's own words.

On voit qu'il faut, de l'expression $\mu\tau\omega\kappa^2$ de l'ancienne théorie, retrancher, pour avoir M quand la section est le carré à angles arrondis et côtés légèrement concaves, une proportion des ·1814. Nous avons vu que, pour le carré rectiligne, il faut prendre $M = \cdot 84346\mu\tau\omega\kappa^2$ ou retrancher une proportion de ·15654 seulement. La légère concavité des côtés a plus influé pour diminuer le moment de torsion (pour même moment d'inertie) que l'arrondissement des quatre angles n'a influé pour l'augmenter.

Pour le carré curviligne à côtés un peu plus concaves et angles aigus, il faut retrancher les ·2217. *Il suffit, comme l'on voit, d'une concavité assez légère des côtés de la base* (1/22 *environ*) *pour diminuer assez notablement le moment de torsion d'un prisme carré.*

Enfin, pour le prisme à côtes saillantes, il faut, de $\mu\tau\omega\kappa^2$, retrancher l'énorme proportion de ·4626, ou prendre seulement ·5374$\mu\tau\omega\kappa^2$ au lieu de $\mu\tau\omega\kappa^2$ que l'on prend pour une section circulaire, ou de ·84346$\mu\tau\omega\kappa^2$ pour une section carrée rectiligne.

Et comme on a, pour une section circulaire, $\kappa^2 = \omega/2\pi$, $M = \mu\tau\omega^2/2\pi$, l'on trouve que les prismes ayant pour bases le carré arrondi, le carré aigu et l'étoile, n'offrent respectivement que les ·867, les ·828, et les ·674 de la résistance élastique à la torsion qu'ils offriraient à égale superficie ω de la section, ou *à égale quantité de matière*, s'ils étaient à base circulaire, bien que les moments d'inertie de leurs sections soient 1^fois ·059, 1^fois ·063, 1^fois ·255 ceux de sections circulaires d'égale superficie.

Ainsi, les quatre saillies qui, malgré leur peu d'épaisseur, ont une influence considérable sur la grandeur du moment d'inertie n'en ont qu'une très-faible sur le moment de torsion. *Les pièces à côtes, employées si utilement contre les flexions, doivent être exclues des parties des constructions où les forces tendent à tordre, ou, du moins, il faut ne compter nullement sur une quote-part des quatre côtes ou saillies dans la résistance* (pp. 439—40).

[38.] Saint-Venant illustrates the inefficiency of projecting parts still more effectually in a footnote to Art. 105, p. 454. He takes a curve of the fourth degree whose equation is given in a footnote, p. 448, and by ascribing a particular value to one of the constants obtains two separate loops. The equation to the contour is:

$$\frac{y^2}{b^2} + \frac{z^2}{c^2} + a\left(\frac{1}{c^2} - \frac{1}{b^2}\right)(y^2 - z^2) - a\,\frac{y^4 - 6y^2z^2 + z^4}{b^2c^2} = 1 - a\,;$$

and the longitudinal shift

$$u = -(1 - 2a)\frac{b^2 - c^2}{b^2 + c^2}\tau yz - \frac{4a\tau}{b^2 + c^2}(y^3z - yz^3).$$

The special value of the constant assumed is $c^2 = -b^2/16$. We have then a figure of the form below and the value of M is only equal to

$\cdot 01857 \mu \tau \omega \kappa^2$, or the torsion of such a pair of cylinders round an intermediate axis is *only one fifty-fourth* of that given by the old

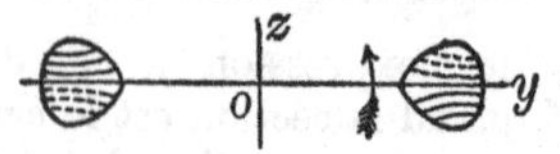

theory:—"Cela ne doit pas étonner, si l'on considère que le glissement est nul aux points $z = 0$, $y = \pm b\sqrt{\frac{23}{32}}$ ou à très-peu près au centre de gravité de chaque orbe."

[39.] Saint-Venant on his pp. 441—9 discusses the contour-lines of the distorted cross-sections of our Art. 37. This he accomplishes by numerical tables in a footnote (pp. 441—3). Then he considers the maximum slides and *fail-points* of the same sections and finally the limiting values of the torsional couples. These values are as follows:

For section (*a*) of Art. 37 $\quad M_0 = \cdot 8269 \dfrac{\omega\kappa^2}{r_0} S_0 = \cdot 7094 \dfrac{\omega^{\frac{3}{2}}}{2\sqrt{\pi}} S_0.$

" (*b*) " $\quad M_0 = \cdot 85514 \dfrac{\omega\kappa^2}{r_0} S_0 = \cdot 6812 \dfrac{\omega^{\frac{3}{2}}}{2\sqrt{\pi}} S_0.$

" (*c*) " $\quad M_0 = \cdot 7285 \dfrac{\omega\kappa^2}{r_0} S_0 = \cdot 5695 \dfrac{\omega^{\frac{3}{2}}}{2\sqrt{\pi}} S_0.$

The reasoning by which Saint-Venant deduces the *fail-points* cannot be considered satisfactory. Indeed the statement as to the 'side of the triangle' and the deduction of the maximum slide on p. 444 are unsound. The same judgment must be passed on the process of p. 447, where the maximum slide for the section (*c*) is shewn to be on the contour. Thus Saint-Venant has *not demonstrated* his very general statement (237) on p. 448. The reader will however find little difficulty in proving the accuracy of Saint-Venant's *results* by casting the expressions on pp. 444 and 447 into other forms or by the ordinary processes of the Differential Calculus. In his edition of the *Leçons de Navier*, our author has recognised the defective reasoning of these pages and replaced them by more accurate arguments. (Cf. his § 31, pp. 308—310 and § 37, pp. 340—1: see our Art. 181 (*e*).)

[40.] In the concluding pages of this chapter Saint-Venant points out how the solutions of a number of other sections can be obtained. Thus we can take solutions like (3) of Art. 36 involving terms of the 12th and 16th degrees and so obtain curves equally symmetrical with regard to the axes of y and z.

Et en y conservant des termes du deuxième, du sixième, du dixième degré à puissances paires de y et z, tels que $b_2r^2 \cos 2\phi = b_2(y^2 - z^2)$, $b_6r^6 \cos 6\phi = \text{etc.}$, l'on aurait une multitude de courbes symétriques par rapport à *chacun* des deux axes de y et z, mais *non égales dans leurs deux sens*, et ayant l'ellipse pour cas particulier (p. 449).

We have referred to an example of this in Art. 38, and another is given by Saint-Venant in a footnote; namely the curve whose equation is

$$\tau \frac{r^2}{2} + b_3 r^3 \cos 3\phi = \text{constant},$$

where $$u = b_3 r^3 \sin 3\phi.$$

By taking $b_3 = -\dfrac{\tau}{6b}$ and the constant $= \frac{2}{3}\tau b^2$, the equation to the contour of the section becomes

$$\left(\frac{r}{2b}\right)^2 - \frac{2}{3}\left(\frac{r}{2b}\right)^3 \cos 3\phi = \frac{1}{3},$$

or $$(y + b)\left(z - \frac{2b - y}{\sqrt{3}}\right)\left(z + \frac{2b - y}{\sqrt{3}}\right) = 0 \quad \text{..............(i)},$$

i.e. an equilateral triangle of height $3b$ and side $= 2b\sqrt{3}$, the axis of y coinciding with a median line. We reproduce Saint-Venant's entire treatment of this case as a good example of his method, and in order in one point to indicate a weakness in his reasoning.

41. We find at once that

$$u = -\frac{\tau}{6b}(3y^2z - z^3) \quad \text{........................(ii)}.$$

Let c be the greatest value of u which, on the side denoted by $y + b = 0$, will be where $z = -b$; then $c = \dfrac{\tau b^2}{3}$, and consequently

$$\frac{u}{c} = -\frac{r^3}{2b^3} \sin 3\phi.$$

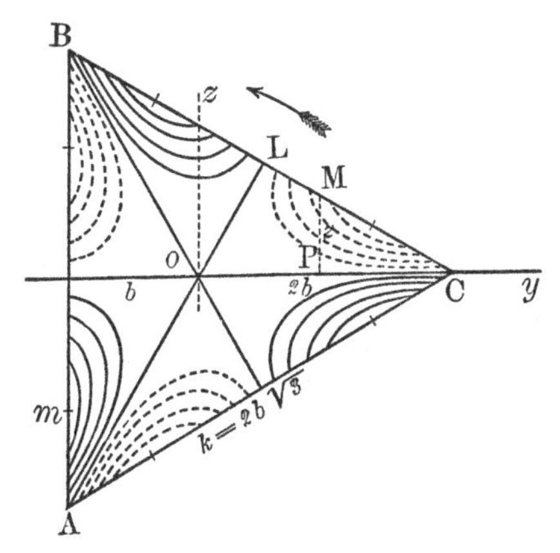

Thus the form of the surface into which the originally plane cross-section becomes changed by torsion is easily understood. In the part between Oy and the perpendicular OL, we have u negative; in the part between OL and OB we have u positive; in the part between OB and yO produced through O we have u negative; in the next piece, which is vertically opposite to the piece between Oy and OL, we have u positive; and so on.

We have as usual the equations

$$\widehat{xy} = \frac{du}{dy} - \tau z, \qquad \widehat{xz} = \frac{du}{dz} + \tau y;$$

these by (ii) of Art. 41 give

$$\widehat{xy} = -\tau\left(z + \frac{yz}{b}\right), \qquad \widehat{xz} = \tau\left(y - \frac{y^2 - z^2}{2b}\right).$$

The moment of torsion by equation (iv) of Art. 17 is

$$M = \mu\tau\left\{\int y^2 d\omega + \int z^2 d\omega + \frac{3}{2b}\int yz^2 d\omega - \frac{1}{2b}\int y^3 d\omega\right\}.$$

All these integrations are easily effected; for here if ξ denote any function of y and z, even in z, we have

$$\int \xi d\omega = 2\int\int \xi dy dz,$$

where we integrate for z from $z = 0$ to $z = \dfrac{2b - y}{\sqrt{3}}$, and for y from $y = -b$ to $y = 2b$. Thus we find that

$$\int z^2 d\omega = \tfrac{3}{2} b^4 \sqrt{3},$$
$$\int y^2 d\omega = \tfrac{3}{2} b^4 \sqrt{3},$$
$$\int yz^2 d\omega = -\tfrac{3}{5} b^5 \sqrt{3},$$
$$\int y^3 d\omega = \tfrac{3}{5} b^5 \sqrt{3}.$$

Then for the moment of inertia round the axis we have

$$\omega\kappa^2 = \int y^2 d\omega + \int z^2 d\omega = 3b^4\sqrt{3} = \frac{\omega^2}{3\sqrt{3}}, \text{ for } \omega = 3b^2\sqrt{3}.$$

Hence
$$M = \tfrac{3}{5}\mu\tau\omega\kappa^2 = \frac{\mu\tau\omega^2}{5\sqrt{3}}.$$

The new theory thus gives a value for M only ·6 of that given by the old.

42. To find the greatest slide, Saint-Venant considers the side which is parallel to the axis of z; then he says that along this side $y + b = 0$, so that $\widehat{xy} = 0$, and $\widehat{xz} = -\dfrac{3b^2 - z^2}{2b}\tau$. Thus the greatest value of $\widehat{xz}$ is when $z = 0$. Hence he tells us that the *fail-point* is on the boundary at the point which is nearest to the axis. The greatest value of the *glissement principal* is then $\dfrac{3b}{2}$; and to ensure safety we must have as before

$$S_0 = \text{or} > \tfrac{3}{2}\mu b\tau.$$

Combining this with $M = \frac{3}{5}\mu\omega\kappa^2\tau$ we have at the limit

$$M_0 = \tfrac{2}{5}\frac{\omega\kappa^2}{b}S_0 = \tfrac{6}{5}S_0 b^3 \sqrt{3} = \frac{2\sqrt[4]{3}}{15}S_0\omega^{\frac{3}{2}}.$$

Thus next to the circular section, the section in the form of an equilateral triangle gives the simplest results.

[The above reasoning involves the assumption that the point of maximum slide *lies on the contour* and is thus unsatisfactory. Saint-Venant has given a thorough investigation of the point in his edition of the *Leçons de Navier*, pp. 287—9.]

[43.] In conclusion we may note that Saint-Venant holds that, among the numerous curves he has considered, one can be found sufficiently close to give practically the laws of torsion for a prism of any given cross-section (pp. 451—2).

[44.] The tenth chapter of the memoir deals with those cases in which the slide-moduli are not the same in the direction of the two transverse axes taken as those of y and z. It occupies pp. 454—70.

Nous y avons aussi été déterminé par le désir de donner sous leur forme la plus simple les seules formules que l'on puisse, jusqu'à présent, appliquer à la pratique; car on n'a pas encore trouvé, par des expériences, le rapport que peuvent avoir entre eux les deux coefficients de glissement transversal μ_1, μ_2 pour diverses matières, et il faut bien les supposer ordinairement égaux (p. 454).

Although well-planned experiments on the possible inequality of μ_1, μ_2 arising either from natural structure or from some process of working are still wanting, yet the inequality in the slide-moduli is not without value as a possible explanation of several minor phenomena of physical elasticity.

[45.] The equations which we have now to solve are those numbered (vi) in our Art. 17. Let us put in those equations $y = \sqrt{\mu_1}\,y'$, $z = \sqrt{\mu_2}\,z'$; they at once reduce to

$$\left.\begin{aligned} d^2u/dy'^2 + d^2u/dz'^2 &= 0 \\ (du/dz' + \tau'y')\,dy' - (du/dy' - \tau'z')\,dz' &= 0 \end{aligned}\right\} \qquad \text{(i)},$$

where $$\tau' = \sqrt{\mu_1\mu_2}\,\tau.$$

In other words our equations remain of exactly the same form provided we write $\tau' = \sqrt{\mu_1\mu_2}\,\tau$ for τ. Hence if we remember that every contour must first be projected by means of the above relation between y, z and y', z', we may make use of all the previous results and equations.

[46.] Thus in the case of the ellipse (pp. 455—8 of memoir), we must write for $\frac{y^2}{b^2} + \frac{z^2}{c^2} = 1$, $\frac{y'^2}{b^2/\mu_1} + \frac{z'^2}{c^2/\mu_2} = 1$. Thus we obtain at once the results:

$$u = -\frac{b^2/\mu_1 - c^2/\mu_2}{b^2/\mu_1 + c^2/\mu_2} \tau' y' z' = -\frac{b^2/\mu_1 - c^2/\mu_2}{b^2/\mu_1 + c^2/\mu_2} \tau y z.$$

Similarly $$\widehat{xy} = -\frac{2\tau z b^2}{b^2/\mu_1 + c^2/\mu_2}, \quad \widehat{xz} = \frac{2\tau y c^2}{b^2/\mu_1 + c^2/\mu_2},$$

and $$M = \frac{\pi b^3 c^3 \tau}{b^2/\mu_1 + c^2/\mu_2} \text{ (see Art. 18).}$$

Saint-Venant remarks that with this inequality, the cross-section of a circular cylinder will be distorted by torsion. The elliptic prism however, for which the ratio of the semi-axes $b/c = \sqrt{\mu_1/\mu_2}$, will retain undistorted cross-sections although under torsion (p. 456). Saint-Venant in the course of the chapter again refers to relations of this kind (p. 462), but it is obvious that such are extremely unlikely to occur in practice.

It must be noted that the 'fail-limit' (*condition de non-rupture*, pp. 456—7) now takes another form, namely that of our Art. 5 (f),

$$1 = \text{or} > \left(\frac{\mu_1 \sigma_{xy}}{S_1}\right)^2 + \left(\frac{\mu_2 \sigma_{xz}}{S_2}\right)^2.$$

From this we find at once

$$(b^2/\mu_1 + c^2/\mu_2)^2 = \text{or} > 4\tau^2 (b^4 z^2/S_1^2 + c^4 y^2/S_2^2).$$

We have then to find the maximum value of the right-hand side. It is easily seen to be on the contour of the cross-section, and at the extremities of the minor or major axis according as b/c is > or < S_1/S_2. In the first case we find that the limiting value of M is given by

$$M_1 = \frac{\pi b c^2}{2} S_1.$$

[47.] Saint-Venant devotes pp. 458—460 to describing the changes which must be made in the general solutions of our Art. 36 in order to adapt them to this case of unequal slide-moduli. They follow easily from our Art. 45. On pp. 460—8 he treats at some length the case of the prism with rectangular cross-section. The results are the same as those of our Art. 27, provided we replace the ratios $\frac{c}{b}$ and $\frac{b}{c}$ where they occur in our formulae by $\frac{c}{b}\sqrt{\frac{\mu_1}{\mu_2}}$ and $\frac{b}{c}\sqrt{\frac{\mu_2}{\mu_1}}$ respectively, and the exponentials

$e^{\pm\frac{(2n-1)\pi}{2}\frac{y}{c}}$ and $e^{\pm\frac{(2n-1)\pi}{2}\frac{z}{b}}$ by $e^{\pm\frac{(2n-1)\pi}{2}\frac{y}{c}\sqrt{\frac{\mu_2}{\mu_1}}}$ and $e^{\pm\frac{(2n-1)\pi}{2}\frac{z}{b}\sqrt{\frac{\mu_1}{\mu_2}}}$ respectively.

The maximum slides still occur at the middle points of the sides, but at the middle of the greater side $2b$ or the lesser side $2c$ according as $b/c >$ or $< \sqrt{\mu_1/\mu_2}$. Saint-Venant gives at the conclusion of the memoir a very useful table, which we reproduce for reference. It serves for equal slide-moduli when we simply put $\mu_1 = \mu_2$. The parameter in the first column is $\frac{b}{c}\sqrt{\frac{\mu_2}{\mu_1}}$ and for it values are taken from 1 to 100 as well as ∞. The second column

TABLE I.

Torsion of Prisms of rectangular Cross-Section.

$\frac{b}{c}\sqrt{\frac{\mu_2}{\mu_1}}$	$(M=\beta\mu_1\tau bc^3)$ β	$\left(\sigma_1=-\gamma_1 c\tau \text{ middle of side } 2b\right)$ γ_1	$\left(\sigma_2=\gamma_2 b\tau \text{ middle of side } 2c\right)$ γ_2	$\left(M_1=\frac{\beta}{\gamma_1}bc^2S_1\right)$ β/γ_1	$\left(M_2=\frac{\beta}{\gamma_2}\frac{c^2}{b^2}\frac{\mu_1}{\mu_2}b^2cS_2\right)$ $\frac{\beta}{\gamma_2}\frac{c^2}{b^2}\frac{\mu_1}{\mu_2}$
1	2·24923	1·35063	1·35063	1·66534	1·66534
1·05	2·35908	1·39651		1·68954	
1·1	2·46374	1·43956		1·71146	
1·15	2·56330	1·47990			
1·2	2·65788	1·51753		1·75363	
1·25	2·74772	1·55268	1·13782	1·76970	1·54556
1·3	2·83306	1·58544		1·7852	
1·35	2·91379	1·61594		1·80316	
1·4	2·99046	1·64430		1·81868	
1·45	3·06319	1·67265			
1·5	3·13217	1·69512	·97075	1·84776	1·43402
1·6	3·25977	1·73889	·91489	1·87463	1·39180
1·7	3·37486	1·77649			
1·75	3·42843	1·79325	·84098	1·91170	1·33107
1·8	3·47890	1·80877		1·92334	
1·9	3·57320	1·83643			
2	3·65891	1·86012	·73945	1·96703	1·15286
2·25	3·84194	1·90546			
2·5	3·98984	1·93614	·59347	2·06072	1·07566
2·75	4·11143	1·95687			
3	4·21307	1·97087		2·13767	
3·333			·44545		
3·5	4·37299	1·98672		2·20111	
4	4·49300	1·99395	·37121	2·25332	·757
4·5	4·58639	1·99724		2·29636	
5	4·66162	1·99874	·29700	2·33200	·628
6	4·77311	1·99974		2·38687	
6·667			·22275		
7	4·85314	1·99995		2·42663	
8	4·91317	1·99999	·18564	2·45660	
9	4·95985	2		2·47993	
10	4·99720	2	·14858	2·49860	
20	5·16527	2	·07341	2·58264	
50	5·26611	2		2·63306	
100	5·29972	2		2·64986	
∞	5·33333	2	0	2·66667	

gives the value of β, where $M=\beta\mu_1\tau bc^3$ is the value of the torsional couple. The third and fourth columns give the maximum slides by means of the coefficients γ_1 and γ_2 where $\sigma_1=-\gamma_1 c\tau$ and $\sigma_2=\gamma_2 b\tau$. The fifth and sixth columns give the maximum value M_0 of M by means of the tabulated values of β/γ_1 and $\frac{\beta}{\gamma_2}\frac{c^2}{b^2}\frac{\mu_1}{\mu_2}$, where $M_1=\left(\frac{\beta}{\gamma_1}\right)bc^2 S_1$ and $M_2=\left(\frac{\beta}{\gamma_2}\frac{c^2}{b^2}\frac{\mu_1}{\mu_2}\right)b^2cS_2$. M_0 is to be taken equal to the lesser of M_1 and M_2.

[48.] Pages 468—9 of this chapter suggest the modifications which must be made in the results obtained for prisms of other cross-sections, when μ_1 differs from μ_2; while on p. 470 we have a simple proof that in this case at corners and angles which project there is no slide, or the intersection of the lateral faces at such corners remains normal to the cross-section.

[49.] Saint-Venant's eleventh chapter deals with the torsion of hollow prisms (pp. 471—6).

In this case we have to satisfy the surface shift-equation

$$\mu_2(du/dz+\tau y)\,dy-\mu_1(du/dy-\tau z)\,dz=0 \quad\text{...........(i)}$$

over two surfaces. If then we form a family of surfaces satisfying this equation and give to the arbitrary constant which appears on the right-hand side two different values we shall obtain the two boundaries of a hollow prism satisfying all the required conditions.

For example:

(a) $$u=-\frac{b^2/\mu_1-c^2/\mu_2}{b^2/\mu_1+c^2/\mu_2}\tau yz$$

satisfies the body shift-equation. Substituting in (i) we have on integration

$$c^2y^2+b^2z^2=\text{constant}.$$

Giving the constant different values we obtain a system of similar and similarly placed ellipses. Thus we find for a hollow elliptic cylinder formed by the ellipses $(2b\times 2c)$ and $(2b'\times 2c')$

$$M=\tau\left\{\frac{\pi b^3c^3}{b^2/\mu_1+c^2/\mu_2}-\frac{\pi b'^3c'^3}{b'^2/\mu_1+c'^2/\mu_2}\right\}=\frac{\tau\pi b^3c^3}{b^2/\mu_1+c^2/\mu_2}\left\{1-\left(\frac{b'}{b}\right)^4\right\}.$$

(b) In the rectangular section

$$u=-\tau yz+\Sigma A_m\sinh(my/\sqrt{\mu_1})\sin(mz/\sqrt{\mu_2}),$$

where $$A_m=\tau\frac{bc}{2}\left(\frac{4}{\pi}\right)^3\frac{c}{b}\sqrt{\frac{\mu_1}{\mu_2}}\frac{(-1)^{n-1}}{(2n-1)^3}\operatorname{sech}(mb/\sqrt{\mu_1})$$

and $$m=\frac{(2n-1)\pi}{2}\frac{\sqrt{\mu_2}}{c}.$$

Substituting in (i) and integrating we find:

$$\frac{32}{\pi^3}\Sigma\frac{(-1)^{n-1}}{(2n-1)^3}\frac{\cosh(my/\sqrt{\mu_1})}{\cosh(mb/\sqrt{\mu_1})}\cos(mz/\sqrt{\mu_2}) = 1-\frac{z^2}{c^2}+C.$$

By variation of C we get possible boundary lines for hollow sections, but since only $C=0$ gives a rectangle, the boundaries will not be similar rectangles. Most of these curves would be extremely difficult to trace; for small values of C, however, we may practically assume we have a hollow cylinder whose cross-section is bounded by two nearly equal rectangles. Saint-Venant finds in curves thus obtained an analogy to the *surfaces isothermes* of Lamé.

(*c*) Lastly we find briefly described the method of dealing with solutions of the form (5) of our Art. 36. The curves are sketched on p. 476 for the double family given by the equation of our Art. 38. Any two of either set might serve as the basis of a hollow prism. Saint-Venant returns in the *Leçons de Navier* (pp. 306, 325–332) to this family and treats a special case of it—*Section en double spatule, analogue à celle d'un rail de chemin de fer*,—at considerable length.

[50.] I now reach Saint-Venant's twelfth chapter which is thus entitled: *Cas où il y a en même temps une torsion, une flexion, des dilatations et des glissements latéraux. Conditions de non-rupture sous leurs influences simultanées* (pp. 476—522). It deals with the all important practical question of combined strain, and may be described as the first scientific treatment of the subject: see our Arts. 1377* and 1571*. The chapter may be looked upon as an extension of the safe-stretch conditions formulated for the first time in the *Cours lithographié*, see our Art. 1567*. In the treatment of the problem to be found there it will be remembered that the slide was dealt with as constant over the cross-section; here the new results with regard to the flexural and torsional distortion of the cross-sections are applied to that extended form of the earlier formula which was cited in our Art. 5 (*d*).

[51.] Before I enter upon an analysis of Saint-Venant's results I may refer to the substance of a footnote given on pp. 477—8 of the memoir. Saint-Venant notes that under torsion the sides and fibres of a prism originally parallel become inclined and helical and so must suffer a stretch. This stretch is, however—if the product of the torsion τ and the distance of the farthest fibre from the axis be small—a small quantity of the *second order*. Wertheim in a memoir to be considered later (see our Chap. XI.) has referred to certain phenomena which he attributes to this stretch.

By a simple analysis Saint-Venant finds its absolute magnitude for a *right-circular* cylinder of radius a. Take a fibre at distance r from the axis and let us consider the element PP' of it between two planes at unit distance. Suppose owing to torsion that the two planes approach each other by a quantity η and let $P'N$ be the perpendicular from the new position of P' on the cross-section through P,

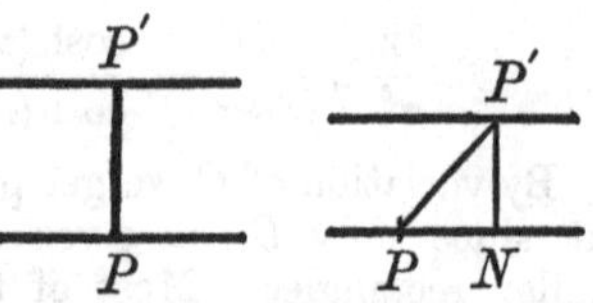

$$PP' = \sqrt{P'N^2 + PN^2} = \sqrt{(1-\eta)^2 + \tau^2 r^2}$$
$$= 1 - \eta + \frac{\tau^2 r^2}{2} \text{ nearly.}$$

Saint-Venant takes for PP' the quantity

$$(1-\eta)\sqrt{1+\tau^2 r^2}$$
$$= 1 - \eta + \frac{\tau^2 r^2}{2},$$

but I do not think he obtains the first expression very rigorously. He has practically the same value in the *Leçons de Navier* (pp. 240—1).

The traction in the fibre will now be given by

$$Ed\omega\left(\frac{\tau^2 r^2}{2} - \eta\right),$$

where E is the longitudinal stretch-modulus. The quantity $-\eta$ must be determined by the condition that the total traction is zero, or

$$\int_0^a 2\pi r dr E\left(\frac{\tau^2 r^2}{2} - \eta\right) \sin P'PN = 0.$$

Since $$\sin P'PN = \frac{1-\eta}{1-\eta+\dfrac{\tau^2 r^2}{2}} = 1 - \frac{\tau^2 r^2}{2}$$

it may be put $= 1$ in the integral.

We find $\dfrac{\tau^2 a^4}{4} = a^2 \eta$, giving $\eta = \dfrac{\tau^2 a^2}{4}$, a result which agrees with Saint-Venant's; our analysis thus proves that η is of the *second order* in τ.

Further we have for the total-moment of these tractions about the axis

$$M = \int_0^a 2\pi r dr E\left(\frac{\tau^2 r^2}{2} - \frac{\tau^2 a^2}{4}\right) \cos P'PN \times r$$
$$= E\pi\tau^3 \int_0^a r^3 dr\left(r^2 - \frac{a^2}{2}\right), \text{ since } \cos P'PN = \tau r;$$
$$= \frac{E\pi a^3}{24}(\tau a)^3.$$

If one takes account of the tractions produced by the lateral squeezes of the fibres, we shall have a similar expression with a change only in the elastic constant. Thus it appears that the effect produced by the stretch of the fibres is of the third order in the torsion and may be legitimately neglected if the torsion be small.

This point—that the stretch only varies as the cube of the torsion—was first stated by Young without proof in his *Lectures on Natural Philosophy*, Vol. I. p. 139. He thence argued that torsional resistance must be due to *detrusion* (slide) and not to stretch. When the torsion τ is considerable, then the quantity M above, due to stretch of the fibres, becomes of importance, as appears from Wertheim's experiments in the memoir referred to: see our Chap. XI.

[52.] Returning to the chief topic of the chapter under consideration we first note with Saint-Venant the linearity of the equations of elasticity, so that it is possible to combine various strains due to different forms of loading by vector-addition and so obtain the total shifts due to a combined load system: see our Art. 1568*. On pp. 479—80 Saint-Venant deduces the shifts for an elliptic prism subject at the same time to traction, flexure and torsion. Use is made of the results obtained on pp. 304 and 455 of the memoir: see our Arts. 12 and 46.

[53.] Saint-Venant now turns to equation (iii) of our Art. 5 (*d*) and after pointing out the difficulties of the general solution by analysis for the case of any prism (p. 482) proceeds to some more special and simple cases when the cubic can be reduced to an equation of the second degree.

Case (1). Let the elasticity be symmetrical about the axis of x, and let the solid be a prism subjected only to a uniform lateral traction, we have

$$s_y = s_z, \quad \bar{s}_y = \bar{s}_z, \quad \bar{\sigma}_{xy} = \bar{\sigma}_{xz} \text{ and } \sigma_{yz} = 0.$$

Hence, if $\sigma_x = \sqrt{\sigma_{xz}^2 + \sigma_{xy}^2}$, we find

$$\left(\frac{s}{\bar{s}} - \frac{s_x}{\bar{s}_x}\right)\left(\frac{s}{\bar{s}} - \frac{s_y}{\bar{s}_y}\right) = \frac{\sigma_x^2}{\bar{\sigma}_x^2},$$

or
$$\frac{s}{\bar{s}} = \frac{1}{2}\left(\frac{s_x}{\bar{s}_x} + \frac{s_y}{\bar{s}_y}\right) \pm \sqrt{\frac{1}{4}\left(\frac{s_x}{\bar{s}_x} - \frac{s_y}{\bar{s}_y}\right)^2 + \left(\frac{\sigma_x}{\bar{\sigma}_x}\right)^2}.$$

In this equation we may put $\bar{s}_x = T_1/E_1$, $\bar{s}_y = T_2/E_2$, $\bar{\sigma}_x = S/\mu$, $\bar{s}_x/\bar{s}_y = \frac{E_2 T_1}{E_1 T_2} = \eta_1/\eta$ and $s_y = -\eta s_x$, where η = ratio of lateral squeeze to longitudinal stretch. Thus we find as safe-stretch limit

$$\text{(i)} \ldots 1 = \text{maximum of } \frac{1-\eta_1}{2}\frac{E_1}{T_1}s_x + \sqrt{\left(\frac{1+\eta_1}{2}\frac{E_1}{T_1}s_x\right)^2 + \left(\frac{\mu\sigma_x}{S}\right)^2}$$

We take the positive sign of the radical, because if $\sigma_x = 0$ we should have the alternative between $\frac{E_1}{T_1}s_x$ and $-\eta_1\frac{E_1}{T_1}s_x = \frac{E_2}{T_2}s_y$, and the former will be considered the greater (Saint-Venant, p. 484).

Case (2). A like equation is obtained, if, without supposing an axis of elasticity, two out of the three slide components vanish at a *fail-point.*

Case (3). This is a case of approximation, Saint-Venant supposes σ_{yz} to be zero; but $-s_y/\bar{s}_y$ and $-s_z/\bar{s}_z$ without being equal to differ but slightly, and he then takes them equal to $\eta_2\frac{s_x}{\bar{s}_x}$ '*une certaine moyenne entre ces deux rapports.*' Thus he replaces $(s/\bar{s} - s_y/\bar{s}_y)\ (s/\bar{s} - s_z/\bar{s}_z)$ by $(s/\bar{s} + \eta_2 s_x/\bar{s}_x)^2$ and divides out all the terms by the same factor. We thus reach the equation

$$\left(\frac{s}{\bar{s}} - \frac{s_x}{\bar{s}_x}\right)\left(\frac{s}{\bar{s}} + \eta_2\frac{s_x}{\bar{s}_x}\right) - \frac{\sigma_{xy}^2}{\bar{\sigma}_{xy}^2} - \frac{\sigma_{xz}^2}{\bar{\sigma}_{xz}^2} = 0,$$

and obtain for the safe-stretch condition

(ii) ... 1 = maximum of

$$\frac{1-\eta_2}{2}\frac{E_1}{T_1}s_x + \sqrt{\left(\frac{1+\eta_2}{2}\frac{E_1}{T_1}s_x\right)^2 + \left(\frac{\mu_1\sigma_{xy}}{S_1}\right)^2 + \left(\frac{\mu_2\sigma_{xz}}{S_2}\right)^2}.$$

Here η_2 is given, I think, most satisfactorily by the arithmetic mean

$$\tfrac{1}{2}\left(\frac{s_y}{\bar{s}_y} + \frac{s_z}{\bar{s}_z}\right) = -\eta_2\frac{s_x}{\bar{s}_x}.$$

Now if $s_y = -\eta s_x$, and $s_z = -\eta' s_x$,

$$\eta_2 = \tfrac{1}{2}\left(\eta\frac{\bar{s}_x}{\bar{s}_y} + \eta'\frac{\bar{s}_x}{\bar{s}_z}\right)$$

$$= 2\left(\eta\frac{1}{\bar{\sigma}_{xy}^2} + \eta'\frac{1}{\bar{\sigma}_{xz}^2}\right)\bar{s}_x^2 : \text{ see our Art. 5 } (d),$$

$$= 2\left\{\eta\left(\frac{\mu_1}{S_1}\right)^2 + \eta'\left(\frac{\mu_2}{S_2}\right)^2\right\}\frac{T_1^2}{E_1^2}.$$

This result gives a constant value for η_2 and appears to agree with Saint-Venant's note on *Clebsch*, p. 275. I do not think the value given for $\epsilon''_1 (= \text{our } \eta_2)$ on p. 485 of the memoir is quite satisfactory.

It will be noted that in all three cases the resulting quadratic is practically of the same form and the condition may for all three be thrown into a somewhat different shape, namely, transposing and squaring we find

$$\text{(iii)} \ldots \left(1 - \frac{E_1 s_x}{T_1}\right)\left(1 + \eta_2\frac{E_1}{T_1}s_x\right) - \left(\frac{\mu_1\sigma_{xy}}{S_1}\right)^2 - \left(\frac{\mu_2\sigma_{xz}}{S_2}\right)^2 = \text{or} > 0.$$

On p. 486 Saint-Venant gives the value of the moduli in terms of the 21 coefficients, and points out the changes which arise when we assume bi- or uni-constant isotropy. On pp. 487—8 we have a direct deduction of the formula of Case (2) on the lines of the *Cours lithographié*: see our Art. 1571*.

[54.] On pp. 488—491 Saint-Venant points out the method by which a general solution for a prism can be worked out. Let the axes of z and y be the principal axes of inertia of the cross-section and P_x, P_y, P_z the load-components parallel to the axes at one terminal and M_x, M_y, M_z the moments round the corresponding axes. Let σ'_{xy}, σ'_{xz} be the slides at any point on the section ω due to the flexure or to M_y, M_z; let σ''_{xy}, σ''_{xz} be the slide-components due to the torsional couple M_x, then

$$s_x = \frac{P_x}{E_1\omega} + \frac{M_y z}{E_1\omega\kappa_y^2} + \frac{M_z y}{E_1\omega\kappa_z^2}.$$

Further the σ' and σ'' components of slide will be known as soon as the section is known and their *sums* must pair and pair be substituted in equation (ii) or (iii) of Art. 53 for σ_{xy} and σ_{xz}.

The equations of equilibrium,

$$\text{(iv)} \quad \left.\begin{aligned} P_y &= \mu_1 \int_0^\omega \sigma'_{xy}\, d\omega \\ P_z &= \mu_2 \int_0^\omega \sigma'_{xz}\, d\omega \end{aligned}\right\}, \qquad M_x = \mu_2 \int_0^\omega \sigma''_{xz}\, y\, d\omega - \mu_1 \int_0^\omega \sigma''_{xy}\, z\, d\omega,$$

will determine the constants in terms of the applied forces.

[55.] In section 125 (pp. 492—4) Saint-Venant treats the exceptional case of a cross-section constrained to remain plane.

Telles sont celles qui sont soumises à ce que M. Vicat appelle un *encastrement complet*, c'est-à-dire qui ne sont pas seulement contenues, mais *scellées* ou *soudées* avec une matière plus rigide; ou bien celles qui se trouvent serrées et sollicitées latéralement *dans leur plan même* par des forces tendant à trancher, comme il arrive aux sections des rivets dans le plan de contact des tôles qu'ils assemblent, ou aux bases des prismes tordus *de longueur nulle* comme dit le même illustre ingénieur (p. 492).

Other such sections occur from the symmetry of load distribution etc.

For such non-distorted sections, we can suppose the 'fibres' formerly perpendicular to become equally inclined, or the slide due to flexure constant, and that due to torsion to follow the old law of Coulomb, i.e.

$$\text{(v)} \quad \begin{cases} \mu_1 \sigma'_{xy} = P_y/\omega, & \mu_2 \sigma'_{xz} = P_z/\omega, \\ \sigma''_{xy} = -\tau z, & \sigma''_{xz} = \tau y, \end{cases}$$

whence by means of equation (iv) of our Art. 54, we can easily express the slides in terms of M_x, P_y and P_z.

The expressions (v) of course are only true for these exceptional sections, which can never occur in pure torsion as sections of danger, while in practical cases of flexure combined with torsion or slide they are frequently found to be specially strengthened (e.g. built-in ends).

[56.] We will now enumerate the examples Saint-Venant gives of the above condition of safety.

Case (1). Consider a rectangular prism (cross-section $2b \times 2c$) subjected only to a force P parallel to the axis of z (or side $2c$). Let the built-in terminal of the prism be so fixed that it can be distorted by flexure. Then if the length of the prism be a, and $2c$ be much greater than $2b$, we have

$$\mu_2 \sigma_{xz} = \frac{3}{2}\frac{P}{\omega}\left(1 - \frac{z^2}{c^2}\right), \qquad P_x = P_y = 0,$$

$$\sigma_{xy} = 0, \qquad Es_x = Paz \Big/ \frac{\omega c^2}{3};$$

so that, *granting uniconstant isotropy*, $S = \frac{4}{5} T$, $\eta = \frac{1}{4}$, and thus the equation (i) of our Art. 53 becomes

$$1 = \text{maximum of } \frac{3Pa}{T\omega c}\left[\frac{3}{8}\frac{z}{c} + \frac{5}{8}\sqrt{\frac{z^2}{c^2} + \left(\frac{c}{a}\right)^2\left(1 - \frac{z^2}{c^2}\right)^2}\,\right].$$

Saint-Venant gives a table of the values of the quantity between square brackets for values of $z/c = 0$ to 1, and for values of $2c/a$ (depth to length) from 3 to 6. From this table the following results may be drawn. So long as $2c/a < 3{\cdot}05$ the *fail-point* lies on the surface of the prism where $z/c = 1$, or at that point *where there is no slide.* If then the ratio of depth to length be $< 3{\cdot}05$, the prism's resistance is just that of flexure without consideration of slide. If on the other hand $2c/a > 3{\cdot}05$ the maximum passes abruptly to the points for which $z/c = {\cdot}2$ about, and approaches more and more to those for which $z = 0$. But this latter point lies on the *neutral-axis*, or it must be *slide* and not *flexure* which produces the failure. When $2c/a = 3{\cdot}2$ we may calculate the resistance either from flexure or transverse slide, but after $2c/a = 4$, it is the slide alone which is of importance. Similar conclusions Saint-Venant tells us may be obtained for a circular section (radius r); in this case the fail-point passes abruptly when $2r/a = 4{\cdot}3$ from $z = r$ to $z = {\cdot}2r$ about.

The reader who bears in mind Vicat's attack upon the mathematical theory of elasticity (see our Arts. 732*—733*) will find that the above remarks satisfactorily explain Vicat's experimental results.

Case (2). This is that of a prism (length $2a$, section $2b \times 2c$) terminally supported and centrally loaded. Here the section of greatest strain suffers no distortion. If the load P be in the direction of the axis of z, we have by equation (v) of our Art. 55, $\sigma_{xy}=0$ and $\sigma_{xz}=P/(\omega\mu)$. Whence supposing uni-constant isotropy we find:

$$1=\frac{3}{8}\frac{3Pa}{4Tbc^2}+\sqrt{\left(\frac{5}{8}\cdot\frac{3Pa}{4Tbc^2}\right)^2+\left(\frac{P}{4Sbc}\right)^2}.$$

Suppose b' and c' to be the values to be given to b and c that the prism might safely withstand a couple Pa producing flexure only, and b'', c'' to be the values to be given to b and c that it might safely withstand a shearing force P applied to the undistorted section. Then we easily find

$$1=\frac{3Pa}{4Tb'c'^2},\text{ and }1=\frac{P}{4Sb''c''}.$$

Hence:

$$1=\frac{3}{8}\frac{b'}{b}\left(\frac{c'}{c}\right)^2+\sqrt{\left(\frac{5}{8}\frac{b'c'^2}{bc^2}\right)^2+\left(\frac{b''c''}{bc}\right)^2}$$

gives the limiting safe values of b and c for the strain in question. Saint-Venant puts first $c'=c''=c$ and so gets

$$b=\tfrac{3}{8}b'+\sqrt{(\tfrac{5}{8}b')^2+b''^2},$$

whence he deduces and tabulates the values of b/b' and b/b'' for various values of b''/b' and b'/b'' respectively, and also the value of

$$\frac{2c}{2a}\left(=\frac{3S}{T}\frac{b''}{b'},\text{ or}=\frac{12}{5}\frac{b''}{b'}\text{ for isotropy}\right).$$

From his table it appears that when

$\dfrac{2c}{2a}=\dfrac{\text{depth}}{\text{length}}=\text{or}>\tfrac{1}{2}$ the slide begins to influence sensibly the result,

$\dfrac{2c}{2a}=\text{or}<10$ the flexure begins to influence sensibly the result.

Between $2c/2a=\frac{1}{2}$ and 10 we are compelled to take both into account.

Case (3). This is the treatment of a cylinder on a circular base subjected at the same time to flexure, torsion and extension. Saint-Venant neglects the flexural slides and ultimately the extension. He obtains an equation similar in character to that of the preceding case and tabulates the values of the radius of safety in terms of the radius of safety in the case of flexure alone for different values of the elastic constant η_1. He remarks (p. 503) that it is not necessary to consider values of $\eta_1>\frac{1}{2}$ for then a stretch would not produce a positive dilatation, '*ce qui n'est point supposable.*' This remark is omitted in the *Leçons de Navier* where a number of values of $\eta_1>\frac{1}{2}$ are dealt with. I may add that the problem is far more completely treated in that work (pp. 414—21). Saint-Venant's tables shew that the results obtained are for values of η_1 between 1/5 and 1/3 very much the same, or we may adopt generally without fear of error the uni-constant hypothesis $\eta_1=1/4$. This hypothesis Saint-Venant tells us is amply verified by the experiments of M. Gouin (see page 486 of the memoir).

I shall have something to say of these experiments when dealing with Morin's *Résistance des matériaux*, 1853: see our Chap. XI.

Case (4). This case gives the calculation of the 'solid of equal resistance' for a bar built-in at one end and acted upon at the other by a non-central load perpendicular to its axis, i.e. combined flexure and torsion. Saint-Venant supposes uni-constant isotropy and neglects the flexural slides. His final equation is

$$2\frac{T\pi r^3}{P} = 3x + 5\sqrt{x^2 + k^2}.$$

Here P is the load acting on an arm k, and r is the sectional radius at distance x from the loaded terminal. (p. 504.)

Case (5). An axle terminally supported has weight Π and carries two heavy wheels (ϖ and ϖ') upon which act forces, whose moments about the axle are equal and whose directions are perpendicular to the axle. We have thus another case of combined flexure and torsion, which is dealt with as before.

[57.] The next case treated by Saint-Venant is of greater complexity; it occupies pp. 507—18 of the memoir. It is the investigation of combined flexural and torsional strain in rectangular prisms ($2b \times 2c$), and possesses considerable theoretical interest. In practice also the non-central loading of beams of rectangular section must be a not infrequent occurrence.

Case (6). Saint-Venant in his treatment does not suppose the elasticity round the prismatic axis to be isotropic, but takes the general case of two slide-moduli, supposing, however, that $b\sqrt{\mu_2} > c\sqrt{\mu_1}$.

He neglects also the flexural slide-components. Let the torsional slide-components be given by $\sigma_1 = -\gamma_y c\tau$ and $\sigma_2 = \gamma_z b\tau$ for $z/c = 1$ and $y/b = 1$ respectively. τ must be eliminated by means of the relation $M'' = \beta\mu_1\tau bc^3$. If ϕ be the angle the plane of the flexural load makes with the plane through the prismatic axis and the axis of y, and M' the flexural moment at section x, we easily obtain for the stretch s_x the value

$$s_x = \frac{3M'}{4Ebc}\left(\frac{z\cos\phi}{c^2} + \frac{y\sin\phi}{b^2}\right)$$

$$= (\text{for } z = c)\ \frac{3M'}{4bc^2E}\left(\cos\phi + \frac{c}{b}\frac{y}{b}\sin\phi\right)$$

$$= (\text{for } y = b)\ \frac{3M'}{4bc^2E}\left(\frac{z}{c}\cos\phi + \frac{c}{b}\sin\phi\right).$$

Let us substitute these values in equation (ii) of our Art. 53. Taking these expressions alternately for the sides $2b$ and $2c$ we obtain:

$$1 = \text{maximum}\ \frac{1-\eta_2}{2T}\frac{3M'}{4bc^2}\left(\cos\phi + \frac{c}{b}\frac{y}{b}\sin\phi\right)$$
$$+\sqrt{\left[\frac{1+\eta_2}{2T}\frac{3M'}{4bc^2}\left(\cos\phi + \frac{c}{b}\frac{y}{b}\sin\phi\right)\right]^2 + \left(\frac{\gamma_y}{S_1}\frac{M''}{\beta bc^2}\right)^2},$$

$$1 = \text{maximum } \frac{1-\eta_2}{2T}\frac{3M'}{4bc^2}\left(\frac{z}{c}\cos\phi + \frac{c}{b}\sin\phi\right)$$
$$+\sqrt{\left[\frac{1+\eta_2}{2T}\frac{3M'}{4bc^2}\left(\frac{z}{c}\cos\phi + \frac{c}{b}\sin\phi\right)\right]^2 + \left(\frac{\gamma_z}{S_2}\frac{b\mu_2}{c\mu_1}\frac{M''}{\beta bc^2}\right)^2}.$$

By means of the Table II. below and Table I. on our p. 39 all the terms of these expressions can be calculated; for γ_y/γ_1 and γ_z/γ_2 are given for values of $\frac{b\sqrt{\mu_2}}{c\sqrt{\mu_1}}$ and also for values of y/b and z/c respectively. Hence so soon as ϕ and the section of danger, i.e. where M' is greatest, are known we can solve the problem by equating to unity the greater of the two maxima written down above and so determine bc^2 for the section.

Saint-Venant by using b', c', b'', c'' with similar meanings to those of our Art. 56, *Case* (2), throws the equation into a somewhat different form.

If the section for which M' is greatest be so built-in or symmetrically situated that no distortion is possible the values of the slides must be those of equations (v) of our Art. 55 and not σ_1, σ_2 as taken above.

TABLE II.

Slides at points of the contour of the Cross-Section of a Prism on rectangular base subjected to Torsion.

$\sigma_1 = -\gamma_y c\tau$ (for $z=c$, or along the sides $2b$)					$\sigma_2 = \gamma_z b\tau$ (for $y=b$, or along the sides $2c$)			
	Value of ratio γ_y/γ_1					Value of ratio γ_z/γ_2		
For y/b	$\frac{b\sqrt{\mu_2}}{c\sqrt{\mu_1}}=1$	$=1{\cdot}5$	$=2$	$=4$	For z/c	$\frac{b\sqrt{\mu_2}}{c\sqrt{\mu_1}}=1$	$=2$	$=4$
0	1·0000	1·0000	1·0000	1·0000	0	1·0000	1·0000	1·0000
·1	·9932	·9949	·9962	·9991	·1	·9932	·9932	·9933
·2	·9750	·9795	·9846	·9973	·2	·9750	·9729	·9729
·3	·9429	·9526	·9639	·9928	·3	·9429	·9384	·9383
·4	·8963	·9127	·9321	·9842	·4	·8963	·8887	·8885
·5	·8333	·8572	·8857	·9678	·5	·8333	·8224	·8220
·6	·7510	·7820	·8196	·9371	·6	·7510	·7369	·7363
·7	·6447	·6811	·7260	·8793	·7	·6447	·6282	·6278
·8	·5063	·5441	·5916	·7695	·8	·5063	·4892	·4885
·9	·3185	·3497	·3896	·5540	·9	·3185	·3044	·3040
1	·0000	·0000	·0000	·0000	1·0	·0000	·0000	·0000

This Table gives γ_y, γ_z in terms of the principal slides γ_1, γ_2 at the centre of the corresponding sides $2b$ and $2c$; the values of γ_1, γ_2 are given in Table I. p. 39.

[58.] Saint-Venant treats with numerical tables the following special cases:

(1) $\phi = 0$ and $c < b$ (pp. 511—2).

(2) c so much less than b that $c/b . \tan\phi$ may be neglected as compared with 1, i.e. the case of a 'plate' (pp. 511—2).

(3) Prism on square base, when $\tan\phi = 0$, $= \frac{1}{2}$, $= 1$, and = anything whatever when there is a non-distorted section for section of least safety (pp. 512—4). The fail-points are also determined.

(4) Prism on rectangular base for which $b = 2c$, when $\tan\phi = 0$, $= \frac{1}{2}$, $= 1$, $= 2$, $= \infty$, and = anything whatever when there is a non-distorted section for that of least safety (pp. 514—518). The fail-points are also determined.

[59.] On pp. 518—22 we have the treatment of a prism on elliptic base subjected at the same time to flexure and torsion. Saint-Venant only works this out numerically for the case of uni-constant isotropy and when $\tan\phi = \infty$.

It is found that after a certain value of the ratio of torsional to flexural couple, the fail-point leaves the end of the major axis (through which the flexural load-plane passes[1]) and traverses the quadrant of the ellipse till it reaches the end of the minor axis (p. 522).

[60.] We now turn to Saint-Venant's final chapter (pp. 522—558). This consists of three parts: § 135 *Résumé général*; § 136 *Récapitulation des formules et règles pratiques* and § 137 *Exemples d'applications numériques.*

In the first article there is little to be noted. A reference is made on p. 528 to the models of M. Bardin shewing the *gauchissement* of the cross-section to which we have previously referred. Saint-Venant also mentions the visible distortion of the cross-sections obtained by marking them on a prism of caoutchouc and then subjecting it to torsion.

In the general recapitulation of formulae we have some results not in the body of the memoir, as on p. 536 (d_1) where the flexural slides for the prism whose base is the curve $\left(\frac{y}{b}\right)^4 + \left(\frac{z}{c}\right)^2 = 1$ are cited from the memoir on flexure: see our Art. 90. So again on p. 546 for the flexural slides of other cross-sections. The best *résumé*, however, of formulae as well as numbers for both flexure and torsion is undoubtedly to be found in Saint-Venant's *Leçons de Navier* to which we shall refer later. The last section § 137 contains some instructive numerical examples of Saint-Venant's treatment of combined strain.

[1] Saint-Venant terms this *sollicité de champ*. When the load-plane is perpendicular to this the prism is *sollicité à plat.*

The memoir concludes with the tables for rectangular prisms which we have in part reproduced on pp. 39 and 49.

[61.] We here bring to a close our review of this great memoir. Since Poisson's fundamental essay of 1828 (see our Art. 434*) no other single memoir has really been so epoch-making in the science of elasticity. It is indeed not a memoir, but a classical treatise on those branches of elasticity which are of first-class technical importance. Written by an engineer who has kept ever before him practical needs, it is none the less replete with investigations and methods of the greatest theoretical interest. Many of its suggestions we shall find have been worked out in fuller detail by Saint-Venant himself, not a few remain to this day unexhausted mines demanding further research.

SECTION II.

Memoirs of 1854 *to* 1864.

Flexure, Distribution of Elasticity, etc.

[62.] *Comptes rendus*, T. XXXIX. pp. 1027—1031, 1854. *Mémoire sur la flexion des prismes élastiques, sur les glissements qui l'accompagnent lorsqu'elle ne s'opère pas uniformément ou en arc de cercle, et sur la forme courbe affectée alors par leurs sections transversales primitivement planes.* This is a *résumé* of the results of the later memoir on flexure (see our Arts. 69 and 93). It cites the general equations for flexure, and the particular results for the case of a rectangular cross-section.

[63.] *L'Institut*, Vol. 22, 1854, pp. 61—63. *Solution du problème du choc transversal et de la résistance vive des barres élastiques appuyées aux extrémités.* This is an account of Saint-Venant's memoir presented to the *Société Philomathique.* It contains only matter given in the *Comptes rendus*, and afterwards more completely in the annotated *Clebsch*: see our Art. 104.

[64.] In the same volume of the same Journal, pp. 220—1, are particulars of the memoir on the Flexure of Prisms communicated to the *Société Philomathique.*

[65.] In the same volume of this Journal, pp. 396—398, is another communication of Saint-Venant's to the *Société Philomathique* (July 8, 1854). This deals with the formulae for the flexure of prisms and for their strength, when the cross-section does not possess inertial isotropy. It gives the general equations and treats specially the case of a rectangular cross-section: see the *Leçons de Navier*, pp. 52—58 and our Arts. 1581*, 14 and 171.

A final paragraph to the paper points out that the resistance to torsion varies more nearly inversely than directly as the axial moment of inertia: see our Art. 290.

[66.] On pp. 428—31 of the same volume of the same Journal Saint-Venant communicates to the *Société Philomathique* (July 8 and October 21, 1854) the results obtained from the stretch-condition of strength. These results were afterwards published in the memoir on Torsion: see our Arts. 53 *et seq.*

[67.] Volume 23 of the same Journal, pp. 248—50. Further results of the memoir on *Torsion* communicated to the *Société Philomathique* (April 12 and May 12, 1855), notably the case of a prism on an equal-sided triangular base: see our Arts 40—2.

[68.] The same volume of the same Journal, pp. 440—442. *Diverses considérations sur l'élasticité des corps, sur les actions entre leurs molécules, sur leurs mouvements vibratoires atomiques, et sur leur dilatation par la chaleur.* An account of a memoir presented October 20, 1855, to the *Société Philomathique* containing general remarks on the rari-constant theory of intermolecular action. The expression for the velocity of sound on p. 441 *b* should be $\sqrt{\frac{3G-p}{\rho}}$ and not $\sqrt{\frac{3G+p}{\rho}}$: see *L'Institut*, Vol. 24, p. 215. Saint-Venant refers to the labours of Newton, Ampère and others on this subject: see our Art. 102. He points out that in order to explain heat by translational vibrations, the second differential of the function which expresses the law of intermolecular force must be *positive*: see our Arts. 268 and 273.

The method, however, of dealing with the velocity of sound by means of an initial stress in an isotropic medium is unsatisfactory. This was recognised by Saint-Venant himself, and he cancelled the entire paragraphs on p. 441, beginning *Newton va même* and *Quelque différents*, of 42 and 10 lines respectively: see *Comptes rendus*, 1876, Vol. 82, p. 34.

[69.] *Mémoire sur la flexion des prismes, sur les glissements transversaux et longitudinaux qui l'accompagnent lorsqu'elle ne s'opère pas uniformément ou en arc de cercle, et sur la forme courbe affectée alors par leurs sections transversales primitivement planes. Journal de Mathématiques de Liouville*, Deuxième Série, T. I. 1856, pp. 89—189.

This is Saint-Venant's classical memoir on flexure; extracts from it will be found in the *Comptes rendus*, T. XXXIX. 1854, p. 1027 and T. XLI. 1855, p. 143.

Certain portions are reproduced in the *Leçons de Navier*, pp. 389—414, but the analytical work does not seem yet to have passed into the text-books.

[70.] Sections 1, 2 (pp. 89—98) are occupied with a history of the old theories and an account of the Bernoulli-Eulerian hypothesis as generally accepted at the date of the memoir. Saint-Venant refers to the labours of Galilei (see our Art. 3*), Mariotte (Art. 10*), Hooke (Art. 7*), James Bernoulli (Art. 18*), Coulomb (Art. 117*), Leibniz (Art. 11*), Duleau (Art. 227*), Barlow (Art. 189*), Hodgkinson (Art. 232*), Tredgold (Art. 197*), Girard (Art. 127*), Navier (Art. 254*), Young (Art. 134*), Robison (Art. 146*), Dupin (Art. 162*) for the theory of beams, and to those of Cauchy, Poisson, Lamé and Clapeyron for the general theory of elasticity. His remarks are reproduced at greater length in the *Historique Abrégé*, and as the reader of our first volume is already acquainted with the researches of these scientists we pass over these pages of the memoir.

In the second section Saint-Venant points out the falseness of the Bernoulli-Eulerian theory, and refers to the corrections and criticisms of Vicat, Persy and himself: see our Arts. 721*, 726*, 811* and 1571*.

As we have already pointed out Saint-Venant in the memoir

on *Torsion* had given the outlines of the true theory of flexure: see our Arts. 9—13.

[71.] The third section (pp. 98—101) is entitled: *Objet et sommaire de ce memoire.* Saint-Venant here indicates that he intends to use the semi-inverse method (see our Art. 3) to test how far the Bernoulli-Eulerian formulae:

$$\begin{cases} \text{Traction} = Ez/\rho, \\ \text{Bending moment} = E\omega\kappa^2/\rho, \\ \int z d\omega = 0, \end{cases}$$

(see our Arts. 20*, 65*, 75*, etc.)

are correct, when consideration is paid to the influence of slide. There is also a succinct account of the contents of Sections 4—32 of the memoir.

[72.] Sections 4—12 (pp. 101—120) contain an elementary sketch of the general theory of elasticity. Saint-Venant wrote three other such sketches, namely (i) in the memoir on *Torsion* (see our Art. 4); (ii) in the *Leçons de Navier* (see our Art. 190); and (iii) for Moigno's *Statique* (see our Arts. 224—9). This sketch falls between (i) and (ii). It adopts rari-constancy and bases it upon intermolecular action being central and a function of central distance only. This rari-constancy Saint-Venant holds to be without doubt true for bodies of 'confused crystallisation' such as are used for the materials of construction (p. 108). At the same time for the sake of the 'weaker brethren,' and as it does not increase the difficulty of solving the elastic equations, he adopts multi-constant formulae.

[73.] As a specimen of the mode of treatment, we reproduce his proof of the equality of the cross-stretch and direct-slide coefficients, i.e. in our notation $|xxyy| = |xyxy|$[1].

We have to shew that the coefficient of s_y in $\widehat{xx}$ = the coefficient of σ_{xy} in $\widehat{xy}$.

Suppose all the strain-components zero except s_y and σ_{xy} and these to be constant for all points of the body. Suppose the central distance of two molecules m', m'' to have length r, and projections x, y, z on the coordinate axes before strain. After strain x and z remain unchanged, but y will be increased by ys_y owing to the stretch and $x\sigma_{xy}$ owing to

[1] See the footnote to our Art. 116.

the slide. Thus the distance r between the molecules will be increased by the quantity

$$\delta r = (ys_y + x\sigma_{xy})\frac{y}{r}.$$

A mutual action

$$f(r) \cdot (ys_y + x\sigma_{xy})\frac{y}{r}$$

will thus be developed between the molecules by the displacement, where $f(r)$ is some function of r.

If these molecules m', m'' form part of two groups situated at either side of an elementary area ω taken perpendicular to the axis of x, we shall have $\omega \,.\, \widehat{xx}$ and $\omega \,.\, \widehat{xy}$ for the stresses obtained by resolving such mutual actions as the above along the axes of x and y respectively and summing them for all actions which cross the area ω. (See our Art. 1563*.)

Thus we have

$$\omega \,.\, \widehat{xx} = \Sigma f(r)\,(ys_y + x\sigma_{xy})\frac{y}{r}\cdot\frac{x}{r},$$

$$\omega \,.\, \widehat{xy} = \Sigma f(r)\,(ys_y + x\sigma_{xy})\frac{y}{r}\cdot\frac{y}{r},$$

or,

$$\widehat{xx} = \frac{s_y}{\omega}\,\Sigma\frac{f(r)}{r^2}\,xy^2 + \frac{\sigma_{xy}}{\omega}\,\Sigma\frac{f(r)}{r^2}\,x^2y,$$

$$\widehat{xy} = \frac{s_y}{\omega}\,\Sigma\frac{f(r)}{r^2}\,y^3 + \frac{\sigma_{xy}}{\omega}\,\Sigma\frac{f(r)}{r^2}\,xy^2.$$

The form of these expressions thus proves the identity of the cross-stretch and direct-slide coefficients on the rari-constant hypothesis.

[74.] In Section 12 (pp. 117—120) Saint-Venant applies the general formulae of elasticity to the simple case of a prism under pure traction. He then deduces the stretch-modulus in terms of the elastic constants for various kinds of elastic bodies.

In a footnote to p. 120 he supposes the body to have weight and to be vertically stretched. He obtains with the notation of our Art. 1070* the following results :

$$\left.\begin{matrix} u \\ v \end{matrix}\right\} = -\begin{Bmatrix} \eta \\ \eta' \end{Bmatrix}\frac{1}{E}\left(\frac{F'}{\omega} - \frac{Wz}{\omega l}\right)\begin{Bmatrix} x \\ y \end{Bmatrix},$$

$$w = \frac{1}{E}\left(\frac{F'z}{\omega} - \frac{W}{\omega}\frac{z^2}{2l} - \frac{W}{\omega}\,\frac{\eta y^2 + \eta' z^2}{2l}\right).$$

These results agree with those of our Art. 1070*, if we take $\eta = \eta'$, or suppose isotropy in the cross-section. Here η, η' are the stretch-squeeze ratios in the directions z, x and z, y respectively.

I had not noticed this footnote when commenting in the first volume on Lamé's treatment of the problem.

[75.] Section 13 (pp. 121—123) deals with Poisson and Cauchy's method of treating the problem of flexure by expanding the stresses as positive integral algebraic functions of the co-ordinates of the point on the cross-section referred to axes in the cross-section: see our Arts. 466* and 618* (footnote). This method Saint-Venant admits had served for the departure of his own researches (p. 99), and he deals more gently with it here (p. 124) than he does in his later work. The assumption of the possibility of the expansion in a convergent series is a very dangerous one, and leads in the case of torsion to very erroneous results: see our Arts. 1626* and 191 (or *Leçons de Navier*, footnote pp. 621—7).

[76.] In §§ 14—17 (pp. 125—36) Saint-Venant gives the general solution of the problem of flexure, carefully stating his assumptions and once integrating his equations. He reduces the solution to the determination of a single function F, which can be chosen to suit a great variety of cross-sections. I will reproduce as briefly as possible the matter of these sections.

[77.] Taking a portion of a weightless prism between two cross-sections Saint-Venant proposes to determine its state of equilibrium after it has been subjected to flexure on the following suppositions:

(i) The character of a certain portion of the shifts and strains is assumed; namely, the axis of the prism, or the right line joining the centroids of the cross-sections, is supposed to become a plane curve (*elastic line* here one with the *neutral line*), and further the stretches of the longitudinal 'fibres' vary in a uniform manner with their distances from each other measured parallel to the plane of the elastic line.

Let x be the direction of the line of centroids before flexure and let the origin be its fixed extremity (see (iii)), and let xz be the plane of flexure (or of the elastic line), then the above condition is analytically represented by

$$s_x = Cz + C' \dots\dots\dots\dots (1),$$

where C and C' are constants for the cross-section.

(ii) The character of a certain portion of the stresses is assumed; namely, it is supposed that the fibres exercise no mutual *traction* upon each other, or that their mutual action is solely of the nature of shear. Further, on the terminal cross-sections there is supposed to be no tractive loading.

These assumptions may be expressed analytically by

$$\widehat{yy} = \widehat{zz} = \widehat{yz} = 0 \text{ .... (2)},$$

$$\int \widehat{xx}\, d\omega = 0 \text{ for a terminal cross-section..............(3)}.$$

Further, it is supposed that although the mode of application and distribution of the load is unknown, yet the resultant load and its moment (M) for each cross-section ω at distance x from the origin are known.

It follows that

$$M = \int \widehat{xx}\, z d\omega \text{ for each section .... (4)}.$$

Further, to simplify the equations of unnecessary elements all motion of rotation, or translation of the prism as a whole, all stretching of the central axis or torsion of the prism are excluded. The latter elements by the principle of superposition of strains can afterwards be added.

(iii) One extremity of the central axis, the central elementary area of the cross-section at that extremity and an elementary strip along the trace of the plane of flexure on the cross-section remain fixed.

Analytically this gives us the conditions:

$$u = v = w = 0,\ du/dz = 0, \text{ when } x = y = z = 0 \text{...........(5)},$$

$$v = 0,\ dv/dz = 0, \text{ when } y = z = 0 \text{ for all values of } x \text{.........(6)}.$$

[78.] Let us adopt the following additional notation: l, $\omega\kappa^2$ and ρ are the length, cross-sectional moment of inertia ($= \int z^2 d\omega$) and radius of curvature at any point of elastic line of the prism. Let us further suppose that the material is such that the cross-sections of the prism are planes of elastic symmetry, it follows easily that the stress-strain relations will be of the form

$$\left.\begin{aligned}
\widehat{xx} &= a s_x + f' s_y + e'' s_z + h\sigma_{yz}\\
\widehat{yy} &= f'' s_x + b s_y + d' s_z + k\sigma_{yz}\\
\widehat{zz} &= e' s_x + d'' s_y + c s_z + n\sigma_{yz}\\
\widehat{yz} &= h' s_x + k' s_y + n' s_z + d\sigma_{yz}\\
\widehat{zx} &= e\sigma_{zx} + h''\sigma_{xy}\\
\widehat{xy} &= h'''\sigma_{zx} + f\sigma_{xy}
\end{aligned}\right\} \text{ .... (7)}.$$

See the annotated *Clebsch*, pp. 75, 6.

Since $\widehat{yy} = \widehat{zz} = \widehat{yz} = 0$ we can determine from the first four equations $\widehat{xx}$, s_y, s_z and σ_{yz} in terms of s_x, we may thus write:

$$\widehat{xx} = E s_x, \quad s_y = -\eta_1 s_x, \quad s_z = -\eta_2 s_x, \quad \sigma_{yz} = \epsilon s_x \text{............ (8)}.$$

[79.] Considering the portion of the prism between the cross-section ω at distance x and the cross-section at the origin we have by (3) and (4):—

$$0 = \int \widehat{xx}\, d\omega = \int E\,(Cz + C')\, d\omega,$$

$$M = \int \widehat{xx}\, z d\omega = \int E\,(Cz + C')\, z d\omega,$$

whence

$$C' = 0 \text{ and } C = M/E\omega\kappa^2 \text{........................(9)}.$$

It follows that

$$\widehat{xx} = zM/\omega\kappa^2 \text{............................(10)}.$$

If we now turn to the body stress-equations we find they reduce to

$$\left.\begin{array}{c}\dfrac{d\widehat{xy}}{dy}+\dfrac{d\widehat{xz}}{dz}=-\dfrac{z}{\omega\kappa^2}\dfrac{dM}{dx}\\ \dfrac{d\widehat{xy}}{dx}=0,\quad \dfrac{d\widehat{xz}}{dx}=0\end{array}\right\}\ldots\ldots\ldots\ldots\ldots\ldots(11),$$

while the surface stress-equations reduce to the single one

$$\widehat{xz}dy-\widehat{xy}dz=0\ldots\ldots\ldots\ldots\ldots\ldots(12).$$

The last two equations of (11) lead us by means of the last two of (7) to the conditions[1]

$$\frac{d\sigma_{xy}}{dx}=0,\quad \frac{d\sigma_{zx}}{dx}=0,$$

or, to

$$\frac{d^2u}{dxdy}+\frac{d^2v}{dx^2}=0,\quad \frac{d^2u}{dxdz}+\frac{d^2w}{dx^2}=0.$$

Hence, putting for $du/dx=s_x$ its value $zM/E\omega\kappa^2$, we have, *since M is supposed a function of x only,*

$$\frac{d^2v}{dx^2}=0,\qquad -\frac{d^2w}{dx^2}=M/E\omega\kappa^2\ldots\ldots\ldots\ldots\ldots\ldots(13).$$

The first equation tells us that there is no curvature in the direction of y after flexure, the second that the curvature $\left(1/\rho=-\dfrac{d^2w}{dx^2}\right.$ for small shifts$\left.\right)$ in direction of z is equal to $M/E\omega\kappa^2$.

We thus obtain

$$M=E\omega\kappa^2/\rho,\quad s_x=z/\rho,\quad \widehat{xx}=Ez/\rho\ldots\ldots\ldots\ldots(14),$$

the formulae of the Bernoulli-Eulerian theory, here deduced without its invalid assumptions (i.e. that the cross-sections remain plane and normal to the strained fibres).

[80.] The first equation of (11) shews that if M is variable or in other words the curvature changes, the stresses $\widehat{xy}$, $\widehat{xz}$ and therefore the slides σ_{xy}, σ_{xz} cannot be zero, or it involves the contradiction of the Bernoulli-Eulerian assumptions.

Further differentiating the same equation with regard to x, we deduce by the second and third equations of (11) the result

$$\frac{d^2M}{dx^2}=0\ldots\ldots\ldots\ldots\ldots\ldots\ldots\ldots(15),$$

or M must be of the linear form in x,

$$=P(a-x)\ldots\ldots\ldots\ldots\ldots\ldots\ldots\ldots(16),$$

[1] Provided the relation $e/h''=h'''/f$ does not hold between the elastic constants.

if we suppose Pa to be the value of M when $x = 0$. In many cases $a = l$, the length of the prism.

This result (obtained on p. 130 of the memoir) is extremely important, and does not seem to me to have been sufficiently regarded. I remark that it is obtained without any consideration of the surface condition (12). *It thus follows that the assumptions* $s_x = Cz + C'$, $\widehat{zz} = \widehat{yy} = \widehat{yz} = 0$ *are not legitimate, if* M *is other than a linear function of the length of the prism. In other words all the important practical cases of continuous loading are excluded from Saint-Venant's theory of flexure, and it remains yet to be shewn that for such cases the Bernoulli-Eulerian hypothesis of* (14) *gives even an approximation to the truth.*

[81.] With regard to the quantity P of the previous Article, we obviously have $-P$ equal to the resultant, in the direction of z of the load, or to the total shear across each section, that is

$$\int_0^\omega \widehat{xz}\, d\omega = -P \quad\text{.........................(17)}$$

for all sections.

Thus we see that Saint-Venant's theory, even without the limitation of equation (12), excludes the possibility of any discontinuous change in the shear, or the transverse load. He supposes the resultant of the whole external load to act either at the extreme section $(x = l)$ or beyond it in the central axis produced. This again narrows down very much the number of practical cases for which the Bernoulli-Eulerian equations have been shewn to be applicable.

[82.] Saint-Venant now proceeds to a first integration of his equations and deduces the following results (p. 131):

$$\left.\begin{aligned} u &= P\frac{2ax - x^2}{2E\omega\kappa^2}z + F(y, z), \quad v = -P\frac{a-x}{2E\omega\kappa^2}(2\eta_1 yz - \epsilon z^2), \\ w &= \sigma_0 x + P\frac{a-x}{2E\omega\kappa^2}(\eta_1 y^2 - \eta_2 z^2) - P\frac{3ax^2 - x^3}{6E\omega\kappa^2} \end{aligned}\right\} \text{...(18)},$$

where σ_0 is a constant, representing the value of σ_{xz} at the origin, and $F(y, z)$ is a function to be determined by the conditions

$$\left.\begin{aligned} &F = 0, \quad dF/dz = 0, \text{ when } y = z = 0; \\ &f\frac{d^2F}{dy^2} + (h'' + h''')\frac{d^2F}{dydz} + e\frac{d^2F}{dz^2} \\ &\qquad = P\frac{E - \eta_1 f - \eta_2 e + \epsilon h''}{E\omega\kappa^2}z + \eta_1\frac{h''' - h''}{E\omega\kappa^2}Py, \\ &\text{for all points of the cross-section; and,} \\ &\left(h'' - f\frac{dz}{dy}\right)\left(\frac{dF}{dy} + P\frac{2\eta_1 yz - \epsilon z^2}{2E\omega\kappa^2}\right) + \left(e - h'''\frac{dz}{dy}\right) \\ &\qquad\times\left(\frac{dF}{dz} + \sigma_0 + P\frac{\eta_2 z^2 - \eta_1 y^2}{2E\omega\kappa^2}\right) = 0 \\ &\text{for all points of the contour of the cross-section.} \end{aligned}\right\} \text{...(19)}.$$

These results follow by simple analytical work if we start with the value of u obtained from the equation $s_x = z/\rho = P(a-x)\,z/E\omega\kappa^2$ and then proceed to those of v and w given by the second and third equations of (8), the values so found being made to satisfy (11), etc.

[83.] Saint-Venant, however, does not deal in his special examples with this general case of elastic distribution; he assumes the material to have planes of elastic symmetry perpendicular to y and z, as well as perpendicular to x. We then have $h'' = h''' = h = k = n = h' = k' = n' = 0$, and clearly $\epsilon = 0$.

Further,

$$\widehat{xy} = f\sigma_{xy}, \quad \widehat{xz} = e\sigma_{xz}, \quad \sigma_{yz} = 0 \qquad (20).$$

The equations (18) and (19) now become, if we take[1]

$$f = \mu_1, \quad e = \mu_2,$$

$$\frac{\eta_1 f}{E} = \gamma_1, \quad \frac{\eta_2 e}{E} = \gamma_2.$$

$$\left.\begin{aligned} u &= P\frac{2ax - x^2}{2E\omega\kappa^2}z + F(y,z), \qquad v = -\gamma_1 P\frac{(a-x)}{\mu_1\omega\kappa^2}yz, \\ w &= \sigma_0 x + \frac{P(a-x)}{2\omega\kappa^2}\left(\frac{\gamma_1 y^2}{\mu_1} - \frac{\gamma_2 z^2}{\mu_2}\right) - P\frac{3ax^2 - x^3}{6E\omega\kappa^2} \end{aligned}\right\} \qquad (18'),$$

$$\left.\begin{aligned} &\mu_1\frac{d^2F}{dy^2} + \mu_2\frac{d^2F}{dz^2} = P\frac{1-\gamma_1-\gamma_2}{\omega\kappa^2}z \text{ throughout the section;} \\ &F = 0, \quad dF/dz = 0, \text{ when } y = z = 0; \\ &-\mu_1\left\{\frac{dF}{dy} + \gamma_1\frac{Pyz}{\mu_1\omega\kappa^2}\right\}dz + \mu_2\left\{\frac{dF}{dz} + \sigma_0 + \frac{P}{2\omega\kappa^2}\left(\frac{\gamma_2 z^2}{\mu_2} - \frac{\gamma_1 y^2}{\mu_1}\right)\right\}dy = 0 \\ &\text{over the contour of the cross-section.} \end{aligned}\right\} \qquad (19').$$

[84.] The last section of general treatment (pp. 133—6) gives formulae for various quantities used for the special cases afterwards dealt with. Thus we note:

First, the values of the stresses:

$$\left.\begin{aligned} \widehat{xx} &= P\frac{(a-x)}{\omega\kappa^2}z, \qquad \widehat{xy} = \mu_1\left(\frac{dF}{dy} + \gamma_1\frac{Pyz}{\mu_1\omega\kappa^2}\right), \\ \widehat{xz} &= \mu_2\left\{\frac{dF}{dz} + \sigma_0 + \frac{P}{2\omega\kappa^2}\left(\frac{\gamma_2 z^2}{\mu_2} + \frac{\gamma_1 y^2}{\mu_1}\right)\right\} \end{aligned}\right\} \qquad (20').$$

It follows that

$$\sigma_{xz} = \widehat{xz}/\mu_2, \text{ or } = \sigma_0 \text{ for } y = z = 0,$$

that is the inclinations of all the cross-sections at their centres to the axis is the same and equals σ_0.

[1] I have altered Saint-Venant's notation to correspond with that of our History, he puts for our $\begin{Bmatrix}\mu_1, & \mu_2, & \gamma_1, & \gamma_2, & \omega\kappa^2, & \eta_1, & \eta_2, & \epsilon, & \sigma_0 \\ G, & G', & \eta, & \eta', & I, & \epsilon, & \epsilon', & \epsilon'', & g_0\end{Bmatrix}$:

Secondly, the equation to the curved surface taken by the cross-section, on neglecting small quantities of the second order, is shewn to be

$$x' = \sigma_0 z' + F(y', z') \ldots\ldots\ldots\ldots\ldots\ldots\ldots (21),$$

where the origin is the centroid of the cross-section, the axis of x' is the tangent there to the elastic line, that of y' is parallel to y and the plane $y'z'$ is the tangent plane to the cross-section at the origin.

It is obvious that x' is not a function of x, or the cross-sections all assume the same distorted form. Hence we see why it is that the different fibres are stretched precisely as they would be, were the cross-sections to remain plane.

Thirdly, the total deflection δ (*la flèche de flexion*) is obtained by putting $y = z = 0$ and $x = l$ in the value of w in (18′), or,

$$\delta = -\sigma_0 l + \frac{P(3al^2 - l^3)}{6E\omega\kappa^2} \ldots\ldots\ldots\ldots\ldots\ldots\ldots (22).$$

Saint-Venant assumes the resultant load to be applied at the terminal, or that $a = l$, thus still further limiting his solution. In this case

$$\delta = -\sigma_0 l + Pl^3/3E\omega\kappa^2 \ldots\ldots\ldots\ldots\ldots\ldots\ldots (22').$$

[85.] The next twelve sections (18—29), pp. 136—68, deal with the determination of σ_0 and F for various forms of cross-section.

In the first place Saint-Venant assumes F is to be a positive integral algebraic function of y, z. In this case it must be of the form

$$F(y, z) = A_0 y + A\left(y^2 - \frac{\mu_1}{\mu_2} z^2\right) + A'yz + B'\left(y^2 z - \frac{\mu_1}{3\mu_2} z^3\right) + B''\left(yz^2 - \frac{\mu_2}{3\mu_1} y^3\right)$$
$$+ P\frac{1 - \gamma_1 - \gamma_2}{6\mu_2\omega\kappa^2} z^3 + C''\left(y^4 - 6\frac{\mu_1}{\mu_2} y^2 z^2 + \frac{\mu_1^2}{\mu_2^2} z^4\right) + C'''\left(yz^3 - \frac{\mu_2}{\mu_1} y^3 z\right) + \ldots (23),$$

in order to satisfy the first of equations (19′).

If this value be substituted in the third equation of (19′) we obtain the differential equation to the corresponding contour-curve.

[86.] Saint-Venant deals however only with the special case, in which the terms in $y^2 z$ and z^3 are alone retained. He puts

$$m = 1 - \gamma_1 - \frac{2\mu_1\omega\kappa^2}{P} B',$$

and thus throws F into the form

$$F(y, z) = P\frac{m - \gamma_2}{6\mu_2\omega\kappa^2} z^3 + P\frac{1 - m - \gamma_1}{2\mu_1\omega\kappa^2} y^2 z \ldots\ldots\ldots (24).$$

After some reductions and an integration he finds for the contour from the third equation of (19′):

$$Cy^{\frac{m}{1-m}} + \frac{\mu_2}{\mu_1}\frac{1 - 2\gamma_1 - m}{3m - 2} y^2 + z^2 = -\frac{2\mu_2\omega\kappa^2}{mP}\sigma_0 \ldots\ldots\ldots\ldots (25),$$

where C is a constant.

If $C=0$ this represents a family of ellipses. If C be finite and we give various values to m we have curves symmetrical with regard to the axis of y, and symmetrical or not with regard to the axis of z according as $m/(1-m)$ is even or odd. Equation (25) can be thrown into a somewhat different form by assuming c to be the semi-axis of the curve in the direction of $\pm z$, and b the semi-axis in the direction y. Thus $y=0$, $z=\pm c$, but for $z=0$, $y=+b$ always, $=-b$ also if $m/(1-m)$ be treated as even.

In putting $y=0$, $z^2=c^2$ we find,

$$\sigma_0 = -\frac{mPc^2}{2\mu_2\omega\kappa^2} \quad \text{.......................} \quad (26).$$

Equation (25) now becomes:

$$\left(1-\frac{1-2\gamma_1-m}{3m-2}\frac{\mu_2 b^2}{\mu_1 c^2}\right)\left(\frac{y}{b}\right)^{\frac{m}{1-m}} + \frac{1-2\gamma_1-m}{3m-2}\frac{\mu_2 b^2}{\mu_1 c^2}\frac{y^2}{b^2} + \frac{z^2}{c^2} = 1.$$

Saint-Venant now proceeds (pp. 138—143) to discuss the various forms that can be taken by this system of curves. This discussion seems to me perhaps a little too brief. Thus, he says: Supposing $m/(1-m)$ to be treated as even, then it is sufficient and necessary in order that the curve may be closed,—and so capable of serving as a contour for a cross-section,—that z/c have a real value when $y/b = 1-\chi$, χ being an extremely small positive quantity. This leads him to the condition that m must lie between

$$\frac{1-2\gamma_1}{1+\mu_1 c^2/(\mu_2 b^2)} \text{ and } 1.$$

[87.] We may note the following cases:

The *ellipse* ($2b \times 2c$) is obtained (not by putting $m/(1-m)=2$ which leads to a logarithmic curve owing to the appearance of indeterminate forms, but) by making the coefficient of $y^{m/(1-m)}$ vanish. Thus we have

$$m = \frac{2\mu_1 c^2 + (1-2\gamma_1)\,\mu_2 b^2}{3\mu_1 c^2 + \mu_2 b^2}$$

The *circle* (radius b) is obtained by putting $b=c$ or

$$m = \frac{2\mu_1 + (1-2\gamma_1)\,\mu_2}{3\mu_1 + \mu_2}.$$

The *false ellipse*, $\frac{y^4}{b^4} + \frac{z^2}{c^2} = 1$, is obtained by putting $m/(1-m) = 4$ in the case of isotropic material for which uni-constancy holds, or

$$\mu_2 = \mu_1,\ E/\mu_1 = 5/2 \text{ and } \gamma_1 = 1/10.$$

More generally we must take $m = 2\gamma_1 - 1$, for a similar curve in bodies with tri-planar elastic symmetry.

[88.] On pp. 139—40 Saint-Venant deals with and figures the

various curves which arise in the case of isotropy, when m is given different values, especially for the cases $c = b$ and $c = 2b$.

On pp. 141—3 he refers to the case of $m/(1 - m)$ being odd, and shews that not only are the limits for m narrower than in the previous case, but that the ratio b/c must remain within certain limits determined by those for m.

The case of $m = 5/7$ and $\mu_1 = \mu_2$ is fully treated and it is shewn that the equations represent for four values of b/c:

des *ovales* ou courbes ovoïdes *dont un des bouts est plus gros que l'autre.* Le *petit bout* dégénère en pointe pour la première et pour la dernière.

L'axe des z ne passe qu'exceptionnellement par le centre de gravité des sections terminées par ces contours non symétriques ; mais peu importe, car comme les fibres restent toutes dans les plans, tout ce qui précède est également vrai si l'on prend pour axe des x l'une quelconque des fibres qui ne varieront pas de longueur. (p. 143.)

[89.] We will next write down in a form corresponding to equation (25), the values of the three stresses; these we easily obtain from equations (20). They are:

$$\left.\begin{aligned}\widehat{xx} = \frac{P(a-x)z}{\omega\kappa^2}, \quad \widehat{xy} = \frac{P(1-m)yz}{\omega\kappa^2} \\ \widehat{xz} = -\frac{mP(c^2-z^2)}{2\omega\kappa^2} + \frac{\mu_2}{\mu_1}\frac{P(1-2\gamma_1-m)}{2\omega\kappa^2}y^2\end{aligned}\right\} \text{.........} \quad (27).$$

As one terminal cross-section usually corresponds to $x = l = a$, we see that $\widehat{xx} = 0$ across it, or the total external force exhibits itself as a shearing load, the resultant of which $-P$ is distributed according to a paraboloidal law.

Saint-Venant adds to these results that for the total deflection δ from equations (22) and (26); thus we have (see his p. 148):

$$\delta = \frac{Pl^3}{3E\omega\kappa^2}\left(1 + \frac{3m}{2}\frac{E}{\mu_2}\frac{c^2}{l^2}\right) \text{.................} \quad (28).$$

The form of the distorted cross-section deduced from equation (21) is:

$$x' = -\frac{P}{\omega\kappa^2}\left(\frac{m}{2\mu_2}c^2z - \frac{m-\gamma_2}{6\mu_2}z^3 - \frac{1-\gamma_1-m}{2\mu_1}y^2z\right) \text{......} \quad (29).$$

If $x'_0 = \dfrac{2m+\gamma_2}{3}\dfrac{Pc^3}{2\mu_2\omega\kappa^2}$ be the value of x' when $y = 0$, $z = -c$, this may be written:

$$\frac{x'}{x'_0} = -\frac{3m}{2m+\gamma_2}\frac{z}{c} + \frac{m-\gamma_2}{2m+\gamma_2}\left(\frac{z}{c}\right)^3 + 3\frac{\mu_2}{\mu_1}\frac{1-\gamma_1-m}{2m+\gamma_2}\left(\frac{y}{c}\right)^2\frac{z}{c} \ldots (29').$$

[90.] Saint-Venant specialises the results of the previous Article on pp. 144—148 for definite values of m. Thus he takes the case of the *false ellipse* for a uni-constant isotropic material ($m=4/5$, $\gamma_1=\gamma_2=1/10$); of the curve $m=9/10$ (or 18/20, considered as even), also for a uni-constant isotropic material—this curve approaches a rectangle of which the angles have been rounded off and the top and bottom hollowed out; and of $m=1$ ($=2/2$), the contour is here a quadrilateral formed by four curved lines. Then he proceeds to cases which have for practical purposes more definite contours, namely:

(i) *The ellipse.* Here, if $q=\mu_1 c^2/\mu_2 b^2$, we have:

$$m=\frac{1+2q-2\gamma_1}{1+3q};$$

$$\sigma_0=-\frac{1+2q-2\gamma_1}{1+3q}\frac{2P}{\mu_2\omega},\ \left[=-\frac{5c^2+2b^2}{3c^2+b^2}\frac{4P}{5\mu\omega},\ \text{for uni-constant isotropy};\right]$$

$$\widehat{xy}=\frac{4P}{\omega}\frac{q+2\gamma_1}{1+3q}\frac{yz}{c^2},\ \left[=\frac{4P}{5\omega}\frac{5c^2+b^2}{3c^2+b^2}\frac{yz}{c^2},\ \text{for uni-constant isotropy};\right]$$

$$\widehat{xz}=-\frac{2P}{\omega}\frac{1+2q-2\gamma_1}{1+3q}\left(1-\frac{z^2}{c^2}\right)+\frac{2P}{\omega}\frac{1-6\gamma_1}{1+3q}\frac{y^2}{b^2},$$

$$\left[=-\frac{4P}{5\omega}\frac{5c^2+2b^2}{3c^2+b^2}\left(1-\frac{z^2}{c^2}\right)+\frac{4P}{5\omega}\frac{y^2}{3c^2+b^2},\ \text{for uni-constant isotropy};\right]$$

$$\delta=\frac{4Pl^3}{3E\omega c^2}\left\{1+\frac{3(1+2q-2\gamma_1)}{2(1+3q)}\frac{E}{\mu_2}\frac{c^2}{l^2}\right\},$$

$$\left[=\frac{4Pl^3}{3E\omega c^2}\left\{1+\frac{15c^2+6b^2}{6c^2+2b^2}\frac{c^2}{l^2}\right\},\ \text{for uni-constant isotropy.}\right]$$

(ii) *The circle.* We have only to put $b=c$ in the above results.

[91.] We may note that the term to be added to the deflection owing to shear is generally about $3\left(\frac{c}{l}\right)^2$ of that due to bending, if we deal with a uni-constant isotropic material (i.e. for circle $\frac{21}{8}\left(\frac{c}{l}\right)^2$, for false ellipse $3\left(\frac{c}{l}\right)^2$, for rectangle with flattened angles $\frac{27}{8}\left(\frac{c}{l}\right)^2$, etc.). This represents the amount neglected in the ordinary theory. If in practice we may safely neglect an error of 1/100 in the deflection, it follows that the ordinary theory will give sufficiently close practical results so long as the length of the beam is 8 or 9 times its diameter.

[92.] On pp. 148—156 Saint-Venant goes through some most interesting work to trace the form of the distorted cross-sections. He traces these surfaces by means of level or contour lines for different ratios of x'/x'_0 [see equation (29')], that is by the trace of the surfaces on planes parallel to the tangent plane at the origin.

The form of these families of curves may be roughly described as follows:

The critical member ($x' = 0$) of the family is an ellipse (or in special cases a circle) and its diameter (the neutral axis). The critical member divides the family into two—for x'/x'_0 a positive fraction, we have a loop below the neutral axis and a 'snake' passing outside and *above* the critical ellipse with the neutral axis for its asymptote; —for x'/x'_0 a negative fraction we have curves congruent to these only the loop is above and the 'snake' below the neutral axis.

The contour of the section itself falls almost entirely within the critical ellipse and so gives a surface cutting the loops, the 'snakes' only apply for the distorted cross-section ideally produced.

The traces of the section made by planes parallel to the plane of flexure are cubical parabolas and are hatched in Saint-Venant's figures. It appears from them that the slide σ_{xz} has its maximum value at the centre. Saint-Venant draws attention to a noteworthy point on p. 152: Since b does not occur in the equation (29′) the contour-lines are the same for all sections having the same m, c and x'_0. The constancy of x'_0 involves $P/\omega\kappa^2$ remaining the same, except in the case of the *false-ellipse* where the term involving $\frac{y}{c}$ disappears from the equation to the contour; thus such ellipses are all orthogonal projections of each other.

We have reproduced three figures giving the form of the distorted sections on the frontispiece to this volume.

Only in Fig. (i) the 'snakes,' which are contour-lines falling outside the real section, are given. The contour-lines for elevations above the tangent plane are given by whole lines, those depressed below it by dotted lines. The traces by planes parallel to the plane of flexure are shaded. The figure corresponds to a circular cross-section when the material has uni-constant isotropy.

It gives very approximately the surface for elliptic cross-sections when b is $< 1{\cdot}5c$.

In Fig. (ii) we have the contour-lines for a *false ellipse.*

In Fig. (iii) for the rectangle with rounded angles and hollowed top and bottom referred to in our Art. 90 ($m = 9/10$). We see that the contour-lines become *straight.*

In calculating and plotting out both Figs. (ii) and (iii) Saint-Venant has supposed uni-constant isotropy.

It may be remarked that the conception of these surfaces is much assisted by plaster-models, which exist for the case of the circular and square cross-sections (see below Art. 111).

[93.] Saint-Venant now passes to the discussion of the flexure of a beam of rectangular cross-section. This occupies pp. 156—168.

By the assumption

$$F(y, z) = \chi(y, z) + \frac{P(1-\gamma_2) z^3}{6\mu_2 \omega \kappa^2} - \frac{\gamma_1 P}{2\mu_1 \omega \kappa^2} y^2 z \text{........(30)},$$

Saint-Venant reduces the equations of condition (19′) for $F(y, z)$ to

$$\left.\begin{array}{l} \mu_1 \dfrac{d^2\chi}{dy^2} + \mu_2 \dfrac{d^2\chi}{dz^2} = 0 \text{ for all values of } y \text{ and } z, \\ \chi(-y, z) = \chi(y, z) \text{ everywhere}, \\ \chi = 0 \text{ and } d\chi/dz = 0 \text{ for } y = z = 0, \\ \dfrac{d\chi}{dz} = -\sigma_0 - \dfrac{Pc^2}{2\mu_2 \omega \kappa^2} + \dfrac{\gamma_1 P}{\mu_1 \omega \kappa^2} y^2 \text{ for } z = \pm c \text{ and } y \text{ between } \pm b, \\ \dfrac{d\chi}{dy} = 0 \text{ for } y = \pm b \text{ and } z \text{ between } \pm c. \end{array}\right\} \text{...(31).}$$

Here $2b$ and $2c$ are the horizontal and vertical (flexure plane) sides of the rectangle.

The first equation of (31) is satisfied by taking

$$\chi = \Sigma e^{qz} \left\{ A_q \cos \sqrt{\frac{\mu_2}{\mu_1}}\, qy + A'_q \sin \sqrt{\frac{\mu_2}{\mu_1}}\, qy \right\} \text{........(32).}$$

The sines must however disappear in virtue of the second equation, and since $\chi = 0$ when $y = z = 0$, we must have $A_q = -A_{-q}$, or,

$$\chi = \Sigma A_q (e^{qz} - e^{-qz}) \cos \sqrt{\frac{\mu_2}{\mu_1}}\, qy.$$

The condition $d\chi/dz = 0$ for $y = 0$, $z = 0$, shews us that a certain relation must hold among the coefficients A_q; it will serve later to determine σ_0.

The condition $d\chi/dy = 0$ for $y = \pm b$ will be satisfied if

$$q = \frac{n\pi}{b} \sqrt{\frac{\mu_1}{\mu_2}},$$

n being any whole number, and obviously it will be sufficient to deal only with positive whole numbers. For $n = 0$, we must introduce a term $A_0 (e^{0.z} - e^{-0.z})$ which gives us a quantity Kz.

Hence finally we may write:

$$\chi = Kz + \sum_1^\infty 2A_n \sinh \frac{n\pi z}{b} \sqrt{\frac{\mu_1}{\mu_2}} . \cos n\pi y/b.$$

The fifth condition of (31) then gives us the following equation to determine A_n by Fourier's method

$$K+\sum_1^\infty 2A_n\frac{n\pi}{b}\sqrt{\frac{\mu_1}{\mu_2}}\cosh\frac{n\pi c}{b}\sqrt{\frac{\mu_1}{\mu_2}}\cos n\pi y/b$$
$$= -\sigma_0-\frac{Pc^2}{2\mu_2\omega\kappa^2}+\frac{\gamma_1 P}{\mu_1\omega\kappa^2}y^2 \ldots (33).$$

Saint-Venant indicates in a foot-note (p. 159) that the form (32) is the most general form which will satisfy all the conditions of the problem.

[94.] Equation (33) easily gives us the following results:

$$K=-\sigma_0-\frac{Pc^2}{2\mu_2\omega\kappa^2}+\frac{\gamma_1 Pb^2}{3\mu_1\omega\kappa^2},$$

$$2A_n=-\frac{4b^3}{\pi^3}\sqrt{\frac{\mu_2}{\mu_1}}\frac{\gamma_1 P}{\mu_1\omega\kappa^2}\frac{(-1)^{n-1}}{n^3}\operatorname{sech}\left(\frac{n\pi c}{b}\sqrt{\frac{\mu_1}{\mu_2}}\right).$$

We are thus able to write down the complete value of χ, namely:

$$\chi=\left(-\sigma_0-\frac{Pc^2}{2\mu_2\omega\kappa^2}+\frac{\gamma_1 Pb^2}{3\mu_1\omega\kappa^2}\right)z$$
$$-\frac{\gamma_1 Pb^3}{\mu_1\omega\kappa^2}\sqrt{\frac{\mu_2}{\mu_1}}\frac{4}{\pi^3}\sum_1^\infty\frac{(-1)^{n-1}}{n^3}\frac{\sinh\left(\frac{n\pi z}{b}\sqrt{\frac{\mu_1}{\mu_2}}\right)}{\cosh\left(\frac{n\pi c}{b}\sqrt{\frac{\mu_1}{\mu_2}}\right)}\cos\frac{n\pi y}{b}\ldots(34).$$

In order finally to fulfil the condition $\frac{d\chi}{dz}=0$ for $y=z=0$ we must take

$$\sigma_0=-\frac{Pc^2}{2\mu_2\omega\kappa^2}\left\{1-\frac{\mu_2}{\mu_1}\cdot\frac{2\gamma_1 b^2}{3c^2}\left[1-\frac{12}{\pi^2}\sum_1^\infty\frac{(-1)^{n-1}}{n^2}\operatorname{sech}\frac{n\pi c}{b}\sqrt{\frac{\mu_1}{\mu_2}}\right]\right\}\ldots(35).$$

We have thus the complete determination of all the constants of the problem.

[95.] In the following pp. 162—3, Saint-Venant deduces from (18'), (20), (30), (34) and (35) the values of the three shifts and the three stresses; we tabulate them for reference.

$$u=P\frac{2lx-x^2}{2E\omega\kappa^2}z$$
$$+(1-\gamma_2)\frac{Pz^3}{6\mu_2\omega\kappa^2}-\gamma_1\frac{Py^2z}{2\mu_1\omega\kappa^2}+\gamma_1\frac{Pb^2z}{\mu_1\omega\kappa^2}\frac{4}{\pi^2}\sum_1^\infty\frac{(-1)^{n-1}}{n^2}\operatorname{sech}\left(\frac{n\pi c}{b}\sqrt{\frac{\mu_1}{\mu_2}}\right)$$
$$-\gamma_1\frac{Pb^3}{\mu_1\omega\kappa^2}\sqrt{\frac{\mu_2}{\mu_1}}\cdot\frac{4}{\pi^3}\sum_1^\infty\frac{(-1)^{n-1}}{n^3}\frac{\sinh\left(\frac{n\pi z}{b}\sqrt{\frac{\mu_1}{\mu_2}}\right)\cos\frac{n\pi y}{b}}{\cosh\left(\frac{n\pi c}{b}\sqrt{\frac{\mu_1}{\mu_2}}\right)},$$

$$v = -\gamma_1 \frac{P(l-x)}{\mu_1 \omega \kappa^2} yz,$$

$$w = -\frac{Pc^2}{2\mu_2 \omega \kappa^2} \left\{1 - \gamma_1 \frac{\mu_2}{\mu_1} \frac{2b^2}{3c^2} \left[1 - \frac{12}{\pi^2} \sum_1^\infty \frac{(-1)^{n-1}}{n^2} \operatorname{sech}\left(\frac{n\pi c}{b}\sqrt{\frac{\mu_1}{\mu_2}}\right)\right]\right\} x$$
$$- \frac{P}{2E\omega\kappa^2} \frac{3lx^2 - x^3}{3} + P \frac{l-x}{2\omega\kappa^2}\left(\frac{\gamma_1 y^2}{\mu_1} - \frac{\gamma_2 z^2}{\mu_2}\right),$$

$$\widehat{xy} = \frac{\gamma_1 P}{\omega\kappa^2} \frac{4b^2}{\pi^2} \sqrt{\frac{\mu_2}{\mu_1}} \sum_1^\infty \frac{(-1)^{n-1}}{n^2} \frac{\sinh\left(\frac{n\pi z}{b}\sqrt{\frac{\mu_1}{\mu_2}}\right) \sin \frac{n\pi y}{b}}{\cosh\left(\frac{n\pi c}{b}\sqrt{\frac{\mu_1}{\mu_2}}\right)},$$

$$\widehat{xz} = -\frac{Pc^2}{2\omega\kappa^2}\left(1 - \frac{z^2}{c^2}\right)$$
$$+ \gamma_1 \frac{Pb^2}{3\omega\kappa^2} \frac{\mu_2}{\mu_1} \left\{1 - \frac{3y^2}{b^2} - \frac{12}{\pi^2} \sum_1^\infty \frac{(-1)^{n-1}}{n^2} \frac{\cosh\left(\frac{n\pi z}{b}\sqrt{\frac{\mu_1}{\mu_2}}\right) \cos \frac{n\pi y}{b}}{\cosh\left(\frac{n\pi c}{b}\sqrt{\frac{\mu_1}{\mu_2}}\right)}\right\}.$$

Saint-Venant verifies these results by shewing that they satisfy the boundary equation $\widehat{xz}\, dy - \widehat{xy}\, dz = 0$ and the load-conditions $\int \widehat{xz}\, d\omega = -P$, $\int \widehat{xy}\, d\omega = 0$.

[96.] The next two sections 28 and 29 (pp. 164—8) are occupied with numerical, graphical and simpler algebraic expressions for the quantities which occur in the previous sections.

For σ_0 Saint-Venant obtains the following results *when* $\gamma_1 = \frac{1}{10}$:

When $\frac{c}{b}\sqrt{\frac{\mu_1}{\mu_2}} =$	$\frac{1}{4}$	$\frac{1}{2}$	$\frac{3}{4}$	1	1·25	1·5	2	2·5	3
$-\sigma_0 = \frac{Pc^2}{2\mu_2\omega\kappa^2} \times$	·67624	·84918	·90729	·94031	·96177	·97101	·98341	·98934	·99259

It is shewn that for all values of $c\sqrt{\mu_1} > b\sqrt{\mu_2}$ the sum-term in equation (35) may be omitted, or we can write

$$-\sigma_0 = \frac{3P}{2\mu_2\omega} - \frac{\gamma_1 b^2}{c^2} \frac{P}{\mu_1\omega}.$$

Further the deflection δ is then given by:

$$\delta = \frac{Pl^3}{3E\omega\kappa^2}\left(1 + \frac{3E}{2\mu_2}\frac{c^2}{l^2} - \eta_1 \frac{b^2}{l^2}\right)$$

since
$$\gamma_1 = \eta_1 \frac{\mu_1}{E}.$$

For the case of isotropy: $\eta_1 = \frac{1}{4}$, $E/\mu_1 = 5/2$, or

$$\delta = \frac{Pl^3}{3E\omega\kappa^2}\left(1 + \frac{15c^2 - b^2}{4l^2}\right).$$

The Bernoulli-Eulerian theory takes no account of the second term within the brackets.

[97.] Saint-Venant devotes his next few pages to a calculation of the value of x' which gives (see our equation (21)) the form of the distorted surface. He treats especially the case of $b = c$, and uni-constant isotropy (i.e. $\gamma_1 = \gamma_2 = 1/10$, $\mu_1 = \mu_2$).

I have reproduced his diagram of the contour-lines, as Fig. iv. of the frontispiece; the hatched lines as before denoting sections of the surface by planes parallel to that of flexure. The contour-lines are drawn for $x'_0 = 0$ to ± 1 by steps of $\cdot 2$.

The trigonometrical terms in x' have little importance when $b < c$, so that in that case we can practically take

$$x' = -\frac{Pc^3}{2\mu_2\omega\kappa^2}\left\{\frac{z}{c} - \frac{1-\gamma_2}{3}\left(\frac{z}{c}\right)^3\right\}$$

$$= -x_0'\frac{3}{2+\gamma_2}\left\{\frac{z}{c} - \frac{1-\gamma_2}{3}\left(\frac{z}{c}\right)^3\right\}.$$

This is equivalent to neglecting terms in the expression for x' involving the factor b/c. It is obvious that the contour-lines now become straight lines.

The above value of x' is obtained by Saint-Venant from very simple considerations in a foot-note on pp. 184—5. It had already been given in the memoir on *Torsion* (see our Art. 12) without the term γ_2 (circa 1/10); a similar proof of the formula is given in the *Leçons de Navier*: see our Art. 183 (*a*).

[98.] Saint-Venant's thirtieth section (pp. 168—171) is entitled: *Sections de forme quelconque.* This amounts to little more than the statement that, a solution having been found for the equations (19) with regard to certain cross-sections we may infer that a solution exists for all cross-sections. The inference is strengthened by reference to a corresponding problem in the conduction of heat.

[99.] Section 31 (pp. 171—187) is termed: *Démonstration directe et sans analyse des formules connues de la flexion des prismes due à leurs seules dilatations longitudinales.* This investigation can be easily followed by those who have grasped the analytical calculations, but it seems to me very doubtful if it would be of value for elementary teaching (e.g. of engineering students). Saint-Venant did not reproduce it in his *Leçons de Navier.*

[100.] The final section of the memoir (§ 32, pp. 187—9) is entitled: *Conclusion. Observation générale pour le cas où le mode d'application et de distribution des forces extérieures vers les extrémités est différent de celui qui rend tout à fait exactes les formules auxquelles conduit la méthode mixte.*

This reiterates the principle of the practical equivalence in elastic effect of two surface distributions of load which are statically equivalent: see our Arts. 8 and 9.

101. *Sur les conséquences de la théorie de l'élasticité en ce qui regarde la théorie de la lumière. L'Institut*, Vol. 24, 1856, 32—34. The article adopts the view that much remains to be done to render the theory of Physical Optics satisfactory; it supports the views of Cauchy, especially with regard to the existence of a *third* ray as obtained by him in his discussion of what is termed *double* refraction. The article concludes thus:

Quoi qu'il puisse être de ces explications, que nous devons nous borner à soumettre aux physiciens et aux physiologistes, et bien que l'on puisse continuer sans doute de regarder le mouvement de la lumière dans les cristaux comme représenté *approximativement* par la surface d'onde du quatrième degré de Fresnel, nous pensons qu'il convient de ne plus passer sous silence les composantes longitudinales des vibrations pour éluder quelques difficultés dont elles sont le sujet, et que, pour rendre la théorie de la lumière exempte d'inexactitude logique, et provoquer pour l'avenir des recherches qui seront peut-être suivies d'importantes découvertes, il y a lieu de ne plus présenter les vibrations de l'éther, dans les milieux biréfringents, comme étant tout à fait parallèles aux divers plans tangents à la surface des ondes lumineuses qui s'y propagent.

102. *Sur la vitesse du son. L'Institut*, Vol. 24, 1856, 212—216. Newton obtained a certain expression for the velocity of sound which gives a result much smaller than that found by experiment. Laplace modified the formula, and thus obtained a result agreeing with experiment: see our Arts. 310* and 68. Saint-Venant is not satisfied with any investigation which has been given, even with the aid of the formulae of the theory of elasticity. He says

On voit toujours, par ce qui précède, qu'il reste encore bien des choses à savoir sur la théorie du son, objet des recherches d'hommes tels que Newton, Lagrange, Euler, Laplace, Poisson et Dulong; qu'on ne doit pas s'étonner de trouver des différences entre les résultats de l'observation et ceux de la formule de vitesse la plus généralement adoptée jusqu'ici $\sqrt{\frac{c}{c'}\frac{p}{\rho'}}$, ni se hâter de déduire de cette formule, probablement fausse, des valeurs du rapport c/c', comme l'ont fait plusieurs physiciens éminents; enfin que ce qu'il paraîtrait y avoir de mieux à faire dans l'enseignement, jusqu'à éclaircissement, serait de démontrer la formule newtonienne et d'énoncer simplement les raisons qui rendent son résultat trop faible (pp. 115—6).

Saint-Venant's article contains valuable references to preceding writers on the subject. See too *Die Fortschritte der Physik im Jahre* 1856, pp. 159—164.

103. *Sur la résistance des solides. L'Institut,* Vol. 24, 1856, pp. 457—459. This article relates to the moments of inertia and the situation of the principal axes of plane figures; the results given are useful in connexion with the resistance of beams to flexure, and are accompanied by various numerical calculations. Two formulae are given with respect to the moment of inertia of a triangle which may have been new at the time, but which now are particular cases of a known general proposition, namely that the moment of inertia of a triangle of mass M about any axis is the same as that of three particles of mass $\frac{1}{12}M$ at the angular points, and a particle of mass $\frac{3}{4}M$ at the centre of gravity. From this may be easily deduced another formula which Saint-Venant gives: the moment of inertia of a trapezium of mass M about one of the non-parallel sides is $\frac{1}{6}M\ (y^2 + y'^2)$, where y and y' are the perpendiculars from the two opposite angles on this side. Again we have a formula respecting the *product of inertia* for a right-angled triangle. Let M be the mass, and a, b the lengths of the sides. Then if the origin be at the angular point, and the axes coincide with the sides, the value as found by an obvious integration is $\frac{1}{12}Mab$. Hence if the origin be at the centre of gravity and the axes parallel to the sides, the value is $\frac{1}{12}Mab - \frac{1}{9}Mab$, that is $-\frac{1}{36}Mab$. This will hold also if the origin is on either of the straight lines through the centre of gravity parallel to the sides, the axes remaining always parallel to their original position.

[104.] *Sur l'Impulsion transversale et la Résistance vive des barres élastiques appuyées aux extrémités. Comptes rendus,* T. XLV.

1857, pp. 204—8. This memoir was presented on August 10, 1857. It was referred to Poncelet, Lamé, Bertrand and Hermite. An extract by the author is given in the *Comptes rendus.* Some of the results of this memoir were communicated to the *Société Philomathique*, November 5, 1853 and January 21, 1854, and partially published in *L'Institut*, T. 22, 1854, pp. 61—3, under the title: *Solution du problème du choc transversal et de la résistance vive des barres élastiques appuyées aux extrémités.* This is a special case of the resilience problem experimentally investigated by Hodgkinson and theoretically by Cox: see our Arts. 939*, 942*, 999* and 1434—7*. Saint-Venant, however, does not like Cox neglect the vibrations of the bar, or assume that its form will be that of the elastic line for a beam which centrally loaded has the same central deflection. In the *Comptes rendus*, Saint-Venant gives some account of the history of both transverse and longitudinal impact problems, but Cox's memoir seems to have escaped him.

The following result is given in the *Comptes rendus*, p. 206:

$$y = V\tau\Sigma \frac{4}{m^3} \frac{\dfrac{\sin mx/l}{\cos m} - \dfrac{\sinh mx/l}{\cosh m}}{\sec^2 m - \operatorname{sech}^2 m + \dfrac{2}{m^2}\dfrac{P}{Q}} \sin(m^2 t/\tau),$$

where the Σ refers to all the real and positive roots m of the equation

$$m(\tan m - \tanh m) = 2P/Q,$$

and the following is the notation used:

$2l$ = length of bar, P its weight, Q that of body striking the bar horizontally with velocity V at its mid-section, y is the horizontal displacement at distance x from one end and $\tau = \sqrt{Pl^3/(2gE\omega\kappa^2)}$.

[105.] Saint-Venant makes the following remark:

Du calcul tant numérique que graphique d'une suite de ces valeurs du déplacement y, on peut déduire la suite des formes très-variées prises par la barre heurtée; ce qui permet de modeler un relief en plâtre donnant la surface que décrirait cette barre supposée emportée transversalement d'un mouvement rapide, perpendiculaire au sens où elle oscille. Cette surface est très-ondulée à cause des oscillations provenant des second et troisième termes surtout de la série Σ (p. 206).

This surface in plaster of Paris was actually prepared under Saint-Venant's directions; and I have found a copy of it very useful for lecture purposes.

When P/Q does not exceed 3, the deflection obtained is very approximately that given by Cox in his memoir: see our Art. 1437*. It is not directly upon the *deflection*, however, but upon the *greatest curvature* that the maximum resistance of the bar depends, and this when $P/Q = 2$ is about 1·5 as great when obtained from the true transcendental formula as when obtained from statical considerations in Cox's manner. (See also *Notice* II. p. 20, under 2°.)

[106.] If the transverse blow be vertical, we must add to the above value of y the statical deflection and replace $V\tau \sin (m^2 t/\tau)$ by the expression $V\tau \sin (m^2 t/\tau) - (g\tau^2/m^2) \cos (m^2 t/\tau)$.

[107.] Saint-Venant compares his results with the numbers obtained by Hodgkinson: see our Arts. 1409*—10*. He finds that the values of the stretch-modulus so obtained agree among themselves, but differ from the statical values obtained from pure traction-experiments. He attributes this to thermal differences, such as had been considered by Duhamel and Wertheim: see our Arts. 889* and 1301*. On p. 207 there is a brief reference to some results for longitudinal impact.

[108.] The memoir itself appears never to have been published but its results together with many extensions and developments are given in the *Note finale du* § 61 of the annotated *Clebsch* pp. 490—596. Just *thirty years* after their discovery! We shall consider them in detail when dealing with that work, as the problem is an extremely important one in the theory of structures. See in particular *Notice* I. pp. 36—41 and *Notice* II. pp. 19—20.

[109.] *Établissement élémentaire des Formules de la torsion des prismes élastiques. Comptes rendus,* T. XLVI. pp. 34—8, 1858. The formulae in question are those of our Art. 17 but they are obtained only for the torsion of isotropic bodies. Saint-Venant's object is to deduce the results of the memoir on Torsion in an elementary fashion for the use of technical schools and practical men. The method does not seem to me entirely clear and satisfactory, and it is not at once obvious why the reasoning only applies to an *isotropic* body. Special proofs of various portions of the theory of elasticity may be now and then of service, but it cannot be denied that they, by tending to obscure the broad lines and general principles of the subject, may do more harm than good to the student.

The fairly elementary treatment of the *Leçons de Navier* seems to me more advantageous (pp. 245—250). The treatment of the present paper is also reproduced in § 7 (pp. 250—2) of the same work.

[110.] *L'Institut*, Vol. 26, 1858, pp. 178—9. Further results on Torsion communicated to the *Société Philomathique* (April 24 and May 15, 1858) and afterwards incorporated in the *Leçons de Navier* (pp. 305—6, 273—4). They relate to cross-sections in the form of doubly symmetrical quartic curves and to torsion about an external axis: see our Arts. 49 (*c*), 182 (*b*), 181 (*d*), and 182 (*a*).

[111.] Vol. 27, 1860, of same Journal, pp. 21—2. Saint-Venant presents to the *Société Philomathique* the model *de la surface décrite par une corde vibrante transportée d'un mouvement rapide perpendiculaire à son plan de vibration.* Copies of this as well as some other of Saint-Venant's models may still be obtained of M. Delagrave in Paris and are of considerable value for class-lectures on the vibration of elastic bodies.

[112.] Vol. 28, 1861, of same Journal, pp. 294—5. This gives an account of a paper of Saint-Venant's read before the *Société Philomathique* (July 28, 1860). In this he deduces the *conditions of compatibility*, or the six differential relations of the types:

$$2\frac{d^2 s_x}{dy\,dz} = \frac{d}{dx}\left(\frac{d\sigma_{xz}}{dy} + \frac{d\sigma_{xy}}{dz} - \frac{d\sigma_{yz}}{dx}\right)$$

$$\frac{d^2\sigma_{yz}}{dy\,dz} = \frac{d^2 s_y}{dz^2} + \frac{d^2 s_z}{dy^2}$$

which must be satisfied by the strain-components. These conditions enable us in many cases to dispense with the consideration of the shifts. A proof of these conditions by Boussinesq will be found in the *Journal de Liouville*, Vol. 16, 1871, pp. 132—4. At the same meeting Saint-Venant extended his results on torsion to: (1) prisms on any base with at each point only one plane of symmetry perpendicular to the sides, (2) prisms on an elliptic base with or without any plane of symmetry whatever; see our Art. 190 (*d*).

[113.] *Sur le Nombre des Coefficients inégaux des formules donnant les composantes des pressions dans l'intérieur des solides élastiques. Comptes rendus,* T. LIII. 1861, pp. 1107—1112. This paper gives very meagrely the outlines of Appendix V. to the *Leçons de Navier*: see our Arts. 192 to 195. Cf. also Moigno's *Statique*, Art. 270 and Stokes' *Report on Double-Refraction*, p. 260.

[114.] *Sur les divers genres d'homogénéité des corps solides et principalement sur l'homogénéité semi-polaire ou cylindrique, et sur les homogénéités polaire ou sphériconique et sphérique.* This paper was read to the Academy on May 21, 1860 and published in Liouville's *Journal de Mathématiques*, 1865, pp. 297—349. An abstract appeared in the *Comptes rendus*, T. L. 1860, pp. 930—4. See also *Notice* II. p. 23 and Moigno's *Statique*, p. 668.

This memoir is important as the first attempt to explain various results of experiment inconsistent with uni-constant isotropy by an extended conception of homogeneity applied to aeolotropic bodies. Cauchy had defined *homogeneity* as consisting in the elasticity of a body being the same *for the same directions* at all points. Saint-Venant alters the latter words and thus defines homogeneity:

> *Un corps est homogène lorsque l'un quelconque de ses éléments imperceptibles est identique à tout élément du même corps pris ailleurs ayant même volume et même forme, mais orienté d'une certaine manière qui peut changer d'un endroit à l'autre.* Il l'est même encore lorsque cette identité de deux éléments, pris n'importe où et convenablement orientés, souffre exception pour certains points isolés ou *ombilicaux* (tels que sont ceux de l'intersection commune des plans des cercles de longitude de la sphère dont on vient de parler...).
>
> Le mode d'orientation des éléments, ou la direction relative de leurs lignes homologues, détermine le genre de l'homogénéité, genre dont chacun admet, comme nous verrons au no. 3, des *sous-genres* où les orientations possibles en chaque point sont multiples. (p. 299.)

Let us take any two lines of the elastic system at right angles and arrange all lines homologous to the first along the normals to a given surface, the second system of lines may then be arranged according to any law we please, e.g. as tangents to any system of curves we please to draw on the surface. If the given surface be of the nth order, we have an *n-ic distribution of elastic homogeneity*; the curves on the surface to which the second system of homologous lines are tangents determine the *sous-genre* or *sub-class*.

[115.] The following paragraphs describe the quadric distributions of elasticity with which Saint-Venant proposes to deal.

After describing the *amorphic* body or body of *confused-crystallisation,* such as a rolled metal plate, the elasticity of which varies in length, breadth and depth,—Saint-Venant continues:

Qu'on enroule en tuyau cylindrique cette plaque homogène rectangulaire non isotrope supposée mince, en dirigeant, par exemple, les génératrices dans le sens de sa longueur. *Elle ne cessera pas d'être homogène;* mais l'égalité d'élasticité aux divers points n'aura pas lieu pour les directions parallèles entre elles. Il y aura égale élasticité suivant les rayons qui vont tous couper perpendiculairement l'axe du cylindre : ce sera l'élasticité dans le sens de l'épaisseur. Il y aura égale élasticité suivant les diverses tangentes aux cercles ayant leur centre sur cet axe. Il n'y aura que les élasticités égales suivant la longueur qui auront conservé des directions parallèles entre elles. (p. 298.)

We shall term this a cylindrical distribution of elastic homogeneity.

The following describes a spherical distribution:

Qu'on imagine maintenant une sphère solide pleine ou creuse, ou un corps de forme quelconque divisible en couches sphériques concentriques. Si la résistance ou la réaction élastique, pour mêmes déplacements de ses points, est partout égale dans le sens des rayons, et partout égale aussi dans certains sens perpendiculaires entre eux et aux rayons, ceux par exemple où se comptent les latitudes et les longitudes pour un équateur donné, la matière est homogène, mais *polairement,* ou d'une manière que nous pouvons appeler *sphériconique* vu le rôle qu'y jouent les *cônes de latitude* ayant un axe déterminé, le même pour tous. (p. 298.)

Such distributions of elasticity are, Saint-Venant asserts,—and I hold him to be entirely right—the true explanation of the anomalies which occur in experiments on a variety of cast, rolled and forged bodies. Even granted that isotropy is bi-constant, it is certainly not scientific to seek by means of two constants to account for the divergency between uni-constant formulae and experimental results on wires, plates, or cylindrical and spherical bodies. Physically it is obvious that the *working* of such bodies really produces in them varied distributions of elastic homogeneity, which bi-constant formulae only serve to mask. The 'isotropic boilers' treated of by Lamé (see our Art. 1038*) or his 'isotropic piezometers' (see our Art. 1358*) have practically no existence (see our Arts. 332* and 1357*), and all elasticians can adopt Saint-

Venant's formulae with entire approval although they may not accept his view of the equations of uni-constant isotropy:

> Formules qui sont les conséquences obligées et rigoureuses de la loi des actions moléculaires *que tout le monde invoque ouvertement ou tacitement*, et même sans laquelle tout établissement de formules mathématiques d'élasticité est illusoire. (p. 300.)

[116.] Saint-Venant on pp. 301—3 makes some remarks on the elastic coefficients, and on the subject of multi-constancy; for the purpose of the memoir, however, he adopts the 21 constants of Green[1].

If the stress be given by formulae of the type

$$p_{xx} = |xxxx|\, s_x + |xxyy|\, s_y + |xxzz|\, s_z + |xxyz|\, \sigma_{yz} + |xxzx|\, \sigma_{zx} + |xxxy|\, \sigma_{xy},$$
$$p_{yz} = |yzxx|\, s_x + |yzyy|\, s_y + |yzzz|\, s_z + |yzyz|\, \sigma_{yz} + |yzzx|\, \sigma_{zx} + |yzxy|\, \sigma_{xy},$$

then the coefficients can only be treated as constants when we suppose the axes-system to vary in direction from point to point of the material. This granted, the above expressions for the stresses will be given in terms of constant coefficients.

[117.] In section 3 (pp. 303—6) after some general remarks as to homogeneity and its various sub-classes, Saint-Venant supposes the distribution of elasticity to be symmetrical with regard to

[1] He refers to Rankine's terminology, which we may here throw into a form brief enough for convenience:

$|xxxx|$ = direct stretch coefficient = the coefficient of direct elasticity of Rankine.
$|xxyy|$ = cross stretch coefficient = the coefficient of lateral elasticity of Rankine.
$|xyxy|$ = direct slide coefficient = the coefficient of tangential elasticity of Rankine.

$|xyyz|$ = cross slide coefficient
$|xxxz|$ = direct slide-stretch coefficient
$|xxyz|$ = cross slide-stretch coefficient
$|xyxx|$ = direct stretch-slide coefficient
$|xyzz|$ = cross stretch-slide coefficient
} = coefficients of asymmetrical elasticity of Rankine.

All elasticians agree that the slide-stretch coefficients whether direct or cross are equal to the corresponding stretch-slide coefficients; further that the cross stretch and cross slide coefficients are equal for the pair of faces involved in the cross. This amounts to saying that we may interchange the first and second pairs of subscripts. We have thus the fifteen relations of Green. For a body with three planes of elastic symmetry all the *asymmetrical* coefficients vanish. The rari-constant elasticians assert that the cross stretch coefficients are equal to the direct slide coefficients, when the cross is made for the two directions involved in the slide (i.e. $|xxyy| = |xyxy|$), and further that the cross slide-stretch coefficients are equal to the cross slide coefficients when the direction of the stretch is involved in both the slides which are crossed (i.e. $|xxyz| = |xyxz|$). This gives the six additional relations of Poisson, or we may interchange between the first and second pair of subscripts.

three planes, or all the asymmetrical coefficients to vanish. In this case the types of traction and shear are:

$$(a)\quad \begin{aligned} \widehat{xx} &= as_x + f's_y + e's_z & \widehat{yz} &= d\sigma_{yz},\\ \widehat{yy} &= f's_x + bs_y + d's_z & \widehat{zx} &= e\sigma_{zx},\\ \widehat{zz} &= e's_x + d's_y + cs_z & \widehat{xy} &= f\sigma_{xy}. \end{aligned}$$

(See our Art. 78.)

(b) If the normal to the distribution-surface be the axis of x and the elasticity be isotropic in the tangent plane, we have also:

$$b = c,\ e = f,\ e' = f' \text{ and } b = 2d + d'.$$

(c) If the material be *amorphic*, there is an *ellipsoidal* distribution of direct-stretch coefficients (see our Arts. 139 and 142), and we have

$$2d + d' = \sqrt{bc},\ 2e + e' = \sqrt{ca},\ 2f + f' = \sqrt{ab}.$$

(d) In the case of *rari-constant* elasticity, the dashed and undashed letters are equal. Thus for the amorphic body we have:

$$\begin{aligned} \widehat{xx} &= 3\frac{ef}{d}s_x + fs_y + es_z & \widehat{yz} &= d\sigma_{yz},\\ \widehat{yy} &= fs_x + 3\frac{fd}{e}s_y + ds_z & \widehat{zx} &= e\sigma_{zx},\\ \widehat{zz} &= es_x + ds_y + 3\frac{de}{f}s_z & \widehat{xy} &= f\sigma_{xy}. \end{aligned}$$

(See, however, our Art. 313.)

[118.] Before we can apply these formulae to any given distribution of elasticity determined by curvilinear coordinates, it is necessary to find:

(1) Expressions for the above strain-components (s_x, s_y, s_z, σ_{yz}, σ_{zx}, σ_{xy}) corresponding to the elements of the three rectangular surface normals or intersection-traces in terms of the curvilinear coordinates.

(2) To express the body-stress equations in terms of curvilinear coordinates. Saint-Venant indicates in § 4 (pp. 306—12) two methods of attacking this problem, and compares them with Lamé's method (in the *Leçons*, 1852, § 77) which he terms "un procédé en quelque sorte *mixte*." The analysis of the problem does not probably admit of much simplification, and for practical purposes the general results of Lamé's treatise on Curvilinear Coordinates may well be assumed: see our Arts. 1150*—3*.

In § 5 (pp. 312—18) and in § 9 (pp. 333—9) Saint-Venant obtains expressions for the strains and the body-stress equations in terms of cylindrical and spherical coordinates respectively. These agree with those of Lamé[1]: see our Arts. 1087* and 1093*. The relations between stress and strain are then given by the formulae of the preceding article.

[119.] The novelty of the present memoir consists in the solution of the elastic equations for cylindrical and spherical shells subjected internally and externally to uniform tractive loads, when the material of these shells is amorphic and has cylindrical or spherical distribution of elasticity. By means of the solutions given, we see that the difficulties encountered by Regnault and others can be more naturally met by presupposing aeolotropy, than by assuming bi-constant isotropy.

[120.] Saint-Venant takes first (§ 7) the case of a long cylindrical shell subjected to internal tractive load $-p_0$ and external $-p_1$. As in Lamé's problem, we may suppose it closed by flat ends in such a manner that the transverse sections are not distorted. Supposing $dw/dz = \gamma$, we easily deduce (see footnote) the equation $\frac{d\widehat{rr}}{dr} + \frac{\widehat{rr} - \widehat{\phi\phi}}{r} = 0$, or substituting the stress-values from formulae (a) of Art. 117 expressed in terms of the strains given in the footnote we have, if $a_r = du/dr$

$$a\,(u_{rr} + u_r/r) - bu/r^2 + (e' - d')/(\gamma r) = 0.$$

[1] As in this volume we shall have frequent occasion to refer to these formulae I tabulate them here for reference—the notation will readily explain itself:

	Cylinder		Sphere (ϕ = co-latitude)
Body Stress-Equations	$\frac{d\widehat{rr}}{dr} + \frac{d\widehat{r\phi}}{rd\phi} + \frac{d\widehat{rz}}{dz} + \frac{\widehat{rr} - \widehat{\phi\phi}}{r} + \rho R = 0$ $\frac{d\widehat{\phi r}}{dr} + \frac{d\widehat{\phi\phi}}{rd\phi} + \frac{d\widehat{\phi z}}{dz} + \frac{2\widehat{r\phi}}{r} + \rho\Phi = 0$ $\frac{d\widehat{zr}}{dr} + \frac{d\widehat{z\phi}}{rd\phi} + \frac{d\widehat{zz}}{dz} + \frac{\widehat{zr}}{r} + \rho Z = 0$		$\frac{d\widehat{rr}}{dr} + \frac{d(\widehat{r\phi}\cos\phi)}{r\cos\phi d\phi} + \frac{d\widehat{r\psi}}{r\cos\phi d\psi} + \frac{2^{rr} - \widehat{\phi\phi} - \widehat{\psi\psi}}{r} + \rho R = 0$ $\frac{d\widehat{\phi r}}{dr} + \frac{d(\widehat{\phi\phi}\cos\phi)}{r\cos\phi d\phi} + \frac{d\widehat{\phi\psi}}{r\cos\phi d\psi} + \frac{2\widehat{\phi r} + \widehat{\psi\psi}\tan\phi}{r} + \rho\Phi = 0$ $\frac{d\widehat{\psi r}}{dr} + \frac{d(\widehat{\psi\phi}\cos\phi)}{r\cos\phi d\phi} + \frac{d\widehat{\psi\psi}}{r\cos\phi d\psi} + \frac{3\widehat{\psi r} - \widehat{\psi\phi}\tan\phi}{r} + \rho\Psi = 0$
s_r	u_r	s_r	u_r
s_ϕ	$v_\phi/r + u/r$	s_ϕ	$v_\phi/r + u/r$
s_z	w_z	s_ψ	$w_\psi/(r\cos\phi) + u/r - v/r\,.\tan\phi$
$\sigma_{\phi z}$	$v_z + w_\phi/r$	$\sigma_{\phi\psi}$	$v_\psi/(r\cos\phi) + w_\phi/r + w/r\,.\tan\phi$
σ_{zr}	$w_r + u_z$	$\sigma_{\psi r}$	$w_r + u_\psi/(r\cos\phi) - w/r$
$\sigma_{r\phi}$	$u_\phi/r + v_r - v/r$	$\sigma_{r\phi}$	$u_\phi/r + v_r - v/r$

Hence we find for the shifts

$$u = Cr^{\sqrt{\frac{b}{a}}} + C'r^{-\sqrt{\frac{b}{a}}} + \frac{d' - e'}{a - b}\gamma r, \quad v = 0, \quad w = \gamma z.$$

The stresses and strains can be at once deduced; they will contain constant terms in γ and powers of r of order $\pm\sqrt{\frac{b}{a}} - 1$. The constants C, C' and γ are to be determined from the surface conditions and the relation $p_0\pi r_0^2 - p_1\pi r_1^2 = 2\pi \int_{r_0}^{r_1} \widehat{zz} r dr$ for total terminal tractive stress.

[121.] Saint-Venant considers various special cases:

(1) $r_1 - r_0$ is a small thickness ϵ. (pp. 324—5.)

(2) $a = b$. Here the solution changes its form, we have (p. 326):

$$u = C_1 r + C_1'/r + \frac{d' - e'}{2a}\gamma r \log r.$$

If $d' = e'$ the solution becomes that found by Lamé and Clapeyron, and applied by Lamé to Regnault's piezometers: see our Arts. 1012* and 1358*

(3) When there is an ellipsoidal distribution of elasticity and rari-constancy is assumed, i.e. when $a = 3ef/d$, $b = 3fd/e$, $c = 3de/f$. In this case $u = Cr^{d/e} + C'r^{-d/e} - d\gamma r/\{3f(d/e + 1)\}$.

The values of the stresses are then easily determined, as well as those of C, C' and γ (p. 329).

The results contain *three independent elastic constants*, and they differ in the *form* of the r-index from those found for the case of isotropy. Hence we can explain by means of them as well as or better than by biconstant formulae the divergencies remarked by Regnault in his piezometer experiments.

[122.] A result is given on p. 331, which is worth citing. The constants d, e, f of the ellipsoidal distribution are not easy to determine by direct experiment. Let E_r, E_ϕ, E_z however be the three stretch moduli in directions r, ϕ, z, then we easily find that:

$$d = \tfrac{2}{5}\sqrt{E_\phi E_z}, \quad e = \tfrac{2}{5}\sqrt{E_z E_r}, \quad f = \tfrac{2}{5}\sqrt{E_r E_\phi}.$$

From equations 50 (p. 332) Saint-Venant might have deduced the criterion for failure arising first by lateral or first by longitudinal stretch. These equations are:

$$s_z = \tfrac{1}{4}\,\frac{2\sqrt{E_\phi} - \sqrt{E_z}}{E_z\sqrt{E_\phi}}\,\frac{p_0 - p_1}{\epsilon}\,r', \quad s_\phi = \tfrac{1}{8}\,\frac{8\sqrt{E_z} - \sqrt{E_\phi}}{E_\phi\sqrt{E_z}}\,\frac{p_0 - p_1}{\epsilon}\,r',$$

where $r_0 = r' - \frac{\epsilon}{2}$ and $r_1 = r' + \frac{\epsilon}{2}$, so that $r' = \frac{r_0 + r_1}{2}$, $r_1 - r_0 = \epsilon$.

So long as $E_z > \cdot 4195\, E_\phi$, s_ϕ is $> s_z$ and failure will occur by lateral stretch. If the absolute strengths R_z and R_ϕ were, as some writers have supposed, proportional to the moduli, and rupture took place in the same manner as failure of linear elasticity, we should say the cylinder would burst across a cross-section or open up longitudinally according as the longitudinal absolute strength R_z was $<$ or $>$ than $\cdot 4195$ times the transverse absolute strength R_ϕ.

A footnote on pp. 331—3 criticises with hardly sufficient severity a memoir of Virgile to which we shall refer later.

[123.] Saint-Venant (pp. 339—47) obtains similar results for the case of a spherical shell. He seeks first to find a solution of the equations (footnote p. 79 and stress-strain relations (a) of Art. 117) by taking $v = 0$, $w = 0$ and $u_\phi = 0$. This gives three equations to be satisfied which are inconsistent unless a certain relation is satisfied by the constants. Now $v = w = 0$ must for the case of uniform internal and external tractive loads be a necessary condition for change in size without distortion. Hence the equation (74) arrived at by Saint-Venant must be the condition for such a strain; it is:

$$\text{(i)} \quad \left(\frac{b-c}{e'-f'}\right)^2 + \frac{b-c}{e'-f'} = \frac{b+c+2d'-e'-f'}{a} \qquad \text{(p. 340).}$$

In this case the solution is simply

$$\text{(ii)} \quad u = Cr^{\frac{b-c}{e'-f'}}.$$

The condition (i) is however not sufficient; we find also from the surface equations that we must have

$$p_0/p_1 = (r_0/r_1)^{\frac{b-c}{e'-f'}-1} \qquad \text{(p. 342).}$$

It will be seen that without elastic isotropy in the tangent plane, it is only very special surface loads which will not produce distortion.

[124.] In § 11 (pp. 342—8) the problem of isotropy for all directions in the tangent plane is dealt with. In this case $e' = f'$, $b = c$, and stresses and constants are easily obtained by aid of the solution:

$$v = w = 0, \quad u = Cr^{n-\frac{1}{2}} + C'r^{-n-\frac{1}{2}},$$

where

$$n = \tfrac{1}{2}\sqrt{1 + 8\,\frac{b+d'-e'}{a}},$$

the body-shift equations being now reduced to the single one:

$$au_{rr} + 2au_r/r - 2\,(b + d' - e')\,u/r^2 = 0.$$

By evaluating the constants Saint-Venant obtains the following expression for u:

$$u = \frac{1}{r_1^{2n} - r_0^{2n}}\left\{\frac{p_0 r_0^{n+\frac{3}{2}} - p_1 r_1^{n+\frac{3}{2}}}{(n-\frac{1}{2})\,a + 2e'}\, r^{n-\frac{1}{2}} + \frac{p_0 r_0^{\frac{3}{2}-n} - p_1 r_1^{\frac{3}{2}-n}}{(n+\frac{1}{2})\,a - 2e'}\,(r_0 r_1)^{2n} r^{-n-\frac{1}{2}}\right\},$$

which gives the lateral stretches $s_\phi = s_\psi = u/r$ at once.

The important point in the piezometer problem is the dilatation of the spherical cavity. This is equal to $\frac{3u_0}{r_0}$

$$= \frac{3r_0^{n-\frac{3}{2}}}{r_1^{2n} - r_0^{2n}} \left\{ \frac{p_0 r_0^{n+\frac{3}{2}} - p_1 r_1^{n+\frac{3}{2}}}{(n-\frac{1}{2})\, a + 2e'} + \frac{p_0 r_0^{\frac{3}{2}-n} - p_1 r_1^{\frac{3}{2}-n}}{(n+\frac{1}{2})\, a - 2e'}\, r_1^{2n} \right\}$$

We see that it involves *three* elastic coefficients, and is thus, even as an *empirical* formula, better adapted to satisfy *numerically* Regnault's experiments than Lamé's bi-constant isotropic formula obtained by putting $d' = e'$, $b = a$ and $n = 3/2$.[1] On the other hand it is physically more plausible. The constants reduce to *two*, if we suppose the body *amorphic* and of *rari-constant* elasticity ellipsoidally distributed. If we take $r_0 = r' - \epsilon/2$, $r_1 = r' + \epsilon/2$, we easily find for the mid-sphere of radius r':

$$s_\phi = s_\psi = \frac{a}{a(b + d') - 2e'^2} \frac{(p_0 - p)\, r'}{2\epsilon},$$

or in the case just mentioned

$$s_\phi = s_\psi = \frac{3}{20d} \frac{(p_0 - p_1)\, r'}{\epsilon}$$

But $E_\phi = \frac{5}{2}\frac{fd}{e}, = \frac{5d}{2}$ by Art. 117 (6) if there be tangential isotropy. Hence finally:

$$s_\phi = s_\psi = \frac{3}{8E_\phi} \frac{(p_0 - p_1)\, r'}{\epsilon}$$

[125.] The final section of the memoir is entitled: *Vase cylindrique terminé par deux calottes sphériques* (pp. 347—9). This treats a problem similar to that dealt with by Lamé in his *Note* of 1850: see our Art. 1038* The mean lateral expansion of the spherical ends is made to take the same value as that of the cylindrical body by equating the expressions for s_ϕ obtained in our Arts. 122 and 124. Saint-Venant thus reaches a more general rule than that given by Lamé as a result of bi-constant isotropy. We have:

$$\frac{1}{8} \frac{8\sqrt{E_z} - \sqrt{E_\phi}}{E_\phi \sqrt{E_z}} \frac{r'}{\epsilon} = \frac{3}{8} \frac{1}{E_{\phi 1}} \frac{r_1'}{\epsilon_1}$$

where the subscript $_1$ refers to the spherical portions of the surface.

Hence

$$\frac{\epsilon}{\epsilon_1} = \frac{r'}{r_1'}, \times \frac{8/E_\phi - 1/\sqrt{E_z E_\phi}}{3/E_{\phi 1}}.$$

In the case of the two portions being of the same *isotropic* material, we have $E_\phi = E_z = E_{\phi 1}$, or

$$\frac{\epsilon}{\epsilon_1} = \frac{7}{3} \frac{r'}{r'}.$$

[1] In Lamé's notation $a = \lambda + 2\mu$ and $e' = \lambda$: see our Art. 1093*.

This agrees with Lamé's result: see our Vol. I. p. 564. If the thick nesses are equal, the radii ought to be as 3 : 7:

ce qui est la règle indiquée par M. Lamé pour les fonds sphériques compensateurs, élevant en quelque sorte, dit-il, le système des chaudières cylindriques au rang des formes naturelles ou des solides d'égale résistance. (p. 349.)

[126.] *Sur la distribution des élasticités autour de chaque point d'un solide ou d'un milieu de contexture quelconque, particulièrement lorsqu'il est amorphe sans être isotrope; Comptes rendus,* T. LVI. 1863, pp. 475—479, p. 804. This is an abstract of the memoir published in Liouville's *Journal* in 1863: see the following article.

[127.] *Mémoire sur la distribution des élasticités autour de chaque point d'un solide ou d'un milieu de contexture quelconque, particulièrement lorsqu'il est amorphe sans être isotrope.* This memoir was presented to the Academy, March 16, 1863, and some account of it appeared in the *Comptes rendus*, see preceding article. It is printed at length in Liouville's *Journal de mathématiques*, Vol. VIII. 1863, pp. 257—95 and 353—430

[128.] The opening pages of the memoir (257—9) as well as the concluding (425—30) entitled respectively: *Objet* and *Résumé et conclusions pratiques*, give an account of the purpose and results of the memoir. As these will sufficiently appear in our treatment of the intervening five sections (four, according to Saint Venant, but III occurs twice by mistake), we shall not reproduce here any part of these preliminary and final remarks.

[129.] The second section is entitled: *Formules diverses où entrent les coefficients dont l'élasticité dépend. Établissement, de plusieurs manières, d'une partie souvent omise, où figurent six constantes complémentaires, qui sont les composantes des pressions pouvant exister antérieurement aux déplacements des points* (pp. 260—286).

The aim of this section may be thus expressed: Let there be an initial system of stress given by $\widehat{xx}_0$, $\widehat{yy}_0$, $\widehat{zz}_0$, $\widehat{yz}_0$, $\widehat{zx}_0$, $\widehat{xy}_0$, and let the elastic nature of the body be given by thirty-six constants $|xxxx|$, $|xyxy|$, $|xyyy|$, etc. Green has decisively determined that these thirty-six can be reduced to twenty-one by the law of energy: see the footnote to our Art. 117. It is desirable to obtain a proof of the elastic formulae due to Cauchy without appealing to the principle of inter-molecular action being central and a function *only* of the distance.

Subscript letters attached to the shifts u, v, w denoting fluxions, the formulae are given by the types:

$$\left.\begin{array}{l}\widehat{xx} = \widehat{xx}_0\,(1 + u_x - v_y - w_z) + 2\,\widehat{xy}_0\,u_y + 2\,\widehat{zx}_0\,u_z + \widehat{xx}_1 \\ \widehat{yz} = \widehat{yz}_0\,(1 - u_x) + \widehat{yy}_0\,w_y + \widehat{zz}_0\,v_z + \widehat{zx}_0\,v_x + \widehat{xy}_0\,w_x + \widehat{yz}_1\end{array}\right\} \quad \ldots\ldots\ldots\text{(i)},$$

where

$$\left.\begin{array}{l}\widehat{xx}_1 = |xxxx|\,s_x + |xxyy|\,s_y + |xxzz|\,s_z + |xxyz|\,\sigma_{yz} + |xxzx|\,\sigma_{zx} + |xxxy|\,\sigma_{xy} \\ \widehat{yz}_1 = |yzxx|\,s_x + |yzyy|\,s_y + |yzzz|\,s_z + |yzyz|\,\sigma_{yz} + |yzzx|\,\sigma_{zx} + |yzxy|\,\sigma_{xy}\end{array}\right\} \quad \text{(ii)},$$

while the type of resulting body-shift equation is:

$$\left.\begin{array}{l}\quad -\rho X = \widehat{xx}_0 u_{xx} + \widehat{yy}_0 u_{yy} + \widehat{zz}_0 u_{zz} + 2\,\widehat{yz}_0 u_{yz} + 2\widehat{zx}_0 u_{zx} + 2\,\widehat{xy}_0 u_{xy} \\ + |xxxx|\,u_{xx} + |xyxy|\,u_{yy} + |xzxz|\,u_{zz} \\ + 2\,|zxxy|\,u_{yz} + 2\,|xxzx|\,u_{zx} + 2\,|xxxy|\,u_{xy} \\ + |xxxy|\,v_{xx} + |xyyy|\,v_{yy} + |zxyz|\,v_{zz} \\ + \{|xyyz| + |zxyy|\}\,v_{yz} + \{|xxyz| + |zxxy|\}\,v_{zx} + \{|xxyy| + |xyxy|\}\,v_{xy} \\ + |xxzx|\,w_{xx} + |xyyz|\,w_{yy} + |zxzz|\,w_{zz} \\ + \{|zxyz| + |xyzz|\}\,w_{yz} + \{|xxzz| + |zxzx|\}\,w_{zx} + \{|xxyz| + |xyzx|\}\,w_{xy}\end{array}\right\} \quad \text{(iii)}.$$

These results representing the most general equations of elasticity for small strains were originally given by Cauchy, as is implied in our Arts. 615*, 616*, 662*, 666* He obtained them by calculating the stresses as the sums of intermolecular actions on the rari-constant hypothesis. Saint Venant in this section proposes to deduce them from the principle of energy (by Green's method) in a manner which will satisfy multi-constant elasticians.

[130.] The proof attempted by Saint-Venant is not legitimate, because in the expression he takes for the work the linear term

$$\widehat{xx}_0\,s_x + \widehat{yy}_0\,s_y + \widehat{zz}_0\,s_z + \widehat{yz}_0\,\sigma_{yz} + \widehat{zx}_0\,\sigma_{zx} + \widehat{xy}_0\,\sigma_{xy}$$

occurs where s_x, s_y, s_z, σ_{yz}, σ_{zx}, σ_{xy} are stretches and slides. Assuming this term correct, which it is not, these ought to be expressed to the second power of the shift-fluxions as in our Art. 1622*, for we want the work to the second power. This Saint-Venant does not do, but treats the strains s and σ as if they were the quantities ϵ_x, ϵ_y, ϵ_z, η_{yz}, η_{zx}, η_{xy} of our Art. 1619*. This mistake was pointed out by Brill and Boussinesq and is acknowledged by Saint-Venant in a memoir of 1871: see our Art. 237. The formulae (i)—(iii) of the preceding article can thus only be considered as valid, when we accept the rari-constant hypothesis and deduce them after the manner of Cauchy. We shall see this point more clearly when dealing with the memoir of 1871. Green gets over the difficulty by expanding his work-function in powers of the ϵ's and η's; he thus gets a linear term, whose constants vanish with the initial stresses, but are not determined as functions of the initial stresses, still less does he show what functions, if any, the remaining constants are of the initial stresses.

[131.] In the course of this section Saint-Venant gives a proof of Cauchy's formulae (i) to (iii) above on the rari-constant hypothesis (footnote, pp. 273—5); he refers to the memoir of C. Neumann (*Zur Theorie der Elasticität,* Crelle, LVII, 1860, p. 281: see our Chap. XI.), where a similar method to his own is used for the case of isotropy (footnote, pp. 275—80), and to the memoir of Haughton (see our Art. 1505*) for a treatment which generalised leads to the same formulae on the rari-constant theory (p. 280 and footnote). Finally we may refer to his footnote (pp. 284—5) for a process by which the body-shift equations (iii) are deduced by means of the rari-constant hypothesis, without a previous investigation of the stresses[1].

[132.] The third section of the memoir (pp. 286—95) is entitled: *Formule symbolique générale fournissant, en fonction des coefficients d'élasticité pour des axes donnés, ceux qui sont relatifs à d'autres axes aussi donnés et rectangulaires, et, aussi, les coefficients qui doivent entrer dans l'expression d'une composante quelconque de pression même oblique.*

Saint-Venant adopts a symbolic representation of the stresses, strains and coefficients in order to express the relations among them. He thus describes this method:

> On abrége singulièrement le calcul et l'on arrive à quelque chose de fort simple au moyen de notations symboliques comme celles que plusieurs auteurs anglais appellent *Sylvestrian umbrae,* parce que M. Sylvester, qui les a employées avec succès, appelle *ombres de quantités* (shadows of quantities) ces sortes de notations dont se sont servis précédemment, au reste, Cauchy et d'autres analystes (p. 290).

[1] There is a wrong reference to Rankine's paper (p. 269, footnote), it should be Vol. VI. (p. 63), not Vol. v. of the *Camb. and Dublin Math. Journal.*

There is a footnote referring to Sylvester's papers in *Camb. and Dublin Math. Journal,* Vol. VII. 1852, p. 76, and *Phil. Trans.* 1853, p. 543.

[133.] Suppose *symbolically* $\iota_{jklm} = \iota_{jk}\iota_{lm} = \iota_j\iota_k\iota_l\iota_m$ to represent $|jklm|$, where j, k, l, m are any of the letters xyz, $x'y'z'$ etc. Further $\widehat{r}\ \widehat{r}$ to represent the stress $\widehat{rr}$, and $\epsilon_r\epsilon_r$ or ϵ^2_r to represent s_r, and finally $2\epsilon_r\epsilon_{r'}$ to represent $\sigma_{rr'}$. Let $c_{rr'}$ denote the cosine of the angle between the directions r, r'.

We are now able to reproduce in symbolic form the following well-known typical relations:

$$\widehat{rr'} = \widehat{xx}\, c_{rx}\, c_{r'x} + \widehat{yy}\, c_{ry}\, c_{r'y} + \widehat{zz}\, c_{rz}\, c_{r'z}$$
$$+ \widehat{yz}\,(c_{ry}\, c_{r'z} + c_{rz}\, c_{r'y}) + \widehat{zx}\,(c_{rz}\, c_{r'x} + c_{rx}\, c_{r'z}) + \widehat{xy}\,(c_{rx}\, c_{r'y} + c_{ry}\, c_{r'x}) \ldots \text{(iv)},$$

$$s_x = s_{x'}\, c^2_{xx'} + s_{y'}\, c^2_{xy'} + s_{z'}\, c^2_{xz'} + \sigma_{y'z'}\, c_{xy'}\, c_{xz'} + \sigma_{z'x'}\, c_{xz'}\, c_{xx'} + \sigma_{x'y'}\, c_{xx'}\, c_{xy'} \ldots \text{(v)},$$

$$\sigma_{yz} = 2s_{x'}\, c_{yx'}\, c_{zx'} + 2s_{y'}\, c_{yy'}\, c_{zy'} + 2s_{z'}\, c_{yz'}\, c_{zz'}$$
$$+ \sigma_{y'z'}\,(c_{yy'}\, c_{zz'} + c_{yz'}\, c_{zy'}) + \sigma_{z'x'}\,(c_{yz'}\, c_{zx'} + c_{yx'}\, c_{zz'}) + \sigma_{x'y'}\,(c_{yx'}\, c_{zy'} + c_{yy'}\, c_{zx'}) \ldots \text{(vi)}.$$

See our Arts. 659* and 663*.

(The last two are most readily obtained from the stretch-quadric of Art. 612* for axes $x'y'z'$, namely:

$$s_{x'}\, x'^2 + s_{y'}\, y'^2 + s_{z'}\, z'^2 + \sigma_{y'z'}\, y'z' + \sigma_{z'x'}\, z'x' + \sigma_{x'y'}\, x'y' = \pm 1.$$

Substitute for x' its equivalent $xc_{xx'} + yc_{yx'} + zc_{zx'}$ and similar quantities for x' and y', then the coefficients of x^2 and yz will be s_x and σ_{yz} as given above.)

The symbolical forms are:

$$\widehat{xx} \text{ or } \widehat{yz} = \iota_{xx} \text{ or } \iota_{yz} \times (\iota_x\epsilon_x + \iota_y\epsilon_y + \iota_z\epsilon_z)^2 \ldots\ldots\ldots\ldots \text{(vii)},$$

whence it follows from (iv) that

$$\widehat{rr'} = (\iota_x c_{rx} + \iota_y c_{ry} + \iota_z c_{rz})\,(\iota_x c_{r'x} + \iota_y c_{r'y} + \iota_z c_{r'z}) \times (\iota_x\epsilon_x + \iota_y\epsilon_y + \iota_z\epsilon_z)^2 \ldots \text{(viii)}.$$

Further we have from (v):

$$\epsilon_x\epsilon_x = (\epsilon_{x'}c_{xx'} + \epsilon_{y'}c_{xy'} + \epsilon_{z'}c_{xz'})^2$$
$$\epsilon_y\epsilon_z = (\epsilon_{x'}c_{yx'} + \epsilon_{y'}c_{yy'} + \epsilon_{z'}c_{yz'})\,(\epsilon_{x'}c_{zx'} + \epsilon_{y'}c_{zy'} + \epsilon_{z'}c_{zz'}),$$

whence we can take

$$\epsilon_j = \epsilon_{x'}c_{jx'} + \epsilon_{y'}c_{jy'} + \epsilon_{z'}c_{jz'} \ldots\ldots\ldots\ldots\ldots\ldots\ldots \text{(ix)}.$$

Put $j = x$, y, z successively and substitute for ϵ_x, ϵ_y, ϵ_z in (viii), we have

$$\widehat{rr'} = (\iota_x c_{rx} + \iota_y c_{ry} + \iota_z c_{rz})\,(\iota_x c_{r'x} + \iota_y c_{r'y} + \iota_z c_{r'z}) \times$$
$$\{(\iota_x c_{xx'} + \iota_y c_{yx'} + \iota_z c_{zx'})\,\epsilon_{x'} + (\iota_x c_{xy'} + \iota_y c_{yy'} + \iota_z c_{zy'})\,\epsilon_{y'}$$
$$+ (\iota_x c_{xz'} + \iota_y c_{yz'} + \iota_z c_{zz'})\,\epsilon_{z'}\}^2 \ldots\ldots\ldots\ldots\ldots\ldots \text{(x)}.$$

But we may obviously also express $\widehat{rr'}$ in the form

$$\widehat{rr'} = |rr'x'x'|\, s_{x'} + \ldots\ldots + \ldots\ldots$$
$$+ |rr'y'z'|\, \sigma_{y'z'} + \ldots\ldots + \ldots\ldots$$
$$= \iota_r\iota_{r'}\,(\iota_{x'}\,\epsilon_{x'} + \iota_{y'}\,\epsilon_{y'} + \iota_{z'}\,\epsilon_{z'})^2 \ldots\ldots\ldots\ldots\ldots \text{(xi)}.$$

Comparing (x) and (xi) which must on development give the same result, we see that it is necessary to take:

$$\iota_n = \iota_x c_{nx} + \iota_y c_{ny} + \iota_z c_{nz} \ldots\ldots\ldots\ldots\ldots\ldots\ldots\text{(xii)},$$

or,

$$|rr'x'y'| = \iota_r \iota_{r'} . \iota_{x'} \iota_{y'} \ldots\ldots\ldots\ldots\ldots\ldots\ldots\text{(xiii)},$$

where ι_n is given by (xii), and x', y' are any two of the three new axial directions, (x', y', z'), r, r' any two directions we please, and n any arbi trary direction.

Thus we have any coefficient of one set of axes expressed in terms of those obtained for another set. The product $\iota_r \iota_{r'} . \iota_{x'} \iota_{y'}$ ought to be made in the order indicated, except that the first pair and the last pair may have their members interchanged *in themselves*. If r, r' are both axial directions (i.e. chosen from x', y', z') then the first pair may be interchanged as a whole with the last pair. If we accept the rari-constant hypothesis, however, for axial directions all interchanges of the order of the ι's will be permissible.

[134.] Saint-Venant notes one or two other symbolical results. Thus, if ϕ be Green's work-function and we suppose no initial stresses:

$$\phi = \tfrac{1}{2}\left[(\iota_x \epsilon_x + \iota_y \epsilon_y + \iota_z \epsilon_z)^2\right]^2 \ldots\ldots\ldots\ldots\ldots\text{(xiv)}.$$

Further the types of stress and of the general body-shift equations (i) to (iii) of our Art. 129 become on the rari constant hypothesis:

$$\begin{aligned}
\begin{Bmatrix} \widehat{xx} \\ \widehat{yz} \end{Bmatrix} &= \begin{Bmatrix} \widehat{xx}_0 \\ \widehat{yz}_0 \end{Bmatrix} (1 - u_x - v_y - w_z) \\
&+ \left(\widehat{x}_0 \frac{d}{dx} + \widehat{y}_0 \frac{d}{dy} + \widehat{z}_0 \frac{d}{dz}\right) \begin{Bmatrix} 2\widehat{x}_0 u \\ \widehat{y}_0 w + \widehat{z}_0 v \end{Bmatrix} \\
&+ \begin{Bmatrix} \iota_{xx} \\ \iota_{yz} \end{Bmatrix} (\iota_x \epsilon_x + \iota_y \epsilon_y + \iota_z \epsilon_z)^2 \ldots\ldots\ldots\ldots\ldots\ldots\text{(xv)},
\end{aligned}$$

$$-\rho X = \left(\widehat{x}_0 \frac{d}{dx} + \widehat{y}_0 \frac{d}{dy} + \widehat{z}_0 \frac{d}{dz}\right)^2 u$$

$$+ \iota_x \left(\iota_x \frac{d}{dx} + \iota_y \frac{d}{dy} + \iota_z \frac{d}{dz}\right)\left(\iota_x \frac{d}{dx} + \iota_y \frac{d}{dy} + \iota_z \frac{d}{dz}\right)(\iota_x u + \iota_y v + \iota_z w) \text{ (xvi)}.$$

[135.] The next section, III *bis* (pp. 353—380), contains some very interesting and important matter. It is entitled: *Surfaces donnant la distribution des élasticités autour d'un même point.—Maxima et minima—Distribution ellipsoïdale des élasticités directes.—Solides ou milieux amorphes.—Intégrabilité des équations.*

Some of the results had already been given by Rankine in his memoir: *On Axes of Elasticity and Crystalline Forms*, *Phil. Trans.* 1856, pp. 261—85, but there is much that is new and the method is very good.

[136.] The relation (xiii) gives for $|rrrr|$ the value

$$|rrrr| = [(\iota_x c_{rx} + \iota_y c_{ry} + \iota_z c_{rz})^2]^2 \text{..................(xvii)},$$

or the direct stretch coefficient in direction r, (c_{rx}, c_{ry}, c_{rz}), in terms of the system of elastic coefficients for the axes x, y, z.

If we put

$$x = c_{rx}/\sqrt[4]{|rrrr|}, \quad y = c_{ry}/\sqrt[4]{|rrrr|}, \quad z = c_{rz}/\sqrt[4]{|rrrr|},$$

and substitute, we obtain the surface

$$1 = \{(\iota_x x + \iota_y y + \iota_z z)^2\}^2,$$

which expanded gives us Rankine's *tasinomic* quartic:

$$\left.\begin{aligned}1 = {} & |xxxx|\, x^4 + |yyyy|\, y^4 + |zzzz|\, z^4 \\ & + 2\{|yyzz| + 2\,|yzyz|\}\, y^2z^2 + 2\{|zzxx| + 2\,|zxzx|\}\, z^2x^2 + 2\{|xxyy| \\ & \qquad + 2\,|xyxy|\}\, x^2y^2 \\ & + 4\{|xxyz| + 2\,|zxxy|\}\, x^2yz + 4\{|yyzx| + 2\,|xyyz|\}\, y^2zx \\ & \qquad + 4\{|zzxy| + 2\,|yzzx|\}\, z^2xy \\ & + 4\,|yyyz|\, y^3z + 4\,|zzzy|\, z^3y + 4\,|zzzx|\, z^3x + 4\,|xxxz|\, x^3z \\ & \qquad + 4\,|xxxy|\, x^3y + 4\,|yyyx|\, y^3x\end{aligned}\right\} \text{(xviii)}.$$

This equation with its fifteen *homotatic* coefficients was first given by Haughton in his memoir of 1846. These 15 coefficients are the 15 coefficients of *rari-constancy* multiplied by the numbers 1, 6, 12 or 4, so that the expressions for the work, stresses etc., can on that hypothesis be given in terms of the coefficients of this equation.

Its fundamental property is that the direct-stretch coefficient in any direction varies inversely as the fourth power of the corresponding ray.

[137.] Paragraphs 10 and 11 together with the footnote pp. 359—62 reproduce results of Rankine and Haughton with regard to the nature of the elastic coefficients. Thus it is pointed out:

(i) That there are sixteen directions real or imaginary for which $|rrrr|$ is a maximum or minimum. These directions cut the tasinomic surface at right-angles, and possess the peculiarity that any stretch in their direction produces a *traction* only across a plane normal to their direction (pp. 356—7).

(ii) That if we take

$$|x'x'x'x'| + |x'x'y'y'| + |x'x'z'z'| = S_{x'},$$

or $S_{x'}$ equal to the sum of direct- and cross-stretch coefficients for the direction x', then

$$S_{x'} = (\iota_x c_{x'x} + \iota_y c_{x'y} + \iota_z c_{x'z})^2\,(\iota_{xx} + \iota_{yy} + \iota_{zz}).$$

Thus $S_{x'}$ varies inversely as the square of the ray of the ellipsoidal surface:

$$1 = (\iota_{xx} + \iota_{yy} + \iota_{zz}) (\iota_x x + \iota_y y + \iota_z z)^2,$$

which developed gives us:

$$1 = S_x x^2 + S_y y^2 + S_z z^2 + 2R_{yz} yz + 2R_{zx} zx + 2R_{xy} xy,$$

where $$R_{nm} = \iota_{xxnm} + \iota_{yynm} + \iota_{zznm} \qquad \text{(xix)}.$$

This is the ellipsoid discovered by Haughton in 1846 and termed by Rankine *orthotatic*. It shews us that by a suitable change of axes we can put $R_{yz} = R_{zx} = R_{xy} = 0$, which give three inter-constant relations, and so reduce the 21 (or 15) elastic constants to 18 (or 12).

(iii) That if an equal stretch s be given in the three orthotatic directions {i.e. those of the axes of the ellipsoid (xix)} this stretch system will produce no shear, for if x_1, y_1, z_1 be these orthotatic directions:

$$\widehat{y_1 z_1} = |y_1 z_1 x_1 x_1|\, s + |y_1 z_1 y_1 y_1|\, s + |y_1 z_1 z_1 z_1|\, s \text{ (from Equation (ii) of Art. 129)}.$$
$$= R_{y_1 z_1}\, s = 0.$$

The orthotatic directions are thus those for which the sum of the corresponding (direct and cross) slide-stretch coefficients vanish.

(iv) That a body may possess orthotatic isotropy, or $R_{x'y'} = 0$ for all rectangular systems x', y', z'. The orthotatic surface now becomes a sphere or $S_x = S_y = S_z$. Such a body however does not possess complete elastic isotropy.

(v) That there exists a surface which measures the difference D between a cross stretch and direct-slide coefficient, i.e.

$$D = |y'y'z'z'| - |y'z'y'z'|.$$

This is Rankine's *heterotatic* surface, and is given by

$$\left.\begin{aligned} D = \{|yyzz| - |yzyz|\}\, c^2_{xx'} + \{|zzxx| - |zxzx|\}\, c^2_{yx'} + \{|xxyy| - |xyxy|\}\, c^2_{zx'} \\ + 2\{|xxyz| - |zxxy|\}\, c_{yx'} c_{zx'} + 2\{|yyzx| - |xyyz|\}\, c_{zx'} c_{xx'} \\ + 2\{|xxyz| - |yzzx|\}\, c_{xx'} c_{yx'} \end{aligned}\right\} \text{(xx)}.$$

The thorough-going rari-constant elastician will fail to observe the existence of this surface, at least the Ossa of his multi-constant colleague will appear to him a wart.

(vi) Finally that there exist nine axes at each point of a body for which

$$|y_1 y_1 y_1 z_1| = |z_1 z_1 z_1 y_1|,$$

or the two direct-slide-stretch coefficients are equal. These directions Rankine terms *metatatic*. The condition for the metatatic isotropy of a body, or for metatatism in all pairs of rectangular directions, is

$$|y'y'z'z'| + 2\,|y'z'y'z'| = \tfrac{1}{2}\{|y'y'y'y'| + |z'z'z'z'|\} \qquad \text{(xxi)}.$$

Such a body, however, is not elastically isotropic[1].

[1] I have here introduced some portion of Rankine's work as given with great clearness by Saint-Venant in order that it may be the more easy to refer to these results in later articles.

[138.] Saint-Venant in the twelfth paragraph of this section of his memoir (pp. 360—5) treats the case in which the elastic material has three rectangular planes of symmetry. This reduces the 21 coefficients to nine, for all the stretch-slide coefficients and cross-slide coefficients (i.e. Rankine's *asymmetrical elasticities*) must now vanish.

$$\left.\begin{array}{lll} \text{Let } a, b, c & \text{be the} & \text{direct-stretch} \\ \quad d, e, f & \quad ,, & \text{direct-slide} \\ \quad d', e', f' & \quad ,, & \text{cross-stretch} \end{array}\right\} \text{coefficients.}$$

Then the tasinomic surface (xviii) becomes:

$$1 = ax^4 + by^4 + cz^4 + 2(2d + d')y^2z^2 + 2(2e + e')z^2x^2 + 2(2f + f')x^2y^2 \ldots \text{(xxii)}.$$

The maximum-minimum values of $|rrrr|$ are now sought and are found to lie in the three axial directions x, y, z, and in pairs of others lying in each plane yz, zx, xy, or 9 in all. The first three solutions are always real; the second six will be imaginary, since the ratio of their direction-cosines become imaginary, when

$$\left.\begin{array}{l} 2d + d' \\ 2e + e' \\ 2f + f' \end{array}\right\} \text{lie between} \left\{\begin{array}{l} b \text{ and } c \\ c \text{ and } a \\ a \text{ and } b \end{array}\right\} \text{respectively} \ldots\ldots \text{(xxiii)}$$

Saint-Venant remarks that the conditions (xxiii) are those for the *gradual variation in one sense* of the stretch-coefficients in the three principal planes of elastic symmetry—a physical characteristic, he holds, probably possessed by all natural bodies.

[139.] In the following section we have the statement of the conditions for *ellipsoidal* elasticity, i.e. that the first three quantities of (xxiii) be respectively equal: (i) to the arithmetic, or (ii) to the geometric mean of the corresponding second three quantities of (xxiii). In either case the direct-stretch coefficient $|rrrr|$ can be represented by the ray of an ellipsoid. In the first case the direct-stretch coefficient varies as the inverse square of the ray of the ellipsoid:

$$1 = ax^2 + by^2 + cz^2;$$

and in the second case as the inverse fourth power of the ray of the ellipsoid:

$$1 = x^2\sqrt{a} + y^2\sqrt{b} + z^2\sqrt{c}.$$

The practical application of this ellipsoidal distribution has been discussed by Saint-Venant in the annotated *Clebsch*: see our analysis of that work in Arts. 307 to 313.

[140.] The next two paragraphs (pp. 367—72) are occupied with an extension of Lamé's solution of the equations of elastic equilibrium by means of *potential functions*: see our Arts. 1061*—3*.

On the rari-constant hypothesis we should have $d = d'$, $e = e'$ and $f = f'$. As a sop to Cerberus Saint-Venant assumes that

$$d'/d = e'/e = f'/f = i \text{.....................(xxiv)}.$$

We may, however, doubt whether Cerberus would accept this sop; for, while supposing the constants unequal, it yet assumes their inequality isotropic in character. If multi-constancy really does exist, the relations (xxiv) are still probably very approximately satisfied for many bodies: see our Arts. 149 and 310.

Writing
$$a/(2 + i) = \mathrm{a}^2,$$
$$b/(2 + i) = \mathrm{b}^2,$$
$$c/(2 + i) = \mathrm{c}^2,$$

and supposing ellipsoidal distribution of the second kind, Saint-Venant finds

$$f = f'/i = \mathrm{ab}, \quad d = d'/i = \mathrm{bc}, \quad e = e'/i = \mathrm{ca}.$$

This enables him to reduce his body-shift equations to the type

$$\mathrm{a}u_{xx} + \mathrm{b}u_{yy} + \mathrm{c}u_{zz} + (1 + i)\frac{d\phi}{dx} = 0,$$

where
$$\phi = \mathrm{a}u_x + \mathrm{b}v_y + \mathrm{c}w_z \text{......................(xxv)}.$$

A very straightforward analysis then leads him to the result:

$$\phi = \iiint f(\alpha, \beta, \gamma)\left\{\frac{(x-\alpha)^2}{\mathrm{a}} + \frac{(y-\beta)^2}{\mathrm{b}} + \frac{(z-\gamma)^2}{\mathrm{c}}\right\}^{-\frac{1}{2}} d\alpha d\beta d\gamma \text{...(xxvi)}.$$

He also obtains (p. 371) the shift-type:

$$u = \iiint \chi_1(\alpha, \beta, \gamma)\left\{\frac{(x-\alpha)^2}{\mathrm{a}} + \frac{(y-\beta)^2}{\mathrm{b}} + \frac{(z-\gamma)^2}{\mathrm{c}}\right\}^{\frac{1}{2}} d\alpha d\beta d\gamma \text{.... (xxvii)},$$

where v and w will have other arbitrary functions χ_2, χ_3.

These arbitrary functions χ_1, χ_2, χ_3 do not seem to me so arbitrary as the reader might assume from Saint-Venant's words. We have so to choose χ_1, χ_2, χ_3 that the value of ϕ obtained from (xxv) by means of (xxvii) shall be the same as that obtained for ϕ from (xxvi).

It appears to me that u, v, w ought to be the x-, y-, z-fluxions respectively of a quantity

$$\psi = \tfrac{1}{2}\iiint f(\alpha, \beta, \gamma)\sqrt{\frac{(x-\alpha)^2}{\mathrm{a}} + \frac{(y-\beta)^2}{\mathrm{b}} + \frac{(z-\gamma)^2}{\mathrm{c}}}\, d\alpha d\beta d\gamma \text{... (xxviii)}.$$

In addition we might add to them certain expressions arising from the twists and giving a zero value for ϕ.

[141.] In the following paragraph Saint-Venant shows that the ellipsoidal conditions of the type $(2d + d') = \sqrt{bc}$ are necessary if a solution in terms of *direct and inverse potentials* is obtainable (pp. 372—4).

[142.] Hitherto the set of ellipsoidal conditions of the type

$$2d + d' = \sqrt{bc}$$

has been seen as one only of the number which satisfies the relations (xxiii). Saint-Venant now attempts to give it a far more important and special physical meaning. Namely, he proceeds to show that these relations hold exactly or very closely for bodies which originally isotropic have afterwards received a permanent strain unequal in different directions. He describes the bodies in question in the following terms:

En effet, dans les corps à cristallisation confuse tels que les métaux, etc., employés dans les constructions, où les molécules affectent indistinctement toutes les orientations, si les élasticités sont égales dans trois directions rectangulaires, elles doivent l'être en tous sens, car on ne voit aucune raison pour qu'elles soient plus grandes ou moindres dans les autres directions. Si les élasticités y sont inégales, cela ne peut tenir qu'à des rapprochements moléculaires plus grands dans certains sens que dans d'autres, par suite du forgeage, de l'étirage, du laminage, etc. ou des circonstances de la solidification. Calculons les grandeurs nouvelles que doivent prendre les coefficients d'élasticité dans un corps primitivement isotrope ainsi modifié (p. 374).

Bodies with 'confused crystallisation' Saint-Venant terms *amorphic solids*, and he now proceeds to show that within certain limits of aeolotropy, they possess an ellipsoidal distribution of elasticity. He assumes that the bodies have *rari-constant* elasticity.

[143.] Let s, s', s'' be the *principal stretches* of the permanent set given to the body, let ρ_0, r_0, x_0, y_0, z_0 be the density, distance between two elements, and its projections on the directions of the principal stretches before the isotropy is altered. Then if ρ, r, x, y, z be the value of these quantities after aeolotropy is produced, we have

$$\left.\begin{array}{c} x = x_0(1+s), \quad y = y_0(1+s'), \quad z = z_0(1+s'') \\ \rho = \rho_0 / \{\overline{1+s} \,.\, \overline{1+s'} \,.\, \overline{1+s''}\} \end{array}\right\} \text{.......} \quad \text{(xxix)}.$$

Let $f(r)$ be the law of intermolecular action, and $F(r) = \frac{1}{r}\frac{d}{dr}\left\{\frac{f(r)}{r}\right\}$ then we have, m being the mass of a molecule

$$\left\{\begin{array}{c} |xxxx| \\ |yyyy| \\ |xyxy| \end{array}\right\} = \frac{\rho}{2}\, \Sigma\, mF(r) \left\{\begin{array}{c} x^4 \\ y^4 \\ x^2y^2 \end{array}\right\} \text{..............} \quad \text{(xxx)}.$$

These results flow at once from the definition of stress on the rari-constant hypothesis and had been given by Cauchy in 1829 (see for example the annotated *Leçons de Navier*, p. 570, footnote and our Art. 615*).

Further if $r - r_0$ be small, we have:

$$F(r) = F(r_0) + (r - r_0) F'(r_0),$$

$$r - r_0 = \frac{x_0^2}{r_0} s + \frac{y_0^2}{r_0} s' + \frac{z_0^2}{r_0} s''$$

In the case of primitive isotropy we have

$$\left.\begin{aligned} &\frac{\rho_0}{2} \Sigma m F(r_0) x_0^4 = \frac{\rho_0}{2} \Sigma m F(r_0) y_0^4 = c_4, \text{ say,} \\ &\frac{\rho_0}{2} \Sigma m \frac{F'(r_0)}{r_0} x_0^6 = \frac{\rho_0}{2} \Sigma m \frac{F'(r_0)}{r_0} y_0^6 = c_6, \text{ say,} \\ &\frac{\rho_0}{2} \Sigma m \frac{F'(r_0)}{r_0} \{x_0^4 y_0^2, \text{ or } x_0^4 z_0^2, \text{ or } y_0^4 z_0^2, \text{ or } y_0^4 x_0^2\} \text{ are all equal} \\ &\qquad = c_{4,2}, \text{ say.} \end{aligned}\right\} \text{(xxxi)}$$

We will also put

$$\frac{\rho_0}{2} \Sigma m F(r_0) x_0^2 y_0^2 = c_{2,2},$$

$$\frac{\rho_0}{2} \Sigma m \frac{F'(r_0)}{r_0} x_0^2 y_0^2 z_0^2 = c_{2,2,2}.$$

Now substitute from (xxix) in (xxx) and using these values, we find

$$\left.\begin{aligned} |xxxx| &= \frac{(1+s)^3}{(1+s')(1+s'')} c_4 + c_6 s + c_{4,2} s' + c_{4,2} s'', \\ |yyyy| &= \frac{(1+s')^3}{(1+s)(1+s'')} c_4 + c_{4,2} s + c_6 s' + c_{4,2} s'', \\ |xyxy| &= \frac{(1+s)(1+s')}{1+s''} c_{2,2} + c_{4,2} s + c_{4,2} s' + c_{2,2,2} s''. \end{aligned}\right\} \ldots\text{(xxxii)}.$$

Now there are certain relations holding between the constants c, which are easily found thus: Change the axis of x by linear transformation:

$$x_0' = \alpha x_0 + \beta y_0 + \gamma z_0, \text{ where } \alpha^2 + \beta^2 + \gamma^2 = 1,$$

then from the initial isotropy we have

$$\Sigma m \chi(r_0) x_0^4 = \Sigma m \chi(r_0) x_0'^4,$$

and

$$\Sigma m \chi(r_0) x_0^6 = \Sigma m \chi(r_0) x_0'^6,$$

where $\chi(r_0)$ is any function of r_0 and α, β, γ may be any direction-cosines we please; it follows that:

$$\left.\begin{aligned} (\alpha^2 + \beta^2 + \gamma^2)^2 \Sigma m \chi(r_0) x_0^4 &= \Sigma m \chi(r_0) (\alpha x_0 + \beta y_0 + \gamma z_0)^4 \\ (\alpha^2 + \beta^2 + \gamma^2)^3 \Sigma m \chi(r_0) x_0^6 &= \Sigma m \chi(r_0) (\alpha x_0 + \beta y_0 + \gamma z_0)^6 \end{aligned}\right\} \ldots\text{(xxxiii)}.$$

These must be identities as they are true for all values of α, β, γ.

Hence we may equate like powers of α, β, γ on both sides. In the first relation by equating the coefficients of β^4 and again of $\alpha^2\beta^2$ we deduce the first of relations (xxxi) and also

$$\Sigma m \chi(r_0) x_0^4 = 3 \Sigma m \chi(r_0) x_0^2 y_0^2,$$

or

$$c_4 = 3 c_{2,2} \text{ ....(xxxiv).}$$

In the second relation by equating the coefficients of $\alpha^4\beta^2$, $\alpha^4\gamma^2$, $\beta^4\gamma^2$ and $\beta^4\alpha^2$ we obtain the third of relations (xxxi), as well as the new one

$$c_6 = 5c_{4,2} \text{ ....(xxxv).}$$

By equating the coefficients of β^6 we reach the second of relations (xxxi); and by equating those of $\alpha^2\beta^2\gamma^2$ the new one:

$$c_6 = 15c_{2,2,2} \text{ ....(xxxvi).}$$

From these relations[1] among the c's we have by multiplying out the first two expressions of (xxxii) *and neglecting the products of* s, s', s'',

$$|xxxx| \times |yyyy| = 9 \left\{ \frac{(1+s)^2(1+s')^2}{(1+s'')^2} c_{2,2}^2 + c_{2,2} c_{2,2,2} (6s + 6s' + 2s'') \right\}$$
$$= 9 \, |xyxy|^2.$$

This is the required type of relation on the hypotheses of *rari-constancy* and *small* permanent strain.

[144.] With regard to the latter assumption Saint-Venant remarks that the terms neglected can only produce very small errors:

> ...si l'on considère que les écrouissages et la trempe, qui changent très-sensiblement la ténacité et les coefficients d'élasticité, altèrent à peine la densité des corps. On peut d ailleurs s'assurer, par un calcul, que les portions ainsi négligées de l'expression de $3|xyxy|$ sont constamment comprises entre les portions correspondantes de celles de $|xxxx|$ et $|yyyy|$, en sorte qu'en supposant même qu'elles altèrent légèrement les valeurs absolues de ces trois coefficients, elles n'altèreront pas sensiblement pour cela la relation de moyenne proportionnalité de $3|xyxy|$ entre $|xxxx|$ et $|yyyy|$, donnée par les termes du premier ordre en s, s', s'' (p. 379).

The calculation mentioned is made by Saint-Venant in a foot note pp. 379—81.

The other assumption that rari-constancy holds for isotropy seems very approximately, if indeed not absolutely, true in the

[1] Saint-Venant obtains these relations among the c's by appealing to a general principle given by Cauchy in his *Nouveaux Exercices*, Prague, 1835, p. 35. It amounts to replacing 4 or 6 in (xxxiii) by the general index $2n$ and then equating general terms.

case of metals. We may then, I think, very legitimately adopt the ellipsoidal distribution indicated by the relations

$$2d + d' = \sqrt{bc},\ \ 2e + e' = \sqrt{ca},\ \ 2f + f' = \sqrt{ab} \ldots\text{(xxxvii)}$$

together with rari-constancy $d/d' = e/e' = f/f' = 1$ for most cases of worked metal such as is used in constructions.

[145.] The fourth section of the memoir (pp. 381—414) is entitled: *Conséquence, en ce qui regarde la théorie du mouvement de la lumière dans les milieux non isotropes, en tenant compte des pressions antérieures aux vibrations excitées.*

This section more properly belongs to the history of physical optics, and I shall content myself here with referring to its chief points without reproducing the analysis.

[146.] In the first place Saint-Venant refers to Green's memoir of 1839 (see our Arts. 917—18*), and states the conditions Green thinks needful in the optical medium which doubly refracts. These conditions in our notation are:

$$\left.\begin{aligned} |xxxx| = |yyyy| = |zzzz| = 2\,|yzyz| + |yyzz| &= 2\,|zxzx| + |zzxx| \\ &= 2\,|xyxy| + |xxyy| \\ |yyyz| = |zzzy| = |zzzx| = |xxxz| = |xxxy| = |yyyx| &= 0 \\ |xxyz| + 2\,|zxxy| = |yyzx| + 2\,|xyyz| = |zzxy| + 2\,|yzzx| &= 0 \end{aligned}\right\}\text{(xxxviii)}$$

They are obtained on the hypotheses of multi-constancy, of what Green terms extraneous pressures,—but Saint-Venant better initial stresses (*pressions antérieures*),—and finally of transverse vibrations being always accurately in the front of the wave. These conditions are practically identical with those obtained by Lamé: see our Art. 1106*.

[147.] Saint-Venant asserts that these conditions involve the isotropy of the medium in question, and therefore destroy the possibility of double refraction. If we suppose rari-constancy they are of course the conditions for isotropy,—does this however remain true in the case of multi-constancy?

Glazebrook in his *Report on Optical Theories* (*British Association Report,* 1885), p. 171, holds that Saint-Venant's criticism fails to reach Green. Let us endeavour briefly to indicate the lines of Saint-Venant's attack.

On pp. 384—393 he shews that Green's conditions flow from the hypotheses with which he has started. He then proves

(i) From the tasinomic relation that the stretch-coefficient is the same for every direction or

$$|x'x'x'x'| = |xxxx|.$$

Thus an equal stretch always produces the same element in the traction whatever its direction.

(ii) That the second set of Green's conditions are fulfilled for all axes, i.e.

$$|y'y'y'z'| = |z'z'z'y'| = 0, \text{ etc.}$$

(iii) That the conditions whose type is

$$|z'z'z'z'| = 2\,|y'z'y'z'| + |y'y'z'z'|$$

are true for any change of rectangular axes.

(iv) That the third set of conditions of the type

$$|x'x'y'z'| + 2\,|z'x'x'y'| = 0$$

are also true for any change of rectangular axes.

(v) That the reciprocal theorems are true, i.e. if any one of the relations in (i) to (iv) hold for all rectangular axes, then Green's fourteen conditions follow.

It will thus be noted that Green's conditions are not based upon any conception of direction in the body, if fulfilled for one set of rectangular axes they are fulfilled for all. So far as these conditions are concerned the body possesses isotropy of direction, i.e. there is nothing of the nature of crystalline axes, or *the peculiarity of the medium has no relation to direction in space.* This seems to me the element of isotropy in Green's conditions which Glazebrook misses, and which Saint-Venant overstates when he identifies it with absolute elastic isotropy. Glazebrook well points out that if we give a stretch s_x only we have the following system of stresses[1]:

$$\widehat{xx} = |xxxx|\,s_x, \qquad \widehat{yz} = |yzxx|\,s_x,$$
$$\widehat{xy} = |xxyy|\,s_x, \qquad \widehat{zx} = 0,$$
$$\widehat{zz} = |xxzz|\,s_x, \qquad \widehat{xy} = 0.$$

Here we are at liberty to take the stretch in the direction of the axis

[1] By choosing as our axes the *orthotatic axes* we can reduce the stress-strain relations as given by Green to the following types:

$$\widehat{xx} = a\theta - 2fs_y - 2es_z,$$
$$\widehat{yz} = d\sigma_{yz},$$

where $a = |xxxx| =$ same for all directions

$$\left.\begin{aligned} d &= |yzyz| \\ e &= |zxzx| \\ f &= |xyxy| \end{aligned}\right\} = \text{values for orthotatic axes of direct-slide coefficients,}$$

of x, because of the directional isotropy of Green's conditions. It follows that such a stretch produces no shear on a face perpendicular to its direction. Glazebrook notes that it does produce a shear $\widehat{yz}$, and that this shear together with the tractions $\widehat{yy}$, $\widehat{zz}$ may be functions of the direction, since Green's conditions do not involve

$$|xxyy|, \ |xxzz|, \text{ and } |yzxx|$$

being the same for all systems of rectangular axes.

But is this the system of stresses we should expect to find in the ether in a crystallised medium? It seems to me physically very improbable, but it is best to let Saint-Venant speak himself, only remarking that the reader will do well to understand by Saint-Venant's use of the word isotropy, the independence of Green's conditions of all sense of direction, as explained above:

Il en résulte que l'exacte *transversalité* des mouvements moléculaires, ou leur parallélisme à des ondes de toutes les directions dans un milieu transparent, exige une foule de conditions qu'on ne voit remplies que dans les corps isotropes. On remarque, surtout, que non-seulement *une dilatation $s_{x'}$ ne produit qu'une pression exactement normale $\widehat{x'x'}$*, ou aucune composante tangentielle de pression sur une face qui est perpendiculaire à sa direction ($|x'y'x'x'| = |x'z'x'x'| = 0$) et, aussi, *qu'un glissement sur une face n'y engendre jamais que des composantes tangentielles* ($|x'x'x'y'| = 0$), mais encore qu'*en tout sens*, ou quelle que soit la direction x' dans ce milieu, une égale dilatation $s_{x'}$ y produit une pression d'égale intensité $\widehat{x'x'}$ ($|x'x'x'x'|$ constant).

Or une pareille égalité est contraire à toutes les idées qu'on peut se former, d'après les faits, des corps doués de la double réfraction. Ils sont cristallisés sous des formes polyédriques non régulières et variées; ils offrent des clivages suivant certaines directions; ils sont, en un mot, d'une contexture essentiellement inégale dans les divers sens, et qui doit, tout porte à le faire présumer, rendre inégaux les rapports $\widehat{x'x'}/s_{x'} = |x'x'x'x'|$ des pressions dans l'éther dont ils sont imprégnés, aux petites dilatations qui les engendrent, et rendre les pressions obliques aux dilatations, excepté pour certains sens principaux Cette présomption est changée en certitude, si l'on considère la biréfringence artificiellement produite *par une compression* donnée dans un seul sens, ou inégalement dans plusieurs, à un corps amorphe primitivement isotrope et uni-réfringent, tel que le verre. On a en effet calculé, au no. XXXII (equation of our Art. 143), l'inégalité des coefficients $|xxxx|$, $|yyyy|$ due à l'inégalité des rapprochements moléculaires dans les sens x et y. Ce calcul était fondé, il est vrai, sur les expressions (equation XXX) assignées aux deux coefficients par l'analyse des actions s'exerçant entre les points matériels suivant leurs lignes de jonction deux à deux, et proportionnellement à une fonction de leur distance. Mais quelque motif qu'on puisse s'alléguer de révoquer en doute cette grande loi qui ne préjuge pourtant rien quant à la forme de la fonction, et quelque chose qu'on puisse concevoir à sa place, il est impossible de ne point convenir que l'inégal rapprochement moléculaire en divers sens doit influer sur la grandeur des *élasticités directes* $|xxxx| = \widehat{xx}/s_x$ comme elle influe bien certainement sur celle des autres élasticités, dites *latérales*, $|xxyy| = \widehat{xx}/s_y$, ou tangentielles, $|xyxy| = \widehat{xy}/\sigma_{xy}$, etc., puisque sans les inégalités au moins de celles-ci en divers sens, les formules ne donneraient pas de double réfraction. Un milieu ne peut être élastique et vibrant si ses parties n'agissent pas les unes sur les autres, et quelque soit le mode de leur action, il n'est pas possible d'imaginer qu'elles engendrent des élasticités directes parfaitement égales, lorsqu'il y a une inégalité de contexture qui rend inégales les élasticités latérales ou tangentielles. (pp. 396—8.)

This argument seems to me of great weight (see, however, a point raised in our Art 193 (1)), and would incline me to reject Green's conditions (especially when we remember that Green himself supposed the ether-density to vary in refracting media), even were there no other grounds for questioning his hypotheses.

[148.] Saint-Venant now proceeds to deduce the *exact* wave-surface of Fresnel on the supposition that the vibrations are not accurately in the wave-front. He does this on the lines of Cauchy's memoir of 1830, but he does not assume rari-constancy and in many respects his method is an improvement on Cauchy's. This leads him to the following inter-constant conditions; the structure of the ether being supposed to have three planes of symmetry and thus its elasticity to be represented by the nine constants of our Art. 117 (a):

$$\left.\begin{aligned}&(b-d)(c-d)=(d+d')^2,\ (c-e)(a-e)=(e+e')^2\\&(a-f)(b-f)=(f+f')^2\\&(a-e)(b-f)(c-d)+(a-f)(b-d)(c-e)\\&\qquad\qquad=2(d+d')(e+e')(f+f')\end{aligned}\right\}\ldots\text{(xxxix)}$$

If the relations (xxxix) are satisfied we shall have Fresnel's wave-surface. If we make $a=b=c$ we shall reduce these conditions to Green's, which are thus only a particular case of those of Cauchy and Saint-Venant. (pp. 398—406.)

[149.] On pp. 406—411 Saint-Venant demonstrates that the relations (xxxix) give practically the same results as the ellipsoidal distribution of (xxxvii). He supposes $d/d' = i$ and then solves the first equation of both sets (xxxix) and (xxxvii) for d; let the values so obtained be respectively d_1 and d_2. Then by a numerical calculation we reach the following results:

If b/c varies from 1·1 to 1·5, then for values of ι between $\frac{1}{2}$ and 2 the ratio of d_1/d_2 always lies between ·98641 and ·99962. In other words whatever the multi-constant i is between these limits, the relations (xxxix) and (xxxvii) give practically the same value of d.

Thus the Cauchy-Saint-Venant conditions correspond closely to the ellipsoidal distribution, which is the distribution we should expect in a body like the ether originally isotropic, but, owing to its presence in the doubly-refracting medium, subjected to an initial state of strain.

The fourth condition of (xxxix) is shewn to be very nearly true if the first three are satisfied (pp. 409—411).

[150.] The objections to Saint-Venant's theory are given by Glazebrook (*op. cit.* pp. 172—3). They consist in: the difficulty of reconciling the theories of double refraction and reflexion so long as we suppose the latter to depend "on difference of density and not of rigidity in the two media," and the existence of the "quasi-normal wave." The latter objection is met by Saint-Venant with the arguments of Cauchy (see his pp. 411—13), and it does not seem insuperable; the former is in some respects serious, and is not discussed by Saint-Venant. At the same time we must observe that the ellipsoid-distribution to which the Cauchy-Saint-Venant conditions approximate does suppose a change in the elastic constant $|yzyz|$ owing to the isotropic ether being rendered aeolotropic in the doubly-refracting medium: see our Art. 143, equation xxxii.

The whole subject is of peculiar interest apart from its bearing on the theory of light, as tending to introduce us by means of the elastic constants into the molecular laboratory of nature—indeed this is the transcendent merit of rari-constancy, if it were only once satisfactorily established!

[151.] Saint-Venant's fifth section (pp. 414—425) is entitled: *Distribution, en divers sens, des modules ou coefficients d'élasticite définis à la manière de Young et de Navier.* This is the determination of the stretch-modulus quartic as first given by Neumann (see our Art. 799*). It is shewn how this may be determined for multi-constancy, but it is pointed out that in the most general case there will be a denominator of 720 terms in the constants, and Saint-Venant wisely contents himself with the case of three planes of symmetry and a 9-constant medium.

The conclusions drawn as to the nature of the quartic and its special reduction to an ellipsoid, are all treated with somewhat fuller detail in the annotated *Clebsch,* and we have accordingly discussed them in our analysis of that work: see our Arts. 308 to 310.

[152.] We may note that Saint-Venant (pp. 424—5) attempts to apply the ellipsoidal distribution of elasticity, which leads to the ellipsoidal distribution of stretch-modulus, i.e.

$$\frac{1}{\sqrt{E_r}}=\frac{c^2_{xr}}{\sqrt{E_x}}+\frac{c^2_{yr}}{\sqrt{E_y}}+\frac{c^2_{zr}}{\sqrt{E_z}},$$

to the case of *wood.* He appeals to Hagen's results (see our Art. 1229*) and compares Hagen's empirical formula

$$\frac{1}{E_r}=\frac{c^3_{xr}}{E_x}+\frac{c^3_{yr}}{E_y} \quad\ldots\ldots\ldots\ldots\ldots\ldots\ldots\ldots(\alpha)$$

with that given by the ellipsoidal distribution

$$\frac{1}{\sqrt{E_r}}=\frac{c^2_{xr}}{\sqrt{E_x}}+\frac{c^2_{yr}}{\sqrt{E_y}} \quad\ldots\ldots\ldots\ldots\ldots\ldots(\beta).$$

He shews the theoretical impossibility of Hagen's formula, arising from the fact that if $E_x = E_y$, E_r is not equal to them, and endeavours to shew by curves that (β) and (α) coincide within the limits of experimental error. By graphical representation of the curves it is seen that only the ellipsoidal distribution gives anything like a satisfactory theoretical as well as practical figure, and Saint-Venant concludes that, although proved for a different kind of medium (see our Arts. 142 and 144), it may be practically of use in the case of fibrous material like wood. Later Saint-Venant saw occasion to alter this opinion; he treats this important material very fully in the *Leçons de Navier* (pp. 817—25) and in the annotated *Clebsch* (pp. 98—110). Under the latter heading we shall discuss his more complete treatment of the subject: see our Arts. 308—310. The memoir ends with the *résumé* to which we have before referred.

[153.] *Sur la détermination de l'état d'équilibre des tiges élastiques à double courbure. Les Mondes,* Tome 3, 1863, pp. 568—575. This note was a contribution to the *Société Philomathique,* August 8, 1863; see also *L'Institut,* 1863, pp. 324—5.

Consider a rod of double curvature; let M_t, M_n, M_ρ be the moments of the applied forces about the tangent to the central axis, the normal to the osculating plane and the principal radius of curvature. Let I, I' be the moments of inertia about the principal axes of the cross-section, and let e be the angle the radius of curvature ρ makes in the unstrained state with the axis of I'; then Saint-Venant gives the two following formulae, where ϵ is the increment in e and δs is an element of central axis:

$$\sin \epsilon = \frac{\rho}{E}\left\{M_\rho\left(\frac{\cos^2 e}{I} + \frac{\sin^2 e}{I'}\right) - M_n \sin e \cos e\left(\frac{1}{I} - \frac{1}{I'}\right)\right\},$$

$$\frac{dM_t}{ds} = \frac{M_\rho}{\rho}.$$

Hence when $e = 0$ or $\pi/2$, or $I = I$, ϵ depends only on M_ρ the moment of the forces round the radius of curvature.

The second equation shews that the moment of torsion M_t is only constant when $M_\rho = 0$ along the whole length of the wire.

Saint-Venant refers to the work of Poisson, Wantzel and Binet: see our Arts. 1599*—1607* He also reproduces the example of the *Comptes rendus*: see our Art. 155, and that of the horizontal semi-circular bar of rectangular cross-section built-in at both terminals and loaded at its mid-point used in the *Leçons de Navier*, p. cxxxiv, which bring out clearly the need of taking into consideration the angle ϵ.

Saint-Venant refers to Bresse: *Cours de mécanique appliquée: Résistance des matériaux*, 1859, p. 86, for a good investigation of the general formulae for elastic wires of double-curvature when the shifts are small.

[154.] *Sur la théorie de la double réfraction: Comptes rendus*, T. 57, 1863, pp. 387—391.

This is a note on a memoir by Galopin, and points out that there is no need to put the initial stresses zero in the ether in order to obtain Cauchy's conditions for double refraction see our Art. 148. The contents of this note are practically involved in the memoir of 1863: see our Art. 127, and concern properly the historian of the undulatory theory of light.

[155.] *Sur les flexions et torsions que peuvent éprouver les tiges courbes sans qu'il y ait aucun changement dans la première ni dans la seconde courbure de leur axe ou fibre moyenne* *Comptes rendus*, T. 56, 1863, pp. 1150—54. See also *L'Institut*, Vol. 31, 1863, pp. 195—6.

This memoir draws attention to the point considered by Saint-Venant in his memoirs of 1843 and 1844; see our Arts. 1598* and 1603*; namely the importance of taking into consideration the 'angle of torsion' or angle between new and old osculating planes in dealing with the elastic equilibrium of wires of double-curvature. Saint-Venant brings out the importance by a good example, namely a curved wire turned upon itself so as to have the same curvature at each point of the central axis, but so that the naturally longest and shortest 'fibres' interchange places.

He points out that the stretch in a fibre distant z from the central axis is:

$$z\sqrt{1/\rho^2 - 2/\rho\rho_0 \,.\, \cos\epsilon + 1/\rho_0^2},$$

where ρ, ρ_0 are the new and the primitive radii of curvature and ϵ the angle the new and old radii of curvature make with each other. In the example above referred to $\rho = \rho_0$ and $\epsilon = \pi$, so that the stretch becomes

$$2z/\rho_0.$$

Generally when $\rho = \rho_0$, the stretch equals

$$2z/\rho_0 \ \sin\tfrac{1}{2}\epsilon.$$

In conclusion Saint-Venant refers to the contributions of Lagrange, Poisson, Binet, Wantzel and himself to the subject: see our Art. 1602* for references.

[156.] *Mémoire sur les contractions d'une tige dont une extrémité a un mouvement obligatoire; et application au frottement de roulement sur un terrain uni et élastique: Comptes rendus*, T. 58, 1864, pp. 455—8.

This memoir was written in 1845, and is an attempt to apply the theory of elasticity to the phenomena of rolling friction. The chief results were published in the *Bulletin de la Société Philomathique* of June 21, 1845. The following conclusions are given in the *résumé* in the *Comptes rendus*:

On en déduit que le frottement de roulement sur un pareil sol est: 1° proportionnel à la pression; 2° en raison inverse du rayon du cylindre; 3° indépendant de sa longueur (ou de la largeur de jante, si c'est une roue); 4° proportionnel à la vitesse; 5° d'autant moindre que le terrain élastique est plus roide ou moins compressible.

Saint-Venant remarks:

Ces résultats sont d'accord avec un certain nombre d'expériences de Coulomb et de M. Morin. (p. 457.)

There is a general indication of the method of treatment adopted in the original memoir, but it is not sufficient to replace its analysis. The memoir itself appears never to have been published.

157. *Travail ou potentiel de torsion. Manière nouvelle d'établir les équations qui régissent cette sorte de déformation des prismes élastiques. Comptes rendus*, T. 59, 1864, pp. 806—809. Translated in the *Philosophical Magazine*, January, 1865, pp. 61—64.

In his memoir on Torsion Saint-Venant used one equation which holds at every point within a body, and one which holds at every point of the convex surface: see equations (vi) of our Art. 17 on that memoir. In the present paper Saint Venant undertakes to obtain these equations simultaneously by the aid of the principle of *Work*.

The potential of elasticity, that is to say the molecular work ϕ which a deformed element is capable of furnishing, is thus expressed for the unit of volume of the element:

$$\phi = \tfrac{1}{2}\widehat{xx}s_x + \tfrac{1}{2}\widehat{yy}s_y + \tfrac{1}{2}\widehat{zz}s_z + \tfrac{1}{2}\widehat{yz}\sigma_{yz} + \tfrac{1}{2}\widehat{zx}\sigma_{zx} + \tfrac{1}{2}\widehat{xy}\sigma_{xy}.$$

Now the values of the component stresses $\widehat{xx}$, $\widehat{yy}$,..... can, we know, be expressed as linear functions of the six strains s_x, s_y, s_z, σ_{yz}, σ_{zx}, σ_{xy}; substitute these values in ϕ, and we obtain an expression of the second degree in the strains, consisting of twenty-one terms. In the case of torsion which we are considering, the strains reduce to the two σ_{xy} and σ_{xz}, so that we have

$$\phi = \tfrac{1}{2}\mu_1 (\sigma_{xy})^2 + \tfrac{1}{2}\mu_2 (\sigma_{xz})^2,$$

where μ_1 and μ_2 are the slide-moduli in the directions of y and z: see Art. 17 of our account of the memoir on Torsion.

Now let M denote the *moment of torsion* so that

$$M = \int dy\, dz\, (\widehat{xz}\, y - \widehat{xy}\, z).$$

Thus if the moment of torsion is measured by an angle τ we have $M\,\dfrac{\tau}{2}$ for the molecular work; so that by equating the two expressions for this work we obtain

$$\tfrac{1}{2}\iint dy\, dz\, (\mu_1\sigma^2_{xy} + \mu_2\sigma^2_{xz}) = \tfrac{1}{2}\tau\iint dy\, dz\, (\widehat{xz}y - \widehat{xy}z) \quad \text{........} \quad (1).$$

Now we assume that the body has three planes of symmetry perpendicular to the axes of x, y, z respectively; so that

$$\widehat{xy} = \mu_1\sigma_{xy}, \qquad \widehat{xz} = \mu_2\sigma_{xz};$$

also
$$\sigma_{xy} = \frac{du}{dy} - \tau z, \qquad \sigma_{xz} = \frac{du}{dz} + \tau y,$$

by equation (iii) of our Art. 17.

Substitute in the above equation (1) and we obtain

$$\iint dy\,dz\left\{\mu_1\frac{du}{dy}\left(\frac{du}{dy}-\tau z\right)+\mu_2\frac{du}{dz}\left(\frac{du}{dz}+\tau y\right)\right\}=0.$$

Integrate this equation by parts in the usual way, and it becomes

$$\int u\left[\mu_1\left(\frac{du}{dy}-\tau z\right)\cos(ny)+\mu_2\left(\frac{du}{dz}+\tau y\right)\cos(nz)\right]ds$$
$$-\tau\iint u\left[\mu_1\frac{d^2u}{dy^2}+\mu_2\frac{d^2u}{dz^2}\right]dy\,dz=0 \ldots\ldots\ldots\ldots\ldots (2);$$

here (ny) and (nz) denote the angles which the normal to the surface at the point (x, y, z) makes with the axes of y and z respectively; and ds is an element of the curve of intersection of the body by a plane at right angles to the axis of x.

If we equate to zero the term in brackets in the double integral we obtain the equation which must hold at every point of the interior; and if we equate to zero the term in brackets in the single integral we obtain the equation which must hold at every point of the surface.

But Saint-Venant does not explain why we must equate these terms separately to zero; that is, he does not explain why he breaks up equation (2) into *two* equations. Moreover the whole process borrows so much from the memoir on Torsion that it has not the merit of being an independent investigation.

Saint-Venant says:

Or la deuxième et la première parenthèse carrée, égalées séparément à zéro... :

by this he means the terms contained within the square brackets in (2). The English translation has very strangely "Now the squares of the second, and of the first parenthesis, each equated to zero,..."

[158.] A remark of Saint-Venant's on p. 809 may be cited:

Le calcul du potentiel de torsion a aussi, en lui-même, une valeur pratique; car les ressorts en hélice, qu'on oppose souvent à divers chocs, travaillent *presque* entièrement par la torsion de leurs fils, ainsi que je l'ai montré en 1843, et que l'ont ermarqué, au reste, Binet dès 1814, M. Giulio en 1840, et récemment des ingénieurs des chemins de fer.

See our Arts. 175*, 1220*, 1382* and 1593—5*. The 1814 and the *récemment* (1864) mark the wide interval which too often separates theory from practice!

[159.] *Théorie de l'élasticité des corps, ou cinématique de leurs déformations. Les Mondes*, Tome 6, 1864, pp. 607 and 608. If a body is deformed any small portion originally spherical becomes an ellipsoid: see our Art. 617* In the present paper Saint-Venant undertakes to establish this proposition by simple general reasoning; the process does not seem very satisfactory.

SECTION III.

Researches in Technical Elasticity.

[160.] *Résumé des Leçons...sur l'application de la mécanique à l'établissement des constructions et des machines....Première section. De la Résistance des corps solides, par Navier....Troisième Édition avec des Notes et des Appendices par M. Barré de Saint-Venant.* The title-page bears the imprint, Paris, 1864. A foot-note, however, on p. 1 tells us that pp. 1—224 appeared in 1857, pp. 225—336 in 1858, pp. 337—496 in 1859, pp. 497—688 in 1860, pp. 689—849 in 1863, while the *Notices et l Historique*, pp. i—cccxi, were finally added in 1864. Thus the whole work of more than 1100 pages occupied some seven years in the production, and thus necessarily lacks somewhat of the unity which is to be met with in other treatises. Under the form of notes to a few sections of Navier's original work (see our Art. 279*), Saint-Venant has given us a complete text-book of elasticity from the practical standpoint. At the same time, by additional notes and appendices, he has rendered his text-book of surpassing historical value and physical suggestiveness. The leading characteristics of the book are simplicity of analysis and copiousness of reference. See *Notice* I., pp. 41—2 and *Notice* II., pp. 28—9.

[161.] The cccxi. pages of introductory matter are occupied with the following subjects: Table of Contents, pp. i—xxxviii; *Notice biographique sur Navier* by de Prony extracted from the *Annales des ponts et chaussées* (1837, 1er semestre, p. 1), pp. xxxix—

li; the funeral discourses on Navier by Emmery and Girard, pp. li—liv: a bibliography of the works of Navier with copious remarks due to Saint-Venant, pp. lv—lxxxiii; the original prefaces to the editions of Navier's *Leçons* published in 1826 and 1833; pp. lxxxiv—xc; and finally Saint-Venant's *Historique abrégé des recherches sur la résistance et sur l'élasticité des corps solides*, pp. xc—cccxi

[162.] The *Historique abrégé* is practically the only brief account of the chief stages of our science extant. Girard had written what was for his day a fair sketch of the *incunabula* (see our Art. 123*), but it remained for Saint-Venant, without entering into the analysis of the more important memoirs, to describe their purport and relationship. It fulfils a different purpose to our own history—for it makes no attempt to replace the more inaccessible memoirs—but as a model of how mathematical history should be written, we hold it to be unsurpassed, and can only regret that a recent French historian has not better profited by the example thus set[1]

We would especially recommend to the student of Saint-Venant's memoirs pp. clxxiii—cxcii, which treat of the relation of his own researches by means of the *semi-inverse* method to the work of his predecessors. The point we have referred to in our Arts. 3, 6, 8 and 9 is well brought out in relation to Lamé's problem of the right-six-face.

We will note one or two further points of the *Historique* in the following five articles.

[163.] On p. cxcviii in the footnote Saint-Venant gives the expression for the work-function in terms of the stresses when there is an *ellipsoidal* distribution of elasticity: see our Art. 144. He finds

$$W = \frac{1+i}{2(2+3i)}\left(\frac{\widehat{xx}}{a} + \frac{\widehat{yy}}{b} + \frac{\widehat{zz}}{c}\right)^2 + \frac{\widehat{yz}^2 - \widehat{yy}\,\widehat{zz}}{2bc} + \frac{\widehat{zx}^2 - \widehat{zz}\,\widehat{xx}}{2ca} + \frac{\widehat{xy}^2 - \widehat{xx}\,\widehat{yy}}{2ab},$$

where for isotropy $i = \lambda/\mu$ and $a^2 = b^2 = c^2 = \mu$.

[1] The essential feature of scientific history is the recognition of growth, the interdependence of successive stages of discovery. This evolution is excellently summarised in Saint-Venant's *Historique*. Our own 'history is only a bibliographical repertorium of the mathematical processes and physical phenomena which form the science of elasticity, as a rule for the purpose of convenience chronologically grouped. M. Marie's *Histoire des sciences mathématiques* is a chronological biography, without completeness as bibliography or repertorium. Excellent fragments there are in it, but the conception of evolutionary dependence is wanting.

Generally:

$$|xxxx| = (2+i)\,a^2, \qquad |yzyz| = bc, \qquad |yyzz| = ibc,$$
$$|yyyy| = (2+i)\,b^2, \qquad |zxzx| = ca, \qquad |zzxx| = ica,$$
$$|zzzz| = (2+i)\,c^2, \qquad |xyxy| = ab, \qquad |xxyy| = iab.$$

[164.] Pages cxcix—ccix deal with the history of the problem of rupture. According to Saint-Venant, two kinds of rupture may be distinguished: *rupture prochaine* and *rupture éloignée.* The former falls outside the theory of 'perfectly elastic' bodies, the latter he thinks may be deduced from the hypothesis that when the limit of mathematical elasticity is passed,—i.e. when the stretch is greater than the limit at which stretch remains wholly elastic and proportional to traction,—then the body will ultimately be ruptured if it has to sustain the same load. The reader who has followed our analysis of the *state of ease* and the *defect in Hooke's Law* given in the appendix to Vol. I. and also our Arts 4 (γ) and 5 (a) in the present volume will recognise that this hypothesis has only a small field of application. What we have really obtained is a *limit to linear elasticity.* It is the more important to notice this because Saint-Venant argues that we must take as our limit the maximum *positive stretch,* for, as Poncelet has asserted: "*que le rapprochement moléculaire ne peut être une cause de désagrégation*" (p. cci). It is probably true that rupture can only be produced by stretch, but *squeeze* can surely produce failure of linear elasticity when the body is so loaded that no transverse stretch is possible. Hence when Saint-Venant introduces the stretch and slide-moduli into his condition for safe loading and so makes it a question of *linear elasticity,* it seems to me that he ought at the same time to alter his statement as to the *greatest positive stretch* being the only quantity we are in search of. Indeed, his condition seems partly based upon an idea associated with *rupture,* and is then applied to constants and equations deduced from the principle of linear elasticity (see his p. ccviii, § XLVIII.). The limitations to which his theory is subjected were, however, partially recognised by Saint-Venant himself (see his pp. ccv—vii). Thus he writes:

Nous ne prétendons pas, au reste, qu'une théorie subordonnant uniquement le danger de rupture d'un solide à la grandeur qu'atteint une dilatation linéaire n'importe dans quelle de ses parties, et indépendamment des autres circonstances où il se trouve en même temps, soit le dernier mot de la science et de l'art.

He refers on this point to the experiments of Easton and Amos: see our Art. 1474*.

[165.] Pages ccxiv—xxiv deal with the problems of resilience and impact.

In the footnote p. ccxvii, there is an error in the integral of the equation $\frac{d^2z}{dt^2} = g \cos \alpha - \frac{g}{f} z$ there given. It should be

$$z = f \cos \alpha + V \sqrt{\frac{f}{g}} \sin \sqrt{\frac{g}{f}}\, t - f \cos \alpha \cos \sqrt{\frac{g}{f}}\, t.$$

The error was noted by Saint-Venant himself in a letter to the Editor of this History, August, 1885.

On p. ccxxii and footnote there should have been a reference to Homersham Cox with regard to the factor $k = 17/35$. His memoir of 1849 (see our Art. 1434*) seems to have escaped Saint-Venant's attention.

A further consideration of the effect of impact on bars when the vibrations are taken into account occurs on pp. ccxxxii—viii, and then follows (pp. ccxxxix—xlix) an account of Stokes' problem of the travelling load (see our Art. 1276*). Saint-Venant refers to the researches of Phillips and Renaudot, but his account wants bringing up to date by reference to more recent researches.

[166.] On pp. ccxlix—ccliii Saint-Venant refers to the rupture conditions given by Lamé and Clapeyron and again by Lamé for cylindrical and spherical vessels. It seems to me that he has not noticed here that these conditions are, on his own hypothesis of a *stretch* and not a *traction* limit, erroneous: see the footnotes to our Arts. 1013* and 1016*.

[167.] After an excellent and succinct account of the course of the investigations of Euler, Germain, Poisson, Kirchhoff &c. with regard to the vibrations of elastic plates (pp. ccliii—cclxxi) the *Historique* closes with two sections LXI. and LXII. (pp. cclxxi—cccxi) on the experiments made by technologists and physicists previously to 1864 on the elasticity and strength of materials. Good as these pages are, they are insufficient to-day in the light of the innumerable experiments of first-class importance made during the last twenty years.

[168.] In considering Saint-Venant's edition of Navier we shall leave the original text out of consideration, and note only those points of Saint-Venant's additions (ten-fold as copious as the original text) which present novelty of treatment or result. We put aside all matters already discussed in the memoirs on Torsion and Flexure. Those memoirs are here to a great extent embodied, their processes simplified and their results extended.

[169.] (*a*) On pp. 2—3 we find Saint-Venant basing the theory of elasticity on the principle of a central inter-molecular action which is a function of the distance.

(*b*) On p. 4, § 6 we have *écrouissage* and *énervation* defined. These definitions are rather theoretical than practical. Thus Saint-Venant defines as *écrouissage* the arrangements taken by the molecules of a body when they pass by changes which are persistent from a less to a more stable condition of equilibrium, as *énervation* the arrangements when they pass to a less stable condition. It will be noted that the physical characteristics of set, yield-point and plasticity are not clearly brought out by these definitions.

(*c*) Pp. 5—14 treat of rupture by compression. Saint-Venant rejects the theory of Coulomb (see our Art. 120*) as giving a stress not a stretch limit. He adopts that of Poncelet, who in 1839 in a course given at Paris, ascribed rupture by compression to the transverse stretch which accompanies longitudinal squeeze (pp. 6 and 10, and compare with footnote p. 381) That short prisms of cast iron, cement &c. often take 8 to 10 times as great a load to rupture them by negative as by positive traction and not the 4 times of the uni-constant theory, is attributed not to bi-constant isotropy but to terminal friction which hinders the lateral expansion, or to want of isotropy (pp. 10 and 12). Such *rupture*, however, really lies at present outside theory.

(*d*) On pp. 15—19 we have the generalised Hooke's Law and the definition of the stretch-modulus (E) and the stretch-squeeze ratio (η). Saint-Venant remarks, that theoretically $\eta = \frac{1}{4}$ (i.e. on the uni-constant hypothesis), that Wertheim finds it differs little from $\frac{1}{3}$, and that it can never be $> \frac{1}{2}$ as otherwise a traction would diminish the volume of a prism of the given substance, "*ce qui n'est pas supposable.*" There is no further reason given why we cannot suppose the volume to diminish We may, however, look at the matter thus:

Let ξ = the ratio of the slide-modulus to the dilatation coefficient ($=\mu/\lambda$), then (Vol. I. p. 885):

$$\xi = \frac{3 - E/\mu}{E/\mu - 2}$$

Hence, since ξ is necessarily positive, we must have $E/\mu > 2$ and < 3 (the mean of these gives the uni-constant hypothesis $E/\mu = 5/2$). But $\eta = \frac{E}{2\mu} - 1$, or η can only have values from 0 to $\frac{1}{2}$.

This proof holds only for an isotropic material. In the case of an aeolotropic material it does not seem obvious why a longitudinal stretch should not produce a negative dilatation. The ratio of dilatation to stretch

$$= \frac{s_x + s_y + s_z}{s_x} = 1 - \eta_1 - \eta_2,$$

and in the case of wood the values obtained for η_1, η_2 would seem to give this a negative value, for they are $> \frac{1}{2}$. Saint-Venant admits later this possibility: see his pp. 821—2. Hence any set of experiments which give values for $\eta > \frac{1}{2}$ may be taken to denote that the material in question is not isotropic and homogeneous.

(*e*) On pp. 20—21 it is suggested that for some substances it is advisable to consider the stretch-modulus E as varying over the cross-section of a prism. Saint-Venant refers to the experiments on this point of Collet-Meygret and Desplaces: see our Chapter XI. He also regards Hodgkinson's experiments as leading to a like conclusion notwithstanding a special experiment to the contrary: see our Arts. 952* (iii), 1484* and references there. We thus have the formula

$$P_x = s_x \int E_x d\omega$$

put forward by Bresse, where P_x is the total traction in a prism stretched s_x in the direction of its axis x, and $(\int E_x d\omega)/\omega$ is the mean value of the stretch-modulus over the cross-section ω. For metals *coulés ou laminés*, where on the lateral faces there is a surface or *skin* change of elasticity, Saint-Venant would take:

$$P_x = s_x (E_0\omega + \epsilon\chi),$$

χ étant le périmètre de la section supposée diminuée d'un à deux millimètres tout autour, afin de représenter le développement moyen de la croûte douée généralement de plus de roideur et de nerf que le reste; et E_0 et ϵ étant deux coefficients à déterminer par les méthodes connues de compensations d'anomalies en faisant des expériences d'extension sur des barres ayant des grosseurs ou des formes sensiblement différentes (p. 21).

(*f*) Saint-Venant returns to this same point on pp. 42—44, and pp. 115—118 when treating of the problem of flexure. In the former passage, Saint-Venant gives reasons for adopting in the case of *metal* a *skin* change only in the elastic-modulus. He proposes the formula

$$M = \frac{E_0 I + ei}{\rho},$$

for the bending-moment, $1/\rho$ being the curvature, I the moment of inertia of the section, and i that of its contour, or rather of the mean line of the skin zone (ligne qu'on peut placer à 1 ou à 2 millimètres à l'intérieur). E_0 and e are to be determined by experiments on the flexure of bars of the given material but sensibly different in size and form.

In the case of wood, Saint-Venant, referring to the experiments of Wertheim and Chevandier (see our Art 1312*), adopts a parabolic law for the variation of the stretch-modulus. Let E_0 and E_1 be the moduli in the direction of the fibre at the centre ($r = 0$) and circumference ($r = r_1$) of the tree, then at any other point (r) we have

$$E = E_0 - (E_0 - E_1)\, r^2/r_1^2.$$

Saint-Venant determines the value of $\int Ey^2 d\omega$—i.e. the 'rigidity'—for a bar of rectangular cross-section ($b \times c$) whose centre of gravity was, before it was hewn, distant r_0 from the centre of the tree (p. 44).

In the second passage to which I have referred the rupture condition (rather the failure of linear elasticity) is deduced from the like hypothesis of skin-change. Saint-Venant obtains a formula

$$M_0 = \frac{T_0}{Ey}(E_0 I + ei),$$

where M_0 is the maximum bending moment which will not cause the elasticity of a 'fibre' at distance y from the neutral axis (where the stretch-modulus $= E$) to fail by giving it a greater stretch than T_0/E. We have then to find the fibre for which T_0/Ey is smallest.

Si l'on a des raisons de penser que c'est la fibre la plus dilatée comme quand la matière est homogène, ou que la contexture hétérogène est telle que le rapport T_0/E varie moins que y, l'équation sera, en désignant comme à l'ordinaire par y' la grandeur de l'ordonnée de cette fibre, et par E', T'_0, les valeurs correspondantes de E, T_0;

$$M_0 = \frac{E_0 T'_0 I}{E'y'} + \frac{e T'_0 i}{E'y'},$$

ou bien, C et c désignant deux constantes dépendant comme E_0 et e de la nature de la matière et de son mode de forgeage ou de fusion,

$$M_0 = C\, I/y' + c\, i/y'.$$

Saint-Venant calculates the value of M_0 for a rectangular section, and also deals with a similar expression for the case of the *wood* prism referred to above; see his pp. 117—8.

(*g*) In §§ 8—12 (pp. 22—26) the reader will find some account of the behaviour of a material under stress continued even to rupture. This account was doubtless for the time succinct and good, but there are several points which could only be accepted now-a-days with many reservations. For example the statement (§ 11): *Le calcul théorique est toujours applicable pour limiter les dilatations et établir les conditions de résistance à la rupture éloignée*—is one which requires much reservation. We have seen in Vol. I. p. 891 that a material may be in a state of ease and yet not possess linear elasticity for strains such as often occur in practice. Further that even when there is linear elasticity its limit can often be raised without *enervation* almost up to the yield-point, where one exists. Hence when Saint Venant takes s_0 to be the stretch at which material ceases *de s'écrouir et commence à s'énerver, ce qui se manifeste par la marche des allongements persistants*, and puts $P_0 =$ or $< E\omega s_0$ as the safe tractive load—where E is the stretch-modulus and ω the sectional area—we find some difficulty in ascertaining what limit s_0 really represents. In most cases before enervation begins, *linear* elasticity will be long gone, and all the formula really can tell us is the stage at which *linear elasticity fails*; this fail-limit may be very far from the yield-point, and in some materials very far indeed from the elastic limit.

Saint-Venant refers to the 'fatigue' of a material due to repeated loading and to the question whether vibrations can change the molecular structure from fibrous to crystalline (see our Arts. 1429*, 1463* and 1464*). These are points on which we know to-day a good deal more than was accessible in 1857.

[170.] Article III. is devoted to the flexure of prisms and commences with a criticism of the Bernoulli-Eulerian hypothesis as expounded by Navier. Saint-Venant shews with the simplest analysis that the cross-sections neither retain their original contour (not even in the simple case of 'circular' flexure, § 3, p. 34) nor their original planeness (§ 4, pp. 36—9). To § 6, p. 40—2, we have already referred when dealing with the question of equipollent load-systems in Art. 8 of our account of the memoir on *Torsion.*

[171.] Pages 52—58 of this Article reproduce with some important additions the formulae of Art. 14 of our account of the memoir on *Torsion.* Saint-Venant proves the following results for the case when the load plane is not a plane of inertial symmetry:

(*a*) *The neutral line is the diameter of the ellipse of inertia conjugate to the trace of the load-plane on the cross-section.* (This theorem was given by Saint-Venant and Bresse about the same time: see our Arts. 1581* and 14.)

(*b*) *The 'deviation' or angle between the load- and flexure-planes is a maximum when the former has for trace on the cross-section a diagonal of the rectangle formed by the tangents at the extremities of the principal axes of the ellipse of inertia.*

A good illustration of a simple kind shewing the deviation is given in § 7, p. 57.

[172.] The notes on pp. 73—85 deal with the elastic line when the flexure is *not so small* that we may neglect the square of the slope which the elastic line makes with the unstrained position of the central axis. The results here given express the maximum deflection and terminal slope in series ascending according to powers of $\dfrac{\text{load} \times (\text{span})^2}{\text{rigidity}}$, further the load and maximum stretch in series of ascending powers of $\dfrac{\text{max. deflection}}{\text{span}}$, and finally the stretch-modulus in terms of max. deflection, span and load. Saint-Venant in *Notice* I. (p. 42) claims some originality for these results. This I think can only refer to the convenient form into which he has thrown them: see our Art. 908*.

[173.] Article IV. (pp. 86—186) is entitled: *Rupture par Flexion.*

This practically deals with the formula for the maximum moment

$$M_0 = \frac{T_0 \omega \kappa^2}{h},$$

where h is the distance of the 'fibre' most stretched from the neutral axis[1] and $\omega\kappa^2$ the sectional-moment of inertia about that axis. The question then arises: what is T_0? Saint-Venant holds that if T_0 be the stress at which *enervation* commences, we have in reality a condition for the safety of a permanent structure. This involves *the enervation-point being very close to the limit of linear elasticity.* In many materials this is certainly not the case, even were it possible to define exactly this enervation-point. We must treat the results of this article as applying only to the *fail-limit,* i.e. the failure of linear elasticity (p. 91). Saint-Venant indeed fully recognises that the formula does not give any condition for *immediate* rupture, and that no argument *against* the mathematical theory of 'perfect elasticity' can be drawn from experiments on absolute strength. He states clearly enough that for beams of various sections, for which $\omega\kappa^2/h$ retains the same value, T_0 varies with the form of the section and is greater than, even to the double of, the value obtained from pure traction experiments (this is the well-known 'crux' which the technicists raise against the mathematicians): see his pp. 90, 91. Yet it seems to me that even the extent to which he adopts the formula is not valid. It only gives the fail-limit, which in some cases, perhaps, may indicate *rupture éloignée.*

[174.] On pp. 95—101 our author treats of 'Emerson's paradox' or the existence of 'useless fibres In other words, the expression $\omega\kappa^2/h$ can be occasionally increased by cutting away projecting portions of ω.

We have the cases of beams of square, triangular and circular cross-sections fully treated, as well as that of the *croix d'équerre.*

[1] We use 'neutral *axis*' for the trace of the plane of unstrained 'fibres' on the cross section, while we retain 'neutral *line*' for the succession of points in the plane of flexure through which pass real or imaginary elements of unstretched fibre. It will only coincide with the 'elastic line' or distorted central axis when there is no thrust.

The elastic failure of such outer fibres does not however denote that the truncated section possesses greater strength than the complete section, as Emerson argued from the formula, Rennie confirmed and Hodgkinson refuted by experiment: see our Arts. 187* and 952* (ii). Saint-Venant very aptly terms them *fibres inutiles.* We may indeed calculate the maximum elastic efficiency of such sections by supposing them truncated till $\omega\kappa^2/h$ is a maximum, but the difference is generally so small as not to repay the labour of calculation, albeit it suggests a method of economising material.

[175.] Pages 103—105 treat of the obscure point of how to determine the value of T_0 in the formula of Art. 173, so that there shall be no danger of *rupture éloignée.* Saint-Venant apparently recognises that the exact point at which *enervation* begins is difficult to discover experimentally, especially when the duration and repetition of loads have to be taken into account (p. 105).

Let T_0, T_0' be the stresses which in positive and negative traction respectively mark the limit of *rupture éloignée* ; let T_1, T_1' be the corresponding easily discovered stresses which mark *cohésion instantanée.* Then Saint-Venant observes that we may learn from previous constructions and from our experience of structures submitted to long use what fraction T_0 is of T_1, and that we are justified in taking for the *same* kind of material, even in its several varieties, a constant ratio between T_0 and T_1, e.g. $T'_0 = \frac{1}{8} T_1$.

On n'aura pas pour cela la dilatation limite $s_0 = T_0/E$ égale au 1/8 de la dilatation finale positive ou négative, puisque la proportionalité des efforts aux effets cesse longtemps avant. Mais on aura un certain rapport aussi à peu près constant entre ces deux dilatations (p. 106).

Saint-Venant even suggests (p. 107) that T_0 may be taken proportional to the T obtained from the formula $P = T \cdot \frac{4}{l} \cdot \frac{\omega\kappa^2}{h}$ where P is the concentrated mid load which will rupture immediately a bar of length l terminally supported. As the T obtained from this formula when used for rupture is found to be a function of the section, this suggestion seems to me a dangerous one.

[176.] On p. 109 (§ 13) a formula is given for finding T_0' when T_0 is known. Suppose that the material is a prism with longitudinal stretch-modulus E, and that E_t is the same modulus for *all* directions transverse to the axis; let $T_{0,t}$ and $T_{1,t}$ be the limiting elastic and the rupture stresses when the material sustains a tractive load in the transverse sense, η the stretch-squeeze ratio. Then:

$\frac{T_{0,t}}{E_t}$ = stretch in transverse direction due to $T_{0,t}$,

$\frac{1}{\eta}\frac{T_{0t}}{E_t}$ = squeeze in longitudinal direction...,

$E\frac{1}{\eta}\frac{T_{0,t}}{E_t}$ = safe limit to negative traction in longitudinal direction.

Thus we must have:

$$T'_0 = E\frac{1}{\eta}\frac{T_{0,t}}{E_t},$$

hence

$$\frac{T'_0}{T_0} = \frac{1}{\eta}\frac{E}{E_t}\frac{T_{0,t}}{T_0}.$$

Now by what precedes, Saint-Venant holds that we can legitimately replace $T_{0,t}/T_0$ by $T_{1,t}/T_1$, a ratio easily found from rupture experiments, thus:

$$\frac{T'_0}{T_0} = \frac{1}{\eta} \cdot \frac{E}{E_t} \cdot \frac{T_{1,t}}{T_1}.$$

In the case of isotropy $T_{1,t} = T_1$, $E = E_t$, and thus on the uni-constant hypothesis we should have $T'_0/T_0 = 1/\eta = 4$.

Saint-Venant finds from experiments of Wertheim and Chevandier, that for oak $T'_0/T_0 = 1{\cdot}21$ or $1{\cdot}08$; for cast-metals he suggests 3, for stone 8 to 10, and for wrought iron 2. He holds the value 6 as obtained by Hodgkinson for cast-iron much too large to be prudently adopted, and discusses at some length Hodgkinson's experiments on the beam of strongest section: see our Art. 243*.

Finally we may note that on p. 115, he states that for different varieties of the same material it is more legitimate to take T_0 proportional to T of the formula of Art. 175, than to the stretch-modulus as some writers have done.

[177.] Pp. 122—171 are occupied with what is generally known as the comparative strength of beams of various sections—in reality it is the failure of linear elasticity and not *strength* with which we are dealing.

(*a*) On pp. 123—5 we have the fail-limit determined for cases of loading in planes of inertial asymmetry. The formula of our Art. 14 namely:

$$M_0 = \text{minimum of } \frac{T_0\varpi}{\frac{z\cos\phi}{\kappa_y^2} + \frac{y\sin\phi}{\kappa_z^2}},$$

we find repeated.

When, as in the case of a rectangular section, z, y have values independent of ϕ corresponding to a maximum of the denominator, we find at once

$$M = T_0\omega \Big/ \sqrt{\frac{z^2}{\kappa_y^4} + \frac{y^2}{\kappa_z^4}}.$$

Saint-Venant applies these results to rectangular and elliptic sections.

(*b*) On pp. 143—156 we have a very full investigation of the I-section with special reference to Hodgkinson's section of greatest strength. Although Hodgkinson's experiments were made on *absolute* strength, Saint-Venant finds that his results are true for the fail-limit (*rupture éloignée*). The general conclusions given on p. 155 are: (1) When T'_0 is sensibly greater than T_0 the I-section with unequal flanges has a higher fail-limit, but a less resistance to flexure, than one with equal flanges, provided the squeeze of the smaller flange is not accompanied by lateral stretches more dangerous than the longitudinal in the larger flange, nor the smaller flange receive lateral flexure (buckle) owing to its compression. (2) When the height of the section is increased by ·4 to ·7 of itself we obtain for the same area a I-section of equal flanges with a higher fail-limit than one of unequal flanges and the lesser height; at the same time the resistance to flexure is largely increased. Such increase of height, however, increases the possibility of *déversement* being produced by a slightly oblique load and facilitates the lateral flexure of the squeezed flange.

(*c*) On pp. 156—163 we have a discussion of the fail-limit of *feathered axes*. Saint-Venant shews that their advantages are not so great as has been frequently supposed, while as we have seen (Art. 37) in the case of torsion they give no increased resistance worth mentioning.

[178.] The next point we have to notice is one of considerable interest and has recently been again attracting the attention of the technicists[1]. It is the calculation of the absolute strength from an empirical relation between stress and strain supposed to hold nearly up to rupture. That strain increases more rapidly than stress after the beginning of set even up to rupture had been long noticed by experimentalists, and various modifications of Hooke's Law had been suggested by Varignon, Parent, Bülfinger and Hodgkinson: see our Arts 13*, 29 β*, 234* and 1411*. There has been, however, considerable obscurity about the various empirical formulae suggested, and they have only been applied to the old Bernoulli-Eulerian theory of flexure with its unchanged

[1] See the discussion and references in *Stabilité des Constructions: Résistance des Matériaux* by M. Flamant, pp. 322—9, and also in the *Engineer*, Vol. LXII., 1886, pp. 351, 392, 407.

cross-sections. To begin with, they can hardly be taken as approximate for any material having a distinct yield-point; nor in the second place is it clearly stated how far they represent stress-strain relations for bodies whose elasticity is non-linear, or how far elastic-strain and set are to be treated as coexistent.

Saint-Venant after citing Hodgkinson s formulae (see our Art. 1411*) takes by preference the following for the positive and negative tractions p_1, p_2 at distances y_1, y_2 from the neutral axis of a beam under flexure:

$$p_1 = P_1\left\{1 - \left(1 - \frac{y_1}{Y_1}\right)^{m_1}\right\}, \quad p_2 = P_2\left\{1 - \left(1 - \frac{y_2}{Y_2}\right)^{m_2}\right\},$$

where P_1, P_2 are the tractions at distances Y_1, Y_2 from the axis, and m_1, m_2 are constants. On p. 177 traces of the curves for p in terms of y are given for values of m from 1 to 10, and they are compared with the curves obtained from Hodgkinson's formula.

It will be observed that the difficulty of stating exactly the physical relation between stress, elastic-strain and set is avoided by an assumption of this kind. There is, however, another assumption of Saint-Venant's which does not seem wholly satisfactory He states it in the following words:

Observons d'abord que lorsque la dilatation d'une fibre a atteint sa limite, comme une faible augmentation qu'on lui fait subir produit la rupture ou bien fait décroître très-rapidement sa force de tension, il est naturel de regarder la courbe des tensions comme ayant à l'instant de la rupture sa tangente verticale ou parallèle à l'axe coordonné des y, d'autant plus que cet instant a été précédé d'une énervation graduelle (pp. 180—1).

This paragraph assumes that for the material dealt with the rupture stress is an *absolute maximum*, but in several automatically drawn stress-strain relations which I have examined this does not appear to be the case (see Vol. I. p. 891), and at any rate in some materials it could only refer to the maximum stress *before stricture* and not to the rupture-stress.

On pp. 178—184 the case of a rectangle is treated at some length. Saint-Venant obtains general formulae on the supposition that the curves for negative and positive traction coincide at the origin, i.e. on the supposition that the stretch- and squeeze-moduli *for very small strains* are equal $(m_1P_1/Y_1 = m_2P_2/Y_2)$. The limiting value of the bending moment is then calculated.

In § 3 various values are assumed for m_1 and m_2; in particular if $m_1 = m_2 = 1$, it is shewn that to make the initial stretch- and squeeze-moduli unequal is to *increase the resistance to rupture by flexure.*

In the case of $m_1 = m_2$, $P_1 = P_2$, $Y_1 = Y_2$ we easily find for a rectangular cross-section $(b \times c)$:

$$M_0 = R_0 \cdot \frac{bc^2}{6} \cdot \frac{3m(m+3)}{2(m+1)(m+2)},$$

which increases from $R_0 \frac{bc^2}{6}$ to $R_0 \frac{bc^2}{4}$ as m increases from 0 to ∞.

If we take $m_2 = 1$ and m_1 any value, we obtain a more complex value for M_0, which increases with m_1 from $R_0 \frac{bc^2}{6}$ to $R_0 \frac{bc^2}{2}$ Thus in all cases the value lies between those given by Galilei's theory and by the ordinary Bernoulli-Eulerian hypothesis.

Saint-Venant does not venture into the analysis required to determine how the constant n given by $M_0 = n R_0 bc^2/6$ varies with the shape of the section, which must be the true test of any theory of this kind, i.e. the constant m must be found to have the same value for all sections.

[179.] Saint-Venant gives on pp. 186—204 an excellent elementary discussion of slide and shear; on pp. 206—214 a like discussion of the effect of slide in changing the contour and shape of the cross-sections of a beam under flexure. The method of treatment is very simple, and by the consideration of a special case the action of the slide is well brought out.

[180.] Pages 216—237 are devoted to combined strain, flexure, stretch due to pure traction and slide The fail-limit is determined by simple geometrical considerations, and the examples, chosen from those of Chapters XII. and XIII. of the memoir on *Torsion* (see our Arts. 50 to 60), are treated with considerable numerical detail. The example on the combined flexure and slide exhibited by the strained axis of a pulley is new (p. 234).

[181.] On pp. 239—271 the general equations of torsion are deduced. The treatment is in some respects better than in the memoir of 1853. We may note a few points:

(*a*) Pages 240—242 give a fuller discussion of the resistance to torsion due to longitudinal stretch of the 'fibres : see our Art. 51.

(*b*) Pages 244—5 (§ 4). Elementary proof that the cross-sections of all prisms, except the right-circular cylinder, are distorted by torsion.

(*c*) Pages 261—2. The expressions $\int \widehat{xy}\, d\omega$ and $\int \widehat{xz}\, d\omega = 0$ for every section of a prism under torsion. This is true whether or not the axis of torsion passes through the centre of the section, supposing it to have one. Saint-Venant had only treated of this matter in the case of the elliptic section (§ 59 of the memoir on *Torsion*: see our Art. 22). A general proof is here given in a footnote.

(*d*) In § 15, pp. 264—7, we have a fuller treatment than occurs in the memoir on *Torsion* of eccentric torsion, or torsion about *any* axis parallel to the prismatic sides. Taking the equations of torsion for an isotropic material (equations vi. of Art. 17):

$$u_{yy} + u_{zz} = 0,$$

$$(u_z + \tau y)\, dy - (u_y - \tau z)\, dz = 0,$$

for which the origin lies on the axis of torsion, let us put $y' = y + \eta$, $z' = z + \zeta$ we find η and ζ being constants:

$$u_{y'y'} + u_{z'z'} = 0,$$

$$\{u_{z'} + \tau (y' - \eta)\}\, dy' - \{u_{y'} - \tau (z' - \zeta)\}\, dz' = 0.$$

These equations have for solution

$$u = u' - \tau (\zeta y' - \eta z'),$$

where u' is the value of u when $\eta = \zeta = 0$, or in other words the shift when the torsion operates round an axis through the new origin. The shifts u' and u giving the distortions in the two cases differ only by

$$\tau (\zeta y' - \eta z') = \tau (\zeta y - \eta z),$$

or the two distorted surfaces are superposable by rotating the one through small angles $\tau\eta$ and $-\tau\zeta$ round the axes of y and z respectively.

Further, $\begin{Bmatrix} u_z + \tau y = u'_{z'} + \tau y' \\ u_y - \tau z = u'_{y'} - \tau z' \end{Bmatrix}$ or the slides determined for either axis are equal for the same points. Thus it follows that the torsional couple will in both cases be the same.

Saint-Venant then shews how by placing two prisms of equal cross-sections with corresponding lines parallel, and fixing their terminal faces so as to remain parallel after torsion about a mid-axis, we can obtain eccentric torsion. The torsional couple will be just double of that obtained from the simple torsion of either. Their axes it is true will be bent into helices, but the bending introduced is a small quantity of the *second order* in the torsion.

(*e*) In § 17 (pp. 268—71) we have an investigation of the maximum-slide and the fail-points. We cite the following passage:

Si σ_x^2 [σ_x= le plus grand *glissement principal*] croissait toujours de l'intérieur à l'extérieur de la section pour chaque direction, ce serait constamment sur son contour qu'il faudrait chercher les points dangereux. Mais nous savons qu'il y a souvent des points du contour où le glissement est nul, et il peut y avoir, dans l'intérieur, quelque point de maximum absolu de σ_x^2 (quoique cela ne se soit présenté dans aucun des exemples ci-après traités); et il n'est pas impossible que ce maximum excède toutes les valeurs de σ_x^2 relatives aux points du contour (p. 269).

We have then an analytical investigation of the fail-points, which suggests a general method of investigation adopted in the sequel for the special cases. This method avoids the ambiguities of some of the paragraphs on this subject in the memoir of *Torsion*: see our Arts. 39 and 42.

[182.] Pages 271—372 treat very thoroughly of the torsion-problem. They reproduce to a great extent the formulae and tables of the memoir on *Torsion*, but at the same time make frequent additions and improvements. We may note the following:

(*a*) Eccentric torsion of a right-circular cylinder. The coordinates of the centre referred to the axis of torsion being η, ζ, we find with the notation of our Art 181 (*d*), a being the radius:—

$$u = \tau (\zeta y - \eta z),$$
$$\sigma_{xy} = -\tau (z - \zeta), \quad \sigma_{xz} = \tau (y - \eta),$$

while $M = \mu\tau \int_0^a r^2 d\omega = \mu\tau\omega \cdot \frac{a^2}{2}$, as in the case of central torsion.

(*b*) A fuller treatment of the prisms whose cross-sections are included in the equation:

$$\frac{r^2}{2} + a_2 r^2 \cos 2\phi + a_4 r^4 \cos 4\phi = \text{const.} \quad \text{(See our Art. 49 } (c).)$$

The most interesting of the cross-sections included in this equation is entitled by Saint-Venant: *Section en double spatule analogue à celle d'un rail de chemin de fer* (p. 365). It has the shape given in the accompanying figure.

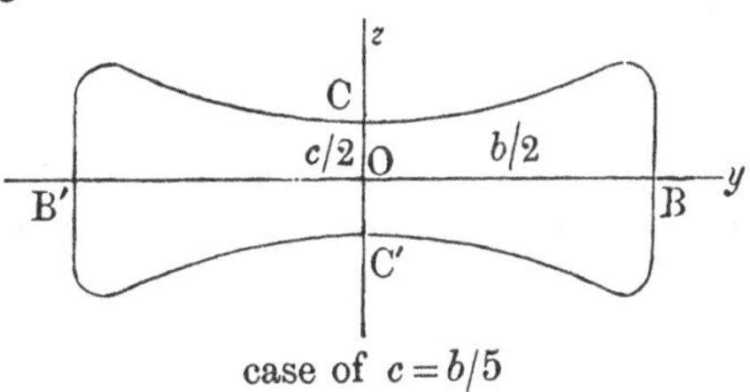

case of $c = b/5$

See pp. 305—307, 312—317, 325—335.

(*c*) The accurate investigation of the fail-points for the bi-symmetrical curves of the 4th and 8th order; see pp. 308—312, 339—341. Cf. our Arts. 37 and 39.

(*d*) In a foot-note to p. 335 Saint-Venant treats a special case of the curve of the fourth degree

$$\frac{y^2 + z^2}{2} + a_2 (y^2 - z^2) + a_4 (y^4 - 6y^2z^2 + z^4) = \text{const.}$$

By taking $a_2 = -1/\sqrt{2}$, $a_4 = 2(\sqrt{2}-1)/b^2$ and the constant $= 0$, we obtain an isosceles triangle having for base a portion of the hyperbola $y^2 = b^2/4 + (\sqrt{2}-1)^2 z^2$ and for sides lines making with the bisector of the base angles whose tangent $= \sqrt{2}-1$. The length of the bisector from vertex to hyperbolic base is then $b/2$. The torsion takes place round an axis through the *vertex*. Saint-Venant finds approximately,

$$M = .56702\, \mu\tau\omega\kappa^2.$$

This value agrees very closely with that of the equilateral triangle: see our Art. 41.

[183.] Pages 372—460 deal with the conditions for resistance to *rupture éloignée* under simultaneous torsion and flexure. Most of this matter had already been given in Chapters XII. or XIII. of the memoir on *Torsion* or in the memoir on *Flexure*: see our Arts. 50—60 and 90—8. One or two points may be noticed:

(*a*) In the memoir on *Torsion* Saint-Venant when seeking for the fail-limit neglects as a rule the flexural slides (see our Art. 56, Case (iii) etc.). Here he commences with an investigation of the values of these slides. The approximate methods of Jouravski and Bresse for obtaining the slide in a beam of small breadth are considered (see our Chapter XI.), and are applied to the rectangle, ellipse and ⌶-cross-sections. A footnote gives the value of the slide in the same approximate manner for an isosceles triangle. See pp. 391—8. But the expressions thus obtained are not exact, and in a considerable number of cases differ sensibly from the real values, especially when the section has a measurable breadth perpendicular to the load plane. The expressions found by Jouravski and Bresse give the *total* shear upon a strip of unit breadth taken on a section of the beam perpendicular to both the cross-section and the load plane, but they do not determine how such shear is transversely distributed, still less the magnitude of the maximum slide on the cross-section. Saint-Venant then proceeds as in the memoir on *Flexure* to deduce exact expressions for the flexural slides (pp. 399—414). The notation used differs from that in the original memoir. The reader will find the two notations placed side by side in the footnote, p. 405. The treatment in the *Leçons de Navier* is shorter and not nearly so complete as in the memoir. The diagrams reproduced in our frontispiece are given in a footnote on pp. 410—12: see our Arts. 92 and 97.

(*b*) Pages 414—60 are occupied with combined flexure and torsion in those cases where we may neglect the flexural slides. They reproduce with some modifications and extensions the results of Chapter XII. of the *Torsion* memoir. There is a good summary on pp. 453—9.

[184.] On pp. 461—9 Saint-Venant treats of rupture (*rupture immédiate*) by torsion.

(*a*) He shews that the moments capable of producing rupture are for similar sections as the *cubes* of their homologous dimensions. A footnote (p. 463) refers to Vicat's experiments which apparently contradict this result; see our Art. 731* Saint-Venant attributes this divergence to flexure having taken place in the short prisms of *plâtre* and *brique crue* used by Vicat.

(*b*) In § 61 (p. 464) Saint-Venant endeavours to find the absolute strength of a circular prism (radius a) under torsion by the assumption of an empirical formula, similar to that of our Art. 178, for the shear q at distance r from the axis of torsion. Namely:

$$q = Q\left[1 - \left(1 - \frac{r}{b}\right)^m\right]$$

where Q is the shear at distance b, and m is a constant.

We are only told in favour of this formula, (1) that for small values of r and for very small shears q is proportional to r and thus to the slide, (2) that q increases less rapidly than r, or the slide, when the slide becomes greater.

If S_1 be the rupture shear and correspond to $r = a$, we have

$$Q = S_1/\{1 - (1 - a/b)^m\}.$$

Then, introducing the same sort of questionable condition as in our Art. 178, namely that $dq/dr = 0$ when $r = a$, we have further

$$a = b \text{ and } S_1 = Q.$$

This leads us to a rupture couple $M_1 = \int_0^\omega rq\,d\omega$,

$$= 2\pi a^3 S_1\left(\frac{1}{3} - \frac{2}{(m+1)(m+2)(m+3)}\right).$$

Or, as m changes from 1 to ∞, M_1 changes from $\frac{1}{2}$ to $\frac{2}{3}$ of $\pi a^3 S_1$ (p. 466)[1].

(*c*) Saint-Venant then attacks the problem of the prism of rectan-

[1] Saint-Venant's result seems to be $\frac{1}{4}$ of the real value, owing to the displace ment of a factor 2.

gular cross section ($b \times c$) for which b is much greater than c. Here the approximate values of the slides before the linear limit is passed are:

$$\sigma_{xy} = -2\tau z, \quad \sigma_{xz} = \frac{2\, c^2 \mu_1}{b^2 \mu_2} \tau y.$$

These results may be deduced from Art. 46 by replacing the elongated rectangle by its inscribed ellipse and neglecting c^2/μ_2 as compared with b^2/μ_1. See also Table I. p. 39, and Art. 47.

He assumes that after the linear limit is passed:

$$\widehat{xy} = -Q'\left\{1 - \left(1 - \frac{z}{h}\right)^m\right\}, \quad \widehat{xz} = Q''\left\{1 - \left(1 - \frac{y}{k}\right)^n\right\}.$$

Hence, since for small slides or small values of z and y, $\widehat{xy} = \mu_1 \sigma_{xy}$ and $\widehat{xz} = \mu_2 \sigma_{xz}$, we must have:

$$-\mu_1 2\tau = -\frac{Q'm}{h}; \quad \mu_2 \frac{2\, c^2 \mu_1}{b^2 \mu_2} = \frac{Q''n}{k}$$

These give, $$Q''_2 = Q'_1 \frac{m}{n} \frac{c^2}{b^2} \frac{k}{h}.$$

Further, since the fail-points are the mid-points of the much longer side b, the rupture points are taken there also. Thus it is necessary that:

$$d\widehat{xy}/dz = 0, \quad \widehat{xy} = S' \text{ when } z = c/2.$$

It follows that $h = c/2$ and $Q' = S'$, the *absolute* shearing strength in direction of y.

To proceed further Saint-Venant assumes that the slide σ_{xz} always remains much less than the slide σ_{xy}, so that for the former it is sufficient to retain the linear strain form, we have thus

$$\widehat{xz} = Q''ny/k = S'm \frac{c}{b} \cdot \frac{2y}{b},$$

together with $$\widehat{xy} = -S'\left\{1 - \left(1 - \frac{2z}{c}\right)^m\right\}.$$

It easily follows that

$$M_1 = bc^2 S'\left(\frac{m}{6} + \frac{2^{m-1} m}{(m+1)(m+2)}\right).$$

Cases (b) and (c) confirm the law of the cube stated in (a). Such formulae, although by no means satisfactory from the theoretical standpoint, are yet useful as suggesting lines for future experiment.

[185.] Pages 469—77 (§ 62) contain a useful discussion of the various methods of determining the elastic and fail-point constants, especially in the case of prisms whose material is transversely aeolotropic. Saint-Venant (p. 471) adopts the result given in Art. 5. d. of our account of the memoir on *Torsion*, $\overline{\sigma}_{yz} = 2\sqrt{\overline{s}_y \overline{s}_z}$, to obtain a plausible relation between shear fail-limit (S_0) and tractive fail-limits T_0 and T_0' We have thus the formula $S_0/G = 2\sqrt{T_0/E \times T_0'/E'}$.

[186.] Pages 477—480 deal with the problem of the torsion of circular cylinders (radius a) having a cylindrical distribution of elastic homogeneity. In this case μ is a function of the axial distance r. There will be no distortion of cross-sections. Saint-Venant supposes μ to remain constant from the axis up to a radius $a - \zeta$, and then at distances r from $a - \zeta$ to a to follow the law

$$\mu = \mu_0 + (\mu_1 - \mu_0)\left(\frac{z}{\zeta}\right)^n \text{ where } z = r - (a - \zeta).$$

He easily deduces the following formula for M,

$$M = \mu_0\tau\frac{\pi a^4}{2} + 2\pi(\mu_1 - \mu_0)\tau\left\{\frac{(a-\zeta)^3\zeta}{n+1} + \frac{3(a-\zeta)^2\zeta^2}{n+2} + \frac{3(a-\zeta)\zeta^3}{n+3} + \frac{\zeta^4}{n+4}\right\}.$$

Special cases are:

(1) Wooden cylinder whose axis is about the same as that of the tree out of which it has been cut; here we may put $\zeta = a$, and we have:

$$M = \frac{n\mu_0 + 4\mu_1}{n+4}\,\omega\kappa^2\tau.$$

(2) Forged or cast iron cylinder with skin change of elasticity:

$$M = \tau\,(\mu_0\omega\kappa^2 + \gamma 2\pi a^3) \text{ where } \gamma = \frac{\mu_1 - \mu_0}{n+1}\zeta.$$

Supposing the fail-point to be on the surface, we have $S_0 = \mu_1\tau a$, and eliminating τ:

$$M_0 = A\pi a^3 + B\pi a^2,$$

where A and B are two constants depending only on the elastic nature of the material. Thus the fail-couple depends partly on the cube, partly on the square of the radius of the cylinder.

[187.] The text of the work concludes with numerical examples such as are given on pp. 551—8 of the memoir on *Torsion*. The remainder of the volume is filled with five appendices and an *Appendice complémentaire* occupying pp. 510—849, which from their historical and physical aspects are perhaps the most interesting portions of the work.

[188.] Appendix I. (pp. 512—19) contains certain elementary proofs due to Poncelet as to the curvature, deflection etc. of the elastic line. A point on p. 518 on the question of built-in terminals (*encastrements*) may be noted. Poncelet remarks that for a cantilever we may suppose two forces, whose resultant is equal and opposite to that of the load, to act at the built-in end. These forces—whose points of application are very close, one on the upper and one on the lower surface of the beam—are very great and alter the surfaces of the built-in beam and the surrounding material, so that the elastic line at this end is not horizontal, but takes a certain inclination varying as the terminal moment directly and inversely as the *profondeur de l'encastrement*. Small as this inclination is, it affects sensibly the experimental accuracy of the theoretical results based on the perfect horizontality of the elastic line at the built-in end. This was noted by Vicat: see our Art. 733*. Saint-Venant holds that careful experiments ought to be made to determine its influence.

[189.] Appendix II. is entitled: *Sur les conditions de l'exactitude mathématique des formules tant anciennes que nouvelles d'extension, de torsion, de flexion avec ou sans glissement.—Démonstration synthétique de ces formules quand on suppose ces conditions remplies.* This appendix contains first an easy refutation of Lamé's ill-judged sneer at the *procédés hybrides, mi-analytiques, et mi-empiriques ne servant qu'à masquer les abords de la véritable science:* see our Arts. 1162* and 3. Saint-Venant shews that his methods have precisely the same validity as those adopted in the cases of simple traction, of the old theories of flexure, and of torsion for a circular cylinder. In the sequel he demonstrates afresh the torsion and flexure equations. He starts from an axiom and definitions involving the hypothesis of central intermolecular action as a function of the central distance only. The appendix occupies pp. 520—541.

[190.] Appendix III. contains a complete theory of elasticity for aeolotropic bodies so far as the establishment of the general equations of elasticity and the usual formulae of stress and strain are concerned. It occupies pp. 541—617. Proceeding from central intermolecular action, Saint-Venant on pp. 556—9 reduces the 36 constants of the stress-strain relations to 15. We may note one or two points of interest:

(*a*) § 23 (pp. 562—74) with its long footnote is specially worthy of the reader's attention. Saint-Venant obtains expressions for the stresses on the hypothesis that initial stress has produced *considerable* initial strain in the body. In this case the strain developed by the initial stress sensibly influences the effect of the later strain. We can no longer add initial and secondary stresses as independent factors of total stress.

Let the initial stresses be $\widehat{xx}_0$, $\widehat{xy}_0$ etc...., the secondary stresses $\widehat{xx}_1$, $\widehat{xy}_1$ etc., and the total stresses $\widehat{xx}$, $\widehat{xy}$, etc.

Let x_1, y_1, z_1 be the directions of three right-lines slightly oblique to each other which were initially the rectangular set x, y, z; let x', y', z' be three other lines rectangular or slightly oblique among themselves, taken close to the former (x, y, z) and normal to the three planes by which we determine the six stress-components. Then, if c_{rs} be the cosine of the angle between the lines r and s we have as stress types:

$$\left.\begin{aligned}\widehat{x'x'} &= \widehat{xx}_0\,(1 + s_x - s_y - s_z) + 2\,\widehat{xy}_0 c_{y_1x'} + 2\,\widehat{zx}_0 c_{z_1x'} + \widehat{xx}_1,\\ \widehat{y'z'} &= \widehat{yz}_0\,(1 - s_x) + \widehat{yy}_0 c_{y_1z'} + \widehat{zz}_0 c_{z_1y'} + \widehat{zx}_0 c_{x_1y'} + \widehat{xy}_0 c_{x_1z'} + \widehat{yz}_1.\end{aligned}\right\}\ldots\text{(i)}.$$

To these we must add the purely geometrical relations of the type:

$$c_{y_1z'} + c_{z_1y} = \sigma_{yz} + c_{y'z'}\,\ldots\ldots\ldots\ldots\ldots\ldots\ldots\ldots\text{(ii)},$$

which reduce if x', y', z' are taken rectangular to the type:

$$c_{y_1z'} + c_{z_1y'} = \sigma_{yz}\,\ldots\ldots\ldots\ldots\ldots\ldots\ldots\ldots\ \text{(iii)}.$$

When, however, the initial stress is not such that the shears are zero or can be neglected when multiplied by small strains, we may simplify equations (i) by a proper choice of x', y', z'. Thus if x', y', z' be taken perpendicular to y_1, z_1, x_1 or z_1, x_1, y_1 respectively, which is compatible with their rectangularity, then either $c_{z_1y'} = c_{x'_1z'} = c_{y_1x'} = 0$, or, $c_{y_1z'} = c_{z_1x'} = c_{x_1y'} = 0$, and we can replace the remaining cosines in (i) by the slides $\sigma_{yz}, \sigma_{zx}, \sigma_{xy}$. By taking x', y', z' bisectors of the angles between the lines $x_1 . y_1, z_1$ and the perpendiculars to the three slightly oblique planes y_1z_1, z_1x_1, x_1y_1, i.e. the closest rectangular system to x_1, y_1, z_1, we obtain:

$$c_{y_1z'} = c_{z_1y'} = \tfrac{1}{2}\,\sigma_{yz}\,\ldots\ldots\ldots\ldots\ldots\ldots\ldots\ldots\text{(iv)}$$

as the type of equation (iii).

In the case (*le seul qui ait été supposé par les divers auteurs de mécanique moléculaire*, ftn. p. 571) in which the shifts are very small and consequently the directions x_1, y_1, z_1 almost coincident with x, y, z, we can take the latter for the rectangular system x', y', z' and we thus find:

$$\begin{aligned}c_{y_1z} &= dw/dy, & c_{z_1y} &= dv/dz, & c_{x_1y} &= dv/dx,\\ c_{y_1x} &= du/dy, & c_{z_1x} &= du/dz, & c_{x_1z} &= dw/dx,\end{aligned}$$

and reach the equations (i) of our Art. 129.

These again reduce to the relations of our Art. 666*, if we put $\widehat{xx}_0 = \widehat{yy}_0 = a$, $\widehat{zz}_0 = c$, and $\widehat{yz}_0 = \widehat{zx}_0 = \widehat{xy}_0 = 0$, and give the proper values to the secondary stresses.

Saint-Venant proves equations (i) by the molecular method in the footnote before referred to. He makes some remarks on the Navier-Poisson controversy, and refers to a paper of his own published in 1844 on Boscovich's system: see our Arts. 527* and 1613*.

(*b*) On p. 587 the remark is made that the stress-strain relations, the body stress-equations and the body strain-equations remain true whatever be the amount of the shifts in space provided the *relative* shifts of adjacent parts or the strain-components are small. In this case, however, the values to be given to the strains in terms of the shifts are those of our Art. 1618*. The ordinary shift-equations of elasticity hold only for small portions of an elastic body, when the total shifts are not small. Hence they cannot be directly applied to large torsional or flexural shifts. The whole treatment on pp. 587—92 is good, and better than that of the memoir of 1847: see our Art 1618*

(*c*) Saint-Venant points out that it is not sufficient to find values of the stress-components which satisfy the body and surface stress-equations. There are also certain *conditions of compatibility* between the strain components deduced from these stresses which also must be satisfied: see our Art. 112.

These equations hold for *all values of the shifts*, provided the strains remain small, i.e. if they take the forms given in our Art. 1618*.

(*d*) Pp. 603—17 contain a direct investigation of Saint-Venant's torsion and flexure equations from the general equations of elasticity. In both cases the method adopted assumes a given distribution of stress and deduces the corresponding shift-equations.

In dealing with torsion Saint-Venant supposes a single plane of elastic symmetry perpendicular to the axis of torsion, and starts from formulae for the shears of the form

$$\widehat{xy} = f\sigma_{xy} + h\sigma_{zx}, \quad \widehat{zx} = e\sigma_{zx} + h'\sigma_{xy},$$

where h and h' are supposed unequal. See our Art. 4 (θ) on the memoir on *Torsion*. He deduces the general torsional equations, which now contain four constants, and solves them for the case of the ellipse. The discussion does not seem to me of much value, as all elasticians, multi- or rari-constant, would agree that $h=h'$, in which case by a change of axes we can take $h=h'=0$: see the same Article. In the case of an elliptic contour a direct analysis gives:

$$M = \frac{4\tau\omega}{1/\mu_1\kappa_1^2 + 1/\mu_2\kappa_2^2 + (1/\kappa_1^2 - 1/\kappa_2^2)(1/\mu_2 - 1/\mu_1)\sin^2\alpha},$$

where α is the angle between the direction in which the slide-modulus is μ_1 and the axis of the ellipse about which the swing-radius is κ_1 The reader must note that μ_1 and μ_2 are not the same constants as in Art. 46 of our discussion of the memoir on *Torsion*, where we supposed 'the principal axes of elasticity' to coincide with the principal axes of the elliptic section.

[191.] The fourth Appendix occupies pp. 617—45 and contains a careful comparison of Saint-Venant's theory of Torsion with the experimental results of Wertheim, Duleau and Savart: see our Arts. 1339* and 31. It is followed by some discussion of torsional vibrations. This appendix is practically directed against Wertheim's memoir on *Torsion* of 1857: see our Art. 1343*. It will be remembered that Wertheim had asserted the theoretical accuracy of Cauchy's erroneous torsion formula (see our Art. 661*), had persisted in retaining the value for the squeeze-stretch ratio which he had deduced by a fallacious theory in 1848 (see our Art. 1319*), and finally had exhibited complete ignorance of Saint-Venant's results for the elliptic cylinder. Saint-Venant easily shews the insufficiency of Wertheim's criticism, and how the mean results of Savart and Duleau for rectangular prisms, and of Wertheim himself for elliptic prisms confirm the new theory: see our Arts. 31 and 35.

In the discussion on torsional vibrations, Saint-Venant reproduces the matter of his memoir of 1849: see our Art. 1628* He regards of course the theory given as only *approximate* (p. 633), but sufficiently so for all practical purposes, as indeed appears from the comparison of theory and experiment (p. 643).

[192.] The fifth Appendix, devoted to the elastic-constant controversy, occupies pp. 645—762. It is an excellent piece of scientific criticism, to which some multi-constant elasticians have insufficiently replied by squeezing caoutchouc or loading pianoforte wires. The difficulty of critical experiments lies first in obtaining a purely isotropic material free from all initial stress and without any superficial elastic variation, and then in assuring the extreme nicety required to determine successfully the stretch-squeeze-ratio. In our first volume we have referred to the leading features of the controversy (see our Arts. 921*—932*) and the chief of the earlier experiments in this field (see our Arts. 470* 1034*, 686*—90*, 1358*). We shall find other remarkable experiments as well as theoretical conclusions have sprung from the controversy in the last 40 years; these will lead us on more than one occasion to examine the validity of Saint-Venant's arguments. Meanwhile we may refer to one or two points brought forward in the present essay.

(*a*) Saint Venant's criticism seems to me unanswerable, when he attacks the validity of the method by which Poisson, Cauchy, Green, or

Lamé have deduced the linearity of the stress-strain relation without any appeal to experiment or any statement of physical fact or any axiom of intermolecular action: see Saint Venant's pp. 660—5 and our Arts. 553*, 614*, 928*, 1051* and 1164* footnote.

(*b*) On pp. 665—676 we have a long and careful numerical examination of the experiments of Regnault on piezometers of copper and brass. In general they accord with the uni-constant theory, or at least better with that than with Wertheim's (see our Art. 1319*). This is followed by some remarks on Wertheim's and Clapeyron's experiments on caoutchouc. The former found $\frac{\mu}{\lambda} = -\frac{1}{4}$ to $\frac{3}{4}$, and the latter $\frac{\mu}{\lambda} = \frac{1}{2202}$ see our Art. 1322* and Chapter XI. Results so discordant as these lead Saint-Venant to remark that neither uni- nor bi-constant isotropy, nor

même des formules linéaires quelconques, ne sont pas applicables au caoutchouc, liquide coagulé ou épaissi plutôt que solidifié, et d'une nature en quelque sorte intermédiaire entre les fluides et les solides (p. 678).

(*c*) Pages 679—89 are occupied with a criticism of Wertheim's hypothesis, that $2\mu = \lambda$, and with the results of his experiments. Saint Venant points out the great probability of a want both of homogeneity and of isotropy in the cylinders used by Wertheim (see our Art. 1343*) and he examines analytically the ratio of longitudinal to transverse stretch-moduli, when such isotropy is not presupposed. We shall return to some of Saint-Venant's arguments when examining Wertheim s later memoirs.

(*d*) On pp. 689—705 we have a consideration of Cauchy's hypothesis of 1851: see our Art. 681*, namely, that it is possible if a body be crystalline that:

les coefficients des déplacements et de leurs dérivées dans les équations d'équilibre intérieur ne sont plus des quantités constantes, mais deviennent des fonctions périodiques des coordonnées (p. 689).

In other words we arrive at stress-strain relations in which the 36 constants are *not* connected by 21 relations. Saint-Venant conducts a new investigation (pp. 697—706) with fairly simple analysis. The turning point of rari- or multi-constancy for such regularly crystallised bodies is then seen to lie in the legitimacy of bringing stretches like s'_x outside certain summations of the form

$$\Sigma_x R \cos^3 (rx) \,.\, s'_x, \quad \Sigma_x R \cos (rx) \cos^2 (rz) \,.\, s'_z,$$

and replacing them by their mean values s_x, s_z. Here s_x is the mean value of s'_x for all the atoms under consideration, and we may replace s'_x by s_x if the body is *isotropic* or possesses *confused crystallisation*. On the other hand in regularly crystallised bodies, there may be terms in s'_x periodic in the coordinates and we cannot replace s'_x by s_x and bring the mean stretch outside the summation. Hence we have not the 21 relations between the coefficients fulfilled. Saint-Venant holds, however, that even if this periodicity be true for regularly crystallised bodies, it can only introduce small differences into the otherwise equal constants. But further, if it does exist

cette altération ne peut regarder que *certains* cristaux réguliers. Elle n'est jamais relative aux corps à cristallisation confuse, comme sont tous les matériaux de construction, et comme sont aussi tous les corps isotropes. Il n'y a donc aucune raison de changer les formules trouvées depuis un tiers de siècle pour les pressions dans ces sortes de corps (p. 705).

[193.] On pp. 706—742 we have an analysis and criticism of the various methods which English and German elasticians have adopted in order to obtain the fundamental equations of elasticity; there is also a *résumé* of their views on the elastic-constant controversy. Here the memoirs of Green, Neumann Haughton, Clebsch, Clausius, Thomson, Kirchhoff, Maxwell and Stokes are briefly considered. Saint-Venant devotes special attention (pp. 721—32) to the value of Green's results as bearing on double refraction and the disappearance of the stretch-wave. This discussion is only of importance to us in its bearing on the elastic-constant controversy Green's treatment of the ether demands the independence of the 21 constants, but we may question whether his results are the only possible ones, nay, even whether they are so satisfactory, as to stand *per se* as a justification of multi constant formulæ.

In order that the vibrations may be exactly parallel to the wave-face Green finds the relations xxxviii. of our Art. 146.

If to these 14 conditions of Green we were to add the six additional conditions of rari-constancy, namely:

$$|yyzz| = |yzyz|, \quad |xxyz| = |zxxy|, \text{ etc.} \quad \text{(i)}$$

we should then have:

$$|xxxx| = |yyyy| = |zzzz| = 3\,|yzyz| = 3\,|zxzx| = 3\,|xyxy| = 3\,|yyzz| = 3\,|zzxx| = 3\,|xxyy| \quad \text{(ii)}$$

and all the other constants zero.

Thus the condition for exact parallelism would be *isotropy*, or this parallelism would be incompatible with double refraction. Now are Green's conditions so extremely probable that we ought to reject the six molecular conditions (i) which render them nugatory? Saint-Venant argues that they are not, chiefly for the reason that they involve: $|x'x'x'x'| = |xxxx|$.

This is proved in the footnote p. 726. It denotes physically that, in whatever direction we take x', the same stretch $s_{x'}$ will produce a traction $\widehat{x'x'}$ of the same intensity. Such an equality seems opposed to our ideas on the nature of bodies endowed with double refraction. The arguments used to support the improbability of this relation are identical with those of the memoir of 1863 and have been cited in our Art. 147.

[194.] While recognising the weight of Saint-Venant's reasoning in this Appendix and in the memoir of 1863, and admitting the difficulty of conceiving a double-refracting medium to obey such conditions as those given by Green, we have yet to notice a point with regard to the arguments Saint-Venant advances. A distinction must be drawn between an isotropic body held by external pressures in an aeolotropic state of *elastic strain*, and a body also primitively isotropic which has received *set* of different intensity in different directions. In the former case the initial stresses may enter into the elastic constants (as in our Art. 129) and so affect the elasticity in different directions. In the latter it would *appear* as if the molecules must be brought in some directions nearer together and so the direct stretch coefficients be affected and varied. But is this experimentally the fact? If a bar of metal be taken and stretched beyond the elastic limit, so that it receives set, it is found that its *stretch-modulus*, which is certainly a function of the direct stretch-coefficients remains nearly constant. Now this set may be of two kinds, first: a set occurring far below the yield-point, which is often little more than a removal of an initial state of strain due to the working: and secondly, a set which denotes a large change in the relative molecular positions and can occur after the yield-point has been reached. If it can be shewn that the stretch-modulus remains nearly constant notwithstanding one or both of these sets, it would be interesting to investigate experimentally whether such is also true for the slide-moduli and the cross-stretch coefficients before we condemn Green entirely.

Experiments on simple traction and torsion of *large bars* before and after very sensible set would throw light on this matter.

[195.] Saint-Venant further remarks that Green's conditions are not *necessary* in order that we may obtain exactly Fresnel's wave-surface. Saint-Venant in a foot-note gives a fairly easy analysis leading to Cauchy's four conditions which are compatible with rari-constancy (see the memoir of 1830, *Exercices mathématiques* 5e année). These conditions are given in our Art. 148 as Equations xxxix.

Les *quatre* relations ou conditions (xxxix)......n'ont rien d'arbitraire ni de bizarre, bien qu'elles soient d'une forme moins simple à coup sûr que les *cinq* conditions de Green (xxxviii of our Art. 146) qui n'en sont qu'un cas particulier....En effet lorsque les trois coefficients d'élasticité directes a, b, c, entre lesquels elles permettent telle inégalité qu'on veut, ont des rapports mutuels n'excédant pas $1\frac{1}{2}$ ou 2, il est facile de s'assurer par des calculs qu'elles sont, numériquement, presque identiques aux relations $2d+d'=\sqrt{bc}$, $2e+e'=\sqrt{ca}$, $2f+f'=\sqrt{ab}$ que nous verrons être celles qui donnent la distribution la plus simple des élasticités autour de chaque point dans les corps hétérotropes, et appartenir, au moins avec une grande approximation, aux corps dont l'isotropie primitive a été altérée par de simples compressions ou dilatations inégales, c'est-à-dire généralement aux corps amorphes ou à cristallisation confuse. Or tous les physiciens admettent que c'est seulement à cet état d'inégal rapprochement moléculaire en divers sens que se trouve l'éther dans les cristaux dont la forme n'est pas un polyèdre régulier (p. 731, foot-note).

We have ventured so far from our subject into that of Light, only to shew that Saint-Venant brings forward strong reasons why, even if we dogmatically assert the elastic jelly character of the ether, it is not necessary to summarily reject the rari-constant hypothesis.

[196.] Pages 732—42 are occupied with an excellent discussion of Stokes' memoir of 1845: see our Arts. 925*—6* and 1264*. There are also a few remarks upon Maxwell's memoir of 1850: see our Art. 1536*. Saint-Venant states that Thomson and Kirchhoff while adopting multi-constancy have not added any additional reasons for its validity. This at the present time is hardly true. I may note Kirchhoff's memoir of 1859: see *Poggendorff's Annalen*, Bd. 108, p. 369, and Thomson's of May, 1865: see *Proceedings of Royal Society* for that date. Saint-Venant's objections to those arguments of Stokes which are drawn from the 'doctrine of continuity,'—practically from the equivalence of the plasticity of metals and the viscosity of fluids—seem to me very forcible and should be read by all scientists interested in the ultimate molecular constitution of bodies. In the question of rari- or multi-constancy are involved, not merely points of technical expediency, but principles going to the base of our knowledge of matter,—such as our *proofs* of the equation of energy and the application of the laws of motion to inter-molecular action.

[197.] Pages 742—46 are occupied with a review of Clausius' memoir of 1849: see our Art. 1398*. It is only necessary to remark here that recent experiments would, we think, have removed Saint-Venant's doubt as to the existence of elastic after-strain in metals (p. 745). The appendix concludes with a *résumé* of all the arguments brought forward in favour of rari-constancy (pp. 746—62).

[198.] The *Appendice complémentaire* is chiefly occupied with an examination of the elastical researches of Rankine, Clebsch and Kirchhoff, which Saint-Venant tells us had not then been properly studied in France. We note one or two points:

(*a*) In § 78 (pp. 764—7) Saint-Venant cites experiments of Morin to prove the linearity of the stress-strain relation. These experiments are really not conclusive, and I especially distrust the results cited for cast-iron. For elastic strains of such magnitude as occur in structures, the stress-strain relation for this material is certainly *not* linear. Nor again can arguments drawn from wires reduced to a state of ease serve the purpose Saint-Venant has in view of demonstrating the linearity and perfect elasticity of all materials for small strains.

(*b*) § 80 (pp. 771—4) treats of what Saint-Venant terms *l'état dit naturel ou primitif*. This is the state of no internal stress. It is used as a means of deducing the uniqueness of the solution of the elastic equations. If there be no body force or surface load the internal stresses are all zero, and *vice-versâ*. I have already had occasion to remark on the caution with which this principle must be accepted: see our Arts. 6 and 10. The arguments of this section do not seem to me very convincing.

(*c*) On pp. 783—86 the reader will find some interesting notes and valuable historical references on the origin of the terms *potential* and *potential function*.

(*d*) § 84 (pp. 789—96) reproduces the erroneous method of the memoir of 1863, for finding the stresses when there is an *initial state of stress*. C. Neumann (see our Chapter XI.) had previously obtained similar results for the case when the initial stress is given by an *uniform traction*: see our Arts. 129—31.

(*e*) Pages 801—25 are occupied with an important discussion of the distribution of elasticity in aeolotropic bodies. Saint-Venant using the symbolic method of Rankine arrives at some of the results of his memoir of 1863: see our Arts. 135—7.

The investigation of the *tasinomic* equation for particular cases, of the distribution of the stretch-moduli, and of the ellipsoidal distribution of elasticity in amorphic solids or cases of *confused crystallisation* follow the lines of the memoir of 1863: see our Arts. 136 and 151. They are accompanied by a discussion of the experimental results of Hagen, Chevandier and Wertheim, as bearing upon this theoretical distribution of elasticity. We shall return to this point when treating of the annotated *Clebsch*: see our Arts. 306—13.

(*f*) The remaining pages of the volume (825—49) are occupied with a sketch of Clebsch's treatment of the problem of torsion and flexure (see his *Theorie der Elasticität* § 23) and Kirchhoff's memoir on rods (see *Crelles Journal*, T. 56, p. 285, *Ueber das Gleichgewicht und die Bewegung eines unendlich-dünnen elastischen Stabes*). Saint-Venant shews how they are in agreement with his treatment of the problem, but does not contribute any additional matter.

[199.] Our analysis of Saint-Venant's edition of the *Leçons de Navier* will, we hope, have gone some way to convince the reader of the thorough study which this work deserves. Taken in conjunction with the annotated *Clebsch* (see our Art. 297) it forms the best introduction to the wide subjects of elasticity and the strength of materials yet published.

SECTION IV.

Memoirs of 1864—1882.

Impulse, Plasticity, etc.

[200.] *Compléments au Mémoire lu le* 10 *août* 1857 *sur l'impulsion transversale et la résistance vive des barres, verges ou poutres élastiques.*

Comptes rendus, T. LX. 1865, pp. 42—47 and pp. 732—35, T. LXI. 1865, pp. 33—37 and T. LXII. 1866, pp. 130—134. These extracts of additions to the memoir of 1857 (see our Arts. 104—8) are all more fully developed in the annotated *Clebsch*: see our Arts. 342 *et seq.*

[201.] *Note sur les pertes apparentes de force vive dans le choc des pièces extensibles et flexibles, et sur un moyen de calculer élémentairement l'extension ou la flexion dynamique de celles-ci: Comptes rendus*, T. LXII. 1866, pp. 1195—99.

This note suggests the application of the principle of virtual displacements and of the hypothesis that dynamical strain is of the same form as statical strain to the problem of impact. Saint-Venant apparently considers that in his papers of 1865—66 he had been the first to adopt this method, but as we have seen it is really due to Cox: see our Art. 1434*. The discussion in this *Note* appears in a more consistent form in the annotated *Clebsch*: see our Art. 368. It is Saint-Venant's great service to have shewn that the accurate and approximate methods agree fairly closely, and *why* they agree. Cox's method gives a result which is almost the same as that given by taking the term involving the principal vibration only. This point is well brought out in the concluding paragraphs of the *Note*, pp. 1198—9.

[202.] *Démonstration élémentaire:* (1°) *de l'expression de la vitesse de propagation du son dans une barre élastique;* (2°) *des formules nouvelles données, dans une communication précédente, pour le choc longitudinal de deux barres: Comptes rendus*, Tome

LXIV. 1867, pp. 1192—5. This is an extract from a memoir afterwards published in the *Journal de Liouville*: see our Arts. 203—20. Other parts of the same memoir are extracted in *Comptes rendus*, T. LXIII. 1866, pp. 1108—1111, and T. LXIV. 1867, pp. 1009—1013.

[203.] *Sur le choc longitudinal de deux barres élastiques de grosseurs et de matières semblables ou différentes, et sur la proportion de leur force vive qui est perdue pour la translation ultérieure; ...Et généralement sur le mouvement longitudinal d'un système de deux ou plusieurs prismes élastiques: Journal de Liouville*, T. XII. 1867, pp. 237—376, (the last two pages containing errata).

This is a long and theoretically very interesting memoir on the longitudinal impact of rods. It is the first complete treatment of the subject published. German writers have made some claim in this respect for Franz Neumann, who in his Königsberg lectures of 1857—8 dealt with the problem in somewhat the same fashion. But Neumann's investigations as first published in the *Vorlesungen über die Theorie der Elasticität*, 1885, pp. 340—346, are very insufficient and incomplete as compared with Saint-Venant's. Experimental investigations have been made by Boltzmann, W. Voigt, Hausmaninger and Hamburger with a view to testing the theory. Their results are not in full accordance with Saint-Venant's formulae. I shall refer to certain points of difference in discussing the present memoir, but the articles devoted to their memoirs must be consulted for fuller details.

[204.] The memoir is divided into two parts, the first treats of the impact of two rods of the same material and of equal cross-section. It is divided into seven articles. The first of these (pp. 237—244) deals with the history of the problem. At the invitation of Coriolis in 1827 Cauchy had investigated the influence of the vibrations produced by impact in altering the translational energy of two rods; Coriolis having recognised that these vibrations must be a source of loss in visible energy. Cauchy accordingly presented on February 19, 1827, a short note to the Academy, which was printed in the *Bulletin...de la Société Philomathique*, December 1826, pp. 180—182, and afterwards in the *Mémoires de l'Institut*. Cauchy treated only of the longitudinal impact of two rods of the same material and section. He concluded that the impulse terminated whenever the two bars had not the same speed at their impellent terminals. This, as we shall see, is not true, and the conclusion vitiated some of Cauchy's results, the analysis of which does not appear to have been published.

Poisson in the second edition of the *Traité de Mécanique* (1833, Vol. II. pp. 331—47) also attacked the problem supposing his rods of the same material and cross-section. He used a double condition for separation, namely, not only that the bar which precedes shall have a greater speed at the impelled terminal than that which follows, but that the squeeze in both at the impellent terminals shall be simultaneously zero. This condition led Poisson to the singular conclusion that two unequal bars would never separate. He had forgotten that physically they can never sustain a stretch at the impellent terminals. In fact Cauchy's condition of excess of speed in the preceding bar is insufficient, and Poisson's additional one of no squeeze is superabundant. The true condition is clearly excess of speed at a time when there is zero squeeze at the impellent terminals, which can never sustain a stretch. It will also be necessary to shew that the bars thus separated are separated for good, and do not, owing to their vibrations, come again into contact.

[205.] Saint-Venant's method of treatment is to investigate the vibrations of a bar, of which the initial condition is given by zero stretch throughout, and by speeds constant for each of the several parts into which the rod may be supposed divided. The first instant at which a zero stretch at the section between any two of these parts is accompanied by an excess speed in the terminal of the preceding section marks a disunion if the parts are not those of a continuous rod. In this manner Saint-Venant shews that if two bars of the same section and material are in impact the shorter takes ultimately and uniformly, while losing all strain, the initial speed of the longer.

This result was stated by Cauchy in 1826. Saint-Venant refers to the elementary proof of it given by Thomson and Tait in §§ 302—304 of their *Treatise on Natural Philosophy* which in 1867 was in the press. His notice had been drawn to this proof by an article in *The Engineer* (February 15, 1867) due to Rankine who, reviewing the extract in the *Comptes rendus* of Saint-Venant's memoir, had also given an elementary proof of one of his results for rods of different materials and cross-sections.

[206.] The second paragraph of the memoir (pp. 244—251) gives the general solution in *finite terms* of the equation for the longitudinal vibrations of a rod, when the initial speed and stretch of each point are given. The third paragraph deals with the special case of this when a rod of length $a = a_1 + a_2 + a_3 + \ldots$ has these parts initially subjected to uniform speeds V_1, V_2, $V_3 \ldots$ and uniform *squeezes* J_1, J_2, $J_3 \ldots$ etc. respectively (pp. 252—259). On pp. 254 and 258 we have diagrams which exhibit graphically in the special cases of two or three parts the speed and squeeze at each point of the rod during the motion. These diagrams are extremely instructive, and a similar method might be used with advantage in other cases of vibratory motions solved by arbitrary functions.

[207.] The fourth paragraph is entitled: *Problème du choc longitudinal de deux barres de longueur* a_1, a_2 *parfaitement élastiques, de même matière et de même section, animées primitivement de vitesses uniformes* V_1, V_2 *sans compression initiale* (pp. 259—262). This applies the results of the preceding paragraph to the simple case of impulse above stated, taking $V_1 > V_2$ and $a_1 < a_2$. Diagrams are given for the values of the speed and squeeze up to the time t given by $kt = 2a_1 + 2a_2$ for the two cases $2a_1 < a_2$ and $a_1 < a_2 < 2a_1$. Here k = velocity of sound $(= \sqrt{E/\rho})$. I have reproduced these diagrams reduced in scale on p. 140. Along the horizontal axis the values of kt are laid down, and along the vertical we have the various points of the combined rods, $OA_1 = a_1$, $A_1A = a_2$. In each area is placed the value of the speed and squeeze for that area, so that by means of the coordinates kt and x we can find the speed and squeeze of any point of the rod at any time. We see from this that at time $t = 2a_1/k$ the contiguous terminals will be moving with unequal velocities V_2 and $\frac{1}{2}(V_1 + V_2)$ but that this is *only for the instant*, and as there is no stretch at those terminals, the bars will not separate. They afterwards move till $t = 2a_2/k$ with the same velocity at the contiguous terminals and no squeeze. *The impulse is terminated, but the bars do not yet separate.*

Unequal speeds occur again when $t = 2a_2/k$, and now the upper bar has a negative squeeze, $j = -(V_1 - V_2)/2k$ at the

CASE $2a_1 < a_2$.

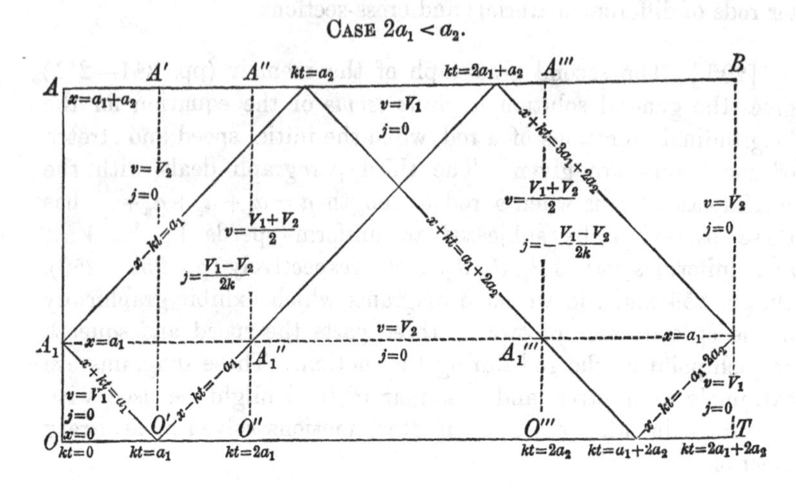

CASE $a_1 < a_2 < 2a_1$.

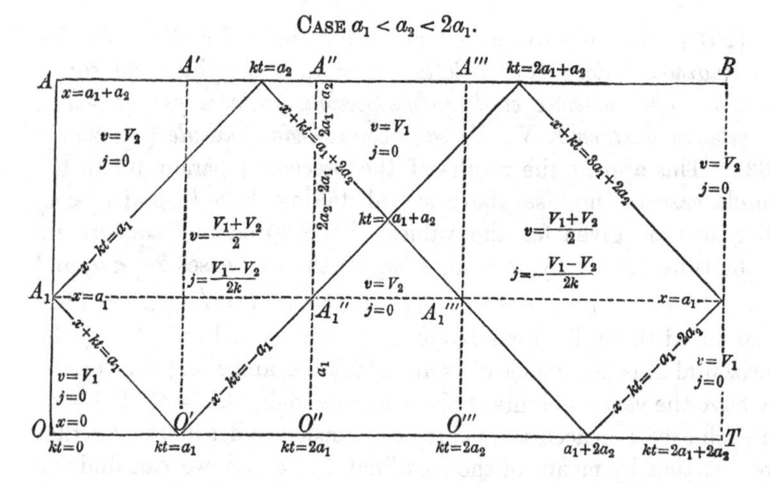

impelled terminal. Hence the solution no longer holds, and we have to treat each bar from this epoch as a distinct one. The bar a_1 moves obviously without strain and with the speed V_2 which the bar a_2 initially had. To deal with the bar a_2, we have to distinguish two cases. Let us suppose:

(1) $2a_1 < a_2$. We have to enquire how a bar of which a

portion $2a_1$ has initially a speed $v=\frac{1}{2}(V_1+V_2)$ and a negative squeeze $j=-\frac{1}{2}(V_1-V_2)/k$, and a portion a_2-2a_1 a speed $v=V_2$ and a squeeze $j=0$ subsequently moves. This has been ascertained in the second paragraph of the memoir and is represented by Saint-Venant in the accompanying diagram.

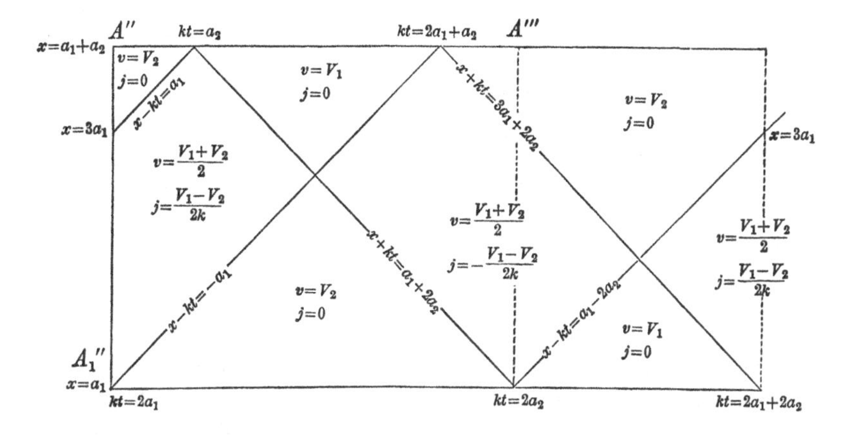

We see at once that after $t=2a_2/k$ the terminal moves with speed V_1 and therefore separates from the terminal of a_1 with speed V_1-V_2. This lasts till $t=2(a_1+a_2)/k$, when what happened at time $t=2a_1/k$ repeats itself and the terminal moves with speed V_2, i.e. with the same speed as the terminal of a_1. Thus it alternately moves with greater and equal speed, or the two terminals never again come into contact.

(2) $a_1<a_2<2a_1$. We have to enquire how a bar of which a portion $2a_2-2a_1$ has initially a speed $v=\frac{1}{2}(V_1+V_2)$ and negative squeeze $j=-\frac{1}{2}(V_1-V_2)/k$, and a portion $2a_1-a_2$, a speed $v=V_1$ and squeeze $j=0$ subsequently moves.

The motion is represented in the first diagram on p. 142, and we see that after the time $t=2a_2/k$ these bars never again come into contact.

[208.] The second diagram on p. 142 represents the whole motion of the two bars supposing them to be endowed with a uniform velocity perpendicular to their lengths during and subsequent to the impact. The full lines give the paths of various points of the rods, the dotted lines give the points at which the speed or squeeze of the rods changes abruptly. They corre-

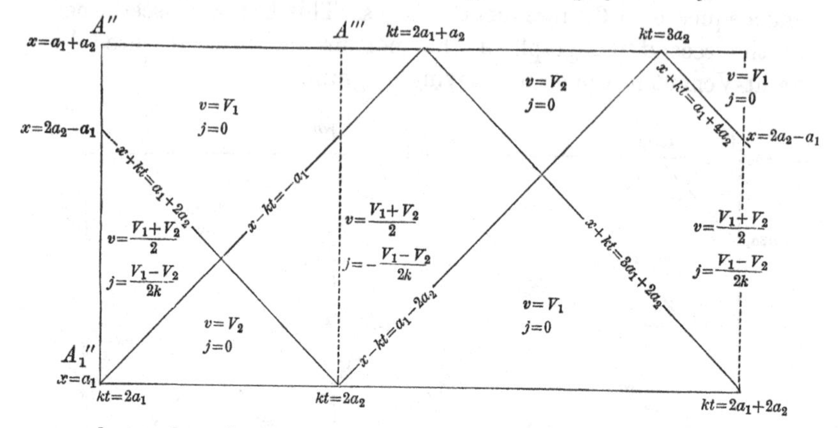

spond to the sloping lines of the previous diagrams. Saint-Venant calls the points at which velocity and squeeze change abruptly *points d'ébranlement.* It is hardly necessary to add that the stretch and squeeze of the rods are for diagrammatic purposes enormously exaggerated.

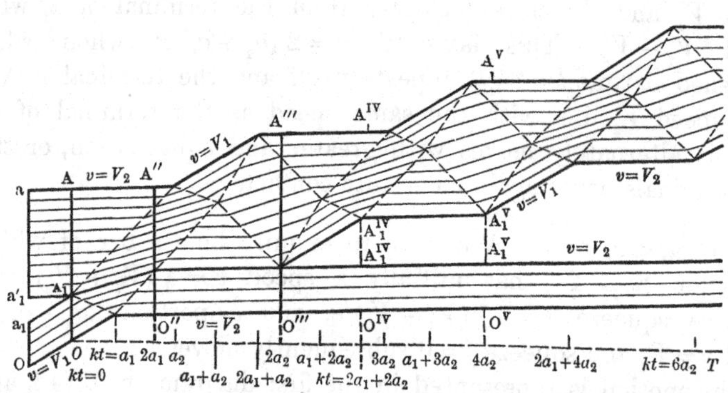

The separation of the two rods is discussed in Saint-Venant's sixth paragraph, the fifth having been devoted to a verification by means of the solution in trigonometrical series of the general results of the fourth paragraph: see pp. 262—269 of the memoir.

[209.] The seventh paragraph (pp. 278—86) is entitled; *Conséquences.—Force vive translatoire perdue dans le choc des deux barres élastiques de même grosseur et de même matière.—Vitesses de translation après le choc.* Let U_1, U_2 be the *centroidal* speeds after the impulse, i.e. at time $t = 2a_1/k$; then, as we have seen on p. 140, $U_1 = V_2$. To obtain U_2 we have only to make use of the principle of conservation of momentum, or

$$a_1 U_1 + a_2 U_2 = a_1 V_1 + a_2 V_2,$$

whence we find

$$\left.\begin{aligned} U_2 &= V_2 + \frac{a_1}{a_2}(V_1 - V_2) \\ \text{together with} \quad U_1 &= V_2. \end{aligned}\right\} \quad \text{..................... (i).}$$

We easily deduce

$$\frac{m}{2}(a_1 V_1^2 + a_2 V_2^2) - \frac{ma_1}{2} U_2^2 - \frac{ma_2^2}{2} U_1^2 = \frac{ma_1}{2}\left(1 - \frac{a_1}{a_2}\right)(V_1 - V_2)^2.$$

Or, the loss of kinetic energy of translation

$$= \frac{ma_1}{2}\left(1 - \frac{a_1}{a_2}\right)(V_1 - V_2)^2 \quad \text{................. (ii).}$$

Writing $M_1 = ma_1$, $M_2 = ma_2$, we see the following differences between Saint-Venant's theory and the ordinary theory of the impact of *perfectly elastic* bodies:

	Saint-Venant's theory	Ordinary theory
$U_1 =$	V_2	$V_1 - \frac{2M_2}{M_1 + M_2}(V_1 - V_2)$
$U_2 =$	$V_2 + \frac{M_1}{M_2}(V_1 - V_2)$	$V_2 + \frac{2M_1}{M_1 + M_2}(V_1 - V_2)$
Loss of Energy	$\frac{1}{2} M_1 \left(1 - \frac{M_1}{M_2}\right)(V_1 - V_2)^2$	0

Comparison with the so-called *inelastic* bodies of the ordinary theory gives no better agreement.

[210.] It may be noted here that Voigt's results for rods of equal cross-sections do not agree with Saint-Venant's theory when the shorter is the *impelling* rod. (*Annalen der Physik*, Bd. XIX. 1883, p. 51.) Further Saint-Venant makes the duration of the impulse $= 2a_1/k$, or $= 2a_2/k$ if we take it till the instant when the rods actually separate. In either case the duration of the impulse is proportional to the length of one of the rods and independent of the area of the cross-section. These results do not agree with Hamburger's experiments (*Untersuchungen über die Zeitdauer des Stosses elastischer cylindrischer Stäbe: Inaugural-Dissertation*, Breslau 1885, pp. 23—27). Hamburger finds that the duration is a function of the velocity of impact, which contradicts Saint-Venant's results.

[211.] The second part of Saint-Venant's memoir is entitled: *Choc de deux barres dont les sections et les matières sont différentes.*

The first paragraph (§ 8, pp. 286—98) gives in a double form the solution of the problem of the motion of two contiguous rods:

1° in trigonometrical series. This result Saint-Venant had obtained in an earlier memoir: see our Arts. 107 and 200. He adds the solution for beams in the form of truncated cones as given in the *Comptes rendus*, LXVI.; see our Art. 223. He remarks of these solutions:

> Au mémoire cité, complément de ceux que j'ai présentés depuis 1857 et qui vont être imprimés au *Journal de l'École Polytechnique*, on trouvera le développement de cette solution, à laquelle il convient de recourir quelquefois même pour les barres prismatiques, comme nous verrons plus loin, notamment quand une des deux parties a une section relativement fort grande, une longueur fort petite ou une résistance élastique considérable; suppositions qui poussées plus loin encore, permettent de réduire l'une des deux parties ou barres à une masse étrangère parfaitement dure, pouvant être venue heurter l'autre barre supposée libre aussi, ce qui constitue un problème dont la solution directe, a été présentée en 1865 (*Comptes rendus*, T. LXI., p. 33: see our Arts. 200 and 221).

2° in finite terms. This solution is somewhat lengthy, but is accompanied by diagrammatic representations of speed and squeeze of the same character as in the simpler case when the bars have equal cross-sections and sound-velocities. It is of a more complex nature, however, in particular the sloping lines become more numerous and change their slope abruptly at the horizontal line which marks the contiguous terminals: see p. 297 of the memoir.

[212.] The general solution is applied to the special case of the impact of two rods where initially the squeeze is zero throughout and the velocities are respectively V_1, V_2: see the ninth and tenth paragraphs. The results are again of a somewhat complex nature, but are rendered more intelligible by the aid of diagrams. They occupy pp. 299—326 of the memoir.

[213.] The eleventh paragraph is entitled: *Conséquences, en ce qui regarde le mouvement des deux barres après l'instant de leur choc, leur séparation, et les vitesses à l'instant où elle s'opère* (pp. 327—336).

Let $M_1 (= m_1 a_1)$, a_1, k_1, E_1, V_1, v_1, j_1 be the mass, length, velocity of sound, stretch-modulus, initial velocity, and velocity and squeeze of any point at any time of the first bar; similar quantities with the subscript 2 will refer to the second bar. Let $r = m_2 k_2/(m_1 k_1)$, and $\tau_1 = a_1/k_1$, $\tau_2 = a_2/k_2$. We shall suppose $\tau_1 < \tau_2$ or that sound traverses the following in less time than it does the preceding bar; this supposition is allowable as we can choose arbitrarily which sense of the velocity shall be considered positive. In discussing the results of the investigation we have to consider three possible cases:

$$r = 1, \quad r > 1, \quad \text{and} \quad r < 1.$$

Case (i). $\qquad r = 1$, or $m_2 k_2 = m_1 k_1$.

The impulse ends when $t = 2\tau_1$, but the bars do not separate until $t = 2\tau_2$. We have for the centroid-velocities after impact,

$$U_1 = V_2, \quad U_2 = V_2 + \frac{M_1}{M_2}(V_1 - V_2).$$

Thus the two rods behave in this case exactly like bars of the same material and of equal cross-section.

Case (ii). $\qquad r > 1$, or $m_2 k_2 > m_1 k_1$.

The impulse ends and the bars separate when $t = 2\tau_1$.

In this case:

$$U_1 = V_2 - 2\frac{m_2 k_2}{m_1 k_1 + m_2 k_2}(V_1 - V_2),$$

$$U_2 = V_2 + 2\frac{M_1}{M_2}\frac{m_2 k_2}{m_1 k_1 + m_2 k_2}(V_1 - V_2).$$

Case (iii). $\qquad r < 1$ or $m_2 k_2 < m_1 k_1$.

The bars no longer separate when $t = 2\tau_1$, but at the instant given by $t = 2\tau_2$.

If n be a whole number such that

$$n\tau_1 < \tau_2 < (n+1)\tau_1,$$

then:

$$U_1 = V_2 + \left(\frac{m_1k_1 - m_2k_2}{m_1k_1 + m_2k_2}\right)^n \left\{1 - \frac{2m_2k_2}{m_1k_1 + m_2k_2}\left(\frac{\tau_2}{\tau_1} - n\right)\right\}(V_1 - V_2),$$

$$U_2 = V_2 + \frac{M_1}{M_2}(V_1 - U_1).$$

where the value for U_1 on the right of the value for U_2 must be substituted from the first expression.

[214.] It will be observed that these formulae are again widely removed from those of the ordinary theory. They have been tested by Voigt for the velocities U_1 and U_2 of rebound, and by Hamburger for the duration of the impact. Neither find a really sufficient experimental accordance. Voigt attributes the discrepancy to the hypothesis adopted for the contiguous terminals, and considers that the rods cannot, while the contiguous terminals are in contact, be replaced by a single rod. He proposes a new theory, which introduces an elastic couch of some indefinite material (*Zwischenschicht*) between the terminals. This in a limiting case reduces the expressions for U_1 and U_2 to those of the ordinary theory, which in the same case agrees fairly with the results of experiment. In the general case, however, he has neither sufficiently specialised his hypotheses nor worked out his analytical results, so that we are unable to form any but the vaguest comparison of theory and experiment. His constants are unknown functions of material and of cross-section, and there seems no means of determining their form: see our discussion of his memoir later. A good test of Saint-Venant's theory might be made by experimenting in a vacuum and so removing a portion of Voigt's couch. I am inclined to think the discrepancy has more to do with thermal effect than with the couch of air, and that we ought to seek for results corresponding to those of the ordinary theory not when the coefficient of elastic impact is taken as unity, or the 'elasticity perfect,' but when it has a value differing from unity and so allowing for a loss of energy by heat. The problem ought not to be impossible with the aid of Duhamel's thermo-elastic equations.

[215.] In the twelfth paragraph (pp. 336—342) it is shewn that the bars after separating at time $t = 2\tau_1$, or $= 2\tau_2$ as the case may be, do not again come into contact. The thirteenth paragraph represents by diagrams similar to the figure on our p. 142 the motion of the two bars before, during and after the impact. These diagrams bring out very clearly the time of separation, and in *Case* (iii.), $r < 1$, shew how both bars retain a portion of the energy in the vibrational form, while in the previous case one bar only has any vibrational energy: see pp. 342—7 of the memoir, especially the diagram p. 345.

[216.] The following or fourteenth paragraph (pp. 347—50) is entitled: *Condition générale de séparation des barres à un instant donné quelconque, exprimée en fonction des vitesses et des compressions de leurs extrémités jointives à cet instant.*

Let V_2', J_2' be the velocity and squeeze of the bar a_2, supposed to be the impelled or preceding bar, at the point of contact.

Let V_1', J_1' be the like quantities for the impelling or following bar.

Saint-Venant deduces the necessary and sufficient condition for separation as follows:

Supposons en premier lieu, ce qui est permis, qu'elles se séparent pendant un temps *infiniment petit*. Le diagramme (23) du no. 3 relatif aux barres se mouvant isolément, ou le théorème qu'on en déduit, énoncé à la fin de ce même numéro, montre que leurs vitesses, au point de leur jonction, deviendront immédiatement après:

$$V_2' - k_2 J_2' \text{ pour } a_2,$$

$$V_1' + k_1 J_1' \text{ pour } a_1.$$

Cette soustraction $-k_2 J_2'$ et cette addition $k_1 J_1'$, faites à leurs vitesses positives, viennent, comme on a dit alors, de la *détente* de compressions J_1', J_2'. Si la nouvelle vitesse de a_2 excède la nouvelle vitesse de a_1, elles s'éloignent alors l'une de l'autre.

La condition de séparation ou d'éloignement est donc

$$V_2' - k_2 J_2' - V_1 - k_1 J_1' > 0.$$

This arises from the fact that a wave of squeeze j is propagated along the rod with the velocity k of sound; $\pm kj$ is then the velocity at which a cross-section is shifted (*vitesse de détente*), and if the whole of the rod were moving with velocity v, the rate of transfer of the section through space would be $v \pm kj$. But in the case of a free terminal section this must denote its absolute velocity, where v now becomes the velocity through space of the element at the end of the rod: cf. pp. 357—8 of the memoir with p. 347.

[217.] The fifteenth paragraph treats of the loss of kinetic energy, or the energy of translation transformed into energy of vibration. All the formulae of our Art. 213 may be included in the forms

$$U_1 = V_1 - \alpha (V_1 - V_2), \quad U_2 = V_2 + \alpha \frac{M_1}{M_2} (V_1 - V_2).$$

The energy lost is then represented by

$$M_1\{2\alpha-(1+M_1/M_2)\,\alpha^2\}\,\tfrac{1}{2}\,(V_1-V_2)^2.$$

Since there must always be a loss of energy, it is necessary that

$$\alpha<\frac{2M_2}{M_1+M_2}.$$

Saint-Venant shews from the values of α in the various cases referred to in Art. 213 that this is always true (pp. 361—355).

The coefficient of dynamic elasticity e as investigated by Newton (*Principia, Ed. Princeps,* p. 22) has probably relation to the energy lost not only in vibrations, but also in the form of heat. To make Newton's formula agree with the above, it is necessary to take $\alpha=\dfrac{M_2}{M_1+M_2}(1+e)$, supposing for a moment Newton's laws to hold for rods and that the energy lost is *principally vibrational, not thermal.* This gives us, for example in *Case* (ii) of Art. 213,

$$1+e=2\,\frac{m_1a_1+m_2a_2}{m_1k_1+m_2k_2}\cdot\frac{k_2}{a_2}.$$

Thus if the rods were of different materials, it is difficult to see how e could be independent of *their masses,* which Newton proved for the impact of spherical bodies. Further in the case of equal rods of the same material e would always equal unity. This again is not true for most bodies. Hence we are driven to conclude either that the amount of thermal energy generated is generally of importance or that the conditions at the surface of impact adopted by Saint-Venant are not satisfactory. It would be interesting to make experiments for a material for which e is nearly unity, the rods being of equal cross-section and the same material, and then endeavour to ascertain by varying their masses whether there was any change in e. Haughton's experiments seem to indicate that e is not constant but a function of the velocity of impact; this does not suggest Saint-Venant's form, but it is interesting as pointing out a want of constancy in this coefficient: see our Arts. 1523* and also 941*, 1183*.

[218.] In his sixteenth paragraph (pp. 355—373) Saint-Venant proceeds to give an elementary proof of the formulae of Art. 213. This proof does not involve differential or integral processes, but it seems to me that, while luminous and suggestive to the reader of the previous analysis, it would not in the more complex cases be of equal value to the student who approached in this manner for the first time the problem of the impact of bars. Similar proofs for the *simpler* cases have been given by Thomson and Tait (§§ 302—305 of their *Natural Philosophy*), and by Rankine (*The Engineer*, February, 1867, p. 133).

[219.] The elementary discussion opens with a deduction of the value of the velocity of longitudinal sound vibrations in a rod $(=\sqrt{E/\rho})$. At that time Saint-Venant thought it novel, believing that no elementary proof had been offered since Newton's rather obscure demonstration of the velocity of sound. In a Note in the *Comptes rendus*, LXXI. 1867, p. 186, Saint-Venant acknowledges the priority of Babinet, who had given the proof in oral lectures 40 years previously and published it in his *Exercices sur la Physique*, Second Edn, 1862. In the same Note Saint-Venant gives in a footnote an elementary demonstration of the velocity of slide waves $(=\sqrt{\mu/\rho})$.

[220.] We shall not reproduce any of Saint-Venant's elementary treatment, but merely refer the curious reader to the sixteenth paragraph of his memoir. We conclude with a short extract on this point from the *résumé* of his memoir which he gives in the seventeenth paragraph:

J'aurais pu borner mon travail à ces sortes de démonstrations. Mais les solutions analytiques, telles que celles qui m'ont conduit aux résultats présentés, portent leur genre de conviction comme les solutions synthétiques, et ce n'est pas trop du concours de deux genres de recherches et de raisonnements pour établir complètement des résultats tout nouveaux et controversés. Et puis, il eut manqué quelque chose, savoir la preuve que les deux barres, après s'être séparées pendant un temps fini, ne se rejoindront pas en vibrant (p. 374).

[221.] *Choc longitudinal de deux barres élastiques, dont l'une est extrêmement courte ou extrêmement roide par rapport à l'autre: Comptes rendus*, LXVI. 1868, pp. 650—3.

This may be looked upon as a supplement to the memoir in the *Journal de Liouville*: see our Art. 203. Saint-Venant had treated this case in that memoir by expressions involving trigonometrical series; he now proposes to give its solution in finite terms.

If a_1, a_2 be the lengths of the two bars, k_1, k_2 the corresponding velocities of sound, M_1, M_2 the masses, V_1, V_2 the initial velocities, U_1, U_2 the final *mean* velocities of the impelling and impelled bars, then Saint-Venant had obtained in that memoir the following results for the case in which a_2/k_2, or the time sound takes to traverse the second bar, is an exact multiple n of the time a_1/k_1 it takes to traverse the first bar:

$$\left.\begin{aligned} U_1 &= V_2 + \left(\frac{1-r}{1+r}\right)^n (V_1 - V_2), \\ U_2 &= V_2 + \frac{M_1}{M_2}\left\{1 - \left(\frac{1-r}{1+r}\right)^n\right\} (V_1 - V_2), \\ \text{where,}\quad r &= \frac{M_2}{M_1}\frac{a_1 k_2}{a_2 k_1} < 1. \end{aligned}\right\} \text{............ (i),}$$

Now if the impelling bar is infinitely short or infinitely hard (if $a_1 = 0$ or $k_1 = \infty$), the number $n \left(= \frac{M_2}{M_1}\frac{1}{r}\right)$ will be infinitely great, hence it follows that:

$$\left(\frac{1-r}{1+r}\right)^{1/r} = e^{-2},$$

and the formulae (i) become:

$$U_1 = V_2 + (V_1 - V_2)\, e^{-2M_2/M_1},$$
$$U_2 = V_2 + M_1/M_2 \,.\, (1 - e^{-2M_2/M_1})\,(V_1 - V_2).$$

[222.] Saint-Venant also shews in this memoir how to obtain from the results of his previous memoir the velocity and squeeze of each bar at each instant of the impact. Thus:

(1°) For the impelling bar. From $t = 0$ to $2a_2/k_2$,

$$\text{velocity} = V_2 + (V_1 - V_2)\, e^{-M_2/M_1 \,.\, k_2 t/a_2},$$
$$\text{squeeze} = 0.$$

(2°) For the impelled bar. First from $t = 0$ to a_2/k_2:

From $x = 0$ to $k_2 t$, $\begin{cases} \text{velocity} = V_2 + (V_1 - V_2)\, e^{-M_2/M_1 \,.\, (k_2 t - x)/a_2}, \\ \text{squeeze} = (V_1 - V_2)/k_2 \,.\, e^{-M_2/M_1 \,.\, (k_2 t - x)/a_2}. \end{cases}$

From $x = k_2 t$ to a_2, $\begin{cases} \text{velocity} = V_2, \\ \text{squeeze} = 0. \end{cases}$

Secondly from $t = a_2/k_2$ to $2a_2/k_2$:

From $x = 0$ to $2a_2 - k_2 t$ the velocity and squeeze have the same values as previously from $x = 0$ to $k_2 t$.

From $x = 2a_2 - k_2 t$ to a_2,

$$\begin{cases} \text{velocity} = V_2 + (V_1 - V_2)\,\{e^{-M_2/M_1 \,.\, (k_2 t - x)/a_2} + e^{-M_2/M_1 \,.\, (k_2 t + x - 2a_2)/a_2}\}, \\ \text{squeeze} = (V_1 - V_2)/k_2 \,.\, \{e^{-M_2/M_1 \,.\, (k_2 t - x)/a_2} - e^{-M_2/M_1 \,.\, (k_2 t + x - 2a_2)/a_2}\}. \end{cases}$$

This gives the whole state of the bars up to the end of the impact or until $t = 2a_2/k_2$.

Saint-Venant tests these results: 1° by the principle of conservation of momentum, 2° by that of conservation of energy, 3° by comparing the above finite forms with the solutions in trigonometrical series. He finds them verified in all cases. In the concluding paragraph he promises in a future communication to deal with the case of a bar with one terminal fixed and the other terminal struck by a load represented by an infinitely short second bar. This is a fundamental problem in suspension bridge bars, and solutions in trigonometrical series had been given by Navier and Poncelet: see our Arts. 272* and 991*. Saint-Venant promises one in *finite* terms.

[223.] *Solution, en termes finis, du problème du choc longitudinal de deux barres élastiques en forme de tronc de cône ou de pyramide: Comptes rendus*, LXVI. 1868, pp. 877—81.

This is again a complement to the memoir in the *Journal de Liouville* (see our Art. 213). It gives in finite terms a solution for a case in that memoir, which Saint-Venant had only solved in trigonometrical series. Namely the case when the bars instead of being prismatic are truncated cones or pyramids.

The equations for the vibrations are in this case of the form:

$$\frac{d\,(E_1\Omega_1\,du_1/dx_1)}{dx_1}=\Omega_1\rho_1\,\frac{d^2u_1}{dt^2},$$

where ρ_1 is the density and Ω_1 the cross-section $=\omega_1\,(1+x_1/h_1)^2$, ω_1 and h_1 being constants. If we put $E_1/\rho_1=k_1^2$, we have an integral of the form:

$$u_1=\frac{f_1\,(x+h_1+k_1t)+F_1\,(x_1+h_1-k_1t)}{x_1+h_1}.$$

Similarly there will be two arbitrary functions f_2, F_2 for the second bar. The problem is to determine these four functions by the initial conditions $du_1/dt=V_1$ from 0 to a_1, $du_2/dt=-V_2$ from 0 to a_2, while the initial squeeze is zero throughout the bars. The terminal conditions have also to be satisfied throughout the motion. The forms of the functions are given on pp. 879—80 of the memoir, and the general treatment of the problem indicated, without, however, any numerical details for special cases.

La solution s'étendrait même à plusieurs barres juxtaposées bout à bout, et par conséquent au choc de deux solides allongés quelconques à axe rectiligne, car ces solides peuvent toujours être approximativement décomposés en troncs de pyramide à base quelconque (p. 881).

[224.] *Leçons de mécanique analytique*, par M. l'Abbé Moigno. *Statique*. Paris, 1868. The last two *Leçons* of this work, the twenty-first and twenty-second, pp. 616—723, contain a general theory of elasticity by Saint-Venant. This is the fourth such general theory that we have from his pen, the former three being respectively in the memoir on *Torsion*, in that on *Flexure*, and in the *Leçons de Navier*: see our Arts. 4, 72, and 190. Saint-Venant's treatment is in the main a modified and improved form of that of the second, third and fourth years of Cauchy's *Exercices de mathématiques*; that is to say it starts from the molecular definition of stress (p. 617). After a very full analysis of stress and strain we reach the general elastic equations. The hundred odd pages form one of the best introductions to the subject of elasticity, though they naturally contain no new results. We may refer to one or two points.

[225.] Saint-Venant rejects like Lamé that definition of stress across a plane, which considers stress as the force necessary to retain the plane in equilibrium if it were to become rigid (footnote p. 619). This apparently simple definition conveys, he holds, no exact notion and its simplicity is a pure delusion. In other words he insists upon the importance of the molecular-definition of stress: see Lamé's *Leçons sur l'élasticité*, § 5, and our Arts. 1051* and 1164*.

[226.] The well-known theorems of Cauchy and the equations to his ellipsoids are reproduced with short proofs: see our Arts. 603*—12*. We may note also on p. 630 a demonstration of Hopkins' theorem: see our Art. 1368*. Relations for change of direction of stretch and slide, such as those of our Art. 133, are given on pp. 644—5. Saint-Venant remarks that these relations were first given by Lamé in 1851, but that he assumes that the *shifts are small*; the proof given by Saint-Venant holds for any shifts, provided the relative shifts, i.e. the local strains, are small.

[227.] On pp. 652—3 Saint-Venant states as a Lemma and proves the principle of linearity of the stress-strain relations, i.e. the generalised Hooke's Law. The proof appeals to the rari-constant hypothesis. The reader will remember that there is an unjustifiable assumption often made in the proof of the generalised Hooke's law by Green's method: see our Art. 928*. We may note here how Saint-Venant as a rari-constant elastician proves his Lemma. After stating that the stresses must be functions of the strains he continues:

Et elles en sont fonctions linéaires ou du premier degré; car, comme les actions réciproques entre molécules sont fonctions continues de leurs distances mutuelles r, celles que développent de *très-petites* augmentations rs_r des distances leur sont proportionnelles; et les changements très-petits des inclinaisons mutuelles de ces actions à composer ensemble pour avoir les pressions sont proportionnelles aussi à des augmentations rs_r de distances. Or ces petites augmentations positives ou négatives:

$$rs_r = rc^2_{rx}s_x + rc^2_{ry}s_y + rc^2_{rz}s_z + rc_{ry}c_{rz}\sigma_{yz} + rc_{rz}c_{rx}\sigma_{zx} + rc_{rx}c_{ry}\sigma_{xy},$$

[see our Art. 547*]

sont sommes de produits des premières puissances des dilatations et glissements s, σ par des quantités $rc^2_{rx}, \ldots rc_{rx}c_{ry}$ qui ne dépendent que de l'état antérieur aux déformations......Les composantes $\widehat{xx} \ldots \widehat{xy}$ des pressions sont donc fonctions du premier degré des mêmes six quantités très-petites s et σ, ce qui est le lemme énoncé.

It will be noted that a clear reason is here given for the legitimacy of Taylor's theorem and the retention of the first powers. *It depends on the rari-constant hypothesis.* A slight discussion of this point with a reference to the *Appendice V.* of the *Leçons de Navier* will be found on pp. 654—6: see our Arts. 192 and 298. There is a footnote on the arbitrary assumption of the stress-strain relations for isotropic bodies by Cauchy and Maxwell: see our Arts. 614* and 1537*.

[228.] On p. 670 there is a footnote citing the values of the stretches and slides for large shifts. This requires modifying in the sense of my remarks in Art. 1619*—22*.

There is an excellent proof on rari-constant lines following Cauchy of the most general elastic equations with initial stresses on pp. 673—689. It is followed on pp. 694—7 by some useful remarks on the difficulties which occur in the treatment of stresses as the sums of intermolecular actions: see our Arts. 443* and 1400* The pitfalls into which Poisson, Navier and others have fallen are well brought out.

[229.] This discussion on elasticity concludes with a deduction of the expression for the strain-energy (Green's function) by means of Lagrange's process and the rari-constant hypothesis (p. 717). The method is similar to that used by C. Neumann in his memoir of 1859: see our Chap. XI. It is pointed out that if Navier's error of taking $(r_1 - r)f'(r)$ instead of $f(r) + (r_1 - r)f'(r)$ for $f(r_1)$ be avoided, and if the summations be not replaced by integrals, then Poisson's objection to the application of the Calculus of Variations to molecular problems falls to the ground (p. 719): see our Arts. 266* and 446*. Finally there is an account of Green's process and an unfavourable criticism of his theory of double refraction (pp. 719—23): see our Arts. 147 and 193.

[230.] *Formules de l'élasticité des corps amorphes que des compressions permanentes et inégales ont rendus hétérotropes. Journal des mathématiques,* Tome XIII. 1868, pp. 242—254. In his memoir of 1863 Saint-Venant has shewn on the rari-constant hypothesis that the ellipsoidal distribution of elasticity holds for aeolotropic, but amorphic bodies, i.e. bodies such as the metals, whose primitive isotropy has been altered by a permanent strain, which has not converted their elements into crystals; such a permanent strain for instance as would be produced by the processes of rolling, forging, etc. This ellipsoidal distribution he has applied to explain the phenomena of double refraction, without adopting exact transversality of vibration, but obtaining without approximation Fresnel's wave-surface. The ellipsoidal conditions are of two kinds:

$$\left.\begin{array}{ll} \text{(i) a group of the type} & 2d + d' = \sqrt{bc} \\ \text{or,}\quad \text{(ii)}\quad ,, \quad ,, \quad ,, & 2d + d' = \tfrac{1}{2}(b + c) \end{array}\right\} \ldots\ldots\ldots \text{(i)}.$$

If the differences of the direct-stretch coefficients $(b - c, c - a, a - b)$ are so small that their squares may be neglected, these two groups of conditions are identical; this is probably the case in the metals used for construction, and in doubly-refracting media: see our Arts. 142—7. The conditions by which Saint-Venant would replace Green's relations—the Cauchy-Saint-Venant conditions as we have termed them—amount to an ellipsoidal distribution of elasticity (see our Art. 149), but this distribution Saint-Venant has only discussed on the basis of rari-constant equations. Boussinesq in a memoir entitled: *Mémoire sur les ondes dans les milieux isotropes déformés,* which immediately precedes the present memoir (pp. 209—241 of the same volume) has deduced the Cauchy-Saint-Venant conditions for double refraction on the basis of the ellipsoidal distribution without any appeal to rari-constancy. The ellipsoidal distribution is proved by Boussinesq for amorphic bodies on the multi-constant hypothesis,—provided we assume the elastic coefficients to be themselves *linear* functions of three small quantities corresponding respectively to the three principal rectangular directions of the permanent strain given to the initially isotropic material. Saint-Venant proposes to give a new proof of Boussinesq's result, so that the ellipsoidal distribution may be accepted for the amorphic bodies in question even by multi-constant elasticians.

[231.] Suppose the body initially isotropic to be permanently strained in such manner that at each point there are three planes of elastic symmetry, then the stress-strain relations are of the form:

$$\left.\begin{array}{ll} \widehat{xx} = as_x + f's_y + e's_z, & \widehat{yz} = d\sigma_{yz}, \\ \widehat{yy} = f's_x + bs_y + d's_z, & \widehat{zx} = e\sigma_{zx}, \\ \widehat{zz} = e's_x + d's_y + cs_z, & \widehat{xy} = f\sigma_{xy}. \end{array}\right\} \text{.................. (ii).}$$

Let ϵ, ϵ', ϵ'' be the three small quantities corresponding to the three rectangular directions x, y, z of which the elastic constants are, according to hypothesis, to be functions, or let the types be

$$a = \alpha + l_1\epsilon + m_1\epsilon' + n_1\epsilon'',$$
$$d' = \delta' + p_1\epsilon + q_1\epsilon' + k_1\epsilon'',$$
$$d = \delta + r_1\epsilon + s_1\epsilon' + h_1\epsilon''.$$

Then since the original condition is isotropy, a must be related to ϵ' and ϵ'' in the same way, and further in the same way to ϵ as b to ϵ' and c to ϵ''. Thus $l_1 = m_2 = n_3$, and $m_1 = n_1 = l_2 = n_2 = l_3 = m_3$. Similar relations hold for the constants of d and d'. Thus we may write as types:

$$a = \alpha + l\epsilon + m(\epsilon' + \epsilon''),$$
$$b = \alpha + l\epsilon' + m(\epsilon + \epsilon''),$$
$$d' = \delta' + p\epsilon + q(\epsilon' + \epsilon''),$$
$$e' = \delta' + p\epsilon' + q(\epsilon + \epsilon''),$$
$$d = \delta + r\epsilon + s(\epsilon' + \epsilon''),$$
$$e = \delta + r\epsilon' + s(\epsilon + \epsilon'')\text{...etc.}$$

Now if we take $\epsilon' = \epsilon''$, or the stretch the same all round the direction x, we ought to have not only $b = c$, $e = f$, $e' = f'$, which easily follows, but in addition the values of the constants ought not to be affected by a rotation of the axes round that of x. This however is easily shewn to involve

$$b = 2d + d',$$

or what is the same thing

$$\alpha + m\epsilon + (l + m)\epsilon' = 2\delta + \delta' + (2r + p)\epsilon + (4s + 2q)\epsilon'.$$

This involves, as an identity true for all values of ϵ and ϵ', the further results

$$a = 2\delta + \delta', \quad m = 2r + p, \quad l + m = 4s + 2q.$$

Whence we easily find generally:

$$\begin{aligned} b + c &= 2a + (\epsilon' + \epsilon'')(l + m) + 2m\epsilon, \\ &= 2(2\delta + \delta') + 2(\epsilon' + \epsilon'')(2s + q) + 2(2r + p)\epsilon, \\ &= 2(2d + d'), \end{aligned}$$

or

$$2d + d' = \tfrac{1}{2}(b + c),$$

the type of ellipsoidal condition for the second group. It will be identical with the group of type $(2d + d') = \sqrt{bc}$, when we may neglect the squares of the differences of a, b, c, or quantities like $(l - m)^2 (\epsilon - \epsilon')^2$. Hence the ellipsoidal conditions have been deduced on a hypothesis very probable in character and not opposed to multi-constancy.

[232.] The memoir concludes by noting that to the stress-strain relations (ii) subject to the inter-constant relations (i), we must add terms of the type:

$$\widehat{xx}_0 (1 + u_x - v_y - w_z) \text{ to } \widehat{xx},$$

$$\widehat{zz}_0 v_z + \widehat{yy}_0 w_y \qquad \text{to } \widehat{yz},$$

if there be an initial stress $\widehat{xx}_0$, $\widehat{yy}_0$, $\widehat{zz}_0$ symmetrical with regard to the planes of symmetry of the primitive strain. Saint-Venant appeals for these to his memoir of 1863, but as we have seen he has really only proved them there for rari-constancy (see our Art. 129).

[233.] *Calcul du mouvement des divers points d'un bloc ductile, de forme cylindrique, pendant qu'il s'écoule sous une forte pression par un orifice circulaire; vues sur les moyens d'en rapprocher les résultats de ceux de l'expérience: Comptes rendus*, LXVI. 1868, pp. 1311—24. This memoir deals only with the motion of the parts of a ductile mass, and does not take into consideration the stresses which produce those motions. Its methods thus approach those of hydrodynamics rather than of elasticity; it belongs as Tresca's own theory, to which it refers, to the pure kinematics of deformation. A report drawn up by Saint-Venant on Tresca's communications to the Academy immediately precedes the above memoir (pp. 1305–11). It deals with and criticises Tresca's pure kinematic theory.

Memoirs by Saint-Venant treating of the flow of a ductile solid or of a liquid out of a vessel will be found in the *Comptes rendus*, LXVII. 1868, pp. 131—7, 203—211, 278—282 and LXVIII. 1869, pp. 221—237, 290—301. They cannot be considered to fall in any way under the title of the elasticity or even the *strength* of materials.

[234.] *Note sur les valeurs que prennent les pressions dans un solide élastique isotrope lorsque l'on tient compte des dérivées d'ordre supérieur des déplacements très-petits que leurs points ont éprouvés: Comptes rendus*, LXVIII. 1869, pp. 569—571. This note gives without proof expressions for the traction and shear at any point of an elastic solid. when we do not neglect the squares of the shift-fluxions. Saint-Venant says that his results have been obtained from rari-constant considerations. He finds:

$$\widehat{xx} = \epsilon_0 \left(\theta + 2\frac{du}{dx}\right) + \epsilon_1 \left\{2\frac{d^2\theta}{dx^2} + \nabla^2\left(\theta + 2\frac{du}{dx}\right)\right\}$$

$$+ \epsilon_2 \left\{4\nabla^2\frac{d^2\theta}{dx^2} + \nabla^2\nabla^2\left(\theta + 2\frac{du}{dx}\right)\right\}$$

$$+ \epsilon_3 \left\{6\nabla^2\nabla^2\frac{d^2\theta}{dx^2} + \nabla^2\nabla^2\nabla^2\left(\theta + 2\frac{du}{dx}\right)\right\} + \dots$$

$$\widehat{yz} = \epsilon_0 \left(\frac{dv}{dz} + \frac{dw}{dy}\right) + \epsilon_1 \left\{2\frac{d^2\theta}{dydz} + \nabla^2\left(\frac{dv}{dz} + \frac{dw}{dy}\right)\right\}$$

$$+ \epsilon_2 \left\{4\nabla^2\frac{d^2\theta}{dydz} + \nabla^2\nabla^2\left(\frac{dv}{dz} + \frac{dw}{dy}\right)\right\}$$

$$+ \epsilon_3 \left\{6\nabla^2\nabla^2\frac{d^2\theta}{dydz} + \nabla^2\nabla^2\nabla^2\left(\frac{dv}{dz} + \frac{dw}{dy}\right)\right\} + \dots$$

Here θ is as usual the dilatation, ∇^2 is the Laplacian $\frac{d^2}{dx^2} + \frac{d^2}{dy^2} + \frac{d^2}{dz^2}$, and $\epsilon_0, \epsilon_1, \epsilon_2, \epsilon_3 \dots$ are constants depending on the elastic nature of the body.

Saint-Venant concludes his note with the remark:

Ces formules serviront peut-être à expliquer des faits relatifs à certaines substances élastiques pour lesquelles le rapport entre les efforts et les effets varie plus rapidement lorsqu'on les comprime que lorsqu'on les étend, en sorte que les vibrations qui y seraient excitées augmenteraient leurs dimensions comme fait la chaleur, dont les effets de dilatation peuvent être attribués, comme j'ai eu l'occasion de le faire remarquer (*Société Philomathique*, October 20, 1855: see our Art. 68), à ce que les actions entre les derniers atomes suivraient une loi analogue, (p. 571).

[235.] *Sur un potentiel de deuxième espèce, qui résout l'équation aux différences partielles du quatrième ordre exprimant l'équilibre intérieur des solides élastiques amorphes non isotropes: Comptes rendus*, LXIX. 1869, pp. 1107—1110.

This note merely refers to E. Mathieu's discussion of the potential of the second kind

$$\phi = \iiint f(\alpha, \beta, \gamma) \sqrt{(x-\alpha)^2 + (y-\beta)^2 + (z-\gamma)^2}\, d\alpha d\beta d\gamma,$$

by means of which the equation $\nabla^2\nabla^2\phi = 0$ can be solved. This equation occurs in the treatment of an isotropic solid. Saint-Venant notices the form

$$\phi = \iiint f(\alpha, \beta, \gamma) \sqrt{\frac{(x-\alpha)^2}{A} + \frac{(y-\beta)^2}{B} + \frac{(z-\gamma)^2}{C}}\, d\alpha d\beta d\gamma,$$

which solves the equations of elasticity when there is an ellipsoidal distribution of elasticity: see our Arts. 140–1.

Saint-Venant speaks highly of Cornu's memoir of 1869 and its bearing on the constant-controversy: see our Articles below on that physicist's work.

[236.] *Preuve théorique de l'égalité des deux coefficients de résistance au cisaillement et à l'extension ou à la compression dans le mouvement continu de déformation des solides ductiles au delà des limites de leur élasticité: Comptes rendus,* LXX. 1870, pp. 309—11.

The object of this note is to prove the equality between the coefficient of resistance to slide and the coefficient of resistance to stretch or squeeze, when both slide and stretch are plastic.

Saint-Venant takes a right six-face of edges $a, b, c,$ and supposes the two faces $a \times b$ to be subjected to shearing forces in direction of a which produce a plastic slide-set $\sigma \times c$, so that the limit of elasticity is passed. If K' be the force necessary per unit of area, the work expended in producing this set is

$$K'ab \times \sigma \times c,$$

or, it equals $K'\sigma$ *per* unit of volume.

Now this same slide-set could have been produced by diagonal stretch and squeeze of magnitude $\sigma/2$: see our Art. 1570*. Let us take the right six-face abc and divide it up into others of the same breadth b, but of length a' and height c' making angles of 45° with a and c and having their end-faces $a' \times c'$ in the faces $a \times c$. In order to produce set-stretch it is necessary to apply to the faces bc' a traction given by Kbc' and to the faces ba' a negative traction given by Kba', where K is the coefficient of resistance to *both* stretch and squeeze. Hence to produce a stretch of $\sigma/2$ and a squeeze of $\sigma/2$ parallel to a' and c' respectively, we require work equal to

$$Kbc' \,.\, \frac{\sigma}{2}\, a' \text{ and } Kba' \,.\, \frac{\sigma}{2}\, c',$$

or, *per* unit volume of the little prism $a'bc'$, we require work equa to

$$K\sigma.$$

But this quantity must equal the previous $K'\sigma$ or

$$K' = K,$$

the result experimentally ascertained by Tresca.

Saint-Venant concludes the note as follows:

> Ce raisonnement me paraît, aussi, justifier l'hypothèse, hardie au premier aperçu, mais, en y réfléchissant, très-rationnelle, de l'égalité des résistances à l'extension et à la compression permanente, par unité superficielle des bases des prismes qu'on y soumet; bien entendu, sous la condition générale, que tout ceci suppose remplie, de mouvements excessivement lents, ou tels que leur vitesse n'entre pour rien dans les résistances aux déformations qu'ils produisent.

In a footnote he refers to a method by which the flow-lines of a plastic material might be obtained experimentally.

It must be noted that the proof assumes the coefficients K_1, K_2 of resistance to squeeze- and stretch-set to be equal, otherwise we should have

$$K_1 + K_2 = 2K'.$$

The reader may compare Coulomb's results on shearing and tractive *strength* referred to on p. 877 of our first volume.

[237.] *Formules des augmentations que de petites déformations d'un solide apportent aux pressions ou forces élastiques, supposées considérables, qui déjà étaient en jeu dans son intérieur.—Complément et modification du préambule du mémoire: Distribution des élasticités autour de chaque point, etc. qui a été inséré en* 1863 *au Journal de Mathématiques,* (see our Arts. 127—152). This memoir is published in the *Journal de Mathématiques,* Tome XVI. 1871, pp. 275—307, and is divided into two parts; the *Première Partie* (pp. 275—291) is occupied with correcting an error which Brill and Boussinesq had pointed out in the memoir of 1863 (see our Art. 130); the *Deuxième Partie* deals with the relations between the elastic constants $|xxxx|$, etc. and the six components of initial strain. It occupies pp. 291—307 and forms the subject of a note on pp. 355 and 391 of the *Comptes rendus,* T. LXXII. 1871

[238.] The error in question was really indicated in our first volume (see Art. 1619*), namely that the true relations between the strains, s_x, σ'_{yz} and the shift-fluxions are in their most general form of the types[1]:

$$\left.\begin{aligned} s_x + \tfrac{1}{2} s_x^2 &= u_x + \tfrac{1}{2}(u_x^2 + v_x^2 + w_x^2) \\ \sigma'_{yz}(1 + s_y)(1 + s_z) &= v_z + w_y + u_y u_z + v_y v_z + w_y w_z \end{aligned}\right\} \ldots\text{(i)},$$

but that these are not the values taken by Saint-Venant in his memoirs of 1847 and 1863: see our Arts. 1622* and 130. Accordingly Saint-Venant's attempt to deduce Cauchy's equations from a multi-constant hypothesis is erroneous.

The full value of the potential energy is

$$\left.\begin{aligned} \phi = \phi_0 &+ \widehat{xx}_0 (s_x + \tfrac{1}{2} s_x^2) + \ldots\ldots + \ldots\ldots \\ &+ \widehat{yz}_0 \sigma'_{yz}(1 + s_y)(1 + s_z) + \ldots\ldots + \ldots\ldots \\ &+ \phi_1 \end{aligned}\right\} \ldots\ldots\ldots\ldots\text{(ii)},$$

as Boussinesq had pointed out, and not

$$\phi = \phi_0 + \widehat{xx}_0 s_x + \ldots\ldots + \ldots\ldots + \widehat{yz}_0 \sigma'_{yz} + \ldots\ldots + \ldots\ldots + \phi_1,$$

as assumed in the memoir of 1863 (see our Art. 130). But the expression (ii) *has been deduced only from molecular considerations on the rari-constant hypothesis.* The fact is that we can on the multi-constant hypothesis expand ϕ in linear and quadratic terms of the strain-components ϵ_x, ϵ_y, ϵ_z, η_{yz}, η_{zx}, η_{xy} of our Art. 1619*, as Green in fact did (*Collected Papers*, pp. 298–9), but we cannot determine to what extent the resulting coefficients are functions of the initial stress-components. This apparently requires us also to make some molecular assumption.

[239.] Starting with expression (ii) for the potential energy, we should arrive at the equations of Cauchy (as Saint-Venant had done in his memoir of 1863 by a double self-correcting error), but we must renounce the hope of arriving at (ii) on the simple assumption of a generalised Hooke's Law. We may note one or two further points in the first part of the memoir:

(*a*) To the second order of small quantities,

$$\left.\begin{aligned} s_x &= u_x + \tfrac{1}{2}(v_x^2 + w_x^2) \\ \sigma_{yz} &= v_z + w_y + u_y u_z - v_y w_y - v_z w_z \end{aligned}\right\} \ldots\ldots\ldots\ldots\ldots\text{(iii)}.$$

This was first noticed by Brill: see p. 279 of Saint-Venant's memoir.

[1] σ'_{yz} differs from the σ_{yz} of our Art. 1621*, it being the *cosine* and not the cotangent of the slide-angle. See Saint-Venant's definition of slide in Art. 1564*.

(*b*) If we assume that the work-function may be expanded in powers of s_x, s_y, s_z, σ_{zy}, σ_{xz}, σ_{yx}, and write

$$\left.\begin{aligned}\phi = \phi_0 &+ \widehat{xx}_0\, s_x + \widehat{yy}_0\, s_y + \widehat{zz}_0\, s_z \\ &+ \widehat{zy}_0\, \sigma_{zy} + \widehat{xz}_0\, \sigma_{xz} + \widehat{yx}_0\, \sigma_{yx} \\ &+ \phi_2\end{aligned}\right\}\dots\dots\dots\dots\dots\dots\text{(iv)},$$

then we are throwing a portion of ϕ involving initial stresses into ϕ_2, which thus differs from the ϕ_1 of (ii). We thus obtain for the stresses the types:

$$\left.\begin{aligned}\widehat{xx} &= \widehat{xx}_0\,(1 - v_y - w_z) - \widehat{xy}_0\,(v_x - u_y) + \widehat{zx}_0\,(u_z - w_x) + \widehat{xx}_2 \\ \widehat{yz} &= \widehat{yz}_0\,(1 - u_x - v_y - w_z) + \widehat{yy}_0\, w_y + \widehat{zz}_0\, v_z + \widehat{zx}_0\, v_x + \widehat{xy}_0\, w_x + \widehat{yz}_2\end{aligned}\right\}\dots\text{(v)}.$$

But $\widehat{xx}_2$ and $\widehat{yz}_2$ while being of the same form as Cauchy's $\widehat{xx}_1$, $\widehat{yz}_1$ [see our Art. 129, (ii)], will in reality have constants increased by the corresponding initial stresses, as is shewn by the rari-constant investigation. Thus:

$$\left.\begin{aligned}|xxxx|_2 &= |xxxx| + \widehat{xx}_0 \\ |yyyz|_2 &= |yyyz| + \widehat{yz}_0 \\ |zzyz|_2 &= |zzyz| + \widehat{yz}_0\end{aligned}\right\}\dots\dots\dots\dots\dots\dots\dots\dots\text{(vi)}.$$

It is the impossibility of determining on the multi-constant theory how these initial stresses occur in the changed values of the constants, which throws us back on rari-constancy for a proof of (ii). Results (vi) combined with (v) convert the latter into Cauchy's formulae: see our Art. 129, (i).

[240.] The second part of the memoir deals with the following problem: *If* $|xxxx|$, $|xxxy|$, $|xxyz|$, *etc. are the elastic constants when there is an initial state of stress* $\widehat{xx}_0$, $\widehat{xy}_0$, *etc. it is required to determine these constants in terms of* $|xxxx|_0$, $|xxxy|_0$, $|xxyz|_0$, *etc. the elastic constants before this initial state of stress.*

Saint-Venant deals with the problem on rari-constant lines. We have, with abbreviated symbols (see our Art. 143):

$$|x^4| \text{ or } |y^2z^2| \text{ or } |y^3z| \text{ or } |x^2yz| = \frac{\rho}{2}\,\Sigma m\,\frac{d}{rdr}\,\frac{f(r)}{r}\,\{x^4 \text{ or } y^2z^2 \text{ or } y^3z \text{ or } x^2yz\}\dots\text{(vii)}.$$

Further we have, if x_0, y_0, z_0 be the position of the molecule m relative to a second before the initial strain, u_0, v_0, w_0 its shift due to that strain, and x, y, z the relative position after the strain,

$$x = x_0 + x_0\frac{du_0}{dx_0} + y_0\,\frac{du_0}{dy_0} + z_0\,\frac{du_0}{dz_0} + \frac{1}{1\,.\,2}\left(\frac{d^2u_0}{dx_0^2}\,x_0^2 + \dots\right),$$

$$r - r_0 = \frac{1}{r_0}\left[x_0^2\,\frac{du_0}{dx_0} + \dots\dots + \dots\dots \right.$$

$$\left. + y_0z_0\left(\frac{dv_0}{dz_0} + \frac{dw_0}{dy_0}\right) + \dots\dots + \dots\dots\right]$$

$$+\frac{1}{2r_0^2}\left[\left(\frac{du_0}{dx_0}x_0+\frac{du_0}{dy_0}y_0+\frac{du_0}{dz_0}z_0\right)^2 + \left(\frac{dv_0}{dx_0}x_0+\ldots\ldots+\ldots\ldots\right)^2 + \left(\frac{dw_0}{dx_0}x_0+\ldots\ldots+\ldots\ldots\right)^2\right]$$

$$-\frac{1}{2r_0^2}\left[x_0^2\frac{du_0}{dx_0}+y_0^2\frac{dv_0}{dy_0}+\ldots\ldots+x_0y_0\left(\frac{du_0}{dy_0}+\frac{dv_0}{dx_0}\right)\right]^2 + \frac{1}{2r_0}\left(x_0^3\frac{d^2u_0}{dx_0^2}+\ldots\ldots\right),$$

$$\frac{d}{rdr}\left(\frac{f(r)}{r}\right)=\frac{d}{r_0dr_0}\left(\frac{f(r_0)}{r_0}\right)+(r-r_0)\frac{d}{dr_0}\frac{1}{r_0}\frac{d}{dr_0}\left(\frac{f(r_0)}{r_0}\right)+\ldots\ldots,$$

$$\rho=\frac{\rho_0}{\left(1+\frac{du_0}{dx_0}\right)\left(1+\frac{dv_0}{dy_0}\right)\left(1+\frac{dw_0}{dz_0}\right)-\frac{dv_0}{dz_0}\frac{dw_0}{dy_0}-\frac{dw_0}{dx_0}\frac{du_0}{dz_0}-\frac{du_0}{dy_0}\frac{dv_0}{dx_0}+\ldots}$$

$$=\rho_0(1-s_{x_0}-s_{y_0}-s_{z_0}),$$

if we suppress squares and products.

Substituting in Equation (vii) and remembering that

$$|x^4|_0 \text{ or } |y^2z^2|_0 \text{ or } |y^3z|_0 \text{ or } |x^2yz|_0=\frac{\rho_0}{2}\Sigma m\frac{d}{r_0dr_0}\left\{\frac{f(r_0)}{r_0}\right\}\{x^4_0 \text{ or } y^2_0z^2_0 \text{ or } y^3_0z_0 \text{ or } x^2_0y_0z_0,\}$$

we obtain the typical results:

$$|x^4| = |x^4|_0(1-3u_{x_0}-v_{y_0}-w_{z_0})+4(|x^3y|_0u_{y_0}+|x^3z|_0u_{z_0}),$$

$$|y^2z^2| = |y^2z^2|_0(1-u_{x_0}+v_{y_0}+w_{z_0})+2(|yz^3|_0v_{z_0}+|xyz^2|_0v_{x_0}+|xy^2z|_0w_{x_0}+|y^3z|_0w_{y_0}),$$

$$|y^3z| = |y^3z|_0(1-u_{x_0}+2v_{y_0})+3(|y^2z^2|_0v_{z_0}+|xy^2z|_0v_{x_0})+(|xy^3|_0w_{x_0}+|y^4|_0w_{y_0}),$$

$$|x^2yz| = |x^2yz|_0(1+u_{x_0})+2(|xy^2z|_0u_{y_0}+|xyz^2|_0u_{z_0})+|x^3z|_0v_{x_0}+|x^2z^2|_0v_{z_0}+|x^3y|_0w_{x_0}+|x^2y^2|_0w_{y_0}.$$

Here $u_{x_0},\ldots$ denote $du_0/dx_0\ldots$, and since the stresses $\widehat{xx}_0$, $\widehat{yz}_0$ are given functions of u_{x_0} $v_{y_0}\ldots u_{z_0}\ldots$etc., we can express the new coefficients $|x^4|\ldots$ in terms of the old $|x^4|_0\ldots$ and the initial stresses. These results are obviously only a more general case of the formulae of our Art. 616*. The following pages 297—304 are concerned with other modes of looking at these results or expressing the stresses in terms of them.

[241.] Let us take as a special case that of a bar of primitively isotropic material subjected to a traction $\widehat{xx}$, there being an initial traction $\widehat{xx}_0$. We have

$$s_{x_0}=\widehat{xx}_0/E_0,\quad s_{y_0}=s_{z_0}=-s_{x_0}/4.$$

Further, if $|x^2y^2|_0=\lambda=\mu$, then $|x^4|_0=3\lambda$ and $E_0=5\lambda/2$.

Thus, $$|x^4| = 3\lambda\,(1 - \tfrac{5}{2}s_{x_0}), \quad |y^4| = |y^4|_0 = 3\lambda,$$
$$|x^2y^2| = |x^2z^2| = \lambda\,(1 + s_{x_0}),$$
$$|y^2z^2| = \lambda\,(1 - \tfrac{3}{2}\,s_{x_0}),$$
$$|y^3z| = |x^2yz| = \text{etc.} = 0.$$

Substituting in the traction-type as given by Cauchy's formula, Eqn. (i) Art. 129, we have

$$\widehat{xx} = \widehat{xx}_0\,\{1 + s_x - 2s_y\} + 3\lambda\,(1 - \tfrac{5}{2}\,s_{x_0})\,s_x + 2\lambda\,(1 + s_{x_0})\,s_y,$$
$$\widehat{yy} = 0 = 3\lambda s_y + \lambda\,(1 + s_{x_0})\,s_x + \lambda\,(1 - \tfrac{3}{2}\,s_{x_0})\,s_y.$$

Whence we find from the second equation:

$$s_y\,(4 - \tfrac{3}{2}\,s_{x_0}) = -\,(1 + s_{x_0})\,s_x,$$

or $$s_y = -\tfrac{1}{4}\,s_x\,\{1 + \tfrac{11}{8}\,s_{x_0}\},\ \text{neglecting } s^2_{x_0}.$$

Substituting in the first we easily deduce

$$\widehat{xx} = \widehat{xx}_0 + \frac{5\lambda}{2}\,s_{x_0} \times \tfrac{3}{2}s_x + \lambda s_x\,\{\tfrac{5}{2} - \tfrac{139}{16}\,s_{x_0}\},$$
$$= \widehat{xx}_0 + \frac{5\lambda}{2}\,s_x\,(1 - \tfrac{79}{40}\,s_{x_0}),$$

or $$\frac{\widehat{xx} - \widehat{xx}_0}{s_x} = E_0 - \frac{79}{40}\,\widehat{xx}_0.$$

Thus if E be the new stretch-modulus, we have $E = E_0 - \tfrac{79}{40}\,\widehat{xx}_0$.

This shows that a large initial traction can alter to some extent the value of the stretch-modulus. It slightly *decreases it.* Saint-Venant obtains in our notation

$$E = E_0 + \tfrac{11}{2}\,\widehat{xx}_0,$$

but I do not think this result is correct. It would denote an *increase* of the stretch-modulus. Saint-Venant in fact puts the stretch-squeeze ratio after the initial stress $= \tfrac{1}{4}$, (thus on p. 305 he writes $s_z = s_y = -\tfrac{1}{4}\,s_x$), but it seems to me that this ratio

$$= -\,(1 + s_{x_0})/(4 - \tfrac{3}{2}\,s_{x_0}) = -\tfrac{1}{4}\,(1 + \tfrac{11}{8}\,s_{x_0}),$$

and is only $= -1/4$ when $s_{x_0} = \widehat{xx}_0/E_0 = 0$, or, when there is no initial stress.

The matter is one of theoretical rather than practical interest, for supposing E were 30,000,000 lbs. per sq. inch, it is unlikely that $\widehat{xx}_0$ could be at most more than 40,000 to 60,000 lbs. per sq. inch; hence the change in E would not amount to more than 140,000 to 200,000 lbs., or at most to 1/150 of E, which with the want of uniformity in any material is in practice almost within the limits of experimental error.

[242.] In Tome XV. of the *Journal de Liouville*, 1870, there are two articles by Saint-Venant, but they refer to a matter which I have thought it well to treat as lying outside our field, namely the stability of masses of loose earth. The history of the memoirs in question may be briefly referred to. Maurice Lévy in 1867 had presented to the Academy a memoir entitled: *Essai sur une théorie rationnelle de l'équilibre des terres fraîchement remuées, et ses applications au calcul de la stabilité des murs de soutènement* (published in the *Journal de Liouville* T. XVIII. 1873, pp. 241—300). This memoir had been referred to a committee including Saint-Venant for report. The report appeared in the *Comptes rendus*, T. LXX. 1870, pp. 217—28, and was reprinted in Vol. XV. of the *Journal*, pp. 237—249. Lévy as well as the committee appear to have been ignorant of Rankine's memoir: *On the Stability of Loose Earth* (*Phil. Trans.* 1857, pp. 9—27) which had contained most of Lévy's results. Lévy had started from Cauchy's stress-theorems (see our Arts. 606* and 610*), and arrived at certain general equations. Saint-Venant in his first note solves to a first approximation Lévy's equation (pp. 250—63 of Tome XVIII.) and hopes some mathematician will proceed further. This was done by Boussinesq, who proceeded to a second approximation in a memoir occupying pp. 267—70 of Tome XV. of the *Journal*. Saint-Venant then reconsidered the whole matter in a second memoir, which occupies the following pp. 271—80. In a footnote he recognises Rankine's priority of research. The memoirs of Saint-Venant and Boussinesq appear also in the *Comptes rendus*. T. LXX. 1870, pp. 217—28, 717—24, 751—4 and 894—7.

[243.] *Rapport sur un mémoire de Maurice Lévy: Comptes rendus*, T. LXXIII. 1871, pp. 86—91. This is a report by Saint-Venant and others on Lévy's memoir establishing the general body-stress equations of plasticity in three dimensions: see our Art. 250. The *Rapport* speaks well of Lévy's memoir as advancing the new branch of mechanics, "pour laquelle l'un de nous a hasardé, sans le préconiser comme le meilleur, le terme *d'hydrostéréo-dynamique*." This branch of research has been called later *plastico-dynamics*, a better word, and we shall refer to it simply as *plasticity*.

[244.] *Sur la mécanique des corps ductiles: Comptes rendus*,

T. LXXIII. 1871, pp. 1181—1184. Saint-Venant here replaces his first name—hydrostereo-dynamics—by *plastico-dynamics*. He refers to the *Complément* to his memoirs on this subject in the *Journal de Liouville*: see our Art. 245, (iii), and to the two examples of the plasticity of a cylinder under torsion and of a prism under circular flexure dealt with there. The object of this note is to show that a formula obtained by Tresca for the torsion of a semi-plastic cylinder contributes no more than Saint-Venant's formula of the above-mentioned *Complément*, while it is at the same time obtained in a semi-empirical fashion. While Tresca's formula involves a new constant K', Saint-Venant depends only on the elastic slide-modulus μ and the plastic-modulus K. Saint-Venant distinguishes in his cylinder only two zones, an elastic and a plastic one, Tresca supposes a mid-zone in which elasticity alters to plasticity or, as Tresca terms it, fluidity. Saint-Venant's discussion has the theoretical advantage, but it seems not improbable that physically something corresponding to Tresca's mid-zone has an existence.

[245.] We have next to turn to a series of interesting and important memoirs by Saint-Venant in which he deals with the *plastic* equations. These are:

(i) *Mémoire sur l'établissement des équations différentielles des mouvements intérieurs opérés dans les corps solides ductiles au delà des limites où l'élasticité pourrait les ramener à leur premier état. Journal de Mathématiques.* Tome XVI. 1871, pp. 308—316. [See also *Comptes rendus*, T. LXX. 1870, p. 473.]

(ii) *Extrait du mémoire sur les équations générales des mouvements intérieurs des corps solides ductiles au delà des limites où l'élasticité pourrait les ramener à leur premier état. Par M. Maurice Lévy. Ibid.* pp. 369—372. [See also *Comptes rendus*, T. LXX. p. 1323, and Saint-Venant's correction referred to in our Art. 263. Some account of the memoir itself will be given under the year 1870.]

(iii) *Complément aux mémoires du* 7 *mars* 1870 *de M. de Saint-Venant et du* 19 *juin* 1870 *de M. Lévy sur les équations différentielles 'indéfinies' du mouvement intérieur des solides ductiles etc.;...Equations 'définies' ou relatives aux limites de ces corps;*—Applications, *Ibid.* pp. 373—382.

[246.] The first paper begins with an interesting account of the history of the theory of plasticity. It refers to Tresca's memoirs and to the attempts of Tresca and Saint-Venant himself to obtain solutions by means of pure kinematics. It is pointed out that the problem is essentially mechanical as well as kinematical and involves a consideration of stress as well as of mere continuity.

In the first place the ordinary equations of fluid-motion must be replaced by others involving inequality of pressure in different directions. Thus the well-known type of hydrodynamic equation:

$$\frac{dp}{dx} = \rho \left(X - \frac{du}{dt} - u\frac{du}{dx} - v\frac{du}{dy} - w\frac{du}{dz} \right),$$

becomes the plastico-dynamic type:

$$\frac{d\widehat{xx}}{dx} + \frac{d\widehat{xy}}{dy} + \frac{d\widehat{xz}}{dz} = -\rho \left(X - \frac{du}{dt} - u\frac{du}{dx} - v\frac{du}{dy} - w\frac{du}{dz} \right) \text{.........(i).}$$

The change of sign is due to change from pressure to *traction*. To this we must add the equation of continuity:

$$\frac{du}{dx} + \frac{dv}{dy} + \frac{dw}{dz} = 0 \text{..........................(ii).}$$

The four equations given by (i) and (ii) represent the relation between the flow (velocity-components u, v, w) of the material and the stress-components. The material in the plastic state is treated as incompressible.

[247.] Now Tresca has demonstrated that, if a material is in the plastic stage, the maximum shear across any face must have a constant value K, which he has ascertained experimentally for a variety of materials. This constant resistance to maximum slide we shall term in future the *plastic modulus*. Hence to obtain the plastico-dynamic equations we must express the fact that

the maximum shear across any face $= K$.........(iii).

Again, Tresca has demonstrated that the direction of the maximum shear is also that of the maximum velocity of slide. This forms then our last condition:

maximum shear and maximum slide-velocity are co-directional }.........(iv).

Equations (i) and (ii), with conditions (iii) and (iv) should give the complete plastico-dynamic equations.

[248.] Saint-Venant only treats the case of what we may term uniplanar plasticity, or the motion the same in all planes parallel to that of x, z. Thus the co-ordinate y disappears from his results.

Let x', z' be two rectangular axes making an angle α with those of x, z, then it easily follows from the first formula in our Art. 1368* that,

$$\widehat{x'z'} = \frac{\widehat{zz} - \widehat{xx}}{2} \sin 2\alpha + \widehat{zx} \cos 2\alpha.$$

This takes its maximum value for

$$\tan 2\alpha = \frac{\widehat{zz} - \widehat{xx}}{2\widehat{xz}} \quad \text{.........................(v)},$$

and is then of the intensity

$$\tfrac{1}{2}\sqrt{4\widehat{zx}^2 + (\widehat{zz} - \widehat{xx})^2}.$$

Thus condition (iii) becomes

$$\widehat{zx}^2 + \left(\frac{\widehat{zz} - \widehat{xx}}{2}\right)^2 = K^2 \quad \text{.......................(vi)}.$$

Further the slide-velocity is easily found to be given by

$$\frac{dw'}{dx'} + \frac{du'}{dz'} = \left(\frac{dw}{dz} - \frac{du}{dx}\right) \sin 2\alpha + \left(\frac{dw}{dx} + \frac{du}{dz}\right) \cos 2\alpha,$$

and therefore takes its maximum when

$$\tan 2\alpha = \frac{\dfrac{dw}{dz} - \dfrac{du}{dx}}{\dfrac{dw}{dx} + \dfrac{du}{dz}}.$$

Hence condition (iv) becomes

$$\frac{\widehat{zz} - \widehat{xx}}{2\widehat{xz}} = \left(\frac{dw}{dz} - \frac{du}{dx}\right) \Big/ \left(\frac{dw}{dx} + \frac{du}{dz}\right) \quad \text{.............(vii)}.$$

Finally equations (i) and (ii) take in this case the simpler forms:

$$\left.\begin{aligned} \frac{d\widehat{xx}}{dx} + \frac{d\widehat{xz}}{dz} &= -\rho\left(X - \frac{du}{dt} - u\frac{du}{dx} - w\frac{du}{dz}\right) \\ \frac{d\widehat{xz}}{dx} + \frac{d\widehat{zz}}{dz} &= -\rho\left(Z - \frac{dw}{dt} - u\frac{dw}{dx} - w\frac{dw}{dz}\right) \\ \frac{du}{dx} + \frac{dw}{dz} &= 0 \end{aligned}\right\} \quad \text{........(viii)}.$$

Equations (vi), (vii), and (viii) are those for uniplanar plasticity.

[249.] Saint-Venant remarks that even these equations will be difficult to solve for any except the simplest cases. He suggests, however, that those for a cylindrical plastic flow would not be difficult to obtain.

In a final paragraph (p. 316) to the first paper Saint-Venant remarks:

Je ferai seulement une dernière remarque: c'est que si, aux six composantes de pressions ci-dessus, $\widehat{xx}, \ldots\ldots \widehat{xy}$, l'on ajoute respectivement les termes: $2\epsilon u_x, 2\epsilon v_y, 2\epsilon w_z, . \epsilon(v_z + w_y), \epsilon(w_x + u_z), \epsilon(u_y + v_x)$, représentant, comme on sait, ce qui vient du frottement dynamique dû aux vitesses de glissement relatif dans les fluides non visqueux se mouvant avec régularité, les équations des solides plastiques, ainsi complétées, s'étendront au cas où les vitesses avec lesquelles leur déformation s'opère, sans être considérables, ne seraient plus excessivement petites, et pourraient engendrer ces résistances particulières, ordinairement négligeables, dont on a parlé au No. 3. Les mêmes équations, avec tous ces termes, seraient propres, aussi, à exprimer les mouvements réguliers (c'est-à-dire pas assez prompts pour devenir tournoyants et tumultueux) des *fluides visqueux*, où il doit y avoir des composantes tangentielles de deux sortes, les unes variables avec les vitesses u, v, w, et mesurées par les produits de ϵ et de leurs dérivées, les autres indépendantes de ces grandeurs des vitesses, ou les mêmes quelle que soit la lenteur du mouvement, et attribuables *à la viscosité*, dont K *représenterait alors le coefficient spécifique.*

[250.] In the second paper to which we have referred in our Art. 245, Maurice Lévy establishes two sets of results. In the first place he obtains the general equations of plasticity; in the next he considers the special case of a cylindrical plastic flow.

We cite the general equations here, but refer to our later discussion of Lévy's memoir for remarks on his method of obtaining them.

The general equations (i) and (ii) hold for this case. The condition (iii) becomes:

$$4(K^2 + q)(4K^2 + q) + 27r^2 = 0 \quad\ldots\ldots\ldots\ldots\ldots\ldots\text{(ix)},$$

where

$$q = \Delta_y\Delta_z + \Delta_z\Delta_x + \Delta_x\Delta_y - \widehat{yz}^2 - \widehat{zx}^2 - \widehat{xy}^2,$$

$$r = \Delta_x\widehat{yz}^2 + \Delta_y\widehat{zx}^2 + \Delta_z\widehat{xy}^2 - \Delta_x\Delta_y\Delta_z - 2\,\widehat{yz}\,\widehat{zx}\,\widehat{xy},$$

and

$$\widehat{xx} - \Delta_x = \widehat{yy} - \Delta_y = \widehat{zz} - \Delta_z = \tfrac{1}{3}(\widehat{xx} + \widehat{yy} + \widehat{zz}).$$

The condition (iv) becomes

$$\frac{\widehat{yz}}{v_z + w_y} = \frac{\widehat{zx}}{w_x + u_z} = \frac{\widehat{xy}}{u_y + v_x} = \frac{\widehat{yy} - \widehat{zz}}{2(v_y - w_z)} = \frac{\widehat{zz} - \widehat{xx}}{2(w_z - u_x)} \ldots\ldots\text{(x)}.$$

Thus (i), (ii), (ix) and (x) are the requisite equations.

[251.] On p. 371 Lévy remarks that Saint-Venant in the case of uniplanar plasticity has not considered the stress $\widehat{yy}$. From equation (x) since $v_y = 0$, and therefore $w_z + u_x = 0$ from (ii), we have

$$\frac{\widehat{yy} - \widehat{zz}}{2u_x} = \frac{\widehat{zz} - \widehat{xx}}{-4u_x},$$

or,

$$\widehat{yy} = \tfrac{1}{2}(\widehat{zz} + \widehat{xx}) \qquad \text{(xi)}.$$

[252.] On p. 372 we have the equations for a cylindrical plastic flow. If z be the axial, r the radial directions, ϕ the meridian angle, u, w radial and axial velocities, they take the form:

$$\left.\begin{aligned} \frac{d\widehat{rr}}{dr} + \frac{d\widehat{rz}}{dz} + \frac{\widehat{rr} - \widehat{\phi\phi}}{r} &= -\rho\left(R_0 - \frac{du}{dt} - u\frac{du}{dr} - w\frac{du}{dz}\right) \\ \frac{d\widehat{rz}}{dr} + \frac{d\widehat{zz}}{dz} + \frac{\widehat{rz}}{r} &= -\rho\left(Z_0 - \frac{dw}{dt} - u\frac{dw}{dr} - w\frac{dw}{dz}\right) \end{aligned}\right\} \quad \text{(xii)},$$

$$\frac{du}{dr} + \frac{u}{r} + \frac{dw}{dz} = 0 \qquad \text{(xiii)},$$

$$4\widehat{rz}^2 + (\widehat{rr} - \widehat{zz})^2 = 4K^2 \qquad \text{(xiv)},$$

$$\frac{\widehat{rz}}{w_r + u_z} = \frac{\widehat{rr} - \widehat{zz}}{2(u_r - w_z)} = \frac{\widehat{rr} - \widehat{\phi\phi}}{2(u_r - u/r)} \qquad \text{(xv)}.$$

We shall see later that the condition (xiv) is not sufficient nor always correct: see our Art. 263.

As a rule when the plastic movements are very small and the effects of gravity can be neglected, we may put the right-hand sides of equations (i) and (xii) equal to zero.

[253.] In the third paper whose title is given in our Art. 245 Saint-Venant first makes the remark that if the velocities be neglected the equations of uniplanar plasticity reduce to the discovery of an unknown auxiliary ψ, where:

$$\widehat{xx} = \frac{d^2\psi}{dz^2}, \quad \widehat{zz} = \frac{d^2\psi}{dx^2},$$

$$\widehat{xz} = -\frac{d^2\psi}{dz\,dx},$$

and

$$4\left(\frac{d^2\psi}{dx\,dz}\right)^2 + \left(\frac{d^2\psi}{dx^2} - \frac{d^2\psi}{dz^2}\right)^2 = 4K^2 \qquad \text{(xvi)}.$$

He suggests that this equation might be solved by approximation.

[254.] Saint-Venant next passes to the treatment of the limiting or surface conditions of plasticity, i.e. the conditions which hold at the boundary of the portion of the material in a plastic condition. He terms them the *équations définies ou déterminées;* the previous equations being called the *équations indéfinies.*

These conditions are of various kinds. A certain portion of the block of matter alone is plastic (called by Tresca the *zône d'activité*), other portions may remain elastic, or after passing through a plastic condition return to elasticity (e.g. a jet of metal after passing an orifice).

The conditions break up into three classes:

1st. Those which relate to the surface of the material at points which have retained or resumed their elasticity. Let such a surface be exposed to a traction T_e and let the elastic stresses be $\widehat{xx}_e \ldots\ldots \widehat{xy}_e$, the suffix e merely referring to their elastic character. The type of surface condition will be

$$\widehat{xx}_e \cos(nx) + \widehat{xy}_e \cos(ny) + \widehat{xz}_e \cos(nz) = T_e \cos(lx) \ldots\ldots \text{(xvii)},$$

where n is the direction of the surface-normal and l that of the applied traction T_e.

2nd. The material is in a plastic stage at the bounding surface, T_p being the traction: the type of equation, if $\widehat{xx}_p \ldots\ldots \widehat{xy}_p$ denote the plastic stresses, is:

$$\widehat{xx}_p \cos(nx) + \widehat{xy}_p \cos(ny) + \widehat{xz}_p \cos(nz) = T_p \cos(lx) \ldots\ldots \text{(xviii)}.$$

3rd. Equations which must hold at the surface at which the material changes from plasticity to elasticity. These are of the type:

$$(\widehat{xx}_e - \widehat{xx}_p) \cos(nx) + (\widehat{xy}_e - \widehat{xy}_p) \cos(ny) + (\widehat{xz}_e - \widehat{xz}_p) \cos(nz) = 0 \ldots \text{(xix)}.$$

In the equations (xvii)—(xix) the elastic stresses and plastic stresses must be obtained from the general equations of elasticity and of plasticity respectively.

[255.] On pp. 378—380, Saint-Venant treats the special case of a right circular cylinder of radius r subjected to torsion till plasticity commences in the outer zone from r_0 to r. He easily finds if M be the torsional couple, μ the slide-modulus and τ the torsional angle:

$$M = 2\pi \left[\mu\tau \frac{r_0^4}{4} + K\left(\frac{r^3}{3} - \frac{r_0^3}{3}\right)\right],$$

while at the surface of elasticity and plasticity we must have

$$\mu\tau r_0 = K.$$

There will be no plasticity then so long as

$$\tau < \frac{K}{\mu r}, \text{ or } M < \frac{\pi r^3}{2} K.$$

If τ be greater than this we have:

$$M = \pi K \left(\frac{2}{3} r^3 - \frac{1}{6} \frac{K^3}{\mu^3 \tau^3}\right).$$

[256.] On pp. 380—381 we have the case of plasticity produced by the equal or 'circular' flexion of a prism of rectangular section.

Let $2c$ be the height in the plane of flexure, $2b$ the breadth of the section, $2c_0$ the height of the middle portion which remains elastic, and $1/\rho$ the uniform curvature. Then it is easy to see that the bending moment M is given by:

$$M = \frac{4}{3}\frac{E}{\rho} bc_0^3 + 4Kb(c^2 - c_0^2).$$

At the surface of separation of the plastic and elastic parts:

$$\frac{Ec_0}{\rho} = 2K.$$

Whence we find:

$$M = 4Kb\left(c^2 - \frac{4}{3}\frac{K^2\rho^2}{E^2}\right),$$

where we must have $\rho < \dfrac{Ec}{2K}$ or the prism will remain elastic.

[257.] Saint-Venant in conclusion indicates that only after first ascertaining *experimentally* the general form taken by the flow in special cases will it be possible to attempt approximate solutions of the equations of plasticity.

I may remark that Saint-Venant assumes that elasticity and plasticity are continuous. This does not seem to me at all borne out by experiment, the stresses have long ceased to be proportional to the strains before plasticity commences: see the diagram on p. 890 of our Vol. I. and my remark in Art. 244.

[258.] Two memoirs by Saint-Venant on *plastico-dynamics* or plasticity occur in Vol. LXXIV. 1872, of the *Comptes rendus.* They are entitled:

(1) *Sur l'intensité des forces capables de déformer avec continuité des blocs ductiles, cylindriques, pleins ou évidés, et placés dans diverses circonstances* (pp. 1009—1015 with footnotes to p. 1017).

(2) *Sur un complément à donner à une des équations présentées par M. Lévy pour les mouvements plastiques qui sont symétriques autour d'un même axe* (pp. 1083—1087).

These memoirs may be looked upon as supplements to those of Saint-Venant and Lévy in the *Journal de Liouville:* see our Arts. 245—57.

[259.] The general principle, Saint-Venant tells us in his first memoir, of plastic deformation is that the greatest shear at each point shall be equal to a specific constant (denoted by K in Tresca's memoir of 1869). It follows by Hopkins' theorem that at each point the greatest difference between the tractions across different faces ought to equal $2K$: see our Art. 1368*.

Saint-Venant treats two special cases, and a third by approximation. We will devote the following three articles to their discussion.

[260.] The first is that of a right six-face of ductile metal.

If the axes of coordinates be taken parallel to its edges, and its faces be subjected to uniform tractions $\widehat{xx}$, $\widehat{yy}$, $\widehat{zz}$, then these tractions will be the principal tractions at any point of the material, and it will be necessary if $\widehat{xx} \sim \widehat{zz}$ be the greatest difference that:

$$\widehat{xx} \sim \widehat{zz} = 2K \qquad \text{(i)}.$$

This condition is fulfilled if

$$\widehat{xx} = -\widehat{yy} = -\widehat{zz} = K,$$

or if

$$\widehat{xx} = -\widehat{yy} = -\widehat{zz} = -K.$$

Of this Saint-Venant remarks:

C'est dans ce sens qu'il faut entendre, avec M. Tresca, que la résistance, soit à l'allongement, soit à l'accourcissement du solide plastique, est constante, et égale à sa résistance au cisaillement (p. 1010).

An extension of this case is that of a cylinder on any base, for which $\widehat{yy} = \widehat{zz}$ without being equal to K, that is to say the transverse or radial tractions which we will denote by $\widehat{rr}$ are all equal and the longitudinal tractions $\widehat{xx}$ are greater than them. We have then for the condition of plasticity:

$$\widehat{xx} = 2K + \widehat{rr} \qquad \text{(ii)}.$$

If the radial tractions are greater than the longitudinal we have:

$$\widehat{rr} = 2K + \widehat{xx} \qquad \text{(iii)}.$$

Either equation (ii) or (iii) gives us by variation

$$\delta\widehat{xx} = \delta\widehat{rr},$$

or, any increment of longitudinal, is accompanied by an equal increment of transverse traction. This is Tresca's principle that *in plastic solids pressure transmits itself as in fluids*, although he proves it by the principle of work.

[261.] The second case dealt with by Saint-Venant is that of a hollow right circular cylinder placed between two rigid fixed planes perpendicular to its axis. The external face being submitted to a pressure p, we require the internal pressure p_1, necessary to reduce the material to a plastic condition.

This problem can be solved by introducing the velocities, here solely radial, of the points of the material. The principles which determine these velocities for a plastic material are: 1st, that there is no change in the volume of the element; and 2nd, that on each elementary area in the material the direction for which the shear is zero, must be that for which the slide-velocity is zero. The latter principle involves the ratios of the half-differences of the tractions to the corresponding stretch-velocities being equal two and two.

Let r be the radial distance from the axis of any element of the material, R and R_1 the external and internal radii of the cylinder; V the radial velocity of the element at r, and $\widehat{rr}$, $\widehat{\phi\phi}$, $\widehat{zz}$ the tractions along the radius, in the meridian plane and parallel to the axis at the same element. Then for the equilibrium and conservation of volume of an elementary annulus $2\pi r dr dz$, it is necessary that:

$$\frac{d\widehat{rr}}{dr} + \frac{\widehat{rr} - \widehat{\phi\phi}}{r} = 0, \quad \frac{dV}{dr} + \frac{V}{r} = 0 \quad \text{(iv)}.$$

Further from the second principle it follows that:

$$\frac{\widehat{rr} - \widehat{zz}}{dV/dr} = \frac{\widehat{rr} - \widehat{\phi\phi}}{dV/dr - V/r} \quad \text{(v)}.$$

Eliminating dV/dr between the second equation of (iv) and (v) we have

$$\widehat{rr} - \widehat{\phi\phi} = 2(\widehat{rr} - \widehat{zz}) = 2(\widehat{zz} - \widehat{\phi\phi}),$$

whence it results that $\widehat{rr} - \widehat{\phi\phi}$ is the greatest difference, and therefore by Eqn. (i)

$$\widehat{rr} - \widehat{\phi\phi} = -2K \quad \text{(vi)}.$$

It follows from the first equation of (iv) that

$$d\widehat{rr}/dr = 2K/r.$$

Or, integrating $$\widehat{rr} = -p_1 + 2K \log(r/R_1) \quad \text{(vii)}.$$

Hence from (vi) we deduce:

$$\left.\begin{aligned} \widehat{\phi\phi} &= -p_1 + 2K + 2K \log(r/R_1), \\ \widehat{zz} &= -p_1 + K + 2K \log(r/R_1). \end{aligned}\right\} \quad \text{(viii)}.$$

We see from these equations that: (a) the pressure on the rigid faces is not uniformly distributed over the surface of the material in contact with them, (b) the meridian traction will increase and generally change from a negative to a positive value as we pass outwards.

If we make $r = R$, we have $\widehat{rr} = -p$,

or, $p_1 = p + 2K \log (R/R_1)$.........................(ix).

If the pressure applied p_1 has a less value than this, the 'annular fibres' near to the inside face can very well acquire stretches exceeding the limit of elasticity and even that of cohesion for isolated straight fibres; but as the fibres in the neighbourhood of the external face remain elastic, there will not be rupture, nor sensible deformation. Saint-Venant refers to the well-known experiment of Easton and Amos: see our Art. 1474*.

In the last Section 5 of the *Note* Saint-Venant refers to Tresca's somewhat unsatisfactory proof of the formula (ix).

[262.] In a foot-note pp. 1015—1017, Saint-Venant deals approximately with the following case: the outer surface of a right circular hollow cylinder (radii R, R_1) is supposed to rest on a rigid envelope, the internal surface is then subjected to great pressure which diminishes the thickness $(R - R_1)$, but increases the height (h), to determine the pressure which will produce this plastic effect. Tresca had obtained a solution of this problem on two hypotheses, which cannot be considered as entirely satisfactory. The general equations of plasticity are indeed too complex to offer much hope of an exact solution for this case. Saint-Venant gives a solution involving only the acceptance of Tresca's second hypothesis namely: that the upper base of the cylinder and all the plane-sections parallel to it remain plane and perpendicular to the axis of the block, and that lines parallel to the axis preserve their parallelism. It is obvious that this hypothesis is only approximately true; but Saint-Venant's investigation is an interesting one, as it deals with one of those cases, in which the maximum difference of the principal tractions is not given by the same pair for all values of the radial distance. This breaks up the solution into two parts corresponding to $3r^2 <$ or $> R^2$, and the case itself into two sub-cases corresponding to $3R_1^2 <$ or $> R^2$. Saint-Venant's results are not in accordance with Tresca's.

[263.] *Sur un complément à donner à une des équations présentées par M. Lévy pour les mouvements plastiques qui sont symétriques autour d'un même axe: Comptes rendus*, T. LXXIV. 1872, pp. 1083—7.

Saint-Venant refers to Lévy's third equation for plasticity with axial symmetry. This equation is (see our Art. 252, Eqn. xiv.):

$$4\widehat{rz}^2 + (\widehat{rr} - \widehat{zz})^2 = 4K^2.$$

He remarks that this equation is only the true condition for plastic motion, when the greatest and least of the *negative* tractions (*pressures*) are in the meridian plane of the point considered. This is not always true and Lévy's third condition requires to be replaced by the following one:

$2K$ = the greatest in absolute value of the three quantities:

$$\left.\begin{array}{c} 2\sqrt{\widehat{rz}^2 + \frac{1}{4}(\widehat{rr} - \widehat{zz})^2} \\[2ex] \widehat{\phi\phi} - \dfrac{\widehat{rr} + \widehat{zz}}{2} - \sqrt{\widehat{rz}^2 + \frac{1}{4}(\widehat{rr} - \widehat{zz})^2} \\[2ex] \widehat{\phi\phi} - \dfrac{\widehat{rr} + \widehat{zz}}{2} + \sqrt{\widehat{rz} + \frac{1}{4}(\widehat{rr} - \widehat{zz})^2} \end{array}\right\}$$

This follows at once from the consideration that the discriminating cubic for the principal tractions is:

$$(T - \widehat{xx})(T - \widehat{yy})(T - \widehat{zz}) - \widehat{yz}^2(T - \widehat{xx}) - \widehat{zx}^2(T - \widehat{yy}) - \widehat{xy}^2(T - \widehat{zz}) - 2\,\widehat{yz}\,\widehat{zx}\,\widehat{xy} = 0,$$

and this becomes when we put:

$$\widehat{rr},\ \widehat{\phi\phi},\ \widehat{zz},\ 0,\ \widehat{rz},\ 0 \text{ for } \widehat{xx},\ \widehat{yy},\ \widehat{zz},\ \widehat{yz},\ \widehat{zx},\ \widehat{xy},$$

respectively

$$(T - \widehat{\phi\phi})\left(T - \frac{\widehat{rr} + \widehat{zz}}{2} - \sqrt{rz^2 + \tfrac{1}{4}(\widehat{rr} - \widehat{zz})^2}\right) \left(T - \frac{\widehat{rr} + \widehat{zz}}{2} + \sqrt{\widehat{rz}^2 + \tfrac{1}{4}(\widehat{rr} - \widehat{zz})^2}\right) = 0.$$

Lévy appears to have divided out by $T - \widehat{\phi\phi}$ and neglected this root.

[264.] Saint-Venant remarks that $\widehat{\phi\phi}$ is, however, sometimes the greatest or least of the three principal tractions, as for example in the problem of our Art. 261, for in that case

$$\widehat{zz} = \frac{\widehat{rr} + \widehat{\phi\phi}}{2}.$$

In the approximate solution of our Art. 262, the traction $\widehat{\phi\phi}$ is involved also in the maximum difference when $3r^2 < R^2$. Thus Lévy's memoir requires to be corrected so far as this equation is concerned.

In a foot-note Saint-Venant points out that his solutions (see our Arts. 261—2), are really obtained by the *semi-inverse* method and he suggests that the same method might be used to solve other plastic problems.

[265.] *Sur les diverses manières de présenter la théorie des ondes lumineuses. Annales de Chimie et de Physique*, 4e série, T. XXV. 1872, pp. 335—381. This memoir was also separately published by Gauthier-Villars in the same year.

The contents belong essentially to the history of the undulatory theory of light. Saint-Venant considers at considerable length the researches of Cauchy, Briot, and Sarrau in this field and points out the defects in the various theories which they have propounded. Finally he deals with Boussinesq's method of obtaining from a general type of equation the special differential equations which fulfil the conditions necessary for explaining the various phenomena of light. Saint-Venant praises highly Boussinesq's hypothesis, and considers that his theory:

> qui offre à la fois plus de simplicité, d'unité, de probabilité, et je crois aussi, de rigueur que les autres (quel que soit le remarquable talent avec lequel ont été présentés ces autres essais, qui ont toujours avancé les questions), mérite d'être enseignée de préférence (pp. 380—1).

I must remark, however, that convenient as Boussinesq's hypotheses may be as a grouping together of analytical results under one primitive formula, it cannot be held as sufficient till we understand the reasons why and how the molecular shifts are functions of the ether-shifts and their space and time fluxions, and are able to deduce the *form* of these functions from some more definite physical hypothesis.

§§ 1—2 treat of the early history of elasticity. As in the memoir of 1863 (see our Art. 146—7) Saint-Venant holds that the conditions presented by Green for exact parallelism and those suggested by Lamé for double refraction are only consistent with isotropy.

> Aussi Lamé et Green ne sont pas compris dans l'analyse que je fais des recherches de divers auteurs sur la lumière. Il importe que des hommes de talent ne s'égarent plus, en pareille matière, sur les errements des deux illustres auteurs de tant d'autres travaux plus dignes d'eux. (Footnote, p. 341.) See our Arts. 920*, 1108*, 146 and 193.

[266.] *Rapport sur un Mémoire de M. Lefort présenté le 2 août* 1875. This report is by Tresca, Resal and Saint-Venant (*rapporteur*) and will be found in the *Comptes rendus*, T. LXXXI. 1875, pp. 459—464. It speaks favourably of the memoir, which deals with the problem of finding the bending moment at the several sections of simple and continuous beams traversed by moving loads. We shall refer to the memoir under Lefort.

[267.] *De la suite qu'il serait nécessaire de donner aux recherches expérimentales de Plastico-dynamique: Comptes rendus*, T. LXXXI. 1875, pp. 115—122.

This note refers to the need of new plastico-dynamic experiments with a view of extending the number of solutions hitherto obtained and also the basis of the existing plastico-dynamic theory. Saint-Venant points out the insufficiency of Tresca's method of dividing the plastic solid into separate portions and applying to these the laws of fluid-continuity; he refers to his own researches in this purely *kinematic* direction: see our Art. 233, and then to his later theory and equations, as supplemented by Lévy, and based on Tresca's law of the equality of the stretch and slide coefficients of resistance: see our Arts. 236, 245 and 258. He points out that to develop this theory, what we want is not the form taken by jets of plastic material, but the *absolute paths of the elements in the material.* He suggests how this might be ascertained by allowing the same load to act in the same manner but for different periods on a number of like plastic blocks, in which a series of points had been previously marked by a three-dimensional wire netting placed in the molten metal. He notes also other methods likely to give the same result. In the course of the note he refers to the simple cases of plasticity solved by Lévy, Boussinesq and himself: see our Arts. 255—61. At the end are a few lines from Tresca, who recognises the importance of the experiments proposed by Saint-Venant, which, I believe, he did not live to undertake.

[268.] *Sur la manière dont les vibrations calorifiques peuvent dilater les corps, et sur le coefficient des dilatations; Comptes rendus*, 1876, T. LXXXII., pp. 33—38.

This is an attempt to represent thermal effects by the change produced by thermal vibrations directly in intermolecular distance rather than indirectly by their influence in altering the constants of molecular attraction. Saint-Venant deals with two molecules only and supposes one fixed.

Let r_0 be the intermolecular distance in equilibrium, $r = r_0 + v$ the displaced distance and $f(r)$ the law of intermolecular action, then we easily find for our equation of vibration:

$$m\frac{d^2v}{dt^2}=f(r)=f(r_0+v),$$

$$=vf'(r_0)+\frac{v^2}{2}f''(r_0)+\frac{v^3}{6}f'''(r_0)+\dots$$

If $dv/dt=\dot{v}_0$, for v and $t=0$, we have as a first approximation

$$v=\frac{\dot{v}_0}{a}\sin at,\ \text{where}\ f'(r_0)/m=-a^2.$$

For a second approximation:

$$v=\frac{\dot{v}_0}{a}\sin at+\frac{b^2\dot{v}_0^2}{6a^4r_0}(1-\cos at)^2,\ \text{where}\ \frac{r_0f''(r_0)}{m}=b^2.$$

Let us find the mean value v_m of v from $t=0$ to $2\pi/a$; we have:

$$v_m=\frac{b^2\dot{v}_0^2}{4a^4r_0}.$$

Hence the stretch due to the thermal vibration

$$=\frac{v_m}{r_0}=\frac{m\dot{v}_0^2}{2}\frac{1}{2r_0}\frac{f''(r_0)}{\{f'(r_0)\}^2}.$$

Thus we see that the stretch is proportional to the kinetic energy $m\dot{v}_0^2/2$, which is generally regarded as a measure of the absolute temperature, and will be positive if $f''(r_0)$ is positive.

Saint-Venant states that these conclusions will still hold, if the two molecules be replaced by a system. The thermal effect would thus depend on *the derivatives of the second order of the function* $f(r)$.

If there should be a point of inflexion in the curve which represents the law of intermolecular action plotted out to distance, we should have a case in which increase of temperature reduced the volume, as occurs in certain exceptional substances. Saint-Venant suggests the form of the figure below for the curve $y=f(r)$; OD being the distance and Oy the force axis.

Here $Ok=r_0$ marks the point at which the action changes from repulsion to attraction; if the axes Oy, OD are asymptotic in character, we have the infinitely great force and infinitely small force at infinitely small and infinitely great distances respectively well marked. pM marks the maximum attractive force between the molecules, and any force greater than this, if maintained, will produce rupture. It corresponds to a distance Op, which defines that of rupture. Great thermal vibrations which impose such a velocity on the molecule that the intermolecular distance exceeds Op may perhaps, indicate liquefaction by heat. The point i corresponds to a point of inflexion, and to a contraction due to heating the substance in the liquid state.

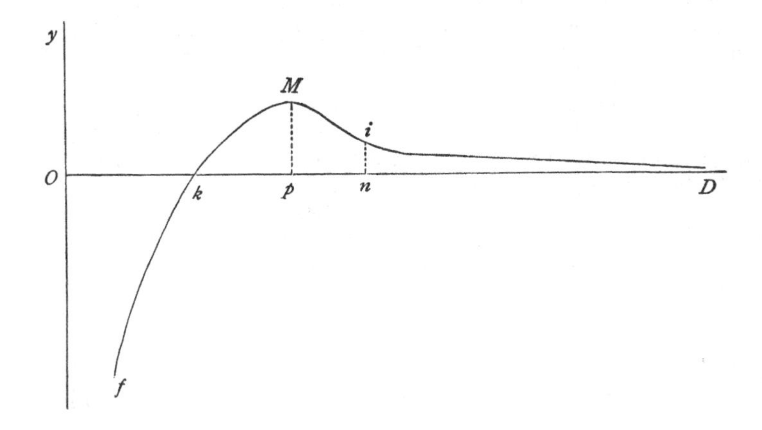

The discussion, if not very conclusive, is interesting especially in its bearings on rari-constancy. See our Arts. 439* and 977*.

[269.] *Sur la constitution atomique des corps: Comptes rendus*, T. LXXXII. 1876, pp. 1223—26.

Saint-Venant in this note refers to a remark of Berthelot on the paradox involved in the indivisibility of an atom supposed to be endowed with matter and therefore of necessity extended. He refers to his memoir of 1844 (see our Art. 1613*), and declares that he considers partly for metaphysical, partly for physico-mathematical reasons, *continuous* extension to belong neither to bodies nor to their component atoms. The point which alone concerns us here is his reference to the rari-constant hypothesis:

A cette occasion je ferai une remarque. Plusieurs auteurs, soit anglais, soit allemands, dans des œuvres qui sont du reste d'une haute portée, voulant étendre à des substances élastiques celluleuses, ou spongieuses, ou demi-fluides, telles que le liége, les gelées, les moelles végétales, le caoutchouc, les formules d'élasticité des solides, découvertes et établies en France de 1821 à 1828 par Navier, Cauchy, Poisson, Lamé et Clapeyron, et ayant besoin, pour une pareille extension, d'augmenter en nombre ou de rendre indépendants les uns des autres des coefficients de ces formules, se sont pris à condamner vivement, sous le nom de *théorie de Boscovich*, non pas son idée capitale de réduction des atomes à des centres d'action de forces, mais la loi même, la loi physique générale des actions fonctions des distances mutuelles des particules qui les exercent réciproquement les unes sur les autres. Et ils attribuent ainsi au célèbre religieux *l'erreur grave* où sont tombés, suivant eux, Navier, Poisson et nos autres savants, créateurs, il y a un demi-siècle, de la Mécanique moléculaire ou interne. Or cette loi blamée, cette loi qui a été mise en œuvre aussi par Laplace, etc. et prise par Coriolis et Poncelet pour base de la Mécanique physique, n'est autre que celle de Newton lui-même, comme on le voit non seulement dans son grand et principal ouvrage, mais dans le Scholie général de sa non moins immortelle Optique. L'usage fait de cette grande loi n'est point une erreur; et les formules d'élasticité à coefficients réduits ou, pour mieux dire, *déterminés*, où elle conduit pour les corps réellement solides, tels que le fer et le cuivre, sont conformes aux résultats bien discutés et interprétés d'expériences faites sur ces métaux (Appendice V. des *Leçons de Navier*: see our Art. 195), expériences au nombre desquelles il y en a de fort concluantes, récemment dues à M. Cornu (p. 1225).

That Boscovich deprived an atom of its extension, but that Newton treated intermolecular force as central, is a point which deserves to be recalled to mind: see our Art. 26*.

[270.] *Sur la plus grande des composantes tangentielles de tension intérieure en chaque point d'un solide, et sur la direction des faces de ses ruptures. Comptes rendus*, 1878, T. LXXXVII., pp. 89—92.

Potier had given the following formulae for the shear $\widehat{rs}$ across a face whose normal r makes angles α, β, γ with the directions of the principle tractions T_1, T_2, T_3:

$$\widehat{rs} = (T_1 - T_2)\cos^2\alpha\cos^2\beta + (T_2 - T_3)\cos^2\beta\cos^2\gamma + (T_3 - T_1)\cos^2\gamma\cos^2\alpha,$$

maximum value of $\widehat{rs} = \frac{1}{2}$ (difference of greatest and least principal tractions).

He had then proceeded to apply these formulae to the conditions of rupture. Saint-Venant notices that these results had been given by Kleitz in 1866, by Lévy in 1870, and by himself in 1864. He might also have added by Hopkins in 1847. The note then points out that rupture in the direction of maximum shear is hardly confirmed by experiments, which point rather to rupture in the direction of maximum stretch. Saint-Venant finally considers the results of some then recent experiments, but remarks on the need for further research in this direction.

[271.] *Sur la dilatation des corps échauffés et sur les pressions qu'ils exercent.* *Comptes rendus*, 1878, T. LXXXVII., pp. 713—18.

This memoir should be read in conjunction with that of 1876: see our Art. 268. It shews us how the phenomena of heat may possibly be accounted for by the law of intermolecular force as assumed by rari-constant elasticians. The assumption made by Saint-Venant is that the vibrations of the molecules, to which the phenomena of heat are due, are translational vibrations, and not of the nature of surface pulsations. This does not seem to me very probable, because in a highly rarified gas, it would denote the absence of any thermal vibration; for, there seems no reason why a molecule should have a *periodic* translational vibration when its fellow-molecules exercise little or no influence upon it.

The bright line spectra of such gases appear indeed to contradict the assumption, and it seems probable that if the thermal vibrations are pulsatory in the case of gaseous molecules, they will be of a like nature in the case of liquids and solids.

[272.] Saint-Venant commences his article with some account of his earlier memoirs, namely the communication made to the *Société Philomathique* in 1855 (see our Art. 68), and the first memoir of 1876 (see our Art. 268). He deduces by similar analysis to that of the latter memoir the same result

$$v = \frac{\dot{v}_0}{a} \sin at \text{.........................(i)},$$

shewing that it is necessary to take into account the terms of the second order, if we are to deal on these lines with thermal phenomena.

[273.] In addition, however, he here proceeds to consider the effect that translational vibrations would have on the pressure exerted by a system of molecules on a surrounding envelope. To obtain some idea of this he supposes a free molecule placed between two fixed ones at a distance $2r_0$ from each other. He easily obtains for the vibrations of the free molecule the equation:

$$m \frac{d^2 v}{dt^2} = 2vf'(r_0) + \frac{v^3}{3} f'''(r_0) + \text{.................(ii)}.$$

If we put $2f'(r_0) = -ma'^2$, and neglect *only the cubes*, we find

$$v = \frac{\dot{v}_0}{a'} \sin a't.$$

As the other two molecules are fixed, there is no question here of dilatation. To find the reaction on either molecule we have to substitute this value of v in $f(r_0+v)$ and we obtain

$$f(r_0+v)=f(r_0)+\frac{f'(r_0)}{a'}\dot{v}_0\sin a't+\frac{f''(r_0)}{2a'^2}\frac{\dot{v}_0^2}{2}\,2\sin^2 a't+\ldots\ldots\ldots\text{(iii)}.$$

Thus the mean value of p, the pressure upon the envelope of the vibrating elementary mass, would be

$$=f(r_0)-\frac{f''(r_0)}{4f'(r_0)}\frac{\dot{v}_0^2}{2}m\ldots\ldots\ldots\ldots\ldots\ldots\ldots\ldots\text{(iv)}.$$

Saint-Venant remarks that as $f'(r_0)$ is obviously negative $(=-ma'^2/2)$, we have only to suppose $f''(r_0)$ negative in order that this may connote an increase of pressure due to the vibration.

Referring to the value of the pressure as given by Eqn. (iv) he suggests in a footnote:

Cette sorte de considération, avec mise en compte, comme il est fait ici, des *dérivées* du second ordre $f''(r)$ des actions, n'est-elle pas propre à remplacer, avec avantage, ces chocs brusques des molécules des gaz contre les parois de leurs récipients, avec réflexions multiples et répétées, que des savants distingués de nos jours ont inventés ou revivifiés, dans la vue de rendre compte mathématiquement des pressions exercées sur ces parois, etc.? (p. 717.)

[274.] Saint-Venant in his fourth paragraph (p. 717) asks whether we can extend the results here found for two or three molecules to a multitude of molecules. He replies, yes, because it is easy to see that the new terms of the second degree due to the first derivatives $f'(r)$ will *add* to the second derivatives in $f''(r)$. On this point he refers to a footnote on p. 281 of his memoir in the *Journal de Liouville*, 1863 (see our Art. 127), and to one by Boussinesq in the same *Journal*, 1873, pp. 305—61.

Saint-Venant concludes therefore that when on the *rari-constant* hypothesis, we calculate the stresses by means of the linear terms only for the shifts, we destroy all dilatation and all stress due to increase of temperature; *we annul in fact all thermodynamics.* According to his theory then thermal effect is entirely due to the second derivatives of the intermolecular action expressed as a function of intermolecular distance. The point is obviously important in its bearing on the rari-constant hypothesis. Do the constants of $f(r)$, the law of intermolecular reaction, vary with the temperature—as would be the case if the "strength of the intermolecular reaction" were to vary with the energy of pulsational vibrations,—or, does heat only affect the mean distance of the molecules by producing molecular translational vibrations, so that $f(r)$ is no direct function of the thermal state of the body?

[275.] *De la Constitution des Atomes.* This paper was contributed to the *Annales de la Société scientifique de Bruxelles*, 2e année, 1878. No copy of this Journal is to be found in the British Museum, the Royal Society Library, or the Cambridge Libraries, and my references will therefore be to the pages of an off-print (*Hayez, Bruxelles*) for which I am indebted to the kindness of M. Raoul de Saint-Venant. The off-print contains 78 pages, and deals—with considerable historical, philosophical and scientific detail—with the continuity of matter and Boscovich's theory of atoms. It may be considered as Saint-Venant's final *résumé* of the arguments brought forward in the memoirs of 1844 and 1876: see our Arts. 1613*, 268 and 269[1]

[276.] The theoretical basis of the theory of elasticity and the strength of materials must be ultimately sought for in the law of molecular cohesion; the discovery of that law will revolutionize our subject as the discovery of gravitation revolutionized physical astronomy. Hence it is that the elastician looks for aid to the atomic physicist, who in his turn will find much that is suggestive for the theory of molecular structure in experiments on the constants of elastic and plastic materials. Bearing this in mind, a great deal that is profitable may be obtained by a perusal of the above memoir, although many scientists would disapprove of much of the method and of several of the conclusions of the author.

In order to place clearly before the reader the scope of the memoir, I preface my discussion of it with one or two remarks. We may legitimately question whether the laws of motion as based upon our experience of sensible bodies really apply to those elementary entities which form the basis of the kinetic properties of sensible bodies[2]. It is, however, most advisable to investigate

[1] In a footnote (p. 1) Saint-Venant remarks from hearsay that the memoir of 1844 (of which I have only seen the extracts in *l'Institut*), appeared in full in a Belgian Journal *Le Catholique* in 1852.

[2] For example the *Second Law of Motion* depends on the masses of the reacting bodies A and B not being influenced by the presence of a third body C, but it is conceivable that the 'apparent mass' of an atom is a function of its internal vibratory velocities, and that these are themselves dependent upon the configuration of surrounding atoms (see Arts. 51 and 52 of a paper in the *Camb. Phil. Trans.* Vol. XIV., p. 110).

what results must flow from applying the principles of dynamics to atoms and throwing back the origin of those principles on some still more simple entity. There is much that would induce us to believe (e.g. bright line spectrum of elementary gas at small pressure and not too high temperature) that an atom has an independent motion of its parts, and this suggests that we should try the effect of applying the principles of dynamics not only to the action of one atom upon another, but also to the mutual action of an atom's parts. If multi-constancy be experimentally demonstrated, then we must suppose either (i) the law of intermolecular action is a function of *aspect*, or (ii) the action of the element A upon another B is not independent of the configuration of surrounding elements (*Hypothesis of Modified Action:* see our Vol. I., p. 814). There may be other possibilities, but these, as the most probable, deserve at least early investigation. If the law of intermolecular action is a function of *aspect*, then we should expect to find that intermolecular distance is commensurable with molecular dimensions. According to Ampère and Becquerel the former is immensely greater than the latter; according to Babinet, they are in the ratio of at least 1800 : 1 (see § 13 of Saint-Venant's memoir). It is difficult to understand under these circumstances how aspect could be of influence, it would be sufficient to treat each molecule as a mere point or centre of action, which is practically Boscovich's hypothesis. According to the more recent researches of Sir William Thomson, who deals with a molecule as an extended, material body, the mean distance between two contiguous molecules of a solid is less than the $\frac{1}{100000000}$ of a centimetre while the diameter of a *gaseous* molecule is greater than $\frac{1}{500000000}$ of a centimetre (*Natural Philosophy*, Part II., p. 502). Thus intermolecular distance would be less than five times molecular dimensions. In this case it would seem probable that the law of action between parts of two molecules must be the same as the law of action between parts of the same molecule, for it is difficult, although, perhaps, not impossible to understand how one could begin and the other cease to be of importance at such small relative distances as 5 to 1. Resistance alike to positive and negative traction shews that the mean intermolecular distance cannot differ much from that at which intermolecular action changes its sign; further the capacity of the molecule itself to vibrate or suffer relative motion of its parts must point to a further change of sign in the action between parts of the molecule, or if this action be really intermolecular action, we are compelled, on the hypothesis of the elementary parts of a substance having extension, to presuppose a law of mutual action capable of *thrice* changing its sign within very narrow limits indeed. An analytical expression for such a law may not be hard to discover, but it would probably be difficult to conceive any mechanical system which could give rise to such an expression. On the other hand, it is perhaps impossible to conceive "aspect" as a factor of a centre of action according to Boscovich. Nor is it easy to picture the latter centre as the source of a vibration such as seems required by the bright line gas spectrum, such a vibration, on the other hand, being easily explained as the free vibration of an extended material molecule. If we turn, however, to the hypothesis of action modified by surrounding elements, there seems no reason why we should not apply it to the Boscovichian centre just as well as to the materially extended molecule of Thomson. The essential characteristic of the theory of Boscovich is the non-extension of the ultimate source of action, not the hypothesis that intermolecular action is a function of the individual molecular distance only. Rari-constancy is not then a necessity of the fundamental portion of Boscovich's doctrine, the two do not stand or fall together as some writers have assumed. Thus Saint-Venant's supposition as to the constitution of atoms in the present memoir is essentially Boscovichian, but he writes:

> La supposition dont nous parlons entraîne celle que l'intensité de chaque action entre deux particules très-proches soit généralement fonction non-seulement de leur distance mutuelle propre, mais encore, à un certain degré, de leurs distances aux particules environnantes, et même des distances de celles-ci entre elles (p. 17, § 7).

This hypothesis of modified action leading to multi-constancy[1]

[1] The *Hypothesis of Modified Action* leads to results akin to those I have referred to in the second footnote to p. 183, and which are expressed by Saint-Venant in the following sentences on p. 17:

> On remarquera qu'elle entraîne aussi que la force *totale* sollicitant une particule n'est pas exactement la résultante géométrique, composée par la règle statique du parallélogramme ou du polygone que l'on connaît, de toutes les forces avec lesquelles la solliciteraient séparément les autres particules si chacune existait seule avec elle, comme on l'a cru jusqu'à nos jours; cette règle ne serait plus vraie que pour les actions à des distances perceptibles, dont l'intensité, réciproque aux carrés des

is presupposed by Saint-Venant throughout the memoir, although, as he remarks, he does not agree with it. It forms indeed the essential difference between this memoir and that of 1844: see his § 7, p. 18.

[277.] Saint-Venant's arguments in favour of the ultimate atom being without extension are of a threefold character:

(i) arguments from the known physical properties of atoms;

(ii) metaphysical arguments;

(iii) theological arguments.

We will briefly refer to some points with regard to these in the following three articles.

[278.] §§ 3—21 deal with more purely scientific arguments based on known properties of atoms. In § 3 we have arguments from the theory of elasticity with special reference to the controversy between Navier and Poisson: see our Arts. 527*—534*. Saint-Venant points out how the continuity of matter is related to the possibility of replacing atomic summations by definite integrals. He proves with great clearness in a footnote pp. 12—15, on the hypothesis of continuity, the following propositions, which are really involved in the result of Poisson's memoirs of 1828 and 1829 and Cauchy's memoir of 1827 (see especially *Journal de l'École polytechnique,* 1831, p. 52, and the *Exercices de mathématiques,* 1828, p. 321, comparing our Arts. 443*, 548* and 616*):

1°. The stress across an elementary plane in a solid body will like that of a liquid at rest have no shearing component.

2°. The traction at any point varies as the square of the density.

3°. If there were no *initial stresses,* no state of strain would produce stress.

Thus *on the rari-constant hypothesis,* we reach impossible physical results or it follows that matter cannot be continuous. This applies also to the ether which could not propagate slide

distances, est celle de la pesanteur universelle, toujours négligeable vis-à-vis des actions à des distances imperceptibles qui produisent l'élasticité, la capillarité, les chocs, les pressions et les vibrations; et ces dernières et énergiques actions se soustrairaient à la règle statique dont nous parlons.

vibrations if continuous. Elementary proofs of the same propositions apparently not involving the hypothesis of rari-constancy are given in §§ 9 and 10.

The following §§ 11—17 contain various arguments against the continuity as well as against the extension of the ultimate elements of matter; they are certainly not conclusive, but they are extremely suggestive, especially with regard to the difficulty I have indicated on pp. 184—5 of the rapid changes in sign which must be attributed on the hypothesis of extension to the law of action. §§ 18—21 consider the explanation of various phenomena —e.g. crystallisation and inertia—on the Boscovichian hypothesis, while a footnote pp. 36—7 deals with a possible form for the law of action and some results of it: compare our Arts. 268 and 273.

[279.] §§ 22—39 deal with what Saint-Venant terms the metaphysical objections, which he says have been the only ones raised by those to whom he has communicated his theory. Some light on Saint-Venant's method of treatment may be gained from his remark on p. 9 :

> Je soumets d'avance, du reste, ce que j'énoncerai, dans les cas surtout où je serai forcé de me placer plus avant sur le terrain métaphysique, à toute autorité ayant pouvoir pour prononcer sur ce qui serait faux, et condamner ou *signaler* ce qui pourrait être dangereux.

It is beyond our province to enter into a discussion of the metaphysical arguments propounded, or the very wide range of philosophical reading evidenced by these sections[1]. It must suffice to say that Saint-Venant shews a decided preference for the scholastic writers, and an occasional tendency to imitate late-scholastic quibbles, as for example the arithmetical paradox on p. 55 by which

> Sans être donc dans les secrets du Créateur nous pouvons prononcer ...qu'il n'a composé ni les corps perceptibles ni leurs dernières parties, d'un nombre infini de points de matière.

[280.] A consideration of the theological arguments up to which the metaphysical lead would be out of place here, as they are, I venture to think, out of place in the pages of a scientific

[1] There is a good criticism of the antinomy of Kant (*du terrible penseur*) with regard to the divisibility of matter on pp. 37—9.

journal. Those who are anxious to determine the real source of cohesion will not be hindered from adopting the principle of extended material atoms, if it agrees best with the facts of observation, by the assertion that if they accept and comprehend thoroughly the system of Boscovich it will preserve them from the

deux principales et plus funestes aberrations philosophiques de notre temps et des temps anciens, le panthéisme et le matérialisme (p. 74).

Notwithstanding that many readers will find themselves unable to approve either the method or conclusions of the latter portion of the memoir, the whole should certainly be read for the interesting questions it raises with regard to the physics of elasticity.

[281.] *Des paramètres d'élasticité des solides et de leur détermination expérimentale. Comptes rendus*, T. LXXXVI, 1878, pp. 781—5.

This is a good *résumé* of the relations holding between the various elastic coefficients and moduli in the case of a body possessing three planes of elastic symmetry, and of the experimental methods of finding their values.

[282.] (1) The stress-strain relations will be those of our Art. 117 (*a*). The coefficients are now nine in number; namely, the three direct stretch-coefficients, a, b, c the three direct slide-coefficients d, e, f and the three cross-stretch-coefficients d', e', f'. We have the following special cases:

(2) Elastic isotropy in planes perpendicular to the axis of x:

$$e=f,\ e'=f',\ b=c=2d+d'.$$

Saint-Venant states the conditions erroneously and says they reduce the *nine* constants to *six*, a, b, d, e, d', e', but d' is known in terms of b and d, or we reduce them to *five*.

(3) Complete elastic isotropy, or as Saint-Venant puts it, isotropy in two of the axial planes:

$$a=b=c=2d+d'=2e+e'=2f+f',\ \text{and}\ d=e=f.$$

This reduces the nine coefficients to *two*, namely $d'=\lambda$ the dilatation coefficient, and $d=\mu$ the slide modulus. Saint-Venant has forgotten to state the relations $d=e=f$.

(4) Que si, sans vouloir (ce qui n'a aucune utilité) étendre l'applicabilité de ces formules aux déformations perceptibles de corps spongieux stratifiés, comme est le liége, ou de mélanges celluleux de solides et de

liquides, tels que sont les *gelées*, et même le caoutchouc, on se borne aux *vrais* solides, et si l'on admet que chacune des actions mutuelles entre deux molécules, dont les $\widehat{xx}\ldots\widehat{xy}$ sont les sommes de composantes, est *fonction d'une seule distance*, savoir celle des deux molécules qui l'exercent l'une sur l'autre, on peut prouver très-facilement (sans user de ces *intégrations autour d'un point* que Lamé a désapprouvées en 1852) que l'on a

$$d' = d, \quad e = e', \quad f = f'.$$

This reduces the coefficients in cases (1), (2) and (3) to six, three and one, respectively (p. 782).

In the second case Saint-Venant says *four*, but this is an error.

With regard to these rari-constant conditions the memoir continues:

Et ces égalités peuvent être admises; car, outre la presque évidence de leur principe, l'unité de paramètre ($\lambda = \mu$ ou $d' = d$) dans tout corps réellement isotrope se trouve prouvée par des faits nombreux, dont les derniers sont fournis par les ingénieuses expériences de 1869 de M. Cornu (p. 782).

In a footnote Saint-Venant refers to the experiments cited by Sir W. Thomson in the *Philosophical Magazine*, Jan. 1878, p. 18: see our Chapter devoted to that scientist. He holds that the discordant results there given for copper, prove either a fault in the experimental method adopted, or aeolotropy in each specimen of a diverse kind...*probablement écroui de manière à rendre, dans plusieurs d'entre elles, E_x, beaucoup plus grand que E_y ou E_z.*

The results for flint-glass and iron are he considers sufficiently near the rari-constant values, while those for cork and caoutchouc may be dismissed as proving nothing either way.

Turning to the stretch-modulus we easily find:

(5) in case (2),

$$E_x = a - 2\eta_{xy} e', \text{ and } \eta_{xy} = \eta_{xz} = \tfrac{1}{2}\,\frac{e'}{d + d'};$$

(6) in case (3),

$$E_x = 2d\,(1 + \eta_{xy}), \quad \eta_{xy} = \eta_{xz} = \tfrac{1}{2}\,\frac{d'}{d + d'};$$

(7) in case (4),

$$E_x = \tfrac{5}{2} d \,(= \tfrac{5}{2}\mu), \quad \eta_{xy} = \eta_{xz}\,(= \eta) = \tfrac{1}{4}.$$

(8) For *amorphic* materials, or bodies without regular crystallisation, such as drawn or rolled metals, stratified stone, wood etc., the aeolotropy of which can be regarded as due to unequal initial

stresses in three directions, or to a fibrous formation, three relations of the type:

$$a = \frac{(2e+e')(2f+f')}{2d+d'} \text{ ....(i),}$$

will sensibly hold, provided E_x, E_y, E_z have not ratios exceeding $\frac{3}{2}$ or at most 2 among themselves. This is the ellipsoidal distribution of elasticity: see our Arts. 138 and 142.

For the case of rari-constant isotropy we have:

$$a = \frac{3ef}{d}, \quad b = \frac{3fd}{e}, \quad c = \frac{3de}{f} \text{ ....(ii),}$$

relations admissible in general for the metals.

(9) For wood, where the ratio of E_x to E_y (the axis of x having the sense of the fibres) can amount to 10, 20, 40 and more, we can only take two of the above relations, namely:

$$b = \frac{3fd}{e}, \quad c = \frac{3de}{f} \text{ ....(i),}$$

which give:

$$E_x = a - \frac{ef}{2d}, \quad E_y = \frac{fd}{e}\frac{8ad - 4ef}{3ad - ef}, \quad E_z = \frac{e^2}{f^2}E_y \text{ ....(ii),}$$

$$\eta_{xy} = \frac{1}{4}\frac{e}{d}, \quad \eta_{xz} = \frac{1}{4}\frac{f}{d}.$$

For a modification of the statements in (8) and (9) with regard to *wood*: see our Arts. 308, 312 and 313.

[283.] Saint-Venant now proceeds to indicate experimental methods of arriving at the values of the following moduli and coefficients.

(1) To find the three direct slide-coefficients, or the slide-moduli d, e, f.

Case (a). If there be isotropy in the plane perpendicular to axis of x ($e = f$). We experiment on the torsion of a right circular cylinder.

Case (b). If e be not equal to f, we use the formula of Art. 29 (modified by Art. 47 and Table I) for the torsion of a prism on rectangular base. Let the base be $2b' \times 2c'$ and let b' be much $> c'$,

$$M_x = \frac{16\alpha}{l} f \frac{b'c'^3}{3}, \text{ sensibly.}$$

If c' be much $> b'$:

$$M_x = \frac{16\alpha}{l} e \frac{b'^3 c'}{3}, \text{ sensibly,}$$

where α is the *total angle* of torsion ($= l\tau$). These give the values of e, f, and similar experiments with prisms whose axes are parallel to y and z give d, f, and d, e, so controlling the former results.

(2) To find the three direct stretch-coefficients a, b, c,

(i) They are given in cases (4), (7) and (8) Eqn. (ii) of the previous article, so soon as we know, d, e, f.

(ii) In case (9) we know b and c, while a will be given from equation (ii), or $a = E_x + \frac{ef}{2d}$, so soon as we have by pure tractional, or better, flexural experiments, obtained the value of E_x; the values of E_y and E_z will then be known.

[284.] We may cite the following from Saint-Venant's concluding remarks (p. 785):

Au reste, si l'expérimentateur possède des moyens d'observation assez délicats pour mesurer aussi η_{xy}, η_{xz}, et par des extensions ou des *flexions* de petits prismes taillés transversalement, pour mesurer même

$$E_y,\ E_z,\ \eta_{yz},\ \eta_{yx},\ \eta_{zx},\ \eta_{zy},$$

les expressions en $a, b, c, d, .. f'$ qu'on peut tirer de ces diverses quantités en résolvant les équations (i.e. those with nine coefficients: see our Art. 307) à second membre trinôme, en annulant deux à deux leurs premiers membres, donneront des moyens de contrôle des mesurages opérés, et même des suppositions (4), (8), (9) (of Art. 282), qui ne sont pas admises par tout le monde. C'est un contrôle de ce dernier genre qu'opère la principale expérience de 1869 de M. Cornu.......(See our discussion of his memoir *infra.*)

On n'a pas besoin d'ajouter qu'aux mesurages statiques des dilatations, flexions et torsions, on pourra substituer au besoin, comme ont fait MM. Wertheim et Chevandier, des observations des sons rendus par des vibrations longitudinales, transversales et tournantes.

Saint-Venant has forgotten to add that the kinetic values of the elastic coefficients thus obtained will probably differ from the statical values: see our Arts. 1301* (3) and 1404*

[285.] *Sur la torsion des prismes à base mixtiligne, et sur une singularité que peuvent offrir certains emplois de la coordonnée logarithmique du système cylindrique isotherme de Lamé.* *Comptes rendus*, T. LXXXVII. 1878, pp. 849—54 and 893—9. There are additions in the off-print. This memoir was read on the 2nd and 9th of December.

Its object is explained in § 2 (pp. 850—1), after the solutions given in the memoir on *Torsion* (see our Art. 36, Nos. 4 and 5) have been cited:

Clebsch a remarqué, en 1862, qu'on obtient une variété de contours plus grande encore en se servant des coordonnées curvilignes isothermes orthogonales de Lamé (i.e. *conjugate functions*); et MM. Thomson et Tait dans leur beau livre *A Treatise on Natural Philosophy*, 1867, ont indiqué, sans le développer, leur emploi pour étendre les solutions telles que (3) {= (1) of our Art. 36}, relatives aux rectangles rectilignes, à des contours rectangulaires mixtilignes se composant d'un arc de cercle ou de deux arcs concentriques et des deux rayons qui les limitent, "ce qui est" disent-ils, "très-intéressant en théorie et d'une réelle utilité en Mécanique pratique."

Il m'a paru que la solution relative à ces sortes de sections pouvait être obtenue d'une manière simple et directe, sans substituer préalablement une certaine inconnue auxiliaire à l'inconnue géométrique u, et en s'en tenant aux coordonnées polaires ordinaires r, ϕ.

[286.] In § 3, Saint-Venant obtains the required solution in cylindrical coordinates. The fundamental equations (see our Art. 17, Eqn. vi.) become

$$\left.\begin{array}{r} u_{rr} + u_r/r + u_{\phi\phi}/r^2 = 0 \\ \tau r dr + u_\phi dr/r - r u_r\, d\phi = 0 \end{array}\right\} \quad \ldots\ldots\ldots\ldots\ldots\ldots (\text{i}).$$

If γ be the angle of the annular sector, r_0 and $r_1\ (> r_0)$ its radii, then the second or surface equation reduces to the following conditions when the median line is taken as initial line:

$$\left\{\begin{array}{l} \tau r^2 = -u_\phi \text{ for values of } r > r_0 < r_1, \text{ when } \phi = \pm\gamma/2, \\ u_r = 0 \text{ when } r = r_0 \text{ or } r_1, \text{ for all values of } \phi \text{ between } \pm\gamma/2 \end{array}\right\} \ldots (\text{ii}).$$

These conditions are found to be satisfied by the following value of u;

$$u = -\frac{\tau r^2}{2}\frac{\sin 2\phi}{\cos\gamma} - \frac{2\tau}{\pi}\frac{(-1)^n}{2n+1}\sum_0^\infty \frac{(r_1^{m+2} - r_0^{m+2})\, r^m - (r_0 r_1)^{m+2}(r_1^{m-2} - r_0^{m-2})\, r^{-m}}{r_1^{2m} - r_0^{2m}} \times \frac{\sin m\phi}{1 - m^2/4},$$

where

$$m = \frac{2n+1}{\gamma}\pi.$$

This result is practically obtained by assuming u to be of the form

$$Cr^2 \sin 2\phi + \Sigma\,(Ar^m + A'r^{-m}) \sin m\phi,$$

and determining the constants by the surface conditions (ii).

[287.] In the following section of the memoir, § 4, Saint-Venant treats precisely the same problem by the aid of the conjugate functions,

$$\beta = \tan^{-1}(z/y), \qquad \alpha = \log\sqrt{\frac{z^2+y^2}{a^2}}$$

He obtains two solutions in terms of α, β for a function V,—related to u by the equations $V_z = u_y$, $V_y = -u_z$.

The first contains two infinite summations and is similar in character to those given by Lamé in the *Onzième Leçon* of his work on Curvilinear Coordinates (see his p. 184). The second is that of Thomson and Tait, (see § 707, p. 252, Part II. of the second edition of their treatise).

He remarks, however, (§ 5) that although the value of the function u, obtained from V, is quite determinate when $r_0 = 0$, yet that of V becomes *indeterminate*. In fact the series for V cease to be *convergent*, and at least for the case of $r_0 = 0$, we have reached the value of u by means of an expression for V, which has ceased to have any meaning. We are thus thrown back in this case on the value of u determined by the process indicated in our Art. 286. See on this point the footnote on p. 143 of the memoir of January, 1879, considered in our Art. 291.

[288.] In § 6 Saint-Venant expresses analytically the value of the torsional moment M and the slides, and in the following sections gives the results of numerical calculations made with these formulae.

We may cite the following for the torsional moment M:

(1) Full Sectors:

$\gamma=$	45°	60°	90°	120°	180°	270°	300°	360°
$\frac{M}{\mathfrak{M}}$	·0923	·1333	·2096	·2754	·3776	·4486	·5253	·5589
$\frac{M}{\mathfrak{M}'}$	·5921	·7036	·7499	·7023	·5902	·4876	·5429	·5589

Here $\mathfrak{M} = \mu\tau \times \frac{r_1^2\gamma}{2} \times \frac{r_1^2}{2}$, or the torsional moment about the *centre* of the sector on the old Coulomb theory; $\mathfrak{M}' = \mathfrak{M} \times \left(1 - \frac{16}{9}\frac{1-\cos\gamma}{\gamma^2}\right)$ = torsional moment about the *centroid* of the sector on the old Coulomb theory. As is well known (see our Art. 181, (*d*)) Saint-Venant's theory makes *both torsional moments equal.* It will be seen at once that for bodies of this kind the results of the old theory are most erroneous and very dangerous in practice. The reduction of the torsional resistance for a split section is well brought out by the result $M/\mathfrak{M} = \cdot 5589$ for $\gamma = 360^\circ$.

(2) Annular sectors when $r_1 = 2r_0$:

$\gamma=$	60°	120°	180°
$\frac{M}{\mathfrak{M}}$	·0800	·1068	·1160
$\frac{M}{\mathfrak{M}'}$	·6812	·3160	·1909

We see again that the errors, when the old theory is used, are simply enormous.

[289.] In §§ 9—10 Saint-Venant determines the points of the full sectors, $\gamma = 60^\circ$ and $\gamma = 120^\circ$, where the slide is zero. These points are at distances from the centre differing from those of the centroids by a small amount only. He then gives the values of the shift u for various points of the same two sectors:

Les plus grandes valeurs de u sont aux points de rencontre de l'arc avec les deux côtés rectilignes. La médiane $\phi = 0$ reste immobile, et les éléments de l'arc prennent, sur le plan primitif de la section, des inclinaisons croissantes avec les distances où ils sont du milieu de cet arc.

[290.] In § 11 Saint-Venant states the value and position of the maximum slides for the same two sectors, i.e. he finds the fail-points (*points dangereux*). In both cases the maxima lie upon the contour, but the maximum of the maxima upon the rectilinear sides.

For $\gamma = 60^\circ$ the fail-point is distant $\cdot 5622 r_1$ from the centre and $\sigma = \cdot 4900\ \tau r_1$,

" $\gamma = 120^\circ$ the fail-point is distant $\cdot 3671 r_1$ from the centre and $\sigma = \cdot 6525\ \tau r_1$.

In a footnote Saint-Venant refers to the remark of Thomson and Tait (see their § 710) that for $\gamma > \pi$ the slide becomes infinite at the centre (i.e. when $r = 0$). This does not necessarily connote rupture, but only that the strain is greater than that to which we can apply the equations of mathematical elasticity. It suggests, however, the advisibility in practice of rounding off re-entering angles.

[291.] *Sur une formule donnant approximativement le moment de torsion.* *Comptes rendus,* T. LXXXVIII. pp. 142—7, 1879. This note was read on January 27, 1879.

This memoir has considerable practical value; it gives an *empirical* formula which embraces within narrow limits all Saint-Venant's torsional results; full sectors with re-entering angles alone excluded.

Starting with the formula for an elliptic section (see our Art. 18)

$$M = \frac{\pi b^3 c^3}{b^2 + c^2} \mu\tau,$$

we may write it

$$M = \kappa \frac{\alpha^4}{I_0} \mu\tau,$$

where I_0 is the moment of inertia of the cross-section about an axis perpendicular to the section through the *centroid* and α is the area. The quantity

$$\kappa = \frac{1}{4\pi^2} = \cdot 025330 = \frac{1}{39 \cdot 48}.$$

Now Saint-Venant finds that for the chief sections he has treated in his various memoirs κ varies only from $\cdot 0228$ to $\cdot 026$, while its mean value is very nearly $\cdot 025 = \frac{1}{40}$.

Hence we have very approximately for all sections the formula:

$$M = \frac{1}{40} \frac{\alpha^4}{I_0} \mu\tau.$$

It will be noted that the torsional moment varies *inversely* as the moment of inertia and not directly as in the old theory. Saint-Venant adds:

En y réfléchissant, on comprend qu'il en doit être généralement ainsi, car les sections allongées qui, à égale surface, ont le plus grand moment d'inertie polaire, sont aussi celles auxquelles la torsion fait prendre le plus de cette incurvation, de ce *gauchissement*, qui diminue l'inclinaison prise par les fibres sur les normales à leurs éléments, surtout aux points les plus éloignés du centre, et par conséquent, sont celles sur lesquelles les réactions élastiques développées ont le moment total M le plus petit (p. 142).

The final section of the memoir § 3 (pp. 143—7), is occupied with some general observations on the elasticity of rods whose axes are curves of double curvature. Their only relation to the preceding formula for torsion is the remark that the coefficient of torsional resistance used by some writers, namely $\mu\tau I_0$, must be replaced by $\frac{1}{40}\mu\tau\alpha^4/I_0$. Saint-Venant compares the results of his memoir of 1843 (see our Art. 1584*) with the more recent researches of Bresse and Resal: see our discussion of their memoirs below. There is nothing of importance to note; the footnote p. 145 should be cancelled.

[292.] *Analyse succincte des travaux de M. Boussinesq, professeur à la Faculté des sciences de Lille, faite par M. de Saint-Venant*, 1880. This report consists of 23 *lithographed* pages.

In April, 1880, Boussinesq had printed and presented to the members of the Academy a notice of his scientific writings. (Danel, Lille, in 4°.) Saint-Venant then drew up the above analysis, strongly recommending Boussinesq for membership of the Academy. Pp. 12—17 (§§ 6—9) treat of his contributions to the theory of elasticity ('*Les travaux de M. Boussinesq sur les corps solides et leur élasticité ne sont pas moins originaux et importants*'). Pp. 17—20 deal with his various mechanical and philosophical papers; pp. 20—23 with his contributions to the undulatory theory of light. We shall have occasion to return to Saint-Venant's essay when discussing Boussinesq's memoirs.

[293.] A second paper of Saint-Venant's dealing with the elastical researches of a contemporary may be noted here. It is entitled: *Sur le but théorique des principaux travaux de Henri Tresca. Comptes rendus*, T. CI., 1885, p. 119—22.

The influence on theory of Tresca's researches and the origin of the science of plasticity are sketched. The writer attributes to Tresca a keen appreciation of theory; he was no mere empiricist, as many have erroneously believed:

Il importe de montrer, dans l'intérêt de sa mémoire comme dans celui de la vérité scientifique, que Tresca fut un esprit plus large, un homme de vraie Science et par conséquent de *théorie* dans la meilleure et la plus saine acceptation de ce mot si souvent mal compris, si fréquemment accusé, par légèreté ou en haine systématique de la Science, de n'exprimer que des chimères (p. 119).

[294.] Géométrie cinématique:—*Sur celle des déformations des corps soit élastiques, soit plastiques, soit fluides: Comptes rendus*, 1880, T. XC., pp. 53—56.

Saint-Venant draws attention to the importance of *pure kinematics* and notes how far it is possible to advance in physical problems without the aid of force or stress considerations. Saint-Venant may be legitimately looked upon as one of the forerunners of that reduction of all dynamics to kinematics, or the exclusion of the idea of force from physics, which is now probably only a matter of time. In a lithographed course of lectures given in 1851 (*Principes de Mécanique fondés sur la Cinématique*, delivered at Versailles to engineer-students) he had treated of great portions of mechanics on kinematic principles. In this direction he had been preceded by Grassmann and followed by Resal (*Cinématique pure*, 1862, and *Mécanique générale*, 1873). The present article points out how far we can advance in the geometry of strain or displacement without the conception of stress. Saint-Venant adduces the theorem of the distortion of a sphere into an ellipsoid, and speaks as if it were only true for small strains. That it is true for all strains was pointed out by Tissot (see a supplementary *Note*, p. 209 of same volume of *Comptes rendus*) who had given a demonstration of it in the *Nouvelles Annales de Mathématiques*, 1878, p. 152. Saint-Venant points out in this *Note* that his own proof of 1864 (*L'Institut*, No. 1614, p. 389) did not really introduce this restriction. The kinematics of strain had, moreover, been thoroughly considered in 1867 by Thomson and Tait in their *Treatise on Natural Philosophy*, pp. 98—124.

[295.] *Du choc longitudinal d'une barre élastique libre contre une barre élastique d'autre matière ou d'autre grosseur, fixée au bout non heurté; considération du cas extrême où la barre heurtante est très raide et très courte: Comptes rendus*, T. XCV., 1882, pp. 359—365, *Errata*, p. 422.

This is only an abstract of the memoir. It gives a solution in trigonometrical series for the case of one bar striking longitudinally a second with one end fixed.

If V be the initial uniform speed of the impelling bar, a_2 its length, a_1 that of the fixed bar, P_2, P_1 the weights of the two bars, x the abscissa measured along the common axis of the two bars from the end of the fixed bar, then the shifts u_2 and u_1 of either bar at any point x *during the impact* are:

$$u_2 = P_2 V \Sigma \frac{2\cos\{m\tau_2 (a_1 + a_2 - x)/a_2\} \sin mt}{m \cos m\tau_2 \left(\dfrac{P_1}{\sin^2 m\tau_1} + \dfrac{P_2}{\cos^2 m\tau_2}\right)},$$

$$u_1 = P_2 V \Sigma \frac{2\sin\{m\tau_1 x/a_1\} \sin mt}{m \sin m\tau_1 \left(\dfrac{P_1}{\sin^2 m\tau_1} + \dfrac{P_2}{\cos^2 m\tau_2}\right)},$$

where m is a root of the equation:

$$\frac{P_1}{\tau_1} \cot m\tau_1 - \frac{P_2}{\tau_2} \tan m\tau_2 = 0,$$

and $\tau_1 = a_1/k_1$, $\tau_2 = a_2/k_2$; k_1 and k_2 being the velocities of sound in the two bars.

[296.] Saint-Venant then considers the case when τ_2 is very small as compared with τ_1, and so deduces Navier and Poncelet's expression for the vibrations of a bar struck by a weight on its free terminal: see our Arts. 273*, and 991*. Saint-Venant does not enter into the question of the *time and manner* in which the bars separate. He goes on to remark that in the case of two *free* bars we may express the result in finite terms, as also in the case of one free bar and a weight moving with a definite velocity and striking it longitudinally on one terminal. The case of a bar fixed at one terminal and struck by a moving weight at the other, he does not in this memoir attempt to solve in finite terms. This, however, he proceeded to do in an article in the same volume of the *Comptes rendus*, on pp. 423—427, entitled:

[297.] *Solution, en termes finis et simples, du problème du choc longitudinal, par un corps quelconque, d'une barre élastique fixée à son extrémité non heurtée.*

This solution is very similar to the full treatment of the problem by Boussinesq referred to in our Art. 341. But it fails to determine the instant of separation, and so does not completely solve the problem. After Boussinesq had given his solution Saint-Venant with the aid of Flamant concluded the whole subject with a graphical investigation of the successive states of the bar and the impelling load for the whole duration of the impact: see our Arts. 401—7.

SECTION V.

The Annotated Clebsch.

[298.] *Théorie de l'élasticité des corps solides de Clebsch. Traduite par MM. Barré de Saint-Venant et Flamant, avec des Notes étendues de M. de Saint-Venant. Paris*, 1883, *pp.* 1—900 (but by means of subscripts the number of pages is much greater than thus appears, e.g. 480. *a*—480. *gg*).

This is Saint-Venant's last great and, we may say, most complete contribution to the theory of elasticity. By means of footnotes, section-notes and appendices he has almost trebled the matter given by Clebsch, and the result is a treatise on the theory of elasticity from the mathematico-physical standpoint which will long remain the standard work on this subject.

> Au moyen de ces explications et annexes, auxquelles nous aurions pu donner plus d'étendue en rapportant d'autres résultats inédits de nos recherches déjà anciennes, nous espérons, si l'on veut bien y donner quelque attention, que la traduction offerte par nous aura une réelle utilité et que la belle et intéressante branche de physique mathématique ayant, avec l'art des constructions, des rapports si intimes, pourra être de mieux en mieux comprise, étudiée et appliquée (p. xxi).

With Clebsch's contributions to elasticity we shall busy ourselves later; so far as the text of his work is concerned, we have only to note here that his isotropic formulae are everywhere replaced by those for suitable distributions of homogeneity (see our Art. 114), and that various obscurities in his treatment are explained or corrected in copious footnotes. We shall occupy ourselves in the following articles with an analysis only of Saint-Venant's contributions to the volume.

[299.] Saint-Venant's first important note occurs on pp. 39—42. It is headed: *La preuve de la forme linéaire des expressions des composantes de tensions ne peut pas être purement mathématique.* This deals with the same matter as pp. 662—5 of the *Leçons de Navier*: see our Arts. 192 (*a*) and 928*, namely the futility of all purely mathematical deductions of the linearity of the stress-strain relations. Such deductions have been given by

Green, Clebsch, Thomson and others: see our Art. 928* and the footnote Vol. I., p. 625.

Généralement et philosophiquement aucune considération purement *mathématique* ne saurait révéler le mode de la dépendance mutuelle des forces agissant sur les éléments des corps, et des changements géométriques qui s'y opèrent, tels que ceux des longueurs et des angles de leurs côtés: la connaissance de ce mode ne peut être dérivée que des faits, ou de quelque loi *physique* exprimant un ensemble de faits constatés (p. 39).

Saint-Venant appeals to experiment and cites Stokes' adduction of the isochronism of sound vibrations with approval: see our Art. 928*. We have remarked elsewhere that the stress-strain relation cannot, however, be treated as linear for the slight elastic strains in many of the materials of practical structures: see Note D of our Vol. I., p. 891.

[300.] But Saint-Venant is not satisfied with appeal to experiment and observation; these give Keplerian laws, without the backbone of Newtonian gravitation:

En général, pour convaincre nos esprits, l'empirisme, qui ne rend compte de rien, ne suffit pas: il nous faut encore une explication, une raison scientifique, où la preuve que les formules qu'on nous propose dépendent de quelque loi assez générale, assez *grandiose*, c'est-à-dire simple, pour que nous puissions en raisonnant, comme faisait Leibnitz, quand ce ne serait que d'une manière instinctive, la regarder comme pouvant être celle à laquelle le souverain Législateur a soumis les phénomènes intimes dont les formules en question représentent et mesurent les manifestations extérieures (pp. 40–1).

Saint-Venant finds this *loi assez générale, assez grandiose* in the law of intermolecular central action, as a function only of the distance, and cites its acceptance by the leading physical mathematicians from Newton to Clausius. He then refers to Green and his followers, who, as we know, appealed to Taylor's Theorem, as a *loi assez grandiose.* Now behind this appeal for 21 independent constants to Taylor's Theorem, although unrecognised by Green, was the important conception that possibly intermolecular action depends not only on the individual molecules, but on the position of each pair of them in the universe relative to other molecules. For example, if intermolecular action arises from molecular pulsations in a fluid ether, we find intermolecular force is a function of molecular surface energy, which surface energy is itself a function of position relative to the totality of other molecules. It is true that the law of intermolecular force thus resulting is not *simple*, although with the knowledge we have of thermal and optical phenomena, it may tend to coordinate far better than any simpler law the total physical universe. Saint-Venant does not appear here to strengthen the arguments of the *Appendice* V (see our Art. 192 (*a*)) by the introduction of a *souverain Législateur*, for whom a *loi assez grandiose* must necessarily be *assez simple*. The assumption is, indeed, anthropomorphical in the extreme. When we regard thermal and optical phenomena,—and note the probable vibration of molecules and the existence of an ether—we may be quite certain that the law of intermolecular action whatever be its nature is far from being primary in the universe; it must be a result of the structure of molecule and ether; *grandiose* it certainly may be, but the addition *c'est-à-dire simple* is an anthropomorphical dogma, which recalls to our minds the *mundi fabrica est perfectissima* of Euler.

[301.] We must next consider the *Note finale du* § 16 which occupies pp. 63—111.

§§ 1—12 of the Note (pp. 65—75) are again concerned with the coefficient controversy, but take up a different line of argument from that of the *Appendice* V: see our Arts. 192—5. Saint-Venant here enquires how far Green's appeal to the principle of work and the impossibility of perpetual motion in itself involves the reduction of the elastic constants to 15.

[302.] He starts from the equation

$$\Sigma m V^2/2 + \psi(x, y, z, x', y', z', x'', y'', z''\ldots) = \text{some constant } C\ldots(a),$$

where V is the translational velocity of the molecule m, whose centroidal position is x, y, z, and the dashed letters give the positions of other molecules m', m'', etc. In other words he makes the total *translational* energy of the system a function of molecular position. He omits:

(1) from the kinetic energy a possible internal vibratory motion of the molecule due to pulsations in its atoms or to change in the relative motion of the atoms of the same molecule;

(2) possible factors in the potential energy due to strains in the molecule itself or to changes in its *aspect* with regard to other molecules.

Is he justified in thus making the translational energy of the molecular centroids a function solely of their position? He seems to think that both the omissions (1) and (2) are legitimate provided that there are no such changes of *temperature* as produce violent atomic vibrations, and that we take the mean of large numbers (see his § 12). But is it not within the bounds of possibility that the mean internal potential energy of the molecules may be changed by an elastic strain, although the mean internal kinetic energy on which the temperature may be supposed to depend remains unchanged? This change in the potential energy of the molecule will be a function of the relative molecular position, but it may be one of *aspect* as well as of centroidal position. If we accept, however, with both Green[1] and Saint-Venant that the former can only depend on the latter, we are thrown back, supposing no sensible thermal changes, on Equation (a).

[303.] Saint-Venant in § 5 proceeds to question whether the Equation (a) can give the form of ψ required by Green. He says that we can replace it by an equation of the form:

$$\Sigma m V^2/2 + \Psi_1 (r, r', r'' \ldots) = C \ldots\ldots\ldots\ldots\ldots\ldots (b),$$

où Ψ_1 est une nouvelle fonction dont il importe peu que les variables r, r', r'' &c. soient ou ne soient pas, en partie, dépendantes les unes des autres, ...r, r', r''...étant les distances des molécules du système tant entre elles qu'avec les centres d'action *fixes* extérieurs (p. 68).

Is this change legitimate? The form (a) retains the possibility of intermolecular action being a function of *aspect*. Is this lost in (b)?—It does not appear to be so if some of the variables r are the distances from *fixed external points*. From this equation we easily deduce for any molecule m, the typical equation:

$$m\ddot{x} = \Sigma \frac{d\Psi_1}{dr} \cos (rx) \ldots\ldots\ldots\ldots\ldots\ldots\ldots\ldots (c),$$

where Σ denotes a summation with regard to all values of r.

[1] Both Green and Sir William Thomson make the potential energy of the element a function only of the *change in shape*, i.e. of the relative position of molecular centroids. I think this assumes that the internal potential energy of the molecule can only be a function of centroidal position. It may, however, be that the internal potential energy of (either the molecule or) the element is a function of the relative *motion* of (the atoms or) the elements, in which case the velocities would appear in Ψ_1, and we should obtain by the Hamiltonian process totally different equations to those of Green for elasticity. These *generalised equations of elasticity* leading to the *Dissipative Function* etc., I propose to discuss elsewhere.

The rest of the investigation now turns upon the question whether $\frac{d\Psi_1}{dr}, \frac{d\Psi_1}{dr'}, \frac{d\Psi_1}{dr''} \ldots$ are solely functions of $r, r', r'' \ldots$ respectively. If they are, then the 36 coefficients reduce to 15. If they are not, then the action of one molecule on a second can depend: (1) upon mutual aspect, (2) upon the position of other molecules. The dependence of the mutual action of each molecular pair solely on their centroidal distance is the hypothesis, as Saint-Venant remarks, upon which most writers on mechanics have based their proofs of the conservation of energy (e.g. Helmholtz). At the same time it does not seem necessary to assume it for more than the atoms, and for the molecules *aspect* may really be important.

[304.] Saint-Venant now proceeds to investigate what consequences flow from rejecting this hypothesis. He remarks that the action between two molecules will now be a function of their distances from *other* molecules, and not only of their mutual distance. It appears to me that the action does not necessarily depend solely on their distances from other molecules, but perhaps also on their distances from imaginary molecules or fixed centres, which give the *aspect* influence. Saint-Venant tries to prove in the first place that the work done in a complete cycle cannot generally be zero, if the intermolecular force is a function of more than the single intermolecular distance. It is quite true, as he observes, that if we move two molecules from a mutual distance r, where the action is R_1 and bring them again to a mutual distance r, the action R_2 need not be equal to R_1, and so the elements of work $R_1 dr$ and $-R_2 dr$ need not be equal and opposite, provided the other intermolecular distances are not the same in the two positions. It is only necessary that the positive work created by one pair of molecules, shall be exactly equal to the negative work created by the action of the remaining pairs of molecules. Is there anything improbable in this? Saint-Venant seems to think so:

Or, quelle que soit la loi imaginable à laquelle on soumette les intensités des actions entre deux molécules, et leur mode de dépendance de la *simple présence* d'autres molécules, si une juste compensation, comme celle dont nous parlons, s'observe ainsi entre deux moitiés de certains systèmes parcourant certains cycles, elle cessera de s'observer en ajoutant à ces systèmes d'autres systèmes pouvant être pris infiniment variés, et en ajoutant aux parcours d'autres parcours quelconques arbitrairement choisis.

La nullité du travail total produit par un cycle ne peut donc être générale qu'autant qu'elle a lieu *pour chaque action individuelle* ; ce qui oblige à admettre que la force que nous avons appelée R soit fonction *de la seule distance* que nous avons appelée r (p. 71).

I do not understand the argument which follows the words: *elle cessera de s'observer en ajoutant.* Suppose the molecules represented by electro-magnets then the total action during any motion of one such magnet A on another B would depend not only on the initial and final *relative* positions of A and B, but owing to the induced currents on the paths and positions of A and B with regard to the other bodies in the field. It seems to me that Saint-Venant's argument would compel us to assert that by introducing other magnets into the field or by moving them about in a proper manner, we could obtain perpetual motion.

[305.] Saint-Venant's second argument is of the following kind (see his § 9). If the intermolecular force depends on more than the particular centroidal distance, then the distances between astral molecules will affect the action between terrestrial. Here to start with, we have somewhat of an assumption; the action of A upon B may depend on the distance of both from C and D but not necessarily on the distance of C from D. For example such might be the case when we treat of *aspect* influence, as given by means of fixed centres having reference only to A and B. Saint-Venant continues: the influence of an astral intermolecular distance on a terrestrial must be absolutely insensible, for even when we are dealing with a small portion of terrestrial matter, the action of its molecules is sensibly independent of the state of other matter even at a visible distance.

Hence the form of Ψ_1, $(r, r', r''...)$ ought to be such that for any small system $d\Psi_1/dr$ depends sensibly only on the molecules in the immediate neighbourhood of m. This condition of exclusion can be easily fulfilled for molecules at *sensible* distances by making Ψ_1 a function of the inverse powers of r, r', r''.... We will now cite Saint-Venant's actual words:

Mais cette ressource d'exclusion sensible est impuissante à l'égard des distances mutuelles de molécules appartenant en particulier à chacun de ces systèmes ou éléments non proches de celui dont on s'occupe.

Les distances mutuelles insensibles entre les molécules composant même chaque étoile auront une influence du même ordre sur la grandeur de $d\Psi_1/dr$, ou sur l'intensité de l'action mutuelle des deux molécules m, m' d'un corps terrestre que les petites distances des molécules qui les avoisinent dans le même corps, tant qu'on n'aura pas imposé à la forme de la fonction $\Psi_1 (r, r', r'' \dots)$ une restriction ou particularisation plus grande (p. 72).

The reader will indeed find it difficult to discover a form of function in which the influence of A upon B, shall be affected by the distance between C and D, and yet shall vanish when C and D are both distant from A and B. Its discovery, however, does not seem impossible, and when we regard the ether as producing the action between A and B by its state of stress, it seems by no means improbable that the approach of C and D may affect the action of A on B.

If, however, we suppose that it is only the distances of A and B from C and D which influence the action of A on B, there is less difficulty in the matter. This case, of special importance, seems to have escaped Saint-Venant's notice. Thus let $\phi'(r)$ be a law of intermolecular action, which gives a zero action for sensibly large values of r, and a strong repulsive action for all values of r less than β, so that r is usually $> \beta$ and β/r a small quantity. Let $f(z_1, z_2, z_3, \dots)$ be a function of the variables z_1, z_2, $z_3, \dots$ which is practically independent of z_r, when z_r is small. Then the following form of Ψ_1 is suitable:

$$\Psi_1 = \Sigma \phi_{pq} (r_{pq}) \{m_p m_q + \beta_{pq} f_{pq} (\beta_{pn}/r_{pn}, \ \beta_{sq}/r_{sq} \dots)\},$$

where in the variables of the function f_{pq} n and s are to take all values except p and q; finally we must sum the expression for all different values of p and q. Since $(\beta/r)^2$ is negligible, f' will not occur and thus $d\Psi_1/dr_{pq}$ will be independent of r_{ns} when n and s are both different from p and q; so that Saint-Venant's objection falls to the ground.

But we are not even compelled to suppose the action of A, B independent of the position of C, D. Let us take q_1, q_2, $q_3 \dots$ as either aspect or internal position coordinates of the molecules, for the purposes of illustration one for each molecule will suffice. Then it seems extremely probable that the potential energy of the system,—as a result of the stress in the ether—involves the generalized velocities $\dot{q}_1$, $\dot{q}_2$, $\dot{q}_3$, etc., so that we must write for Ψ_1 a function of $\dot{q}_1, \dot{q}_2, \dot{q}_3 \ldots r, r', r'' \ldots$ In this case our equation will be of the form:

$$\frac{\Sigma m V^2}{2} + \frac{\Sigma \alpha \dot{q}^2}{2} + \frac{\Sigma \gamma_{sn} \dot{q}_s \dot{q}_n}{2} + \Psi_1 (\dot{q}_1, \dot{q}_2, \dot{q}_3, \ldots r, r', r'' \ldots) = \text{const.} \ldots (d),$$

where α and γ are certain constants. We should have to apply the general dynamical equations to determine the V's and $\dot{q}$'s. Thus the intermolecular force between m_1 and m_2 might be a function of q_3, which in its turn might be found from the dynamical equations as a function of r' and r'', etc., distances, let us say, between m_3, m_4 etc., while r', r'' would have no *direct* influence on the action between m_1 and m_2: see Arts. 931*, 1529*.

The point is of very great physical interest, as it really concerns the *direct* application of the Second Law of Motion to the ultimate particles of bodies. Can we or can we not superpose the action of C on A to that of B on A, or does the action of C on A, affect that of B on A? See the footnotes to our pp. 183 and 185.

[306.] The strong points of the rari-constant argument seem to me to lie in: (i) the *probable* insignificance of the indirect action of C as compared with the direct action of A on B; (ii) the insufficiency of most of the experiments yet brought to bear against rari-constancy.

Be this as it may, I still feel it impossible to accept the following statements of Saint-Venant as satisfactory:

j'affirme hardiment, et tout le monde, j'en suis convaincu, pensera comme moi, qu'il faudra absolument adopter la forme ou la particularisation indiquée ci-dessus:

$$\Psi_1 (r, r', r'' \ldots) = f(r) + f_1(r') + f_2(r'') + \ldots\ldots$$

...Elle fait revenir à l'adoption, comme voulue ainsi par l'expérience même, de la loi des actions *fonctions des seules distances où elles s'exercent*, et non des autres distances; loi que le simple bon sens, aidé d'une observation générale des faits, a fait accepter pendant plus d'un siècle et demi. Et je suis convaincu que Green lui-même y croyait sans s'en rendre compte. Je ne peux, en effet, interpréter d'une autre manière cet instinct de physicien et de géomètre, ce sentiment "que les forces, dans l'univers, sont disposées *de manière à faire, du mouvement perpétuel, une naturelle impossibilité.*" Green, sans aucun doute, refusait ainsi, *à chaque action moléculaire mutuelle en particulier*, la possibilité contraire...(pp. 72 and 73).

I doubt whether Green had thoroughly seen the important

physical consequences which flow from multi-constancy, but I do not see why he should have objected to two molecules having done work on their return to the same distance at a different point of the field. In § 12 (pp. 74–5) Saint-Venant recognises a distinction between atomic actions and their resultant, or molecular action. At the same time, however, he holds that if the latter be indeed a function of *aspect*, it will not produce on the principle of averages any great inequality in the coefficients of type $|xxyy|$ and $|xyxy|$. Notwithstanding these rari-constant views, he wisely adopts in the *Clebsch* for the equal coefficients of rari-constancy letters distinguished by a dash.

[307.] On pp. 75—84 (§§ 13—16) Saint-Venant reproduces the results of the memoirs of 1863 and 1878, or of the *Leçons de Navier*, p. 808 *et seq.*: see our Arts. 151, and 198 (*e*). The results given in § 15 are precisely those obtained by Neumann in 1834: see our Art. 796*. In the notation of our work, if a, b, c are the direct-stretch, d, e, f the direct-slide and d', e', f' the cross-stretch coefficients, for a material with three planes of elastic symmetry, then:

$$\frac{(bc-d'^2)}{1/E_x}=\frac{(ca-e'^2)}{1/E_y}=\frac{ab-f'^2}{1/E_z}=\frac{ad'-e'f'}{1/F_x}=\frac{be'-f'd'}{1/F_y}$$
$$=\frac{cf'-d'e'}{1/F_z}=\Delta=\begin{vmatrix} a & e' & f' \\ e' & b & d' \\ f' & d' & c \end{vmatrix}.$$

Further as a typical strain-stress equation we have:

$$s_x=\widehat{xx}/E_x-\widehat{yy}/F_z-\widehat{zz}/F_y,$$

so that $1/E_x$, $-1/F_z$, $-1/F_y$ etc., are Rankine's *thlipsinomic* coefficients: see our Chapter XI.

In addition we have for the stretch-squeeze ratios equations of the type:

$$\eta_{yz}/E_y=\eta_{zy}/E_z=1/F_x.$$

[308.] In § 17 Saint-Venant deals with *amorphic* bodies, or those for which the following relations hold:

$$2d+d'=\sqrt{bc},\quad 2e+e'=\sqrt{ca},\quad 2f+f'=\sqrt{ab}\ldots\ldots(\text{i}).$$

If the quantities $\frac{1}{2}(\sqrt{b}-\sqrt{c})^2$, $\frac{1}{2}(\sqrt{c}-\sqrt{a})^2$, $\frac{1}{2}(\sqrt{a}-\sqrt{b})^2$ are small we may write these relations:

$$2d+d'=\frac{b+c}{2},\quad 2e+e'=\frac{c+a}{2},\quad 2f+f'=\frac{a+b}{2}\ldots(\text{ii}).$$

See the memoirs of 1863 and 1868; or our Arts. 139, 142—4 and 281.

Saint-Venant holds that for a feeble degree of aeolotropy produced by permanent compressions as, for example, in drawn or rolled metal and in some kinds of stone the relations (i) or (ii) suffice. For wood however some other conditions must hold. For let us suppose:

(*a*) The relations (ii) to hold with equal transverse elasticity (or $a=b$) and rari-constancy then:

$$3f=a, \quad 6e=c+a \text{ and } c=6e-3f.$$

We easily find from the formulae of Art. 307, that:

$$\eta_{zx}=\tfrac{1}{4}e/f, \quad E_z/E_x=\tfrac{1}{8}\left(18\, e/f-\frac{e^2}{f^2}-9\right),$$

whence $$\eta_{zx}=9/4-\tfrac{1}{2}\sqrt{18-2\,E_z/E_x},$$

or, in order that the stretch-squeeze ratio be real we must have $E_z/E_x<9$.

This result is contradicted by Hagen's experiments (see our Art. 1229*). Hagen found:

$$E_z/E_x=\begin{cases}15 & \text{for oak,}\\ 22{\cdot}5 & \text{for beech,}\\ 48 & \text{for pine,}\\ 83 & \text{for fir.}\end{cases}$$

(*b*) The relations (i) to hold together with $a=b$, $d=e=d'=e'$.

It follows that $\eta_{zx}=\tfrac{1}{4}\,e/f$, $E_z/E_x=e^2/f^2$,

whence $$\eta_{zx}=\tfrac{1}{4}\sqrt{E_z/E_x}, \quad E_z/G=\tfrac{5}{2}\sqrt{E_z/E_x}.$$

These expressions are never imaginary and give reasonable values for η_{zx} up to $E_z/E_x=4$. After this η_{zx} begins to take unsuitable values till for $E_z/E_x=80$, we have η_{zx} so large as $2{\cdot}236$.

Clebsch (p. 8, § 2) and at one time Saint-Venant (see our Art. 169 (*d*)) had held that η must necessarily be $<\tfrac{1}{2}$. This error the latter had recognised in the *Appendice complémentaire* to the *Leçons de Navier*, and he now adds:

Cette opinion n'est fondée sur aucun fait; il ne l'exprime même que *pour les corps isotropes*, et quelques expériences de Wertheim ont montré qu'aux approches de la rupture d'une tige métallique, c'est-à-dire au moment où sa matière est arrivée à un état très fibreux, comparable à celui des bois, une extension de plus diminue le volume; en sorte que, sans pouvoir aller jusqu'à $\eta=2{\cdot}236$, rien n'empêcherait de porter η jusqu'à 1 pour les bois tendres (p. 89).

Saint-Venant now seeks some correction of the amorphic formulae (i) which will give better results than this for η_{zx} when E_z/E_x is large.

[309.] He first proceeds on pp. 89—95 to determine Neumann's stretch-modulus quartic; he obtains it in the form:

$$\frac{1}{E_r}=\frac{c_x^4}{E_x}+\frac{c_y^4}{E_y}+\frac{c_z^4}{E_z}+2\,\frac{c_y^2c_z^2}{F_1}+2\,\frac{c_z^2c_x^2}{F_2}+2\,\frac{c_x^2c_y^2}{F_3}\;\ldots\ldots\ldots\text{(iii)},$$

where c_x, c_y, c_z are the direction-cosines of the line r, and

$$\left.\begin{array}{l} 1/F_1 = 1/(2d) - 1/F_x \\ 1/F_2 = 1/(2e) - 1/F_y \\ 1/F_3 = 1/(2f) - 1/F_z \end{array}\right\}.$$

This agrees with Neumann's result (see our Art. 799*) if we note that his N_a, N_c, M_b, M_c, P_a, P_b are really *cross*-stretches and therefore of negative sign[1].

By taking $x = c_x \sqrt[4]{E_r}$, $y = c_y \sqrt[4]{E_r}$, $z = c_z \sqrt[4]{E_r}$,
we have a surface of the fourth order, whose ray measures $\sqrt[4]{E_r}$ in the same direction.

[310.] In § 21 (pp. 95—8), Saint-Venant enters upon a lengthy calculation of the maxima and minima values of E for different directions. If three relations of the type

$$F_3 = \sqrt{E_x E_y} \quad\text{.............................(iv)}$$

hold, then (iii) reduces to an ellipsoid and we have the ellipsoidal distribution of elasticity. This gives only three maxima and minima for E_r. Saint-Venant seeks conditions under which there shall only be three maxima for the surface (iii) when the relations of type (iv) are not fulfilled; in other words, he seeks when there will be, as he expresses it, a *variation simple et graduelle des élasticités*.

The conditions are

(1) that F_1 lie between E_y and E_z, and two others of the same type;

(2) that the three expressions whose type is

$$(1/E_y - 1/F_3)(1/E_z - 1/F_2) + (1/F_3 - 1/F_1)(1/F_1 - 1/F_2),$$

shall not all be of the same sign.

[311.] In § 22 Saint-Venant shows that the three ellipsoidal conditions of type $F_3 = \sqrt{E_x E_y}$ are identical with the three of type $2f + f' = \sqrt{ab}$, provided either $\frac{d}{d'} = \frac{e}{e'} = \frac{f}{f'}$, or again that rari-constancy is assumed to hold.

[312.] He next seeks for some non-ellipsoidal distribution which shall satisfy the conditions for *variation simple* of our Art. 310. He takes as a probable solution: (1) rari-constancy, and (2) two of the ellipsoidal relations, i.e. he writes:

$$a = 3ef/d, \quad b = 3fd/e,$$

and searches for a value of n, where

$$c = 3de/(fn),$$

[1] Unfortunately the wrong signs are given in Art. 796* to all the quantities M, N, P. If these are corrected, a negative sign must be inserted in the second table of Art. 795* before the $1/F$'s. The value of $1/E_r$ in Art. 799* is then accurate. I regret that this slip of Neumann's escaped me.

which shall satisfy those conditions. After some rather complex analysis the necessary and sufficient conditions are found to reduce to

$$\frac{9+12\,E_z/E_x-\sqrt{81+144\,E_z/E_x}}{2+4\,E_z/E_x}$$

$$>n>\frac{12+9\,E_z/E_x-\sqrt{144\,E_z/E_x+81\,(E_z/E_x)^2}}{4+2\,E_z/E_x},$$

where we suppose $E_z > E_x > E_y$.

Saint-Venant then gives a table of the limiting values of n and of $\eta_{zx}=\frac{1}{4}\frac{d}{f}=\sqrt{\frac{n}{18-2n}\cdot\frac{E_z}{E_x}}$, for various values of E_z/E_x from 1 to 80 and also for ∞.

The values of η_{zx} are now found to be possible, provided a suitable value of n be chosen. What shall this be?

[313.] The empirical formula for n

$$1/n=1+\frac{1}{\gamma}\,(E_z/E_x-1)\;\ldots\ldots\ldots\ldots\ldots\ldots\ldots\ldots(\text{v}),$$

is suggested on p. 104. On p. 105 Saint-Venant tabulates the values of n and η_{zx} for the parameter E_z/E_x (= 1 to 80) when γ has the numerical values 9 and 22·22. These values for γ are chosen because, for $E_z/E_x = 80$, they give respectively η_{zx} = about 2/3 and 1. The Table also contains the corresponding values of E_z/e ($=E/\mu$ with transverse isotropy). These values vary on the first supposition ($\gamma = 9$) from 2·5 to 78·2, and on the second ($\gamma = 22{\cdot}22$) from 2·5 to 52·67. The ratio E/μ can thus be very great, but for E_z/E_x very great, this does not seem at all improbable, at least we have at present no experiments to contradict it. As for the value of γ we need not confine it to 9 or 22·22, but in general we may take it from 7 or 8 to 30 (p. 108). *Nous pensons qu'on ne courra guère risque de se tromper en faisant* $\gamma = 16$ (p. 108).

As Saint-Venant observes there is a great need of new experiments to determine E_z and E_x (by flexure), μ (by torsion) and η ($=-s_x/s_z$, by delicate measurements of the transverse dimensions of bars under traction).

[314.] In default of experiment we may finally adopt as formulae most probably sufficient for elastic problems concerning amorphic aeolotropic solids, such as stone, wood, and the metals employed in the construction of bridges and machines:

$$\left.\begin{array}{ll}\widehat{xx}=\frac{3ef}{d}\,s_x+fs_y+es_z, & \widehat{yz}=d\sigma_{yz}\\ \widehat{yy}=fs_x+\frac{3fd}{e}\,s_y+ds_z, & \widehat{zx}=e\sigma_{zx}\\ \widehat{zz}=es_x+ds_y+\frac{3de}{nf}\,s_z, & \widehat{xy}=f\sigma_{xy}\end{array}\right\}\;\ldots\ldots\ldots\ldots(\text{vi}),$$

where at each point yz, zx, xy are three rectangular planes of elastic symmetry and z is the direction of greatest elastic resistance, generally 'longitudinal,' that is, in the case of wood in the direction of the fibre, or in a metal bar in the direction of the prismatic axis. In a metal plate it will be perpendicular generally to the plane of the plate.

The quantity n is to be determined by Equation (v), where γ may be taken $=16$, when we have no further experimental data to suggest a better value.

Since $E_z = \frac{de}{f}\frac{6-n}{2n}$, it is obvious that three torsional experiments and one tractional experiment will give d, e, f and n, or all the constants of the stress-strain relations (vi).

Indeed we may write the value of $\widehat{zz}$

$$\widehat{zz} = es_x + ds_y + \left(E_z + \frac{1}{2}\frac{de}{f}\right)s_z,$$

and so get rid of n altogether.

For the case of transverse isotropy, if $E_z = E$, $d = e = \mu$, $f = \mu'$, we have:

$$\left.\begin{array}{ll} \widehat{xx} = \mu'(3s_x + s_y) + \mu s_z & \widehat{yz} = \mu\sigma_{yz} \\ \widehat{yy} = \mu'(s_x + 3s_y) + \mu s_z & \widehat{zx} = \mu\sigma_{zx} \\ \widehat{zz} = \mu(s_x + s_y) + \left(E + \dfrac{\mu^2}{2\mu'}\right)s_z & \widehat{xy} = \mu'\sigma_{xy} \end{array}\right\} \ldots\ldots\ldots\ldots (\text{vii}).$$

Here μ and E are easy to determine experimentally, but μ' far more difficult.

Saint-Venant gives the following empirical formula for μ' which he considers very probably exact enough in practice:

$$\frac{\mu'}{\mu} = 1 - \beta\left(\frac{2}{5} - \frac{\mu}{E}\right) \ldots\ldots\ldots\ldots\ldots\ldots\ldots\ldots (\text{viii}).$$

When γ of (v) is taken $= 9$, then $\beta = \frac{5}{3}$, or $\frac{\mu'}{\mu} = \frac{1}{3} + \frac{5}{3}\frac{\mu}{E}$,

" " " " $= 22{\cdot}22$, then $\beta = 2$, or $\frac{\mu'}{\mu} = \frac{1}{5} + 2\frac{\mu}{E}$.

For these values of β, the corresponding values of μ'/μ and μ/E differ by only 1/16, from those obtained from equation (v).

We have reproduced these results because they supply, although to some extent empirically, the most probable formulae yet suggested for technical materials. Such formulae have been much needed, and Saint-Venant, as usual, has been the first to recognise the wants of practice.

[315.] A note of Saint-Venant to § 22 (see pp. 142—5) deals briefly with the history of the flexure and torsion of prisms. It contributes nothing to the section on the same subject in the *Historique Abrégé*. We pass on to the longer note attached to § 28 which occupies pp. 174—90.

[316.] This note is concerned with the applicability of Saint-Venant's torsion and flexure solutions to such cases as occur in practice. The first four sections (pp. 174—7) reproduce arguments already given in the memoir on *Torsion* or the *Leçons de Navier* for the approximate elastic equivalence of statically equipollent loads: see our Arts. 8, 9 and 170. The remaining sections (§§ 5—17) seek arguments in favour of the legitimacy of the assumptions

$$\widehat{xx} = \widehat{yy} = \widehat{xy} = 0 \ldots\ldots\ldots\ldots\ldots\ldots\ldots\ldots (a),$$

taken by Saint-Venant as the basis of his solutions. In other words, is it legitimate to assume that for all practical loadings there is little or no mutual action *parallel* to the prismatic cross-section between adjacent longitudinal fibres?

After referring to the labours of Poisson and Cauchy on the subject of rods (see our Arts. 466* and 618*) as involving arbitrary assumptions only true for rods of length great as compared with the linear dimensions of the cross-section, Saint-Venant enquires whether the investigations of Kirchhoff give any better validity to the assumptions (*a*). He points out that Kirchhoff proves only the possibility, not the necessity of these questionable relations (p. 181): see my footnote, p. 266.

[317]. Saint-Venant next turns to Boussinesq's memoirs of 1871 and 1879: see later our discussion of that author's researches. Saint-Venant applies the method of those memoirs to the simple case of a bar of homogeneous material with three planes of elastic symmetry.

Instead of setting out from the assumptions (*a*) our author supposes the following conditions to hold, z being the direction of the prismatic axis:

$$\frac{d^2 s_x}{dz^2} = \frac{d^2 s_y}{dz^2} = \frac{d^2 s_z}{dz^2} = \frac{d^2 \sigma_{xy}}{dz^2} = \frac{d\sigma_{zx}}{dz} = \frac{d\sigma_{zy}}{dz} = 0 \ldots\ldots\ldots (b).$$

These are described as *fort approchées, quand elles ne sont pas rigoureuses.*

From the conditions (*b*) the conditions (*a*) are deduced by the principle of elastic work. The proof holds only for *rods*, i.e. prisms the length of which is great as compared with the linear dimensions of the cross-section; the cross-section may, however, be supposed to vary slightly, and the terminal load as well as the distribution of body force are perfectly general, provided only the body force on any element of length of the rod does not exceed the surface stresses or the loads on the terminal cross-sections of the element.

[318.] We may ask: whether the conditions (*b*) do not assume as much as conditions (*a*)? We reproduce the arguments by which Saint-Venant reaches (*b*). It does not seem to me that the condition $d^2s_z/dz^2=0$ would be true when the flexure was due to *buckling*, which in the case of a long rod does not seem excluded by the load distributions referred to: see our Art. 911*.

Prenons pour axe des z, en chaque endroit, la ligne des centres de gravité des sections transversales, et les axes des x, y, rectangulaires entre eux et à cette ligne sur une des sections. Dans une quelconque des portions dont nous parlons, que nous appelons longues parce qu'elles sont supposées l'être beaucoup par rapport aux dimensions transversales, il est facile de reconnaître que les composantes de tension et les dilatations ou glissements s, σ, varient d'une manière incomparablement moins rapide dans le sens longitudinal z que dans les sens x et y; de sorte que, si nous exceptons de petites portions de tige avoisinant les extrémités, où se trouvent les points d'application des forces locales ou discontinues, les dérivées de ces déformations s, σ, par rapport à z seront, de nécessité, considérablement moindres que ce que sont ou peuvent être leurs dérivées par rapport à x et à y. En effet, pour σ_{zx}, par exemple, $d\sigma_{zx}/dz$ sera de l'ordre de grandeur du quotient, par la longueur de la tige ou de la longue portion de la tige considérée, de cette déformation σ_{zx}, ou de la différence des valeurs qu'elle a aux extrémités; tandis que $d\sigma_{zx}/dx$ pourra être de l'ordre de grandeur du quotient de σ_{zx} par la demi-épaisseur, qui n'est, disons-nous, qu'une fort petite fraction de la longueur. Autrement dit, si pour fixer les idées nous divisons la tige, par la pensée, en tronçons dont la longueur soit de l'ordre de grandeur de la dimension transversale moyenne, les s, σ auront des valeurs extrêmement peu différentes en deux points homologues des bases de chaque tronçon, tandis qu'ils pourront avoir, du centre au périmètre des sections, des différences de valeur aussi considérables que d'une extrémité à l'autre de la tige. Nous pouvons donc comme approximation, déterminer la loi de variation des déformations s, σ, transversalement, ou en fonction de x et y, *comme si leurs dérivées par rapport à z étaient nulles*. Cette hypothèse, ou ce point de départ, n'est que comme une traduction analytique de l'énoncé de la question même qui nous occupe, et qui est de déterminer ce qui se passe dans une tige allongée et très mince sollicitée de la manière continue que nous venons de supposer (pp. 184—5).

This reasoning does not appear to me wholly satisfactory, and can at best only apply to *rods* and not the prisms of Saint-Venant's problems. It may, however, still be the method

la meilleure et la plus complète qui en ait été théoriquement donnée (p. 190).

Perhaps on the whole the appeal to experiment referred to in our Arts. 8—10 is more satisfactory.

[319.] In a note pp. 195—7 Saint-Venant proves for the case of flexure the results

$$\int \widehat{zx}\, d\omega = \frac{d}{dz}\int \widehat{zz}\, x d\omega\, ; \quad \int \widehat{zy}\, d\omega = \frac{d}{dz}\int \widehat{zz}\, y d\omega,$$

where z is an axis in direction of the prismatic axis, and x, y are any rectangular axes in the cross-section of which $d\omega$ is an element of area. These formulae express analytically:

ce théorème connu et très utile, que l'*effort tranchant*, pour une section quelconque, ou la force tangentielle totale dans une direction transversale *aussi quelconque*, est égale à la dérivée, par rapport à la coordonnée longitudinale, du *moment de flexion* autour d'une droite tracée sur la section perpendiculairement à cette direction (p. 197).

See pp. 389—9 etc. of the *Leçons de Navier*.

[320.] The following *Note*, pp. 210—20, reproduces only portions of the great or the subsidiary memoirs on *Torsion*: see our Arts. 1, 285 and 291; and the *Note*, pp. 240—2, some results from Chapter XI. of the *Torsion*: see our Art. 49.

The *Note finale du* § 37 (pp. 252—82) corrects Clebsch's erroneous assumption of a stress-limit by the proper stretch-conditions. Its contents are extracted from the memoir on *Torsion* and the *Leçons de Navier*: see our Arts. 5, (*b*)—(*f*), and 180.

[321.] We may refer to one or two points in this last *Note*:

(*a*) Saint-Venant takes two simple cases for an isotropic material and compares the stress and stretch-conditions for safe loading. First take the case when only the stresses $\widehat{xx}$, $\widehat{xz}$, $\widehat{xy}$ have values differing from zero, we easily find from the equation of our Art. 53, Case (i), that we must have

$$T_0 = \text{or} > (1-\eta)\,\widehat{xx}/2 + (1+\eta)\sqrt{\widehat{xx}^2/4 + \widehat{xy}^2 + \widehat{xz}^2},$$

while Clebsch obtains from the stress condition

$$T_0 = \text{or} > \widehat{xx}/2 + \sqrt{\widehat{xx}^2/4 + \widehat{xy}^2 + \widehat{xz}^2}.$$

In the second case suppose the traction $\widehat{xx}$ zero, then we have:

from the stretch condition,

$$\sqrt{\widehat{xy}^2 + \widehat{xz}^2} = \text{or} < T_0/(1+\eta),$$

from the stress condition,

$$\sqrt{\widehat{xy}^2 + \widehat{xz}^2} = \text{or} < T_0.$$

When the shears are zero the conditions agree. As a rule *safety is on the side of the stretch-condition.*

(*b*) Some remarks confirmatory of Poncelet's theory of rupture (better *elastic failure*) under compression by transverse stretch are given on p. 270 and may be cited[1]. The theory leads, as we have seen, in isotropic material to the relation $T_0/T_0' = 1/\eta$: see our Arts. 164, and 175.

1°. Les petits prismes de pierre dure, lors de leur écrasement, se séparent d'abord en aiguilles verticales, ce qui prouve bien une extension dans le sens transversal.

2°. Lors de l'écrasement des bois par compression dans le sens de leurs fibres, celles-ci se séparent d'abord, et ensuite ploient sans résistance.

3°. Les petits cylindres de fonte douce ou malléables, écrasés, se gercent sur les bords de manière à former une rosette, ce qui prouve qu'il y a eu, tout autour, rupture par dilatation transversale vers la circonférence.

4°. Dans beaucoup d'expériences de rupture de pièces de fonte par flexion, il s'est détaché latéralement une sorte de coin du côté devenu concave ou comprimé.

5°. La puissante machine de M. Blanchard, de Boston, à courber les pièces de bois, contenues de manière à ne pouvoir se dilater du côté convexe ni se boursoufler latéralement du côté concave, comprime violemment ce dernier côté sans le désorganiser aucunement.

6°. Le rapport des coefficients T_1' et T_1 de rupture immédiate par écrasement et par traction, ou des forces capables de produire, pour une base $=1$, ces deux sortes d'effet, a été trouvé le plus souvent, pour la fonte, entre 4 : 1 et 6 : 1 ; et il devait, en effet, excéder $1/\eta$ qui est 4 pour les corps isotropes. Car lorsqu'on opère la compression d'un prisme court, entre deux plans durs où ses bases s'appliquent, celles-ci sont empêchées de se dilater, en sorte que le renflement latéral n'acquiert toute sa grandeur que vers le milieu de la hauteur du prisme.

Saint-Venant remarks that the limits T_0, T_0' must be based directly on experiment; but experiment only gives such limits as

[1] Professor A. B. W. Kennedy has kindly made some experiments for me on lateral stretch in which *three* short cast-iron prisms placed end to end were subjected to contractive load. The load terminals of the outer prisms were found to have expanded somewhat, but not to the same extent as their other terminal sections or those of the mid-prism. Rupture took place by portions of the end prisms shearing off. The mid-prism was then cut open longitudinally and acid applied to the face, the openings thus brought to sight were more or less longitudinal, but not very definite. Indeed the condition marked rather a plastic than a ruptural change.

the T_1 and T_1' of immediate rupture. A constant ratio between T_1 and T_0 is usually assumed:

rapport qu'on prend généralement d'un dixième en France, d'après l'exemple des colonnes légères d'une ancienne église d'Angers, mais que des ingénieurs anglais portent à un sixième (p. 271).

(c) We may note that Saint-Venant on pp. 274—5 in repeating case 3° of Art. 122 of the *Torsion*: see our Art. 53, Case (iii), now replaces the $s_y/\bar{s}_y$ and $s_z/\bar{s}_z$ of the notation of that article by *their mean*, so that he appears to have been dissatisfied with the value adopted in the memoir. He does not, however, work out the value of η_2 of our Art. 53, Case (ii) ($=\eta_1$ of his notation).

(d) A very good example of Saint-Venant's fail-point method is given on pp. 279—82 (§ 17). It brings out well the influence which want of isotropy and slide have on the condition for safety.

Let us take the case of a beam of length l, of cross-section ω, and of transverse elastic isotropy denoted by E, μ and η. Suppose it built-in at one end and loaded with P at the other, or of length $2l$ with a load $2P$ in the centre. Then if κ be the swing-radius of the section about the neutral axis and h the distance from that axis of the farthest 'fibre', we see that the fail-point will be at the built-in section which remains plane. Here the maximum stretch and the uniform slide are given by:

$$s = Plh/(E\omega\kappa^2), \quad \sigma = P/(\mu\omega).$$

Whence the condition of our Art. 53, (i), becomes with slightly modified notation:

$$T_0 = \text{or} > \frac{1-\eta}{2}\frac{Plh}{\omega\kappa^2} + \sqrt{\left(\frac{1+\eta}{2}\right)^2\left(\frac{Plh}{\omega\kappa^2}\right)^2 + \left(\frac{P}{\omega}\right)^2\left(\frac{T_0}{S_0}\right)^2},$$

$$= \text{or} > \frac{Plh}{\omega\kappa^2}\left\{\frac{1-\eta}{2} + \frac{1+\eta}{2}\sqrt{1+\left(\frac{E}{2(1+\eta)\mu}\right)^2\left(\frac{2\kappa^2}{lh}\right)^2}\right\}\ldots\text{(i)},$$

since $S_0/\mu = 2T_0/E$ by our Art. 5, (*d*).

In the case of the rectangular cross section $b\times c$, with c parallel to the load-plane, we have $\kappa^2 = c^2/12$, $\omega = bc$, $h = c/2$ and the condition becomes:

$$T_0 = \text{or} > \frac{6Pl}{bc^2}\left\{\frac{1-\eta}{2} + \frac{1+\eta}{2}\sqrt{1+\left(\frac{E}{2(1+\eta)\mu}\right)^2\left(\frac{c}{3l}\right)^2}\right\}\ldots\text{(ii)},$$

or,

$$T_0 = \text{or} > \frac{6Pl}{bc^2}\left\{1 + \frac{1}{1+\eta}\left(\frac{E}{12\mu}\right)^2\left(\frac{c}{l}\right)^2\right\}\ldots\ldots\ldots\ldots\text{(iii)},$$

if the second term under the radical is, as usual, small.

Saint-Venant now introduces the following suggestive table determined by the method of our Arts. 312–4, z being the direction of the prismatic axis:

For	$E_z/E_x=1$	1·5	2	5	10	15	20	40	80
	$\eta=\cdot 25$	·30	·34	·48	·60	·65	·70	·80	·85
	$E/\mu=2\cdot 5$	3	3·8	6·6	11	15	18	36	66
Whence	$\left\{\frac{E}{2(1+\eta)\mu}\right\}^2=1$	1·331	2·011	4·973	11·816	20·657	28·026	100	318·27
	$\frac{1}{1+\eta}\left(\frac{E}{12\mu}\right)^2=\cdot 035$	·048	·075	·204	·525	·947	1·323	5	16·35

Now the value given by the old theory was

$$T_0 = \text{or} > \frac{6Pl}{bc^2}.$$

Whence we see that for certain kinds of wood when $E_z/E_x = 80$, for short lengths only double of the diameter, the value of P obtained from the old theory may be *double* what is given by the true theory. Indeed these numbers are most suggestive and valuable for the problem of flexure.

[322.] To Clebsch's §§ 39—46 dealing with thick plates, Saint-Venant contributes two long notes. The first occupies pp. 337—367 and treats of various *rigorous* solutions for the bending of plates by an analysis which for simplicity compares favorably with that of Clebsch. Indeed we have here the most complete account yet given of the bending of *thick circular* plates, and as usual Saint-Venant keeps in view practical cases. The results are all given in terms of the 5-constant formulae (see our Art. 282, (2)), or for a material with transverse isotropy on the multi-constant hypothesis. Many of the results are new and the method seems to me novel; some of the formulae are apparently due to Saint-Venant's old pupil M. Boussinesq, who investigated the matter at his request. The following problems are investigated: (i) case of simple cylindrical flexure; (ii) case of combined cylindrical flexure; (iii) cases of shearing load on the lateral sides of a plate; (iv) general case of circular plate with a great variety of special cases of contour and load conditions.

[323.] *Case of simple cylindrical flexure.* Let 2ϵ be the thickness of the plate; let the axis of z be perpendicular to the initial plane of the plate; and let those of x, y, lie in the plane of the plate. Suppose the plate to be infinite in the direction of y, but of any length in the direction of x. Then consider the following shifts, where ρ is a constant:

$$u = -\frac{xz}{\rho}, \quad v = 0, \quad w = \frac{1}{2\rho}\left(x^2 + \frac{d'}{c}z^2\right) \text{..............(i).}$$

We find at once:

$$\left.\begin{array}{l} s_x = -z/\rho, \quad s_y = 0, \quad s_z = \dfrac{d'}{c}\dfrac{z}{\rho}, \\ \sigma_{yz} = \sigma_{zx} = \sigma_{xy} = 0 \end{array}\right\} \dots\dots\dots\dots\dots\dots\dots\dots (ii).$$

Substitute in the formulae of Art. 117, (*a*) and (*b*), and we have:

$$\left.\begin{array}{l} \widehat{yz} = \widehat{zx} = \widehat{xy} = 0; \quad \widehat{zz} = 0; \\ \widehat{xx} = \left\{\dfrac{d'^2}{c} - (2f + f')\right\}\dfrac{z}{\rho}; \quad \widehat{yy} = \left\{\dfrac{d'^2}{c} - f'\right\}\dfrac{z}{\rho} \end{array}\right\} \dots\dots\dots\dots (iii).$$

Here the quantity $2f + f' - d'^2/c$ corresponds for the case of plates to the stretch-modulus in the simple flexure of a bar. We shall denote it by H, where in the case of isotropy, $H = \dfrac{4\mu(\mu + \lambda)}{2\mu + \lambda}$.

We easily see that (iii) satisfy the body-stress equations.

The load reduces to

$$\widehat{xx} = -\frac{Hz}{\rho}$$

over the sides perpendicular to x, and we can see that this gives a couple round the axis of y for each element $2\epsilon\delta y$ of the side $= M_y \delta y$, where

$$M_y = \int_{-\epsilon}^{+\epsilon} \widehat{xx} \,.\, z\,dz = -2H\epsilon^3/(3\rho).$$

We can cut away a portion of the plate by planes perpendicular to the axis of y if we impose a load at each point of the new sides given by

$$\widehat{yy} = -(H - 2f)\, z/\rho.$$

Obviously $1/\rho$ must be very small, and the plate then takes a cylindrical curvature of radius ρ.

[324.] *Case of two combined cylindrical flexures.* In § 3 Saint-Venant first combines two solutions such as that of our Art. 323, the value of ρ being the same for both. He transfers to cylindrical coordinates r, ϕ, and thus obtains with the notation of our p. 79 the results:

$$\left.\begin{array}{l} u = -rz/\rho, \quad v = 0, \quad w = \left(r^2/2 + \dfrac{d'}{c} z^2\right) \Big/ \rho \\ \widehat{zz} = 0, \quad \widehat{rr} = \widehat{\phi\phi} = -2(H - f)\, z/\rho \end{array}\right\} \dots\dots\dots\dots (iv).$$

This is the case of *spherical* curvature. The proper distribution of side load must be obtained by compounding $\widehat{rr}$ and $\widehat{\phi\phi}$, the shears being all zero. The corresponding total couples are

$$M_r = M_\phi = -\frac{4\epsilon^3}{3}\frac{H - f}{\rho} \dots\dots\dots\dots\dots\dots\dots (v).$$

Saint-Venant remarks:

Ils ont un intérêt pratique bien que l'application, au contour, de forces normales distribuées comme l'exigent les expressions ci-dessus $\widehat{rr}$, $\widehat{\phi\phi}$ soit irréalisable; car si à leur place, il y a [see our Arts. 8 and 170] tout auprès des bords d'une plaque mince, d'autres forces appliquées par exemple sur les faces supérieure et inférieure de manière à n'avoir pas de résultante et à produire des couples dont les moments fléchissants aient par unité de longueur la valeur (v), la plaque soit rectangle, soit circulaire, éprouvera *très approximativement* la déformation sphérique indiquée, *partout* sauf de très petites zones auprès des bords, par les raisons que nous avons données précédemment en traitant des tiges (p. 343).

[325.] The second case of combined flexure given by Saint-Venant is obtained by taking for u and v two expressions like that given for simple cylindrical flexure, with ρ different; we have at once:

$$u = -xz/\rho, \quad v = -yz/\rho', \quad w = x^2/(2\rho) + y^2/(2\rho') + (1/\rho + 1/\rho')\frac{d'}{2c}z^2,$$

$$\widehat{zz} = \widehat{yz} = \widehat{zx} = \widehat{xy} = 0, \quad \widehat{xx} = -\left(\frac{H}{\rho} + \frac{H-2f}{\rho'}\right)z, \quad \widehat{yy} = -\left(\frac{H-2f}{\rho} + \frac{H}{\rho'}\right)z.$$

Here the curvature is elliptic or hyperbolic according as ρ and ρ' are of the same or different signs. If $\rho = -\rho'$:

le feuillet moyen devient une de ces surfaces à courbures principales égales et opposées, appelées *anticlastiques* par MM. Thomson et Tait dans leur grand *A Treatise of Natural Philosophy*, de 1867, dont un seul exemplaire existe en France, et dont il n'a encore été réédité que le premier volume (p. 344).

As is well-known the distinguished scientists gave up in their second edition the idea of proceeding further. How Saint-Venant formed his conclusion as to the existence of a *seul exemplaire*, we cannot say, as with few exceptions French scientists refrain when citing from giving exact references to the sources of their information.

[326.] *Plates subjected laterally to shearing load.* Saint-Venant first takes the case of a rectangular plate *infinitely* long in the direction of y but bounded in the direction of x by the planes $x = \pm a$.

Let $P\delta y$ be the total shearing-load parallel to z, on the strip $2\epsilon dy$, then we have for a section of the plate by a plane at distance x from the origin:

$$\int_{-\epsilon}^{+\epsilon} \widehat{zx}\, dz = P; \quad \int_{-\epsilon}^{+\epsilon} \widehat{xx}\, z dz = P\,(a - x).$$

Boussinesq had found at Saint-Venant's request the following suitable values for the shifts:

$$\left.\begin{aligned} u &= \frac{3P}{2H\epsilon^3}\left[-z\left(ax - \frac{x^2}{2}\right) + \frac{d'}{c}\frac{z^3}{6}\right] + \frac{3P}{4e}\left(\frac{z}{\epsilon} - \frac{z^3}{3\epsilon^3}\right), \\ v &= 0, \quad u = \frac{3P}{2H\epsilon^3}\left[\frac{ax^2}{2} - \frac{x^3}{6} + \frac{d'}{c}(a - x)\frac{z^2}{2}\right]. \end{aligned}\right\} \text{......(vi)}.$$

Hence we find for the stresses:

$$\left.\begin{aligned} &\widehat{zz} = \widehat{yz} = \widehat{xy} = 0, \\ &\widehat{xx} = -\frac{3Pz}{2\epsilon^3}(a-x), \quad \widehat{xz} = \frac{3P}{4\epsilon}\left(1-\frac{z^2}{\epsilon^2}\right), \\ &\widehat{yy} = -\frac{H-2f}{H}\,\frac{3Pz}{2\epsilon^3}(a-x) \end{aligned}\right\} \text{..........(vii).}$$

The deflection of the central plane is given by the cubical parabola

$$w_0 = \frac{Pa^3}{2H\epsilon^3}\left(\frac{3}{2}\frac{x^2}{a^2} - \frac{1}{2}\frac{x^3}{a^3}\right) \text{....................(viii).}$$

This agrees with the case of a rod of length $2a$ and depth 2ϵ, terminally supported and loaded with $2P$ at the centre if the plate-modulus H be replaced by the stretch-modulus E.

[327.] We can cut out a definite portion of the plate by planes perpendicular to y, if we impose the tractive loads given by $\widehat{yy}$ of equations (vii).

Suppose we try to combine two sets of solutions such as (vi) of the previous Article, giving the plate now a flexure parallel to y. Then we find, if Q corresponding to P, and b to a, from (viii):

$$w_0 = \frac{3}{2H\epsilon^3}\left(\frac{Pax^2 + Qby^2}{2} - \frac{Px^3 + Qy^3}{6}\right).$$

Hence although we combine this with a solution of the form given in Art. 325, we can make only the square not the cubic terms in x and y vanish. In other words for $x = \pm a$, together with $y =$ any value from b to $-b$, and for $y = \pm b$, together with $x =$ any value from a to $-a$, we cannot make $w_0 = 0$. Thus the contour of the mid-plane of the rectangular plate cannot be treated as *fixed*.

Le problème de la flexion de la plaque rectangulaire posée de niveau tout autour ne peut probablement recevoir que des solutions approximatives.... (p. 346).

[328.] *Problem of the thick circular plate.* This can be solved accurately for flexure whatever the thickness, if the plate be *symmetrically* loaded in all directions round its axis of figure by forces applied to its cylindrical boundary. Just as in the case of torsion or flexure, these forces will be supposed distributed in a definite manner, but the resultant shearing force and couple about the tangent to the contour of the mid plane will be arbitrary. In practical applications we must appeal to the principle of the elastic equivalence of statically equipollent load-systems: see our Art. 8. We shall suppose that there is no tendency to extension in the plate and that it is bounded by two coaxial cylinders of radii r_1 and r_0 ($r_1 > r_0$).

We shall find that the magnitude of the central shift can be determined for any load whatever, not necessarily symmetrical.

[329.] *The general solution.* Let 2ϵ be the thickness; P the shearing load parallel to the axis per unit of length of contour of the plate; $Q = 2\pi a P$ the total shearing load on the whole lateral area $2\pi a \times 2\epsilon$ of the plate; M_r the moment of the couple, per unit of length, on a vertical strip of the cylindrical surface of radius r about the tangent to the contour of the mid-plane; M_{r_1} will then denote the corresponding load couple on the outer bounding cylinder. We shall suppose the mid-circle of the inner cylindrical boundary fixed.

The strains are given in the footnote to our p. 79, except that on account of the symmetry we put $v = 0$, and the variation with regard to ϕ zero for all quantities. The stresses then become on the hypothesis of elastic isotropy in the plane of the plate [see Art. 117 (*b*)]:

$$\left.\begin{aligned} \widehat{rr} &= (2f+f')\, u_r + f'u/r + d'w_z, & \widehat{rz} &= e\,(u_z + w_r) \\ \widehat{\phi\phi} &= f'u_r + (2f+f')\, u/r + d'w_z, & \widehat{r\phi} &= \widehat{z\phi} = 0 \\ \widehat{zz} &= d'\,(u_r + u/r) + cw_z & & \end{aligned}\right\} \quad \text{.........(i).}$$

Further we have $M_r = \int_{-\epsilon}^{+\epsilon} \widehat{rr}\, z dz.$

The body stress-equations reduce to:

$$\frac{d\widehat{rr}}{dr} + \frac{d\widehat{rz}}{dz} + \frac{\widehat{rr} - \widehat{\phi\phi}}{r} = 0; \quad \frac{d\widehat{rz}}{dr} + \frac{d\widehat{zz}}{dz} + \frac{\widehat{zr}}{r} = 0 \quad \text{........ (ii).}$$

The surface or load conditions are:

$$\left.\begin{aligned} &\text{for } z = \pm\epsilon, \quad \widehat{zz} = \widehat{rz} = 0 \text{ for all values of } r, \\ &\text{for } r = r_1, \quad \int_{-\epsilon}^{+\epsilon} \widehat{rr}\, dz = 0, \quad \int_{-\epsilon}^{+\epsilon} \widehat{rz}\, dz = P, \\ &\qquad M_r = M_{r_1} \end{aligned}\right\} \quad \text{..................... (iii).}$$

[330.] Saint-Venant's mode of solution is the following. He assumes $\widehat{rz}$ to be of the form $\frac{3P}{4\epsilon^3} \cdot \frac{f(r)}{f(r_1)} (\epsilon^2 - z^2)$, and also that, $\widehat{zz} = 0$ throughout the plate. He thus satisfies the load conditions.

These assumptions of the *semi-inverse* method were undoubtedly suggested by equations (vii) of our Art. 326.

The second body-stress equation at once gives us $f(r) = \frac{\text{constant}}{r}$;

so that
$$\widehat{rz} = \frac{3P}{4\epsilon^3} \frac{r_1}{r} (\epsilon^2 - z^2) \quad \text{........................ (iv).}$$

Straight-forward substitution, remembering $\widehat{zz} = 0$, or $w_z = -\frac{d'}{c}\frac{1}{r}\frac{d(ru)}{dr}$, leads to the following form of the first body-stress equation (ii):

$$\frac{d}{dr}\left(\frac{1}{r}\frac{d(ru)}{dr}\right) = \frac{4z}{Ir},$$

where
$$I = \frac{8H\epsilon^3}{3Pr_1}.$$

Integrating we find, if A be an arbitrary constant:

$$\frac{1}{r}\frac{d(ru)}{dr} = z\left(\frac{4}{I}\log(r/r_1) - \frac{2}{A}\right) = -\frac{c}{d'}w_z \text{ .... (v).}$$

Integrating again we have:

$$\left.\begin{aligned} u &= \frac{z}{I}\{2r\log(r/r_1) - r\} + \frac{Bz}{r} - \frac{rz}{A} + \frac{\phi(z)}{r} \\ w &= -\frac{d'}{c}\left\{\frac{2z^2}{I}\log(r/r_1) - \frac{z^2}{A}\right\} + \chi(r) \end{aligned}\right\} \text{ ....(vi).}$$

Here B is another arbitrary constant and χ, ϕ arbitrary functions of r and z respectively.

Now we have $\widehat{rz} = e(u_z + w_r) = \frac{3P}{4\epsilon^3}\frac{r_1}{r}(\epsilon^2 - z^2)$; substituting for u and w from (vi) we find the following relation between χ and ϕ:

$$\frac{2r}{I}\log\frac{r}{r_1} - \frac{r}{I} + \frac{B}{r} - \frac{r}{A} + \frac{d\chi}{dr} = \frac{1}{r}\left\{\frac{2}{I}\frac{d'}{c}z^2 + \frac{2}{I}\frac{H}{e}(\epsilon^2 - z^2) - \frac{d\phi}{dz}\right\}.$$

Saint-Venant remarks that we can satisfy this relation in several ways (p. 350), but the proper method seems to me to equate either side multiplied by r to the same constant. He *takes this constant to be zero.* If this constant be retained, however, it only alters the value of the constant B in the expressions for the shifts we are about to give, and so may be neglected. We ought to add a constant C' to the value of $\chi(r)$; but this leads to a term in $u = C'/r$, or in $\widehat{rr} = -2fC'/r^2$, which, not containing an odd power of z, would prevent us from fulfilling the condition

$$\int_{-\epsilon}^{+\epsilon} \widehat{rr}\, dz = 0 \text{ for } r = r_1.$$

Substituting the values obtained by integration for ϕ and χ in (vi), we have:

$$\left.\begin{aligned} u &= -\frac{rz}{A} + \frac{1}{I}\left\{2rz\log\frac{r}{r_1} - rz + \frac{2d'}{c}\frac{z^3}{3r} + \frac{2}{r}\frac{H}{e}\left(\epsilon^2 z - \frac{z^3}{3}\right)\right\} + \frac{Bz}{r} \\ w &= \frac{1}{A}\left(\frac{r^2}{2} + \frac{d'z^2}{c}\right) + \frac{1}{I}\left\{r^2 - \left(r^2 + \frac{2d'z^2}{c}\right)\log\frac{r}{r_1}\right\} - C - B\log\frac{r}{r_1} \end{aligned}\right\} \text{...(vii).}$$

The values of $\widehat{rr}$ and M_r may then be easily deduced. Saint-Venant gives expressions for them on pp. 351—2. By putting $r = r_1$, we obtain:

$$M_{r_1} = -\frac{4\epsilon^3}{3}\frac{H - f}{A} + \frac{4\epsilon^3}{3}\frac{f - H\gamma^2}{I} - \frac{4\epsilon^3}{3}f\frac{B}{r_1^2} \text{ .... (viii),}$$

where γ^2 is given by:

$$\frac{2}{5}\frac{f}{H}\left(\frac{d'}{c} + 4\frac{H}{e}\right)\frac{\epsilon^2}{r_1^2} = \gamma^2 \text{ .... (ix),}$$

and may be neglected when ϵ/r_1 is small.

If P or $1/I = 0$ and we put $B = 0$, this value of M_{r_1}, agrees with that of M_r in equation (v) of our Art. 324. We shall then write for simplification

$$\frac{1}{\rho} = -\frac{3}{4\epsilon^3}\frac{M_{r_1}}{H-f},$$

and we find
$$\frac{1}{A} = \frac{1}{\rho} + \frac{1}{I}\frac{f-H\gamma^2}{H-f} - \frac{f}{H-f}\frac{B}{r_1^2} \quad\text{..................(x).}$$

Substituting this value of $\frac{1}{A}$ in equations (vii) we note the following final results given on p. 354 and attributed by Saint-Venant to Boussinesq ('*que M. Boussinesq a cherchées et trouvées à ma prière*'):

$$\left.\begin{aligned}
u &= -\frac{rz}{\rho} + \frac{1}{I}\left\{rz\left(2\log\frac{r}{r_1} - H\frac{1-\gamma^2}{H-f}\right) + \frac{2}{r}\frac{d'}{c}\frac{z^3}{3} + \frac{2}{r}\frac{H}{e}\left(\epsilon^2 z - \frac{z^3}{3}\right)\right\} \\
&\qquad + \frac{Bz}{r} + \frac{f}{H-f}\frac{Brz}{r_1^2}, \\
w &= \frac{1}{\rho}\left(\frac{r^2}{2} + \frac{d'}{c}z^2\right) + \frac{1}{I}\left\{r^2 + \left(\frac{r^2}{2} + \frac{d'}{c}z^2\right)\left(\frac{f-H\gamma^2}{H-f} - 2\log\frac{r}{r_1}\right)\right\} \\
&\qquad - C - B\log\frac{r}{r_1} - \frac{f}{H-f}\cdot\frac{B}{r_1^2}\left(\frac{r^2}{2} + \frac{d'}{c}z^2\right), \\
M_r &= \frac{4\epsilon^3}{3}\left\{-\frac{H-f}{\rho} + \frac{1}{I}\left[2(H-f)\log\frac{r}{r_1} - H\gamma^2\left(\frac{r_1^2}{r^2} - 1\right)\right]\right. \\
&\qquad \left. - fB\left(\frac{1}{r^2} - \frac{1}{r_1^2}\right)\right\}; \\
\text{where}\quad \frac{1}{I} &= \frac{3Pr_1}{8H\epsilon^3},\quad \frac{1}{\rho} = -\frac{3}{4\epsilon^3}\frac{M_{r_1}}{H-f},\quad \gamma^2 = \frac{2}{5}\frac{f}{H}\left(\frac{d'}{c} + 4\frac{H}{e}\right)\frac{\epsilon^2}{r_1^2}
\end{aligned}\right\}\quad\text{...(xi).}$$

For the vertical shift and shift-fluxion of the mid-plane we have when $B = C = 0$:

$$\left.\begin{aligned}
w_0 &= \frac{r^2}{2\rho} + \frac{1}{I}\left(r^2 + \frac{r^2}{2}\frac{f-H\gamma^2}{H-f} - r^2\log\frac{r}{r_1}\right) \\
\frac{dw_0}{dr} &= \frac{r}{\rho} + \frac{1}{I}\left(H\frac{1-\gamma^2}{H-f}r - 2r\log\frac{r}{r_1}\right)
\end{aligned}\right\}\quad\text{..........(xii).}$$

These very important results can be applied to a great number of special examples. They include the solutions of Poisson given for *thin* circular plates, and various other particular cases (as of isotropy, etc.) treated by diverse writers: see our Arts. 494*—504*[1].

[331.] *Special cases.*

(*a*) Suppose the plate not to be *annular*, but to rest on the rim of a disc of radius r_0 in such a manner that its bending is not interfered

[1] To obtain Saint-Venant's notation we must replace, u, w by capitals, H by a_1, I by H, r_1 by a, and ρ by R.

with (§ 14). The plate may now be dealt with as consisting of an 'inner disc' and 'outer annulus.' Then evidently $dw_0/dr = 0$ when $r = 0$ because the tangent plane to the mid-section at $z = 0$, $r = 0$, must be horizontal; further round the ring $r = r_0$ the shearing stress must vanish for the inner disc which can thus only be acted upon by couples and will take a spherical curvature $(1/\rho_0)$ as in our Art. 324. Thus for the inner disc

$$u = -\frac{rz}{\rho_0}, \quad w = \frac{1}{\rho_0}\left(\frac{r^2 - r_0^2}{2} + \frac{d'}{c}z^2\right), \quad M_r = -\frac{4\epsilon^3}{3}\frac{H-f}{\rho_0},$$

and for the conditions at $r = r_0$

$$w_0 = 0, \quad \frac{dw_0}{dr} = \frac{r_0}{\rho_0}, \quad M_{r_0} = -\frac{4\epsilon^3}{3}\frac{H-f}{\rho_0}.$$

Three equations to determine the three constants ρ_0, B and C (ρ is known from M_{r_1}) of the problem are then obtainable by putting $r = r_0$ in the equations (xi) which hold for the outer annulus. Saint-Venant finds:

$$\left.\begin{aligned} B &= \frac{r_0 - \gamma^2 r_1^2}{I}, \quad C = \frac{r_0^2}{2\rho} + \frac{1}{I}\left\{r_0^2\left(1 - \frac{\gamma^2}{2}\right) + \frac{f}{H-f}\frac{r_0^2}{2}\left(1 - \frac{r_0^2}{r_1^2}\right)\right. \\ &\qquad\qquad \left. - (2r_0^2 - \gamma^2 r_1^2)\log\frac{r_0}{r_1}\right\}, \\ \frac{1}{\rho_0} &= \frac{1}{\rho} + \frac{1}{I}\left\{\left(\frac{f}{H-f} + \frac{\gamma^2 r_1^2}{r_0^2}\right)\left(1 - \frac{r_0^2}{r_1^2}\right) - 2\log\frac{r_0}{r_1}\right\} \end{aligned}\right\} \dots \text{(xiii)},$$

whence the values of u, and w for $r > r_0 < r_1$, can be at once found.

The solutions obtained by Saint-Venant in this first case are, as he himself observes, hardly satisfactory except for the case of a very *thin* plate. What he does is to make the vertical shifts of the mid-plane zero for the disc and the annulus when $r = r_0$; then the slopes of the tangent planes for both are equated, and finally the total couples along the same circle $r = r_0$. In the solutions he gives for the shifts the u and w for the annulus are not equal to the u and w for the disc when $r = r_0$, except for the mid-plane. In particular u when $r = r_0$ is a function of z only for the disc, but of z^3 as well for the annulus. In other words we have theoretical separation of the material at $r = r_0$. Thus the solutions are at best only approximate, and cannot be considered to hold at all in the neighbourhood of the rim itself. But shall we assume they hold accurately at points not in the neighbourhood of this rim? If the stresses acting at this rim were really confined to a line, they would certainly produce permanent alterations in the material; are we then justified in assuming that equating the vertical shifts and the tangent plane slopes (w and dw/dr) for

$r = r_0$ will give us the best values of the constants? I am inclined to doubt at least the presumed equality of the tangent-plane slopes: see our Art. 1572*, and p. 23 of the *Leçons de Navier*. The results become of course exact when we may neglect ϵ^2.

[332.] (*b*) Suppose the centre of the plate to rest upon a fixed circle of very small radius (p. 357). Then equations (xii) give the total deflection δ by putting $r = r_1$:

$$\delta = \frac{r_1^2}{2\rho} + \frac{3Qr_1^2}{16\pi H\epsilon^3} \cdot \frac{2H - f - H\gamma^2}{2H - 2f} \quad \text{..............(xiv)},$$

where $$Q = 2\pi r_1 P, \quad 1/\rho = -\frac{3}{4\epsilon^3}\frac{M_{r_1}}{H-f}.$$

Sub-cases are:

(i) $M_{r_1} = 0$, or $1/\rho = 0$; this is the simple case of only shearing load on the cylindrical sides. Such might happen if the mid-plane contour were fixed to a ring.

(ii) The cylindrical faces of the plate are fixed (see our footnote p. 231) and a normal load Q applied at the centre by means of a circle of very small radius. Here dw_0/dr of equation (xii) must be zero for $r = r_1$ or:

$$\frac{1}{\rho} = -\frac{H}{I}\frac{1-\gamma^2}{H-f}.$$

This gives $$M_{r_1} = \frac{Pr_1}{2}(1-\gamma^2),$$

and $$\delta = \frac{3Qr_1^2}{32\pi H\epsilon^3}.$$

(iii) Elastic isotropy (p. 358, § 16). We have only to put

$$f = e = \mu, \quad d' = f' = \lambda, \quad c = 2\mu + \lambda,$$

$$H = 2f + f' - \frac{d'^2}{c} = \frac{4(\mu+\lambda)\mu}{2\mu+\lambda},$$

$$\gamma^2 = \frac{2}{5}\frac{f}{H}\left(\frac{d'}{c} + \frac{4H}{e}\right)\frac{\epsilon^2}{r_1^2} = \frac{16\mu + 17\lambda}{10(\lambda+\mu)}\frac{\epsilon^2}{r_1^2}.$$

[333.] (*c*) Saint-Venant now returns to the case of a complete plate resting on a circular rim (of radius r_0) as given in our Art. 331, (*a*), and determines the deflections when the contour (of radius r_1) is (i) fixed, (ii) built-in (see his p. 360).

[334.] (*d*) In § 18 we have the remark that the force exerted on the ring $r = r_0$ must be equal and opposite to the force exerted on the ring $r = r_1$, or it must equal Pr_1/r_0 per unit of length of the arc. Thus the solutions of (*c*) are applicable to the case of a plate either fixed or built-in at its contour and loaded with Q uniformly distributed round the ring $r = r_0$. The deflections obtained by Saint-Venant are (p. 362):

(i) Mid-plane contour simply supported or fixed

$$\delta = \frac{3Qr_1^2}{32\pi H\epsilon^3}\left\{\frac{2H-f}{H-f}\left(1-\frac{r_0^2}{r_1^2}\right)+\left(\frac{r_0^2}{r_1^2}-\gamma^2\right)\log\frac{r_0^2}{r_1^2}\right\}.$$

(ii) Cylindrical face built-in

$$\delta' = \frac{3Qr_1^2}{32\pi H\epsilon^3}\left\{1-\frac{r_0^2}{r_1^2}+\left(\frac{r_0^2}{r_1^2}-\gamma^2\right)\log\frac{r_0^2}{r_1^2}\right\}.$$

For the reasons given in my Art. 331, I am doubtful as to the validity of these results except in the case when we may neglect γ^2.

[335.] (*e*) In § 19, p. 362, Saint-Venant explains how we may treat the problem of a thick circular plate subjected to any symmetrical load continuous or discontinuous on a plane face. We have in the case of a continuous load to substitute $\phi(r_0)\,2\pi r_0 dr_0$ for Q in the equations of (*d*) and integrate between the limits 0 and r_0, to find the total deflection. If we integrate from 0 to r_0, we shall obtain the deflection of the centre below any ring r_0 and so the form of the surface taken by the mid-plane. Saint-Venant seems to think this process more rigorous than that for thin plates dependent on Lagrange's equation and used by Poisson: see our Arts. 284*, 496*—504*. But I cannot get over the difficulty suggested in my Art. 331. The results are not true for the ring in consideration unless γ^2 may be neglected, but Saint-Venant practically divides his whole plate up into such rings, when thus integrating. It appears to me possible that he may thus be really introducing an important sum of small errors.

In § 21, p. 365, he treats by this method the case of a thick plate uniformly loaded and finds from the results in (*d*):

$$\delta = \frac{3Qr_1^2}{128\pi H\epsilon^3}\left(\frac{3H-f}{H-f}+4\gamma^2\right),$$

$$\delta' = \frac{3Qr_1^2}{128\pi H\epsilon^3}(1+4\gamma^2),$$

where Q is the total load.

These results, first given by Boussinesq, agree in the case of uni-constant isotropy and neglect of γ^2 with those of Poisson: see our Art. 502*.

[336.] (*f*) This case is the most general possible and is thus stated by Saint-Venant:

Mais, lorsqu'on se propose *d'avoir seulement la flèche centrale*, sans chercher la forme que prend la plaque en ses divers points, une remarque bien simple montre que les expressions en r_1 et r_0 suffisent au calcul de cette flèche pour toutes les distributions possibles, même non symétriques, même discontinues et irrégulières, des charges que supporte la plaque soutenue en haut (p. 363).

We note that if we have a single load P at any point of a rim-supported plate, it must produce the same central deflection as if it were at *any other point at the same distance from the centre*. Hence by the principle of super-position of displacements in the case of elastic strain, a load P at an isolated point distant r from the centre, must produce the same central deflection as if it were uniformly distributed round the ring of radius r. Thus the formulae of (d) hold if the load Q be concentrated at a distance r_0 from the centre. This result seems first to have been stated by Lévy in a memoir of 1877, although it was involved in the results of § 76 of Clebsch's treatise.

[337.] Saint-Venant concludes this *Note* on thick plates with the following words:

Nous avons démontré, dans la présente Note, comme on a vu, nos formules d'une manière rigoureuse, ou sans annulations de termes. Leur parfaite rigueur est subordonnée, il est vrai, comme est celle de toutes les formules ci-dessus d'extension, flexion, torsion des tiges, à ce que les forces ou les réactions d'appuis et d'encastrements agissent exclusivement sur une certaine surface qui est, pour les plaques, leur cylindre contournant, en s'y distribuant des manières qui sont exprimées en z par les formules du deuxième et du troisième degré donnant $\widehat{rz}$ et $\widehat{rr}$, et spécifiées pour $r = r_1$. Mais, ainsi que nous avons eu bien des fois occasion de le dire, elles donnent des résultats très suffisamment approchés quel que soit le mode d'application et de distribution si la plaque est peu épaisse; et, en tous cas, notre analyse actuelle, outre qu'elle tient compte de termes (ceux en γ^2 ou ϵ^2/r_1^2) dont il n'est nulle question dans l'analyse connue, a l'avantage de ne donner que les résultats où tout a été mis en compte dès le commencement ou sans suppressions faites de prime abord, et dont on n'aperçoit pas *a priori* la portée et le degré d'influence sur les résultats lorsqu'on en opère de ce genre (p. 367).

But does this paragraph explain all the assumptions? I think not: see our Art. 331.

[338.] The second *Note* inserted by Saint-Venant in Clebsch's third chapter is due to Boussinesq. It is a résumé of the results obtained by the latter in a series of memoirs during the years 1878—9, and afterwards published separately under the title: *Recherches sur l'application des potentiels à la théorie de l'équilibre intérieur des solides élastiques*; see our detailed account of this important work below. The *Note* itself is entitled: *Sur l'équilibre des corps massifs sollicités en un point superficiel ou intérieur.* It occupies pp. 374—405; an addition occupies pp. 405^a—407^a; while some consideration, also due to Boussinesq, of Cerruti's

Memoir of 1882 on the same subject, will be found on pp. 881—8 (*Complément à la Note finale du* § 46). As these contributions are not due to Saint-Venant we postpone the discussion of their contents until we are dealing with the special researches of Boussinesq and Cerruti.

[339.] The next important addition of Saint-Venant is the *Note finale du* § 60. It is entitled: *Théorie de l'impulsion longitudinale d'une barre élastique par un corps massif qui vient heurter une de ses deux extrémités ; et de la résistance de la matière de la barre à un pareil choc*; it occupies pp. 480 a—480 gg. The numerical results of this note together with their graphical representation will be considered in our account of the Memoir of 1883: see our Arts. 401—7.

[340.] The first seven sections (pp. 480 a—480 k) give an account of the various tentative stages in the history of the theory. We have first two theorems of Young, which as first approximations may be cited. Let the bar be of weight P, density ρ, section ω, length l and stretch-modulus E; let Q be the weight and V the velocity of the body which strikes it at the free end, the other end being fixed.

Then if u be the total shift of the free end, g gravitational acceleration, and we suppose the stretch uniformly distributed, we have from the principle of work:

(i) *Bar horizontal:*

$$\frac{E\omega u^2}{2l} = \frac{Q}{g}\frac{V^2}{2}, \quad \text{or if } u_0 = \frac{Ql}{E\omega} \text{ be the statical shift,}$$

$$u = V\sqrt{u_0/g}.$$

(ii) *Bar vertical:*

$$\frac{E\omega u^2}{2l} = \frac{Q}{g}\frac{V^2}{2} + Qu \quad \therefore \; u = u_0 + \sqrt{u_0^2 + V^2\frac{u_0}{g}}.$$

Let $u/l = T_0/E$ the greatest safe stretch within the elastic limit, then in Case (i):

$$\frac{E\omega l}{2}\left(\frac{T_0}{E}\right)^2 = \frac{QV^2}{2g}.$$

Now $\frac{QV^2}{2g}$ is the work necessary to destroy the efficiency of the bar, or its resilience. Hence the resilience of a bar varies as its volume ωl, multiplied by $\frac{T_0^2}{2E}$, a quantity depending only on its elasticity. In this form of Young's theorem, the quantity T_0^2/E has been termed by Tredgold the *modulus of resilience:* see our Arts. 999* and 982* and Vol. I., p. 875.

[341.] The next stage in the history of longitudinal impact was due to Navier (see our Arts. 272*—4*). He expressed the complete analytical solution of the problem for the case of the horizontal bar in a Fourier's series. Poncelet added to this solution the effect of gravity and the statical action of the weight, supposed to strike the bar in a vertical position: see our Art. 990*. Neither Navier nor Poncelet developed this analytical solution, except for the special case of P/Q being very small when the results agree with those of the preceding article. Saint-Venant undertook this development, so far as ascertaining the *shift* is concerned, in 1865 and 1868 (see our Arts. 200 and 201) for certain common values of P/Q, i.e. $\frac{1}{4}$, $\frac{1}{2}$, 1, 2, 4. He found it possible to determine the shift of the end struck, but the series gave no prospect, however far the numerical calculations were carried, of ascertaining the maximum stretch or squeeze (p. 480 g). It became necessary then to find a solution in finite terms. The form of these finite terms seems to have been suggested by Saint-Venant's course of memoirs lasting from 1865—1882 on the impact of two bars: see our Arts. 203 and 221. The next stage was Boussinesq's solution in terms of a single exponential for the shift at a time not greater than $2l/a$: see our Art. 403 and the account later of his paper of 1882. Later in the same year two officers of the French marine artillery, Sébert and Hugoniot, obtained an exponential solution in finite terms for a vibrating bar fixed at one end and subjected at the other to a force varying with the time. This solution really covers that of Boussinesq, who hearing only of the method of Sébert and Hugoniot, sent to Saint-Venant in the summer vacation of 1882 a direct and complete solution of the problem of longitudinal impact. Judging from the communications of M. Hugoniot to Saint-Venant (see pp. 480 j—480 k) the merit of the solution must be divided between the two naval officers and the professor of Lille.

The reader will find an account of Boussinesq's solution in the chapter devoted to that elastician.

[342.] The next insertion of Saint-Venant is the *Note finale du* § 61. It occupies no less than 138 pages (pp. 490—627) and contains the complete theory of the transverse impulse of bars, including results of Saint-Venant's not hitherto published: see our Arts. 104–5, 200–1, and *Notice* II. p. 20, 2°. The *Note* is entitled: *De l'Impulsion transversale des barres élastiques, et de leur vibration avec le corps qui les aura mises en mouvement. Détermination de leur flexion ainsi que des conditions de leur résistance vive ou dynamique.*

[343.] The first 51 sections (pp. 490—597) are devoted to the analytical and numerical solution of various problems of bars vibrating transversely with a load attached:

ces pièces sont supposées *vibrer* non pas *seules* comme le supposent les solutions données par Clebsch, mais *unies avec le corps étranger* dont l'impulsion, ou brusque, ou graduée, les a fait sortir de leur état d'équilibre; car c'est pendant cette union, ne durât-elle que le temps d'une demi-période oscillatoire, que les déplacements relatifs des parties de ces pièces atteignent leur maximum et qu'elles courent le plus grand danger de rupture ou d'énervation dont les calculs de résistance ont pour objet de les sauver (p. 490).

Saint-Venant's method is simply to solve in 'normal' functions or coordinates the equation:

$$\frac{d^2}{dz^2}\left(E\omega\kappa^2\frac{d^2u}{dz^2}\right)+\frac{p}{g}\left(\frac{d^2u}{dt^2}-\mathfrak{g}\right)=0\ldots\ldots\ldots\ldots\ldots\text{(i)},$$

where u is the transverse shift of the point in the axis at distance z from one end of the bar, p/g is the mass per unit length of the bar and of any permanent load at the same point, $\mathfrak{g}$ the body acceleration (usually only gravity) on the same length, and $E\omega\kappa^2$ with our usual notation the rigidity, which may vary from point to point. The bar is supposed to be loaded and to receive displacement in a plane which passes through a principal axis of each cross-section. The terminal and initial conditions determine the constants of the normal functions while the conditions at the impelled point select the normal functions required and determine the notes.

[344.] The process of solution and the calculation of the dynamical deflection are generally long, even if we keep only one term of the series, but:

cette expression simple de la flèche dynamique peut, comme je l'ai reconnu dans une multitude d'exemples, être identiquement obtenue sans poser d'équations différentielles, en s'aidant d'une hypothèse plausible sur les rapports mutuels des déplacements, et en y appliquant d'une manière tout élémentaire, le théorème des vitesses virtuelles ou celui des pertes brusques de force vive; en sorte que rien n'empêchera d'introduire dans les cours, même industriels, cette méthode que j'appelle de *deuxième approximation*, tenant suffisamment compte de l'*inertie* des systèmes heurtés, et d'en substituer l'enseignement général à celui qui y est quelquefois donné, pour deux cas particuliers, de la méthode dans laquelle, en abstrayant tout à fait ou en supposant infiniment petite la masse de ces systèmes, on s'éloigne généralement beaucoup de la réalité et des faits (p. 491).

The *hypothèse plausible* which Saint-Venant makes is precisely that of Cox (see his p. 584, § 46) and his results, pp. 584—597, are those of Cox (see our Arts. 1435–7*), or those I had obtained by Cox's method before examining Saint-Venant's work (see Vol. I. pp. 894—6). Thus the merit of this elementary treatment of the problem is entirely Cox's, but Saint-Venant's work, taking first into account the vibratory terms is really the justification of the hypothesis. I am somewhat surprised that Cox's paper escaped Saint-Venant, as he is usually very careful in his historical notices, and he had certainly read Stokes' papers in the volumes of the *Cambridge Transactions*.

The Note terminates with a consideration of Willis's problem and a discussion of the numerical results of the *Iron Commissioners' Report*: see our Arts. 1276*, 1406* and 1417*.

[345.] I propose to describe in one case Saint-Venant's method of solution, and then to record the other problems with which he has dealt in this Note. The following conditions are easily seen to hold:

(i) at a free end:

$$\text{Bending moment} = E\omega\kappa^2 \frac{d^2u}{dz^2} = 0\,;\ \text{shear} = -\frac{d}{dz}\left(E\omega\kappa^2 \frac{d^2u}{dz^2}\right) = 0.$$

(ii) at a fixed end[1]: $u = 0$, $E\omega\kappa^2 \dfrac{d^2u}{dz^2} = 0$. (We retain the $E\omega\kappa^2$ in both cases as $\omega\kappa^2$ may be the vanishing factor in certain systems.)

(iii) at a built-in end: $u = 0$, $\dfrac{du}{dz} = 0$.

[1] At a *fixed* end the terminal direction is free; the word *supported* should also be interpreted as equivalent to *fixed*, i.e. allowing only of shearing force, but this in either sense: see footnote Vol. I., p. 52.

(iv) At the join of two bars:

$$u = u_1, \quad \frac{du}{dz} = \frac{du_1}{dz}, \quad E\omega\kappa^2 \frac{d^2u}{dz^2} = E_1\omega_1\kappa_1{}^2 \frac{d^2u_1}{dz^2}.$$

(v) At the join of two bars where there is a weight of mass Q/g we must have:

$$\frac{Q}{g}\frac{d^2u}{dt^2} - \frac{d}{dz}\left(E\omega\kappa^2 \frac{d^2u}{dz^2}\right) + \frac{d}{dz}\left(E_1\omega_1\kappa_1{}^2 \frac{d^2u_1}{dz^2}\right) = 0,$$

together with the relations (iv): see Saint-Venant's pp. 494-5.

[346.] Let us apply these results to the simple case of a prismatic bar supported terminally and struck by a weight Q with velocity V at its mid-point. Let the length of the bar be $2l$, its weight P, and $\tau^2 = Pl^3/(2gE\omega\kappa^2)$. We shall suppose the bar so placed that the impact is horizontal, or g may be put zero. Equation (i) of Art. 343, thus becomes:

$$\tau^2 \frac{d^2u}{dt^2} + l^4 \frac{d^4u}{dz^4} = 0 \quad\text{.........................(i),}$$

together with the conditions:

$$u = 0, \quad \frac{d^2u}{dz^2} = 0, \text{ when } z = 0 \quad\text{....................(ii),}$$

$$\tau^2 \frac{d^2u}{dt^2} = \frac{Pl^3}{Q}\frac{d^3u}{dz^3}, \quad \frac{du}{dz} = 0, \text{ when } z = l \quad\text{..............(iii).}$$

Take as a particular integral:

$$z = Z_m \left\{A'_m \frac{\tau}{m^2} \sin\frac{m^2t}{\tau} + B'_m \cos\frac{m^2t}{\tau}\right\}.$$

We find $$m^4 Z_m = l^4 \frac{d^4 Z_m}{dz^4}.$$

The solution of this equation takes the well-known form first given by Euler, (see our Art. 52*):

$$Z_m = C \sin\frac{mz}{l} + C_1 \cos\frac{mz}{l} + C_2 \sinh\frac{mz}{l} + C_3 \cosh\frac{mz}{l}.$$

To satisfy (ii) and the second of (iii) we must take

$$C_1 = C_3 = 0, \quad C_2 = -C\frac{\cos m}{\cosh m}.$$

Further $u = 0$ when $t = 0$, therefore we have finally u of the form

$$\left.\begin{aligned} u &= \Sigma \frac{\tau}{m^2} A_m Z_m \sin\frac{m^2t}{\tau} \\ \text{where}\quad Z_m &= \frac{\sin(mz/l)}{\cos m} - \frac{\sinh(mz/l)}{\cosh m} \end{aligned}\right\} \quad\text{....................(iv),}$$

and the first of conditions (iii) gives us the equation for m

$$m(\tan m - \tanh m) = 2P/Q \quad\text{.................(v)},$$

which may be termed the *characteristic* transcendental equation for m. Of this equation m, $-m$, $m\sqrt{-1}$, $-m\sqrt{-1}$ are all roots if m is a root, but they give rise to the same Z_m, so that we need only take the real and positive roots.

[347.] It remains to determine A_m. When $t=0$, let $\dot{u} = \psi(z)$, then

$$\Sigma A_m Z_m = \psi(z) \quad\text{..........................(i)}.$$

Multiply both sides by $Z_{m'}$ and we have;

$$\Sigma A_m Z_m Z_{m'} = \psi(z)\, Z_{m'} \quad\text{.....................(ii)}.$$

Put $z=l$, multiply by Q and add this to the integral of equation (ii) above with regard to $\frac{P}{2l}dz$ from $z=0$ to $2l$, and we find:

$$\Sigma A_m \left\{2\frac{P}{2l}\int_0^l Z_m Z_{m'}\, dz + QZ_m(l)\, Z_{m'}(l)\right\} = 2\frac{P}{2l}\int_0^l Z_{m'}\psi(z)\, dz + QZ_{m'}(l)\times\psi(l).$$

As Saint-Venant remarks this is really an integration of equation (ii) with regard to $d\mathfrak{q}$, where $\mathfrak{q}$ = total weight of beam and load $= P+Q$. Now Saint-Venant shews by straightforward integration that

$$\int_0^l Z_m Z_{m'}\, dz = -\frac{4l}{mm'}\frac{P}{Q} \quad\text{.....................(iii)},$$

or remembering the values of $Z_m(l)$, $Z_{m'}(l)$, we have, save when $m=m'$:

$$2\frac{P}{2l}\int_0^l Z_m Z_{m'}\, dz + QZ_m(l) Z_{m'}(l) = \int_0^{2l} Z_m Z_{m'}\, d\mathfrak{q} = 0.$$

Thus we find:

$$A_m = \frac{2\frac{P}{2l}\int_0^l Z_m\psi(z)dz + QZ_m(l)\psi(l)}{2\frac{P}{2l}\int_0^l Z^2{}_m dz + QZ^2{}_m(l)} \quad\text{............(iv)},$$

which may be written in the form:

$$A_m = \frac{\int_0^{2l} Z_m\psi(z)d\mathfrak{q}}{\int_0^{2l} Z^2{}_m d\mathfrak{q}} \quad\text{.....................(v)}.$$

[348.] Now arises a question as to the value we ought to give to $\psi(z)$. Saint-Venant puts $\psi(z)=0$ except $z=l$, when $\psi(l)=V$. Thus he obtains after some obvious reductions:

$$A_m = \frac{QVZ_m(l)}{\frac{P}{l}\int_0^l Z^2{}_m dz + QZ^2{}_m(l)}.$$

On evaluating this expression we find

$$A_m = \frac{4V}{m(\sec^2 m - \operatorname{sech}^2 m) + \dfrac{2P}{Qm}} \quad \ldots\ldots\ldots\ldots\ldots (\text{vi}).$$

Equations (vi) of this Article with (iv) and (v) of Art. 346 give the complete solution.

Is now this choice of initial velocities a proper one? Saint-Venant has defended it in the *Comptes rendus*, T. LXI. 1865, p. 43, against an objection raised to it. He says that if a small portion of the bar receive an initial velocity, Z_m will be nearly constant for this portion; accordingly equation (v) of the preceding Article gives us for the numerator of A_m, the expression $Z_m(l) \int \psi(z) d\mathfrak{q}$ where $\psi(z)$ is zero except over this small portion, where it has a value slightly less than V. But he remarks that the momentum possessed by this small portion and the weight Q ought to be exactly $\frac{Q}{g} V$, or $\int \psi(z) \frac{d\mathfrak{q}}{g} = \frac{Q}{g} V$.

[349.] We will now indicate the various problems which are dealt with analytically by Saint-Venant.

(*a*) In §§ 7–14 he treats as a general problem the cases when the bar is not prismatic (i.e. the rigidity $E\omega\kappa^2$ varies), when its ends are fixed in different fashions, when there are various bars or when one bar with a varying load forms the complete system. He shews that supposing the functions $Z_{\mathbf{m}}$ can be found which satisfy the equation:

$$\frac{d^2}{dz^2}\left\{E\omega\kappa^2 \frac{d^2 Z_{\mathbf{m}}}{dz^2}\right\} = \mathbf{m}^4 \frac{p}{g} Z_{\mathbf{m}},$$

then the integral $\qquad \int Z_{\mathbf{m}} Z_{\mathbf{m}'} d\mathfrak{q} = 0,$

where $\mathfrak{q}$ represents the total weight of the system and the integration extends from one end to the other of the system; $\mathbf{m}$ and $\mathbf{m}'$ are two unequal roots of the characteristic equation in $\mathbf{m}$ which arises from the terminal and load conditions (p. 506).

The coefficients of the time function $A_{\mathbf{m}} \mathbf{m}^{-2} \sin \mathbf{m}^2 t + B_{\mathbf{m}} \cos \mathbf{m}^2 t$ will be determined by equations similar to (v) of our Art. 347; $A_{\mathbf{m}}$ depending only on the initial velocities, $B_{\mathbf{m}}$ only on the initial displacements (p. 507).

The value of the denominator of these coefficients, i.e. $\int Z^2_{\mathbf{m}} d\mathfrak{q}$, can be obtained by the *differentiation* with regard to $\mathbf{m}$ of a certain function of $Z_{\mathbf{m}}$ and its fluxions with regard to z (p. 508). Compare Lord Rayleigh's *Theory of Sound*, Vol. I., pp. 209–10.

[350.] (*b*) The next special example given by Saint-Venant is that of a doubly built-in beam struck at the mid-point. He finds:

$$u = V\tau\Sigma\frac{2}{m^2}\cdot\frac{(1-\cos m\cosh m)(\cos m-\cosh m)(\sin m+\sinh m)}{(1-\cos m\cosh m)^2+\frac{P}{Q}(\cos m-\cosh m)^2}Z_m\sin\frac{m^2t}{\tau},$$

where $$Z_m = \frac{\sinh\frac{mz}{l}-\sin\frac{mz}{l}}{\cosh m-\cos m} - \frac{\cosh\frac{mz}{l}-\cos\frac{mz}{l}}{\sinh m+\sin m},$$

and the characteristic equation is:

$$\frac{m(1-\cos m\cosh m)}{\sin m\cosh m+\cos m\sinh m} = \frac{P}{Q}.$$

See pp. 511—3.

[351.] (*c*) When one end of the bar, for this particular case supposed of length l, is built-in and the other is struck we have:

$$u = V\tau'\Sigma\frac{2}{m^2}\frac{(\sin m\cosh m-\cos m\sinh m)(\sin m+\sinh m)(\cos m+\cosh m)}{(\sin m\cosh m-\cos m\sinh m)^2+\frac{P}{Q}(\sin m+\sinh m)^2}\times Z_m\sin\frac{m^2t}{\tau'},$$

where $$Z_m = \frac{\cosh\frac{mz}{l}-\cos\frac{mz}{l}}{\cosh m+\cos m} - \frac{\sinh\frac{mz}{l}-\sin\frac{mz}{l}}{\sinh m+\sin m},$$

and for this case: $$\tau'^2 = \frac{Pl^3}{gE\omega\kappa^2},$$

and the characteristic equation is:

$$m\frac{\sin m\cosh m-\cos m\sinh m}{1+\cos m\cosh m} = \frac{P}{Q}.$$

See pp. 513—4.

[352.] (*d*) Suppose the bar of length $2l = a + b$ and the blow to be given at a distance a from one end, then if τ^2 be as in Art. 346:

$$u = V\tau\Sigma\frac{A_mZ_m}{m^2}\sin\frac{m^2t}{\tau}, \text{ from } z=0 \text{ to } a,$$

where $$Z_m = \frac{\sin\frac{mb}{l}\sin\frac{mz}{l}}{\sin m\cos m} - \frac{\sinh\frac{mb}{l}\sinh\frac{mz}{l}}{\sinh m\cosh m};$$

and, $$u' = V\tau\Sigma\frac{A_mZ'_m}{m^2}\sin\frac{m^2t}{\tau}, \text{ from } z'=0 \text{ to } b,$$

where $$Z'_m = \frac{\sin\frac{ma}{l}\sin\frac{mz'}{l}}{\sin m\cos m} - \frac{\sinh\frac{ma}{l}\sinh\frac{mz'}{l}}{\sinh m\cosh m}.$$

In both cases

$$A_m = \frac{4}{m}\frac{1}{\frac{dZ_m(a)}{dm} + \frac{2P}{m^2Q}}, \text{ (where obviously } Z_m(a) = Z'_m(b)),$$

and the characteristic equation is:

$$mZ_m(a) = \frac{2P}{Q}.$$

See pp. 514—7[1].

The results given in our Arts. 349—352 correspond with the *introduction of vibratory terms* to the solutions obtained by Cox's method in Art. 1437* and pp. 894—895, c (i), c (ii) and (*b*) respectively of our first volume. I have gone through Saint-Venant's analysis but not worked out independently his results.

[353.] We have next several cases in which the bar would not be immoveable if it were rigid, *i.e.* the bar is free or pivoted. Here the solution will have an algebraic part as well as a transcendental. This part can sometimes be obtained by retaining the root $m = 0$, which has been divided out of the *characteristic* equation; but as a rule it is better to treat it separately as arising from the kinetic conditions of the problem and determine it by general dynamical principles such as the principle of momentum. I will briefly indicate Saint-Venant's treatment in the following example:

(*e*). A prismatic bar is struck transversely at its two terminals by bodies of weight q and Q moving with velocities v and V respectively. The length of the bar is l and its weight P; the origin is taken at the end at which q strikes the bar. As before let us take

$$q + P + Q = \mathfrak{q}, \quad \frac{Pl^3}{gE\omega\kappa^2} = \tau'^2, \text{ as in Case } (c) \text{ Art. } 351.$$

We have then: $$\tau'^2\frac{d^2u}{dt^2} + l^4\frac{d^4u}{dz^4} = 0 \text{.........................} (i).$$

For the free ends, $$\frac{d^2u}{dz^2} = 0, \text{ when } z = 0 \text{ or } l \text{.................} (ii);$$

$$\left.\begin{aligned} \tau'^2\frac{d^2u}{dt^2} + \frac{Pl^3}{q}\left(\frac{d^3u}{dz^3}\right) &= 0, \text{ when } z = 0 \\ \tau'^2\frac{d^2u}{dt^2} - \frac{Pl^3}{Q}\left(\frac{d^3u}{dz^3}\right) &= 0, \text{ when } z = l \end{aligned}\right\} \text{...........} (iii);$$

$u = 0$, for $t = 0$; $du/dt = v$ when $z = 0$, $= V$ when $z = l$, and equal zero for all other values of z at the epoch $t = 0$.

[1] Saint-Venant has a for our l, b for our a and b_1 for our b, EI for our $E\omega\kappa^2$, with other slight differences. I have altered his notation to agree with that of our first volume.

Saint-Venant assumes a solution of the form:

$$u = \left(C + C' \frac{z}{l}\right) t + \Sigma Z_m A_m \frac{\tau'}{m^2} \sin \frac{m^2 t}{\tau'} .$$

By the principle of linear momentum we have:

$$qv + QV = q \left(\frac{du}{dt}\right)_{z=0} + Q \left(\frac{du}{dt}\right)_{z=l} + \int_0^l \frac{du}{dt} \frac{P}{l} dz \ldots\ldots (\text{iv}),$$

and by that of the moment of momentum about the point $z = 0$,

$$QVl = Ql \left(\frac{du}{dt}\right)_{z=l} + \int_0^l \frac{du}{dt} z \frac{P}{l} dz \ldots\ldots\ldots\ldots\ldots (\text{v}),$$

as equations connecting the velocities before and after the blow. Now Saint-Venant so chooses C and C' *that the algebraic part of* u, *namely* $\left(C + C' \frac{z}{l}\right) t$, *shall satisfy equations* (iv) *and* (v) *independently of the trigonometrical terms.* We easily find:

$$C \Big/ \left\{-1 + \frac{2qv}{QV}\left(1 + 3\frac{Q}{P}\right)\right\} = \frac{C'}{3} \Big/ \left\{1 + 2\frac{q}{P} - \frac{qv}{QV}\left(1 + 2\frac{Q}{P}\right)\right\}$$

$$= \frac{2QV}{P} \Big/ \left\{1 + \frac{4(Q+q)}{P} + 12\frac{qQ}{P^2}\right\} \ldots\ldots\ldots\ldots (\text{vi}).$$

We can now determine the function Z_m from the equations (i) to (iii) as if the algebraic portion of u had no existence, for the latter disappears entirely from these equations. Saint-Venant finds:

$$Z_m = (\cosh m - \cos m)\left(\sinh\frac{mz}{l} + \sin\frac{mz}{l}\right) - (\sinh m - \sin m)\left(\cosh\frac{mz}{l} + \cos\frac{mz}{l}\right)$$

$$+ \frac{2mq}{P}\left(\sin m \sinh\frac{mz}{l} + \sinh m \sin\frac{mz}{l}\right) \ldots\ldots (\text{vii}),$$

with the characteristic equation:

$$1 - \cos m \cosh m + m \frac{Q+q}{P} (\sin m \cosh m - \cos m \sinh m)$$

$$+ \frac{2Qq}{P^2} m^2 \sin m \sinh m = 0 \ldots\ldots\ldots\ldots\ldots (\text{viii}).$$

Initially $\qquad C + C' \frac{z}{l} + \Sigma A_m Z_m = \psi(z)$, say,

multiplying by $Z_m d\mathfrak{q}$ we have as the coefficients of C and C' respectively,

$$\left.\begin{aligned} \int Z_m\, d\mathfrak{q} &\equiv qZ_m(0) + QZ_m(l) + \frac{P}{l}\int_0^l Z_m\, dz \\ \int Z_m \frac{z}{l}\, d\mathfrak{q} &\equiv QZ_m(l) + \frac{P}{l}\int_0^l Z_m \frac{z}{l}\, dz \end{aligned}\right\} \ldots\ldots\ldots (\text{ix}).$$

By straightforward integration we can shew that both these expressions $= \frac{2P}{m} \times$ {the function of m to the left of equation (viii)}, and therefore both $= 0$.

We have thus $\int \left(C + C' \frac{z}{l}\right) Z_m d\mathfrak{q} = 0$, or, *the algebraic part* has no influence on the determination of the value of A_m, which accordingly equals:

$$\frac{qvZ_m(0) + QVZ_m(l)}{qZ^2_m(0) + QZ^2_m(l) + \frac{P}{l}\int_0^l Z^2_m dz},$$

as before. Saint-Venant gives on p. 525 the lengthy expressions for the numerator and denominator of this quantity. Equation (vii) gives us easily the numerator and all but the expression $\int_0^l Z^2_m dz$ in the value of the denominator. The value of this integral I find to be:

$$\frac{1}{P}\Big[P^2\left\{\frac{3}{m}(\sin m\cosh m - \cos m\sinh m)(1 - \cos m\cosh m) + (\sinh m - \sin m)^2\right\}$$
$$+ 2Pq\{6\sin m\sinh m(1 - \cos m\cosh m) + m(\cosh m - \cos m)(\sinh m - \sin m)\}$$
$$+ q^2\{2m^2(\sinh^2 m - \sin^2 m) + 6m\sin m\sinh m(\sin m\cosh m - \sinh m\cos m)\}\Big].$$

There is one point, however, which we must notice, namely, that equations (iv) and (v) have only been proved for the *algebraic portions* of the solution, but they must hold generally. Substituting the full value of u, we find that these equations will still be satisfied, if:

$$q\Sigma AZ_m(0) + Q\Sigma AZ_m(l) + \Sigma\int_0^l AZ_m \frac{P}{l} dz = 0,$$

$$Ql\Sigma AZ_m(l) + \Sigma\int_0^l AZ_m \frac{Pz}{l} dz = 0.$$

But these equations are satisfied for each Z_m of the sum by reason of equations (ix).

I do not think this point is explicitly brought out by Saint-Venant, although in a long footnote pp. 521—4, he proves a more general proposition, namely:

On peut donc, dans les problèmes de mouvement des barres ou tiges élastiques libres ou pivotantes autour de points ou d'axes fixes, établir *séparément* la partie algébrique ou de solidification, et la partie transcendante ou vibratoire, de leur mouvement. Et même on peut généralement, ce qui est encore mieux, ne s'occuper que de celle-ci, qui seule intéresse le problème de la résistance de la matière, sans craindre que la non-prise en considération de celle-là *soit une cause d'erreur* (p. 524).

That is, the principles of kinetics will hold for the algebraic and transcendental parts of the solution separately as we have seen in the above example.

[354.] Saint-Venant on pp. 525—6 treats two special cases of the problem in Art. 353.

(i) If we put $q = 0$ we get the case of a free bar struck transversely at one end. The solution given in the article referred to easily reduces to:

$$u = 2\frac{3\frac{z}{l} - 1}{4 + \frac{P}{Q}} Vt + 2V\tau'\Sigma\frac{(\sin m \cosh m - \cos m \sinh m)\, Z_m \sin\frac{m^2 t}{\tau'}}{(\sin m \cosh m - \cos m \sinh m)^2 + \frac{P}{Q}(\sinh m - \sin m)^2},$$

where Z_m

$$= (\cosh m - \cos m)\left(\sinh\frac{mz}{l} + \sin\frac{mz}{l}\right) - (\sinh m - \sin m)\left(\cosh\frac{mz}{l} + \cos\frac{mz}{l}\right);$$

the characteristic equation being:

$$m(\sin m \cosh m - \cos m \sinh m) + \frac{P}{Q}(1 - \cos m \cosh m) = 0.$$

(ii) If we put $v = 0$, and $q = \infty$, we have the case of a bar fixed or pivoted at one end and struck at the other.

We find

$$u = \frac{z/l}{1 + P/(3Q)} Vt + 2V\tau'\Sigma\frac{1}{m^2}\frac{\frac{\sin(mz/l)}{\sin m} + \frac{\sinh(mz/l)}{\sinh m}}{1 + \frac{P}{2Q}(\operatorname{cosec}^2 m - \operatorname{cosech}^2 m)}\sin\frac{m^2 t}{\tau'},$$

the characteristic equation being:

$$\cot m - \coth m = 2m\, Q/P.$$

On pp. 526—30 are given various verifications of these results.

[355.] Saint-Venant on p. 530 (§ 24) passes to the consideration of *Impulsions graduelles ou tranquilles.* Under this term he includes problems involving the effect of the weights of both the striking body and the bar during the blow, or involving the constrained movement of a portion of the bar. For example, if a horizontal bar be struck vertically we have to solve the equation (i) of Art. 343, with $\mathfrak{g}$ put $= g$. I will briefly indicate Saint-Venant's method in the following problem: *A horizontal bar terminally supported, to the mid-point of which is attached a weight Q', receives a blow at the same point from a body of weight Q falling vertically with velocity V. To determine the transverse shift.*

[356.] Let $2l$ be the length, P the weight of the bar, $Pl^3/(2gE\omega\kappa^2) = \tau^2$, $P + Q + Q' = \mathfrak{q}$.

We have to solve the equation

$$\tau^2\left(\frac{d^2u}{dt^2}-g\right)+l^4\frac{d^4u}{dz^4}=0 \text{ .... (i)},$$

subject to the terminal conditions:

$$u=0,\quad \frac{d^2u}{dz^2}=0 \text{ when } z=0;\quad \frac{du}{dz}=0 \text{ when } z=l \text{........ (ii)},$$

$$\left.\begin{aligned} Q+Q'-\frac{Q+Q'}{g}\frac{d^2u}{dt^2}&=-2E\omega\kappa^2\frac{d^3u}{dz^3} \\ \text{or,}\qquad \tau^2\left(\frac{d^2u}{dt^2}-g\right)&=\frac{Pl^3}{Q+Q'}\frac{d^3u}{dz^3}\end{aligned}\right\} \text{ when } z=l \text{.........(iii)}.$$

Saint-Venant takes $u=u_1+U$ where u_1 is independent of the time and chosen so that the gravitational terms disappear from equations (i) and (iii) i.e.:

$$-g\tau^2+l^4\frac{d^4u_1}{dz^4}=0;\quad -g\tau^2(Q+Q')=Pl^3\frac{d^3u_1}{dz^3}, \text{ when } z=l.$$

Thus we have $$\frac{d^3u_1}{dz^3}=\frac{g\tau^2}{l^4}(z-l)-g\tau^2\frac{Q+Q'}{Pl^3},$$

and integrating having regard to equations (ii), we find

$$u_1=\frac{Pl^3}{48E\omega\kappa^2}\left\{8\frac{z}{l}-4\left(\frac{z}{l}\right)^3+\left(\frac{z}{l}\right)^4\right\}+\frac{(Q+Q')\,l^3}{6E\omega\kappa^2}\left\{\frac{3}{2}\frac{z}{l}-\frac{1}{2}\left(\frac{z}{l}\right)^3\right\} \text{...(iv)}.$$

The equations for U can now be easily solved, we deduce:

$$U=\Sigma Z_m\left(\frac{\tau}{m^2}A_m\sin\frac{m^2t}{\tau}+B_m\cos\frac{m^2t}{\tau}\right) \text{ where } Z_m=\frac{\sin\frac{mz}{l}}{\cos m}-\frac{\sinh\frac{mz}{l}}{\cosh m},$$

and the characteristic equation is:

$$m(\tan m-\tanh m)=2P/(Q+Q') \text{ .... (v)}.$$

In order to determine B_m and A_m we have, if $\phi(z)$ and $\psi(z)$ be respectively the initial shift and velocity corresponding to U,

$$B_m=\frac{\int\phi(z)\,Z_m\,d\mathfrak{q}}{\int Z^2{}_m\,d\mathfrak{q}},\quad A_m=\frac{\int\psi(z)\,Z_m\,d\mathfrak{q}}{\int Z^2{}_m\,d\mathfrak{q}}.$$

Now, $U_{t=0}=$ the initial value of $u-u_1$,

$\dot{U}_{t=0}=$ the initial value of $\dot{u}-\dot{u}_1$.

Further the initial value of u is the deflection due to the bar's own weight P and the load Q' attached.

Hence we have: $$U_{t=0}=-\frac{Ql^3}{6E\omega\kappa^2}\left\{\frac{3}{2}\frac{z}{l}-\frac{1}{2}\left(\frac{z}{l}\right)^3\right\}.$$

Further $\dot{u}_1=0$, or $\dot{U}_{t=0}=V$ for the weight Q, whose abscissa is l, but for all other points $\dot{U}_{t=0}=0$.

Saint-Venant then proceeds to calculate A_m and B_m, and ultimately finds:

$$u = u_1 + \frac{Q}{Q+Q'} \Sigma \frac{4}{m^3} \left\{ \frac{\dfrac{\sin \dfrac{mz}{l}}{\cos m} - \dfrac{\sinh \dfrac{mz}{l}}{\cosh m}}{\sec^2 m - \operatorname{sech}^2 m + 2P/\{m^2 (Q+Q')\}} \right\}$$

$$\times \left\{ V\tau \sin \frac{m^2 t}{\tau} - \frac{g\tau^2}{m^2} \cos \frac{m^2 t}{\tau} \right\} \text{ .... (vi).}$$

Equations (iv), (v) and (vi) form the complete analytical solution of the problem. See pp. 531—5.

[357.] Saint-Venant now treats other problems of *gradual impulse*, or as I should prefer to term it *non-impulsive resilience*. For example:

(*a*) A vertical bar of weight P terminally fixed and having a weight Q attached to its mid-point, is acted upon at that point by a constant horizontal force q. See pp. 535—8.

(*b*) The same bar is acted upon at its mid-point by a horizontal force $q =$ some function $f(t)$ of the time. Here Saint-Venant for his method of treatment appeals to the memoir of Duhamel cited in our Art. 903*. The solution contains an integral of the form $\int_0^t \cos \frac{m^2}{\tau} (t-\epsilon) f'(\epsilon)\, d\epsilon$. See pp. 538—40.

(*c*) The same bar is subjected to a sudden small but afterwards invariable horizontal displacement α of its mid-point. See pp. 540—2.

(*d*) The same bar is subjected to a small horizontal displacement α of its mid-point which is a function of the time: $\alpha = F(t)$. See pp. 542—3.

On pp. 543—7 Saint-Venant indicates another method of treating problems in non-impulsive resilience. For this he appeals to Phillips' memoir of 1864: see our account of it later.

[358.] The next problem investigated is a more important one and is thus stated: *Balancier de machine à vapeur oscillant autour d'une situation horizontale; sa flexion, sa vibration et sa résistance quand il est soumis à l'action et à l'impulsion graduelle de forces périodiquement variables s'exerçant sur ses extrémités par des bielles restant sensiblement verticales.*

I will indicate the method adopted by Saint-Venant. He supposes the arms of the beam each equal l, and that the forces applied to each extremity may be represented by a periodic term of the form $2Q \cos \Omega t$ which practically acts perpendicular to the beam. He justifies this assumption in the following manner:

Quant à la forme à assigner aux expressions des efforts verticaux q et q_1 exercés par les bielles, observons que si, du côté gauche, l'on appelle $-Q_1$ l'effort opérateur qu'exerce, tangentiellement à sa circonférence, une roue montée sur l'arbre du volant, et d'un rayon égal à la longueur r_1 de la manivelle, effort qui est rendu sensiblement constant lorsque le volant a un moment d'inertie de grandeur suffisante, et si Ω est la vitesse angulaire de la manivelle, on doit prendre, le temps t étant supposé compté à partir de l'instant où celle-ci est horizontale,

$$q_1 = -2Q_1 \cos \Omega t.$$

En effet q_1 devra avoir son maximum négatif pour l'angle $\Omega t = 0$, son maximum positif pour $\Omega t = \pi$: il devra être nul aux *points morts*, où $\Omega t = \pi/2$ et $3\pi/2$; enfin comme l'espace parcouru *verticalement* par le bouton de la manivelle pendant le temps dt est $\Omega r_1 \, dt \cos \Omega t$, le travail de la force q_1 pendant le parcours d'une demi-circonférence est $r_1 \int_{-\pi/2}^{+\pi/2} q_1 \cos \Omega t \, d(\Omega t)$; intégrale qui, si l'on y fait $q_1 = -2Q_1 \cos \Omega t$ est justement égale à $-Q_1 \pi r_1$, c'est-à-dire au travail de la force tangentielle constante $-Q_1$; en sorte que l'expression posée pour q_1 est bien ce qu'il faut pour que cette force verticale entretienne le mouvement du mécanisme en fournissant, à la fin de chaque période, le travail opérateur qui a été dépensé pendant sa durée (p. 549).

If we accept these values for the forces acting on the beam we can easily state the analytical conditions of the problem.

$$\left.\begin{array}{l}\text{For the right arm:} \quad \tau^2\, d^2u/dt^2 + l^4 d^4u/dz^4 = 0, \\ \text{with } \; d^2u/dz^2 = 0 \text{ and } E\omega\kappa^2\, d^3u/dz^3 + q = 0, \text{ when } z = l\end{array}\right\} \ldots\ldots\ldots\ldots\ldots \text{(i)}.$$

$$\left.\begin{array}{l}\text{For the left arm:} \quad \tau^2 d^2u_1/dt^2 + l^4 d^4u_1/dz_1^4 = 0, \\ \text{with } \; d^2u_1/dz_1^2 = 0 \text{ and } E\omega\kappa^2\, d^3u_1/dz_1^3 + q_1 = 0, \text{ when } z_1 = l\end{array}\right\} \ldots\ldots \text{(ii)}.$$

$$\left.\begin{array}{l}\text{When} \quad z = z_1 = 0, \text{ we must have } u = u_1 = 0 \\ du/dz = -du_1/dz_1 \text{ and } d^2u/dz^2 = d^2u_1/dz_1^2\end{array}\right\} \ldots\ldots\ldots\ldots \text{(iii)}.$$

The initial conditions will be of the following kind:

When $t = 0$, $u = \phi(z)$, $du/dt = \psi(z)$, $u_1 = \phi_1(z_1)$, $du_1/dt = \psi_1(z_1)$...(iv).

We put also: $\quad q = 2Q \cos \Omega t, \quad q_1 = 2Q_1 \cos \Omega t,$

$$\Omega = n^2/\tau \text{ where } \tau^2 = Pl^3/(2gE\omega\kappa^2).$$

Now Saint-Venant takes $u = v + U$; $u_1 = v_1 + U_1$, and chooses v and v_1 so that q and q_1 shall disappear from the equations (i) and (ii), and shall separately satisfy all the conditions but (iv). Substituting u and u_1 in the equations (i) to (iii) we find they remain the same with the suppression of q and q_1, that is: d^3U/dz^3 and d^3U_1/dz_1^3 vanish for $z = l$ and $z_1 = l$ respectively.

The solutions for U and U_1 take the usual forms in Z_m functions as coefficients in a series of circular functions of m^2t/τ, the characteristic equation being now

$$1 + \cos m \cosh m = 0 \dots\dots\dots\dots\dots\dots\dots\dots (\text{v}).$$

This is the well-known equation of Bernoulli and Euler: see our Arts. 49*, 64* and the footnote Vol. I. p. 50.

It is obvious that v and v_1 will be single terms in circular functions of Ωt or $\frac{n^2}{\tau} t$ the phase of the forced vibration, while U and U_1 will contain series in terms of $\frac{m^2}{\tau} t$ where m satisfies equation (v), or gives a free vibration. We have then to determine the constants A_m and B_m so as to satisfy the relations (iv), or so that $U = \phi(z) - v$, $dU/dt = \psi(z) - dv/dt$, when $t = 0$, with similar values for the quantities with subscript unity. The solution is thus completed. (pp. 547—553.)

[359.] The reader will remark on examining Saint-Venant's results, that if n be nearly $= m$, or the fly wheel rotate with nearly a natural period of the rod vibration, the displacement due to that natural vibration will become excessive and the danger of the beam breaking will be great. This will occur when

$$\Omega = \frac{m^2}{\tau} = m^2 \sqrt{\frac{2gE\omega\kappa^2}{Pl^3}}.$$

Let $p =$ number of revolutions of the fly wheel per second, $= \Omega/2\pi$. Then there will be great danger when:

$$p = \frac{m^2}{2\pi} \sqrt{\frac{2gE\omega\kappa^2}{Pl^3}} \dots\dots\dots\dots\dots\dots\dots\dots (\text{vi}).$$

Saint-Venant in a footnote gives the following first 8 values of m^2:

3·557, 61·70, 199·8, 417, 712·9, 1088·3, 1542·1, 2075·1;

hence, since slackening speed would be dangerous, if p had a value lying between those obtained from (vi) by substituting any two values of m^2, we have the safe maximum number of rotations per second of the fly wheel given by

$$p < \frac{3{\cdot}557}{2\pi} \sqrt{\frac{2g\,E\omega\kappa^2}{Pl^3}}.$$

This seems to me an important condition[1]. I am not aware

[1] A similar condition ought also to be satisfied between the number of rotations of the fly wheel and the least free period of stretch vibrations in the connecting rod of an ordinary engine.

whether it has been previously noticed, or how far the dimensions of the beams of ordinary beam-engines ensure its fulfilment. We can throw it into a simpler form. Let f = the deflection of either extremity of the beam subject only to its own weight, then

$$f = \frac{Pl^3}{8E\omega\kappa^2},$$

or

$$p < \frac{3{\cdot}557}{4\pi}\sqrt{\frac{g}{f}} < {\cdot}272\sqrt{\frac{g}{f}}.$$

[360.] Saint-Venant does not draw any numerical conclusions from his results, which seem to me to suggest several points of importance, but only remarks finally:

Nous n'insisterons pas sur la solution, dont nous croyons avoir posé les bases, de ce problème complexe et délicat, solution qui, une fois développée, fournira la connaissance des plus grandes dilatations à contenir dans de justes limites, en réglant les dimensions de cet organe de mécanisme, soumis à des forces toujours variables, le faisant fléchir et vibrer alternativement dans deux sens opposés (p. 553).

[361.] We now pass to that portion of Saint-Venant's work which is peculiarly characteristic of the man, namely to the practically important numerical calculation of the results given in the previous articles. This occupies pp. 553—576 (§§ 32—42). The appalling amount of work that lies behind the numbers given can only be appreciated by those who have attempted similar calculations. The graphical representation of the results, although the plates have been long engraved, has not yet been published (see footnote p. 557)[1]. The plaster model referred to in our Art. 105 will be found, however, of considerable service as offering a concise picture of the whole motion in a particular and most important case.

[362.] Saint-Venant treats in §§ 32—5 the problem of the doubly supported bar centrally struck: see our Arts. 104 and

[1] I much regret that it has been settled that these plates shall *not* be published, Saint-Venant at a date later than the footnote of 1883 having expressed an opinion that the curves ought to be plotted out for more frequent values of t/τ and z/l, as well as for a wider range of the ratio P/Q. It is to be hoped, having regard to the practical importance of the problem, that some one will be found willing to undertake the labour of the requisite numerical calculations.

346—8. He begins by tabulating the first seven values of m obtained from the characteristic equation:

$$m(\tan m - \tanh m) = 2P/Q,$$

when P/Q is very small, equals $\frac{1}{10}$, $\frac{1}{4}$, $\frac{1}{2}$, 1, 2, or is very great (p. 554)

Then for the three cases $P/Q = \frac{1}{2}$, 1 and 2 he has calculated up to six terms in m the value of the amplitudes in $u/(V\tau)$ for each component harmonic at the points $z/l = \cdot 2$, $\cdot 4$, $\cdot 6$, $\cdot 8$, and 1. He has thus been able to trace the curves having the several terms of $u/(V\tau)$ for ordinate and t/τ for abscissa. Corresponding ordinates added together gave the total deflection for various values of z/l plotted to a time base. These curves were traced from $t/\tau = \cdot 05$ to $2\cdot 25$, except in the third case ($P/Q = 2$) when they were only taken to $t/\tau = 1\cdot 9$. Unfortunately we have not these curves to examine, but the following remarks of Saint-Venant sufficiently characterise the physical nature of the impulse:

Ces cinq courbes partant du même point ($u = 0$, $t = 0$) ne reviennent, au bout de ces temps, couper l'axe des abscisses $u/(V\tau) = 0$, qu'en des points légèrement différents les uns des autres, ce qui montre qu'à aucun instant la barre ne retourne exactement à son état primitif. Ces courbes, représentant la loi et la suite du mouvement de chacun des cinq points, sont fort sinueuses; cela vient de ce que le mouvement résulte de la superposition de vibrations ayant des durées et des amplitudes de moins en moins grandes, dont chacune a son *rebond* bien avant celui de l'oscillation principale provenant du premier terme du Σ ou de la valeur m_0 de m.

Toutes ces courbes serpentantes sont, pour $t = 0$, ou à l'origine, tangentes à l'axe des abscisses, avec lequel, même, elles se confondent dans de très petites étendues, parce que l'ébranlement ne se transmet pas instantanément du point milieu aux points d'appui.

Il y a exception, bien entendu, pour les courbes relatives à $z = l$. La tangente y fait un angle demi-droit avec l'axe; et cela devait être, car, à l'instant initial, les vitesses ne sont nulles qu'en exceptant le point milieu qui reçoit le choc, et où $du/dt = V$; ce qui donne bien 1 pour la tangente trigonométrique, quotient $d(u/V\tau)$ par $d(t/\tau)$, de l'angle fait avec l'axe des abscisses par le premier élément de la courbe représentative du mouvement du point milieu.

Ces courbes, pour des points proches des appuis, s'élèvent même au-dessus de l'axe $u = 0$ des abscisses, c'est-à-dire que, par une sorte de réaction ou de rebond qui suit de près un affaissement imperceptible, les u sont négatifs. (See pp. 557 and 889.)

The last remark should be compared with that of Stokes' in another case of resilience: see our Art. 1282*.

[363.] In § 33 Saint-Venant describes how he has traced the form taken by the rod at different intervals of time from $t = \cdot 1\tau$ up to $t = 2\cdot 25\tau$. From these curves he has deduced by graphical measurement the maximum curvatures and the times at which they occur. I reproduce some of his results in the accompanying table.

Resilience of a simple-supported beam, struck transversely.

When $P/Q=$	1/4	1/2	1	2	4
Maximum deflection	$1\cdot 091\, V\tau$	$\cdot 739\, V\tau$	$\cdot 477\, V\tau$	$\cdot 297\, V\tau$	$\cdot 167\, V\tau$
at about $t=$	$1\cdot 92\tau$	$1\cdot 34\tau$	$1\cdot 18\tau$	$\cdot 82\tau$	$\cdot 78\tau$
Maximum curvature	—	$2\cdot 60\, V\tau/l^2$	$1\cdot 75\, V\tau/l^2$	$1\cdot 30\, V\tau/l^2$	—
at about $t=$	—	$\begin{Bmatrix}1\cdot 25\tau\\ 1\cdot 65\tau\end{Bmatrix}$	$1\cdot 20\tau$	$\begin{Bmatrix}\cdot 75\tau\\ 1\cdot 1\tau\end{Bmatrix}$	—

It will be noted that the instant of maximum deflection and that of maximum curvature do not coincide. Saint-Venant remarks that at each instant of time the maximum curvature is not central, although the maximum of the maxima for the various times as above tabulated is central.

The maximum stretch at any instant $= h/R$, where h is the distance of the 'outer fibre' from the neutral axis and $1/R$ is the curvature; this must be less than T_0/E, where T_0 gives the fail-limit: see our Art. 173. Hence our condition for non-failure is

$$T_0/E > h\beta\, V\tau/l^2,$$

or

$$V < \frac{1}{\beta}\frac{l^2 T_0}{Eh\tau},$$

where $\tau^2 = \dfrac{Pl^3}{2gE\omega\kappa^2}$, and β must be put equal to $2\cdot 60$, $1\cdot 75$ or $1\cdot 30$ according as P/Q equal $\frac{1}{2}$, 1 or 2.

Saint-Venant throws this into a slightly different form. By substituting τ and squaring we find:

$$\frac{QV^2}{2g} < \epsilon \,.\, \frac{T_0^2}{E}\, 2l\omega\, \frac{\kappa^2}{h^2}$$

Here T_0^2/E is the modulus of resilience, $2l\omega$ is the volume of the beam, κ^2/h^2 is in general a number independent of the linear dimensions of the cross-section, i.e. the same for all similar beams, and $\frac{\epsilon}{3} = \frac{1}{6\beta^2}\frac{Q}{P}$, and so takes the values ·04928, ·05442, and ·04933 for the three cases respectively, so that for an approximation we might take $\epsilon = \cdot 15$ for these cases. This constancy of ϵ would give Young's Theorem which was established by neglecting the inertia of the bar ($\epsilon = \frac{1}{6}$), but, as Saint-Venant rightly observes, sufficient cases have not yet been calculated to allow a safe empirical formula to be proposed.

The reader should note, however, the contents of our Art. 371 (iv) as modifying the above results.

[364.] Some remarks of Saint-Venant on p. 627 bearing on the results of the previous article are so suggestive for directions of further physical research that we cite them here in the hope that some one may ultimately be induced to undertake the needful investigations:

Plusieurs questions, du reste, se présentent, dont l'analyse ne peut encore tirer, des faits actuellement connus, une solution suffisante.

1°. Doit-on (comme ont fait les auteurs qui ont traité les problèmes de résistance vive par première approximation) regarder la limite T_0/E des dilatations statiques ou permanentes non dangereuses des fibres, comme s'appliquant aux dilatations dynamiques ne durant qu'une fraction de seconde, et qu'un même choc ne produit qu'une seule fois dans toute leur grandeur; ou bien peut-on, sans péril, en adopter une moins élevée.

2°. Doit-on, dans le calcul (numérique ou graphique) de la plus grande courbure, ajouter, comme nous avons fait [Art. 363], à ce qui vient de la vibration principale et visible, donnée par le premier terme, en m_0, du Σ, ce que fournissent passagèrement les vibrations secondaires, tertiaires, etc., représentées par les autres termes, et dont la durée périodique est incomparablement plus petite; ou bien peut-on négliger, comme sans danger, les surcroîts de dilatations de fibres qu'elles produisent par instants; ce qui reviendrait à s'en tenir aux valeurs[1] de $1/\rho$, en les affectant, tout au plus, de coefficients de sécurité ou de précaution, étrangers au calcul des vibrations accessoires?

3°. Y a-t-il, de la part des vibrations élastiques de peu de durée et d'amplitude, et vu le seul fait de leur fréquente répétition, une sorte particulière de danger, comme serait celui de détruire le *nerf* du fer forgé ou laminé, en le disposant, comme le feraient de fortes vibrations

[1] Saint-Venant here gives a reference to the equations he has given on p. 626, connecting *statical* curvature with statical deflection.

calorifiques ou une sorte de *fusion*, à revenir de l'état fibreux ou à particules entrelacées, à l'état cristallin ou grenu?

Des expériences, dont il est difficile de tracer le programme, mais où pourra jouer un rôle essentiel le mesurage de ces déformations persistantes regardées comme annonçant des commencements d'énervation et de désagrégation, seront nécessaires pour renseigner là-dessus la théorie qui devra, quels qu'en soient les résultats, se bien garder d'abdiquer son rôle et de renoncer aux considérations et patients calculs dont nous avons, à l'instar de nos maîtres, tâché de donner quelques specimens.

The experiments on repeated load to which we shall refer later in this volume have thrown light on some at least of Saint-Venant's problems.

[365.] Saint-Venant passes in § 35 to the problem of our Art. 355 with $Q'=0$. He remarks that the maximum value of the second part of u (Equation vi) treated as *consisting only of the first term* will be reached when

$$\tan\frac{m^2 t}{\tau} = -\frac{m^2 V\tau}{g\tau^2}.$$

He thus deduces for the time-terms' bracket the value

$$\frac{1}{m^2}\sqrt{(g\tau^2)^2+(m^2V\tau)^2}.$$

Hence the total deflection f produced by the blow is given by:

$f=$ maximum of $u-$ initial deflection due to weight of beam,

$$=\frac{Ql^3}{6E\omega\kappa^2}+\frac{4}{m_0^4+\dfrac{Qm_0^6}{2P}(\sec^2 m_0-\operatorname{sech}^2 m_0)}\sqrt{\left(\frac{Pl^3}{2E\omega\kappa^2}\right)^2+(m_0^2V\tau)^2}.$$

Here m_0 is the first root of the characteristic Equation (v) of Art. 356, or since $Q'=0$, of the Equation of Art. 362. Saint-Venant calculates on p. 562 the value of the coefficient of the radical and finds it has almost exactly for values of $P/Q=\frac{1}{4}, \frac{1}{2}, 1, 2, 4$ the same value, namely $Q/(3P)$, as when P/Q is extremely small.

Hence $$f=f_s+\sqrt{f_s^2+f_D^2},$$

where f_s is the statical deflection and

$$f_D=\frac{Qm_0^2}{3P}V\tau \qquad \text{(i)}$$

is the dynamical deflection of Art. 363 to a first approximation.

If $V=0$, we have non-impulsive resilience, and $f=2f_s$, a theorem of Young's.

[366.] In the next sections (§§ 36—7), Saint-Venant shews that the solution obtained on the hypothesis of Cox, that the form of the beam is at each instant of the impact the same as it would be under the same *statical* central deflection, gives a close approximation to the maximum dynamic deflection. Now Cox has shewn that

$$f_D = \frac{V}{\sqrt{\frac{6gE\omega\kappa^2}{Ql^3}\left(1+\frac{17}{35}\frac{P}{Q}\right)}}$$

(see our Vol. I. p. 894, (*b*) with proper change of notation).

Equating this to $f_D = \frac{Qm_0^2}{3P} V_T$ (see Art. 365, Eqn. (i)), we have

$$m_0^4 = \frac{3P}{Q} \Big/ \left(1+\frac{17}{35}\frac{P}{Q}\right),$$

which is Saint-Venant's second approximation to the value of m_0. It appears from his work that Cox's result for the *central maximum deflection* is accurate when we neglect m_0^8 (p. 568).

On p. 570 we have the maximum deflections calculated for the five typical cases:

	$P/Q=$	1/4	1/2	1	2	4
$\frac{f_D}{V_T}$	by Saint-Venant's series,	1·091	·739	·477	·297	·167
	by Cox's formula, or $f_D/V_T = \frac{1}{\sqrt{\frac{3P}{Q}\left(1+\frac{17}{35}\frac{P}{Q}\right)}}$	1·0904	·7375	·4737	·2908	·1683

We see that at any rate for these cases *Cox's formula gives the deflection with all the accuracy needful in practice.*

[367.] Saint-Venant next proceeds to a second approximation in other cases of resilience, i.e. he investigates the values of γ the *mass-coefficient of resilience*, see our Vol. I., p. 894 (*b*).

His § 40, 2° = our Vol. I., p. 895, c. (i).
" § 40, 3° = " " p. 895, c. (ii).
" § 40, 4° = " " p. 894, Eqn. (i).

For the case referred to in our Art. 355,

$$m_0^4 = \frac{3P}{Q+Q'} \Big/ \left(1+\frac{17}{35}\frac{P}{Q+Q'}\right).$$

The case given in our Vol. I., p. 896, (iii), Saint-Venant does not appear to have considered.

In a footnote he remarks that the second approximation will be far from exact in cases like those of Arts. 353 and 354 where the bar is free or pivoted at one point only.

[368.] Saint-Venant next proceeds to obtain Cox's formula by an elementary method. In a long footnote he gives the history and a proof of the principle of virtual shifts as applied to impulsive forces (pp. 577—82). His method is more general and simpler than Cox's, and as it gives a general expression for the value of the mass-coefficient γ, we indicate it here: see his pp. 578—87:

Let V be the initial impact-velocity of the weight Q; let V_1 be the final impact velocity, or the velocity attained by Q when the beam begins to bend, let v_1 be the velocity of any point of the beam immediately after the impact, so that $v_1 = V_1$ at the mid-point. Take the shifts at the instant when the bending effect begins as the virtual shifts, then:

$$\left(\frac{Q}{g}V - \frac{Q}{g}V_1\right)V_1 dt - \int \frac{dP}{g} v_1 \cdot v_1\, dt = 0,$$

the integral extending along the length of the beam. Dividing by $QV_1^2 \dfrac{dt}{g}$, we have

$$V/V_1 = 1 + \frac{P}{Q}\int \left(\frac{v_1}{V_1}\right)^2 \frac{dP}{P},$$

or

$$\left.\begin{aligned} V_1 &= \frac{V}{1+\gamma P/Q} \\ \gamma &= \int \left(\frac{v_1}{V_1}\right)^2 \frac{dP}{P} \end{aligned}\right\} \text{.......................... (i).}$$

where

The determination of γ thus depends entirely on the relation we choose between v_1 and V_1. Cox's assumption is that: the relation between the statical shifts at the centre and any other point holds continuously during the motion. Thus if $u = U_1\phi(z)$ be the relation,

$$\dot{u} = \dot{U}_1\phi(z), \text{ or } v_1 = V_1\phi(z).$$

This gives

$$\gamma = \int \{\phi(z)\}^2 \frac{dP}{P} \text{.......................... (ii).}$$

Now the total kinetic energy of the system after impact must be

$$= \frac{Q}{g}\frac{V_1^2}{2} + \int \frac{v_1^2}{2}\frac{dP}{g} = \frac{QV_1^2}{2g}\left(1+\gamma\frac{P}{Q}\right) = \frac{QV^2}{2g} \Big/ \left(1+\gamma\frac{P}{Q}\right).$$

This must be equal to the maximum strain-energy of the system, which is always of the form $af_D \times \frac{f_D}{2}$, a being a constant depending on the beam and f_D the maximum deflection. Thus we arrive at

$$f_D = V \sqrt{\frac{Q}{ag(1+\gamma P/Q)}} \text{ ....(iii)}.$$

This is Cox's formula: see our Arts. 1435*—7* and Vol. I., pp. 894—6.

If f_s be the statical deflection due to Q, $Q = af_s$ and

$$f_D = V \sqrt{\frac{f_s}{g(1+\gamma P/Q)}} \text{ ....(iv)}.$$

[369.] Saint-Venant adds to Cox's treatment the consideration of the approximate periodic time.

The body moved has the 'reduced total mass' $\frac{Q+\gamma P}{g}$, and the resistance to motion is au_0, where u_0 is the central shift at time t.

Hence we have $$\frac{Q+\gamma P}{g}\frac{d^2u_0}{dt^2} = -au_0 = -\frac{Q}{f_s}u_0,$$

or $$u_0 = A\sin(\beta t + B), \text{ where } \beta = \sqrt{\frac{g}{f_s}\frac{1}{1+\gamma P/Q}}$$

But when $t = 0$, $u_0 = 0$ and $\dot{u}_0 = V_1$.

Thus finally

$$u_0 = V\sqrt{\frac{f_s}{g}\frac{1}{1+\gamma P/Q}}\sin\left(\sqrt{\frac{g}{f_s}\frac{1}{1+\gamma P/Q}}\cdot t\right) \text{ ....(v)}.$$

[370.] (*a*) On pp. 587—589 the values of γ are obtained by Cox's method for the examples referred to in our Art. 367.

(*b*) On pp. 589—90 we have the case of a beam whose length exceeds the distance between the two points of support symmetrically placed. If P be the weight of beam in the span and P' of the total projecting portions we find

$$\gamma = \tfrac{17}{35} + \tfrac{3}{4}\left(\frac{P'}{P}\right)^2.$$

(*c*) On pp. 590—594 Saint-Venant treats the important case of resilience for the "solid of equal resistance," i.e. when the cross-sections are rectangles of equal breadths and of heights given by parabolic ordinates. He deals with this problem by two methods and finds in both cases that $\gamma = \frac{29}{42}$. He remarks in a footnote that the end sections which are of course in practice not of zero height, must be calculated by the methods of the memoir on flexure: see our Art. 69. But the addition of this material only introduces into γ a term of the order $\left(\frac{\text{height of end-section}}{\text{height of mid-section}}\right)^7$ which is negligible.

(*d*) On pp. 595—597 Saint-Venant deduces the result of our Art. 365 by Cox's method.

[371.] Leaving on one side for a moment Saint-Venant's §§ 52—55 we observe the following points in the concluding pages of this long *Note*:

(i) pp. 620—623. An examination of the results of the *Iron Commissioners' Report* and Hodgkinson's experiments: see our Arts. 943* and 1409—10*. This amounts to little more than the remark that Hodgkinson's $\frac{1}{2}$ is almost equal to the theoretical value $\frac{17}{35}$ of γ, and the statement that the values of the modulus obtained by applying the resilience formulae to 67 experiments agree sufficiently well among themselves.

(ii) Saint-Venant remarks that $f_D = V\sqrt{\frac{f_s}{g}\frac{1}{1+\gamma P/Q}}$ can be applied to a variety of cases of impact, as those of carriage springs, etc.; the value of γ being known $\left\{\text{i.e. } \int\left(\frac{v_1}{V_1}\right)^2 \frac{dP}{P}\right\}$, so soon as we have assumed v_1/V_1 to have the ratio of the corresponding statical deflections (p. 624). At the same time the method of vibrations involving the transcendental series ought to be used to control this result wherever it is possible (p. 625).

(iii) *The values obtained by Cox's method for the maximum curvature and so for the maximum stretch are not sufficiently exact*, and we must have recourse to the transcendental series or the numbers given in our Art. 363. Thus in the case of a simply supported beam centrally struck we should have by Cox's method $1/\rho = 3f_D/l^2$, but the values deduced from the Table in our Art. 363 give

$$3f_D/l^2 \times \begin{Bmatrix} 1{\cdot}183 \\ 1{\cdot}252 \\ 1{\cdot}486 \end{Bmatrix} \text{ according as } P/Q = \begin{Bmatrix} \frac{1}{2} \\ 1 \\ 2 \end{Bmatrix}.$$

Saint-Venant gives an empirical formula for these *three* cases on p. 627, but a better form (error $<\frac{1}{16}$) is given in the *Changements et Additions* p. 895, namely:

$$1/\rho = \sqrt{3}\frac{1}{\sqrt{E\omega\kappa^2 l}}\sqrt{\frac{QV^2}{2g}}.$$

This gives the same condition of resilience as the $\epsilon = \frac{1}{6}$ of our Art. 363.

It is noteworthy that in reality Young's Theorem is much more nearly fulfilled than would appear from the application of Cox's method: see our Vol. I., p. 895.

(iv) A second interesting point is raised in the *Changements et Additions* p. 896. Saint-Venant remarks that the formulae given are based upon the supposition that the disturbance due to the blow has had time to be reflected several times from the points of support before the moment of maximum flexure. They cease to be applicable when the bar is very long, and Q a very small weight with a very great velocity of impact:

En effet, préalablement à toute propagation, une flexion brusquement produite à l'endroit du choc peut engendrer des dilatations dangereuses, dépendant de la seule vitesse V et nullement du poids heurtant Q.

Let Ω = velocity of propagation of sound along the rod, or

$$\Omega^2 = \frac{E}{P/(2l\omega g)}.$$

We easily deduce $\frac{V\tau}{l^2} = \frac{V}{\Omega\kappa}$, where $\tau^2 = Pl^3/(2gE\omega\kappa^2)$.

The corresponding maximum values of the stretches are by Art. 363:

For	$P/Q=\frac{1}{2}$	$P/Q=1$	$P/Q=2$
$s_0=$	$2{\cdot}60\frac{h}{\kappa}\frac{V}{\Omega}$	$1{\cdot}75\frac{h}{\kappa}\frac{V}{\Omega}$	$1{\cdot}30\frac{h}{\kappa}\frac{V}{\Omega}$

Now Boussinesq has shewn that the stretch produced in the element struck at the first instant of the blow has for magnitude, whatever be the relation between P and Q:

$$s_0 = \frac{h}{\kappa}\frac{V}{\Omega}.$$

Hence if we find that value for P/Q (say, n) for which the numerical coefficient of $\frac{h}{\kappa}\frac{V}{\Omega}$ is sensibly unity, we may say that the maximum stretch for that and all other larger values of P/Q is given by the expression $\frac{h}{\kappa}\frac{V}{\Omega}$, and takes place in the first instant of the impact.

See the Note in the *Comptes rendus* 1882, p. 1044, or Boussinesq, *Application des Potentiels à l'étude...du mouvement des solides élastiques*, p. 486.

From some slight calculations I have made I believe this result will be reached when P/Q lies between 2·5 and 3. If this be true, it very much limits the range within which there is any necessity to apply the transcendental series to ascertain the curvature and so the condition of failure. We may then, I think, say that after $P/Q = 2{\cdot}5$, the maximum-stretch is always given by the formula $\frac{V}{\Omega}\frac{h}{\kappa}$ or is independent of the mass of Q.

[372.] We must now return to pp. 597—619 of Saint-Venant's note which we have omitted above. They deal with *Willis' Problem* or the resilience of a horizontal beam subjected to a travelling load: see our Arts. 1417*—1422*. We shall include under our discussion the memoirs of Phillips[1] and Renaudot[2], because these writers have made mistakes in their analysis, which have been rectified by Saint-Venant. With Saint-Venant's additions and rectifications we shall thus be able to give the reader a more complete view of the advance made by the problem since the memoir of Stokes: see our Arts. 1276*—1291*.

[373.] We will first give the equations for the complete problem as propounded by Phillips. Let P be the weight, $2l$ the length, $E\omega\kappa^2$ the rigidity of the beam, u the shift to the right and u_1 to the left of the travelling load Q (distant $x = Vt$ from the right-hand terminal) of points distant z and z_1 from right and left-hand ends of the beam. We shall suppose the beam simply supported.

[1] *Calcul de la résistance des poutres droites telles que les ponts, etc. sous l'action d'une charge en mouvement. Annales des mines*, t. VII., pp. 467—506, 1855.

[2] *Etude de l'influence des charges en mouvement sur la résistance des ponts métalliques à poutres droites. Annales des ponts et chaussées*, t. I. 4e série, pp. 145—204, 1861.

Then for the beam we have

$$\left.\begin{aligned} E\omega\kappa^2 \frac{d^4u}{dz^4} - \frac{P}{2l} &= -\frac{P}{2lg}\frac{d^2u}{dt^2} = -\frac{P}{2l}\frac{V^2}{g}\frac{d^2u}{dx^2} \\ E\omega\kappa^2 \frac{d^4u_1}{dz_1^4} - \frac{P}{2l} &= -\frac{P}{2lg}\frac{d^2u_1}{dt^2} = -\frac{P}{2l}\frac{V^2}{g}\frac{d^2u_1}{dx^2} \end{aligned}\right\} \text{..........(i)},$$

$$\left.\begin{aligned} u = 0, \text{ and } \frac{d^2u}{dz^2} = 0 \text{ when } z = 0 \\ u_1 = 0, \text{ and } \frac{d^2u_1}{dz_1^2} = 0 \text{ when } z_1 = 0 \end{aligned}\right\} \text{................(ii)};$$

and for the conditions at the load:

$$(u)_x - (u_1)_{2l-x} = 0,\ \left(\frac{du}{dz}\right)_x + \left(\frac{du_1}{dz_1}\right)_{2l-x} = 0,\ \left(\frac{d^2u}{dz^2}\right)_x - \left(\frac{d^2u_1}{dz_1^2}\right)_{2l-x} = 0 \text{ ... (iii)},$$

together with:

$$-E\omega\kappa^2\left\{\left(\frac{d^3u}{dz^3}\right)_x + \left(\frac{d^3u_1}{dz_1^3}\right)_{2l-x}\right\} - Q = -\frac{Q}{g}\frac{d^2y}{dt^2}, \text{ where } y = (u)_{z=x=Vt}\text{...(iv)}.$$

No general solution has yet been found for these equations. But omitting the condition of *initial zero velocities* it is possible to satisfy all the other Equations (i) to (iv) by algebraic expressions in z and x, when we neglect in successive approximations successive powers of a certain quantity which is small in all practical applications. Further, it is possible to add vibratory parts to the algebraic solution which satisfy very approximately the initial conditions (pp. 599 and 891).

[374.] Saint-Venant's method of solution differs from that of Stokes and includes the effect of the inertia of the beam. We will indicate its stages.

1*st Approximation*. Let us neglect the terms in V^2 in Equations (i) to (iv), or find only the statical shifts for the load Q at a point x. We have:

$$\left.\begin{aligned} u' &= Q\frac{2l-x}{12lE\omega\kappa^2}\left[(4lx - x^2)z - z^3\right] + P\frac{8l^3z - 4lz^3 + z^4}{48lE\omega\kappa^2} \\ u_1' &= Q\frac{x}{12lE\omega\kappa^2}\left[(4l^2 - x^2)z_1 - z_1^3\right] + P\frac{8l^3z_1 - 4lz_1^3 + z_1^4}{48lE\omega\kappa^2} \end{aligned}\right\} \text{.....(v)}.$$

2*nd Approximation*.

Now let $\frac{1}{\beta} = \frac{2QV^2l}{3gE\omega\kappa^2}$, or β is the same as Stokes' β of our Art. 1278* (where c is written for our present l), then in practice $1/\beta$ is always $< 1/12$ or even than $1/20$ and is the small quantity of our approximations. In the above equations we shall replace

$$\frac{PV^2}{2lgE\omega\kappa^2} \text{ by } \frac{1}{\beta}\frac{3P}{4Ql^2}, \text{ and } \frac{QV^2}{E\omega\kappa^2 g} \text{ by } \frac{1}{\beta}\frac{3}{2l}.$$

Let us assume

$$u = u' + \frac{1}{\beta} U, \quad u_1 = u_1' + \frac{1}{\beta} U_1,$$

substitute and neglect $\left(\frac{1}{\beta}\right)^2$. We find from Equations (i) by dividing out by $1/\beta$:

$$\left.\begin{aligned} \frac{d^4U}{dz^4} &= \frac{3P}{8l^3E\omega\kappa^2}(2l-x)\,z = \frac{3P}{8l^3E\omega\kappa^2}x_1 z, \text{ if } x_1 = 2l - x, \\ \frac{d^4U_1}{dz_1^4} &= \frac{3P}{8l^3E\omega\kappa^2}xz_1 \end{aligned}\right\} \ldots \text{(vi)}.$$

Equations (ii) now become:

$$\left.\begin{aligned} U = 0, \text{ and } \frac{d^2U}{dz^2} = 0 \text{ when } z = 0, \\ U_1 = 0, \text{ and } \frac{d^2U_1}{dz_1^2} = 0 \text{ when } z_1 = 0 \end{aligned}\right\} \ldots\ldots\ldots\ldots\ldots \text{(vii)}.$$

Integrating (vi) we have

$$\left.\begin{aligned} U &= \frac{3Px_1}{8l^3E\omega\kappa^2}\left\{\frac{z^5}{5\,!} + \frac{Cz^3}{3\,!} + Dz\right\} \\ U_1 &= \frac{3Px}{8l^3E\omega\kappa^2}\left\{\frac{z_1^5}{5\,!} + C_1\frac{z_1^3}{3\,!} + D_1 z_1\right\} \end{aligned}\right\} \ldots\ldots\ldots\ldots\ldots \text{(viii)}.$$

These satisfy equations (vii).
It remains to determine C, D, C_1, D_1, by Equations (iii) and (iv).
But they become:

$$\left.\begin{aligned} (U)_{z=x} = (U_1)_{z_1=x_1}, \quad \left(\frac{dU}{dz}\right)_{z=x} + \left(\frac{dU_1}{dz_1}\right)_{z_1=x_1} = 0, \\ \left(\frac{d^2U}{dz^2}\right)_{z=x} = \left(\frac{d^2U_1}{dz_1^2}\right)_{z_1=x_1} \end{aligned}\right\} \ldots\ldots\ldots \text{(ix)}.$$

Further, (iv) may be written

$$\left(\frac{d^3U}{dz^3}\right)_{z=x} + \left(\frac{d^3U_1}{dz_1^3}\right)_{z_1=x_1} = \frac{Q\beta}{E\omega\kappa^2 g}\frac{d^2y}{dt^2} = \frac{3}{2l}\frac{d^2y}{dx^2}.$$

Now $y = u_1$ when $z = x_1$, or after a short reduction

$$y = \frac{Qx^2x_1^2}{6lE\omega\kappa^2} + \frac{Pxx_1(4l^2 + xx_1)}{48lE\omega\kappa^2}.$$

Since $x_1 = 2l - x$, we have: $\dfrac{d(xx_1)}{dx} = 2(l - x) = x_1 - x$, and thus find

$$\frac{3}{2l}\frac{d^2y}{dx^2} = \frac{1}{E\omega\kappa^2}\left\{2Q - \frac{3xx_1}{l^2}\left(Q + \frac{P}{8}\right)\right\}.$$

Hence

$$\left.\begin{aligned}\left(\frac{d^3U}{dz^3}\right)_x+\left(\frac{d^3U_1}{dz_1^{\,3}}\right)_{x_1}&=\frac{1}{E\omega\kappa^2}\left\{2Q-\frac{3xx_1}{l^2}\left(Q+\frac{P}{8}\right)\right\}\\&=\frac{Q}{E\omega\kappa^2l^2}(2l^2-6lx+3x^2)+\frac{3P}{8E\omega\kappa^2l^2}(-2lx+x^2)\end{aligned}\right\}\ldots(\text{x}).$$

This result does not agree with Saint-Venant's on p. 606 (Equation (t_9)) but it will do with that in the Errata, p. 900, if the coefficients of the brackets of the latter are inverted.

From the third Equation of (ix) and from (x) we easily find with the help of (viii) the following equations to determine C, C_1,

$$\frac{x^2}{6}+C=\frac{x_1^{\,2}}{6}+C_1;\quad x_1C+xC_1=-2lxx_1+\frac{8Ql}{3P}(2l^2-3xx_1)\ldots(\text{xi}).$$

This differs in the sign of the bracket in the second equation from Saint-Venant's Equation (x_9) on p. 606.

Solving (xi) we have

$$\left.\begin{aligned}C&=-\frac{x}{6}(x+5x_1)+\frac{4Q}{3P}(2l^2-3xx_1)\\C_1&=-\frac{x_1}{6}(x_1+5x)+\frac{4Q}{3P}(2l^2-3xx_1)\end{aligned}\right\}\ldots\ldots\ldots\ldots(\text{xii}).$$

[375.] We can easily test these results. The bending-moment

$$M=-E\omega\kappa^2\frac{d^2u}{dz^2}=-E\omega\kappa^2\frac{d^2u'}{dz^2}-\frac{E\omega\kappa^2}{\beta}\frac{d^2U}{dz^2}$$

$$=\frac{Qx_1z}{2l}+\frac{P(2l-z)z}{4l}-\frac{3Px_1}{8l^3\beta}\left\{\frac{z^3}{6}-\frac{x}{6}(x+5x_1)z+\frac{4Q}{3P}(2l^2-3xx_1)z\right\}\ldots(\text{xiii}^{\text{a}}).$$

Put $z=l$, and $x=x_1=l$, and we find

$$M_l=\frac{Ql}{2}\left(1+\frac{1}{\beta}\right)+\frac{Pl}{4}\left(1+\frac{5}{4}\frac{1}{\beta}\right)\ldots\ldots\ldots\ldots(\text{xiii}^{\text{b}}).$$

This is Saint-Venant's result (z_9) on p. 607.

It gives the bending moment at the centre when the train is passing that point. If we put $P=0$ or neglect the weight of the beam, we have Stokes' result. Phillips finds by overlooking several terms and by means of a longer analysis $3/(4\beta)$ in the second bracket.

[376.] We are now in a position to find D and D_1 and so determine the deflection at any point. From Equations (ix) I have calculated the following value for D:

$$D=\frac{1}{360}\{x_1x(58x^2+8x_1^{\,2}+92xx_1)+7x^4\}+\frac{2Q}{9P}(3xx_1-2l^2)x(x+2x_1)\ldots(\text{xiii}).$$

Equations (xii) and (xiii) determine C, D, and so U from (viii).

Adding $\frac{1}{\beta}U$ to u' of (v) we obtain the complete solution to this degree of approximation. We may write down the complete value thus obtained, u_1 being obviously given by interchanging x_1 with x and z_1 with z:

$$u = \frac{Qx_1}{12lE\omega\kappa^2}\{x(2x_1+x)z - z^3\} + \frac{P}{48lE\omega\kappa^2}(8l^3z - 4lz^3 + z^4)$$
$$+ \frac{1}{\beta}\frac{3Px_1}{8l^3E\omega\kappa^2}\left[\frac{z^5}{120} + \frac{z^3}{6}\left\{-\frac{x}{6}(x+5x_1) + \frac{4Q}{3P}(2l^2 - 3xx_1)\right\}\right.$$
$$\left. + \frac{z}{360}\{x_1x(58x^2+8x_1^2+92xx_1)+7x^4\} + \frac{2z}{9}\frac{Q}{P}(3xx_1-2l^2)x(x+2x_1)\right] \quad\text{....(xiv}^a\text{)}.$$

This embraces both Saint-Venant's forms (ω) and (ω'), p. 615 g, and I have tested them, and find they agree with this result.

If $x = x_1 = z = l$ we find:

$$u_l = \frac{Ql^3}{6E\omega\kappa^2} + \frac{5Pl^3}{48E\omega\kappa^2} + \frac{1}{\beta}\left\{\frac{Ql^3}{6E\omega\kappa^2} + \frac{9}{80}\frac{Pl^3}{E\omega\kappa^2}\right\}$$
$$= \frac{Ql^3}{6E\omega\kappa^2}\left(1 + \frac{1}{\beta}\right) + \frac{5Pl^3}{48E\omega\kappa^2}\left(1 + \frac{27}{25}\frac{1}{\beta}\right) \quad\text{..........(xiv}^b\text{)}.$$

If we put $P=0$, we obtain Stokes' result: see our Art. 1287*. It will be observed that these expressions for the bending-moment and the deflection have been reached *without any assumption as to the value of the ratio* Q/P.

[377.] We may make some remarks on the above results. Phillips first gave the complete equations for the problem and included the effect of the inertia of the beam (i.e. the terms in P). He obtained erroneous coefficients, however, for the terms in $1/\beta$. The correct values were first obtained by Saint-Venant, and his process is much shorter than Phillips'. In § 54 (pp. 609—612) Saint-Venant gives an elementary proof of the value of the bending moment in our equation (xiii^a). He does not make use of the general differential equations, but calculates and sums the parts of the bending moment due to statical loading, to the 'centrifugal force' of the travelling load $\left(\frac{Q}{g}\frac{V^2}{\rho}\right)$, and to the mass accelerations $\left(\frac{P}{2lg}dz\frac{d^2u}{dt^2}\right)$ of each element dz of the beam. The parts due to the last two influences are of the first order in $1/\beta$ and so we use in them the statical values for $1/\rho$, the curvature, and u.

We may ask whether the expressions in Equations (xiii^b) and (xiv^b) give the maxima values of M and u.

In the value of M the part affected by Q has its maximum when z = its greatest value x; further, the principal portion of the same part $\left\{\frac{Q}{2}\frac{x(2l-x)}{l}\right\}$ has its maximum when $x=l$. Again the principal portion of the part in P, namely $\frac{P}{4} z \frac{2l-z}{l}$, has its maximum for $z=l$. The other parts of the expression for M are always much less and thus will give only an influence of the second order on the maximum values of z and x, i.e. their influence will not be sensible on the value of the maximum moment (p. 607). It is also easy to see that the maximum deflection is the mid-point deflection at the instant of transit of the load over the mid-point.

Throughout his discussion of the problem Saint-Venant does justice to Stokes' memoir; it will be observed that he frequently adopts Stokes' methods, but the extension of the results to any ratio of Q/P is in itself no small advance.

[378.] We shall now shew how the results obtained in (xiv[a]) must be modified in order that the condition for initial zero-velocity in the parts of the beam may be satisfied. This involves the introduction of periodic terms. Stokes had introduced such a periodic term on the assumption that Q/P was small (see our Art. 1289*). Phillips had endeavoured to measure the magnitude of the periodic terms which would enable us to dispose of the finite initial velocities which the above solution pre-supposes; he found that these terms were much smaller than the principal algebraic terms (Saint-Venant, pp. 613—614), but this does not prove that we may neglect them as compared with the terms in $1/\beta$. Saint-Venant adopting Stokes' approximate method, but without his assumption of the smallness of Q/P, introduces a periodic term which allows approximately (to the order $1/\beta$) for the zero initial velocities of the beam.

[379.] It will be remembered that Stokes' method consists in replacing each force acting on the beam by a uniformly distributed force which produces the same mean deflection as would be produced by the actual force taken alone (see our Art. 1288*). By this method he arrives at the following equation:

$$\frac{15E\omega\kappa^2}{Pl^3}\nu + \frac{155}{126}\frac{V^2}{g}\frac{d^2\nu}{dx^2} = \frac{Q}{P}X\left\{1 - \frac{V^2}{g}\frac{d^2(\nu X)}{dx^2}\right\} \quad \text{........(i)},$$

where $$\left.\begin{matrix}X\\Z\end{matrix}\right\} = 5\,\frac{8l^3\begin{Bmatrix}x\\z\end{Bmatrix} - 4l\begin{Bmatrix}x^3\\z^3\end{Bmatrix} + \begin{Bmatrix}x^4\\z^4\end{Bmatrix}}{16l^4},$$

and the deflection at z is given by

$$u = Z\left(\nu + \frac{Pl^3}{15E\omega\kappa^2}\right) \text{ ....(ii)},$$

thus determining what is represented by ν.

The equation (i) shews that ν is of the same order as Q/P, and Stokes solves it on the supposition that Q/P is so small that quantities of the order $(Q/P)\times\nu$ may be neglected, i.e. he omits the last term of the bracket on the right-hand side. Saint-Venant, however, seeks a value of ν by approximations in which powers of $1/\beta$ are neglected, in other words, *he makes no assumptions as to the value of the ratio Q/P except that P/Q is not to be extremely large.* In most practical cases Q and P will not be very far from equality, and the exception is accordingly legitimate. If we take

$$r^2/l^2 = \frac{31}{252}\frac{P}{Q}\frac{1}{\beta}, \text{ where } \frac{1}{\beta} = \frac{2QlV^2}{3gE\omega\kappa^2},$$

we have the small quantities in terms of which Saint-Venant solves the equation (i).

It will be found that $r/2l = 1/q$, where q is the constant of Stokes' investigation: see our Art. 1290*.

{We may note that S_1 of Stokes $= \dfrac{Pl^3}{6E\omega\kappa^2}$,

and f_s' of Saint-Venant $= \dfrac{5Pl^3}{48E\omega\kappa^2}$}.

[380.] The solution found by Saint-Venant is given by:

$$\nu = \frac{Ql^3}{15E\omega\kappa^2}\left(X - r^2\frac{d^2X}{dx^2} + \frac{1}{\beta}X\frac{6lx - 3x^2 - 2l^2}{l^2}\right)$$
$$- \frac{Ql^3}{6E\omega\kappa^2}\left\{1 - \frac{2}{\beta} + 3\left(\frac{r}{l}\right)^2\right\}\frac{r}{l}\sin\frac{x}{r} \text{ .... (iii)}.$$

See his p. 615 e.

Substituting for ν in equation (ii) of Art. 379 we have u. For the central-deflection as the load passes we find

$$\left.\begin{aligned} u_l = \frac{125}{128}\frac{Ql^3}{6E\omega\kappa^2}\left(1+\frac{1}{\beta}\right) + \frac{5Pl^3}{48E\omega\kappa^2}\left(1 + \frac{155}{336}\frac{1}{\beta}\right) \\ - \frac{25l^3}{96E\omega\kappa^2}\left[Q\left(1-\frac{2}{\beta}\right) + \frac{31}{84}\frac{P}{\beta}\right]\frac{r}{l}\sin\frac{l}{r}\end{aligned}\right\}\text{...(iv)}.$$

The algebraic terms as might be supposed owing to the method of approximation, are not exactly the same as in (xiv^b). The factor $\frac{125}{128}$ instead of 1 is not, however, important, while the factor $\frac{155}{336}$ instead of

$\frac{27}{25}$ occurs only in terms involving $1/\beta$. Saint-Venant concludes that the algebraic terms given by the first method are the correct ones, and that we may add to them the expression

$$-\frac{25l^3}{96E\omega\kappa^2}\left[Q\left(1-\frac{2}{\beta}\right)+\frac{31}{84}\frac{P}{\beta}\right]\frac{r}{l}\sin\frac{l}{r},$$

in order to approximately account for the periodic terms. This result and (iv) differ from those of Saint-Venant (α') p. 615 h and (δ') p. 615 i but seem to me to give the correct value of u_l.

The corresponding part to be subtracted from the bending-moment at the centre as the load passes it is

$$-\frac{5}{8}l\left[Q\left(1-\frac{2}{\beta}\right)+\frac{31}{84}\frac{P}{\beta}\right]\frac{r}{l}\sin\frac{l}{r}.$$

This again differs from Saint-Venant's results (β') p. 615h and (ϵ') p. 615 j. By a misprint which has escaped correction he has the fraction $\frac{35}{24}$ where I have $\frac{31}{84}$.

[381.] The last extension of the problem which we shall consider here is that of Renaudot, who does not deal with the case of an isolated load (as a locomotive) but with that of a continuous load (as a train of trucks or carriages) crossing the bridge. Let p be the weight per foot-run of the girder, p' that of the travelling load the head of which is distant $x = Vt$ from the right hand terminal. In this case equations (i) of our Art. 373, are replaced, on the supposition that the train is longer than bridge, by

$$E\omega\kappa^2\frac{d^4u}{dz^4}-(p+p')=-\frac{p}{g}\frac{d^2u}{dt^2}-\frac{p'}{g}\frac{d^2w}{dt^2}\text{ ....(i).}$$

$$E\omega\kappa^2\frac{d^4u_1}{dz_1{}^4}-p=-\frac{p}{g}\frac{d^2u_1}{dt^2}=-\frac{pV^2}{g}\frac{d^2u_1}{dx^2}\text{ ....(ii).}$$

Here w is the shift of the element $(p'/g)\,dz$ of the train on the bridge, and z is to be put in w equal to Vt less the constant distance between the given element and the head of the train. Thus while the z in d^2u/dt^2 is not a function of t, that in d^2w/dt^2 is to be treated as a function of t, or since $x = Vt$ we may write:

$$\frac{d^2w}{dt^2}=V^2\left\{\frac{d^2u}{dx^2}+2\frac{d^2u}{dx\,dz}+\frac{d^2u}{dz^2}\right\}$$

Thus the first equation becomes:

$$E\omega\kappa^2\frac{d^4u}{dz^4}-(p+p')=-\frac{pV^2}{g}\frac{d^2u}{dx^2}-\frac{p'V^2}{g}\left\{\frac{d^2u}{dx^2}+2\frac{d^2u}{dx\,dz}+\frac{d^2u}{dz^2}\right\}\text{ ... (iii).}$$

Starting from equations (ii) and (iii) with the necessary terminal conditions for each portion of the girder, we may proceed as in our Arts. 373—376 to determine first the statical and then the first dynamical approximation. The maximum bending-moment will be greatest when the load just covers the whole girder. It is then given by

$$M_l = \frac{p+p'}{2} l^2 \left(1 + \frac{5}{8}\frac{1}{\beta'}\right), \text{ where } \frac{1}{\beta'} = \frac{p' l^2 V^2}{g E \omega \kappa^2}.$$

Similarly we may deduce the bending-moment when the train is headed by a locomotive of weight Q followed by a train of weight p' per foot-run. In order to obtain the position of the maximum bending-moment, it will, as in Art. 377, be sufficient to find the values of z and x which give the maximum moment for the statical approximation. These are

$$x = 2l - Q/p', \quad z = l\left\{1 + \frac{Q}{2pl} \cdot \frac{Q}{2(p+p')l}\right\},$$

and they must be substituted in the second approximation involving the terms in $1/\beta$ and $1/\beta'$ (see pp. 616—618).

Renaudot neglects the term $2\dfrac{d^2u}{dx\,dz}$ in equation (iii) as of small importance. He arrives at a wrong value, i.e. $\left(1 + \dfrac{2}{3\beta'}\right)$, for the bracket in the value of the bending-moment.

[382.] Saint-Venant remarks that Phillips has also treated the case of a travelling load crossing a beam *doubly built-in*. His solution is, however, erroneous, as has been pointed out both by Bresse and Saint-Venant, nor would it be of much value to correct his results, for built-in ends (*encastrements*) never produce their full effect, and such alternating motions as occur with travelling loads in bridges soon deprive such ends of nearly all their effect (see p. 619 and our Arts. 733* and 188).

There is also a reference on p. 619 to Bresse's exact solution for the case of a bridge across which a very long train is continuously moving with velocity V, so that the bridge takes up a permanent form. In this case equation (iii) of Art. 381 becomes

$$E\omega\kappa^2 \frac{d^4u}{dz^4} - (p+p') = -\frac{pV^2}{g}\frac{d^2u}{dz^2},$$

and we can find an exact solution. It gives for the maximum bending-moment

$$M_l = \frac{p+p'}{2} l^2 . 2\beta'' \{\sec\sqrt{1/\beta''} - 1\}, \text{ where } \frac{1}{\beta''} = \frac{pl^2V^2}{gE\omega\kappa^2},$$

or,

$$M_1 = \frac{p+p'}{2} l^2 \left(1 + \frac{5}{12}\frac{1}{\beta''}\right) \text{ approximately.}$$

This result is less than that of our Art. 381, or we see that the dangerous instant is that in which the train just covers the whole bridge. It is then producing impulsive changes in the elastic line of the bridge, and not a steady form of the elastic line as in the case of a very long train imagined to be continuously crossing.

[383.] We now reach Saint-Venant's last contribution to the annotated Clebsch, namely, the *Note finale du* § 73, pp. 689—752. It is entitled: *Théorie de la flexion et des autres petites déformations des plaques élastiques planes minces, tirée directement des équations différentielles générales de l'équilibre d'élasticité des solides.*

The *Note* consists of four essentially distinct parts: (i) a deduction of the general elastic body-shift equations for thin plates; (ii) a full discussion of the contour conditions, and the controversy with regard to them; (iii) the solutions for statical equilibrium of thin circular plates; and (iv) a reproduction with extensions of Navier's results obtained in the memoir of 1820, and hitherto only published in extract: see our Arts. 258*—64*. I propose to deal somewhat at length with this *Note* as it forms distinctly the best treatment hitherto given for thin plates. Saint-Venant adopts Boussinesq's method (see the memoirs of 1871 and 1879 in the Chapter devoted to that elastician) but with certain important modifications. He describes Clebsch's investigation, notwithstanding that it starts with unnecessary simplifications, as "obscure, indirect and very complex." I think the terms are fully warranted.

[384.] Let the mid-plane of the plate be taken as that of $z = 0$ and let its faces be $z = \pm \epsilon$. We shall endeavour to deduce from the three body-stress equations, a single equation involving only the stresses $\widehat{xx}$, $\widehat{yy}$, $\widehat{xy}$ and given quantities. Let the body-stress equations be

$$\left.\begin{aligned}\frac{d\widehat{xx}}{dx} + \frac{d\widehat{yx}}{dy} + \frac{d\widehat{zx}}{dz} + X = 0\\ \frac{d\widehat{xy}}{dx} + \frac{d\widehat{yy}}{dy} + \frac{d\widehat{zy}}{dz} + Y = 0\\ \frac{d\widehat{xz}}{dx} + \frac{d\widehat{yz}}{dy} + \frac{d\widehat{zz}}{dz} + Z = 0\end{aligned}\right\}\dots\dots\dots\dots(\text{i}).$$

Adding the third of these equations to the differentials of the first two with regard respectively to x and y, such differentials before addition being multiplied by z, we find

$$z\left\{\frac{d^2\widehat{xx}}{dx^2} + 2\frac{d^2\widehat{xy}}{dxdy} + \frac{d^2\widehat{yy}}{dy^2}\right\} + \frac{d}{dz}\left\{\frac{d}{dx}(z\,.\,\widehat{zx}) + \frac{d}{dy}(z\,.\,\widehat{yz}) + \widehat{zz}\right\} + z\left(\frac{dX}{dx} + \frac{dY}{dy}\right) + Z = 0.$$

Integrating this from $z=+\epsilon$ to $z=-\epsilon$,

$$\int_{-\epsilon}^{+\epsilon}\frac{d^2\widehat{xx}}{dx^2}zdz+2\int_{-\epsilon}^{+\epsilon}\frac{d^2\widehat{xy}}{dx\,dy}zdz+\int_{-\epsilon}^{+\epsilon}\frac{d^2\widehat{yy}}{dy^2}zdz+\phi(xy)=0\ldots\ldots\text{(ii)},$$

where

$$\phi(xy)=Z'+(\widehat{zz})_{+\epsilon}-(\widehat{zz})_{-\epsilon}+\frac{dX''}{dx}+\frac{dY''}{dy}+\frac{d}{dx}[\epsilon\{(\widehat{zx})_{+\epsilon}+(\widehat{zx})_{-\epsilon}\}]$$
$$+\frac{d}{dy}[\epsilon\{(\widehat{zy})_{+\epsilon}+(\widehat{zy})_{-\epsilon}\}]\ldots\text{(iii)},$$

and $$Z'=\int_{-\epsilon}^{+\epsilon}Zdz,\quad X''=\int_{-\epsilon}^{+\epsilon}zXdz,\quad Y''=\int_{-\epsilon}^{+\epsilon}zYdz;$$

the subscripts denote as usual that the stresses are to be given their values at the surfaces $z=\pm\epsilon$.

All the terms in the expression $\phi(xy)$ are thus known quantities.

[385.] The question of what further assumptions we shall make now arises. Those usually made are the following:

1°. $\widehat{zz}=0$. (This is made even by Boussinesq and Lévy, the most recent writers on the subject.)

2°. $s_x=z/\rho$, $s_y=z/\rho'$, where ρ and ρ' are the two curvatures of the plate at *its mid-plane* for the point x, y. It follows that:

$$\left.\begin{aligned}s_x&=-z\frac{d^2w_0}{dx^2}\\ s_y&=-z\frac{d^2w_0}{dy^2}\end{aligned}\right\}\ldots\ldots\ldots\ldots\ldots\ldots\ldots\ldots\text{(iv)},$$

where w_0 is the normal shift of the point x, y of the mid-plane.

Using the stress-strain relations for three planes of elastic symmetry (see our Art. 117 (*a*)), we easily find from 1° and 2°:

$$\left.\begin{aligned}\widehat{xx}&=(a-e'^2/c)\,s_x+(f'-d'e'/c)\,s_y\\ \widehat{yy}&=(f'-d'e'/c)\,s_x+(b-d'^2/c)\,s_y\end{aligned}\right\},\text{ and }\frac{d^2\widehat{xy}}{dx\,dy}=-2zf\frac{d^4w_0}{dx^2dy^2}\ldots(\text{iv}^{\text{b}}).$$

Substituting in (ii) and integrating we have the equation:

$$(a-e'^2/c)\frac{d^4w_0}{dx^4}+2(2f+f'-d'e'/c)\frac{d^4w_0}{dx^2dy^2}+(b-d'^2/c)\frac{d^4w_0}{dy^4}=\frac{3}{2\epsilon^3}\phi(xy)\ldots\text{(v)}.$$

This becomes in the case of elastic isotropy parallel to the mid-plane:

$$\left(\frac{d^2}{dx^2}+\frac{d^2}{dy^2}\right)^2w_0=\frac{3}{2H\epsilon^3}\phi(xy)\ \ldots\ldots\ldots\ldots\ldots\text{(vi)},$$

where $H=a-e'^2/c$, the plate-modulus of our Art. 323.

This is the equation obtained by Lagrange, Poisson and Cauchy: see our Arts. 284*, 484* and 640*.

[386.] On pp. 696—700 (§§ 4—5) Saint-Venant considers what are the arguments in favour of the assumptions 1° and 2° of the previous Article. He remarks that owing to the thinness of the plate, the normal or z variations of both the stresses and the strains must be large as compared with the longitudinal variations. Hence as a first approximation, we have the fluxions with regard to x and y of both stress and strain components more and more nearly zero as the plate is taken thinner and thinner. It is sufficient however to assume that those of $\widehat{xx}$, $\widehat{xy}$ and $\widehat{yy}$ are zero or small. The body stress equations then give:

$$\frac{d\widehat{zx}}{dz} + X = 0, \quad \frac{d\widehat{zy}}{dz} + Y = 0.$$

Thus the stresses $\widehat{zx}$, $\widehat{zy}$ on integration will be of the order ϵ, or as Saint-Venant puts it:

Si on les intègre par rapport à la petite coordonnée z on voit que les composantes $\widehat{zx}$, $\widehat{zy}$ n'ont de valeurs, à l'intérieur d'un tronçon ou élément de plaque, que celles qu'elles peuvent avoir sur une des deux bases, plus ce qui vient des forces A, B, agissant sur sa masse. Ces forces locales n'ont qu'une influence insignifiante qui n'est presque rien en comparaison de ce qui vient à la fois de toutes les forces agissant sur le reste de la plaque ainsi que sur ses bords par les réactions des appuis ou autrement, et dont les effets accumulés se transmettent au tronçon à travers ses quatre faces latérales, ce qui s'applique surtout aux composantes agissant horizontalement (pp. 697—8).

The third body stress equation, however, shows that $\widehat{zz}$ is very small as compared with $\widehat{zx}$, $\widehat{zy}$ because these quantities occur with *lateral variation*, hence $\widehat{zz}$ is doubly small as compared with $\widehat{xx}$, $\widehat{xy}$ and $\widehat{yy}$. Thus we may take $\widehat{zz} = 0$ as all writers have hitherto done.

[387.] This argument is not, perhaps, quite convincing. It would seem at first sight better to assume $\widehat{zz}$ *to be very approximately a function of* x, y *only*. The expressions then for $\widehat{xx}$, $\widehat{xy}$, $\widehat{yy}$ would contain together with the terms linear in z, terms not involving z, but functions of x, y only. These terms disappear when we substitute them in equation (ii) and integrate between $z = +\epsilon$ and $-\epsilon$. But here a new difficulty arises; suppose the surface of the plate $z = +\epsilon$ subjected to a load $\widehat{zz} = \chi(x, y)$. This will make no change in the first three terms of equation (ii) of Art. 384 although we cannot suppose $\widehat{zz} = 0$, but it will lead to a difficulty with regard to the expression $\phi(xy)$.

This expression contains terms of the form $(\widehat{zz})_{+\epsilon}$ and $(\widehat{zz})_{-\epsilon}$; the former $= \chi(x, y)$ and the latter is zero. Hence it follows that $\widehat{zz}$ must vary with z from $+\epsilon$ to $-\epsilon$. Saint-Venant (p. 699) says we must take $(\widehat{zz})_{+\epsilon} = \chi(x, y)$ and put $(\widehat{zz})_{-\epsilon} = 0$, but this seems to me to destroy the basis of his approximation. Possibly, following the hint he gives on p. 700, the true method is to consider that, when the dimensions of a body are very small in any sense, then a *surface-load* in the same sense will give the same strains perpendicular to that sense as the *integral of a body-force* also in that sense. Thus the flexure-equations for a beam are deduced on the assumption that there is *no* lateral stress, yet we do not hesitate to use them for beams subject to continuous lateral load[1]. I conclude then that it is best to put $\widehat{zz}$ always *zero* (and not a definite value as Saint-Venant does on p. 699) and assume, when plates have a surface distribution of load, that the result of such load so far as the shifts of the points of the mid-plane are concerned can be represented by a body force, whose integral between the faces is equivalent per unit area to the surface load.

[388.] In § 5 Saint-Venant shews that from the assumptions, or approximate values:

$$\frac{d\,(\sigma_{zx},\ \sigma_{zy})}{dx,\ dy}=0, \qquad \frac{d^2 s_z}{dx^2,\ dxdy,\ dy^2}=0 \ldots\ldots\ldots\ldots\ldots(\alpha),$$

(which are less restrictive than $\sigma_{zx}=\sigma_{zy}=0$, and $\widehat{zz}=0$) we can deduce results embracing those of our Art. 385, 1° and 2°.

Writing the first set of expressions at length we easily find that:

$$\frac{ds_x}{dz}=-\frac{d^2w}{dx^2}, \qquad \frac{ds_y}{dz}=-\frac{d^2w}{dy^2}, \qquad \frac{d\sigma_{xy}}{dz}=-2\frac{d^2w}{dxdy}\ \ldots\ \ldots(\beta).$$

Whence we see by differentiating with regard to z that:

$$\frac{d^2s_x}{dz^2}=-\frac{d^2s_z}{dx^2}=0, \qquad \frac{d^2s_y}{dz^2}=-\frac{d^2s_z}{dy^2}=0, \qquad \frac{d^2\sigma_{xy}}{dz^2}=-2\frac{d^2s_z}{dxdy}=0.$$

Thus the second z fluxions of s_x, s_y, σ_{xy} are zero, or we may write w_0 for w in (β); it follows that we must have:

$$s_x=s^0{}_x-z\frac{d^2w_0}{dx^2}, \qquad s_y=s^0{}_y-z\frac{d^2w_0}{dy^2}, \qquad \sigma_{xy}=\sigma^0{}_{xy}-2z\frac{d^2w_0}{dxdy}\ldots(\gamma),$$

where the zero affixed refers the quantity to the mid-plane.

Saint-Venant remarks of the equations (γ):

Elles montrent, comme conséquence cinématique des égalités posées (α), que les dilatations de petites droites matérielles horizontales de direction donnée varient *linéairement* le long de toutes les lignes primitivement verticales, des divers points desquelles ces petites horizontales auraient été tirées.

Il convient de remarquer en passant que cela n'entraîne nullement, comme conséquence, que ces verticales resteront exactement droites et normales au feuillet moyen devenu courbe, car leurs petits intervalles horizontaux peuvent très bien croître linéairement avec z quoiqu'elles soient devenues courbes, si celles qui sont voisines affectent des inflexions pareilles, ainsi qu'il arrive pour les *sections* voisines, dans les tiges éprouvant *la flexion* dite *inégale* (p. 699: see our Art. 325).

It will be noted that this treatment brings out the real difficulties and assumptions of the problem, better than those which start by

[1] Since writing the above I have obtained the *full* solution for a simply supported beam continuously loaded on its upper surface, I find $\widehat{zz}$ is of the *same order* as $\widehat{zy}$, where x is the direction of the axis of the beam, and z the direction of the load.

assuming the strain-energy to be a function of the curvatures and so deduce by Lagrange's, or other, method the fundamental equation of the plate: see Thomson and Tait, § 639, or Lord Rayleigh's *Sound*, Vol. I., § 214.

I may remark that the equations (ii) and (iii) obtained in Art. 384 still hold if 2ϵ the thickness of the plate changes gradually with x and y.

[389.] Returning to the body-stress equations of Art. 384, let us integrate the first two between the limits $\pm\epsilon$ of z. We note first, however, since:

$$\widehat{xx} = \left(a - \frac{e'^2}{c}\right) s_x + \left(f' - \frac{d'e'}{c}\right) s_y$$
$$= \left(a - \frac{e'^2}{c}\right)\left(s^0_x - z\frac{d^2 w_0}{dx^2}\right) + \left(f' - \frac{d'e'}{c}\right)\left(s^0_y - z\frac{d^2 w_0}{dy^2}\right)$$

by equations (γ) of Art. 388, that

$$\left.\begin{aligned} &\int_{-\epsilon}^{+\epsilon} \widehat{xx}\,dz = 2\epsilon\left\{\left(a - \frac{e'^2}{c}\right) s^0_x + \left(f' - \frac{d'e'}{c}\right) s^0_y\right\} = 2\epsilon\widehat{xx}_0, \\ \text{similarly}\quad &\int_{-\epsilon}^{+\epsilon} \widehat{xy}\,dz = 2\epsilon\,\widehat{xy}_0, \text{ and } \int_{-\epsilon}^{+\epsilon} \widehat{yy}\,dz = 2\epsilon\,\widehat{yy}_0 \end{aligned}\right\}\ldots(\text{i}),$$

where the affixed $_0$ denotes a mid-plane value.

Hence from the integration of the body-stress equations we obtain:

$$\left.\begin{aligned} &2\epsilon\left(\frac{d\widehat{xx}_0}{dx} + \frac{d\widehat{xy}_0}{dy}\right) + (\widehat{zx})_{+\epsilon} - (\widehat{zx})_{-\epsilon} + X' = 0, \\ &2\epsilon\left(\frac{d\widehat{xy}_0}{dx} + \frac{d\widehat{yy}_0}{dy}\right) + (\widehat{zy})_{+\epsilon} - (\widehat{zy})_{-\epsilon} + Y' = 0, \end{aligned}\right\}$$

where X' and Y' are the integrals of X and Y across the plate.

Substitute the values of the mid-plane stresses in terms of the mid-plane shifts u_0, v_0 and we have:

$$\left.\begin{aligned} &2\epsilon\left\{(a - e'^2/c)\frac{d^2u_0}{dx^2} + (f + f' - d'e'/c)\frac{d^2v_0}{dxdy} + f\frac{d^2u_0}{dy^2}\right\} + (\widehat{zx})_{+\epsilon} - (\widehat{zx})_{-\epsilon} + X' = 0 \\ &2\epsilon\left\{(b - d'^2/c)\frac{d^2v_0}{dy^2} + (f + f' - d'e'/c)\frac{d^2u_0}{dxdy} + f\frac{d^2v_0}{dx^2}\right\} + (\widehat{zy})_{+\epsilon} - (\widehat{zy})_{-\epsilon} + Y' = 0 \end{aligned}\right\}\ldots(\text{ii}).$$

These equations reduce in the case of isotropy parallel to the plate to the simpler forms (p. 702):

$$\left.\begin{aligned} &2\epsilon\left\{H\frac{d}{dx}\left(\frac{du_0}{dx} + \frac{dv_0}{dy}\right) + f\frac{d}{dy}\left(\frac{du_0}{dy} - \frac{dv_0}{dx}\right)\right\} + (\widehat{zx})_{+\epsilon} - (\widehat{zx})_{-\epsilon} + X' = 0 \\ &2\epsilon\left\{H\frac{d}{dy}\left(\frac{du_0}{dx} + \frac{dv_0}{dy}\right) - f\frac{d}{dx}\left(\frac{du_0}{dy} - \frac{dv_0}{dx}\right)\right\} + (\widehat{zy})_{+\epsilon} - (\widehat{zy})_{-\epsilon} + Y' = 0 \end{aligned}\right\}\ldots(\text{iii}),$$

where H is the plate-modulus of Art. 323.

It will be noticed that these equations for the shifts u_0, v_0 are independent of that for w_0, or the transverse and longitudinal strain exercise no influence on each other. This has already been remarked by Cauchy and Poisson: see our Arts. 483* and 640*.

[390.] In §§ 8—10 Saint-Venant considers the effect of great stresses parallel to the mid-plane on the normal shift w_0. Thus he obtains what may be called the terms due to the action of the plate as a transverse membrane. He finds that in the function $\phi(xy)$ of equations (v) and (vi) of Art. 385 we must include the expression:

$$2\epsilon\left(\frac{d^2w_0}{dx^2}\widehat{xx}_0 + 2\frac{d^2w_0}{dxdy}\widehat{xy}_0 + \frac{d^2w_0}{dy^2}\widehat{yy}_0\right) - \frac{dw_0}{dx}\{(\widehat{zx})_{+\epsilon} - (\widehat{zx})_{-\epsilon} + X'\}$$

$$- \frac{dw_0}{dy}\{(\widehat{zy})_{+\epsilon} - (\widehat{zy})_{-\epsilon} + Y'\}.$$

From the sum of this expression and $Z' + (\widehat{zz})_{+\epsilon} - (\widehat{zz})_{-\epsilon}$ equated to zero we deduce the equation for the transverse equilibrium of a membrane. In its present form it has been obtained on the supposition that 2ϵ is constant; the alterations for 2ϵ variable are indicated by Saint-Venant in a footnote, p. 704.

[391.] In § 13 Saint-Venant commences his treatment of the contour conditions. Let α be the angle between the normal to the mid-plane contour at any point and the axis of x, let P, Q, R be the components of the applied load parallel to the axes, and ds, dn elements of the arc and normal of the mid-plane contour.

We find at once:

$$P = \widehat{xx}\cos\alpha + \widehat{yx}\sin\alpha,\quad Q = \widehat{xy}\cos\alpha + \widehat{yy}\sin\alpha,\quad R = \widehat{xz}\cos\alpha + \widehat{yz}\sin\alpha.$$

Hence by equations (i) of Art. 389, we have:

$$2\epsilon\,(\widehat{xx}_0\cos\alpha + \widehat{yx}_0\sin\alpha) = \int_{-\epsilon}^{+\epsilon} Pdz = P',$$

$$2\epsilon\,(\widehat{xy}_0\cos\alpha + \widehat{yy}_0\sin\alpha) = \int_{-\epsilon}^{+\epsilon} Qdz = Q'.$$

Substitute for the mid-plane stresses in terms of the shifts and we have:

$$2\epsilon\left\{\left[(a - e'^2/c)\frac{du_0}{dx} + (f'' - d'e'/c)\frac{dv_0}{dy}\right]\cos\alpha + f\left(\frac{du_0}{dy} + \frac{dv_0}{dx}\right)\sin\alpha\right\} = P',$$

$$2\epsilon\left\{\left[(f' - d'e'/c)\frac{du_0}{dx} + (b - d'^2/c)\frac{dv_0}{dy}\right]\sin\alpha + f\left(\frac{du_0}{dy} + \frac{dv_0}{dx}\right)\cos\alpha\right\} = Q'.$$

These are the sufficient and necessary contour conditions for longitudinal strain. When there is elastic isotropy parallel to the mid-plane they reduce to

$$2\epsilon\left\{\left[H\left(\frac{du_0}{dx} + \frac{dv_0}{dy}\right) - 2f\,\frac{dv_0/dy}{du_0/dx}\right]\frac{\cos\alpha}{\sin\alpha} + f\left(\frac{du_0}{dy} + \frac{dv_0}{dx}\right)\frac{\sin\alpha}{\cos\alpha}\right\} = \frac{P'}{Q'}$$

[392.] Saint-Venant next turns to the more controverted conditions involving the normal shift w_0. He proceeds to calculate M_s, M_n and R', the first two symbols representing the moments round tangent and normal respectively to the mid-plane contour for the load applied to the strip $2\epsilon \times ds$, and R' being the total shearing load on the same strip.

Now $$M_s = \int_{-\epsilon}^{+\epsilon} (P \cos \alpha + Q \sin \alpha)\, z dz = P'' \cos \alpha + Q'' \sin \alpha$$
$$= -\frac{2\epsilon^3}{3} \{\widehat{xx}'' \cos^2 \alpha + 2\widehat{xy}'' \sin \alpha \cos \alpha + \widehat{yy}'' \sin^2 \alpha\},$$

where, r and p being any two directions,
$$\widehat{rp}'' \times \frac{2\epsilon^3}{3} = -\int_{-\epsilon}^{+\epsilon} \widehat{rp} z dz,$$
and as before, a single dash on the loads denotes an integration with regard to z from $+\epsilon$ to $-\epsilon$, and a double dash an integration after multiplication by z, between the same limits. Substituting from equations (i) of Art. 389 for the stresses we find:
$$M_s = -\frac{2\epsilon^3}{3} \left\{ \left[(a - e'^2/c) \frac{d^2 w_0}{dx^2} + (f' - d'e'/c) \frac{d^2 w_0}{dy^2} \right] \cos^2 \alpha + 4f \frac{d^2 w_0}{dx dy} \sin \alpha \cos \alpha \right.$$
$$\left. + \left[(b - d'^2/c) \frac{d^2 w_0}{dy^2} + (f' - d'e'/c) \frac{d^2 w_0}{dx^2} \right] \sin^2 \alpha \right\}$$

Or, for elastic isotropy parallel to the mid-plane:
$$M_s = -\frac{2\epsilon^3}{3} \left[(H - 2f) \left(\frac{d^2 w_0}{dx^2} + \frac{d^2 w_0}{dy^2} \right) + 2f \frac{d^2 w_0}{dn^2} \right] \ldots\ldots\ldots\ldots (i).$$

This first condition is not the subject of discussion but has been generally accepted.

[393.] In a similar manner we find:
$$M_n = Q'' \cos \alpha - P'' \sin \alpha,$$
$$= \frac{2\epsilon^3}{3} \{\sin \alpha \cos \alpha\, (\widehat{xx}'' - \widehat{yy}'') - (\cos^2 \alpha - \sin^2 \alpha)\, \widehat{xy}''\},$$
where
$$\widehat{xx}'' - \widehat{yy}'' = \left(a - f' + \frac{d' - e'}{c} e' \right) \frac{d^2 w_0}{dx^2} - \left(b - f' - \frac{d' - e'}{c} d' \right) \frac{d^2 w_0}{dy^2},$$
or, $$= (a - f') \left(\frac{d^2 w_0}{dx^2} - \frac{d^2 w_0}{dy^2} \right)$$ with elastic isotropy parallel to the mid-plane.

And again, $$\widehat{xy}'' = 2f \frac{d^2 w_0}{dx dy}.$$

Further: $$R' = \int_{-\epsilon}^{+\epsilon} R dz = \cos \alpha \int_{-\epsilon}^{+\epsilon} \widehat{zx} dz + \sin \alpha \int_{-\epsilon}^{+\epsilon} \widehat{yz} dz$$
$$= \cos \alpha \left[\epsilon \{ (\widehat{zx})_{+\epsilon} + (\widehat{zx})_{-\epsilon} \} + X'' - \frac{2\epsilon^3}{3} \left(\frac{d\widehat{xx}''}{dx} + \frac{d\widehat{yx}''}{dy} \right) \right]$$
$$+ \sin \alpha \left[\epsilon \{ (\widehat{zy})_{+\epsilon} + (\widehat{zy})_{-\epsilon} \} + Y'' - \frac{2\epsilon^3}{3} \left(\frac{d\widehat{xy}''}{dx} + \frac{d\widehat{yy}''}{dy} \right) \right],$$

where $$\frac{d\widehat{xx}''}{dx}+\frac{d\widehat{xy}''}{dy}=(a-e'^2/c)\frac{d^3w_0}{dx^3}+(2f+f'-d'e'/c)\frac{d^3w_0}{dxdy^2},$$

$$\frac{d\widehat{xy}''}{dx}+\frac{d\widehat{yy}''}{dy}=(b-d'^2/c)\frac{d^3w_0}{dy^3}+(2f+f'-d'e'/c)\frac{d^3w_0}{dx^2dy},$$

reducing respectively to the differentials with regard to x and y of $H\left(\frac{d^2w_0}{dx^2}+\frac{d^2w_0}{dy^2}\right)$, when there is elastic isotropy parallel to the mid-plane.

These results can be easily deduced by integrating $\widehat{zx}$ and $\widehat{yz}$ from expressions of the form:

$$\widehat{xz}=\frac{d\,(z.\widehat{xz})}{dz}-z\frac{d\widehat{xz}}{dz}$$

$$=\frac{d\,(z.\widehat{xz})}{dz}+z\left(\frac{d\widehat{xx}}{dx}+\frac{d\widehat{xy}}{dy}+X\right).$$

I have reproduced the values of M_s, M_n and R', because they are the most complete hitherto given and will be useful for reference hereafter.

[394.] Saint-Venant adopts Thomson and Tait's 'reconciliation' (see our Art. 488*) and replaces the couples M_n by an additional shear $\frac{dM_n}{ds}$ added to R'. In other words he equates the contour load to the couple M_s and the shear $R'+\frac{dM_n}{ds}$

He attributes this method of reconciling Kirchhoff and Poisson to Boussinesq (p. 715). There are two points which arise in this reconciliation which deserve to be noted. The first objection to the replacement of the couples M_n by a distributed shear is that referred to in our Art. 488*, namely that the Kirchhoff contour conditions could not be used for the case of a *discontinuous* distribution of shearing force and normal couple. Saint-Venant replies to this:

S'il y avait des forces extérieures isolées, appliquées en certains points de ce cylindre et faisant couples autour de ses normales, elles seraient capables d'y imprimer à la plaque, entre ces points, des torsions finies. Aucun auteur n'a supposé l'existence de pareilles forces, qui sont capables de produire des altérations permanentes de la contexture de la matière de la plaque si elles agissent avec une certaine intensité sur des portions excessivement petites de sa surface. Tous supposent que les forces se répartissent sur des surfaces d'étendue finie; et nous ne considérerons même, ainsi qu'ils l'ont tous fait, que des forces agissant sur le cylindre contournant d'une manière continue et graduelle (p. 714).

The second objection is that due to M. Lévy (see our Art. 397); he holds that when the couple M_n is due to vertical forces we can replace it by a shear distribution perpendicular to the plate, but that when it is due to horizontal forces this is not allowable. This point has been discussed at length by Boussinesq (see the Chapter we have devoted to that elastician), and Saint-Venant sums up his arguments in the following words:

Si ces couples sont formés par des forces horizontales tangentes au cylindre, agissant en sens opposés, les unes au-dessus, les autres au-dessous de la périphérie moyenne, et si la plaque est supposée avoir une épaisseur comparable aux deux autres dimensions, ces couples pourront conspirer pour produire certains effets d'ensemble dont nous ne nous occupons pas, tels qu'une *inflexion* imprimée à toutes les arêtes, et accompagnée de cette torsion générale autour d'un axe vertical dont il a été traité dans les chapitres relatifs aux tiges. Mais si la plaque est extrêmement mince, ces sortes de déformations sont négligeables. Les couples de forces horizontales dont il s'agit s'exerçant d'une manière continue sur les arêtes successives, ne produiront que ces torsions *locales* dont nous nous occupons ici; et leurs effets seront sensiblement les mêmes que ceux de couples de forces verticales *de même moment*, qu'on leur substituerait en faisant tourner ceux-là de 90 degrés, substitutions qui se font, comme on sait, dans la statique élémentaire des corps solides (p. 714).

We may I think conclude that:

1°. The shift-equation ((vi) of Art. 385) for thin plates is only an approximation and depends upon the assumptions that $\widehat{zz} = 0$ and that s_x, s_y, σ_{xy} contain *only the first power* of z, as in Eqns. (γ) of our Art. 388. These assumptions are, however, probable and the approximation is close when the thickness of the plate is extremely small.

2°. To the same degree of approximation the *two* boundary conditions of Kirchhoff are true for very thin plates.

3°. When the plate has a thickness small but not indefinitely small compared with its other dimensions, the equation of Lagrange can under certain conditions still hold, but it is not then legitimate to replace the normal couple by a distribution of shearing-load.

This latter conclusion is *opposed* to Saint-Venant's opinion on p. 720. He shews that if the following conditions hold:

$$\frac{d^2\widehat{zz}}{dx^2,\, dxdy,\, dy^2} = \frac{d^2(\sigma_{zx},\, \sigma_{zy})}{dx^2,\, dxdy,\, dy^2} = \frac{d^3 s_z}{dx^3,\, d^2xdy,\, d^2ydx,\, dy^3} = 0,$$

we can still deduce Lagrange's equation, but these conditions allow of a definite but small thickness for the plate. He then states that Kirchhoff's contour conditions remain true. Now it seems to me that we can no longer replace normal couples by vertical shearing-loads, for this will cause a difference in the strain of the plate to a distance into its material of the same order as its thickness, and this distance is no longer vanishingly small as compared with the other dimensions of the plate.

[395.] Saint-Venant now proceeds to an interesting summary of other writers' treatment of the problem of thin plates. He notes that Poisson and Cauchy assume that the stresses can be expanded in powers of z giving convergent series. From this assumption Saint-Venant deduces equations (γ) of our Art. 388. He remarks of this assumption that it has never found supporters—*elle n'est pas suffisamment fondée, et peut se trouver souvent en défaut.* I must notice, however, that Saint-Venant's own assumptions of our Art. 385 really lead to the expression of the stresses $\widehat{xx}$, $\widehat{xy}$ and $\widehat{yy}$ as *linear functions* of z, (see equations (iv^b) of Art. 385 and (γ) of Art. 388,) while from the first two body stress-equations we obtain by integration for $\widehat{zx}$ and $\widehat{yz}$ *quadratic functions* of z together with terms $-\int_0^z X dz$ and $-\int_0^z Y dz$ which will in the great majority of cases be linear in z. Thus Poisson's and Cauchy's assumption is only a too general statement of the results reached by Saint-Venant himself.

[396.] Saint-Venant appears to think that the terms $Z' + \frac{dX''}{dx} + \frac{dY''}{dy}$ which occur in the function $\phi(xy)$ of his result (see equation (iii) of our Art. 384) do not agree with the similar terms obtained by Poisson (see equation (9) of our Art. 484*) and Cauchy (see Equation (70) of our Art. 640*). With proper transformation of symbols these are:

$$2\epsilon \left\{ \left[Z + \frac{\epsilon^2}{6} \frac{d^2 Z}{dz^2} \right]_{z=0} + \frac{\epsilon^2}{3} \left[\frac{d}{dz} \left(\frac{dX}{dx} + \frac{dY}{dy} \right) \right]_{z=0} \right\}.$$

Now Poisson and Cauchy assume forms such as:

$$X = X_0 + z \left(\frac{dX}{dz} \right)_0 + \frac{z^2}{2} \left(\frac{d^2 X}{dz^2} \right)_0 + \text{etc.},$$

$$\therefore X'' = \int_{-\epsilon}^{+\epsilon} Xz dz = \frac{2\epsilon^3}{3} \left(\frac{dX}{dz} \right)_0 + \text{terms involving fifth and higher}$$

powers of ϵ.

Similarly:

$$Z' = \int_{-\epsilon}^{+\epsilon} Z dz = 2\epsilon Z_0 + \frac{2\epsilon^3}{6}\left(\frac{d^2Z}{dz^2}\right)_0 + \text{terms involving fifth and higher}$$

powers of ϵ,

$$\text{or} \quad Z' + \frac{dX''}{dx} + \frac{dY''}{dy} = 2\epsilon\left\{\left[Z + \frac{\epsilon^2}{6}\frac{d^2Z}{dz^2}\right]_{z=0} + \frac{\epsilon^2}{3}\left[\frac{d}{dz}\left(\frac{dX}{dx} + \frac{dY}{dy}\right)\right]_{z=0}\right\}$$

+ terms involving fifth and higher powers of ϵ, which may be neglected.

Thus their assumption does not lead to an error in this point as Saint-Venant suggests. It seems to me also that Poisson and Cauchy's hypothesis is more valid than that of Clebsch and other writers who simply put $\widehat{zz} = \widehat{zx} = \widehat{zy} = 0$.

[397.] At this point Saint-Venant notices Maurice Lévy's memoir of 1877 (see *Journal de Liouville*, 1877, pp. 219—306, or our discussion of the memoir later). Lévy investigates what are the possible solutions for a prism with two free faces, when the shifts u, v, w are supposed capable of being expressed in a series of ascending powers of z and the forces acting on the lateral sides of the prism have a given resultant load. It follows that the stresses will now be given in ascending powers of z, and that there is no limitation as to the thickness of the plate (or height of the prism). Lévy finds (1) that the powers of z in u, v, w and in $\widehat{xx}$, $\widehat{xy}$ and $\widehat{yy}$ cannot surpass the third, (2) that the stress $\widehat{zz} = 0$ throughout the prism, and (3) that the stresses $\widehat{xz}$, $\widehat{yz}$ contain only second powers of z, which appear through the factor $(\epsilon^2 - z^2)$.

It will be seen that these results of Lévy give the proper limitation to Cauchy and Poisson's hypothesis, and shew clearly its relation to Saint-Venant's assumptions. Saint-Venant on pp. 726—733 deals with another part of Lévy's memoir; namely, the term he has introduced into the values of the shifts u and v in order that the *three* surface conditions of Poisson may be satisfied for thin plates. This term is periodic in z, but Saint-Venant following Boussinesq rejects it as producing effects only of the same order as those we are neglecting in our approximation. We shall return to this point later when treating of Lévy's memoir and his controversy with Boussinesq.

[398.] Saint-Venant now turns to the concrete application of the thin plate formulae. He first deals with circular plates and obtains the following results:

Let there be a uniform surface load p per unit of area.

(i) *When the contour simply rests on a ring of its own radius a.* Shift of mid-plane at radius r:

$$w_0 = \frac{3pa^4}{128H\epsilon^3}\left[\left(\frac{r}{a}\right)^4 - 1 - \frac{4H-2f}{H-f}\left\{\left(\frac{r}{a}\right)^2 - 1\right\}\right].$$

Central Shift $$f_0 = \frac{3H-f}{H-f}\frac{3pa^4}{128H\epsilon^3}.$$

(ii) *When the contour is built-in.*

$$w'_0 = \frac{3pa^4}{128H\epsilon^3}\left\{1 - 2\left(\frac{r}{a}\right)^2 + \left(\frac{r}{a}\right)^4\right\},$$

$$f'_0 = \frac{3pa^4}{128H\epsilon^3}.$$

Further we find that:

When the plate is simply supported: the line of inflexion, given by $\frac{d^2w_0}{dr^2} = 0$, is determined by $r = a\sqrt{\frac{2H-f}{3(H-f)}}$, or $= \cdot 931\, a$ if $H/f = 8/3$, as in the case of uniconstant isotropy.

When it is built-in: the line of inflexion is determined by $r = \cdot 5773\, a$, and the line of zero-moment (i.e. where $M_s = 0$) by $r = a\sqrt{\frac{H-f}{2H-f}}$, or $= \cdot 6202\, a$ in the case of uniconstant isotropy.

The maximum stretches in the two cases, given by the greatest values of $s_r = -z\frac{d^2w_0}{dr^2}$, are respectively:

$$s_0 = \frac{2H-f}{H-f}\cdot\frac{3pa^2}{32H\epsilon^2}\left(= \frac{117pa^2}{1280\mu\epsilon^2}, \text{ for uniconstant isotropy}\right),$$

$$s'_0 = \frac{3pa^2}{16H\epsilon^2}\left(= \frac{9pa^2}{128\mu\epsilon^2}, \text{ for uniconstant isotropy}\right).$$

The conditions for the fail-limit are thus easily written down. Compare with these results those of Poisson in our Arts. 497*—504*.

[399.] The last pages of Saint-Venant's *Note* are occupied with a reproduction and extension of Navier's memoir on rectangular plates (pp. 740—52).

Let us take the origin at one of the angles of the plate (sides a, b, and $a < b$) the sides being the axes.

We have here to solve the equation (vi) of our Art. 385, namely:

$$\left(\frac{d^2}{dx^2} + \frac{d^2}{dy^2}\right)^2 w_0 = \frac{3}{2H\epsilon^3}\phi(xy),$$

subject in the case of a simply supported edge to the conditions

$$\left.\begin{array}{l} H\dfrac{d^2w_0}{dy^2}+(H-2f)\dfrac{d^2w_0}{dx^2}=0, \text{ when } x=0 \text{ or } a, \text{ for values of } y \text{ from } 0 \text{ to } b, \\ H\dfrac{d^2w_0}{dx^2}+(H-2f)\dfrac{d^2w_0}{dy^2}=0, \text{ when } y=0 \text{ or } b, \text{ for values of } x \text{ from } 0 \text{ to } a, \\ \qquad\qquad w_0=0 \text{ for all points of the contour.} \end{array}\right\}$$

The solution is easily found to be (p. 743):

$$w_0=\frac{3}{2\pi^4 H\epsilon^3}\sum_{m=1}^{m=\infty}\sum_{n=1}^{n=\infty} A_{mn}\frac{\sin\dfrac{m\pi x}{a}\sin\dfrac{n\pi y}{b}}{\left(\dfrac{m^2}{a^2}+\dfrac{n^2}{b^2}\right)^2},$$

where $$A_{mn}=\frac{4}{ab}\int_0^a d\alpha\int_0^b d\beta \sin\frac{m\pi\alpha}{a}\sin\frac{n\pi\beta}{b}\phi(\alpha,\beta).$$

This result is applied to the calculation of the following special examples:

Case (*a*). A uniformly distributed load p per unit of surface area.

Here:

$$w_0=\frac{24p}{\pi^6 H\epsilon^3}\sum_1^\infty\sum_1^\infty\frac{1}{m'n'}\frac{\sin\dfrac{m'\pi x}{a}\sin\dfrac{n'\pi y}{b}}{\left(\dfrac{m'^2}{a^2}+\dfrac{n'^2}{b^2}\right)^2},$$

the summations being for odd numbers m', n' only.

The maximum or central deflection f_0 is very nearly given by the first term of this series with $x=a/2$, $y=b/2$, or we find

$$f_0=\frac{24\,(pab)\,a^3b^3}{\pi^6 H\epsilon^3\,(a^2+b^2)^2}.$$

The second term will, for $a=b$, be only 1/75 of this.

The maximum stretch s_0 is very nearly given by

$$s_0=\frac{24ab^3}{\pi^4 H\epsilon^2\,(a^2+b^2)^2}pab.$$

Case (*b*). An isolated central load $=P$. Here:

$$w'_0=\frac{6P}{\pi^4 Hab\epsilon^3}\sum_1^\infty\sum_1^\infty\frac{(-1)^{\frac{m'-1}{2}}(-1)^{\frac{n'-1}{2}}}{\left(\dfrac{m'^2}{a^2}+\dfrac{n'^2}{b^2}\right)^2}\sin\frac{m'\pi x}{a}\sin\frac{n'\pi y}{b},$$

where m' and n' are odd numbers only.

Here, $$f'_0=\frac{6Pa^3b^3}{\pi^4 H\epsilon^3\,(a^2+b^2)^2},$$

nearly, but not so approximately as in the like result for f_0 in *Case* (*a*), and

$$s'_0 = \frac{6ab^3}{\pi^2 H\epsilon^2 (a^2 + b^2)^2} P.$$

Hence f'_0/f_0 and s'_0/s_0 for the same total loads $= \pi^2/4 = 2{\cdot}5$ nearly: see our Art. 263*.

[400.] We have given a large amount of space to this monumental work because it contains much that is of value to both physicist and technologist, and we would gladly bring home to both the important services which Saint-Venant has rendered to the science of elasticity. His annotated *Clebsch* will long form the standard book of reference on our subject, but it is possible that the results we have here collected will reach some to whom it may not be accessible.

[401.] *Détermination et Représentation graphique des lois du choc longitudinal.* This memoir was presented to the Academy on July 16, 23, 30, and August 6, 1883. It appeared in the *Comptes rendus*, T. XCVII., 1883, pp. 127, 214, 281 and 353. An off-print of it with a note by Boussinesq (*Comptes rendus*, T. XCVII., 1883, p. 154) was afterwards put together and repaged. This off-print was distributed also as an appendix to the annotated *Clebsch*. Our references will be to the *sections* which are the same in the *Comptes rendus* and in the off-print.

The memoir is due to Saint-Venant in conjunction with M. Flamant the co-translator of the *Clebsch* and professor at the Ecole des Ponts et Chaussées, Paris.

[402.] After a short account as in the *Clebsch* (see our Art. 341) of the evolution of the problem the authors refer to the analytical solution given by Boussinesq (see the same article and Boussinesq's *Application des potentiels à l'étude de l'équilibre... des solides élastiques*, p. 508 *et seq.*) and reproduced by them on pp. 480 k—480 gg of the *Clebsch*.

D'après cette Note (du § 60 de Clebsch), le choc longitudinal s'accomplit suivant des lois ayant des expressions analytiques différentes, se succédant l'une l'autre à des intervalles déterminés. Par exemple, les dérivées des déplacements des divers points de la barre varient, d'un instant à l'autre, tantôt avec gradation continue, tantôt par *bonds* considérables donnant aux mouvements une empreinte périodique de l'acquisition brusque de vitesse qui a été faite au premier instant du choc par l'élément heurté.

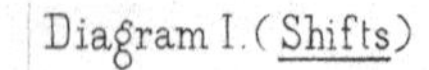

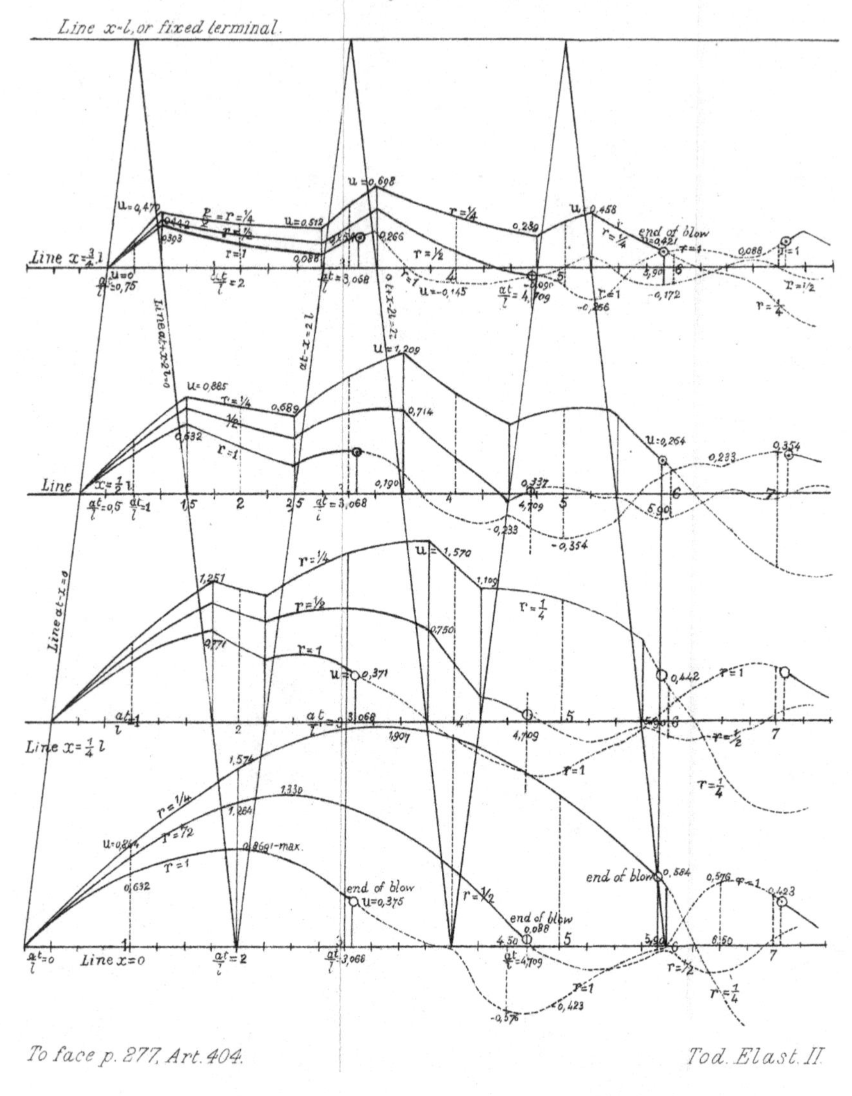

To face p. 277, Art. 404. *Tod. Elast. II.*

The material originally positioned here is too large for reproduction in this reissue. A PDF can be downloaded from the web address given on page iv of this book, by clicking on 'Resources Available'.

Il nous a donc paru utile de présenter ici aux regards, par une suite d'épures ou de diagrammes, une peinture de ces singulières et remarquables lois, afin de les éclairer et d'en faire bien comprendre la nature et les intéressantes conséquences. (§ 1.)

What then our authors accomplish is the *graphical* representation of the results of Boussinesq's analytical solution which was obtained in terms of discontinuous functions. It is one of the many instances in which Saint-Venant has helped to make of practical value the results of most intricate analysis. He was ever conscious that till theoretical formulae are reduced to simple numbers, the task of the mathematician is very far indeed from completion. Only the final diagram or numerical table can fitly crown the analytical labours of the mathematical physicist.

By means of such graphical representation we see at a glance the chief laws of the phenomena investigated, and are able to determine which approximate formulae we may fairly accept, and which we must replace by others better adapted to represent the exact facts of the case.

[403.] I reproduce the more important diagrams of the memoir as their practical value for engineers and technologists seems to me very considerable.

The following notation is adopted[1]:

l = length, ω = cross-section, ρ = density, P = weight, E = stretch-modulus, u = shift (at distance x from impelled end) of the bar, $a = \sqrt{E/\rho}$, or the velocity of sound. One end of the bar is fixed, and we may suppose it placed horizontally and struck horizontally by a mass of weight Q with velocity V. If the bar be vertical the effect produced by its weight must also be taken into account.

At the instant at which the blow ends, $du/dx = s_x = 0$ for $x = 0$ (see our Art. 204) and the following numerical values are obtained:

$Q/P < 1{\cdot}7283$............ the blow ends between the times $t = 2l/a$ and $4l/a$,
$Q/P > 1{\cdot}7283 < 4{\cdot}1511$ .... $= 4l/a$ and $6l/a$,
$Q/P > 4{\cdot}1511 < 7{\cdot}35$ .... $= 6l/a$ and $8l/a$.

[404.] The first diagram which I reproduce gives the shifts u for zero, quarter, half and three-quarter span for times $at/l = 0$ to $at/l = 7{\cdot}5$. Along the horizontal axis at/l is measured, along the

[1] In the memoir the authors use a for our l, σ for our ω, and ω for our a.

vertical axis $u = au/(Vl)$. Three curves are drawn for $r \equiv P/Q = 1$, $\frac{1}{2}$ and $\frac{1}{4}$ respectively, and having for scale 20 mm. for the unit of both at/l and u.

The shifts for the duration of the impulse are denoted by a heavy line ending in a small circle which marks the end of the impulse; the shifts after the impulse are marked by dotted lines, till they begin to repeat themselves when the lines become again heavy.

Whenever $at - x$ or $at + x - 2l$ is a multiple of $2l$ we note sudden changes in the slope (or the *shift-velocity*) of these curves; the points where these changes occur are termed by the authors *points de brisures*.

Les pieds des ordonnées de ces points de brisures sur les lignes horizontales d'abcisses marquées $x = l/4$, $x = l/2$, $x = 3l/4$ se trouvent, ainsi, aux rencontres de ces trois horizontales avec les obliques joignant en deux sens opposés les points $at/l = 0, 2, 4$...de l'horizontale $x = 0$ du bas, avec ceux $at/l = 1, 3, 5$...d'une horizontale $x = l$ tracée au haut. Celles de ces obliques qui montent de gauche à droite ont, en effet, pour équations $at - x = 0, 2l, 4l$...et celles qui descendent ont $at + x - 2l = 0, 2l, 4l$....Ces lignes obliques *figurent, en x et t, la marche de l'onde d'ébranlement*, tant directe que réfléchie aux extrémités de la barre, ou ce que parcourrait la tête de cette onde, si la barre vibrante était emportée perpendiculairement à sa longueur avec une vitesse a/l. Cela montre bien que les *bonds* et les *brisures* sont déterminés *par le passage de cette onde;* et cela donne une raison sensible du binôme et du trinôme $at - x$ et $at + x - 2l$ que M. Boussinesq a fait figurer dans ses formules de déplacements, etc. (§ 8).

We see that in all cases the maximum shift is at the end which receives the impulse.

[405.] The second diagram (Fig. 4 of the memoir) gives graphically the law of squeeze at the terminals and at $\frac{1}{4}$, $\frac{1}{2}$ and $\frac{3}{4}$ span for the same values of P/Q. Here the abscissae are at/l, and the ordinates $d = (-s_x)\, a/V$, or the squeezes reduced in the ratio of a to V. The scale of the abscissae is 20 mm. for $at/l = 1$ and of the ordinates 10 mm. for $d = 1$.

We note that in all cases the maximum squeeze is at the fixed end.

[406.] The third and fourth diagrams (figures 5 and 6 of the memoir) give:

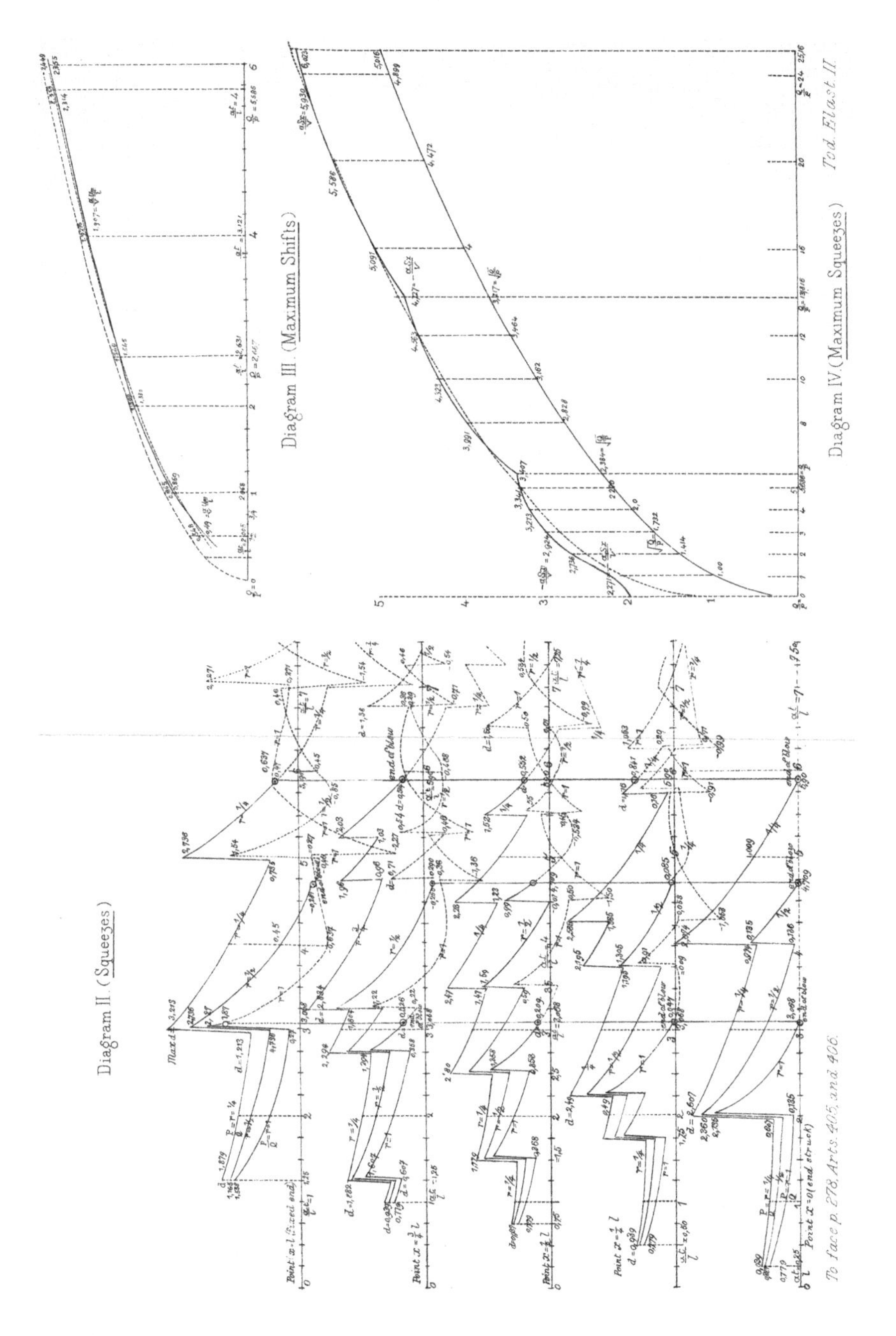

The material originally positioned here is too large for reproduction in this reissue. A PDF can be downloaded from the web address given on page iv of this book, by clicking on 'Resources Available'.

(1) *The maximum shifts at the end struck* (u_m).

Here the abscissa gives Q/P from 0 to 6, and the ordinate $u_m a/(Vl)$, the scale of the unit of both being equal to 20 mm. The heavy line is given by the true theory; the broken line is the parabola given by the first approximation $\frac{a}{V}\frac{u_m}{l} = \sqrt{\frac{Q}{P}}$; the pointed line is the curve given by the second approximation $\frac{a}{V}\frac{u_m}{l} = \sqrt{\frac{Q}{P}\frac{1}{1+\frac{1}{3}\frac{P}{Q}}}$: compare our Arts. 943*, 368, and the *Historique Abrégé, Leçons de Navier*, footnote p. ccxxiii.

We see at once that the *Hodgkinson-Saint-Venant approximation gives the terminal shifts with very considerable accuracy, and may be adopted with safety for all values of* $Q/P > \frac{1}{2}$.

In the course of the calculations the following numerical results not indicated on the diagram are obtained:

The maximum shift u_m is reached if

$$
\begin{array}{lll}
Q/P < 5{\cdot}686 & \text{between } t = 2l/a \text{ and } 4l/a, \\
Q/P > 5{\cdot}686 < 13{\cdot}816 & \quad ,, \quad t = 4l/a \text{ and } 6l/a, \\
Q/P > 13{\cdot}816 < 25{\cdot}16 & \quad ,, \quad t = 6l/a \text{ and } 8l/a.
\end{array}
$$

(2) *The maximum squeezes* $(-s_x)$ *at the fixed end.*

The three curves have for abscissae the values of Q/P from 0 to 25·10; the upper heavy curve has for ordinates the exact values of $(-s_x)\,a/V$ where s_x is given its maximum value, i.e. at the fixed end. The lower heavy curve is the parabola obtained by taking for ordinates

$$d = \frac{a}{V}(-s_x) = \sqrt{\frac{Q}{P}};$$

and the dotted curve by taking for ordinates

$$d = \frac{a}{V}(-s_x) = \sqrt{\frac{Q}{P}} + 1.$$

For the abscissa-scale 5 mm. is taken for $Q/P = 1$, and for the ordinate scale 20 mm. for $d = 1$.

It will be seen that the true values differ immensely from the values given by the old formula $d = \sqrt{Q/P}$. Thus that formula never suffices for finding the maximum squeeze[1].

[1] It may be obtained as follows: Suppose the shift uniformly distributed and its maximum mean value $= u_m$. Then work done $= \frac{1}{2}E\omega l\left(\frac{u_m}{l}\right)^2 = \frac{1}{2}\frac{Q}{g}\frac{V^2}{2}$ when the maximum is reached. Hence

$$(-s_x)_m = \frac{u_m}{l} = \frac{V}{a}\sqrt{\frac{Q}{P}}, \quad \text{or} \quad \frac{a}{V}\left(\frac{u_m}{l}\right) = \sqrt{\frac{Q}{P}}.$$

If, however, we take

$$d = \sqrt{\frac{Q}{P}} + 1,$$

we get the dotted curve of our fourth diagram, which from $Q/P > 5$ approaches closely to the true curve. Saint-Venant gives the following practical rules:

(*a*) For values of $Q/P > 24$ take $d = \sqrt{\frac{Q}{P}} + 1$,

(*b*) For values of $Q/P > 5$ and < 24 take $d = \sqrt{\frac{Q}{P}} + 1{\cdot}10$,

(*c*) For values of Q/P between 0 and 5 take $d = 2\,(1 + e^{-2P/Q})$,

this latter being the exact formula.

[407.] There are one or two other points in § 12 which we may note:

(1) The authors refer to the condition for *cohésion permanente* which is to be obtained from the maximum squeeze given by the results of Art. 406 (2).

Si les chocs ont eu pour tendance de raccourir, et *s'il ne devait en résulter que des compressions*, ces mêmes quantités numériques $-(du/dx)_m$ seraient à égaler à un nombre plus grand T'_0/E, T'_0 étant la limite, toujours très au-dessus de T_0, des forces comprimantes non dangereuses. Mais, comme nous avons vu que, dans la première période *de la détente libre* qui suit le choc, il se produit des dilatations égales aux compressions ayant précédé, le danger de désagrégation de la matière survit à la jonction, et la prudence conseille de traiter les compressions sur le même pied que les dilatations, ou d'égaler leurs valeurs numériques (see our Art. 175) à la même limite T_0/E que si c'étaient des dilatations.

To obtain the true condition for the safety of the structure we must remember that the bar is subjected to a succession of strains approaching the maximum in value. The real limit of T_0, then, *ought probably to be that for a repeated alternately positive and negative strain*, and if we are to give credence to Wöhler's experiments this is not the T_0 for a fail-point in pure tractional experiments. According to Wöhler the former is much less than the latter: see our account of his researches later.

(2) In a footnote (§ 12) the authors remark that the negative traction must be such that it does not cause the bar to buckle. They add that no bar will buckle unless the load is $> \pi^2 E\omega\kappa^2/l^2$, so that they treat the bar as a doubly pivoted strut (see *Corrigenda* to our Vol. I., Art. 959*). It seems just as probable that the bar would have one end built-in, in which case we may take double of the above load. The footnote then continues:

L'on peut admettre analogiquement, et même, ce semble, comme un *a fortiori*, que cette barre d'un poids P, sollicitée par le choc comprimant d'un corps Q tendant à y développer en un seul endroit, et comme maximum, une pression longitudinale égale à $E\omega(-s_m)$, s_m étant la valeur ci-après (see Art. 406 (a)), ne fléchira pas si l'on a

$$E\omega\frac{V}{a}\left(\sqrt{\frac{Q}{P}}+1\right)<\frac{\pi^2 E\omega\kappa^2}{l^2}.$$

This result[1] may be thought a little doubtful, in particular when we take into account the want of accord between the Eulerian theory of struts and experience: see our Arts. 957*—961*, 1255*—1262*. The authors remark of the above condition that it is *presque toujours remplie*, but I should be uneasy with regard to any structure where the above quantities had any approach to equality.

(3) At the end of § 12 it is shewn by a process involving the determination of mean values that the expression given in our Art. 406 (a) is really a close approximation to the true result. This is also proved in Boussinesq's note attached to the memoir: see also p. 544 of the work of his referred to in our Art. 402.

[408.] *Remarques relatives à la Note de M. Berthot sur les actions entre les molécules des corps: Comptes rendus*, 1884, T. XCIX., pp. 5—7.

Berthot in a memoir of 1884 (*Comptes rendus*, T. XCVIII., p. 1570) had suggested the following law of intermolecular force

$$F(r)=Kmm'\frac{r-r_0}{r^3},$$

where m, m' are the masses of two molecules at distance r and K, r_0 are constants. It is obvious that the force changes from attraction to repulsion at $r=r_0$.

Saint-Venant remarks that in 1878 in a footnote to a memoir, *Sur la constitution des atomes* (p. 37: see our Art. 275), he had referred to a law of somewhat like form.

In both cases the force tends to follow the gravitational law when r is much greater than r_0. Saint-Venant refers to the forms given by Poisson and Poncelet for representing intermolecular force (see our Arts. 439* and 977*), but he holds that although such laws are suggestive, it is very unlikely that in the present state of science we shall hit upon the correct one. He

[1] The memoir has $\frac{\sqrt{Q}}{P}$ for $\sqrt{\frac{Q}{P}}$. I may note also the following errata:

In equations (11), (12) and (13) the exponentials following the *curled* brackets should be placed inside them.

In equation (46) for first P/Q read Q/P.

observes that the discovery of its absolute form indeed is unnecessary for the establishment of the formulae of elasticity, hydraulics and electricity.

[409.] *Sur la flexion des prismes. Comptes rendus,* T. CII., 1886, pp. 658—664 and pp. 719—722. This memoir by Resal professes to point out an error in Saint-Venant's memoir on the flexion of prisms of 1856: see our Art. 69. The writer notes that Saint-Venant fixes the direction of a rectilinear element of the first face and not the direction of the first element of the prismatic axis. He then proceeds to assert that Saint-Venant has not taken account of the relation $\int xz d\omega + P = 0$: see our Art. 81. He endeavours to shew that this has led Saint-Venant to erroneous results in the case of the elliptic cross-section, but he himself falls into an error in his algebra, and so gives the colour of an error to Saint-Venant's work.

Boussinesq in a note in the same volume of the *Comptes rendus,* pp. 797—8, entitled: *Observations relatives à une Note récente de M. Resal sur la flexion des prismes,* points out Resal's algebraic error, and remarks that the difference between the terminal conditions of Saint-Venant and those proposed by Resal only produces a small rotation of the coordinate axes, and introduces no change into the expressions for the strains or stresses.

Resal in a few words (p. 799) acknowledges his error.

[410.] *Courbes représentatives des lois du choc longitudinal et du choc transversal d'une barre prismatique, dressées par feu de Saint-Venant, publiées par M. Flamant. Journal de l'École Polytechnique,* LIXe Cahier, pp. 97—128, 1889.

In the case of transverse impulse these curves are those referred to in our Arts. 105 and 361, while results drawn from those for longitudinal impact are mentioned in our Arts. 107 and 341: see also the passages in the *Notices* referred to in our Art. 108. The footnote on p. 244 of the present volume stating that it had been decided that the plates should not be published was printed nearly two years ago, and was made on the authority of M. Flamant. I can only hope that this footnote, however confusing to the reader, may, perhaps, have helped to bring about a reconsideration of the question of publication[1]. The plates were engraved as far back as 1873.

[411.] In the case of longitudinal impact the exact results—calculated from Boussinesq's solution in finite terms—are known and have been discussed by Saint-Venant and Flamant: see our Arts. 401—7. These enable us to compare for this case the approximate graphical and the actual results. While the accordance is fairly good for the maximum shifts, it is not very close for the stretches. In the case of transverse impact we are not yet able to test the accuracy of the graphical values of the curvature, which have been obtained from the shift-curves based on the first few terms of the transcendental series, but the fact that the shift-curves shew *no abrupt changes of slope*, as in the case of longitudinal impact, leads me to believe that far greater accuracy is obtainable by graphical processes for the case of transverse than for that of longitudinal impact. Compare Plates IV.—VI. of the memoir or Diagram I. of our p. 277 with Plates X. and XI. of the memoir.

Flamant himself writes:

Quoi qu'il en soit, le travail de Saint-Venant a un intérêt suffisant pour justifier sa publication: il peut servir d'exemple en montrant comment, grâce à un labeur considérable par l'étendue duquel cet infatigable travailleur ne s'est pas laissé rebuter, les valeurs de ces séries à termes périodiques de périodes décroissantes peuvent être représentées graphiquement; et il donne, tout au moins, sur les grandeurs de ces quantités, une première indication permettant de déduire des conséquences pratiques qui, si elles ne sont pas absolument exactes, n'en sont pas moins précieuses, puisqu'elles constituent tout ce que l'on sait sur ce sujet si important au point de vue de la stabilité des constructions (pp. 98—9).

The text which accompanies the plates is principally extracted from the *Annotated Clebsch* (§§ 60 and 61, see our Arts. 339, 342 *et seq.*), so that the whole may be looked upon as really a work of Saint-Venant which has been carefully edited by Flamant.

[412.] Pages 99—110 deal with longitudinal impact. The Plates I.—III. shew the graphical stages preparatory to drawing the shift-curves for five points of the bar in the case of $P/Q = \frac{1}{4}$ (the notation

[1] The footnote appeared in *The Elastical Researches of Barré de Saint-Venant*, Cambridge, 1889, an off-print of our Chapter x.

being that of our Art. 403). Plate IV. gives these curves, while Plates V. and VI. contain like curves for the cases $P/Q = \frac{1}{2}$, 1, 2, and 4. These curves serve in general the same purposes as those of Diagram I. of our p. 277, but they do not give the same abrupt changes of slope. The slope of these curves measures the stretch (or squeeze) and it is easy to see that its maximum occurs at the fixed end; unfortunately the slope of a tangent to an approximate curve is unlikely to give very good results. Thus for $P/Q = \frac{1}{4}$, Saint-Venant finds the maximum slope to be $2{\cdot}825\ V/a$ and to occur at a time $at/l = 3{\cdot}25$. The accurate values are $3{\cdot}213\ V/a$ and $at/l = 3$ (see our Diagram II. p. 278). The errors are even larger than this in the ratios of the maximum to the mean squeezes (pp. 109—110). But Saint-Venant's graphical values shew at least that one errs greatly in taking, as is usually done in the text-books, the mean for the maximum squeeze.

[413.] The curves for transverse impact we have already discussed in our Arts. 362—3 and 371. It is unfortunate that we have so little means of testing their accuracy. For the reasons given above, however, I am inclined to think the results more accurate than in the previous case. Saint-Venant assumes the form of a small arc at the lowest point of one of the instantaneous positions of the central axis to be a parabola with its axis vertical and so takes the curvature 8 times the subtense divided by the square of the chord (p. 119). I wish it had been possible to reproduce Plates X. and XI., but their scale precluding this, I must content myself by referring the reader to the original memoir.

Flamant makes (pp. 122—4) some interesting comparisons between longitudinal and transverse impact, and shews that if the same body falls with the same velocity first longitudinally and then transversely on a bar, the strain is considerably greater in the latter than in the former case, although the ratio of the strains diminishes as P/Q increases.

[414.] In conclusion Flamant remarks that in both cases the cross-sections are supposed to remain plane and that this is far removed from reality in the case of transverse impact, for the strain will really propagate itself in spherical or ellipsoidal surfaces from the point of impact, and these are not approximately coincident with the plane transverse cross-sections except at distances which are a considerable number of times greater than the depth of the bar. Flamant also refers to Boussinesq's solution (see the Chapter we have devoted to that scientist) but remarks that it leads to formulae so complicated that it does not seem possible to deduce any practical results from them (p. 124). Thus Saint-Venant's graphical calculation of the strain for these special cases is all the theory we can at present use in practical structures subjected to transverse impact.

[415.] Saint-Venant died on January 6, 1886. The President of the Academy on announcing his death at the meeting held on January 11, used the following words:

La vieillesse de notre éminent confrère a été une vieillesse bénie. Il est mort plein de jours, sans infirmités, occupé jusqu'à sa dernière heure des problèmes qui lui étaient chers, appuyé pour le grand passage sur les espérances qui avaient soutenu Pascal et Newton (*Comptes rendus*, T. CII., 1886, p. 73).

Short notices of his life appeared in the *Comptes rendus*, T. CII., pp. 141—7, 1886 by Phillips, and in *Nature*, February 4, 1886 by the Editor of the present volume. A full and excellent account by Boussinesq and Flamant of his life and work, together with a complete bibliography of his contributions to science, was published in the *Annales des Ponts et Chaussées* for November, 1886. In presenting this paper to the Academy, Boussinesq said:

Nous avons tâché d'y rappeler, avec tous les détails que comportait l'étendue matérielle de texte dont nous pouvions disposer, l'existence si bien remplie et les travaux les plus marquants du profond ingénieur-géomètre, notre maître à tous deux, qui a été une des gloires de l'Académie à notre époque et un modèle pour les travailleurs de tous les temps. (*Comptes rendus*, T. CIV., 1887, p. 215.)

A more popular account of Saint-Venant's life based chiefly on the notices in the *Annales* and *Nature* will be found in the *Tablettes biographiques; Dixième Année*, 1888.

[416.] *Summary.* In estimating the value of Saint-Venant's contributions to our subject, we have first of all to note that he is essentially the originator on the theoretical side of modern technical elasticity. In his whole treatment of the flexure, torsion and impact of beams he kept steadily in view the needs of practical engineers, and by means of numerical calculations and graphical representations he presented his results in a form, wherein they could be grasped by minds less accustomed to mathematical analysis. At the same time he was no small master of analytical methods himself, and he undertook in addition purely numerical calculations before which the majority would stand aghast. His memoirs on the distributions of elasticity round a point and of homogeneity in a body opened up new directions for physical investigation, while his numerous discussions on the nature of molecular action have greatly assisted towards clearer conceptions of the points at issue. The hypotheses of modified molecular action and of polar molecular action may either or both be true, or false; but we see now clearly that it is to the investigation of these hypotheses and not to the experiments of Oersted, Regnault etc. nor to the viscous fluid and ether jelly arguments of the first supporters of multi-constancy to which we must turn if we want to investigate the question of rari-constancy[1]. Saint-Venant's foundation, on the basis of Tresca's investigations, of the new branch of theoretical science, which he has termed *plastico-dynamics*, has not only direct value, but shews clearly the fallacy of those who would identify plastic solids and viscous fluids. The fundamental equations in the two cases differ in character; a difference which may be expressed in the words—the plastic solid requires a certain magnitude of stress (shear), the viscous fluid a certain magnitude of time for any stress whatever, to permanently displace their parts.

Not the least merit of Saint-Venant's work is the able band of disciples he collected around him. His influence we shall find strongly felt when investigating the work of Boussinesq, Lévy, Mathieu, Sarrau, Resal and Flamant. He formed the connecting link between the founders of elasticity and its modern school in France.

The vigorous spirit, the striking mental freshness, the perfect fairness of his thought enabled him to penetrate to the basis of things; the depth of his affection, his kindly foresight and consideration, his rare personal devotion attached to him all who came in his way and stimulated them to renewed investigation (Flamant and Boussinesq: *Notice sur la vie et les travaux de B. de St. V.*, p. 27).

[1] This is well brought out by the comparison of Voigt's recent memoir (*Göttinger Abhandlungen*, 1887) with those of the early supporters of multi-constancy.

CHAPTER XI.

MISCELLANEOUS RESEARCHES. 1850-60.

SECTION I.

Mathematical Memoirs[1], *including those of W. J. M. Rankine.*

[417.] W. J. M. Rankine: *On the Centrifugal Theory of Elasticity, and its connection with the Theory of Heat. Edinburgh Royal Society Proceedings,* Vol. III. pp. 86—91, 1851. This paper deals only with the elasticity of fluids and gases.

[418.] W. J. M. Rankine: *Laws of the Elasticity of Solid Bodies.* This paper was read before the British Association at Edinburgh, 1850. It is briefly noticed in the *Report* for that year: see our Art. 1452*. It is published at length in the *Cambridge and Dublin Mathematical Journal,* Vol. VI. 1851, pp. 47—80, with additions, pp. 178—181 and pp. 185—6. It is reprinted on pp. 67—101 of Rankine's *Miscellaneous Scientific Papers* edited by W. J. Millar. The pages of the latter will be briefly referred to as *S. P.* while we are dealing with Rankine's memoirs.

[1] The titles of the separate sections of this chapter refer rather to the method than to the substance of the memoirs. Thus this section discusses researches of Bresse, Phillips, Winkler etc., which are of the first importance to engineers, while the physical and technical sections will be found to contain many papers of great interest to mathematicians. The overwhelming number of memoirs demanded some classification, and the grouping of them broadly into mathematical, physical and technical sections seemed the least objectionable arrangement. In certain cases memoirs have been taken out of their proper section or their chronological order with a view to grouping kindred researches or bringing together the complete work of an individual scientist.

[419.] Section I. of the memoir (pp. 48—54, *S. P.* pp. 68—73) contains reproductions of formulae already given by Cauchy, Lamé and others for expressing the stresses or strains in any three rectangular directions in terms of those in any three other rectangular directions. I may note that Rankine uses *pressures* where I use positive tractions and that he uses symbols T_1, T_2, T_3 for the *halves* of what I have termed the slides, a notation which I am inclined to think would be of value if it had been generally adopted: see our Vol. I. p. 881.

On p. 49 (*S. P.* p. 68) he writes:

It is desirable that some single word should be assigned to denote the state of the particles of a body when displaced from their natural relative positions. Although the word *strain* is used in ordinary language indiscriminately to denote relative molecular displacement, and the force by which it is produced, yet it appears to me that it is well calculated to supply this want. I shall therefore use it, throughout this paper, in the restricted sense of *relative displacement of particles*, whether consisting in dilatation, condensation or distortion.

It is thus to Rankine that we owe the scientific appropriation of the word *strain*.

[420.] In Section II. (pp. 54—63, *S. P.* pp. 73—81) of the paper Rankine restricts his enquiries to

homogenous bodies possessing a certain degree of symmetry in their molecular actions, which consists in this: that the actions upon any given particle of the body of any two equal particles situated at equal distances from it within the sphere of molecular action, in opposite directions, shall be equal and opposite (p. 54, *S. P.* p. 73).

He is thus dealing with a case of what might be termed *central elastic symmetry*.

By a process of rather general reasoning in what is entitled: *Theorem I.* (p. 55) and by the assumption of the linearity of the stress-strain relations Rankine reaches expressions for the stresses which with our system of notation may be given as:

$$\widehat{xx} = as_x + f's_y + e''s_z, \quad \widehat{yz} = d\sigma_{yz},$$
$$\widehat{yy} = f''s_x + bs_y + d's_z, \quad \widehat{zx} = e\sigma_{zx},$$
$$\widehat{zz} = e's_x + d''s_y + cs_z, \quad \widehat{xy} = f\sigma_{xy}.$$

There are thus twelve apparently independent constants. Rankine does not note that the principle of energy requires us to suppose that $d'' = d'$, $e'' = e'$, $f'' = f'$, which reduces these formulae to those of our Art. 117 formulae (a).

[421.] Rankine now proceeds to *Theorem II.* (p. 61, *S. P.* p. 79) which he states as follows: *The coefficient of rigidity is the same for all directions of distortion in a given plane*, or in analytical language he would say that $\widehat{yz} = \widehat{y'z'}$ if $\sigma_{yz} = \sigma_{y'z'}$ whatever rectangular directions lying in the same plane y, z and y', z' may be. This *Theorem II.* does not appear to be correct and Rankine's error seems to have arisen from his supposing that a pure shearing force alone can change the angles of a rhombic prism. He has neglected to take into account the tractions which would have to be distributed over the faces to produce the sort of distortion he is considering, and although the work required to produce the distortion might be the same, however it was produced, yet this equality does not involve the equality of the shears, except when the rhombic angles are right. In the latter case his theorem reduces to the well-known results $\widehat{yz} = \widehat{zy}$ and $\sigma_{yz} = \sigma_{zy}$.

[422.] By means of the erroneous Theorem II., Rankine in Theorem III. (p. 61, *S. P.* p. 80) deduces relations of the type

$$4d = b + c - d' - d'' \qquad \text{(i)},$$

or remembering the real equality of d' and d'',

$$2d + d' = \tfrac{1}{2}(b + c) \qquad \text{(ii)}.$$

But this is the well-known *second* type of relation for bodies with an ellipsoidal distribution of elasticity, or for what Saint-Venant has termed *amorphic bodies:* see our Arts. 230—1, 308. Rankine's results, if true, ought to hold for crystals with three rectangular planes of elastic symmetry, but such bodies do not satisfy the above conditions. Hence: *all the further conclusions of Rankine's paper which depend upon the truth of* (i) *or* (ii) *can be considered to hold only for the limited range of amorphic or other bodies for which the ellipsoidal relations of the second type hold.*

[423.] Section III. (pp. 63—66, *S. P.* pp. 81—4) is entitled: *Results of the Hypothesis of Atomic Centres.* In this by rather vague reasoning Rankine deduces that when molecular force is central and a function only of the distance

$$d = d' = d'', \quad e = e' = e'', \quad f = f' = f'',$$

which are the usual conditions of rari-constancy. He bases his results on what he terms the hypothesis of Boscovich, which he considers not to be true for all solid bodies; he holds, however, that it may be corrected by combining it with an hypothesis of his own, to which we shall return later. We have several times had occasion to point out that the hypothesis of Boscovich does not really involve the conditions of rari-constancy (see our Art. 276), and that Boscovichian systems may be chosen which do not lead to rari-constancy has been recently demonstrated by Sir William Thomson: see *Proceedings of the Royal Society of Edinburgh*, July, 1889, or *Mathematical Papers*, Vol. III. pp. 395—427. Further Rankine's reasoning has been questioned by Sir William Thomson in a note attached to the memoir: see p. 80 (*S. P.* p. 98). The reply of Rankine to this criticism is given on pp. 178—81 (*S. P.* pp. 98—100).

Of course equations (7) and (8) of this section will be erroneous unless the body possesses ellipsoidal elasticity of the second type, and thus we are obliged to reject Rankine's fascinating statement on p. 66 (*S. P.* p. 84) that:

in a body whose elasticity arises wholly from the mutual actions of atomic centres, all the coefficients of elasticity are functions of the three coefficients of rigidity [i.e. the three slide-coefficients, d, e, f]. Rigidity being the distinctive property of solids, a body so constituted is properly termed a *perfect solid*.

[424.] In Section IV. of the memoir (pp. 66—9, *S. P.* pp. 84—6) Rankine applies his *Hypothesis of Molecular Vortices* to the elasticity of solids. He had previously written several papers on this hypothesis dealing with the elasticity of gases and vapours and generally with the mechanical theory of heat. For our present purposes it is sufficient to cite the description Rankine gives of his hypothesis in this memoir (pp. 66—7):

Supposing a body to consist of a continuous fluid, diffused through space with perfect uniformity as to density and all other properties, such a body must be totally destitute of rigidity or elasticity of figure, its parts having no tendency to assume one position as to *direction* rather than another. It may, indeed, possess elasticity of volume to any extent, and display the phenomena of cohesion at its surface and between its parts. Its longitudinal and lateral elasticities will be equal in every direction; and they must be equal to each other by equation (5).

[Here Rankine gives conditions which amount to putting $d = e = f = 0$, $a = b = c = d' = e' = f' = d'' = e'' = f''$ in the stress-strain relations of our Art. 420. He afterwards writes the latter series of quantities $= J$, which he terms the coefficient of *fluid elasticity*.]

If we now suppose this fluid to be partially condensed round a system of centres, there will be forces acting between those centres greater than those between other points of the body. The body will now possess a certain amount of rigidity; but less, in proportion to its longitudinal and lateral elasticities, than the amount proper to the condition of perfect solidity. Its elasticity will, in fact, consist of two parts, one of which, arising from the mutual actions of the centres of condensation, will follow the laws of perfect solidity; while the other will be a mere elasticity of volume, resisting change of bulk equally in all directions.

There is indeed much that is suggestive in Rankine's hypothesis of molecular vortices, as well as in his attempt to separate elasticity into the two factors of perfect solidity and of perfect fluidity, which involve the conceptions of rigidity and bulk elasticity. But from what we have seen above these factors of elasticity do not correspond exactly to fluid and Boscovichian methods of action, and Rankine's *imperfect solid* cannot in general be obtained by superposing on a fluid elasticity the rigidity of a *perfect solid*, i.e. the elasticity of a rari-constant substance.

The expressions given in Rankine's Equation (9) for the direct-stretch and cross-stretch coefficients in terms of the slide modulus and J would only be true for a particular type of amorphic body.

In the case of isotropy relations such as (ii) of our Art. 422 certainly do hold and then we have

$$a = b = c = 3d + J = 3e + J = 3f + J,$$
$$d' = e' = f' = d + J = e + J = f + J.$$

Thus in the ordinary isotropic notation of our *History* the *coefficient of fluidity*, or $J, = \lambda - \mu$ and it vanishes on the Boscovichian or rather uni-constant hypothesis.

Rankine's special error would thus seem to lie in the extension of his results from isotropy to aeolotropy other than that of certain amorphic bodies.

To this Section a *Note* is added (pp. 69—71, *S. P.* pp. 87—9) containing a reference to the researches of Green, MacCullagh, Stokes, Poisson, Navier, Cauchy, Lamé and Wertheim, with a comparison of their notations for the elastic constants with that of Rankine himself. The latter remarks of Wertheim's hypothesis ($\lambda = 2\mu$) that it must be regarded as doubtful. "If the effect of heat is to diminish μ and increase J, there may be some temperature for each substance at which M. Wertheim's equation is verified."

[425.] Section V. (pp. 71—80, *S. P.* pp. 89—98) is entitled: *Coefficients of Pliability, and of Extensibility and Compressibility, Longitudinal, Lateral, and Cubic. Examples of their Experimental Determination.* These are the coefficients which Rankine afterwards termed *Thlipsinomic* (see our Art. 448), or those which express strain in terms of stress. For the stress-strain relations of our Art. 420, the coefficients of pliability are the reciprocals of the coefficients of rigidity, i.e. of the slide-coefficients d, e, f.

For the stretches Rankine has in our notation:

$$s_x = \quad a_1 \widehat{xx} - b_3 \widehat{yy} - b_2 \widehat{zz},$$
$$s_y = - b_3 \widehat{xx} + a_2 \widehat{yy} - b_1 \widehat{zz},$$
$$s_z = - b_2 \widehat{xx} - b_1 \widehat{yy} + a_3 \widehat{zz},$$

and he classifies the coefficients as follows:

a_1, a_2, a_3 are coefficients of *longitudinal extensibility and compressibility*. We may perhaps better term them *direct traction coefficients*, they are coefficients of 'stretchability.' b_1, b_2, b_3 are coefficients of *lateral extensibility and compressibility*. We may perhaps better term them *cross traction coefficients*. Our terminology would thus be in accordance with that which we have adopted for the usual elastic or tasinomic coefficients: see the footnote on our page 77.

Rankine then proceeds to express these six thlipsinomic coefficients in terms of the four constants, d, e, f and J, of which he imagines in the case of central elastic symmetry all the other elastic constants to be functions. His results are rather lengthy and appear to have no application except to the case of a certain type of amorphic body (see our Art. 422). For the special case of isotropy we have

$$a = \frac{2\mu + J}{5\mu^2 + 3\mu J}, \qquad b = \frac{\mu + J}{10\mu^2 + 6\mu J},$$

$$d = \frac{3}{5\mu + 3J} \quad \text{or} \quad \frac{1}{d} = J + \frac{5}{3}\mu.$$

Here d is what Rankine terms the coefficient of cubic compressibility, or $1/d$ what we have termed the *dilatation-modulus* and represented by F: see Vol. I. p. 885. Further $1/a$ is obviously E, the stretch-modulus, and $b = \eta/E$, where η is the stretch-squeeze ratio. There is thus little of importance here beyond the terminology.

Rankine then proceeds to show how the rigidity (μ), fluid elasticity (J), longitudinal elasticity ($\lambda + 2\mu$), lateral elasticity (λ), as well as the thlipsinomic coefficients a, b and d may be experimentally ascertained. He determines them for brass and crystal glass from Wertheim's experiments, and indicates how they might be found for aeolotropic bodies (pp. 75—80).

[426.] We have already referred to Sir W. Thomson's criticism with which the memoir concludes and to Rankine's rejoinder (pp. 178—81, *S. P.* pp. 97—100). An additional *Note* (pp. 185—6, *S. P.* p. 100—1) merely gives the relations between the symbols of the present memoir and those of Clerk-Maxwell's memoir of 1850: see our Art. 1536*.

[427.] W. J. M. Rankine: *On the Laws of Elasticity: Cambridge and Dublin Mathematical Journal*, Vol. VII. 1852, pp. 217—34 (*S. P.* pp. 101—118). This is a sequel to the memoir referred to in our Arts. 418—26, the sections being numbered in continuation. The object of this portion of the memoir is to compare the results and symbols of Haughton and Green with those adopted by the author. Rankine here follows Lagrange's method of investigation and sums up his assumptions in the following postulates:

(i) That the variations of molecular force concerned in producing elasticity are sufficiently small to be represented by functions of the first order of the quantities on which they depend: and,

(ii) That the integral calculus and the calculus of variations are applicable to the theory of molecular action. It is thus apparent that the science of elasticity is, to a great extent, one of deduction *à priori* (p. 230, *S. P.* p. 114).

These do *not* seem to me the only assumptions of the paper, for Rankine again reduces in the case of rari-constancy the stress-strain relations of our Art. 420 to relations having only three independent constants. He obtains the relations of the second ellipsoidal type (see our Art. 422, (ii)) but by a hypothesis very different from that of the earlier part of his paper.

[428.] Section VI. entitled: *On the Application of the Method of Virtual Velocities to the Theory of Elasticity* (pp. 217—24) follows closely the methods of Haughton's memoirs of 1846—9, and contains nothing of special note[1]. We may remark that Rankine endorses Haughton's view of the relation $\widehat{xy} = \widehat{yx}$ cited in our Art. 1517*.

Mr Haughton correctly remarks that this often quoted Theorem of Cauchy is not true for all conceivable media. It is not true, for instance, for a medium such as that which Mr MacCullagh assumed to be the means of transmitting light. It is true, nevertheless, for all molecular pressures which properly fall under the definition of elasticity, if that term be confined to the forces which preserve the figure and volume of bodies (p. 221, *S. P.* p. 105).

Here as in the earlier part of the memoir Rankine insists upon the distinction between the resistances to change of bulk and to change of form, and in the following section he again builds up

[1] See our Arts. 1505*—18*.

an elastic solid of the ordinary type by superposing a fluid elasticity upon a homogeneous body consisting of centres of force only and so having rari-constant equations.

[429.] Section VII. is entitled: *On the Proof of the Laws of Elasticity by the Method of Virtual Velocities* (pp. 224—30, *S. P.* pp. 108—114). The following words exactly reproduce Rankine's position:

The fluid elasticity considered in the last article cannot arise from the mutual actions of centres of force; for such actions would necessarily tend to preserve a certain arrangement amongst those centres, and would therefore resist a change of figure. Fluid elasticity must arise either from the mutual actions of the parts of continuous matter, or from the centrifugal force of molecular motions, or from both those causes combined.

On the other hand it is only by the mutual action of centres of force that resistance to change of figure and molecular arrangement can be explained, that property being inconceivable of a continuous body. The elasticity *peculiar* to solid bodies is, therefore, due to the mutual action of centres of force. Solid bodies may nevertheless possess, in addition, a portion of that species of elasticity which belongs to fluids.

The investigation is simplified by considering in the first place the elasticity of a solid body as arising from the mutual action of centres of force only, and afterwards adding the proper portion of fluid elasticity (p. 224, *S. P.* p. 108).

Rankine deduces by a process, some steps of which I do not grasp[1], the usual rari-constant equations of elasticity. These it will be remembered have 15 independent constants (see our Art. 116 and footnote). To get the most general system of coefficients he adds a constant J to the *rari-constant* direct-stretch and cross-stretch coefficients, i.e. takes

$$|xxxx| + J, \quad |xxyy| + J, \text{ etc.}$$

but he does not add this constant to the coefficients like $|xyxy|$ which in the rari-constant theory are equal to those of type $|xxyy|$. Thus he really supposes the *multi-constant* coefficients to satisfy relations of the type

$$|xxyy| - |xyxy| = |yyzz| - |yzyz| = |zzxx| - |zxzx| \; \{= \text{Rankine's } J\} \ldots\ldots \text{(i)},$$

together with the three purely *rari-constant* conditions:

$$\left.\begin{array}{l} |xxyz| = |xyxz| \\ |yyzx| = |yzyx| \\ |zzxy| = |zxzy| \end{array}\right\} \ldots\ldots\ldots\ldots\ldots\ldots\ldots\ldots\ldots \text{(ii)}.$$

[1] For example, how the expression on p. 226 (*S. P.* p. 110) for the total action of an indefinitely slender pyramid is obtained, supposing $\phi(r)$ to be the law of intermolecular force.

Hence he puts the six rari-constant relations on different footings, the latter three always hold, the former three (obtained by putting $J=0$) do not generally hold and are replaced by the two of type (i) above.

The most general aeolotropy has thus for Rankine only *sixteen* constants. It would be interesting to know how far experimentally Rankine's views are justified. Are any of the inter-constant relations of rari-constancy more generally satisfied than others? Rankine's defective theory can hardly in itself be considered an argument in favour of the reduction of the constants to sixteen. Unfortunately the results (ii) are identically satisfied for all bodies possessing three rectangular planes of elastic symmetry, and thus experiments would have to be made on very complex aeolotropic systems.

[430.] Rankine now proceeds to reduce his sixteen elastic coefficients to *four*, three rari-constant coefficients supplemented by the coefficient of fluidity J. He first reduces the sixteen to seven by putting the coefficients of asymmetrical elasticity (see our footnote p. 77) zero. The exact reasoning by which he reaches this result (p. 228, *S. P.* 112) is far from clear to me, but I presume it does not amount to more than his previous supposition of the central symmetry of the elastic distribution. He apparently supposes that his reasoning is perfectly general.

The next stage is to reduce the six remaining rari-constant coefficients to three by means of the ellipsoidal conditions of the second type: see our Art. 422, (ii). Rankine deduces these conditions by a method totally different from that of the first part of his memoir, and he asserts that they hold for "all known homogeneous substances" (p. 229, *S. P.* p. 112). He proceeds as follows:

Let $\phi(r)$ be the law of central intermolecular force and Σ denote a molecular summation over a cone of elementary solid angle, then he assumes that $R=\Sigma r^2\phi'(r)$ is a function $F(i)$ of i "the *mean interval* between centres of force in a given direction." If the direction-cosines of this direction be l, m, n, and f, g, h, k, be constants, he assumes that referred to the axes of elasticity i will be of the form

$$i=\text{exponential}\,(f+gl^2+hm^2+kn^2).$$

He then continues:

Let us assume as a *Fifth Postulate*, what experience shews to be sensibly true of all known homogeneous substances—viz. that their elasticity varies very little in different directions. Those substances, such as timber whose elasticity in different directions varies much, are not homogeneous, but composed of fibres, layers, and tubes of different substances (p. 229, *S. P.* p. 112).

Thus he deduces that $gl^2 + hm^2 + kn^2$ must be very small as compared with f, or that we may take:

$$R = \psi(f) + \psi'(f)(gl^2 + hm^2 + kn^2),$$

substituting this value of R in the summation expressions[1] for the elastic constants he easily deduces relations of the type

$$\tfrac{1}{2}\{|yyyy| + |zzzz|\} = 3\,|zyzy|$$

or those of the second ellipsoidal type.

[431.] Rankine concludes this section of his memoir by the two postulates I have cited at the beginning of my criticism in Art. 427. Those postulates do not seem to me to involve the reduction of the twenty-one elastic constants to three. The memoir suggestive in parts, seems full of very doubtful reasoning. The results of this memoir are indeed rejected in the one *On Axes of Elasticity...*, discussed in our Arts. 443—52 where Rankine states that "there is now no doubt that the elastic forces in solid bodies are not such as can be analysed into fluid elasticity and mutual attractions between centres simply." But my present point is that, even if they could be, there would be no *necessary* reduction of the constants below sixteen, so that Rankine's reasoning as well as his hypotheses are at fault.

[432.] In a *Note to Sections VI. and VII.* appended to the memoir and entitled: *On the Transformation of the Coefficients of Elasticity by the aid of a Surface of the Fourth Order* (pp. 231—4, *S. P.* pp. 114—8), Rankine gives expressions for the transformation of the coefficients of elasticity from one set of rectangular axes to a second. I believe this to be the first occasion (1852) on which expressions were given for the transformation of the elastic coefficients. The same results, however, were obtained by Saint-Venant in a much simpler symbolic form some years later (1863) and have already been cited in this work: see our Art. 133. Hence I do not propose to reproduce the earlier discussion, merely noting Rankine's undoubted priority, which was fully admitted by Saint-Venant: see our footnote p. 89, and Art. 135, etc.

[1] I am not certain of the accuracy of these summation expressions if $\phi(r)$ be the law of intermolecular force. But I think Rankine's results would follow if $F(r)$ of our equation (xxx.) Art. 143 were made a function of i.

[433.] W. J. M. Rankine: *On the Velocity of Sound in Liquid and Solid Bodies of limited Dimensions, especially along Prismatic Masses of Liquid: Cambridge and Dublin Mathematical Journal*, Vol. VI. 1851, pp. 238—67 (*S. P.* pp. 168—199).

Rankine remarks that if we could ascertain the velocities of transmission of vibratory movements along the axes of elasticity of an *indefinitely extended* mass of any substance we should at once be able to calculate its coefficients of elasticity. As we cannot experiment on such a mass of solid elastic material, the best results, which can be obtained in practice are those based on the transmission of nearly longitudinal vibrations along prismatic or cylindrical bodies. If the vibrations were *solely* longitudinal, we should be able to find the "true longitudinal elasticity," i.e. the direct stretch-coefficient. It is however "impossible to prevent a certain amount of lateral vibration of the particles, the effect of which is to diminish the velocity of transmission in a ratio depending on circumstances in the molecular condition of the superficial particles, which are yet almost entirely unknown." Rankine holds that the supposition that the stretch-modulus is obtained from experiments upon the longitudinal vibrations of a rod or bar is

> inconsistent with the mechanics of vibratory movement; and accordingly, experiment has shown that the elasticity corresponding to the velocity of sound in a rod agrees neither with the modulus of elasticity, nor with the true longitudinal elasticity; although it is in some cases nearly equal to the former of those quantities, and in others to the latter (p. 239, *S. P.* p. 169).

We will briefly indicate the course of Rankine's investigations in the following six articles.

[434.] Pp. 240—6 (*S. P.* pp. 170—6) are entitled: *General Equations of Vibratory Motion in Homogeneous Bodies.* In this paper Rankine integrates the general equations of vibratory motion for a solid having central elastic symmetry, i.e. with nine independent elastic coefficients: see our Art. 117, (*a*).

Rankine adopts as types of solution shifts with factors of the form

$$e^{\frac{2\pi}{\lambda}\{a'x+b'y+c'z\pm\sqrt{-1}(\sqrt{\epsilon}\,.\,t-ax-by-cz)\}},$$

where ϵ has three different values given by the roots of a particular cubic equation. Certain rather complex relations must hold among the constants. Taking a series of such terms Rankine finds for the shifts:

$$\left.\begin{aligned}u=\Sigma\Big[e^{\frac{2\pi}{\lambda}(a'x+b'y+c'z)}\Big\{&l\cos\frac{2\pi}{\lambda}(\sqrt{\epsilon}.t-ax-by-cz)\\&+l'\sin\frac{2\pi}{\lambda}(\sqrt{\epsilon}.t-ax-by-cz)\Big\}\Big]\\v=\Sigma\,&(\text{terms in } m, m' \text{ instead of } l, l')\\w=\Sigma\,&(\text{terms in } n, n' \text{ instead of } l, l')\end{aligned}\right\}\ \ldots\ldots(\text{i}).$$

Thus there are fourteen constants λ, $\sqrt{\epsilon}$, a, b, c, a', b', c', l, l', m, m', n, n' for each set of terms, and these are connected by the equation $a^2+b^2+c^2=1$ and six other equations of condition, or we have seven independent constants.

Such expressions Rankine says "contain the complete representation of the laws of small molecular oscillations in a homogeneous body of any dimensions and figure" (p. 245, *S. P.* p. 175).

The special case of an indefinitely extended medium has been treated by Poisson, Cauchy, Green, MacCullagh, Haughton, Blanchet, and Stokes; see our Arts. 523*, 1166*—78*, 917*—21*, 1519*—22*, and 1268*—75*. Rankine gives the principal results which depend upon the fact that in this case for small oscillations we must have

$$a'=b'=c'=0.$$

See his pp. 246—8 (*S. P.* pp. 176—8).

[435.] Pp. 248—50 (*S. P.* pp. 178—80) deal with the *General Case of a Body of limited Dimensions*. Here the velocity is no longer a function only of the direction-cosines a, b, c of the wave front, but also of a', b', c'. Rankine in these pages gives the shift-speeds, the stretches, the slides and the six stresses as deduced from equations (i). Taking the special case of an isotropic medium (pp. 250—6, *S. P.* pp. 180—184) and the axis of x as direction of propagation, Rankine puts

$$a=1,\ b=0,\ c=0,\ a'=0,$$

and finds for the velocities of propagation in our notation,

$$\sqrt{\epsilon}=\sqrt{\frac{(\lambda+2\mu)}{\rho}}\sqrt{1-b'^2-c'^2},$$

$$\sqrt{\epsilon}=\sqrt{\frac{\mu}{\rho}}\sqrt{1-b'^2-c'^2}\quad\text{(pp. 250—1, \textit{S. P.} p. 181)}.$$

Hence these velocities of propagation are less than in an *unlimited* mass in the ratio $\sqrt{1-b'^2-c'^2}:1$.

[436.] Rankine now remarks that:

It may be shown that the vibrations corresponding to the velocity

$\sqrt{\frac{\mu}{\rho}}\sqrt{1-b'^2-c'^2}$ cannot take place in a body of which the surface is free unless $b'=0$, $c'=0$, in which case they are reduced to ordinary transverse vibrations (p. 251, *S. P.* p. 182).

This is shown in an *Appendix* II. (pp. 265—7, *S. P.* pp. 197—9) entitled: *General Equations of nearly-transverse Vibrations.* Herein Rankine calculates out the surface traction in his prism and shows that if it is to be zero we must have $b'=c'=0$, and the nearly transverse vibrations become accurately transverse.

[437.] The vibrations which have the velocity

$$\sqrt{\frac{\lambda+2\mu}{\rho}}\sqrt{1-b'^2-c'^2}$$

are termed by Rankine *nearly-longitudinal*, for the longitudinal component predominates, and are dealt with by him on pp. 251—4 (*S. P.* pp. 182—4). He finds expressions for the surface stresses and remarks that if we knew "the laws which determine the superficial pressures in vibrating bodies" these expressions would enable us to find b' and c' and so determine the velocity of propagation. "Those laws, however, are as yet a matter of conjecture only." It seems to me that a reasonable hypothesis is that the surface-stress or load vanishes, but even then except in very special cases Rankine's expressions would probably be too complex to afford any manageable solution of the problem (see two memoirs by Chree, *Quarterly Journal of Mathematics*, Vol. XXIII. pp. 317—42 and Vol. XXIV. pp. 340—58).

For a musical note "the velocity of propagation must be the same for all the elementary vibrations into which the motion may be resolved," that is to say $b'^2+c'^2$ must have the same value in all the terms of the expressions for the shifts. This leads Rankine to considerably simplify his equations for the stresses, strains and surface loads in this particular case: see his pp. 254—6 (*S. P.* pp. 184—7). Rankine does not, however, draw any special conclusions from these simplified results.

[438.] Rankine next turns (pp. 256—60) to the relation between the velocities of sound in a rectangular horizontal prism of liquid and in an infinite mass of the same liquid, and he shows that on a certain hypothesis as to the surface conditions, based upon his own theory of molecular vortices, these velocities would be in the ratio of $\sqrt{2}:\sqrt{3}$, as indeed Wertheim found them by experiment: see our Arts. 1349*—51*. Returning to the vibrations of solid rods Rankine, *adopting a hypothesis like that for liquids*, gives results for the cases of rectangular and circular prismatic rods. Not only are the hypotheses here rather vague but the results do not seem very satisfactory. In the case of the rectangular prism Rankine supposes the lateral vibrations of the particles to take place parallel to one pair of faces of the prism only, and he finds that the same relation holds between the velocities of sound in a solid prism and in an infinite mass as for a liquid. The case of the rod of circular cross-section is investigated in an Appendix I. (pp. 262—5, *S. P.* pp. 193—6), and Rankine concludes that the ratio of $\sqrt{1-b'^2-c'^2}:1$ lies between $\sqrt{1}:\sqrt{2}$ and $\sqrt{2}:\sqrt{3}$, approaching the less value as the diameter of the rod diminishes. Comparison with some experiments of Wertheim and Savart does not give very satisfactory results, and Rankine supposes that the freedom of the lateral vibrations is really limited by the means used to fix the rods so that the ratio of the two velocities generally exceeds $\sqrt{2}:\sqrt{3}$ and sometimes approaches equality. See Chree, *Quarterly Journal of Mathematics*, Vol. XXI. p. 295, and Vol. XXIII. pp. 335, 341.

[439.] Rankine concludes generally that:

(i) In liquid and solid bodies of limited dimensions, the freedom of lateral motion possessed by the particles causes vibrations to be propagated less rapidly than in an unlimited mass.

(ii) The symbolical expressions for vibrations in limited bodies are distinguished by containing exponential functions of the coordinates as factors; and the retardation referred to depends on the coefficients of the coordinates in the exponents of those functions, which coefficients depend on the molecular condition of the body's surface—a condition yet imperfectly understood (p. 261, *S. P.* p. 192).

It seems to me that the proper condition at the body's surface is the vanishing of the stress, but that in most cases of longitudinal vibrations this leads to very complex conditions for the determination of the coefficients of the coordinates in the exponentials. Otherwise I think we may safely agree with these conclusions of Rankine's, and he certainly put the matter in a clearer light than it was left by Wertheim: see our Arts. 1349*—51* and the last memoir of Chree's cited, pp. 324—5.

[440.] W. J. M. Rankine: *On the Vibrations of Plane Polarised Light. Philosophical Magazine*, Vol. I. 1851, pp. 441—6 (*S. P.* pp. 150—5). This is an attempt to explain by some rather general reasoning based on Rankine's theory of 'molecular vortices' (or atomic *nuclei* surrounded by elastic atmospheres: see our Art. 424) the phenomena of polarised light. We only refer to it here to cite the following remarks:

For if there is any proposition more certain than others respecting the laws of elasticity, it is this:—that the transverse elasticity of a medium, or the elasticity which resists *distortion* of the particles, depends upon the position of the *plane of distortion*, being the same for all directions of distortion in a given plane. This law is implicitly involved in the researches of Poisson, of M. Cauchy, of Mr Green and others on elasticity (p. 441, *S. P.* p. 150).

This can only refer to Theorem II. of the memoir of 1850: see our Arts. 421—2. As Rankine is here talking of crystalline bodies his statement is erroneous.

The keynote to Rankine's researches is to be found in the hypothesis:

That the medium which transmits light and radiant heat consists of the *nuclei* of the atoms vibrating independently, or almost independently, of their atmospheres; *absorption* being the transference of motion from the nuclei to the atmospheres, and *emission* its transference from the atmospheres to the nuclei (p. 443, *S. P.* p. 152).

The difficulty is then to understand how the ether of space, which must consist of atomic *nuclei* in order to transmit, is still incapable of absorbing.

[441.] W. J. M. Rankine: *General View of an Oscillatory Theory of Light. Philosophical Magazine,* Vol. VI. 1853, pp. 403—14 (*S. P.* pp. 156—67). This paper contains no reference to the theory of elasticity, and is rather difficult to follow owing to the suppression of the "strict mathematical analysis" by which its conclusions were deduced.

[442.] W. J. M. Rankine: *On the General Integrals of the Equations of the Internal Equilibrium of an Elastic Solid.* This is published in the *Proceedings of the Royal Society,* Vol. VII. 1856, pages 196—202. It is an abstract of a memoir which was received December 7, 1854. Judging from the abstract the memoir must have been of a very elaborate character; but it does not seem to have been ever published: see our Arts. 454 and 455.

After some definitions and general statements apparently

reproducing results of the ordinary theory of elasticity, we are told that the "*Second Section* of the paper relates to the problem of the general integration of the equations of the internal equili brium of an Elastic Solid, especially when it is *not isotropic.*" The solution seems to have been in Cartesian coordinates and obtained in some way by expanding the stresses in trigonometrical series of the three coordinates, but it is extremely difficult to follow the account given.

The *Third Section* appears to have dealt with Lamé's problem of the rectangular prismatic solid (see our Arts. 1079*—80*). Apparently the method consisted only in a long series of what, I should imagine, would be very troublesome approximations (p. 201).

The *Fourth Section* dealt with the general integrals of the equations of elasticity for an isotropic solid.

Finally Rankine insists upon the importance for practical purposes of the distinction between the *cone of shear* and the *cone of slide.* By this I judge that he had in the memoir drawn attention to the facts that the directions of maximum stress and strain do not necessarily coincide, and that rupture does not always take place across the direction of maximum stress: see our Arts. 1367*—8*.

[443.] W. J. M. Rankine: *On Axes of Elasticity and Crystalline Forms: Phil. Trans.* 1856, pp. 261—285 (*S. P.* pp. 119—149). This paper was read on June 21, 1855. It is remarkable for the number of new, and not improbably physically important results relating to the twenty-one elastic constants which it states, as well as for the novel nomenclature which it proposes to introduce.

Unfortunately the writer obtains his results by the application of "that branch of the Calculus of Forms which relates to linear transformations, and which has recently been so greatly advanced by the researches of Mr Sylvester, Mr Cayley, and Mr Boole." I say, unfortunately, as it will rarely happen that the elastician will have made a sufficiently wide study of *invariants, covariants, contragredients et hoc genus* to understand the processes of this memoir, while terms such as *umbral matrices* and *contra-ordinates* tend at the best to obscure the simple physical principles which often lie behind the equations. The biquadratics which give the distributions of stretch-modulus and direct-stretch coefficient are no doubt *tasimetric covariants,* but it may well be questioned whether this is the clearest method of approaching their discussion. Luckily Saint-Venant in his memoir of 1863 has given short and direct proofs of most of Rankine's results bringing out in each case their physical bearing: see our Arts. 132—7 which should be compared with Arts. 445—7

[444.] Rankine commences with the statement that:

As originally understood, the term "axes of elasticity" was applied to the intersections of three orthogonal planes at a given point of an elastic medium, with respect to each of which planes the molecular actions causing elasticity were conceived to be symmetrical.

The next two paragraphs (p. 261, *S. P.* p. 119) refer to the peculiar hypothesis of the earlier memoirs see our Arts. 424, 429, etc. The writer states that if the elasticity of solids arose from the action of centres obeying the rari-constant hypothesis or partly from such action and partly from an elasticity like that of a fluid, resisting change of volume only,' then it is easy to prove that three such planes of symmetry exist in every homogeneous solid. It is not obvious that three such planes would exist in a homogeneous aeolotropic solid with 15 constants, unless we could reduce those fifteen constants in the method of the earlier memoirs, a method which we have seen to be erroneous. Rankine further remarks that there is now no doubt that elastic stress is not such as can be accounted for by fluid elasticity and central intermolecular action as a function of the distance. This of course is merely a declaration of his own multi-constant views, which is somewhat obscured by the reference to "fluid-elasticity."

Assuming multi-constancy Rankine conveniently defines an axis of elasticity as any direction with respect to which certain kinds of elastic stresses are symmetrical, or

speaking algebraically, *directions for which certain functions of the coefficients of elasticity are null or infinite* (p. 261, *S. P.* p. 119).

The former seems a clearer statement than the latter.

[445.] We now give a Table of Rankine's nomenclature premising that he adopts θλίψις to denote strain and τάσις to denote stress.

Table of Tasinomic Coefficients or Constants.

Our notation for constant.	Our name for constant.	Rankine's name for Constant.	Rankine's name for Elasticity.
$\lvert xxxx \rvert$	Direct stretch	Orthotatic { Euthytatic	Direct or Longitudinal
$\lvert xxyy \rvert$	Cross stretch	Orthotatic { Platytatic	Lateral
$\lvert xyxy \rvert$	Direct slide	Orthotatic { Goniotatic	Rigidities
All other types	—	Plagiotatic	Unsymmetrical

It will be noted that Rankine's nomenclature does not distinguish by special names between a cross slide and a cross stretch-slide coefficient ($\lvert xyyz \rvert$ and $\lvert xxyz \rvert$).

We have further the following surfaces:

Thlipsimetric Surface = stretch-quadric: see our Art. 612*.

Tasimetric Surface = stress-quadric: see our Art. 610*

Rankine forgets the double sign in writing down these equations: see his p. 264 (*S. P.* p. 122), (3) and (4).

Orthotatic Ellipsoid. If $\iota_x\iota_x\iota_x\iota_x$ symbolically denote $\lvert xxxx \rvert$ its equation is:

$$\{\iota_{xx}(\iota_{xx}+\iota_{yy}+\iota_{zz})\}x^2+\{\iota_{yy}(\iota_{yy}+\iota_{zz}+\iota_{xx})\}y^2+\{\iota_{zz}(\iota_{zz}+\iota_{yy}+\iota_{xx})\}z^2$$
$$+2\{\iota_{yz}(\iota_{xx}+\iota_{yy}+\iota_{zz})\}yz+2\{\iota_{zx}(\iota_{xx}+\iota_{yy}+\iota_{zz})\}zx+2\{\iota_{xy}(\iota_{xx}+\iota_{yy}+\iota_{zz})\}xy$$
$$\equiv(\iota_{xx}+\iota_{yy}+\iota_{zz})(\iota_x x+\iota_y y+\iota_z z)^2=1.$$

Heterotatic Ellipsoid. This has for equation:

$$\{\lvert yyzz \rvert - \lvert yzyz \rvert\}x^2+\{\lvert zzxx \rvert - \lvert zxzx \rvert\}y^2+\{\lvert xxyy \rvert - \lvert xyxy \rvert\}z^2+2\{\lvert zzxy \rvert - \lvert zxyz \rvert\}yz$$
$$+2\{\lvert xyyz \rvert - \lvert yyzx \rvert\}zx+2\{\lvert yzzx \rvert - \lvert zzxy \rvert\}xy=1.$$

The three *orthotatic axes* are the principal axes of the orthotatic ellipsoid, the three *heterotatic axes* those of the heterotatic ellipsoid. For the former three axes equations of the type

$$\lvert yzxx \rvert + \lvert yzyy \rvert + \lvert yzzz \rvert = 0$$

hold, and they possess the physical property which we may state in the following words:

At each point of an elastic solid, there is one position in which a cubical element may be cut out, such, that a uniform dilatation of that element by equal stretches of its three dimensions, shall produce no shear on the faces of the element (p. 266, *S. P.* p. 126).

This physical property is not very obviously conveyed in Rankine's method of looking at the orthotatic ellipsoid. It follows at once from Saint-Venant's treatment of the subject: see our Art. 137, (iii).

For the heterotatic axes we have three relations of the type

$$|zxxy| - |xxyz| = 0,$$

and the physical property which we may thus state:

At each point of an elastic solid, there is one position in which a cubical element may be cut out, such that if there be a distortion of that element round (i.e. a slide perpendicular to) x, and an equal distortion round y, the traction on the faces normal to x arising from the distortion round x shall be equal to the shear round z arising from the distortion round y (p. 267, *S. P.* p. 126).

The coefficients of the heterotatic ellipsoid are termed *heterotatic differences* they vanish on the rari-constant hypothesis. Rankine terms *fluid elasticity* that elasticity for which the heterotatic ellipsoid becomes a sphere; the body is then *heterotatically isotropic.*

A body is *orthotatically isotropic* when the orthotatic ellipsoid becomes a sphere.

A body which is both heterotatically and orthotatically isotropic is *not completely isotropic* as it has still 11 independent constants.

[446.] The next surface dealt with by Rankine is what he terms the *biquadratic tasinomic surface*, or

$$\{(\iota_x x + \iota_y y + \iota_z z)^2\}^2 = 1.$$

It is the biquadratic which gives the distribution of the direct-stretch coefficient.

He terms its coefficients the *homotatic coefficients.* Diameters of this surface which are normal to the tangent planes at their extremities are termed *euthytatic axes* (p. 268, *S. P.* p. 127). Rankine returns later to the consideration of these axes in *Sections* 22—29.

He now proceeds to the dissection of this surface by rectangular linear transformation. By this means it is always possible to make three of the terms with odd exponents or three functions of such terms vanish. Thus Rankine shows we may find three mutually rectangular axes for which three equations of the type

$$|yyyz| = |zzyz|$$

hold. These axes he terms the *principal metatatic axes.* They possess the following property (supposing them to be the axes of x, y, z):

If there be a stretch along y and an equal squeeze along z (or *vice versâ*), no shear will result round x on planes normal to y and z (p. 268, *S. P.* p. 128).

Suppose the axes of coordinates to be any whatever, and let y', z' be any other pair of rectangular axes in the plane of y, z, then it is easy to show (by the method of our Art. 133) that:

$$|y'y'y'z'| - |z'z'y'z'| = \{2\,|yyzz| + 4\,|yzyz| - |yyyy| - |zzzz|\}\,\frac{\sin 4\omega}{4} + \{|yyyz| - |zzyz|\}\cos 4\omega$$

where $\omega = \angle\, yOy'$.

Hence, in each plane in an elastic solid, there is a system of two pairs of axes metatatic for that plane and forming with each other eight equal angles of 45°.

If x, y, z be the principal metatatic axes, then $|yyyz| = |zzyz|$ and we see that $|y'y'y'z'| = |z'z'y'z'|$ when $\omega =$ any multiple of 45°

Or, in each of the three metatatic planes, there is a pair of diagonal metatatic axes, bisecting the right angles formed by the principal metatatic axes (p. 269, *S. P.* p. 128). These six metatatic axes and their productions are perpendicular to the faces of a rhombic dodecahedron.

A solid is *metatatically isotropic* when for a cubical element cut out in any position, a stretch in the direction of one axis and an equal squeeze along another produce no shear on the faces.

Metatatical isotropy involves three relations of the type

$$2\,|yyzz| + 4\,|yzyz| - |yyyy| - |zzzz| = 0$$

for all sets of axes. These expressions are termed *metatatic differences.*

[447.] *Orthotatic Symmetry.* When one and the same set of orthogonal axes are at once orthotatic, heterotatic, metatatic and euthytatic, or the twelve plagiotatic coefficients vanish, the solid is said to possess *orthotatic symmetry.* This reduces the elastic constants to the *nine orthotatic coefficients.*

Cybotatic Symmetry. In addition to orthotatic symmetry let the three direct stretch coefficients be equal to each other, the three direct-slide coefficients and the three cross-stretch coefficients. In this case the coefficients reduce to three and the symmetry is *cybotatic* (p. 270, *S. P.* p. 130).

The metatatic difference will in this case be equal to

$$2\,|yyzz| + 4\,|yzyz| - 2\,|xxxx|$$

and unless this vanishes the body will not be metatatically isotropic. Green's proposed structure for the ether endowed it with cybotatic symmetry: see our Art. 146.

If the metatatic difference vanishes then cybotatic symmetry reduces to bi-constant isotropy, or what Rankine terms *pantatic isotropy* (p. 271, *S. P.* p. 131).

[448.] Rankine next passes to *Thlipsinomic Coefficients* or those which express strain as a linear function of stress. We may express these coefficients as follows, a, b, c denoting the directions of the axes x, y, z:

$$s_x = (aaaa)\,\widehat{xx} + (aabb)\,\widehat{yy} + (aacc)\,\widehat{zz} + (aabc)\,\widehat{yz} + (aaca)\,\widehat{zx} + (aaab)\,\widehat{xy}.$$

If symbolically we put $(aaaa) = \nu_a\nu_a\nu_a\nu_a$ and $\widehat{xx} = \zeta_x\zeta_x$, $\widehat{yz} = 2\zeta_y\zeta_z$, we may throw the strain into the form useful for symbolic operations:

$$= \nu_a\nu_a\,(\nu_a\zeta_x + \nu_b\zeta_y + \nu_c\zeta_z)^2.$$

Rankine gives the following nomenclature:

Our notation for constant.	Our name for constant.	Rankine's name for Constant.	Rankine's name for Strain property.
$(aaaa)$	Direct traction	Orthothliptic { Euthythliptic	Longitudinal Extensibility
$(aabb)$	Cross traction	Orthothliptic { Platythliptic	Lateral Extensibility
$(bcbc)$	Direct shear	Orthothliptic { Goniothliptic	Pliability
All other types		Plagiothliptic	Unsymmetrical Pliability

It is easy to see that all the Thlipsinomic axes coincide with the corresponding systems of Tasinomic axes. As a rule platythliptic (or cross-traction) coefficients are negative (p. 273, *S. P.* p. 134).

[449.] Rankine next proceeds to consider strains and stresses when referred to oblique axes, with a view of dealing more at length with the biquadratic tasinomic surface and the euthytatic axes; see our Art. 446. By transformation to oblique (or rectangular) axes he reduces the equation of this surface to its canonical form, in which it has only nine terms, those in y^3z, yz^3, z^3x, zx^3, x^3y and xy^3 being removed. This involves the vanishing of six plagiotatic coefficients for that system of axes, namely:

$$|yyyz| = |zzyz| = |zzxz| = |xxxz| = |xxxy| = |yyxy| = 0$$

or, as we should say, all the direct stretch slide coefficients are zero. These three axes which always exist but may be oblique or rectangular are termed the *principal euthytatic axes.*

[450.] We have next the following classification with regard to forms of euthytatic distribution:

(i) If a solid has three oblique principal euthytatic axes making equal angles with each other round an axis of symmetry, and if each of these axes has equal systems of homotatic coefficients, i.e. if the biquadratic tasinomic surface reduce to the form

$$|xxxx|\,(x^4+y^4+z^4)+2\,\{|yyzz|+2\,|yzyz|\}\,(y^2z^2+z^2x^2+x^2y^2)$$
$$+\,4\,\{2\,|xzxy|+|xxyz|\}\,xyz\,(x+y+z)=1,$$

then the solid is said to possess *rhombic symmetry*, for the three oblique axes are normals to the faces of one rhombohedron. The axis of symmetry must be a fourth euthytatic axis.

(ii) In the limiting case when the three oblique axes make angles of 120° with each other, they lie in the same plane and are normal to the axis of symmetry and to the faces of a hexagonal prism. This is *hexagonal symmetry*.

Rankine (pp. 278—9, *S. P.* pp. 140—1) proves various properties of this kind of symmetry having regard to the existence of other euthytatic axes.

(iii) If a solid has one euthytatic axis (z) normal to the other two (xy) still oblique, these two having equal sets of homotatic coefficients, it is said to possess *orthorhombic symmetry*, its principal euthytatic axes being normals to the faces of a right rhombic prism. A sub-case of orthorhombic symmetry is the existence of further pairs of euthytatic axes in the planes zx, zy. When such exist they are normals to the faces of an *octohedron with a rhombic base.*

(iv) The three principal euthytatic axes being orthogonal, we have *orthogonal symmetry.* This subdivides itself according to the existence of other euthytatic axes in none, all or two of the principal euthytatic planes into a distribution of euthytatic symmetry marked by a rectangular prism, by an irregular rhombic dodecahedron, or by an octohedron with rectangular base

(v) Orthogonal symmetry with equal sets of homotatic coefficients for each axis is called *cyboïd symmetry.* The three cases corresponding to those of (iv) are marked by a cube, a regular rhombic dodecahedron and a regular octohedron.

(vi) *Monaxal symmetry.* The homotatic coefficients are completely isotropic round one axis. The principal euthytatic axes are the axis of symmetry and all lines perpendicular to it If other euthytatic axes exist they are normal to the surface of a cone (p. 280 *S. P.* p. 143).

(vii) *Complete isotropy* of the homotatic coefficients is the case in which every direction is an euthytatic axis.

[451.] On pp. 280—1 (*S. P.* pp. 143—4) Rankine classifies the several *primitive forms known in crystallography* on the basis of these various distributions of the euthytatic axes. He makes the following statement:

It is probable that the normals to *Planes of Cleavage* are euthytatic axes of minimum elasticity.

He brings no evidence on this point, and it seems to me somewhat doubtful for the following reasons:

(i) *Any* biquadratic surface would give a similar system of symmetrical forms, which might be classified in the same manner. Why should the biquadratic which determines the distribution of the direct-stretch coefficient be chosen? Rankine's euthytatic axes correspond to directions in which this coefficient has a maximum or minimum value, and therefore the planes of cleavage would be perpendicular to directions in which the direct-stretch coefficient has a maximum or minimum value.

(ii) It would seem quite as reasonable, if not more reasonable, to choose as our fundamental biquadratic that which gives the distribution of the stretch-modulus (see our Art. 309). For the directions of the maximum or minimum rays of this figure are those for which a given traction produces a minimum or maximum stretch But even then it is not yet proven that in an *aeolotropic* body rupture will first occur across the directions of greatest stretch.

(iii) If we put all the stresses zero except $\widehat{xx}$, we have

$$s_x = (aaaa)\,\widehat{xx}, \qquad s_y = (bbaa)\,\widehat{xx}, \qquad s_z = (ccaa)\,\widehat{xx},$$
$$\sigma_{yz} = (bcaa)\,\widehat{xx}, \qquad \sigma_{zx} = (caaa)\,\widehat{xx}, \qquad \sigma_{xy} = (abaa)\,\widehat{xx}.$$

The maximum stretch s_x for a given traction $\widehat{xx}$ will thus occur for that direction in which $(aaaa)$ (really $1/E$) is a maximum, but how far will rupture (supposing elasticity to last up to rupture!) be affected by the existence and magnitude of the other components of strain? The magnitude of these depends in each case on the value round the given direction of the platythliptic and plagiothliptic coefficients.

(iv) Thus it would seem to me that if we assume the direction of the greatest stretch for a given traction to determine that of ultimate rupture, then it would be better to form the biquadratic giving the euthythliptic coefficient $(aaaa)$ in any direction, and deduce euthythliptic instead of euthytatic axes as giving the planes of cleavage. The ultimate planes of cleavage thus obtained may coincide with Rankine's, but the conditions would appear in a different form, and the whole process have a more direct physical meaning.

(v) It must be remarked that some geological writers hold that the planes of cleavage are perpendicular to the directions of maximum or minimum traction. These are not necessarily those in which either the stretch-modulus or the direct-stretch coefficient is a maximum or a minimum. Their view would lead to a third method of treating the problem see our Art. 1367*.

[452.] On pp. 282—3 (*S. P.* pp. 145—7) of the memoir are some general remarks. Thus Rankine notes that the 15 homotatic coefficients on which the euthytatic axes depend, may be considered as independent of the six heterotatic differences on which the heterotatic axes depend. In other words, granting an euthytatic classification of crystals, bodies may have the same crystalline form and yet differ materially in the laws of their elasticity. This would not be possible in the case of rari con stant elasticity.

It may be noted that Rankine rejects the hypothesis of the luminiferous ether being a simple elastic medium, as no such medium could give a rotation of the plane of polarisation. He notes also that the refractive action of a crystal on light requires far fewer constants than are supplied by the crystal's elasticity.

The memoir concludes with a note on *Sylvestrian Umbrae* (pp. 284—5, *S. P.* pp. 147—9).

[453.] W. J. M. Rankine: *On the Stability of Loose Earth.* This is published in the *Philosophical Transactions* for 1857, pp.

9—27; it was received June 10 and read June 19, 1856: an abstract of it is given in the *Proceedings of the Royal Society*, Vol. VIII., 1857, pp. 185—7.

The memoir employs some of the elementary formulae of stress in the problem of earthwork. Suppose that the axes of coordinates at a certain point coincide with the principal axes of stress. Let T_1, T_2, T_3 be in descending order of magnitude, and let them denote the *principal tractions*. Take the plane which contains the directions of T_1 and T_3; and in that plane suppose a straight line making an angle ψ with the direction of T_1 consider the stress on the plane at the point which is normal to the straight line. Denote this stress by R; let P be the tractive, Q the shearing component of R; and let θ denote the angle between the directions of P and R. Then put

$$S = \tfrac{1}{2}(T_1 + T_3), \qquad D = \tfrac{1}{2}(T_1 - T_3);$$

and it will be found that the following results are easily deduced from the elementary formulae of stress:

$$P = S + D\cos 2\psi, \qquad Q = D\sin 2\psi,$$

$$\tan\theta = \frac{D\sin 2\psi}{S + D\cos 2\psi};$$

the maximum value of θ is $\sin^{-1} D/S$, and it occurs when

$$\psi = \frac{\pi}{4} + \tfrac{1}{2}\sin^{-1}\frac{D}{V}.$$

[454.] In the *Comptes rendus*, Vol. L., 1860, p. 235, there is a note of the *Grand Prix de mathématiques*. This had been offered for the second time in 1857, for a solution of the following problem:

Trouver les intégrales des équations de l'équilibre intérieur d'un corps solide élastique et homogène dont toutes les dimensions sont finies, par exemple d'un parallélépipède ou d'un cylindre droit en supposant connues les pressions ou tractions inégales exercées aux différents points de sa surface.

The commissioners were Liouville, Lamé, Duhamel and Bertrand. The problem for the right six-face was first proposed by Lamé: see our Arts. 1079*—80*.

Two memoirs were sent in, but as neither of them contained the solution of the question proposed, the prize was not awarded but proposed again for 1861. One of these memoirs was, I believe, due to Rankine. "In 1857......he also sent to the French Academy of Sciences a memoir,—*De l'Équilibre intérieur d'un corps solide, élastique, et homogène.*" See the *Memoir* of Rankine by P. G. Tait prefixed to the *Miscellaneous Scientific Papers*, p. xxiii.

This is probably closely connected with the paper of which an abstract is given in the *Proceedings of the Royal Society*: see our Art. 442. Its non-publication and the failure at Paris suggest that the analysis was probably defective as well as lengthy. A portion only of the Paris paper was afterwards in 1872 communicated to the Royal Society of Edinburgh and is published in the *Transactions* Vol. XXVI.: see our Arts. 455—62.

[455]. W. J. M. Rankine: *On the Decomposition of Forces externally applied to an Elastic Solid. Transactions of the Royal Society of Edinburgh*, Vol. XXVI., 1872, pp. 715—27.

The author writes:

The principles set forth in this paper, though now (with the exception of the first theorem) published for the first time, were communicated to the French Academy of Sciences fifteen years ago, in a memoir entitled: *De l'Equilibre intérieur d'un corps solide, élastique, et homogène*, and marked with the motto, "Obvia conspicimus, nubem pellente Mathesi," the receipt of which is acknowledged in the *Comptes rendus* of the 6th April, 1857.

See our Arts. 442 and 454.

The memoir is like nearly all Rankine's papers, extremely suggestive, and rich in terminology, amounting in this case to very unnecessary verbosity.

[456.] The memoir opens with the statement of the following theorem (which had been given in the *Philosophical Magazine*, Vol. X., p. 400, 1855):

Every self-balanced system of forces applied to a connected system of points is capable of resolution into three rectangular systems of parallel self-balanced forces applied to the same points (p. 715).

The three rectangular axes to which the three systems of self-balanced forces are parallel are termed *isorrhopic axes*. Rankine proves the proposition by appeal to the theory of covariants. But it is easily proved *ab initio*. Let X, Y, Z be the components of force acting on the point x, y, z. Then for equilibrium we must have:

$$\Sigma X = \Sigma Y = \Sigma Z = 0,$$

$$\Sigma (Yz - Zy) = \Sigma (Zx - Xz) = \Sigma (Xy - Yx) = 0.$$

Consider the line drawn through the origin with direction cosines l, m, n, then the force at x, y, z parallel to this line is $lX + mY + nZ = P$ say. Let r be the distance between the origin and x, y, z; let ϕ be the angle between r and (l, m, n). Then the quantity $rP\cos\phi$ is independent of the directions of the coordinate axes and consequently

$$\Sigma rP \cos\phi$$

is (a covariant or) the same in form for all systems of rectangular axes through the origin. But it equals

$$\Sigma (lx + my + nz)(lX + mY + nZ) = l^2\Sigma Xx + m^2\Sigma Yy + n^2\Sigma Zz$$
$$+ mn\Sigma (Yz + Zy) + nl\Sigma (Zx + Xz) + lm\Sigma (Xy + Yx).$$

Putting with Rankine:

$$\Sigma Xx = A,\ \Sigma Yy = B,\ \Sigma Zz = C,$$
$$\Sigma Yz = \Sigma Zy = D,\ \Sigma Zx = \Sigma Xz = E,\ \Sigma Xy = \Sigma Yx = F,$$

we have this equal to

$$Al^2 + Bm^2 + Cn^2 + 2Dmn + 2Enl + 2Flm.$$

But there are three rectangular directions, namely those of the principal axes of the quadratic surface:

$$Ax^2 + By^2 + Cz^2 + 2Dyz + 2Ezx + 2Fxy = 1 \qquad \text{(i)},$$

for which $D = E = F = 0$, in this expression.

Hence there are three directions for which:

$$\Sigma X = 0,\quad \Sigma Xy = 0,\quad \Sigma Xz = 0,$$
$$\Sigma Y = 0,\quad \Sigma Yz = 0,\quad \Sigma Yx = 0,$$
$$\Sigma Z = 0,\quad \Sigma Zx = 0,\quad \Sigma Zy = 0,$$

which proves the theorem.

Rankine terms (i) the *Rhopimetric Surface*; its coefficients the *Rhopimetric Coefficients*; its principal axes are the *Isorrhopic Axes* and the corresponding values of A, B, C the *Principal Rhopimetric Coefficients*. An *Arrhopic System* of forces is defined as one for which *all* the rhopimetric coefficients are zero. Rankine adds that in this case every axis is an isorrhopic axis, but the proper and sufficient conditions for this are that $A = B = C$, while $D = E = F = 0$.

[457.] Rankine next applies his theory of isorrhopic axes to reduce any load system applied to an elastic solid to three separate self-balanced systems of parallel loads and thus the problem of elastic equilibrium to the solution of three separate cases of parallel loading. He justifies this reduction by remarking that although we may not in the treatment of an elastic solid transfer the point of application of a force to *any* point in the line of action of the force, we may still resolve each force at its point of action into components in different directions, or:

When the straining forces to which an elastic solid is subjected are restricted within certain limits, the straining effect of any number of self-balanced systems of forces combined is sensibly equal to the sum of the effects which those systems respectively produce when acting separately (p. 716).

If X, Y, Z be the body-forces at x, y, z, X', Y', Z' the components of load at x', y', z' on the element dS of surface, then the rhopimetric coefficients for an elastic solid will be given by formulae of the type:

$$\left.\begin{aligned} A &= \iiint xX\rho dxdydz + \iint x'X'dS \\ D &= \iiint zY\rho dxdydz + \iint z'Y'dS \\ &= \iiint yZ\rho dxdydz + \iint y'Z'dS \end{aligned}\right\} \quad\text{(ii)},$$

whence the isorrhopic axes can be found (p. 717).

[458.] After reproducing the body- and surface-stress equations in Lamé's notation, Rankine proceeds to remove the terms involving terrestrial gravitation from the body stress-equations. Such gravitation is usually the only body-force which occurs in elastic problems. Take the plane of yz horizontal through the centroid of the body and the axis of x vertically downwards, then by assuming

$$\widehat{xx} = \widehat{xx}' - g\rho x,$$

we cause the body-forces to disappear from the differential equations.

The first surface-stress equation now becomes

$$X' = l\,(\widehat{xx}' - g\rho x') + m\widehat{xy} + n\widehat{xz}.$$

Hence a system of surface tractions given by

$$X' = -g\rho l x', \quad Y' = 0, \quad Z' = 0,$$

would just balance the weight of the body We may thus withdraw the weight of the body from our consideration of the problem, if we take away from the internal stress $\widehat{xx}$ found after removal of the gravita tion terms the quantity $g\rho x'$ and further suppose the surface load increased by the component $g\rho l x'$ parallel to the axis of x.

This system of surface and body-load is according to Rankine arrhopic, for from (ii):

$$B = C = D = E = F = 0.$$

Further:
$$\begin{aligned} A &= \iiint xg\rho dxdydz + \iint x'(-g\rho x'l)\,dS \\ &= g\rho\left\{\iiint x dxdydz - \iint x'^2 l dS\right\}. \end{aligned}$$

Now the first integral vanishes since the plane of yz passes through the centroid and the second term also, Rankine says, if we remember the changes in sign of l. But this seems to me only true if the surface is symmetrical about the plane of yz.

The system of surface-loads which forms with the gravitation of a body an arrhopic system, Rankine terms an *antibarytic* load-system ('antibarytic pressures') and the corresponding body-stresses are *anti barytic stresses.*

The system of stresses left after taking away the antibarytic stresses from the actual stresses at the several elements of a body's surface are termed *abarytic stresses* ('abarytic pressures') (p. 719).

The internal stresses corresponding to an abarytic system of surface loading satisfy equations of the type:

$$\frac{d\widehat{xx}}{dx} + \frac{d\widehat{xy}}{dy} + \frac{d\widehat{xz}}{dz} = 0.$$

[459.] The memoir next proceeds to an analysis of abarytic load systems. Rankine gives the following definition: An abarytic surface load which produces uniform stress throughout an elastic solid is termed *homalotatic.* An abarytic system may be broken up into a homalotatic system and an arrhopic system in the following manner. Calculate the six rhopimetric coefficients and assume the internal stresses to be equal to these coefficients divided by the volume of the solid, i.e. take

$$\widehat{xx} = A/V, \qquad \widehat{yy} = B/V, \qquad \widehat{zz} = C/V,$$
$$\widehat{yz} = D/V, \qquad \widehat{zx} = E/V, \qquad \widehat{xy} = F/V.$$

These satisfy the body-stress equations and give for the surface load

$$X' = \frac{1}{V}(lA + mF + nE),$$

$$Y' = \frac{1}{V}(lF + mB + nD),$$

$$Z' = \frac{1}{V}(lE + mD + nC),$$

or, a homalotatic system of surface-load.

The rhopimetric coefficients for this surface-load are of the type:

$$A_0 = \iint xX'dS$$
$$= \frac{1}{V}\{A\iint xldS + F\iint xmdS + E\iint xndS\}$$
$$= \frac{1}{V}AV = A,$$

for, $$\iint xmdS = \iint xndS = 0.$$

Thus the rhopimetric coefficients of the homalotatic system are equal to those for the complete abarytic system, or if the homalotatic system be subtracted from the abarytic system we must be left with a pure arrhopic system. (pp. 720—1).

Rankine remarks that the above homalotatic system of six uniform stresses really denotes the *mean state of stress* of the whole body. It may be remarked that the axes of principal traction of the homalotatic system are the isorrhopic axes of the complete abarytic system.

[460.] By following out the operations indicated in the above articles we reduce any system of load applied to an elastic body to the solution of a problem in arrhopic loading. Thus the reduction of the load system into three rectangular systems of parallel load can be made for any three rectangular axes; for example, for axes parallel to the axes of figure of a body, which will as a rule considerably simplify the problem. (p. 722).

[461.] The next section of the memoir investigates those cases in which internal stress is independent of the coefficients of elasticity of the solid. Rankine concludes that when the shifts can be expressed by algebraic functions of the coordinates not exceeding the second degree, and consequently the stresses by constants and linear functions of the coordinates, this result will follow. The stresses will then be of the type

$$\widehat{xx} = c_1 + e_1 x + f_1 y + g_1 z,$$

$$\widehat{yz} = c_4 + e_4 x + f_4 y + g_4 z.$$

Rankine gives no general name to stresses of this type[1] but classifies them as follows:

The constant terms c ... c_4 ... etc. correspond to a *homalotatic* load-system.

The coefficients e_1, f_2, g_3 are equivalent to an *antibarytic* load-system.

The coefficients $f_1, g_1, e_2, g_2, e_3, f_3,$ correspond to a *homalocamptic* load-system, or to stresses due to uniform bending.

The coefficients e_4, f_5, g_6 correspond to a *homalostrephic* load-system or to stresses due to uniform twisting (pp. 723—4).

Rankine shows that both homalocamptic and homalostrephic load-systems are arrhopic (pp. 724—5). He does not discuss or give a name to the stresses arising from f_4, g_4, e_5, g_5, e_6 and f_6.

[462.] In conclusion Rankine takes (pp. 725—7) two simple examples of homalocamptic and homalostrephic stresses. The first embraces practically the Euler-Bernoulli theory of flexure, and the second the torsion of an elliptic cylinder allowing for the distortion of the cross-sections.

The latter investigation starts from the assumption that

$$\widehat{xz} = by, \quad \widehat{xy} = cz,$$

where b and c are undetermined constants. Rankine determines them erroneously, for in line 19 of p. 727 he puts $y^2/p^2 + z^2/q^2 = 1$, which

[1] Catching for a moment Rankine's mania for nomenclature we might term all the cases in which the stresses are linear functions of the coordinates, cases of *euthygrammic* stress.

does not hold as his point y, z is not on the perimeter of the elliptic cross-section. Had he noted this he would have found just *double* the values he gives for $\widehat{xz}$ and $\widehat{xy}$ in equation (22), and these would then have been in agreement with the results of Saint Venant as cited in our Art. 18. I do not understand the remark as to Cauchy with which the memoir closes. These two examples are, however, of little importance.

[463.] W. J. M. Rankine: *On the Stability of Factory Chimneys. Proceedings of the Philosophical Society of Glasgow*, Vol. IV pp. 14—18, Glasgow, 1860. This paper treats only of the effects of the wind and of the weight of the chimney, and does not discuss its elastic strength even in the matter of crushing due to the weight of the chimney itself. It is a simple problem in statics which is here dealt with, and can be easily solved by an appeal to the theory of the core: see our Art. 815* and Vol. I., p. 879

[464.] W. J. M. Rankine: *A Manual of Applied Mechanics*. London, 8vo. 1858—1888. The first edition of this work was published in 1858 and the twelfth in 1888 edited by W. J. Millar. The first edition contains xvi + 640 pages and the twelfth xiv + 667 pages. The chief additions made by the Editor are contained in the Appendix. My references will be to the pages of the more readily accessible twelfth edition. The work itself is important in the history of elasticity, for it was among the first to bring the theory of elasticity in a scientific form before engineering students. Rankine himself writes in his preface:

A branch of Mechanics not usually found in elementary treatises is explained in this work, viz., that which relates to the equilibrium of stress, or internal pressure, at a point in a solid mass, and to the general theory of the elasticity of solids. It is the basis of a sound knowledge of the principles of the stability of earth, and of the strength and stiffness of materials; but so far as I know, the only elementary treatise on it that has hitherto been published is that of M. Lamé (p. iii).

We will briefly note the several parts of this work which treat of our subject, commenting on anything which seems to have been novel at the date of its publication.

[465.] Pp. 68—127 deal with stresses in solids and deduce those in liquids as a special case.

(*a*) Rankine, as in his memoir of 1855, reserves the term *stress* for the dynamic aspect of elasticity, strain for its geometrical aspect. A further progress in differentiation of terms is made by defining *shear* as tangential stress, i.e. ceasing to treat it as a name for strain: see our Vol. I., p. 882.

(*b*) Rankine deals with such problems as 'centres of stress' (load points), 'neutral axis,' 'conjugate stresses' and the relation of these quantities to moments of inertia (pp. 71—85). He gives general expressions for the traction and shear across any plane (pp. 92—3) and for the discovery of the principal tractions. He deals with the special case so important in practice of uniplanar stress (pp. 95—112) and with the 'ellipse of stress'. His treatment of this subject is the fullest which, I think, had been given at the time of publication of his work, and his discussion of stress-centres although a little later than that of Bresse (see our Arts. 815* and 516) was probably worked out quite independently. Thus, if the system of stress in a plane be given by $\widehat{xx}$, $\widehat{yy}$, $\widehat{xy}$ referred to rectangular axes in the plane, and n denote the normal to any plane perpendicular to this plane and t the trace of these two planes, Rankine shows that

$$\widehat{nn} = \widehat{xx}\cos^2(xn) + \widehat{yy}\sin^2(xn) + 2\widehat{xy}\cos(xn)\sin(xn),$$

$$\widehat{nt} = \tfrac{1}{2}(\widehat{xx} - \widehat{yy})\sin 2(xn) - \widehat{xy}\cos 2(xn)\,;$$

whence he easily deduces the properties of the principal uniplanar tractions and of the ellipse of stress, and applies them to a variety of special problems. Most of his results have found their way into other text-books and papers sometimes with scanty acknowledgement; I may cite in this matter a dissertation by Kopytowski: *Ueber die inneren Spannungen in einem freiaufliegenden Balken*, pp. 1—17.

(*c*) I may draw special attention to the Problem on pp. 110—12 entitled: *Combined stresses in one plane: Given the normal intensities and directions of any number of simple stresses whose directions are in the same plane; required the directions and intensities of the pair of principal stresses* [tractions] *resulting from their combination.*

Let the principal tractions be T_1 and T_2, and let the first make an angle ϕ with the axis of x, x and y being two arbitrary rectangular axes taken in the plane of the stresses.

Let p, p' denote the normal intensities of any two of the given stresses, then Rankine shows that

$$T_1 + T_2 = \Sigma(p),$$

$$T_1 - T_2 = \sqrt{\Sigma(p^2) + 2\Sigma\{pp'\cos 2(pp')\}},$$

$$\tan 2\phi = \frac{\Sigma\{p\sin 2(xp)\}}{\Sigma\{p\cos 2(xp)\}}.$$

Thus the intensities of the principal tractions can be found without assuming planes of reduction, but to find their directions requires us to do this.

(*d*) On pp. 112—127 we have the body-stress equations deduced and a few special applications to fluids etc.

(*e*) The theory of uniplanar stresses developed in the previous sections is applied on pp. 129—269 to framework, arches, buttresses, earth-pressure, domes, masonry and brick-work of all kinds. This involves a considerable discussion of the properties of the core (see our Art. 815*) which is not however referred to directly by name. The topics discussed fall outside the limits of our present subject.

[466.] Chapter III. entitled: *Strength and Stiffness*, occupies pp. 270—377 and forms for its date an excellent practical treatise on the technical side of elasticity. We can only note a few points:

(*a*) As usual Rankine strives to give scientific definiteness to certain terms which are in wide but rather vague use. Thus for example he proposes the following nomenclature for the fracture associated with characteristic kinds of strain (p. 272):

	Strain.	*Fracture.*
Longitudinal	Extension [Stretch]	Tearing
	Compression [Squeeze]...	Crushing and Cleaving.
Transverse	Distortion [Slide].........	Shearing
	Torsion	Wrenching [Twisting]
	Bending	Breaking across [Snapping].

This analysis of the more usual forms of strains is convenient, but objection might well be taken to some of the words for fracture; thus a *wrenching* fracture is associated in our minds rather with a combination of torsion and pull than with pure torsion[1]. Perhaps the term 'twisting fracture' would be less liable to misinterpretation. 'Breaking across fracture' is also rather cumbersome and might be more briefly termed *snapping*.

(*b*) Rankine's discussion and definitions of perfect and imperfect elasticity and of set (pp. 272—3) are perhaps not wholly satisfactory in the light of more recent knowledge, but his further definitions require notice:

(i) The *Ultimate Strength* of a solid is the stress required to produce fracture in some specified way. [This is now usually termed *absolute strength.*]

[1] I should prefer to retain the name *wrench* for the stress side of the strain combined of a stretch and a torsion (which might perhaps be called a *wring*). We might then scientifically appropriate *sprain* for the set-strain produced by a wrench.

(ii) The *Proof Strength* is the stress required to produce the greatest strain of a specific kind consistent with safety; that is with the retention of the strength of the material unimpaired. A stress exceeding the proof strength of the material, although it may not produce instant fracture, produces fracture eventually by long-continued application and frequent repetition (p. 273).

This definition of proof strength is not scientifically very accurate, for it neither suggests the method nor shows the possibility of its determination. In many cases we may have set without any reduction of absolute strength, and thus proof strength is not by any means measured by the elastic limit. Rankine notes this and remarks (p. 274):

(iii) The determination of proof strength by experiment is now, therefore, a matter of some obscurity; but it may be considered that the best test known is the *not producing an increasing set by repeated application.*

Obviously this is merely a negative test and could only be successful in ascertaining the proof strength of a given piece of material, after the material had been rendered unfit for further use. Rankine defines strength whether ultimate or proof, as the product of two quantities, *Toughness* and *Stiffness.*

(iv) *Toughness*, ultimate or proof, is here used to denote the greatest strain which the body will bear without fracture or without injury as the case may be.

(v) *Stiffness*, which might also be called *hardness*, is used to denote the ratio borne to that strain [toughness] by the stress required to produce it.

Thus while toughness is measured as a strain, stiffness is measured by a tasinomic (or elastic) coefficient of some particular kind. It does not seem correct, however, to identify hardness with this conception of stiffness.

(vi) *Malleable* and *ductile* solids have ultimate toughness greatly exceeding their proof toughness.

(vii) *Brittle* solids have their ultimate and proof toughness nearly equal.

(viii) *Resilience* or *Spring* is the quantity of mechanical work required to produce the proof strain, and is equal to the product of that strain by the *mean stress* in its own direction which takes place during the production of that strain,—such stress being either exactly or nearly equal to one-half of the stress corresponding to the proof-strain p. 273.

It would be better to distinguish between absolute, proof and elastic resilience, and then perhaps to reserve the word *spring* for the latter only: see our Vol. I., p. 875, and Art. 340.

(ix) *Pliability* (Extensibility, Compressibility, Flexibility, to which we might add Shearability and Twistability) is a general term used to denote the inverse of *stiffness*. It is accurately measured by *a thlipsinomic coefficient* (p. 273).

(x) *Working Stress* on the material of a structure is made less than the proof strength in a certain ratio to be determined by practical experience, in order to provide for unforeseen contingencies (p. 274).

Such a ratio is termed a *factor of safety*. The ratios of the ultimate strength to the proof strength and to the working stress are also termed *factors of safety*. There is a table of such factors on p. 274.

[467.] Rankine now turns to the mathematical theory of elasticity, especially to the discussion of strains, strain-energy, and the usual problems of technical elasticity. He considers that the generalised Hooke's Law is "fulfilled in nearly all the cases in which the stresses are within the limits of proof strength—the exceptions being a few substances very pliable, and at the same time very tough, such as caoutchouc" (p. 275). This statement seems practically to identify the proof strength with the limit of linear elasticity—an identity which itself seems to be the exception rather than the rule: see our Arts. 850*—5*, 857*, 1217*, 1296*, and Vol. I., p. 891, Note D.

We may remark that the *Manual* uses *isotropic* and *amorphous* as synonymous terms (p. 278). This is not in accordance with the terminology of the present work: see our Arts. 4 (η), 115, and 142, 230.

[468.] Rankine (pp. 280—3) discusses at some length uniplanar strain and the ellipse of strain. He works out problems of hollow, cylindrical and spherical shells and obtains results corresponding to those of Lamé but he uses only elementary processes. He adopts, however, (pp. 293 and 296) stress limits of strength: see our Arts. 1013*, 1016*, and footnotes. He gives the variation of the cross-section for a doubly built-in heavy beam of 'uniform strength': see his p. 336 and our Art. 5 (*e*), and then passes to shearing-stress and strength (as in rivetted joints of all kinds), to compression and crushing (splitting, shearing, bulging, buckling, cross-breaking), to flexure (bending moment, shear and transverse strength, i.e. snapping), to beams of equal and greatest strength (solids of *equal resistance*, etc.) and to *Lines of Principal Stress in Beams*. These are treated on the supposition that the stress-system of a beam under flexure is uniplanar, but the researches of Saint-Venant have shown this to be incorrect: see our Arts. 99—100. Such lines of stress as are figured by Rankine on p. 342 and are to be found in many practical text-books, are therefore even in the most favourable cases, e.g. the thin web of a girder, only rough approximations. Similarly Rankine's treatment of the influence of slide when combined with flexure in producing deflection is erroneous: see our Art. 556 and our discussion below of Winkler's memoir of 1860. Then follow a number of problems on the elastic line for various beams which do not call for special notice. This section of the chapter concludes with a reference to the 'Hydrostatic Arch' first fully discussed by Yvon-Villarceaux. Its equation may be written

$$Py = E\omega\kappa^2/\rho,$$

where ρ is the radius of curvature at any point of the elastic line, $E\omega\kappa^2$ is the flexural rigidity, P a constant and y the depth of a point on the elastic line below a fixed horizontal. Its full investigation obviously requires elliptic functions: see Rankine's p. 353 and compare his pp. 190—5 for the treatment by elliptic functions.

[469.] Rankine next passes to *Torsion* and *Combined Torsion and Bending* with little to be noted; then to *Crushing by Bending*. Here a formula of the type

$$P = \frac{T_0\omega}{1 + c\dfrac{l^2}{h^2}}$$

is given for the strength P of a pillar or column of length l and least diameter h, cross-section ω and tensile strength T_0; c being an empirical constant depending on the material. Rankine apparently gives T_0 absolute, proof, or working-stress values and considers that corresponding values will thus be obtained for P. He states that this formula was first proposed by Tredgold and afterwards revived by Gordon, who determined the values of c from Hodgkinson's experiments. For pillars with both ends *rounded* instead of built-in we must take $4c$ for c (pp. 361—3).

This part of Rankine's book concludes with a discussion of various kinds of girders and some miscellaneous remarks on strength and stiffness. A considerable number of useful practical tables of elasticity and of strength of various materials will be found in the pages of the work as well as in the *Appendix*[1].

[470.] The last portion of the *Applied Mechanics* which refers to our subject is the fourth chapter of Part V. entitled: *Motions of Pliable Bodies*, pp. 552—65. It treats briefly of bodies attached to light springs the inertia of which is neglected and to a few cases of elastic vibrations. There appears to be no novelty in it.

On the whole Rankine's *Applied Mechanics* may be taken as a book which was a very distinct advance on any work previously

[1] The latter contains also in the later editions a sufficient discussion of the analytical treatment of continuous beams and of Clapeyron's theorem.

published professing to deal with the problems of technical elasticity. Such works as these of Rankine and Weisbach separate very distinctly the first decade of our half-century from the previous thirty years. The step to them from books of the type of Tredgold's is very great and marks the beginning of the era of 'technical education.'

[471.] In Vol. I. of the *Abhandlungen der math.-phys. Classe der Königlich sächsischen Gesellschaft der Wissenschaften*, Leipzig, 1852, pp. 133—168, is a memoir by Seebeck entitled: *Ueber die Querschwingungen gespannter und nicht gespannter elastischer Stäbe.* The memoir itself is due to the year 1849, and belongs essentially to the theory of sound. Let m be the mass per unit length of the bar, $E\omega\kappa^2$ its rigidity and P the longitudinal stress, then the equation for the transverse displacement y at distance x from a terminal is:

$$m\frac{d^2y}{dt^2} = -E\omega\kappa^2\frac{dy^4}{dx^4} + P\frac{d^2y}{dx^2} \quad \text{..................(i).}$$

Seebeck thus omits the effect of the angular rotation of the sections of the rod. His equation may be compared with the fuller equation given by Donkin: *Acoustics*, p. 168. Seebeck first assumes $P = 0$, and finds in this case from the resulting equation the loops, the nodes etc., for the six possible variations among clamped, free and supported terminals. His numerical results are of very considerable value, and have been largely used by later writers on sound.

[472.] The second part of the memoir deals with the vibrations of stretched rods, and the particular point of interest is the modification in tone produced by the stiffness of musical strings. There are two cases which Seebeck deals with, and which have formed the subject of experiments: (i) both ends pivoted, (ii) both ends clamped. (The third case, one end pivoted and one clamped, can of course be deduced from the latter of these by doubling the length of the rod.) In the former case Seebeck shows that

$$n^2 = \frac{i^2\pi^2}{l^2}\left(\frac{P}{m} + \frac{E\omega\kappa^2}{m}\frac{i^2\pi^2}{l^2}\right) = n_1^2\left(1 + i^2\pi^2\frac{E\omega\kappa^2}{Pl^2}\right) \quad \text{........ (ii)}$$

$$= n_1^2 + n_0^2 \quad \text{..(iii),}$$

where $n/2\pi$ is the frequency of vibrations of the stretched rod, $n_0/2\pi$ the frequency without the stretch, $n_1/2\pi$ the frequency for the rod treated as a flexible string under tension P, l the length of the rod and i any integer. This result is the law stated by N. Savart and deduced theoretically by Duhamel, but which *only* holds for pivoted terminals. Savart's experiments were, however, made with rods with *clamped* terminals: see our Art. 1228*.

Obviously the stiffness of a doubly-fixed string destroys the harmonic character of its tones.

Passing to the case of a doubly-clamped rod Seebeck shows that (iii) does not hold and that the determination of the notes is much more complex. For the case, however, of the not too high sub-tone i of a stiff string or flexible rod, he finds

$$n^2 = n_1^2 \left\{1 + 4\left(\frac{E\omega\kappa^2}{Pl^2}\right)^{\frac{1}{2}} + (12 + i^2\pi^2)\frac{E\omega\kappa^2}{Pl^2}\right\} \quad \text{.........(iv)},$$

where the notation is the same as in the previous case (p. 162). Thus in this case we have two different effects, the purity of the harmonics is destroyed by the stiffness and all the notes are raised in pitch.

Both Donkin and Lord Rayleigh refer to Seebeck's memoir, but it is somewhat singular that Donkin misstates the result (iv), and Lord Rayleigh while questioning Donkin's conclusion does not note that Seebeck has really settled the point. Lord Rayleigh possibly had not been able to see Seebeck's memoir and perhaps Donkin, whom he follows, had read it somewhat carelessly. The following are the passages in question:

Donkin gives (iv) without the last term of the curled bracket and after comparing it thus mutilated with (ii) remarks:

We see that they differ essentially, especially in this respect, that, in the case (iv) of *fixed faces* the pitch of all the component tones is raised, by the rigidity, through the same interval, so that they do not cease to form a harmonic series; whereas in the other case (ii) each tone is raised through a greater interval than the next lower one, and the series is therefore no longer strictly harmonic (*Acoustics*, p. 182).

Lord Rayleigh on the other hand, after giving equation (iv) in Donkin's form, remarks:

According to this equation the component tones are all raised in pitch by the same small interval, and therefore the harmonic relation is not disturbed by the rigidity. It would probably be otherwise if terms involving κ^2/l^2 were retained; it does not therefore follow that the harmonic relation is better preserved in spite of rigidity when the ends are clamped than when they are free, but only that there is no additional disturbance in the former case though the absolute alteration of pitch is much greater (*Theory of Sound*, Vol. I. p. 245).

It is to be hoped that this oversight will not lead any one to repeat needlessly Seebeck's investigation.

[473.] Seebeck shows that the correction for stiffness is extremely small in most practical cases (p. 163). For example, on his own lecture room mono-chord, the 27th tone was the first that differed from harmonic purity by as much as a comma ($\frac{81}{80}$).

There is an appendix to the memoir giving an account of some experiments of Seebeck's on the tones of doubly-clamped stiff cords. He considers that experiment and theory (as represented by (iv)) are in close agreement (pp. 164—168).

[474.] A passage in Seebeck's memoir (p. 136 ftn.) refers to the changes in amplitude of vibration produced by causes which are neglected in Equation (i) of our Art. 471. He remarks that one of these causes is elastic after-strain, and refers for a further discussion on this point to the *Programm der technischen Bildungsanstalt zu Dresden*, 1846. This latter contains an excellent little paper by Seebeck on the various methods which have been used for determining the stretch-modulus and the character of the errors to which they are liable. It is entitled: *Ueber Schwingungen, mit besonderer Anwendung auf die Untersuchung der Elasticität fester Körper* (pp. 1—40). Therein will be found lists very complete at that date of the stretch-moduli of various materials obtained by both statical and vibrational methods, as well as a fairly comprehensive list of experimental investigations on this point. One or two statements deserve special notice see my foot-note Vol. I. p. 756.

(*a*) Seebeck discusses (pp. 9—13) the effect of a constant frictional force and of an air-resistance proportional to the velocity in reducing the amplitude of oscillation of an elastic body.

(*b*) He carefully distinguishes between the imperfect elasticity' which arises from set and that which arises from elastic after-strain. He points out that Wertheim's statement that all bodies, even under the feeblest stress, receive set (see our Arts. 1296* and 1301*, 7° and 8°) does not prove anything more than the fact that Wertheim's material had not been reduced to a state of ease (p. 29); and he remarks how absurdly confusing is the term 'perfectly elastic' as used in the text-book theory of the impact of spherical and other bodies (pp. 28 and 31).

(*c*) He attributes the reduction in amplitude of vibration, even in metals, in a great extent to elastic after-strain, at the same time expanding and developing Weber's arguments: see our Art. 712*.

(*d*) He considers that the effect of elastic after-strain must be to render the value of the stretch-modulus as determined by statical measurement smaller than the value obtained from vibrations:

Denn während der kurzen Dauer einer Schwingung kann nur der kleinste Theil der Nachwirkung in Thätigkeit treten, dagegen sie bei der längeren Dauer des statischen Versuchs die gemessene Dehnung merklich vergrössern und daher einen kleineren Modulus geben muss (p. 34).

(e) He holds that the effect of after-strain was mingled with the temperature effect in the experiments of Weber and Wertheim referred to in our Arts. 705* and 1297*. Hence those results must give too great a difference between the specific heats at constant pressure and constant volume. The objection applies perhaps more strongly to Wertheim's than to Weber's mode of experimenting. See the remarks of Clausius referred to in our Art. 1398*—1405*.

(f) Finally I may note a little scrap of historical information bearing on the problem of impact which Seebeck gives on p. 32. He points out that Daniel Bernoulli had attempted to calculate the loss of kinetic energy in the form of elastic vibrations which occurs when a body strikes centrally and transversely a free rod. Bernoulli came to the conclusion that $\frac{5}{9}$ of the total energy before impact would be taken up as elastic vibrations in the rod. His investigation is based upon the assumption that the rod will be bent into the form corresponding to its deepest tone. Bernoulli's memoir is published in the *Novi Commentarii Acad. Petropol.* Tom. xv. p. 361, 1770. It may be taken as the first attempt to treat impact elastically, and the primary step in investigations which have been so ably followed up by Poisson, Cauchy, Saint-Venant, F. Neumann, Boussinesq and Hertz: see our Arts. 203—20, 401—7, 410—14 and subsequent articles in this History.

[475.] Seebeck also contributed papers treating of the theory of the vibrations of elastic bodies to Dove's *Repertorium der Physik*, Bd. VI. pp. 3—100, Berlin, 1842, and Bd. VIII. pp. 1—108 (*Akustik*, separate pagination), Berlin, 1849 These papers deal principally with the theory of sound, and may even yet be read with interest. I would call attention especially to pp. 52—4 of the latter paper wherein Seebeck draws attention to Savart's *etwas künstlichen und nicht einwurffreien Vorstellung* of the mode in which combined longitudinal and transverse vibrations displace the sand on a vibrating rod: see our Art. 327*. These pages are entitled: *Ueber die Sandanhaufungen auf longitudinal-schwingenden Körpern.* Seebeck's theory causes the sand to accumulate at the nodes and not like Savart's at the loops. Although only descriptive, Seebeck's statements are much clearer than Savart's, and they have been reproduced with considerable experimental detail by Terquem: see Section II. of this Chapter.

Seebeck died in 1849.

[476.] Clausen: *Ueber die Form architektonischer Säulen; Bulletin physico-mathématique de l'Académie*, T. IX. pp. 368—79, St Petersburg 1851. Also *Mélanges Mathématiques et Astronomiques*,

Tome I. (1849—53) pp. 279—94, St Petersburg, 1853. Clausen seeks to find the form of a column which for a given buckling load shall have the least volume. This problem as we have remarked is of no very great practical importance, for in the comparatively short columns of architecture, the longitudinal stress produces set long before the buckling load is reached: see our Arts. 1258—9*. Lagrange as we have seen (Art. 113*) obtained the differential equation required for the solution of the problem and showed that the right circular cylinder is *one,* and under certain conditions, the only solution. Clausen has succeeded in solving the general differential equation, and comes to a different result. In the following lines he somewhat misstates Lagrange's conclusions as to the best form of column:

Als Eigenschaft der zweckmässigsten Form wurde angenommen, dass sie bei gleicher Höhe und Tragkraft das kleinste Volumen enthalte. Lagrange wandte zur Auflösung dieser viel schwierigern Aufgabe den von ihm erfundenen Variationscalcul an, und gelangte zuletzt zu dem sehr auffallenden Resultate, dass die Säule von gleicher Dicke die stärkste bei gleichem Volumen sei. Seit dieser Zeit ist diese Aufgabe meines Wissens nicht berührt worden.—Indem ich die Auflösung auf eine andere Art versuchte, gelang es wider Erwarten, die Differentialgleichung, deren allgemeine Integration Lagrange nicht versucht hatte, auf elliptische Transcendenten zu reduciren, wodurch es sich zeigt, dass die zweckmässigste Form vom Cylinder abweicht, und dass das Volumen dieses bei gleicher Höhe und Tragkraft sich zum Volumen jener Form verhält wie 1 : $\sqrt{3/4}$ (p. 368).

[477.] Let ω be the area of the cross-section. Then we have if ds be an element of the axis of the column, volume $= V = \int_0^l \omega ds$, and this is to be a minimum. Further if P be the load and y the deflection, we have upon the Eulerian theory

$$E\omega\kappa^2 \frac{d^2y}{ds^2} = -Py.$$

Now κ^2 varies as ω, if all the cross-sections are similar figures with their centroids in the axis, or $\kappa^2 = \beta\omega$, say. Hence

$$\frac{d^2y}{ds^2} = -\frac{P}{E\beta}\frac{y}{\omega^2}.$$

We have thus to make $\int_0^l \left(y \Big/ \frac{d^2y}{ds^2}\right)^{\frac{1}{2}} ds$ a minimum.

By direct application of the Calculus of Variations I deduce the equation:

$$\frac{d^2}{ds^2}\sqrt{y\Big/\left(\frac{d^2y}{ds^2}\right)^3} = \sqrt{1\Big/y\frac{d^2y}{ds^2}}.$$

This agrees with Clausen's result stated as Equation (3) p. 372 if we introduce a minus sign under both roots. He seems to me to employ an unnecessarily complex process to reach this simple conclusion. Let us write $z = \omega \cdot \sqrt{E\beta/P}$, then we have to solve

$$\frac{d^2y}{ds^2} = -\frac{y}{z^2} = -y^{\frac{1}{3}}u^{-\frac{2}{3}} \quad \text{.......................(1)},$$

$$\frac{d^2u}{ds^2} = -y^{-\frac{2}{3}}u^{\frac{1}{3}} \quad \text{..........................(2)},$$

where $$u = \sqrt{-y\Big/\left(\frac{d^2y}{ds^2}\right)^3}.$$

Multiply these equations by $\frac{du}{ds}$ and $\frac{dy}{ds}$ respectively, add and integrate, and we find:

$$\frac{dy}{ds}\frac{du}{ds} = C - 3(uy)^{\frac{1}{3}} = C - 3z$$

$$= 3(z_0 - z) \quad \text{.............................. (3)},$$

if $z = z_0$ for the point at which

$$\frac{dy}{ds} = 0.$$

Multiply (1) by u, (2) by y, and add their sum to the double of (3) then we have:

$$\frac{d^2(yu)}{ds^2} = \frac{d^2(z^3)}{ds^2} = 6z_0 - 8z,$$

or integrating $$\left(\frac{d(z^3)}{ds}\right)^2 = c^4 + 12z_0z^3 - 12z^4 \ldots \quad \text{................(4)},$$

c^4 being an arbitrary constant.

Whence we deduce

$$ds = \frac{3z^2dz}{\sqrt{c^4 + 12z_0z^3 - 12z^4}} \quad \text{....................(5)}.$$

Thus z, and so the section, is given in terms of the arc s of the axis by means of elliptic functions.

[478.] We have now to determine the value of the constant c^4. According to Lagrange (Art. 112*) we may measure the efficiency of a column of given height by the ratio P/V^2, and P varies as V^2/l^4. Hence if two columns carry the same load we must have $V^2/l^4 = V_0^2/l_0^4$, or the volume of a column which is of unit length and carries the same load is given by

$$V_0 = V/l^2.$$

Clausen takes as his condition for determining c^4 that V_0 must be a minimum. After some rather troublesome analysis he finds $c^4 = 0$. Equation (5) now becomes:

$$ds = \frac{\sqrt{\frac{3}{4}}\sqrt{z}dz}{\sqrt{z_0 - z}} \quad \text{.........................(6)}.$$

Further,

$$dV = \omega ds = \sqrt{\frac{P}{E\beta}}\, z ds = \sqrt{\frac{P}{E\beta}}\sqrt{\frac{3}{4}}\frac{z^{\frac{3}{2}}dz}{\sqrt{z_0 - z}} \quad \text{.........(7)}.$$

Let us put $z = z_0 \cos^2 \theta$,

$$s = z_0 \int \sqrt{3} \cos^2 \theta d\theta = \frac{\sqrt{3}}{4} z_0 [2\theta + \sin 2\theta],$$

$$V = z_0^2 \sqrt{\frac{P}{E\beta}} \int \sqrt{3} \cos^4 \theta d\theta = \frac{\sqrt{3}}{32} z_0^2 \sqrt{\frac{P}{E\beta}} [12\theta + 8 \sin 2\theta + \sin 4\theta].$$

To obtain the total length and volume we must take these expressions between the limits $\pm \frac{1}{2}\pi$ of θ supposing the strut doubly pivoted.

Hence $$l = \frac{\sqrt{3}\pi z_0}{2}, \qquad V = \frac{3\pi\sqrt{3}}{8} z_0^2 \sqrt{\frac{P}{E\beta}} \quad \text{.............(8)},$$

$$V_0 = V/l^2 = \frac{\sqrt{3}}{2\pi}\sqrt{\frac{P}{E\beta}} \quad \text{.......................(9)}.$$

[479.] Now suppose we take a column the cross-section of which is uniform $(\omega = \omega_0)$ but of the same shape as before, then we have

$$\frac{d^2y}{ds^2} = -\frac{P}{E\beta}\frac{y}{\omega_0^2},$$

and $$y = C_1 \sin\left(\sqrt{\frac{P}{E\beta}}\frac{s}{\omega_0} + C_2\right),$$

C_1 and C_2 being constants.

We readily find $\sqrt{\frac{P}{E\beta}}\frac{l}{\omega_0} = \pi$, or $l = \pi\omega_0\sqrt{\frac{E\beta}{P}}$.

Further since the columns are to be of the same height we must have this equal to the l of Equation (8), and it follows that,

$$\omega_0 = \frac{\sqrt{3}}{2}\sqrt{\frac{P}{E\beta}} z_0.$$

We deduce for V', the volume of this uniform column,

$$V' = l\omega_0 = \frac{3}{4}\pi\sqrt{\frac{P}{E\beta}} z_0^2,$$

whence from (8) $$V : V' :: \sqrt{3} : 2,$$

that is, for the same buckling-load and height the volume of the column of variable section is less than that of the column of uniform section in the ratio of $\sqrt{3} : 2$.

[480.] Clausen devotes pp. 770—80 to the consideration of the problem of the column built-in at one end and loaded at the other, instead of the doubly pivoted strut of the previous investigation. He arrives at the conclusion, which he might have foreseen, that the former agrees in shape with either half of the latter. He figures this column and remarks: 'dass die Form, wie mir scheint, eine dem Augen nicht ungefällige ist'—an opinion, I think, which will not be accepted by many.

In the present memoir the form of the cross-sections is left undetermined, they are merely assumed to be similar and similarly placed. Clausen remarks, however, that the circle is not the form which offers the greatest resistance to buckling; he gives no analysis of the point. Since, however, the load carried by the best column is always the same as that of a column of uniform section of $2/\sqrt{3}$ times its volume, we have only to compare the loads carried by the latter for various forms of cross-section to arrive at a variety of comparative results. These loads, if the length and the area of the cross-section of the column remain the same, vary as κ^2. Thus take a rectangular section $2a \times 2b$ and $b < a$ and compare it with a circular section of radius c, the relative efficiencies are as $c^2/4 : b^2/3$ where $\pi c^2 = 4ab$, or they are as $a/\pi : b/3$, therefore the rectangular section will be better than the circular if $b > 3a/\pi$ *i.e.* if the side b lies between $(3/\pi)\,a$ and a.

Thus certain rectangular sections, almost square, are better than circular sections in the matter of buckling. The practical value of the whole of this investigation must, however, be questioned: see our Arts. 146*, 911*, 958* and 1258*.

[481.] E. Segnitz: *Ueber Torsionswiderstand und Torsionsfestigkeit. Journal für die reine und angewandte Mathematik*, Bd. 43, 1852, pp. 340—364.

The author seems quite ignorant of the existence of the slide-modulus and of the shearing resistance of a material; he endeavours to explain torsion by the longitudinal extension of the rod or prism treated as a bundle of fibres. Young (*Natural Philosophy*, Vol. I. p. 139) had already pointed out the insufficiency of this hypothesis. It had also been considered as a corrective factor by Maxwell, Wertheim and Saint-Venant: see our Arts. 1549*, 51, and Wertheim's memoir on *Torsion* in Section II. of this Chapter.

The memoir ought scarcely to have been printed in Crelle's *Journal* in 1852.

[482.] E. Phillips: *Rapport sur un Mémoire de M. Phillips, concernant les ressorts en acier employés dans la construction des véhicules qui circulent sur les chemins de fer. Comptes rendus,* T. 34, 1852, pp. 226—35. This report by Poncelet, Seguier and Combes speaks very fully and favourably of Phillips' results. It will be found useful to those to whom the original memoir in the *Annales des Mines* is not accessible: see our Art. 483. The commissioners remark that:

Le travail de M. Phillips sera fort utile aux ingénieurs et aux constructeurs, qui y trouveront des règles rationnelles et d'une application facile, pour l'établissement des ressorts capables de satisfaire, avec la moindre dépense de matière, à des conditions données de flexibilité et de résistance (p. 235).

They recommend the publication of the memoir in the collection of the *Savants étrangers.*

A portion of the report is printed as a foot-note on the first page of the memoir in the *Annales,* where there is an additional remark by M. Combes that the formulae for springs given somewhat earlier by Blacher (see Section III. of this Chapter) were really due to Clapeyron.

[483.] E. Phillips: *Mémoire sur les ressorts en acier employés dans le matériel des chemins de fer. Annales des Mines,* Tome I. 1852, pp. 195—336. We have already referred to previous notes and memoirs by Phillips on this subject (see our Art. 1504 *), but this memoir is the principal one, indeed it is one of the most important that has ever been published on the theory of laminated springs. It consists of three chapters and a long *Note.* We shall consider these at some length.

[484.] The first chapter is entitled: *Théorie mathématique des ressorts.* It occupies pp. 195—227, and should be taken in conjunction with the *Note* entitled: *Démonstration des formules de la flèche et de la flexion d'un ressort quelconque sous charge,* which is appended to the memoir (pp. 319—36). The theory here developed is very complete and has been carefully verified experimentally by Phillips, the details of his experiments being given in other parts of the memoir.

Les résultats qu'elle donne ont été vérifiés dans les cas les plus divers, par des expériences directes, avec un degré de précision extrême, auquel j'étais loin de m'attendre moi-même, et qui paraît indiquer, de la part de l'acier, un état d'élasticité bien plus parfait que dans le fer ou dans la fonte (p. 196).

[485.] Let a spring be supposed built up of a number of separate laminae eL, e_1L_1, e_2L_2, etc. projecting one beyond the other as in Fig. (i), and let eL be the 'matrix-lamina.' Let the distances of the terminals of these laminae from the mid-plane VV of the spring,—where

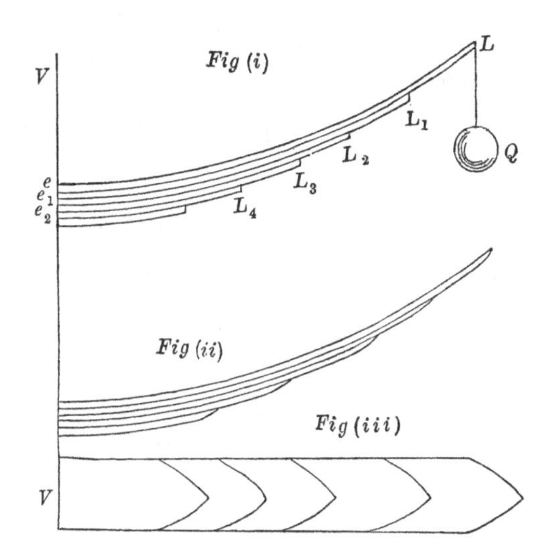

symmetry enables us to treat the spring as *built-in*,—be given respectively by L, L_1, L_2, etc.; let the curvatures of the different laminae at their respective central axes after manufacture be given at a section distant z from VV by $1/r$, $1/r_1$, $1/r_2$, etc.; let a load Q be applied to the terminal L, and $1/\rho$ then be the curvature at z of the matrix-lamina; let ϵ, ϵ_1, ϵ_2, etc. be the distances between the central axis of the matrix-lamina and the central axes of the laminae e_1L_1, e_2L_2, etc.; let M, M_1, M_2, etc. be the flexural rigidities of the successive laminae.

Phillips supposes that the laminae are throughout in contact with each other, and afterwards investigates the conditions for this. Let then p, p_1, p_2, etc. be the pressures per unit length between the first and second, the second and third, laminae etc. at the section distant l from VV. Now Phillips practically *assumes* that the distance between any cross-section and VV is the same whether measured perpendicular to VV or along the central axis of the lamina. This is probably almost true in practice, but such an equation as that at the middle of his p. 201 requires some comment of this kind: see however p. 282 of the memoir and our Art. 488.

Applying the Bernoulli-Eulerian theory of flexure and neglecting the weight of the spring, we easily deduce:

For LL_1, $M\left(\frac{1}{r}-\frac{1}{\rho}\right)=Q(L-z)$, or $1/\rho=\frac{B+Qz}{M}$,

where $B=\frac{M}{r}-QL.$

For L_1L_2 as part of the matrix-lamina,

$$M_1\left(\frac{1}{r}-\frac{1}{\rho}\right)=Q(L-z)-\int_z^{L_1}p(l-z)\,dl \quad \text{...........(i)},$$

and as part of the second lamina,

$$M\left(\frac{1}{r_1}-\frac{1}{\rho+\epsilon}\right)=\int_z^{L_1}p(l-z)\,dl,$$

whence:

$$M\left(\frac{1}{r}-\frac{1}{\rho}\right)+M_1\left(\frac{1}{r_1}-\frac{1}{\rho+\epsilon}\right)=Q(L-z).$$

When ϵ is so small compared with ρ that it may be neglected, as is usually the case, we have:

$$\frac{1}{\rho}=\frac{B_1+Qz}{M+M_1}, \quad \text{where } B_1=\frac{M}{r}+\frac{M_1}{r_1}-QL.$$

Continuing this process we easily find for the curvature at a point on the matrix-lamina lying between L_i and L_{i+1},

$$\left.\begin{aligned}\frac{1}{\rho}&=\frac{B_i+Qz}{M+M_1+M_2+\ldots+M_i}\\ \text{where}\quad B_i&=\frac{M}{r}+\frac{M_1}{r_1}+\frac{M_2}{r_2}+\ldots+\frac{M_i}{r_i}-QL\end{aligned}\right\}\text{..............(ii)}.$$

Calling this $1/\rho_i$, let us find the difference of $1/\rho_{i+1}$ and $1/\rho_i$ where $z=L_{i+1}$, we have

$$\frac{1}{\rho_{i+1}}-\frac{1}{\rho_i}=\frac{M_{i+1}}{\sum_0^i M_q\times\sum_0^{i+1}M_q}\left\{Q(L-L_{i+1})+\sum_0^i M_q\left(\frac{1}{r_{i+1}}-\frac{1}{r_q}\right)\right\}\text{.....(iii)},$$

where $\sum_0^i F(M_q, r_q)$ denotes, if F be any function of the M's and r's,

$$F(M,r)+F(M_1,r_1)+F(M_2,r_2)+\ldots+F(M_i,r_i).$$

Now L is $>L_{i+1}$, and as a rule $r_{i+1}, r_i, r_{i-1}\ldots r_1, r$ must either be equal or in ascending order of magnitude from r_{i+1} to r, if the laminae are to touch, hence $1/\rho_{i+1}-1/\rho_i$ is a finite quantity and there is an abrupt change of curvature at the point where the $(i+1)$th sheet laps the ith. This abrupt change could be got rid of by making $M_{i+1}=0$ at that point, or by trimming and pointing off the end of the lamina as suggested either in our Fig. (ii) or in our Fig. (iii).

Le même raisonnement se continue pour toute l'étendue de la maîtresse feuille, et on vérifie ainsi l'utilité de ce fait pratique que tous les bons ressorts ont les extrémités de leurs feuilles aiguisées et amincies (p. 204).

[486.] The stretch s_z at distance v from the central axis of the matrix-lamina, or at distances v_1, v_2, etc. from the central axes of the second, third laminae, etc. will be given by formulae of the type:

$$s_z = v\left(\frac{1}{r} - \frac{1}{\rho}\right),$$

$$s_z = v_1\left(\frac{1}{r_1} - \frac{1}{\rho + \epsilon}\right), \ldots\ldots s_z = v_i\left(\frac{1}{r_i} - \frac{1}{\rho + \epsilon_{i-1}}\right) \ldots\ldots\ldots (\text{iv}).$$

Whence if e, e_1, e_2, etc. be the successive thicknesses of the laminae we have formulae for the maximum stretches of which the type for the matrix-lamina between L_i and L_{i+1} is:

$$s_z = \frac{e}{2}\left[\frac{\sum_1^i M_q\left(\frac{1}{r} - \frac{1}{r_q}\right) + Q(L-z)}{\sum_0^i M_q}\right] \ldots\ldots\ldots\ldots\ldots (\text{v}).$$

Suppose all the laminae to be of the same curvature before being built-up into the spring, or the r's to be all equal at the same cross-section of the spring, then

$$s_z = \frac{e}{2}\frac{Q(L-z)}{\sum_0^i M_q} \ldots\ldots\ldots\ldots\ldots\ldots\ldots\ldots\ldots (\text{vi}),$$

or, when there is no original difference of curvature in the laminae, the nature of the curve in which the laminae are shaped and their initial curvature have no influence on the stretches in the spring or upon its resistance (pp. 207—8).

[487.] Phillips remarks that the formula (ii) of the previous article enables us to calculate out the value of ρ for a succession of positions on the matrix-lamina for any given spring, and thus to draw a curve of its form under a given load. He gives (p. 206) details of five experiments in which the deflections thus obtained were compared with their experimental values. There is an extremely close accordance between the experimental and theoretical results.

[488.] We next pass to the analytical determination of the deflection, the investigation of which occupies pp. 319 *et seq.* of the *Note*. We have generally from (ii)

$$\frac{1}{\rho} = a + bz \ldots\ldots\ldots\ldots\ldots\ldots\ldots\ldots\ldots (\text{vii}),$$

where
$$a = \frac{\sum\limits_0^i (M/r) - QL}{\sum\limits_0^i M}, \quad b = \frac{Q}{\sum\limits_0^i M}.$$

Phillips has λ where we have z, λ being the distance from VV of a given section measured along the central axis of the matrix-lamina, but as we have already pointed out equation (ii) is accurately true only when we use z and not λ. Phillips by neglecting the difference between $d\lambda/dz$ and unity writes (vii) in the form:

$$\frac{dp}{d\lambda}(1 + p^2)^{-\frac{3}{2}} = a + b\lambda \qquad \text{(viii)},$$

where $p = dy/dz$, or is the slope of the tangent at the central axis to the horizontal, i.e. to the direction perpendicular to that of the load Q. This equation (viii) he integrates on the assumption that a and b are constants along the central axis between the laps, and finds:

$$\frac{p}{\sqrt{1 + p^2}} = C + a\lambda + \frac{b}{2}\lambda^2,$$

but the left-hand side $= dy/d\lambda$, hence integrating again

$$y = C' + C\lambda + \tfrac{1}{2}a\lambda^2 + \frac{b}{6}\lambda^3 \qquad \text{(ix)}.$$

It seems to me that (ix) is only true so far as we may legitimately interchange λ and z. Phillips does not seem to have remarked that he has already supposed this interchange allowable when he puts λ instead of z on the right-hand side of (vii). Thus the true limitation to Phillips' investigations appears to be that any curvatures, however considerable, may be given to the laminae in manufacture, but that when the spring is made up and in the unloaded state it ought to be very approximately flat. This condition is probably satisfied in most springs in practical use.

In our investigations we shall replace Phillips' λ by z, since we use λ in a special technical sense in this work, but we shall suppose z measured indifferently either along the horizontal or along the central axis of the matrix-lamina.

[489.] C and C', a and b will have different values for each separate lap of the spring. Let the spring have $n+1$ laminae and let A_n, B_n be the values of a and b for the portion of the matrix-lamina which covers all the other n laminae, A_{n-1}, B_{n-1} the values of a and b for that portion which covers only $n-1$ laminae, and so on, and let C_n, C'_n, C_{n-1}, C'_{n-1} be the corresponding values of C and C'. As before L_n, L_{n-1}, etc., L will be the semi-lengths of the successive laminae from the lowest upwards. The conditions to be satisfied at the lap of two laminae are that the deflection and slope shall be continuous;

while if deflections be measured from the lowest point of the central axis of the matrix-lamina we must obviously have $C_n = C'_n = 0$.

An easy application of an ordinary method of solving finite difference equations leads to the results:

$$\left.\begin{aligned} C_{n-k} &= \sum_1^k (A_{n-k+1} - A_{n-k}) L_{n-k+1} \\ &\quad + \tfrac{1}{2}\sum_1^k (B_{n-k+1} - B_{n-k}) L^2_{n-k+1} \\ C'_{n-k} &= -\tfrac{1}{2}\sum_1^k (A_{n-k+1} - A_{n-k}) L^2_{n-k+1} \\ &\quad - \tfrac{1}{3}\sum_1^k (B_{n-k+1} - B_{n-k}) L^3_{n-k+1} \end{aligned}\right\} \quad \text{............(x)},$$

the summation being for k; while between L_{n-k+1} and L_{n-k}

$$y = C'_{n-k} + C_{n-k} z + A_{n-k}\frac{z^2}{2} + B_{n-k}\frac{z^3}{6} \quad \text{..............(xi)}.$$

We have thus the deflection at any point of the matrix-lamina. To find the deflection due to the load Q, we must find the value of y when $Q = 0$. Let y_0 be its value and let ${}_0A_{n-k+1}$, ${}_0B_{n-k+1}$, etc. be the corresponding values of A_{n-k+1} and B_{n-k+1}, then we easily see from formulae for a and b that ${}_0B_{n-k+1}$ etc. are all zero, and further that

$$A_{n-k+1} - {}_0A_{n-k+1} = -\frac{QL}{\sum_0^{n-k+1} M}.$$

Thus for $f_z = y_0 - y$ we have the expression:

$$\begin{aligned} f_z = {} & \frac{QL}{2}\sum_1^k \frac{M_{n-k+1} L^2_{n-k+1}}{\sum_0^{n-k+1} M \times \sum_0^{n-k} M} - \frac{Q}{3}\sum_1^k \frac{M_{n-k+1} L^3_{n-k+1}}{\sum_0^{n-k+1} M \times \sum_0^{n-k} M} \\ & - \left\{ QL\sum_1^k \frac{M_{n-k+1} L_{n-k+1}}{\sum_0^{n-k+1} M \times \sum_0^{n-k} M} - \frac{Q}{2}\sum_1^k \frac{M_{n-k+1} L^2_{n-k+1}}{\sum_0^{n-k+1} M \times \sum_0^{n-k} M} \right\} z \\ & + \frac{QL}{2}\frac{z^2}{\sum_0^{n-k} M} - \frac{Q}{6}\frac{z^3}{\sum_0^{n-k} M} \quad \text{......................(xii)}. \end{aligned}$$

[490.] Phillips considers various special cases of the formula (xii) of the preceding article. Thus the total deflection f of the spring due to the load Q will be obtained by putting $z = L$ and $k = n$; we then find

$$f = \frac{QL^3}{3M} + Q\sum_{k=1}^{k=n}\left\{ \frac{M_{n-k+1} L_{n-k+1}}{\sum_0^{n-k+1} M \times \sum_0^{n-k} M}\left(L L_{n-k+1} - \frac{L^2_{n-k+1}}{3} - L^2\right)\right\} \quad \text{...(xiii)}.$$

A special case of this is when all the flexural rigidities are equal; we then have:

$$f = \frac{QL^3}{3M} + \frac{QL_1}{1.2M}\left(LL_1 - \frac{L_1^2}{3} - L^2\right) + \frac{QL_2}{2.3.M}\left(LL_2 - \frac{L_2^2}{3} - L^2\right)$$
$$+ \ldots + \frac{QL_n}{n(n+1)M}\left(LL_n - \frac{L_n^2}{3} - L^2\right) \ldots\ldots\ldots \text{(xiv)}.$$

Phillips still further simplifies this by taking

$$L - L_1 = L_1 - L_2 = L_2 - L_3 = \ldots = l,$$

or supposing the laminae equally spaced out; he then proves that after certain reductions we have:

$$f = \frac{QL^3}{3(n+1)M} + \frac{Ql^3}{3M}\left(\frac{n(n-1)}{2} + \frac{1}{2} + \frac{1}{3} + \frac{1}{4} + \ldots + \frac{1}{n+1}\right) \ldots\ldots \text{(xiv)}'.$$

These results show us that when the flexural rigidity and curvature of each lamina are constant throughout its length and the rigidities the same for all laminae, then the deflection (i) is proportional to the charge, (ii) is independent of the primitive curvature and form of the laminae (pp. 319—329).

[491.] Phillips now proceeds to extend the results just stated by an ingenious process of general analysis to the case in which the primitive curvatures vary in any arbitrary manner. He shows that the deflection is still proportional to the charge and independent of the original form and curvatures of the laminae (pp. 329—33). This independence of the deflection on the primitive form of the laminae seems a result likely to be important in the practical construction of springs.

It is further shown that the change in the sine of the angle which the tangent at any point to the central axis of the matrix-lamina makes with the horizontal is also proportional to the load and independent of the primitive form and curvature of the laminae.

[492.] On pp. 215—19 Phillips calculates the pressures between the various laminae at any section given by z. Suppose the section taken between L_k and L_{k+1} (see fig. (i) in our Art. 485); let ϖ = the pressure per unit length between the matrix-lamina and the first sub-lamina, ϖ_1 between the first and second sub-laminae,... ϖ_{k-1} between the $\overline{k-1}$th and kth laminae, then Phillips easily deduces after the manner of our Art. 485, that:

$$\left.\begin{aligned}
\varpi &= \frac{d^2}{dz^2}\left\{\sum_1^k M_q\left(\frac{1}{r_q}-\frac{1}{\rho+\epsilon_{q-1}}\right)\right\}\\
\varpi_1 &= \frac{d^2}{dz^2}\left\{\sum_2^k M_q\left(\frac{1}{r_q}-\frac{1}{\rho+\epsilon_{q-1}}\right)\right\}\\
&\cdots\cdots\cdots\cdots\cdots\cdots\\
\varpi_{k-1} &= \frac{d^2}{dz^2}\left\{\quad M_k\left(\frac{1}{r_k}-\frac{1}{\rho+\epsilon_{k-1}}\right)\right\}
\end{aligned}\right\}\text{............ (xv).}$$

Thus since ρ is given by (ii) we can find these pressures; they must all be positive if the laminae are to have no tendency to separate.

[493.] The memoir then passes to the effect of vibrations on springs and to their resilience.

(*a*) The case of a weight Q placed upon the centre of a spring is very easily dealt with, if we assume with Phillips that the inertia of the spring may be neglected. The motion is then simple-harmonic and of period $2\pi\sqrt{f/g}$, where f is the statical deflection which Q would produce in the spring.

(*b*) If β be the ratio of load to deflection, so that $Q=\beta f$, the resilience is well known to be $\beta f^2/2$ or $Q^2/(2\beta)$. Now let w be the amount of work due to a blow which will just flatten the spring, and let the statical force required to flatten it be P, then we have

$$w = P^2/(2\beta),$$

or
$$P = \sqrt{2\beta w} = \sqrt{\frac{2w \times Q}{f}}.$$

Phillips gives the result in the form

$$P = \sqrt{\frac{2w}{f}} \times Q$$

on p. 223, which is obviously a misprint.

(*c*) The resilience may also be given another form suggestive of Young's theorem (see our Vol. I. p. 875).

The work required to bend an element dz of a lamina from curvature $1/\rho'$ to $1/\rho$, the sheet having an initial curvature $1/r$ is well known to be

$$\frac{1}{2}M\left\{\left(\frac{1}{r}-\frac{1}{\rho}\right)^2-\left(\frac{1}{r}-\frac{1}{\rho'}\right)^2\right\}dz.$$

Thus the work required to flatten the element from its curvature of manufacture or $1/r$

$$=\frac{1}{2}\frac{M}{r^2}dz.$$

Hence if the length of the lamina be l, and its cross section ω

be rectangular and of height e, the work required to flatten the whole lamina, supposing its stretch-modulus E,

$$= E \frac{\omega l}{24} \frac{e^2}{r^2}.$$

Now the stretch in the lamina has for maximum value $s = \frac{1}{2}e/r$, and if U be the volume of the lamina, the work done

$$= \frac{EUs^2}{6}.$$

Hence the work done in flattening the spring

$$= \Sigma \left(\frac{EUs^2}{6}\right).$$

Supposing all the laminae to have the same final stretch on flattening, then we have, if V be the total volume of the spring:

$$\text{Total resilience } \frac{EVs^2}{6} \quad \text{.....................(xvi).}$$

Cases may arise in which the blow begins to act upon the spring when it is already in a state of strain, i.e. its primitive condition is one of strain. In this case ρ_0, the initial radius of curvature, is not equal to r, but $\frac{1}{r} - \frac{1}{\rho_0} = \frac{2s_0}{e}$, where s_0 is the initial stretch. Hence the work required to flatten the element dz of a lamina is equal to

$$\frac{1}{2} M \left(\frac{1}{r^2} - \frac{4s_0^2}{e^2}\right) dz$$

$$= E\omega \left(\frac{e^2}{24r^2} - \frac{s_0^2}{6}\right) dz,$$

or, for the total work on a lamina we have the expression

$$\frac{E}{6}\left(Us^2 - \int_0^U s_0^2\, dU\right).$$

Hence the total resilience of the spring

$$= \frac{E}{6}\left(Vs^2 - \Sigma \int_0^U s_0^2\, dU\right) \quad \text{..................(xvii).}$$

Of this result Phillips writes:

Le travail se trouve donc diminué toutes les fois que le ressort ne part pas de sa position de fabrication. Or c'est ce qui arrive pour tous les ressorts de choc et de traction qui sont posés avec une certaine bande; mais on voit que la différence sera toujours assez faible quand s_0 ne sera pas très grand, parce que s_0 n'entre que par son quarré. Ainsi, dans les ressorts ordinaires, où s_0 est environ 1/3 de s, on perd environ 1/9 de la puissance du ressort pour résister au choc. On voit, en même temps, qu'il y a avantage à faire en sorte que la bande de pose du ressort qui répond à un effort d'environ 1000 kilogrammes, produise un allongement s_0 le plus faible possible ; par conséquent, sous ce point de vue, il y a avantage, toutes choses égales d'ailleurs, à employer des ressorts un peu roides plutôt que très-flexibles (p. 226).

[494.] *Chapitre Deuxième* entitled: *Des formes les plus convenables à donner aux ressorts et des règles pour les calculer*, occupies pp. 227—301 and contains many points of great interest.

Phillips first draws attention to the fact, referred to in our Art. 491, that the primitive form of the laminae is practically of little importance:

Il y a donc avantage, sous le rapport de la simplicité, à choisir des arcs de cercle, et c'est cette forme que je suppose adoptée (p. 227).

In the second place it is evident that the best sort of spring will be built-up in such a manner that all its parts are equally strained under any load or at least the maximum load (or maximum strain due to any oscillations) which it is designed to bear. As a rule this maximum strain will occur when the spring is completely flattened, and in such state the maximum stretches in all the laminae ought to be equal. The maximum stretch of the matrix-lamina on flattening $= e/(2r)$ and this will be the same for every section of it. If the laminae have initially the same curvature then they will have the same maximum stretch in every cross-section when flattened out. But supposing the laminae have before being formed into the spring initially different curvatures, we have then to ask how they can be spaced out so that the spring can be reduced to approximate flatness, and what conditions must be satisfied in order that the maximum stretches shall be the same for all the laminae.

Let $2P$ be the load which applied to the middle of the spring reduces it to approximate flatness. Then Phillips takes as his condition of flatness that the curvature of the matrix-lamina shall be zero at each lap of a sub-lamina. This gives us from equation (ii) of our Art. 485.

$$B_{i-1} + PL_i = 0, \quad B_i + PL_{i+1} = 0, \text{ etc.},$$

or generally,
$$P(L_i - L_{i+1}) = M_i/r_i,$$

which leads us to
$$L_i - L_{i+1} = M_i/(Pr_i) \quad\text{.................. (xviii)}.$$

(xviii) is the formula which determines the spacing of the laps. If the laminae are all of equal rigidity and initially of equal curvature we have

$$L - L_1 = L_1 - L_2 = \ldots = L_i - L_{i+1} = \ldots = \frac{M}{Pr} \quad\text{........ (xix)},$$

which determines the spacing for this special case.

If any lamina say the kth has a considerable initial strain, then we have $r_k < r$, and therefore if the stretch on flattening is to be the same for the kth lamina and for the matrix-lamina we must have $s_0 = \frac{e_k}{2r_k} = \frac{e}{2r}$, or we must have $e_k < e$. If $l_k = L_k - L_{k+1}$, we have $l_k = M_k/(Pr_k) = Ebe_k^3/(12Pr_k)$, similarly $l = Ebe^3/(12Pr)$, where b is the breadth of the laminae; hence it follows that $e_k/l_k > e/l$, and therefore $l_k < \frac{e_k}{e} l$ and $l_k e_k < \frac{e_k^2}{e^2} le$, but $e_k < e$ so that *à fortiori* we have

$$l_k < l \text{ and } l_k e_k < le.$$

If we flatten the spring out so as easily to calculate its volume, we see that if there is no initial strain and therefore all the depths of the laminae and the spacings equal, the volume, omitting that of the matrix-lamina, will be measured by an isosceles triangle of area L^2e/l less the sum of the little triangles of bases l and height e, or eL. Now if there be initial strain since $e_k/l_k > e/l$, we see that the perimeter of the figure formed by joining the corners of successive laminae falls outside the above isosceles triangle and has therefore a greater area, call it F; we have to subtract from this figure the sum of the little triangles of bases l_k and heights e_k, or the volume of the spring will be measured by $F - \Sigma l_k e_k$, but by what precedes $F > L^2e/l$ and $l_k e_k < le$. Hence the volume of the spring having a considerable initial strain and the same flexibility and absolute resistance which requires a given load to flatten it, is greater than that of a spring with equal heights and spacings for its laminae, and having the same matrix-lamina (pp. 231—3). On the other hand if the thicknesses of the laminae increase from the matrix downwards it may be shewn that the volume of the spring is less than in the case when all the thicknesses are equal (pp. 238—9).

Phillips then proceeds to shew that as a general rule the non-equality of the heights and curvatures of the sub-laminae with those of the matrix-lamina has very little influence upon the deflection of the matrix-lamina. For if $e_k/r_k = e/r$ and $e_k < e$, it follows that $e_k^3/r_k < e^3/r$ or $M_k/r_k < M/r$, or the resistance to initial strain is greater in the matrix-lamina than in any sub-lamina (pp. 233—4).

In the case of a spring with laminae equally curved initially it is easy to prove that the maximum stretches at all the cross-sections in all the laminae will be equal, even if the load be not the maximum or flattening load.

[495.] Hitherto Phillips has only made the curvature for the maximum load P zero at the laps. He now proposes to deduce the proper shaping off of the ends of the laminae in order that the curvature may be zero at all points.

For the matrix lamina itself from L to L_1 we must have

$$\frac{M}{r} - P(L - z) = 0,$$

or, if y be the variable thickness of the lap and $L-z=x$, since M varies as y^3 we have

$$y^3/x = \text{constant} = e^3/(L-L_1),$$

which determines the value of y for each x.

For the first sub-lamina we have:

$$\frac{M}{r}+\frac{M_1}{r_1}-P(L-z)=0,$$

or

$$\frac{M}{r}+\frac{M_1}{r_1}-P(L-L_1)-P(L_1-z)=0,$$

whence, since the matrix lamina has uniformly $M/r=P(L-L_1)$ after $z=L_1$,

$$\frac{M_1}{r_1}=P(L_1-z)\quad\text{..........................(xx)},$$

or, if $L_1-z=x_1$, $\quad y_1^3/x_1=\text{constant}=e_1^3/(L_1-L_2)$.

Thus the thickness at the ends of the first sub-lamina follows the same law as in the case of the matrix lamina, and the like may be shown of the other successive laminae. Instead of tapering off the thickness we might have reduced the breadth, or terminated our laminae in poignard or triangle form (see fig. (iii) of our Art. 485). Phillips states that this latter method is the more wasteful (pp. 237—8).

[496.] A formula is obtained by Phillips on pp. 332—6, which seems of considerable interest and practical value. He finds namely the deflection of a 'complete' or 'incomplete' spring when all the laminae are of the same section except at the laps, where account is taken of their proper shaping. He supposes also equal curvatures of manufacture.

Calling m the flexural rigidity of the kth lamina at the shaped lap, we have by equations of the type (xx),

$$\frac{m}{r}=P(L_{k-1}-z),$$

and by the law of spacings (xix), since the spaces are equal,

$$\frac{M}{r}=Pl.$$

Further $\quad (k-1)\,l+L_{k-1}=L.$

Whence since

$$\{(k-1)\,M+m\}\left(\frac{1}{r}-\frac{1}{\rho}\right)=Q\,(L-z),$$

we easily find:

$$\frac{1}{\rho}=\frac{P-Q}{P}\,\frac{1}{r}\quad\text{..........................(xxi)}.$$

Thus the curvature for the complete portion of the spring or the part which is staged is constant, and thus the matrix lamina takes the form of a circular arc whatever be the load.

Suppose the staging to cease with the nth lamina so that the length $2L_n$ of the spring is neither tapered nor covered by any sub-laminae, then we have

$$nM\left(\frac{1}{r}-\frac{1}{\rho}\right)=Q\,(L-z).$$

But $$\frac{1}{r}=Pl/M=P\,(L-L_n)/(nM).$$

Thus we have

$$\frac{1}{\rho}=a+bz\,\ldots\ldots\ldots\ldots\ldots\ldots\ldots\ldots\ldots\ldots(\text{xxii}),$$

where $$a=\{P\,(L-L_n)-QL\}/(nM),\quad b=Q/(nM),$$

$2L_n$ being the portion of the nth lamina not thinned down.

For the portion of the spring which is complete we have

$$\frac{1}{\rho}=a'\,\ldots\ldots\ldots\ldots\ldots\ldots\ldots\ldots\ldots\ldots(\text{xxiii}),$$

where $$a'=\frac{P-Q}{P}\frac{1}{r}\ \text{by (xxi)}.$$

If (xxii) and (xxiii) be twice integrated and the four constants of integration determined by the vanishing of the deflection and slope when $z=0$, and by the equality of the deflections and slopes when $z=L_n$ as obtained from the two expressions for the curvature of the complete and incomplete portions, then the following expression for f, the droop due to the load Q, is reached after some algebraical reductions:

$$f=\frac{Q}{6nM}\{2L^3+(nl)^3\}\ldots\ldots\ldots\ldots\ldots\ldots\ldots(\text{xxiv}).$$

If the spring is complete, $nl=L$ and

$$f=\frac{QL^3}{2nM}\ldots\ldots\ldots\ldots\ldots\ldots\ldots\ldots\ldots\ldots(\text{xxv}),$$

or 3/2 of the value of the droop of a spring of n equal flat laminae of the same rigidity M and of the same length $2L$.

Phillips gives details (on pp. 214—5 of the memoir) of experiments on the deflection of springs actually in use on various railway wagons and locomotives, and compares the experimental values with those calculated from the formula (xxiv). There is a very remarkable accordance between theory and experiment.

[497.] To calculate the depths and spacings of the laminae of the most general type of spring we must use the formulae:

$$\frac{e}{2r}=\frac{e_1}{2r_1}=\frac{e_2}{2r_2}=\ldots=s_0,$$

where s_0 is the maximum stretch in each lamina when the spring is flattened, and

$$L - L_1 = \frac{M}{Pr}, \quad L_1 - L_2 = \frac{M_1}{Pr_1}, \quad L_2 - L_3 = \frac{M_2}{2r_2}, \quad \text{etc.},$$

P being *half* the central load required to flatten the spring.

Of these results Phillips writes:

S'il arrive que les épaisseurs augmentent de quantités trop petites pour qu'on puisse donner à toutes les feuilles les épaisseurs calculées d'après leurs rayons, on donnera à plusieurs feuilles, en partant du haut, une épaisseur commune égale à la moyenne entre leurs épaisseurs, et un étagement commun égal à la moyenne de leurs étagements; on fera de même pour plusieurs des feuilles suivantes, et ainsi de suite jusqu'à ce que le ressort soit terminé. Quant aux amincissements, ils se calculeront par la règle générale (pp. 240—1).

[498.] There are two special methods of easily designing a laminated spring to which Phillips refers on pp. 238—9:

(*a*) We may suppose all the laminae cut as it were from one and the same hoop of metal, so that all have the same primitive curvature and thickness. When the spring is manufactured there will then be a very slight initial strain in the laminae before the spring is loaded. Such a spring possesses the advantages referred to in our Art. 494.

(*b*) We may suppose the laminae to have no initial strain by describing the laminae from the same centre and with bounding radii increasing by the mean of the thicknesses of adjacent laminae, while the thicknesses themselves increase proportionately to the radii of the central axes, or obey the relation:

$$\frac{e}{2r} = \frac{e_1}{2r_1} = \frac{e_2}{2r_2} = \dots = \frac{e_n}{2r_n} = s_0.$$

This sort of spring besides having no initial strain has also the advantage of a slightly but sensibly less volume than that described in (*a*). This is really the converse of the proposition in our Art. 494, p. 340, because by Art. 497 the ratios of the successive thicknesses to the corresponding spacings vary inversely as the thicknesses and so now decrease.

[499.] On pp. 240—2 of the memoir are given a number of interesting properties of springs, the laminae of which have the same or sensibly the same thickness (Case (*a*) of the previous Article).

If H be the total thickness at the mid-section of such a spring supposed complete, l the equal spacing of the laps, L the half-length and b the breadth of the matrix-lamina, $2P$ the flattening load and V the volume, then we have in the notation of the previous articles:

$$H = \frac{e}{l} L, \text{ nearly}; \quad l = \frac{M}{Pr} = \frac{Ebe^3}{12Pr}; \quad V \doteq HLb, \text{ nearly}.$$

Further, if f be the droop of the spring when unloaded, then since there is little or no initial strain: $r = L^2/2f = \dfrac{e}{2s_0}$ by the first equation of our Art. 498, (*b*). Whence we deduce

$$H = \frac{6Pf}{Es_0^2 bL}.$$

Now let ν equal the *flexibility* of the spring, or, the droop produced when a unit load is put at each extremity, then we must have, supposing stress and strain proportional:

$$f = \nu P,$$

and hence: $$H = \frac{6P^2\nu}{Es_0^2 bL}, \text{ and } V = \frac{6P^2\nu}{Es_0^2}.$$

Thus we find for a given material that:

(*a*) The total thickness of a spring is proportional directly to:

(i) the square of the flattening load,

(ii) the flexibility,

inversely to:

(i) the breadth of the spring,

(ii) its length.

(*b*) The volume of a spring is proportional to:

(i) the square of the flattening load,

(ii) its flexibility,

and further:

(*c*) Springs having the same flexibility and ultimate resistance, $2P$, have also sensibly the same volume.

Since $l/L = e/H = \dfrac{Es_0^3 L^3 b}{6Pf^2}$, and we must have $l < L$, it follows that the length L of the spring ought to be such that:

$$L < \left\{\frac{6Pf^2}{Es_0^3 b}\right\}^{\frac{1}{3}},$$

a condition generally satisfied in practice.

[500.] Phillips next proceeds to apply his theoretical results to the practical calculation of the dimensions of springs, chiefly those of railway wagons. He determines numerically the lengths of the various laminae suitable for springs of various classes. The springs thus calculated were constructed and the experimental deflections agreed very closely with those obtained by theory (pp. 242—52). The data assumed are (i) the flexibility of the spring (ν); (ii) its absolute resistance ($2P$); (iii) the chord of manufacture ($2c$) of the spring; (iv) the normal load ($2Q$); (v) the breadth of the laminae (b). Phillips supposes in addition that the limit of resistance of the spring is reached under the load $2P$ corresponding to flattening. He shows, however, how the details may be calculated when the flattening load corresponds to neither $2P$ nor to $2Q$, and also when the data are otherwise varied. Since the flexibility is known, the droop produced by $2P$ the flattening load, that is the subtense f of manufacture, is known. Phillips then puts without further comment $L = \sqrt{c^2 + f^2}$, or he equates the length of the spring to the chord of half its arc thus tacitly neglecting quantities of the order $\{(L-c)/L\}^2$. This is, however, in accordance with his previous approximations: see our Arts. 484 and 488. He further supposes the laminae to be of equal initial curvature and thickness and neglects any initial strain. Thus he easily deduces that the values e, l of the thickness and the spacing are given respectively by

$$e = \frac{c^2 + f^2}{f} s_0, \quad l = \frac{Ebe^3 f}{6P(c^2 + f^2)},$$

while the number of laminae will be the whole number in the quotient L/l.

For steel Phillips takes $E = 20,000$ kilogrammes per sq. mm. and $s_0 = \cdot 0025$, as a thoroughly safe stretch below the fail-limit for good steel.

On pp. 247—8 he shows that, when the laminae are described about the same centre, the thickness of the kth lamina, its radius of curvature and the corresponding spacing will be found from those of the $(k-1)$th lamina by the formulae

$$e_k = \frac{e_{k-1}(2r_{k-1} + e_{k-1})}{2r_{k-1} - e_{k-1}}, \quad r_k = \frac{e_k}{2s_0}, \quad l_k = \frac{M_k}{Pr_k}.$$

The first formula might for practical purposes be replaced by

$$e_k = e_{k-1}\left(1 + \frac{e_{k-1}}{r_{k-1}}\right).$$

[501.] On pp. 252—5 after discussing the effect of bolting the matrix lamina and under certain conditions several of the sub-laminae of the spring to a rigid frame on which the load is placed, Phillips next turns to the very important practical point of whether adjacent laminae do or do not tend to gape. His consideration of this matter occupies pp. 255—68, and is of great interest. There are three fundamental types of laminated springs to be considered: (*a*) the first type when the curvature of manufacture and the thickness of the laminae are equal for all, (*b*) the second type when the thicknesses decrease from the matrix to the sub-laminae, and (*c*) the third type when they increase. In both (*b*) and (*c*) it is supposed that the thicknesses, radii of curvature and the spacings are calculated by the formulae of our Art. 497, i.e. that they are determined so that the stretch on flattening is the same for all the laminae.

Phillips shows that for the first type of spring each lamina experiences only pressure at its terminals and that each such pressure is half the load, the laminae remain exactly fitted to one another without sensible pressure, but without gaping—'ce qui est conforme à l'expérience' (p. 256). When a spring is of the second type the laminae tend before, but not after flattening to separate. Finally if a spring is of the third type its laminae tend to separate after but not before flattening. In both cases (*b*) and (*c*) there is complete contact right along all the laminae for the load corresponding to flattening. These effects may be somewhat, but only slightly modified at the sections of the spring corresponding to the ends of the laminae. This modification will be very small if the spring under its normal load does its work in a flat condition.

On voit ainsi, en outre, qu'il convient de faire en sorte qu'un ressort travaille habituellement aplati sous la charge qu'il supporte; indépendamment de ce qu'alors les glissements des feuilles, et par suite le travail dû au frottement sont moindres (p. 268).

Phillips deduces the important conclusions we have referred to above from the expressions for the pressure between successive laminae which we have reproduced in our Art. 492.

[502.] Pages 268—93 of the memoir are devoted to what the author terms a *ressort à auxiliaire* or a *reserve spring*. He describes it in the following words:

En principe, on a fait remplir par des appareils différents deux conditions essentiellement distinctes: la flexibilité et la résistance qui n'ont nullement besoin d'être remplies par le même instrument. Le ressort se compose alors des deux parties: l'une, formée de feuilles toutes de même épaisseur, constitue le ressort proprement dit: elle travaille seule ordinairement sous la charge normale; l'autre, placée au-dessous, est plus épaisse et divergente, et ne vient en contact avec elle que sous un excès de charge et successivement. Cette dernière partie qui sert d'auxiliaire est calculée d'après l'excès de résistance propre qu'on désire attribuer au ressort, quelle que soit d'ailleurs cette résistance (p. 269).

The part of a reserve spring which is called into play by the normal load may be termed the *main spring*, the part which is only called into play when the normal load is surpassed the *secondary spring*. In order that a reserve spring may offer a progressive resistance to oscillations beyond the normal load, the secondary spring must be constructed in such a manner as to establish only a gradual contact with the main spring.

If the extreme resistance $2P$ of the spring be reached when both its parts are flattened and $2Q$ be the normal load, then the contact of main spring and secondary spring ought to begin when the load is $2Q$ and go on up to complete coincidence under $2P$. The main spring will generally be formed of a number of laminae of equal thickness spaced in the usual manner and calculated so as to have a given droop i under the normal load $2Q$. The calculation of the main spring under these conditions, especially when the form sought is to involve the least expenditure of material, is a matter of rather troublesome approximation but is discussed very fully by Phillips (pp. 270—5 etc.).

The secondary spring may be made in one of several forms; for example it may be (i) circular,—in this case its radius of manufacture r' ought to be equal to the radius at the centre of the last lamina of the main spring when under load $2Q$, i.e. if there be n laminae of equal rigidity and curvature of manufacture in that part of the spring we must have:

$$1/r' = \frac{\frac{nM}{r} - QL}{nM};$$

or (ii), the shape of the secondary spring may be the elastic line of the last lamina when under the normal load, or better a form a little more curved than this so that the oscillations of the main spring may be carried gradually and not abruptly to the secondary. This case is discussed by Phillips on pp. 286—92.

[503.] He remarks that in most cases it is sufficient to make the secondary spring consist of a single lamina. Its semi-length L' will be that of the last lamina of the main spring diminished by the spacing $M/(Pr)$, and we should then have in case (i) to determine its rigidity M' from the equation

$$M' = Pr'L'.$$

A more complex condition comes in, however, if we take a single lamina for the secondary spring in case (ii), for in this case its rigidity m' (and so the thickness of the lamina) must vary throughout and the maximum stretch must not exceed s_0 when the lamina is flattened. We have then in the notation of our previous articles:

$$m'/r' = P(L' - z),$$

while $m' = \frac{1}{12}Ebe'^3$, and if there be n laminae in the main spring we have

$$1/r' = \frac{\frac{nM}{r} - Q(L-z)}{nM}.$$

Further, $\qquad s_0$ must be $> e'/(2r')$.

Whence we easily deduce

$$(L'-z)\left\{\frac{nM}{r} - Q(L-z)\right\}^2 < \frac{2s_0{}^3Ebn^2M^2}{3P}$$

The left-hand side will be found to be a true maximum for

$$z = \tfrac{1}{3}\left(L + 2L' - \frac{nM}{rQ}\right),$$

and since $\qquad L - L' = nM/(Pr),$

the inequality may be easily reduced to:

$$\frac{2}{9}\frac{Q^2}{r^3}nM\left(\frac{1}{Q} - \frac{1}{P}\right)^3 < \frac{s_0{}^3Eb}{P} < \frac{3}{2}\frac{M}{Pr^3},$$

if we remember the value of M and that $e = 2rs_0$ for the main spring. Hence finally we must have:

$$n < \frac{27}{4} \frac{QP^2}{(P-Q)^3}.$$

For example if $P = 2Q$ we must have $n < 27$. This sets a limit to the number of laminae in the main spring when the secondary spring consists of a single lamina shaped like the form of the main spring under the load $2Q$ (pp. 289—90).

[504.] Phillips on p. 282 draws attention for the first time to the source of error—which may rise to importance, especially in the case of reserve springs,—from the chord and arc of the matrix lamina having been treated as interchangeable in the equations: see our Arts. 485, 488, and 500. He measures the amount of error thus introduced and shows how it may be allowed for. He remarks that the flexure due to a given load is obtained as the difference of two formulae, one of which gives the subtense without load and the other with load. The latter formula he holds to be sufficiently exact in practice when the chord and arc are interchanged, since the normal load approximately flattens the spring; the former must be modified if the difference between the arc and chord gives a sensible difference in the value of the subtense when the two are interchanged. If L and S be semi-chord and semi-arc the quantities $L^2/(2r)$ and $S^2/(2r)$ must be practically equal (pp. 282—4).

[505.] A remark of Phillips on p. 295 is worth citing. It refers to the springs we have classed in our Art. 501 as of the *third* type:

Je ferai remarquer, en passant, que le type déjà décrit des ressorts à feuilles d'épaisseurs croissantes, qui travaillent aplatis sous la charge normale, et dont toutes les feuilles éprouvent dans l'aplatissement un même allongement, rentre réellement dans la classe des ressorts à auxiliaire, car les rayons étant croissants, les feuilles ne viennent en contact que successivement. Ce fait est d'autant plus saillant, que souvent ces ressorts se terminent par une ou deux grosses feuilles. Seulement le propre de ces ressorts est que toutes les feuilles sont jointives sous la charge normale, et qu'alors toutes éprouvent les mêmes allongements.

[506.] The few remaining points in the second chapter of the memoir may be very briefly indicated

On pp. 295—8 Phillips deals more particularly with the calculation of springs intended to resist impact, and gives details of various springs actually constructed for the *Chemin de Fer de l'Ouest.* Phillips describes a novel kind suitable for resisting both impact and steady pressure and offering special advantages for passenger coaches on railways. These springs have secondary springs attached to them consisting of one or more large laminae so arranged that the flexibility is much less after the load has passed a certain limit (e.g. 3000 kilogs.), and thus specially heavy loads or impacts do not tend to vary very greatly the relative heights of the buffers of the carriages.

On pp. 299—301 we have a short résumé of the results of the chapter and an indication of how the theory therein developed may be used for the investigation of new forms of springs. It is followed by a table of numerical details of all the springs which had been constructed according to Phillips' theory before 1851.

[507.] *Chapitre troisième* is entitled: *Expériences sur l'élasticité de l'acier* and occupies pp. 302—18. A description of the apparatus employed is given and long details of experiments on various kinds of steel, tempered, annealed, hammered etc. Phillips concludes that for practical purposes we may take the stretch-modulus at 20,000 kilogs. per sq. mm. and the fail-limit, or that limit which it is not advisable to exceed even for an occasional and exceptional load, as a stretch of from ·004 to ·005 according to the quality of the steel, while for the normal load the stretch should not exceed ·002 to ·003.

Dans les meilleurs ressorts faits jusqu'à présent, l'acier travaille habituellement à environ ·0022 sous la charge normale (p. 317).

In the course of his investigations Phillips notes that to stretch steel for once up to ·005 or ·006 saves it from any sensible set when again subjected to the same strain (p. 316), and further he briefly refers (p. 318) to a result associated with the 'paradox in the theory of beams' as a subject for future study. Thus he states that a stretch of ·0095 (instead of ·005) corresponding to a load of 190 kilogs. per sq. mm. can be reached in *flexure* experiments without danger.

The appended *Note* then follows, the details of which have been given in the course of our analysis of the memoir.

[508.] The memoir just considered is a striking example of how a very simple elastic theory—sufficiently accurate for the range of facts to which it is applied—can be made to yield most valuable results. Phillips' theory of springs such as are employed in the ordinary rolling stock of railways is one of those excellent bits of work which can only be produced by the practical man with a strong theoretical grasp. I have devoted considerable space to its discussion as the Journal in which it appears is not among the most accessible, and so far as I know the only text-book in which extracts have yet found a place is M. Flamant's *Stabilité des constructions, Résistance des matériaux*, Paris, 1886 pp. 574—88.

[509.] Giuseppe Fagnoli: *Riflessioni intorno la teorica delle pressioni che un corpo o sistema di forma invariabile esercita contro appoggi rigidi ed irremovibili dai quali è sostenuto in equilibrio. Mem. dell' Accad. delle Scienze di Bologna*, T. VI., 1852, pp. 109—38.

This memoir, as long-winded as its title, was probably the last attempt to solve without the aid of the theory of elasticity the problem of the reactions upon a body of more than three points of support. It seems to me utterly obscure and involves the strange metaphysical conception of internal reactions in 'perfectly rigid bodies.'

[510.] A. Popoff: *Sur l'intégration des équations relatives aux petites vibrations d'un milieu élastique. Bulletin de la société impériale des naturalistes de Moscou.* T. XXVI., *Première Partie,* pp. 342—56, Moscow, 1853.

This paper deduces by a slightly different method the solutions of the uniconstant elastic equations for small vibrations first obtained by Ostrogradsky and Poisson: see our Arts. 739*—41* and 564*.

There does not seem any particular advantage in the method of Popoff and he draws no new conclusions from his solutions.

[511.] A. Popoff: *Intégration des équations qui se rapportent à l'équilibre des corps élastiques et au mouvement des liquides: Bulletin physico-mathématique de l'Académie...de St Pétersbourg,* T. XIII., 1855, pp. 145—9. This is reprinted (with the title only in Russian) in the *Mélanges mathématiques et astronomiques,* T. II., pp. 284—9.

The paper was received in October 1852.

Adopting the notation of our footnote p. 79, and supposing the elastic body to be under the influence of no body-forces and in equilibrium then we can easily show that the equations of elasticity in cylindrical coordinates are:

$$\left.\begin{aligned} &\nabla^2\theta = 0,\\ &\nabla^2 u - \frac{u}{r^2} - \frac{2}{r^2}\frac{dv}{d\phi} + \frac{\lambda+\mu}{\mu}\frac{d\theta}{dr} = 0,\\ &\nabla^2 v - \frac{v}{r^2} + \frac{2}{r^2}\frac{du}{d\phi} + \frac{\lambda+\mu}{\mu}\frac{d\theta}{rd\phi} = 0,\\ &\nabla^2 w + \frac{\lambda+\mu}{\mu}\frac{d\theta}{dz} = 0, \end{aligned}\right\} \quad \ldots\ldots\ldots\ldots\ldots\text{(i)},$$

where
$$\nabla^2 = \frac{d^2}{dr^2} + \frac{1}{r}\frac{d}{dr} + \frac{1}{r^2}\frac{d^2}{d\phi^2} + \frac{d^2}{dz^2},$$

or is the Laplacian in cylindrical coordinates.

Further, $$\theta = \frac{du}{dr} + \frac{u}{r} + \frac{1}{r}\frac{dv}{d\phi} + \frac{dw}{dz} \quad \text{.....................(ii)}.$$

We can obtain by our Art. 884* the thermo-elastic equations of equilibrium, if we write for θ in (i) θ' and instead of (ii) write

$$\theta' = \frac{du}{dr} + \frac{u}{r} + \frac{1}{r}\frac{dv}{d\phi} + \frac{dw}{dz} - \frac{\beta}{\lambda+\mu} q \quad \text{..............(iii)},$$

where for thermal equilibrium, $\nabla^2 q = 0$ ....(iv), q being the temperature at r, ϕ, z.

It is these thermo-elastic body-shift-equations, which Popoff has solved. He has not considered the surface conditions nor the stresses, and he limits his investigation to cases in which q, θ', u, v and w do not become infinite for $r = 0$[1].

[512.] The solution is really in terms of Bessel's functions, although he expresses them by integrals of the form given in equation (4), Art. 371, of Todhunter's *Functions of Laplace, Lamé and Bessel.* The solution is fairly straight-forward although only the outline of the integrations is given. The results are somewhat too lengthy to be reproduced here, but should be consulted by any one endeavouring to solve the general problem of the strain in a right-circular elastic cylinder subjected to any system of surface-stress. To show the type of solution I cite the value of w:

$$w = \Sigma \left(\left[(Ae^{az} - A'e^{-az}) \cos n\phi + (Be^{az} - B'e^{-az}) \sin n\phi \right] \times (ar)^n \left\{ \epsilon + \frac{\lambda+\mu}{2\mu} (2n\epsilon - ar\zeta) \right\} \right),$$

where $$\epsilon = \int_0^\pi \cos(ar\cos\chi) \sin^{2n}\chi \, d\chi,$$

$$\zeta = \int_0^\pi \sin(ar\cos\chi) \sin^{2n}\chi \cos\chi \, d\chi,$$

and n is an integer to be given all values from 0 to ∞. A, A', B, B', a are constants to be determined by the surface conditions.

The constant a is in practice the most difficult to determine, it appears in each Bessel's function and in each exponential, and even for the simple cases of axial symmetry, we obtain an appalling equation to ascertain its relation to n. The analogy of struts leads us to see that there are many cases in which it must be imaginary.

The values of u and v are still more complex, and it seems to me that really practical progress will hardly be made by attempting to carry this solution in Bessel's functions further. Possibly more might be achieved by solving Laplace's equation in cylindrical coordinates by a definite integral and then attempting to deduce definite integral solutions for the shifts.

[1] In his notation $\theta' = \omega$, $(\lambda+\mu)/\mu = k$, $\beta q/\mu = \Delta\theta$, $u = \delta v$, $v = r\delta\psi$, $w = \delta z$.

[513.] J. B. Phear: *Note on the Internal Pressure at any point within a body at rest. Cambridge and Dublin Mathematical Journal,* Vol. IX., 1854, pp. 1—6. A proof of the existence of Lamé's stress-ellipsoid of no peculiar interest: see our Art. 1059* The author remarks that this representation of stress is "so elegant that it seems to deserve a place in our University mathematics."

[514.] M. Bresse: *Recherches analytiques sur la flexion et la résistance des pièces courbes,* Paris, 1854, 269 pp. and three plates. This treatise consists of five chapters and treats analytically on the Bernoulli-Eulerian hypothesis the flexure of curved ribs, in particular, circular arches. It contains a very complete discussion of the problem, and Bresse's tables are of considerable value in testing any proposed circular arch. At the same time the graphical methods of Eddy are of more general application and would probably be now-a-days adopted, at least as a method of verification and comparison. I proceed to give some account of the contents of this treatise.

[515.] Chapter I., is entitled: *Étude hypothétique de la répartition d'une force sur la section droite d'un prisme.* Pp. 1—43 are occupied with a very full, clear and interesting discussion of the properties of the neutral axis and the load-point (stress-centre) and of their relations to the ellipse of inertia, and applications to the core, the centre of percussion and centre of pressure of a given area or cross-section. After comparing this chapter with the *Cours lithographié* referred to in our Art. 813*, I have no doubt that the *Cours* was due to Bresse, or *that we owe to him the important conception of the core and all that flows from it.* I regret that I was not able to associate his name with this conception in Vol. I.

It is to be noticed that Bresse proves these properties on the assumption that the stretch-modulus varies over the cross-section. He treats it as if it were a variable distribution of surface density over that section.

[516.] Pp. 44—56 of this chapter are entitled: *Répartition d'une charge totale sur la base d'un prisme n'ayant pas d'adhérence avec son appui.*

Suppose a loaded prism to rest on a horizontal base. This base can give pressure but not tension. Suppose farther the resultant vertical load P on the prism to meet this base in the point H. If H lies within the core, the base will be required to give pressure only and the distribution of that pressure will follow the law laid down in our Art. 815*. On the other hand, if H falls outside the core, we cannot make use of the formula in our Art. 815* as it gives in part tensions. The problem considered by Bresse is then: *How must the pressures be distributed over the portion of the prism's base remaining in contact with the plane in order that the resultant of these pressures may be equal and opposite to P?*

Obviously the boundary between the parts of the section remaining and not remaining in contact must be the neutral axis for the part remaining in contact. Otherwise a portion of the section on both sides would give pressure or be in contact. The problem then reduces to the following: To cut a portion off a given area by a straight line, such that the load-point or stress-centre of the area cut off when it has the straight-line as neutral axis may be a given point.

For the general case Bresse only suggests a method of tentative solution. Namely to take: (i) a series of parallel neutral axes and find the load-points of the portions they cut off; the series of points so obtained gives a curve, which we may term the 'load-point curve'; and, (ii) to draw such load-point curves for a variety of directions of the series of parallel neutral axes. Obviously the load-point curve which goes through the given load-point H solves the problem.

On pp. 46—48 Bresse proves an interesting property of the load-point curve, namely that the tangent to this curve at any load-point passes through the centroid of the area cut off by the corresponding neutral axis.

In the particular case when the given load-point lies upon an axis of symmetry of the section of the prism, we have only to draw neutral axes perpendicular to this symmetrical axis, and the required one can often be fairly easily found. Bresse works out the required dividing line in the case of the rectangle, circle, ellipse, etc., in which cases the analysis is not difficult. In particular in the case of a rectangle $2a \times 2b$, when the load-point is at a distance na from the centre $(n > \frac{1}{3})$ on the axis of symmetry parallel to the sides $2a$, the neutral axis lies on the opposite side to the load-point at a distance from the centre equal to $a(2-3n)$, and the maximum stress is in the side of the rectangle parallel to the neutral axis and

$$= \frac{P}{4ab} \frac{4}{3(1-n)}.$$

It is shown on p. 52 that the maximum stress in the case of a circular cross-section increases much more rapidly as the load-point is removed further from the centre than in the case of a rectangular one the side of which is equal to the diameter of the circle.

[517.] Bresse's second chapter is entitled: *Généralités sur la flexion et la résistance des pièces courbes* (pp. 60—67). This chapter gives a very clear account of what the author understands by an arched rib (*pièce courbe*) and the limits he has set to his discussion of the general problem. Thus he neglects slide, he supposes torsion to produce no effect so great that the rib-axis cannot still be dealt with as a plane curve, and he calculates the stress across any section on the assumption that the section is in the unstrained position; he allows, however, for a gradual change of cross-section and for a variation of the stretch-modulus in the cross-section. The 'mean-fibre' of the rib is defined as the locus of the centroids of the cross-sections, when those cross-sections are supposed to have a superficial density at each point equal to the stretch-modulus. He sums up the problems he proposes to deal with as follows:

(i) To find the stress over each cross-section of the rib supposing the loads and reactions given.

(ii) To find the effects of a change of temperature in producing stress and shift.

(iii) To calculate the reactions when the unstrained form and the load are given.

[518.] Chapter III. is entitled: *Flexion et résistance des pièces courbes, lorsque la pièce, dans l'état primitif et dans l'état de flexion, se trouve dans un plan contenant aussi les forces extérieures* (pp. 68—156).

The first section (pp. 68—76) of this chapter deals with problem (i) of the previous article. It shows how to find the stress-centre (load-point) of each cross-section when the reactions and the external forces on the rib are known. Suppose the rib divided up into elements and the corresponding distributed or concentrated loads represented by a single resultant for each element. Now form a vector-polygon of these elementary loads and the two terminal reactions. Choose the meet of the two reactions as ray-pole of this vector-polygon, and draw a corresponding link-polygon[1] for the rib, its first link being the reaction at one of the terminals of the arch. This is the 'line of pressure' of the arch, and it meets each cross-section of the rib in the corresponding stress-centre. The total stress at this stress-centre is measured by the corresponding ray of the vector-polygon. This stress may be resolved in and perpendicular to the cross-section. The component in the plane of the cross-section gives the *total* shearing stress across the section; the component P perpendicular to the plane, if substituted in the formula of our Art. 815* or of p. 879 of Vol. I. gives the distri-

[1] Vector- and link-polygons are the convenient terms by which Clifford generalised the names force- and funicular-polygons.

bution of traction over the cross-section. Bresse is dealing with cases where the plane of flexure is the plane of loading, i.e. the load-plane passes through a principal axis of each cross-section (note our Art. 14), so that the formula takes the simple form

$$T = \frac{P}{\omega}\left(1 + \frac{by}{\kappa^2}\right).$$

Here, if E the stretch-modulus vary, we must put

$$\omega = \frac{\Sigma E\delta\omega}{\Sigma E}.$$

κ will also change if the cross-section be supposed to vary *slightly*. As a rule it will be sufficient to tabulate (or exhibit graphically) T for the extrados and intrados. The quantity b may sometimes be obtained with sufficient accuracy by scaling its value from a carefully drawn line of pressure. It can of course be ascertained for any cross-section by an analytical determination of the resultant of the forces acting on the rib to one side of the cross-section.

In the following section of the chapter (pp. 76—83) Bresse gives two most interesting examples of the calculation of the tractive stress over the cross-sections of arched ribs in the cases of a simple arch due to Tritschler (*Pont de Brest*) and of a combination of ribs forming an arch due to Vergniais. I do not think a more instructive study can be found for an engineering student than to work out for himself with Bresse's data, both analytically and graphically, the stresses in one or both of these two cases.

[519.] § III. (pp. 84—95) is entitled: *Recherche des déformations de la fibre moyenne sous l'action de forces extérieures supposées toutes connues.* Its object is to find expressions for the shifts at each point of the central axis (*la fibre moyenne*) of the arched rib, and for the change in inclination of the cross-section at any point of the central axis. We may obtain Bresse's equations as follows:

Let α be the angle the cross-section at any point of the central axis makes with a given cross-section, measured so that α increases with s the length of arc from the given cross-section, let ϵ be the 'moment of inertia' and e the 'mass' of the cross-section supposing it loaded with a superficial density equal to the stretch-modulus E. Then the change in the angle $\delta\alpha$ due to the strain may be represented by $\Delta\delta\alpha$, and that in the arc δs by $\Delta\delta s$; let ρ be the strained, ρ_0 the unstrained curvature at any point s of the central axis, and N the corresponding total normal stress.

Then we easily deduce for the stretch in a 'fibre' distant z from the line through the centroid of the cross-section perpendicular to the load-plane:

$$\text{stretch} = z\left(\frac{1}{\rho} - \frac{1}{\rho_0}\right) + \left(1 + \frac{z}{\rho}\right)\frac{\Delta\delta s}{\delta s},$$

supposing $(z/\rho)^2$ to be negligible; therefore the bending moment M, taken to increase a, is given by

$$M = \Sigma E z^2 \delta\omega \left(\frac{1}{\rho} - \frac{1}{\rho_0}\right) + \Sigma E \left(1 + \frac{z}{\rho}\right) z \frac{\Delta \delta s}{\delta s} d\omega$$

$$= \epsilon \left(\frac{1}{\rho} - \frac{1}{\rho_0}\right) + \frac{\epsilon}{\rho} \frac{\Delta \delta s}{\delta s}.$$

Putting in for ρ and ρ_0 their values in terms of a and s, we have

$$\frac{M}{\epsilon} = \frac{\Delta \delta a}{\delta s} - \frac{da}{ds} \frac{\Delta \delta s}{\delta s} + \frac{1}{\rho} \frac{\Delta \delta s}{\delta s}$$

$$= \frac{\Delta \delta a}{\partial s} + \left(\frac{1}{\rho} - \frac{1}{\rho_0}\right) \frac{\Delta \delta s}{\delta s};$$

$$\therefore \ \Delta \delta a = \frac{M}{\epsilon} \delta s - \left(\frac{1}{\rho} - \frac{1}{\rho_0}\right) \Delta \delta s.$$

Now the second term on the right-hand side may generally be neglected in arches because it is the product of small differences; hence integrating, it follows that:

$$\Delta a - \Delta a_0 = \sum_{s_0}^{s} \frac{M}{\epsilon} \delta s \text{..........................(i).}$$

This agrees with Bresse's equation (8 *bis*), p. 87. On p. 85 he does not give the second term of the expression above for $\Delta \delta a$, because he appeals to a result on his p. 35, where, however, he has treated the central axis as straight.

We may obtain Bresse's equations (9 *bis*) and (10 *bis*), p. 88, as follows:

$$\partial s \cos a = \partial x, \quad \partial s \sin a = -\partial y.$$

Hence, if u and v be the shifts, and β a coefficient of stretch produced by any cause other than the loads, as for example temperature:

$$\delta u = \Delta \partial s \cos a - \partial s \sin a \Delta a$$

$$= \frac{dx}{ds} \left(\frac{N}{e} + \beta\right) \delta s + \delta y \left(\Delta a_0 + \sum_{s_0}^{s} \frac{M}{\epsilon} \delta s\right).$$

Summing this (the second term on the left by parts), we have

$$u - u_0 = \sum_{s_0}^{s} \frac{N}{e} \frac{dx}{ds} \partial s + \beta (x - x_0) + \Delta a_0 (y - y_0) + y \sum_{s_0}^{s} \frac{M}{\epsilon} \delta s - \sum_{s_0}^{s} y \frac{M}{\epsilon} \partial s.$$

Or rearranging:

$$u - u_0 = \Delta a_0 (y - y_0) + \beta (x - x_0) + \sum_{s_0}^{s} \left[(y - y_1) \frac{M_1}{\epsilon_1} + \frac{N_1}{e_1} \frac{dx_1}{ds_1}\right] \delta s_1 \text{...(ii),}$$

where the summation is to apply only to quantities marked with the subscript $_1$.

Similarly we find:

$$v - v_0 = -\Delta a_0 (x - x_0) + \beta (y - y_0) + \sum_{s_0}^{s} \left[-(x - x_1) \frac{M_1}{\epsilon_1} + \frac{N_1}{e_1} \frac{dy_1}{ds_1}\right] \delta s_1 \text{...(iii).}$$

We have retained the sign of summation as it indicates clearly the method of procedure by quadratures, when, as is most frequently the case, the loads and bending moments are not continuous, and so integration cannot be applied.

[520.] On pp. 90—94 Bresse indicates how the constants $\Delta\alpha_0, u_0, v_0$ can be determined practically. Thus one or more cross-sections will have their directions unchanged, or one or both terminals will be pivoted, or there will be a line of symmetry for the rib; three conditions will always be given which enable us to determine these constants. On pp. 95—105 we have the formulae (i) to (iii) applied to several special examples. Thus Bresse deals with:

(*a*) The case of a uniformly loaded rib of circular form and given span with uniform cross-section. The integration of the equations is easy, though the results are long (see our Arts. 525—6). He considers this case with a uniform load first along the arc and secondly along the chord; the load being in both instances vertical and the chord horizontal.

(*b*) The case of a cast-iron circular rib of the railway viaduct at Tarascon over the Rhône (see our Art. 527). The deflection as obtained by calculation is ·0642 metres, as obtained from the mean of experiments on the rib before and after erection = ·0650 metres. This is an excellent example of the application of theory to practice, and the nearness of the theoretical and experimental results is remarkable, when one remembers the irregularity of the stretch-modulus across the cross-section and even the doubt as to its mean value.

The theoretical result for the deflection due to a change of temperature of 1° centigrade is worked out on the supposition that β the linear dilatation = ·00111. It is ·00159 metres. Experiment gave in the mean ·00135 metres or a difference of about 1/6.

[521.] The following section of the chapter under discussion is entitled: *Recherche des forces inconnues*, and it occupies pp. 105—126. In the examples hitherto considered Bresse has supposed the terminal reactions to be known; this is not generally the case, and we now turn to the problem of discovering the unknown reactions when the primitive form, the nature of the terminal fixings and the superincumbent load are given. We will briefly cite the conditions to be applied to equations (i)—(iii) in order to obtain the unknown reactions:

(*a*) At a *fixed* or *pivoted* terminal we have $u=v=0$ to determine the two components of the unknown reaction at that terminal.

(*b*) At a *built-in* terminal we have $u=v=0$, and $\Delta\alpha=0$ to determine the two components of the reaction and the bending moment at that terminal.

(*c*) When two ribs are fixed or joined together, we have u and v the same for both at that point, which gives two equations to find the components of the mutual reaction.

(*d*) When two ribs are built into each other, we have three equations arising from the equality of the values of u, v and $\Delta\alpha$ for both ribs at that point; these equations suffice to determine the reaction and the bending moment at the point.

(*e*) If a terminal be constrained to move along a smooth curve, we have a relation between u and v for that terminal, which suffices to determine the normal reaction of the curve.

In all these cases there will be three equations of statical equilibrium for each rib, which suffice with the above to determine the constants $\Delta\alpha_0$, u_0, and v_0; thus in each case there will be sufficient equations to determine all the unknowns.

Bresse treats a number of general cases of fixed or built-in terminals etc., or of combinations of ribs, by the principles we have laid down above. His method is, however, sufficiently indicated by our statement; the analysis varies in quantity according to the nature of the structure[1]. Two of the more interesting cases investigated are those of an arched rib with a horizontal tie-bar parallel to but not coincident with the chord, and a system of three mutually built-in pieces such as form the bridge system of Vergniais (pp. 112—122). On pp. 123—5 Bresse shows the sufficiency of the elastic and statical equilibrium equations to determine all the unknown quantities. On p. 125 is a paragraph entitled: *Du calage des arcs.* I do not understand clearly in what this process of *calage* or wedging, used apparently in building-up an arched rib out of its component parts, may consist. According to Bresse it has the effect of increasing the planned length of the central axis, and produces a uniform stretch in the rib and so a pressure upon the buttresses although the rib be not loaded. He proposes to allow for it by adding to the coefficient β a term having a value independent of the temperature and equal to $\dfrac{\text{the sum of the breadths of the wedges}}{\text{the planned length of the central axis}}$

[1] Bresse speaks of a doubly built-in arched rib as having *peu d'importance pratique* (p. 110). This is, however, the type of the remarkable bridge at St Louis, Mass. U. S., which is 518 feet span and formed of doubly built-in steel ribs.

[522.] In § VI. of this chapter (pp. 126—147) we have a very interesting but laborious bit of algebraical work, namely, the application of the results of the previous section to find the mutual actions between the several ribs and the reactions upon the buttresses in the case of a bridge on the Vergniais principle, of which the numerical dimensions are given. It is an excellent application, whose practical suggestiveness is much increased by variations in the treatment according as the ribs are supposed to be either pivoted or built-in to each other and to the buttresses.

[523.] The final section of the chapter is entitled: *Remarques et théorèmes concernant la manière dont les forces extérieures entrent dans les formules de la flexion. Conséquences* (pp. 147—156). The author shows that the shifts as well as the terminal reactions are linear functions of the loads and of the thermal stretch coefficient β. This of course is a result of the general principle of 'perfect elasticity'. It gives us a means, however, of calculating the parts of the shifts or of the reactions due to each individual load and then by adding the parts of ascertaining the totals,—a method which will often be found very convenient. These results depend of course on β being independent of the loads. They would fail:

Par exemple, si la chaleur ne dilatait pas également une barre tendue et une barre comprimée, ce que, à notre connaissance, les physiciens n'ont pas vérifié (p. 149).

The point is of interest. I have only come across Pictet's remark on this subject: see our Art. 876* (3).

[524.] Pp. 153—156 deal with a property of symmetrical arched ribs asymmetrically loaded, and with a special application of it. This property is thus stated by Bresse—it being assumed that the axis of v is that of symmetry and that of u perpendicular to v.

If symmetry be given to the load system:

1° En ajoutant pour chaque force manquant de sa symétrique une force égale et située symétriquement; 2° en supprimant les forces dont les symétriques manqueraient; que dans ces deux hypothèses on détermine soit l'une des variations u, v, $\Delta\alpha$ qui caractérisent la flexion en un point, soit l'une des composantes, parallèlement aux axes, d'une réaction inconnue, soit son moment, la somme des deux quantités ainsi déterminées sera égale à la somme ou à la différence des quantités analogues qui, sous l'action du système primitif des forces, se produisent au point considéré et en son symétrique ;...on doit, de plus, prendre la différence des quantités analogues pour deux points symétriques, lorsque, tout en étant symétriques, elles ont des directions contraires (p. 155).

For example, let an arch have a vertical axis of symmetry and let the load be parallel to this axis. Let Q, Q' be the horizontal thrusts on the terminals, then for any load :

$$Q + Q' = 0.$$

Suppose the load to be made symmetrical, so that Q, Q' become Q_1, $-Q_1$ when we add to make symmetry, and become Q_2, $-Q_2$ when we subtract to make symmetry. Then according to the above principle

$$Q_1 + Q_2 = Q - Q',$$

or

$$Q = \tfrac{1}{2}(Q_1 + Q_2).$$

Thus if we can obtain results for symmetrical loading, we can deduce results for asymmetrical loading.

[525.] Chapter IV. (pp. 157—217) deals with the thrust of arched ribs of uniform cross-section, for which the central axis, originally circular, remains after flexure in one and the same plane. Bresse's method is direct and simple.

He supposes (§ 81) a single isolated load Π at any point acting perpendicular to the span $2a$ of an arched rib. The vertical reactions at the terminals are given by the equations of Statics, the thrust Q is obtained by an application of the principle referred to in our Art. 524, to the equation deduced from constant length of the span : see our Art. 521. Thus Bresse finds :

$$Q = \Pi \frac{\tfrac{1}{2}(\sin^2\phi - \sin^2\theta) + \cos\phi(\cos\theta + \theta\sin\theta - \cos\phi - \phi\sin\phi) - \tfrac{1}{2}\dfrac{G^2}{a^2}\sin^2\phi(\sin^2\phi - \sin^2\theta)}{\phi + 2\phi\cos^2\phi - 3\sin\phi\cos\phi + \dfrac{G^2}{a^2}\sin^2\phi(\phi + \sin\phi\cos\phi)} \quad \text{.........(i)},$$

where

2ϕ = the central angle of arched rib,

θ = the angle the radius to the loaded point makes with the radius to mid-point of rib, and

G = the swing-radius of the cross-section superficially loaded with the stretch-modulus. See our Arts. 1458* and 1573*.

Similarly (§ 82) if there be an isolated load S, at a point determined by θ, acting parallel to the chord of the arch, the terminal thrusts

$$= \tfrac{1}{2}Q_1 + \tfrac{1}{2}S \text{ and } \tfrac{1}{2}Q_1 - \tfrac{1}{2}S,$$

where

$$\tfrac{1}{2}Q_1 = S\frac{\tfrac{1}{2}\theta - \tfrac{1}{2}\theta\sin\theta\cos\theta - \sin\theta\cos\phi + \theta\cos\theta\cos\phi + \tfrac{1}{2}\frac{G^2}{a^2}\sin^2\phi(\theta + \sin\theta\cos\theta)}{\phi + 2\phi\cos^2\phi - 3\sin\phi\cos\phi + \frac{G^2}{a^2}\sin^2\phi\,(\phi + \sin\phi\cos\phi)} \quad \text{.........(ii).}$$

Next (§ 83) if a couple L with its axis perpendicular to the plane of the central-axis be applied to an element of the rib at the point θ

$$Q = -\frac{L}{a}\frac{\sin\phi\,(\sin\theta - \theta\cos\phi)}{\phi + 2\phi\cos^2\phi - 3\sin\phi\cos\phi + \frac{G^2}{a^2}\sin^2\phi\,(\phi + \sin\phi\cos\phi)} \quad \text{......(iii).}$$

Lastly (§ 84), if there be a change in the length of the central axis due to temperature or any other cause and having a stretch-coefficient β (p. 163),

$$Q = \frac{2e\beta\sin^3\phi\,\frac{G^2}{a^2}}{\phi + 2\phi\cos^2\phi - 3\sin\phi\cos\phi + \frac{G^2}{a^2}\sin^2\phi\,(\phi + \sin\phi\cos\phi)} \quad \text{...... (iv),}$$

where e = mass of area of cross-section loaded with the stretch-modulus E.

By applying the principle of superposition of stress we are able from Equations (i) to (iv) to ascertain the thrust due to any conceivable system of isolated loads. Any continuous load may be concentrated over small elements and treated as a system of isolated loads. Or, on the other hand we may replace Π or S by $f(\theta)\,d\theta$ and integrate along the central axis. This is done by Bresse in the following three cases[1]:

(i) Thrust due to $2p\rho\phi$ being the weight of the arch (radius ρ) or a load distributed uniformly along its length (§ 87),

$$Q = 2p\rho\phi x$$

$$\frac{\tfrac{1}{4} - \tfrac{5}{2}\cos^2\phi - \phi\sin\phi\cos\phi + \tfrac{9}{4}\frac{\sin\phi}{\phi}\cos\phi - \tfrac{1}{2}\frac{G^2}{a^2}\sin^2\phi\left(\sin^2\phi - \tfrac{1}{2} + \tfrac{1}{2}\frac{\sin\phi}{\phi}\cos\phi\right)}{\phi + 2\phi\cos^2\phi - 3\sin\phi\cos\phi + \frac{G^2}{a^2}\sin^2\phi\,(\phi + \sin\phi\cos\phi)} \quad \text{.........(v).}$$

(ii) Thrust produced by a load $2p'a$ distributed uniformly along the chord of the arc (§ 88),

$$Q' = 2p'a\frac{-\tfrac{1}{4} + \tfrac{7}{12}\sin^3\phi + \tfrac{1}{4}\frac{\phi}{\sin\phi}\cos\phi - \tfrac{1}{2}\phi\sin\phi\cos\phi - \tfrac{1}{3}\frac{G^2}{a^2}\sin^4\phi}{\phi + 2\phi\cos^2\phi - 3\sin\phi\cos\phi + \frac{G^2}{a^2}\sin^2\phi\,(\phi + \sin\phi\cos\phi)} \quad \text{...(vi).}$$

[1] He also gives results for (i) and (ii) when the uniformly distributed loads do not cover the whole of the arch.

(iii) Thrust produced by a fluid pressure along the extrados of the arch (§ 89). The result is too complex to be cited here.

[526.] The most important case is that represented by Equation (i). Bresse throws it into the form

$$Q = \Pi k_1 \frac{1 - \mathrm{K}\dfrac{G^2}{a^2}}{1 + \mathrm{K}'\dfrac{G'^2}{a^2}},$$

where, if

$$A = \tfrac{1}{2}(\sin^2\phi - \sin^2\theta) + \cos\phi\,(\cos\theta + \theta\sin\theta - \cos\phi - \phi\sin\phi),$$

and $B = \phi + 2\phi\cos^2\phi - 3\sin\phi\cos\phi$,

$$k_1 = A/B,$$

and $$\mathrm{K} = \tfrac{1}{2}\frac{\sin^2\phi\,(\sin^2\phi - \sin^2\theta)}{A}, \qquad \mathrm{K}' = \frac{\sin^2\phi\,(\phi + \sin\phi\cos\phi)}{B}.$$

The quantities k_1, K, K′ are expanded in powers of $2\phi/\pi$ and $r \equiv \theta/\phi$ on pp. 173—191, and their values tabulated in Tables I. to IV. at the end of the volume. The entries give the values of k_1 for values of $2\phi/\pi$ from ·12 to 1 rising by ·01 at first, then by ·02 and ultimately by ·04; and for values of r rising by ·05 from 0 to ·95 (Table I.). The mean values of K and K′ are given (i.e. the mean for all values of θ for any angle ϕ since they vary little with θ) for values of $2\phi/\pi$ from ·12 to 1 (Table III.), and finally the values of $\left(1 - \mathrm{K}\dfrac{G^2}{a^2}\right)\Big/\left(1 + \mathrm{K}'\dfrac{G^2}{a^2}\right)$ for the same range of values of $2\phi/\pi$ and five values of G^2/a^2, namely ·0005 to ·0025 inclusive rising by ·0005 (Table IV.).

Bresse points out on p. 172 that the value of G^2/a^2 varies for seven French bridges between ·000106 and ·000795, and that its maximum value ·0025 in Table IV. is probably seldom approached in practice. As most of the bridges have a value of G^2/a^2 lying between ·0003 and ·0004, the value of Table IV. would have been increased had additional entries been made for values of G^2/a^2 less than ·0005.

[527.] Bresse shows that if a load p be put upon the arched rib per unit length of the *central axis*:

$$Q = 2p\rho\phi \times m_1 \frac{1 - \mathrm{K}G^2/a^2}{1 + \mathrm{K}'G^2/a^2} \quad\text{.................... (v)}',$$

where ρ is the radius of the central axis.

If a load p' be put upon the arched rib per unit length of the *chord*:

$$Q' = 2p'a \times n_1 \frac{1 - \mathrm{K}G^2/a^2}{1 + \mathrm{K}'G^2/a^2} \quad\text{................ (vi)}'.$$

If there be a coefficient of thermal or other stretch β:

$$Q_1 = q_1 \frac{\beta e\, G^2/a^2}{1 + \mathrm{K}'G^2/a^2} \quad\text{........................ (vii)},$$

where $e = \Sigma E\delta\omega$ as before.

The values of m_1, n_1 and q_1 are tabulated for values of $2\phi/\pi$ from 12 to 1 in Table II. Unfortunately q_1 by a printer's error appears as r_1 in that Table and the error is nowhere pointed out.

Bresse's first four tables thus give us a means of ascertaining the thrust in many practically important cases of circular ribs of uniform cross-section. The method of using the Tables is exemplified on pp. 212—217 by their application to the bridges at Brest and Tarascon. The discussion on the former bridge brings out clearly the smallness of the error introduced by concentrating into a series of isolated loads the parts of a continuous load which act upon even considerable portions of the arch.

[528.] We may note one or two other points brought out in the course of this chapter.

(i) On pp. 193—196 it is shown that Equation (vii) may be replaced with sufficient approximation in practice by taking the formula

$$Q_1 = \frac{\beta e\, G'^2}{G'^2 + \frac{8}{15}f^2}$$

where f, the rise of the arch, is measured for the central axis.

(ii) If the same load $(2p\rho\phi = 2ap')$ be distributed uniformly along the central axis or uniformly along the chord, then the ratio of $Q : Q'$ as determined by Equations (v) and (vi) may for most practical purposes be taken as unity. Bresse gives the following values (p. 203):

$2\phi/\pi =$	·12	·2	·3	·4	·5	·6	·7	·8	·9	1
$Q/Q' =$	·997	·993	·984	·971	·953	·930	·900	·863	·814	·750

(iii) If Q'' be the horizontal tension of the cables of a suspension bridge which is of span $2a$, rise f, and loaded with p' lbs. per foot-run, then the ratio of Q' as given by (vi) to Q'' $\left(=\frac{p'a^2}{2f}\right)$ is very nearly unity if G^2/a^2 be small. Thus if G^2/a^2 be less than ·0005 we have sensibly for

$2\phi/\pi =$	·12	·2	·3	·4	·5	·6	·7	·8	·9	1
$Q'/Q'' =$	·999	·996	·992	·985	·975	·962	·946	·922	·893	·849

Hence for most practical problems we may calculate Q' from the tension in the cables of a suspension bridge of the same span and rise. We note that Q' is always *less* than Q'': see our Art. 1459*.

[529.] Chapter V. is entitled: *Résistance d'un arc circulaire à section constante, chargé dans toute sa longueur de poids uniformément répartis suivant l'horizontale* (pp. 218—249). The object of this chapter is to calculate the maximum traction at any point of an arched rib due to a uniform loading of the rib of p' lbs. per foot run of the horizontal chord. This is practically the loading which would occur, if the bridge were tested by a train of locomotives or a uniform pile of iron rails (*l'arc sous l'action de la charge d'épreuve*, pp. 218—9).

Let E be the stretch-modulus of any fibre, N the normal force on a cross-section making an angle α with the central cross-section; e the mass of the cross-section of superficial density E, and ϵ the moment of inertia of the mass of this cross-section about an axis through the central axis perpendicular to the load-plane. Then the traction in a fibre at distance y from that axis is given (pp. 220—2) by:

$$E\left(\frac{N}{e}+\frac{My}{\epsilon}\right)\ldots\ldots\ldots\ldots\ldots\ldots\ldots\ldots\text{(i)}.$$

It is easy to show, ρ being the radius of the arch, that:

$$N=-Q\cos\alpha-p'\rho\sin^2\alpha\ \ldots\ldots\ldots\ldots\ldots\ldots\text{(ii)},$$

$$M=Q\rho(\cos\alpha-\cos\phi)-\tfrac{1}{2}p'\rho^2(\sin^2\phi-\sin^2\alpha)\ldots\ldots\ldots\text{(iii)};$$

or, if $Q=n\times 2p'a$, n being a certain function of ϕ and G^2/a^2 (compare our Art. 527),

$$N=-p'\rho\,(2\cos\alpha\sin\phi+\sin^2\alpha)\ldots\ldots\ldots\ldots\text{(ii)}',$$

$$M=\tfrac{1}{2}p'\rho^2(\cos\alpha-\cos\phi)(4n\sin\phi-\cos\alpha-\cos\phi)\ldots\ldots\text{(iii)}'.$$

We thus know the traction at any point by substituting (ii)′ and (iii)′ in (i).

Now Bresse assumes that in an arched rib it is the pressure or negative traction, which first reaches the elastic limit, he therefore seeks for the greatest negative value of the expression $E(N/e+My/\epsilon)$. I do not think that this is justifiable. What we really want is the greatest positive stretch of the material, and accordingly the proper condition seems to be to find the greatest positive and negative values of $N/e+My/\epsilon$, then to choose the maximum numerical value from either the positive values, or the negative multiplied by η the stretch-squeeze ratio, and equate that maximum to the safe elastic stretch. Bresse really assumes that the elastic limit is reached in compression and extension with the same numerical strain, and therefore as the squeezes are always greater than the stretches, we have only to deal with the former. But we ought I think to investigate whether $\eta\times$ the maximum squeeze is greater than the maximum stretch. If η be, say, $\frac{1}{3}$ or $\frac{1}{4}$, then it by no means follows that Bresse's condition is correct. For example in the results given by him for the *Pont de Brest* and represented graphically in Fig. 23 of Plate II, the maximum positive traction is in the extrados of the arch and very sensibly greater than one-third of the maximum negative traction, which here occurs at the same cross-section in the intrados. Similarly in the stresses for the *Système Vergniais* given in Fig. 26 (B) Plate III, the maximum positive traction in the extrados of the *contrefort* is greater than the maximum negative traction in the same rib, and is about as great as the maximum negative traction in the main arch. Thus in these actually existing bridges, it is obvious that Bresse's method of seeking for the maximum negative traction would be deceptive. The true criterion must in each case be deduced from the situation of the load-point (or stress-centre), *i.e.* whether it lies inside or outside the whorl of the cross-section. Bresse's method only applies if all the load-points lie inside the whorls: see Vol. I. p. 879.

[530.] Pp. 221—230 are occupied with a discussion of the possible magnitudes and positions of the maximum negative traction. These depend largely on the sign of M as given by Equation (iii)′, and Bresse shows that if $n > \frac{1}{2} \cot \phi$, then M vanishes at either four points or two points besides the 'pivoted' terminals: see our Art. 1460*. I will not enter into the details of this investigation, since for the reasons given in the previous article it does not seem to me entirely satisfactory; the graphical construction of curves of thrust and bending-moment, of the line of pressure and of the whorl of the cross-section is the better treatment of the problem, some allowance being made if necessary for the effect of shearing force. Suffice it to add that if E_1 be the stretch-modulus of the 'mean fibre' Bresse reduces the maximum negative traction to the form

$$\zeta \frac{p'a}{e} E_1$$

where ζ is a coefficient depending on $\frac{G^2}{a^2}$, $\frac{2\phi}{\pi}$ and $\frac{h}{a}$, where h is the distance of the central axis from 'the extreme fibre.' The values of ζ are tabulated on pp. 260—269 for a considerable range of values of these arguments, and a horizontal line drawn across Bresse's columns marks whether the maximum negative traction occurs in the extrados or intrados (Table V.).

[531.] After some numerical examples of this Table on pp. 237—8, Bresse concludes his work with a section entitled: *Des circonstances qui peuvent influer sur la résistance d'un arc à section constante, chargé uniformément suivant l'horizontale* (pp. 238—249). This section deals with general theorems (deduced from the numerical results of Table V. and therefore open to the objections of our Art. 529) as to the elastic strength of an arch when we vary: (i) ϕ, or what is the same thing, vary the ratio of rise to span, (ii) the cross-section as determined by the ratios of G and h to $2a$ the span.

If G/a be constant, and we take the mean value of h/a (which does not vary much since G/a is constant) we can find a value of the ratio of rise to span, which gives a minimum of ζ, or a maximum elastic resistance. Thus we find approximately for:

$G^2/a^2=$	·0001	·0002	·0003	·0004	·0005	·0006	·0008	·0010	·0012	·0015
$f/2a=$	·1242	·1495	·1581	·1668	·1756	·1889	·1980	·2117	·2164	·2210

In § 122 Bresse considers a special case of an arched rib of hollow elliptic cross-section and investigates for what values of the ratio of rise to span it is more advantageous to place the cross-section with its major axis horizontal than with it vertical or *vice versâ*.

In § 123 he deals with the problem of the best ratio of the height to the breadth in the cross-section (supposed to be rectangular and of constant area) of an arched rib having a given load, height and span. The laws of ribs with circular central axes differ in respect of relative strength very considerably from those of straight beams.

Although for the reasons stated above, Bresse's results in this section must not be considered as final, still they indicate the existence of numerous very interesting properties varying with the form of the rib. They conclude what is the most thorough investigation hitherto published of the elastic strength of circular arches subjected to uniplanar flexure.

[532.] M. Bresse: *Cours de mécanique appliquée.* The *Première Partie* of this book was published in Paris in 1859 in parts. A second edition of the *Première Partie* appeared in 1866, with, however, few modifications, and a third in 1880. The *Troisième Partie* (*Calcul des moments de flexion dans une poutre à plusieurs travées solidaires*) appeared in 1865. Only the first and third parts deal with topics related to our present subject. The former is entitled: *Résistance des Matériaux et Stabilité des Constructions*, and the chief difference between the first and second editions is that § II. of the third chapter on continuous beams disappears in the later edition, reappearing in a much fuller form in 1865 as the *Troisième Partie* of the work. The *Première Partie* in the second edition from which I cite contains pp. i—xxviii and 1—536. I shall discuss the *Troisième Partie* under the year 1865.

[533.] Chapter I. entitled: *Généralités; Principes fondamentaux. Recherche des tensions dans les diverses parties d'un corps prismatique*, pp. 5—89, is occupied with a discussion of the moment of inertia, the neutral axis, the load-point, the core and the distribution of traction over a cross-section when the line of pressure is known. This follows with some amplifications the treatment of the lithographic course and of the work on arched ribs: see our Arts. 813*—5* and 514—6.

[534.] Chapter II. (pp. 90—149) deals with the general equations for the strain of a rod, whose central axis is not necessarily a straight line. It is an amplification of the treatment in the work on arched ribs and the major portion does not call for special remark. The only part which need be noticed is entitled: *Des mouvements vibratoires dans les pièces élastiques* and occupies pp. 143—9. Bresse deals with the case of the vibrations of a rod, the central axis of which is a plane curve. He supposes this rod to vibrate only in the plane of its own central axis, so that that plane must pass through a principal axis of each cross-section; the cross-section itself is considered to be uniform.

Let u, v be the shifts of the centroid G of any cross-section (distant s along the central axis from any fixed point of the rod) measured along the tangent and normal (outwards) to the central axis at G. Let χ be the variation in the angle which the tangent at G makes with any fixed line, positive when taken clockwise. Let m be the mass of the rod per unit length and T, N the external forces per unit of length of the rod at

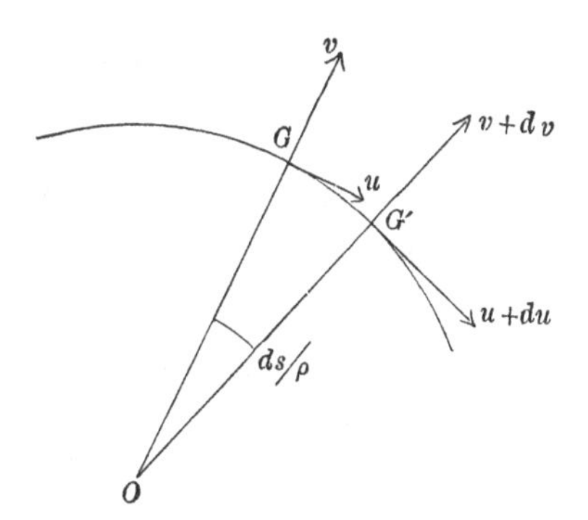

G, M the clockwise couple round G per unit element (this is introduced by Bresse, but it seems to me that in most conceivable cases M would be zero). Let $1/\rho$ be the curvature, κ the swing-radius of the cross-section ω at G round an axis through G perpendicular to the plane of flexure, E the stretch- and μ the slide-modulus, both being supposed uniform for the cross-section. Then Bresse obtains the following equations:

$$m\frac{d^2u}{dt^2} = T + E\omega\frac{d}{ds}\left(\frac{du}{ds} + \frac{v}{\rho}\right) + \frac{\mu}{\rho}\left(\frac{dv}{ds} + \chi - \frac{u}{\rho}\right),$$

$$m\frac{d^2v}{dt^2} = N + \mu\omega\frac{d}{ds}\left(\frac{dv}{ds} + \chi - \frac{u}{\rho}\right) - \frac{E\omega}{\rho}\left(\frac{du}{ds} + \frac{v}{\rho}\right),$$

$$m\kappa^2\frac{d^2\chi}{dt^2} = M + E\omega\kappa^2\frac{d^2\chi}{ds^2} - \mu\omega\left(\frac{dv}{ds} + \chi - \frac{u}{\rho}\right).$$

Here $E\omega\left(\frac{du}{ds}+\frac{v}{\rho}\right)$ is the total traction and $\mu\omega\left(\frac{dv}{ds}+\chi-\frac{u}{\rho}\right)$ the total shear over the cross-section at G.

These three equations suffice theoretically to determine u, v and χ. Bresse makes the following remarks on them:

Pour le moment, nous ne pousserons pas plus loin l'étude de la question; dans les cas pratiques les plus simples, la solution présentera généralement de grandes difficultés, comme on le verra ultérieurement par les exemples que nous indiquerons. Nous n'avons voulu, en donnant les calculs précédents, que compléter la théorie générale de la déformation des pièces élastiques, par l'exposé de la méthode à suivre pour mettre en équation le problème des mouvements vibratoires (pp. 148—9).

[535.] The practical part of the *Cours* begins with Chapter III. which is entitled *Problèmes divers concernant les poutres droites* (pp. 150—224). A good deal of this chapter is not novel, but the methods are very clearly and concisely put, and some interesting problems of continuous beams with large numbers of supports are dealt with on pp. 176—188; these should certainly be studied by any one practically interested in this subject. Slide is considered after the manner of Jouravski (see Section III. of our present Chapter) on pp. 206—9, but there is no reference to the work of Saint-Venant. The chapter concludes with an essentially theoretical treatment of the problem of struts (pp. 210—224).

[536.] Chapter IV. deals with the problem of arched ribs (pp. 225—263) after the manner of the work we have already analysed: see our Arts. 514 to 531. Chapter V. is also a continuation of this subject (pp. 264—338)[1]. It contains, however, a section on the strength of cylindrical vessels (pp. 323—338) which requires some notice on our part. The first problem dealt with is that of a boiler or flue of right-circular cross-section, and the method adopted is the old hydrostatic process, involving no elastic principle: see our Art. 1012*.

[537.] The second case dealt with is novel. It is entitled: *Résistance d'une chaudière à profil faiblement elliptique* (p. 326), and, if we could trust the investigation, this case might be useful in calculating the dimensions of slightly elliptic flues. Bresse however practically treats his elliptic cylinder as if the portion between two

[1] Matter not in the book of 1854 is chiefly confined to some account of the experiments of Desplaces, Collet-Meygret, and Jules Poirée: see Section III. of our present Chapter.

parallel cross-sections at unit-distance could be dealt with as if it were a rod. Thus he takes the product of the flexural rigidity and the change in the curvature as equal to the bending-moment. Let c be the thickness, supposed uniform, of the flue, l its length, M the moment tending to bend the wall of the flue round any longitudinal section, $\Delta\psi$ the change due to the strain in the angle between two tangents to the central line of the flue's cross-section at the ends of an arc δs, then Bresse puts:

$$M = Ec \times l \times \frac{c^2}{12} \times \frac{\Delta\psi}{\delta s}.$$

Now, if it is legitimate to use any formula of this kind at all, it would seem necessary to at least replace the stretch-modulus by the plate-modulus $\{$*i.e.* by $E/(1-\eta^2)\}$, but I must confess to having grave doubts as to the entire method of treatment. To assume the existence of a neutral axis passing through the centroid of a transverse section in the case of a bent *plate* subjected to strain seems in itself a very risky proceeding.

If we may adopt Bresse's assumption we arrive at the following results—in which

p = the internal pressure in the flue; e = the eccentricity of the elliptic cross-section; $2a$ its internal major axis; x the abscissa of any point measured along this major axis from the centre; c the thickness of the flue, supposed small and uniform; l = the length of the flue:

(*a*) The bending-moment of the wall of the flue at points given by x is equal to $\frac{1}{4}pe^2(2x^2 - a^2)$ per unit length of the flue.

(*b*) The maximum traction, which occurs at the ends of the major axis, is given by

$$T = \frac{pa}{c} + \frac{3pa^2e^2}{2c^2}.$$

The first term in the traction is due to the internal pressure supposing the flue to be exactly circular, the second term is due to the flexure produced by the slight ellipticity.

The result (*b*) gives a quadratic to find the proper thickness c for a given value of T and p; the positive root must be taken. Bresse turns this formula into numbers and shows that a very slight value of e^2 ($\cdot 02$) will require the value of c to be increased in the ratio of 5 : 3. Thus the existence of slight ellipticity in a flue is very unfavourable to its strength.

(*c*) There is a decrease in the semi-major axis given by

$$\delta a = -\frac{pa^4e^2}{Ec^3},$$

and an increase in the semi-minor axis of about the same amount (p. 332).

Further the eccentricity e' after strain is given in terms of the eccentricity before strain by

$$e'^2 = e^2 \Big/ \left(1 + \frac{4pa^3}{Ec^3}\right).$$

These results would be, perhaps, slightly improved if E were replaced by $E/(1-\eta^2)$.

(*d*) Bresse next supposes p negative, that is to say that there is an external pressure p. In this case e' will be real or equilibrium possible only if

$$1 > \frac{4pa^3}{Ec^3},$$

or

$$c > a\left(\frac{4p}{E}\right)^{\frac{1}{3}}.$$

This result is independent of the absolute strength T of the material. It is discussed at considerable length by Bresse on pp. 334—7. But the manner in which it has been deduced does not leave an impression of conclusiveness on my mind. Were c even to *approach* this value, e' would become very great, or the strain exceed the elastic limit.

(*e*) Bresse points out that the thickness of the elliptic flue will have to be greater for the case of external pressure than for the case of an equal internal pressure, *supposing that the resistances to compression and extension are equal*, (p. 338.)

[538.] The same problem is considered by J. H. Macalpine in the third of *Three Original Papers* (Glasgow, James Maclehose and Sons) printed in 1889. It is entitled: *On the strength and stiffness of an elliptic cylinder submitted to hydrostatic pressure* (pp. 26—31), and may be referred to here. Macalpine obtains practically the same results as Bresse for the change in the axes, but for the maximum bending moment he finds (instead of $\frac{1}{4}pe^2a^2$) $\frac{1}{4}pe^2a^2\left\{1 + \frac{4p}{E}\left(\frac{a}{c}\right)^3\right\}$ per unit length. The term $\frac{4p}{E}\left(\frac{a}{c}\right)^3$ arises from the *second* approximation and it can obviously in certain cases sensibly modify Bresse's result. To the first approximation, however, I think the two agree. For on p. 30, l. 10 from the bottom, Macalpine finds *in his own notation*:

$$\theta = \frac{pb^4e^2}{4ac}\sin\phi\cos\phi.$$

But M his bending moment $= c\dfrac{d\theta}{ds}$,

or,

$$M = \frac{pb^4e^2}{4a}(2\cos^2\phi - 1)\frac{d\phi}{ds}.$$

Changing to our notation and remembering that, as we are neglecting e^4, we may put $b=a$, $\frac{d\phi}{ds}$ = curvature = $1/a$, and $a \cos \phi = x$; we have :

$$M = \frac{pe^2}{4}(2x^2 - a^2),$$

which agrees with Bresse's result (a). In other respects Macalpine's treatment is neither so full nor so clear as Bresse's. He falls into the same error of treating a bent plate as a rod[1].

[539.] The sixth chapter of Bresse's book is entitled: *Problèmes particuliers sur les poutres vibrantes* and occupies pp. 339—387. This chapter is perhaps the most interesting in the book. It gives a good *resumé* of all the work which had been done in and since the time of Navier and Poncelet on the subject of vibrational stress in bridges.

It opens with a discussion after the manner of Poncelet of the longitudinal vibrations of bars variously loaded (cf. our Arts. 988*—993*), and does not here add much to Poncelet's results. The great defect of Poncelet's work is that it leaves us with complex analytical expressions, which require the patience of a Saint-Venant to reduce them to numerical results of practical value: see our Arts. 401 and 411.

Bresse next passes to the transverse vibrations of a uniform beam simply supported at both terminals and uniformly loaded (pp. 361—374). I think his work here is original, at any rate I have not come across the same results before. It bears of course considerable resemblance to the ordinary theory given in treatises on Sound of the transverse vibrations of a rod.

Bresse obtains an equation of the following type (which may easily be deduced as a special case from those of our Art. 534) for the transverse shift y of the centroid of a section distant x from one terminal of a beam of length l:

$$E\omega\kappa^2 \frac{d^4y}{dx^4} = mg - m\frac{d^2y}{dt^2} + m'\kappa^2 \frac{d^4y}{dx^2dt^2},$$

where $E\omega\kappa^2$ = the flexural rigidity of the beam,

m = combined mass of beam and load per foot-run,

m' = mass of beam only per foot-run,

t = the time from any epoch.

[1] My objections to this method of treatment have been more fully given in Art. 1547*.

Bresse obtains a general solution for any initial shifts and any initial velocities : see his p. 369 ; he also deals with one or two special cases. Thus on pp. 370—1 he shows how the constants may be easily calculated when the shifts and velocities are initially given by integer algebraic functions of x. A further interesting case on p. 372 may be cited here. It practically amounts to an expression for the deflection at any point of a bridge or beam, when a continuous load is suddenly placed upon it.

Bresse finds:

$$\frac{24E\omega\kappa^2}{mg}\, y = x^4 - 2lx^3 + l^3x - \frac{96l^4}{\pi^5}\sum_{i=1}^{i=\infty}\left(\frac{1}{i^5}\sin\frac{i\pi x}{l}\times\cos\frac{i^2\pi^2a^2t}{l\sqrt{i^2\pi^2b^2+l^2}}\right),$$

the summation being for all odd integer values of i.

Now $1/i^5$ = successively 1, $\frac{1}{243}$, $\frac{1}{3125}$, so that for all practical purposes it is unnecessary to go beyond the first or second term of the summation.

Here $a^4 = E\omega\kappa^2/m$, $b^2 = m'\kappa^2/m$, $= \kappa^2$ if the weight of the load be negligible as compared with that of the bridge.

For $x = l/2$, and t = even multiple of $l\sqrt{\pi^2b^2 + l^2}/(\pi a^2)$ the summation is very nearly equal to $\frac{5}{16}\cdot\frac{\pi^5}{96}$, and the maximum central deflection is given by

$$y = \frac{5}{192}\frac{mgl^4}{E\omega\kappa^2}, \text{ nearly.}$$

Thus the maximum deflection is almost twice the statical deflection under the same load, an instance of Poncelet's Theorem : see our Art. 988*.

[540.] The following and last section of this chapter is entitled: *Effet produit sur une poutre par une charge roulante* (pp. 375—87). Bresse begins by analysing Phillips' memoir of 1855 (see our Arts. 372—82 and 552—4) and quoting his results. For the case of a doubly-supported beam, he has not, however, noted Phillips' error: see our Art. 375. For the case of the doubly built-in beam he was, as we have noticed in Art. 382, the first to correct Phillips and he gives this correction on p. 376. With the notation of our Arts. 373—4, where it must be remembered $2l$ is the length of the beam, Bresse finds for the maximum bending moment of a doubly built-in beam subjected to a travelling load:

At the centre:

$$\left(\frac{1}{12}Pl + \frac{1}{4}Ql\right)\left(1 + \frac{3}{8}\frac{1}{\beta}\right),$$

and at the terminals:

$$\frac{1}{6}Pl\left(1 + \frac{3}{16}\frac{1}{\beta}\right) + \frac{1}{4}Ql\left(1 + \frac{3}{8}\frac{1}{\beta}\right).$$

The comparatively small practical value of these results has been pointed out in our Art. 382.

[541.] Bresse then passes on pp. 377—387 to the discussion of his own particular problem in live-load, of which we have already given the statement and chief results in our Art. 382. To the results given there we may add the expression for the central deflection f; in the notation of that article:

$$f=\frac{Pl^3}{2E\omega\kappa^2}\{-\tfrac{1}{2}\beta''+\beta''^2(\sec\sqrt{1/\beta''}-1)\}, \qquad \text{(p. 382.)}$$

Bresse discusses, with numerical values for the limiting speeds, cases of plate-iron railway girders and shows that the speeds obtained are considerably greater than the usual train speeds. The practical value of the investigation is, however, not very great, as the maximum moment is reached (as is pointed out in our Art. 382) just as the train covers the whole bridge, and not after a steady deflection is set up by a very long train.

[542.] The next chapter of the work (*Chapitre septième*) is entitled: *Résultats d'expériences sur l'élasticité des matériaux* and occupies pp. 388—422. This portion of the work was at the time of publication a useful *résumé* of the experiments of Hodgkinson and his contemporaries. It is now somewhat out of date. The remarks, however, on p. 393 as to the ill-founded character of the reproaches against the theory of elasticity, based on the fact that formulae depending on the proportionality of stress and strain will not explain rupture, are still to the point. Were they studied we should hear less of the "paradox in the theory of beams": see our Arts. 178 and 507.

The work concludes with chapters dealing analytically with framework and with the pressure of masses of earth; both topics lie outside the scope of our history.

[543.] M. Painvin: Thèse de Mécanique. *Études sur les états vibratoires d'une couche solide, homogène et d'élasticité constante, comprise entre deux ellipsoïdes homofocaux*, Paris, 1854. This is a thesis presented to the Faculty of Sciences of Paris for the degree of 'docteur ès sciences mathématiques.' The examining commission were Chasles, Lamé, and Delaunay. The memoir contains 46 quarto pages and is, I believe, the first attempt to use the equations of elasticity in curvilinear coordinates for the solution of any problem.

[544.] Lamé in 1841 had published in the *Journal de Liouville* (see our Art. 1037*) the uniconstant equations of elasticity in curvilinear coordinates. It was not till five years after the date of Painvin's memoir that he published the more complete treatment of the subject which is to be found in his *Leçons sur les coordonnées curvilignes* (see our Art. 1149*). Painvin adopts two elastic constants, and puts his body shift-equations into the forms used in Lamé's *Leçons*, but he possibly owes these to Lamé himself. There are also a number of purely analytical propositions proved in the memoir with regard to what would now be called 'ellipsoidal harmonics', which I do not remember to have seen discussed by Lamé and which may possibly be original. At the same time I am not sufficiently acquainted with Lamé's earlier papers on isothermal surfaces to know what is the history of the subject before the publication of Lamé's *Leçons sur les fonctions inverses* in 1857. At any rate Painvin's paper contains some very elegant analysis, although but little which is of value from the standpoint of physical elasticity.

[545.] The memoir consists of the following two distinct parts:

(i) A proof that the equations of elasticity in curvilinear coordinates can be solved for the two cases of longitudinal and transverse vibrations, so soon as solutions can be found of the differential equation: $a^2\nabla^2 F = d^2F/dt^2$, where ∇^2 is the Laplacian expressed in curvilinear coordinates[1], and a^2 a constant.

(ii) An investigation of the vibrations of a shell of isotropic and homogeneous material bounded by two confocal ellipsoids. The shell is surrounded by air and the forces which produce the initial disturbance are applied normally to the surface. Further only the longitudinal vibrations are considered.

I propose to make a few remarks on both these points.

[1] Let ρ_1, ρ_2, ρ_3 be the three curvilinear coordinates, and let v_1, v_2, v_3 be the three shifts as in our Art. 1153*. Then, if

$$h_i^2 = \left(\frac{d\rho_i}{dx}\right)^2 + \left(\frac{d\rho_i}{dy}\right)^2 + \left(\frac{d\rho_i}{dz}\right)^2,$$

we have

$$\nabla^2 F = h_1 h_2 h_3 \left\{\frac{d}{d\rho_1}\left(\frac{h_1}{h_2 h_3}\frac{dF}{d\rho_1}\right) + \frac{d}{d\rho_2}\left(\frac{h_2}{h_3 h_1}\frac{dF}{d\rho_2}\right) + \frac{d}{d\rho_3}\left(\frac{h_3}{h_1 h_2}\frac{dF}{d\rho_3}\right)\right\}$$
$$= \frac{d^2F}{ds_1^2} + \frac{d^2F}{ds_2^2} + \frac{d^2F}{ds_3^2} - \left(\frac{1}{r_2'} + \frac{1}{r_3'}\right)\frac{dF}{ds_1} - \left(\frac{1}{r_1''} + \frac{1}{r_3''}\right)\frac{dF}{ds_2} - \left(\frac{1}{r_1'''} + \frac{1}{r_2'''}\right)\frac{dF}{ds_3},$$

in the notation of our Art. 1150*. This easily follows from the consideration that

$$\frac{ds_i}{d\rho_i} = \frac{1}{h_i}, \text{ and } 1/r_i^{(to\ y\ dashes)} = \frac{h_y}{h_i}\frac{dh_i}{d\rho_y}.$$

[546.] The first theorem is, as Painvin remarks, really obvious, for Lamé has shown that the waves of longitudinal and transverse vibration (dilatational and twist waves) both depend on the solution of an equation of the form:

$$a^2\left(\frac{d^2F}{dx^2}+\frac{d^2F}{dy^2}+\frac{d^2F}{dz^2}\right)=\frac{d^2F}{dt^2};$$

where, as is well known, the quantity ∇^2 is an invariant for all types of coordinates. See our Arts. 526* and 1078*, and compare also Lamé's *Leçons sur la théorie...de l'élasticité...*, pp. 143—6.

The novelty of Painvin's work consists in the types of solution it suggests for the vibrations of bodies when we use curvilinear coordinates. We may indicate his process as follows. Let

$$\tau_1=\frac{h_2h_3}{h_1}\left\{\frac{d}{d\rho_2}\left(\frac{v_3}{h_3}\right)-\frac{d}{d\rho_3}\left(\frac{v_2}{h_2}\right)\right\}\text{ ....(i),}$$

$$=\frac{1}{h_1}\left\{\frac{dv_3}{ds_2}-\frac{dv_2}{ds_3}-\frac{v_3}{r_3''}+\frac{v_2}{r_2'''}\right\},$$

and τ_2 and τ_3 be like quantities obtained by cyclical interchange, then τ_1, τ_2, τ_3 correspond closely to the doubles of the twists in Cartesian coordinates. The body shift-equations are of the type:

$$\frac{d^2v_1}{dt^2}-S_1=\Omega^2h_1\frac{d\theta}{d\rho_1}+\omega^2h_2h_3\left(\frac{d\tau_2}{d\rho_3}-\frac{d\tau_3}{d\rho_2}\right)\text{ .... (ii),}$$

where $\Omega^2=(\lambda+2\mu)/\Delta$, $\omega^2=\mu/\Delta$, $\Delta=$ the density and $\theta=$ the dilatation, which is given by

$$\theta=h_1h_2h_3\left\{\frac{d}{d\rho_1}\left(\frac{v_1}{h_2h_3}\right)+\frac{d}{d\rho_2}\left(\frac{v_2}{h_3h_1}\right)+\frac{d}{d\rho_3}\left(\frac{v_3}{h_1h_2}\right)\right\}\text{.........(iii).}$$

This value is easily seen to be identical with that of our Art. 1153*.

If we suppose the body forces S_1, S_2, S_3 zero, we find that two types of solution for Equation (ii) can be reached. In the first place consider the curvilinear twists τ_1, τ_2, τ_3 zero. This will arise when

$$v_1/h_1=dF/d\rho_1,\quad v_2/h_2=dF/d\rho_2,\quad v_3/h_3=dF/d\rho_3\text{............(iv).}$$

Equations of type (ii) now become of type

$$\frac{1}{h_1}\frac{d^2v_1}{dt^2}=\Omega^2\frac{d\theta}{d\rho_1}\text{.......................... (v).}$$

But (iii), remembering the value of ∇^2 given in the footnote to p. 374, shows us that $\theta=\nabla^2F$, whence we find that F must satisfy the equation

$$\Omega^2\nabla^2F=d^2F/dt^2.$$

Vibrations of the type (v) are those termed *longitudinal* by Lamé and Painvin. I think them best spoken of as *dilatational* vibrations.

[547.] The second type of vibrations depending only on ω are obtained by putting $\theta=0$, and are *pure twist* vibrations.

Painvin obtains a solution of the following type

$$\frac{v_1}{h_2 h_3} = \frac{dX}{d\rho_2}\frac{dx}{d\rho_3} - \frac{dX}{d\rho_3}\frac{dx}{d\rho_2} + \frac{dY}{d\rho_2}\frac{dy}{d\rho_3} - \frac{dY}{d\rho_3}\frac{dy}{d\rho_2} + \frac{dZ}{d\rho_2}\frac{dz}{d\rho_3} - \frac{dZ}{d\rho_3}\frac{dz}{d\rho_2},$$

where X, Y, Z are all solutions of the equation

$$\omega^2 \nabla^2 F = d^2F/dt^2.$$

The twists are shown (pp. 9—17) to be of the type

$$\tau_1 = -\frac{dx}{d\rho_1}\nabla^2 X - \frac{dy}{d\rho_1}\nabla^2 Y - \frac{dz}{d\rho_1}\nabla^2 Z + \frac{d}{d\rho_1}\left(\frac{dX}{dx} + \frac{dY}{dy} + \frac{dZ}{dz}\right).$$

This investigation seems to me unsatisfactory, because the solution is not entirely freed of x, y, z the old Cartesian coordinates.

[548.] In the second portion of his memoir (pp. 17—46) Painvin determines a solution of the equation $\Omega^2\nabla^2 F = d^2F/dt^2$, subject to the following conditions. Let $\rho_1 = a$ and $\rho_1 = a'$ be the parametric values for the confocal ellipsoids, then he supposes:

(i) the shears $\widehat{s_1 s_2}$ and $\widehat{s_1 s_3}$ to be zero for $\rho_1 = a$ and for $\rho_1 = a'$, these he terms the *surface conditions*;

(ii) the values of F and dF/dt for $t = 0$ to be given functions of ρ_1, ρ_2, ρ_3 except as far as the addition of an arbitrary constant is concerned. This is really equivalent to assuming the initial shifts and initial speeds. These are the *initial conditions*.

The supposition (ii) is perfectly straightforward but it is difficult to grasp the physical meaning of (i). The surface traction $\widehat{s_1 s_1}$ is not put zero for $\rho_1 = a$ and a', hence there must be a traction varying with the time exerted over the surfaces of the shell, if it is to vibrate solely *longitudinally*. Painvin does not distinctly say so, but I think he supposes the air, which he refers to as surrounding his shell, capable of giving the necessary traction. This is, of course, impossible (see our Art. 1084*); the traction sometimes will be *positive*, and the air could not even provide anything like as great negative traction (pressure) as would be required for many sound vibrations. Physically the only result of the memoir, assuming its analysis complete, is to show that the vibrations of a free shell bounded by confocal ellipsoids must be partly *twist-vibrations*, for Painvin's solution is evidently impossible. At the same time it involves such very pretty analytical investigations, that we wish it had some real physical value.

[549.] J. Dienger: *Studien zur mathematischen Theorie der elastischen Körper: Grunerts Archiv der Mathematik und Physik*, Theil 23, 1854, pp. 293—359.

This is a treatise on the general theory of elasticity, with applications to the theory of vibrations. The writer proceeds on rari-constant lines, basing his work on that of Navier, Poisson and Lamé. He prefers rari-constancy to Lamé's method as it indicates:

> was die jeweils eingeführten Grössen zu bedeuten haben, wie sie folglich zu berechnen und zu behandeln sind—ein Vortheil, der gewiss nicht zu niedrig anzuschlagen ist. (p. 358.)

He proceeds on the supposition of an initial state of stress, like Cauchy (see our Arts. 615*—6*), but he retains shift-fluxions up to the fourth order; for this he claims originality. The coefficients of the shift-fluxions of the fourth order are given in terms of molecular summations (pp. 300—301). He deals with the relations between these summations on pp. 323—6, and obtains for isotropy body shift-equations of the type:

$$(G+P)\nabla^2 u + 2P\frac{d\theta}{dx} + (A+K)\nabla^2(\nabla^2 u) + 4K\nabla^2\left(\frac{d\theta}{dx}\right) + X = 0,$$

where,

$$G = \tfrac{1}{2}\Sigma m f(r)\, r\cos^2\alpha$$

$$P = \tfrac{1}{2}\Sigma m F(r)\, r^2\cos^2\beta\cos^2\gamma$$

$$A = \tfrac{1}{24}\Sigma m f(r)\, r^3\cos^4\alpha$$

$$K = \tfrac{1}{24}\Sigma m F(r)\, r^4\cos^2\beta\cos^4\gamma = \tfrac{1}{24}\Sigma m F(r)\, r^4\cos^2\lambda\cos^4\beta,$$

where $F(r) = r\dfrac{d}{dr}\left(\dfrac{f(r)}{r}\right)$, $f(r)$ being the law of inter-molecular central action, and α, β, γ the direction angles of the molecule m at distance r. If the body be not subjected to initial stresses, G and A are both zero. In this case the equation above agrees with that which may be deduced from Saint-Venant's values of the stresses: see our Art. 234.

The rest of the discussion—notwithstanding the author's claim on p. 358—does not seem to me to offer any novelty. Dienger concludes with a remark as to Cauchy's explanation of dispersion, which he considers a failure as it would apply to 'empty space'. The promise to explain dispersion in a perfectly natural manner in a later memoir does not seem to have been fulfilled.

[550.] L. F. Ménabréa. *Études sur la théorie des Vibrations. Memorie della Reale Accademia delle Scienze,* T. XV., 1855, pp. 205—329. Turin, 1855. The memoir was read June 12, 1853 and published as an offprint in 1854. It commences with a general discussion of the stability and small oscillations of a slightly disturbed group of particles. This occupies pp. 205—225 and the author draws particular attention to the best mode of integrating the equations which arise. The general discussion is followed by special problems, which introduce elastic bodies as limiting cases. Thus a heavy flexible string is treated as the limiting case of a number of particles united by weightless inextensible strings, or again a rod as the limiting case of a number of heavy particles united by rigid links which resist being displaced about their extremities by forces proportional to the angles the adjacent links make with one another. In this method the limits of finite difference equations become the differential equations for the vibrations of elastic strings, rods, membranes etc. The method is due to Lagrange and is used freely in the *Mécanique Analytique.* Examples of it will be found in Lord Rayleigh's *Theory of Sound* Vol. I. § 120, or in Routh's *Rigid Dynamics* 3rd Edition, § 486. It cannot be considered entirely satisfactory as it often involves somewhat arbitrary hypotheses: as for example, in the case of the transverse vibrations of the rod referred to above.

[551.] The following are the contents of the memoir as far as it relates to special cases:

(*a*) Pp. 226—232: Oscillations of a particle attracted by several fixed centres of force; (*b*) pp. 232—7: Vibratory motion of a flexible string carrying two heavy particles, the string being fixed at one end only; (*c*) pp. 238—73: Vibratory motion of a string fixed at one end and carrying several heavy particles;—this is subdivided into several parts dealing with strings whose parts are not homogeneous, etc.; (*d*) pp. 273—284: Longitudinal vibrations of an elastic rod or string loaded with particles at different points; (*e*) pp. 285—291: Vibrations of a flexible and inextensible string fixed at its terminals and forming a curve under the action of forces distributed along its length; Ménabréa arrives at formulae agreeing with those given by Navier on p. 163 of the work referred to in our Art. 272*; (*f*) pp. 292—297: Vibrations of a funicular polygon formed of flexible and extensible elements; (*g*) pp. 297—306: Transverse vibrations of a rod composed of diverse heterogeneous parts, or having various heavy particles attached to it;

Ménabréa besides obtaining the general equation of the vibrations of a rod with a longitudinal tension T, namely:

$$F\frac{d^4y}{dx^4} = T\frac{d^2y}{dx^2} - m\frac{d^2y}{dt^2} \text{ (see our Art. 471),}$$

(where F is here a constant depending on the material of the rod, which remains undefined owing to the vagueness of the hypothesis adopted), also indicates the solution of the following problem:

A rod is clamped at its upper end; a particle, the weight of which is great relative to that of the rod, is attached to its lower end; to find the motion of the system when set vibrating (pp. 301—2).

The solution is not carried far enough to be of service in dealing with Kupffer's empirical formula for the like case: see Section II. of the present Chapter.

(*h*) Pp. 307—311: Vibrations of a plane rectangular flexible membrane uniformly stretched and composed of two parts of different material; (*i*) pp. 312—22: Radial vibrations of a homogeneous elastic sphere. The results in this case agree with those given by Poisson in his memoir of 1828: see our Art. 449* *et seq.*; (*j*) pp. 323—7: Note on the theory of light; Ménabréa deduces Fresnel's equations from his general theory of a particle oscillating under the action of several fixed centres of force.

The memoir as a whole contains no new results, but there are some interesting and suggestive analytical processes.

[552.] E. Phillips: *Calcul de la résistance des poutres droites, telles que les ponts, les rails, etc., sous l'action d'une charge en mouvement. Annales des Mines*, Tome VII, 1855, pp. 467—506.

This is the important memoir to which we have referred in our Arts. 372—82.

The memoir is divided into three chapters. *Chapitre I.* (pp. 468—87) is entitled: *Des poutres encastrées par leurs deux extrémités,* and it deals with the case of a load crossing with any given velocity a straight beam of uniform cross-section doubly built-in. This problem is not of very much importance, for it is difficult to really build-in the terminals of a girder, and when done there arise several practical disadvantages. Phillips' analysis is only approximate, the deflection being expanded in powers of the distance from a terminal, and the coefficients of these powers being given by rather lengthy series in powers of the time. These series are simplified by the assumption that $m/(E\omega\kappa^2)$ is a small quantity, where m is the mass of the beam per foot-run and $E\omega\kappa^2$ its flexural rigidity; only first powers of this expression are then retained.

I have not verified Phillips' analysis and his results as given on pp. 480—6 are too lengthy for citation.

[553.] *Chapitre II.* entitled: *Des poutres reposant librement sur deux appuis*, deals with the like problem for simply-supported terminals (pp. 487—500). The analysis has for practical purposes been much simplified by Saint-Venant, and as the latter has corrected an error of Phillips we merely refer the reader to our discussion of the problem in Art. 372—6.

In both the cases dealt with in these chapters Phillips does not satisfy the initial condition that the velocity of all parts of the girder shall be zero, before the load comes upon it. In the case of the doubly built-in girder, however, the initial velocity given by the approximate solution is of the order $1/\beta^2 \equiv m/E\omega\kappa^2$ and is therefore very small. For the doubly-supported girder the terms neglected are of the order $Vl/(3\beta)$, where l is the length of the girder and V the velocity of the travelling load.

In order to ascertain the real effect of this initial velocity Phillips supposes the bridge to remain without load and to start from a position of rest with an initial velocity exactly equal to that which must be neglected in his problem. He finds that this initial system of velocities would produce a maximum deflection occurring almost at the centre of the bridge and given with sufficient approximation by

$$\frac{2}{\pi\beta}\left(\frac{l}{\pi}\right)^4 \frac{QV}{E\omega\kappa^2},$$

where Q is the weight of the travelling load.

The ratio of this deflection to the maximum deflection at the centre is very nearly

$$\frac{Vl}{3\beta},$$

while the corresponding curvatures (d^2y/dx^2) have very nearly the ratio

$$\frac{Vl}{4\beta}.$$

Phillips then shows that for four actual bridges with a load moving at 108 kilometres per hour the former of these quantities does not exceed 1/20, and they are thus in practice negligible.

[554.] *Chapitre III.* (pp. 500—6) is entitled: *Conséquences pratiques de la théorie précédente et son extension à d'autres pro-*

blèmes. It deals first with the case of the doubly-supported girder. The conclusions drawn with regard to the curvature and stretch are involved in the results of our Arts. 372—7, where following Saint-Venant we have corrected Phillips' numerical error.

In the latter part of this chapter (pp. 503—6) Phillips deals with the case of the doubly built-in girder. He shows that except just when the load is coming on to or leaving the bridge the maximum curvature at the instant is immediately under the load, and that the maximum maximorum takes place when the load reaches the centre of the beam; we have then[1]:

$$1/\rho = \frac{Ql}{8E\omega\kappa^2}\left(1 + \frac{QV^2l}{8E\omega\kappa^2 g}\right) + \frac{Pl}{24E\omega\kappa^2}\left(1 + \frac{QV^2l}{4E\omega\kappa^2 g}\right).$$

Evidently then the magnitude of the fraction $\dfrac{QV^2l}{8E\omega\kappa^2 g}$ determines the influence of the speed of the travelling load on the deflection. Phillips takes the case of a rail one metre long and for which the rigidity is 197,600 (sq. metre kilogrammes?) and Q is 6000 (kilogrammes? he has kilometres). The value of the fraction is then about ·35 for a velocity of 108 kilometres per hour, and about ·16 for one of 72.

He remarks in conclusion:

> Dans tous les cas de la pratique, ce qu'il y aura de plus simple à faire est ceci. Comme il faut toujours que la poutre puisse supporter la charge au repos, on commencera par calculer les dimensions de cette poutre en conséquence, d'après les règles ordinaires. Puis, l'on vérifiera si la quantité $QV^2l/(3E\omega\kappa^2 g)$ dans le cas de la poutre appuyée librement par ses deux extrémités, ou $QV^2l/(8E\omega\kappa^2 g)$ dans le cas de la poutre encastrée par ses deux bouts, est assez petite. Dans ce dernier cas, et si la charge permanente n'était pas négligeable, il faudrait en outre que $QV^2l/(4E\omega\kappa^2 g)$ fût une petite fraction. Dans le cas où ces diverses fractions ne seraient pas assez petites, on diminuerait l'écartement des points extrêmes ou l'on augmenterait le moment d'élasticité de la poutre jusqu'à ce que la fraction dont il s'agit devienne négligeable (p. 505).

[555.] Kopytowski: *Ueber die inneren Spannungen in einem freiaufliegenden Balken unter Einwirkung beweglicher Belastung.* 4to., Göttingen, 1865. This is an inaugural dissertation for the doctor's degree and contains 88 pages with a plate of figures. Although falling somewhat outside the period we are considering this memoir is so closely related to that of Phillips dealt with in the three previous articles that we may briefly touch upon it here. The author deals with the case of a uniform beam terminally supported which is crossed by a continuous travelling load. He refers in his preface to the labours in this field of Navier, Willis,

[1] I have not verified the analysis by which Phillips reaches this result but have slightly modified the numerical results which follow.

Stokes, Phillips and Renaudot: see our Arts. 1276*, 1417*, and 372—82, 540, 552—4; but he does not note the errors of the last two writers and falls into similar ones himself. He proposes in his preface to extend the results of the last writer, especially by considering the stresses acting at all parts of the beam. He assumes that both the flexure and the resulting system of stresses are uniplanar (pp. 5—7).

[556.] The memoir may be divided into two parts. The first occupies pp. 7—43 and considers the bending moment, total shear on a cross-section and principal tractions at any point of the beam, when its weight is taken into account and the continuous load is supposed merely to act *statically* as it crosses the beam. Thus pp. 7—10 give the usual Bernoulli-Eulerian theory with such results as that the total shear is the slope of the bending-moment curve. Pp. 10—17 give a theory of uniplanar stress which is practically a reproduction of Rankine's treatment of the like problem in his memoir *On the Stability of Loose Earth* or in his *Applied Mechanics*: see our Arts. 453 and 465, (*b*). There are several misprints in the results on p. 17. Pp. 17—23 investigate the principal tractions on the assumption that the stress system in a beam under flexure is uniplanar. Let x be the direction of the axis of the beam, y that of the horizontal neutral axis, and z the vertical in the plane of the cross-section. Then Kopytowski assumes that only the stresses $\widehat{xx}$ and $\widehat{xz}$ have finite values and that these stresses are the same for all points on the cross-section at the same distance from the neutral axis. Thus the whole of his reasoning on p. 19 is fallacious unless $\widehat{xz}$ is uniform along a horizontal parallel to the neutral axis. But Saint-Venant has shown that this is certainly not true for an isolated load, for in that case $\widehat{xz}$ varies right across the section; further the stress $\widehat{yx}$ is not generally negligible as compared with $\widehat{xz}$ but may be of the same order. Like results have been shown by the Editor of the present work to hold for a heavy beam continuously loaded, which is Kopytowski's own case[1]. Thus his application of uniplanar stress to determine the principal tractions in a beam under flexure—a method which is practically identical with Jouravski's—is fallacious both on the ground of the supposed uniformity of $\widehat{xz}$ and also in the neglect of $\widehat{yx}$. The results on p. 21 possess therefore no more exactitude than they would have, if we put $\widehat{xz}$ (or, Kopytowski's Θ_z) $= 0$, or reduced the system to a single principal traction, i.e. the longitudinal traction. The only exception to this seems to be the case of the extremely thin web of a girder of T or I cross-section.

On pp. 23—30 expressions are deduced on Kopytowski's assumptions for the principal tractions in a heavy beam partially covered with a continuous load. I have not investigated these results with

[1] *Quarterly Journal of Mathematics*, vol. xxiv., 1889, pp. 63—110.

the view of recording possible misprints or errors of calculation, as they seem to me for the reasons stated above valueless. The same remark applies to the numerical tables III.—V. on pp. 38—40.

[557.] Besides the principal tractions Kopytowski investigates the values of the bending moment and total shear at any cross-section when a given arbitrary length ξ of the beam is covered by a uniform continuous load. For the maximum bending moment he finds that the beam must be totally covered by the continuous load, and for the maximum shear that the load must cover only the longer portion of the beam from the cross-section to a terminal, both results previously well known. Kopytowski calculates, however, the magnitude and situation of the greatest bending moment, and the value of the total shear at various cross-sections when any given portion ξ of the beam is covered by the continuous load. His results on this point may possess some novelty: see his pp. 23—7 and Tables I—II., pp. 35—6. I have neither tested their accuracy, nor that of Table VI. (p. 41) containing the deflections at the several points of the beam for various positions of the continuous load, because these results seem to me neither of real practical value nor of any special theoretical interest.

[558.] The second part of the memoir pp. 43—88 deals with Renaudot's problem of the influence of a rapidly travelling continuous load on the deflection and stresses in a beam terminally supported. Kopytowski generalises the equations by introducing terms depending on the angular motion of the cross-sections. These terms would in most practical cases be negligible. But our author while introducing these terms drops out another really important term in his Equation (34) on p. 45[1], namely in his notation the term $\dfrac{-2PV}{P+p}\dfrac{d^2y}{dx\,dt}$ on the right-hand side. Thus multiplying up by $P+p$, his equation ought to be:

$$p\frac{d^2y}{dt^2} = g(P+p) - \epsilon g\frac{d^4y}{dx^4} - P\left(V^2\frac{d^2y}{dx^2} + 2V\frac{d^2y}{dx\,dt} + \frac{d^2y}{dt^2}\right) + pr^2\frac{d^4y}{dx^2\,dt^2}.$$

That the term in question does not appear in Kopytowski's equation is due to the singular process by which he deduces it, i.e. he apparently equates the vertical acceleration of a point on the axis of the beam to that of the point of the continuous load *instantaneously* above it. Renaudot introduces this term only to drop it as 'small.' As a matter of fact it is of the same order as the terms retained. Kopytowski solves his equations in the approximate manner suggested by Phillips: see our Art. 552, and follows Phillips very closely in his method of showing that the fact that the initial conditions are not exactly satisfied does not for practical purposes invalidate the solution (pp. 46—63 and pp. 69—72). The whole of his discussion, however, in order to be made of value would require to be modified by the introduction of additional terms depending on the term noted above as

[1] This page abounds with misprints.

omitted in the differential equation. Pp. 63—68 are a reproduction of Bresse's problem (see our Arts. 382 and 540) without, however, any acknowledgment of the source from which the material is drawn.

[559.] Pp. 72—78 return to Renaudot's problem and calculate the bending moment at any point for any position of the load, and also the maximum bending moment. The latter value agrees with Renaudot's, but is wrong. The coefficient of the term $PV^2l^2/(\epsilon g)$ in Equation (65) of p. 78 should be 5/32 and not 1/6.

The expressions for the total shear (pp. 78—80), the terminal reactions, as well as the maximum total shear at the middle of the beam will also be wrong, so far as the numerical coefficients of the terms in $PV^2l^2/(\epsilon g)$ are concerned. I have not, however, recalculated these coefficients. The values of the principal tractions given in Equation (69) of p. 80 are again erroneous for the reasons given in Art. 556, and that for the deflection (pp. 81—2) has also a wrong coefficient. The same remarks apply to the numerical results on pp. 83—5 and p. 88.

[560.] On pp. 85—7, Table VIII., we have the values of $4/\beta'$, where β' is the constant of our Art. 381, calculated for a certain number of actual bridges. This discussion and table might have been of considerable value had not Kopytowski introduced what seems a very doubtful hypothesis into his calculations; he assumes, namely, that in each case the moment of inertia of the cross-section has been designed so as just to carry without failure the weight of the beam and the continuous load considered as acting *statically*. Thus, suppose $2l$ the length of the beam, p the weight per unit-run of the beam and p' that of the load. Then the maximum statical bending moment at the centre when the beam is fully loaded is:

$$M = \tfrac{1}{2}(p' + p)\, l^2.$$

Let h be the vertical diameter of the beam and $E\omega\kappa^2$ its flexural rigidity, then Kopytowski also equates this to

$$\frac{2}{h} T_0 \,\omega\kappa^2,$$

T_0 being the traction which corresponds to the fail-limit of the material. Hence he finds to determine $\omega\kappa^2$:

$$\omega\kappa^2 = \tfrac{1}{4}\frac{p' + p}{T_0}\, l^2 h,$$

or, substituting in $1/\beta'$ of our Art. 381:

$$1/\beta' = \frac{4T_0}{Eg}\, \frac{p'V^2}{(p+p')\,h}.$$

See the memoir pp. 74 and 85.

From this formula Kopytowski calculates $4/\beta'$ for the Britannia and Conway bridges and for bridges near Bordeaux, Bern, St Gallen etc. But the usefulness of his results seems to me vitiated because there is no sufficient reason for supposing that the moment of the cross-section of any of these bridges really has the value which is found by this process.

On p. 87 some remarks occur on the experiments of the *Iron Commissioners* and on Stokes' value for the deflection in the case of an isolated load: see our Arts. 1417* and 1287*. The memoir is a rather more ambitious than satisfactory piece of work.

[561.] H. Resal: *Thèse de Mécanique. Sur les équations polaires de l'élasticité et leur application à l'équilibre d'une croûte planétaire*, Paris, 1855.

This is an academical dissertation occupying 40 quarto pages and dealing with a special case of Lamé's memoir of 1854: see our Art. 1111*. It is reproduced on pp. 395—440 of the *first* edition of Resal's *Traité élémentaire de mécanique céleste* (Paris, 1865) with some of the misprints corrected. As the latter work is more readily accessible than the *Thèse*, our references are to its pages.

Pp. 395—411 are occupied with an investigation of the equations of elasticity in spherical coordinates. Resal adopts uni-constant isotropy, noticing, however, that Wertheim's experiments do not seem to be in complete accordance with the relation $\lambda/\mu = 1$. He rather weakly remarks:

Dans l'incertitude ou nous nous trouvons sur la valeur de ce rapport, dont la connaissance est indispensable pour pouvoir calculer λ et μ en fonctions du *coefficient d'élasticité*, la seule constante que l'on a l'habitude de faire entrer dans les questions de résistance des matériaux, nous avons cru devoir continuer à admettre la relation théorique $\lambda = \mu$, trouvée par MM. Navier, Poisson et Cauchy (footnote, p. 404).

There is no novelty in this part of Resal's investigation except, I think, his application in a footnote (pp. 402—4) of Cauchy's *fonctions isotropes* to determine a relation between the elastic constants in the expressions for the stresses. The method does not seem to present any advantages.

[562.] On p. 411 we have Resal's problem stated: "To determine the elastic equilibrium of the spherical crust of a planet, rotating round a diameter, under the action of the mutual gravitation of its parts and subjected to uniform internal and external normal pressures."

This problem may be termed *Resal's Problem* although as we have seen a portion of it had already been considered by Lamé.

Symmetry shows us at once that the shifts lie entirely in the meridian-plane, or reduce to u and v in the notation of our footnote on p. 79. Now these shifts may be divided into two parts $u' + u''$ and $v' + v''$ where u' and v' are due to the radial surface-forces and body-forces (i.e. pure gravity), while u'' and v'' are due to the so-called 'centrifugal force.' Now it is easily seen that we must have $v' = 0$. The value of u' was determined by Lamé for *bi-constant* isotropy in his *Leçons*: see our Arts. 1094*—5* and compare Arts. 1114*—8*, where it is shown that Lamé made some progress towards the solution of *Resal's Problem.*

[563.] The following are the values at distance r from the centre obtained by Lamé's method for u', for the radial traction $\widehat{rr}'$, and for $\widehat{\phi\phi}'$ ($= \widehat{\psi\psi}'$ in the notation of our p. 79) the meridian traction corresponding to the shift u':

$$u' = \frac{\gamma}{5} r^3 + \frac{r}{3\lambda + 2\mu} \left\{ \frac{r_0^3 p_0 - r_1^3 p_1}{r_1^3 - r_0^3} - (\lambda + \tfrac{6}{5}\mu)\,\gamma \frac{r_1^5 - r_0^5}{r_1^3 - r_0^3} \right\}$$
$$+ \frac{1}{4\mu r^2} \left\{ \frac{r_0^3 r_1^3 (p_0 - p_1)}{r_1^3 - r_0^3} - (\lambda + \tfrac{6}{5}\mu)\,\gamma \frac{r_0^3 r_1^3 (r_1^2 - r_0^2)}{r_1^3 - r_0^3} \right\} \ldots\ldots\ldots \text{(i)},$$

$$\widehat{rr}' = - \frac{r_0^3 p_0 (r_1^3 - r^3) + r_1^3 p_1 (r^3 - r_0^3)}{r^3 (r_1^3 - r_0^3)} + (\lambda + \tfrac{6}{5}\mu) \frac{\Pi\gamma}{r^3 (r_1^3 - r_0^3)} \ldots\ldots\ldots\ldots \text{(ii)},$$

$$\widehat{\phi\phi}' = \widehat{\psi\psi}' = \frac{r_0^3 p_0 - r_1^3 p_1}{r_1^3 - r_0^3} + \frac{r_0^3 r_1^3 (p_0 - p_1)}{2r^3 (r_1^3 - r_0^3)} - \tfrac{4}{5}\mu\gamma r^2$$
$$- \frac{(\lambda + \frac{6}{5}\mu)\,\gamma}{r^3 (r_1^3 - r_0^3)} \{ \tfrac{3}{2} (r_1^2 - r_0^2)\, r_0^3 r_1^3 - \Pi \} \ldots\ldots\ldots \text{(iii)},$$

where: $\Pi = r^5 (r_1^3 - r_0^3) - r^3 (r_1^5 - r_0^5) + r_0^3 r_1^3 (r_1^2 - r_0^2)$ and is divisible by $(r_1 - r_0)(r - r_0)(r - r_1)$,

p_0 and p_1 = the internal and external pressures at the surfaces of the shell of radii r_0 and r_1 respectively,

$\gamma = \dfrac{\rho g}{2(\lambda + 2\mu) r_1}$, ρ being the density, and

g = gravitational acceleration at the outer surface.

The value of $\widehat{\phi\phi}'$ given above does not agree with Lamé's F (p. 216 of the *Leçons*). The coefficient of Π in his expression should be $-2(\lambda + \frac{2}{5}\mu)$ and not $2(\lambda + 2\mu)$ as he has it. The form we have obtained for $\widehat{\phi\phi}'$ is also more convenient for further calculations than Lamé's. Resal obtains a value for $\widehat{rr}'$ agreeing with ours when $\lambda = \mu$, he does not write down the general value of $\widehat{\phi\phi}'$ or u'.

[564.] Both Lamé and Resal proceed to approximations in the special case of a thin crust. We shall examine the true approximations somewhat closely, as it appears that both Lamé and Resal have fallen into error.

Let us put $r_1 = r_0(1+e), \quad r = r_0(1+\epsilon)$

and suppose the squares and products of e and ϵ to be small.

We find : $$\widehat{rr}' = -p_0 + \frac{\epsilon}{e}(p_0 - p_1)\{1 + 2(e-\epsilon)\} \qquad \text{(iv)}.$$

The lowest term containing g as a factor is of the second order and its value is

$$\frac{5\lambda + 6\mu}{2(\lambda+2\mu)} g\rho r_0 \epsilon(\epsilon - e).$$

Further :

$$\widehat{\phi\phi}' = -p_1 + (p_0 - p_1)\left\{\frac{1}{2e} - \frac{\epsilon}{2e} + \frac{e}{3} - \epsilon + \frac{\epsilon^2}{e}\right\}$$
$$- \frac{g\rho r_0}{2}\left\{1 - \frac{3\lambda+2\mu}{\lambda+2\mu}\left(\epsilon - \frac{e}{2}\right)\right\} \qquad \text{(v)}.$$

The value of u' to the same degree of approximation is :

$$u' = (p_0 - p_1) r_0 \left\{\frac{\lambda+2\mu}{4\mu(3\lambda+2\mu)}\frac{1}{e} + \frac{\lambda+2\mu}{2\mu(3\lambda+2\mu)} + \frac{\lambda+2\mu}{6\mu(3\lambda+2\mu)}e\right.$$
$$\left. - \frac{\lambda}{\mu(3\lambda+2\mu)}\epsilon - \frac{\lambda}{2\mu(3\lambda+2\mu)}\frac{\epsilon}{e} + \frac{1}{4\mu}\frac{\epsilon^2}{e}\right\}$$
$$- \frac{g\rho r_0^2}{2}\left\{\frac{\lambda+2\mu}{2\mu(3\lambda+2\mu)} + \frac{e}{4\mu} \frac{\lambda}{\mu(3\lambda+2\mu)}\epsilon\right\}$$
$$- \frac{p_0 r_0(1+\epsilon)}{3\lambda+2\mu} \qquad \text{(vi)}.$$

[565.] If we neglect the products of p_0 or p_1 with e or ϵ, as both Resal and Lamé appear to do, we have on rearranging :

$$u' = \frac{(\lambda+2\mu) r_0}{4\mu(3\lambda+2\mu)}\left(\frac{p_0-p_1}{e} - g\rho r_0\right) + (p_0-p_1) r_0\left(1 + \frac{2\mu}{\lambda} - \frac{\epsilon}{e}\right)\frac{\lambda}{2\mu(3\lambda+2\mu)}$$
$$- \frac{g\rho r_0^2}{2}\left(\frac{e}{4\mu} - \frac{\lambda}{\mu(3\lambda+2\mu)}\epsilon\right) - \frac{p_0 r_0}{3\lambda+2\mu} \qquad \text{(vii)}.$$

Putting successively $\epsilon = 0$ and $\epsilon = e$ we find :

$$u_0' = \frac{(\lambda+2\mu) r_0}{4\mu(3\lambda+2\mu)}\left(\frac{p_0-p_1}{e} - g\rho r_0\right) + \frac{(p_0-p_1) r_0(\lambda+2\mu)}{2\mu(3\lambda+2\mu)}$$
$$- \frac{g\rho r_0^2 e}{8\mu} - \frac{p_0 r_0}{3\lambda+2\mu} \qquad \text{(viii)},$$

$$u_1' = \frac{(\lambda + 2\mu) r_0}{4\mu(3\lambda + 2\mu)}\left(\frac{p_0 - p_1}{e} - g\rho r_0\right) + \frac{(p_0 - p_1) r_0}{3\lambda + 2\mu}$$
$$- \frac{g\rho r_0^2 e}{8\mu}\left(1 - \frac{4\lambda}{3\lambda + 2\mu}\right) - \frac{p_0 r_0}{3\lambda + 2\mu} \quad \text{.........(ix)},$$

and,
$$u_0' = u_1' - \frac{g\rho r_0^2 e}{2}\frac{\lambda}{\mu(3\lambda + 2\mu)} + \frac{(p_0 - p_1) r_0 \lambda}{2\mu(3\lambda + 2\mu)}$$
$$= u_1' + \{(p_0 - p_1) r_0 - g\rho r_0^2 e\}\frac{\lambda}{2\mu(3\lambda + 2\mu)} \quad \text{.........................(x)}.$$

The results (vii)—(ix) differ widely from those given by Lamé and Resal. The equation (x) agrees with one given by the latter author (p. 417) if we put $\lambda = \mu$. The values given by them for the shifts seem to be erroneous.

[566.] Turning to the tractions, our formula (v) gives:

$$\widehat{\phi\phi_1}' = \frac{1}{2e}\{(p_0 - p_1) - g\rho r_0 e\}$$
$$- \frac{p_0 + p_1}{2} + e\left\{\frac{p_0 - p_1}{3} + \frac{3\lambda + 2\mu}{\lambda + 2\mu}\frac{g\rho r_0}{4}\right\} \quad \text{.......(xi)},$$

and:
$$\widehat{\phi\phi_0}' = \widehat{\phi\phi_1}' + \frac{p_0 - p_1}{2} - e\,\frac{3\lambda + 2\mu}{\lambda + 2\mu}\frac{g\rho r_0}{2} \quad \text{..............(xii)}.$$

Lamé (*Leçons*, p. 217) has in our notation the results

$$\widehat{\phi\phi_1}' = \frac{1}{2e}\{p_0 - p_1 - g\rho r_0 e\},$$

$$\widehat{\phi\phi_0}' = \widehat{\phi\phi_1}' - e\,\frac{\lambda}{\lambda + 2\mu} g\rho r_0.$$

It is not obvious without further discussion why in the case of a planetary crust $\frac{1}{2}(p_0 + p_1)$ should be neglected as compared with $\frac{1}{2} g\rho r_0$. The second equation is wrong unless we suppose $e\,\widehat{\phi\phi_1}'$ necessarily negligible.

Resal (*Mécanique Céleste*, p. 417) gives the same value as Lamé for $\widehat{\phi\phi_1}'$, but for $\widehat{\phi\phi_0}'$ he has

$$\widehat{\phi\phi_0}' = \widehat{\phi\phi_1}' - \tfrac{5}{8} g\rho r_0 e.$$

Thus he agrees with the third term on the right of our equation (xii) in the coefficient of $g\rho r_0 e$, since he puts $\lambda = \mu$, but he disagrees with Lamé. Like Lamé he appears to have dropped entirely the term $\frac{1}{2}(p_0 - p_1)$ and I see no reason for this.

Resal (p. 416) gives for $\widehat{rr}'$ the value

$$\widehat{rr}' = -p_0 + \frac{\epsilon}{e}(p_0 - p_1).$$

This neglects the term $2(e - \epsilon)\frac{\epsilon}{e}(p_0 - p_1)$ of equation (iv). If terms involving e, like $g\rho r_0 e$, are retained in $\widehat{\phi\phi}'$ this does not seem legitimate. If Lamé and Resal suppose $p_0 - p_1$ and $g\rho r_0 e$ to be of the same order then this would be allowable, but this would still compel them to retain the term $\frac{1}{2}(p_0 - p_1)$ they have cast out of (xii).

[567.] Both Lamé and Resal apply these results to the structure of the earth; they thus initiated those investigations in terrestrial physics, which have been still further advanced by Sir William Thomson, G. Darwin, Chree and others. Resal closely follows Lamé without, however, so much explanatory statement. Their whole investigation of rupture at the earth's surface is based upon the assumption that rupture takes place where the shear or traction is a maximum. They thus endeavour to explain geological faults. We may note the general drift of their reasoning, modifying it slightly to suit our formulae, as it will be useful for comparison with later work.

(*a*) Lamé remarks (p. 218) that geologists (i.e. those of his day) considered that the thickness of the crust could not be more than $\frac{1}{150}$ of the radius, or $e = \frac{1}{150}$. Hence to a first approximation from (xi),

$$\widehat{\phi\phi}_1' = 75\{(p_0 - p_1) - g\rho r_0 e\}.$$

If therefore $p_0 - p_1$ were not very nearly equal to $g\rho r_0 e$, there would be a very sensible horizontal traction at the surface of the earth. There is nothing to show the existence of such stress and accordingly Lamé supposes $p_0 - p_1 = g\rho r_0 e$ very nearly. This obviously means that the difference of the surface pressures just supports the weight of the crust. If it were exactly true we should have to a first approximation $u_0' = u_1'$, or the earth would retain the original thickness of its crust[1]. If this relation holds we have also from (xii):

$$\widehat{\phi\phi}_0' = -\frac{\lambda}{\lambda + 2\mu} g\rho r_0 e,$$

or the meridian stress at the inner surface of the crust is a pressure.

If $\widehat{\phi\phi}_1'$ is negative, which it must become in the course of time as p_0 diminishes and e increases, then $\widehat{\phi\phi}_0'$ is a still greater pressure.

(*b*) Both Lamé and Resal use the stress-quadric

$$\frac{x^2}{\widehat{rr}'^2} + \frac{y^2 + z^2}{\widehat{\phi\phi}'^2} = 1,$$

and the shear-cone

$$\frac{x^2}{\widehat{rr}'} + \frac{y^2 + z^2}{\widehat{\phi\phi}'} = 0$$

to determine the direction of shearing rupture and the magnitude of the shearing force. They suggest how the magnitude of this force and its direction may be found by experiment and observation of faults.

[1] Lamé (p. 220) puts in this case $u_0' = u_1' = 0$, which arises from the error in his equation corresponding to our (vii).

Their reasoning seems very doubtful; it is not consistent with the more probable hypothesis that the surfaces of rupture are perpendicular to the directions of greatest stretch. It is easy to see that the principal term in the radial stretch (du'/dr) is negative so long as $p_0 - p_1 - g\rho r_0 e$ is positive, and that the meridian stretch (u'/r) is under the same conditions positive. Thus till p_0 gets so small that $p_0 - p_1 - g\rho r_0 e$ becomes negative, rupture will most probably occur in planes vertical to the earth's surface, but afterwards rupture may occur by the crust breaking up into spherical shells. Geological faults possibly arise owing to some inequality of radial pressure after such rupture.

(c) Lamé assumes the value of u' to a first approximation,—namely

$$u' = \frac{\lambda + 2\mu}{4\mu(3\lambda + 2\mu)} \left\{ \frac{p_0 - p_1}{\tau} r_0^2 - g\rho r_0^2 \right\},$$

where $\tau = r_0 e =$ absolute thickness of crust (see equation vii),—to be still true for a spheroidal earth, when we put for r_0 the distance of any point on the crust from the centre of the spheroid. Thus he supposes u_1' and u_2' to be the values of u' corresponding to the values of r_0, r_1 and r_2 say, in Brittany and Sweden. He puts g equal to its values g_1 and g_2 in those two places and neglecting the effect of rotation of the earth on g, he has $g_1/g_2 = r_2^2/r_1^2$ and consequently:

$$u_1' - u_2' = \frac{\lambda + 2\mu}{4\mu(3\lambda + 2\mu)} \frac{p_0 - p_1}{\tau} (r_1^2 - r_2^2).$$

Now r_1 is $> r_2$; hence as p_0 decreases ($p_0 - p_1$ being positive) and τ increases, $u_1' - u_2'$ must diminish, or we should expect the surface of the earth in Brittany to be falling as compared with that in Sweden. This is certainly the case in parts, but whether the method by which the conclusion is reached is valid is another matter. In Brittany there are submarine forests, while recent shells are found in Sweden much above the Baltic level (see Lamé, *Leçons*, p. 221)[1].

[568.] While Lamé in his work merely supposes the effect of centrifugal force to make a slight variation in the value of the gravitational term, Resal has independently investigated this effect. He considers (pp. 419—440) a spherical shell without internal or external pressures rotating about a diameter. By adding the results to those of the preceding articles we can obtain the solution of the most general case. The problem is of course only a special case of *Lamé's Problem* (see our Art. 1111*) but it is one possessing considerable interest for both physicist and geologist.

[1] Captain A. P. Madsen holds that in parts of Jutland the land has risen 20 feet since the Stone Age: see *Nature*, vol. 40, 1889, p. 108. Other northern districts are supposed however to have recently sunk: see Geikie's *Text-book of Geology*, pp. 280, 283—4.

Let u'' as above be the radial shift, and let v'' be the meridional shift towards the pole in latitude ϕ; let ω be the spin about the polar axis, ρ the density of the rotating shell or crust, r_0, r_1 its internal and external radii, and let $r_1 = r_0 + \epsilon$. Then the equations of our Art. 1112* readily give us:

$$\left.\begin{aligned}
&(\lambda + 2\mu)\, r \frac{d\theta}{dr} + \frac{\mu}{r\cos\phi}\frac{d\,(Q\cos\phi)}{d\phi} = -\rho\omega^2 r^2 \cos^2\phi,\\
&(\lambda + 2\mu)\frac{d\theta}{d\phi} - \mu\frac{dQ}{dr} = \rho\omega^2 r^2 \sin\phi\cos\phi,\\
\text{where}\quad & Q = \frac{du}{d\phi} - \frac{d\,(rv)}{dr},\\
\text{and}\quad & \theta = \frac{1}{r^2}\frac{d\,(r^2 u)}{dr} + \frac{1}{r\cos\phi}\frac{d\,(v\cos\phi)}{d\phi}
\end{aligned}\right\}\ldots\ldots\ldots\text{(i)}.$$

Particular solutions of these are given by:

$$\left.\begin{aligned}
&\theta_0 = -\tfrac{7}{2} a\, r^2 \cos^2\phi,\\
&u_0'' = ar^3\left(\tfrac{1}{5} - \cos^2\phi\right),\\
&v_0'' = \tfrac{1}{2} ar^3 \sin\phi\cos\phi,\\
\text{where}\quad & a = \frac{\rho\omega^2}{7\,(\lambda + 2\mu)}
\end{aligned}\right\}\ldots\ldots\ldots\ldots\ldots\ldots\ldots\ldots\text{(ii)}.$$

For the general solution assume:

$$u'' = a_0 r + b_0 r^{-2} + \left(\sin^2\phi - \tfrac{1}{3}\right)\left(a_1 r + b_1 r^{-4} + a_2 r^3 + b_2 r^{-2}\right)\ldots\ldots\text{(iii)}.$$

In order to satisfy (i) we easily find:

$$v'' = \sin\phi\cos\phi\left(a_1 r - \tfrac{2}{3} b_1 r^{-4} + \frac{5\lambda + 7\mu}{3\lambda} a_2 r^3 + \frac{2\mu}{3\lambda + 5\mu} b_2 r^{-2}\right)\ldots\text{(iv)},$$

and

$$\theta = 3a_0 + \left(\sin^2\phi - \tfrac{1}{3}\right)\left(-\frac{7\mu}{\lambda} a_2 r^2 - \frac{6\mu}{3\lambda + 5\mu} b_2 r^{-3}\right)\ldots\ldots\ldots\text{(v)}.$$

Now at the surfaces of the shell we must have $\widehat{\phi r}''$ and $\widehat{rr}''$ zero, i.e. they vanish for all values of ϕ when $r = r_0$ and $r = r_1$.

$$\left.\begin{aligned}
\text{But}\quad \widehat{\phi r}'' = {}& 2\mu\cos\phi\sin\phi\left\{\tfrac{3}{2} ar^2 + a_1 + \tfrac{8}{3} b_1 r^{-5}\right.\\
&\left.\qquad + \frac{8\lambda + 7\mu}{3\lambda} a_2 r^2 + \frac{3\lambda + 2\mu}{3\lambda + 5\mu} b_2 r^{-3}\right\},\\
\widehat{rr}'' = {}& (3\lambda + 2\mu)\, a_0 - 4\mu b_0 r^{-3} - \frac{7\,(5\lambda + 6\mu)}{15} ar^2\\
& + \left(\sin^2\phi - \tfrac{1}{3}\right) 2\mu\left\{\frac{7\lambda + 12\mu}{4\mu} ar^2 + a_1 - 4b_1 r^{-5} - \tfrac{1}{2} a_2 r^2\right.\\
&\left.\qquad - \frac{9\lambda + 10\mu}{3\lambda + 5\mu} b_2 r^{-3}\right\}
\end{aligned}\right\}\ldots\text{(vi)}.$$

In order that these may vanish for all values of ϕ when $r=r_0$ and $r=r_1$ we must have:

$$\left.\begin{aligned} a_0 &= \frac{5\lambda+6\mu}{15(3\lambda+2\mu)(\lambda+2\mu)}\rho\omega^2\frac{r_0^5-r_1^5}{r_0^3-r_1^3}, \\ b_0 &= \frac{5\lambda+6\mu}{60\mu(\lambda+2\mu)}\rho\omega^2\frac{r_0^3r_1^3(r_0^2-r_1^2)}{r_0^3-r_1^3} \end{aligned}\right\} \text{..........(vii)};$$

together with the following four relations to determine a_1, b_1, a_2, b_2:

$$\left.\begin{aligned} \tfrac{3}{2}a\begin{Bmatrix}r_0^2\\ r_1^2\end{Bmatrix} + a_1 + \tfrac{8}{3}b_1\begin{Bmatrix}r_0^{-5}\\ r_1^{-5}\end{Bmatrix} + \frac{8\lambda+7\mu}{3\lambda}a_2\begin{Bmatrix}r_0^2\\ r_1^2\end{Bmatrix} + \frac{3\lambda+2\mu}{3\lambda+5\mu}b_2\begin{Bmatrix}r_0^{-3}\\ r_1^{-3}\end{Bmatrix} &= 0, \\ \frac{7\lambda+12\mu}{4\mu}a\begin{Bmatrix}r_0^2\\ r_1^2\end{Bmatrix} + a_1 - 4b_1\begin{Bmatrix}r_0^{-5}\\ r_1^{-5}\end{Bmatrix} - \tfrac{1}{2}a_2\begin{Bmatrix}r_0^2\\ r_1^2\end{Bmatrix} - \frac{9\lambda+10\mu}{3\lambda+5\mu}b_2\begin{Bmatrix}r_0^{-3}\\ r_1^{-3}\end{Bmatrix} &= 0. \end{aligned}\right\}$$

.........(viii).

[569.] These equations completely solve the problem, but lead to rather lengthy expressions for the constants. Resal confines himself to uni-constant results. Our (vii) corresponds to his (A) p. 434, and our (viii) to the first and third equations in the set at the bottom of his p. 435, except that he has $2A'_2$ in the first, where he ought to have $4A'_2$. He does not, however, obtain the values of the constants even for uni-constancy, but assumes the shell extremely thin and then obtains their values when ϵ/r_0 is negligible. To calculate the constants at least to the first power of the thickness is only laborious not difficult, and would I think be necessary before any conclusions as to the points of maximum strain could be fairly drawn. If we put $\lambda=\mu$ and neglect ϵ/r_0, we have to find a_1, b_1, a_2, b_2 from the first of each set of equations in (viii) and the differentials of those equations with regard to r_0.

Solving the equations so obtained I find:

$$\left.\begin{aligned} a_1 &= -\frac{161}{225}\frac{\rho\omega^2}{\mu}r_0^2, & a_2 &= \frac{37}{350}\frac{\rho\omega^2}{\mu}, \\ b_1 &= \frac{4}{25}\frac{\rho\omega^2}{\mu}r_0^7, & b_2 &= -\frac{112}{225}\frac{\rho\omega^2}{\mu}r_0^5, \end{aligned}\right\} \text{..........(ix).}$$

while (vii) gives:

$$a_0 = \frac{11}{135}\frac{\rho\omega^2}{\mu}r_0^2, \quad b_0 = \frac{11}{270}\frac{\rho\omega^2}{\mu}r_0^5$$

These results agree with those of Resal's p. 436, if proper changes of notation be made, notwithstanding the error I have noted in his equations corresponding to (viii).

Returning to the values of the shifts in (iii) and (iv) we see that r may now be written r_0, so that to Resal's degree of approximation the shifts are constant for each latitude right through the crust and no conclusion can be drawn as to whether points inside or outside the crust are those of maximum strain. We find for the complete values of the shifts to this degree of approximation:

$$u'' = \frac{\rho\omega^2 r_0^3}{10\mu}(4 - 9\sin^2\phi), \qquad v'' = -\frac{\rho\omega^2 r_0^3}{2\mu}\sin\phi\cos\phi \;\ldots\ldots\ldots\text{(x)}.$$

From (iii) we easily find for the mean radial stretch of the crust in the most general case:

$$\frac{u_1'' - u_0''}{r_1 - r_0} = a_0 - b_0\frac{r_0 + r_1}{r_0^2 r_1^2} + (\sin^2\phi - \tfrac{1}{3})\left\{a_1 - b_1\frac{(r_1^2 + r_0^2)(r_1 + r_0)}{r_1^4 r_0^4}\right.$$

$$\left. + a_2(r_1^2 + r_0^2 + r_1 r_0) - b_2\frac{r_0 + r_1}{r_0^2 r_1^2}\right\} + \alpha(r_1^2 + r_0^2 + r_1 r_0)(\tfrac{1}{5} - \cos^2\phi),$$

and therefore, when we neglect $(\epsilon/r_0)^2$, we have after some reductions:

$$\frac{u_1'' - u_0''}{\epsilon} = -\frac{1}{10}\frac{\rho\omega^2 r_0^2}{\mu}\cos^2\phi \;\ldots\ldots\ldots\ldots\ldots\ldots\ldots\text{(xi)}.$$

This agrees with Resal's result p. 437. He appears to deduce it from equation (23) of his p. 436, but he has not proved that the value he there gives for his W is correct even to the terms involving ϵ.

[570.] Resal draws various conclusions from the results (x) and (xi) of the previous article on his pages 437—40. Thus he remarks that:

(i) The flattening at the poles is 5/4 of the bulge at the equator.

(ii) The radius is not changed for the latitude

$$\phi = \sin^{-1} 2/3 \text{ or } = 41^\circ\, 48'\, 37''.$$

(iii) The thickness of the crust remains unchanged at the poles and decreases gradually towards the equator.

(iv) The meridional displacements are towards the equator and are maxima in the latitude 45°.

(v) The meridional curve is approximately an ellipse with the semi-axes:

$$r_0\left(1 + \frac{4\rho\omega^2 r_0^2}{10\mu}\right), \quad r_0\left(1 - \frac{5\rho\omega^2 r_0^2}{10\mu}\right).$$

(vi) Some geologists consider the flattening at the poles of the earth to have arisen from the rotation after solidification. In this case we find that the stretch-modulus for the material of the crust supposed thin, homogeneous and uni-constant would have to be about two and half times that of wrought iron; the mean modulus of the predominant kinds of rock of which the terrestrial crust is built-up probably differs very widely from this value.

(vii) The dilatation

$$\theta = \frac{1}{5} \frac{\rho\omega^2 r_0^2}{\mu} \cos^2 \phi.$$

Thus it is zero at the poles and a maximum at the equator.

Resal deals with the principal tractions and the stress-condition of rupture, but for oft cited reasons (see our Arts. 5 (*c*), 321, etc.) we do not consider this treatment satisfactory.

It is clear that Resal advanced considerably the problem first dealt with by Lamé, and both really laid the foundation of work afterwards done *de novo* by Sir W. Thomson, G. Darwin, Chree and others in applying the theory of elasticity to solve problems connected with the earth's crust.

[571.] E. Lamarle: *Note sur un moyen très-simple d'augmenter, dans une proportion notable, la résistance d'une pièce prismatique chargée uniformément. Bulletin de l'Académie Royale...de Belgique*, Tom. XXII. 1re Partie, pp. 232—52, 503—25. Bruxelles, 1855.

L'objet de cette note est de signaler à l'attention des constructeurs une disposition très-simple qui permet, *en certains cas*, d'augmenter, dans une proportion considérable, la résistance des pièces soumises à la flexion. Cette disposition, que je n'ai vue indiquée nulle part et que je crois nouvelle, consiste essentiellement, soit à remplacer par des encastrements obliques les encastrements horizontaux, soit, plus généralement encore, à établir certaines inégalités de hauteur entre les divers supports d'une même pièce, au lieu de placer tous ces supports à un même niveau, comme on le fait habituellement (p. 232).

[572.] The first part of the memoir deals with the *Cas général de deux supports*, i.e. with simple beams. Lamarle supposes the beam of length l to be uniformly loaded with p lbs. per unit-run and to bend in the plane of loading. It is supported at two points A and B, of which B is not necessarily on the same level as A. Suppose the horizontal through A taken as axis of x and the axis of y taken vertically downwards, then Lamarle shows that:

$$y = \frac{pl^4}{2E\omega\kappa^2}\left[\frac{1}{12}\left(\frac{x}{l}\right)^4 - \frac{a}{6}\left(\frac{x}{l}\right)^3 + \frac{b}{2}\left(\frac{x}{l}\right)^2\right] + mx \text{ .... (i)},$$

where, m and m' being the values of dy/dx at A and B and f the value of y at B,

$$\left.\begin{aligned} a &= 1 + \frac{12E\omega\kappa^2}{pl^4}\{2f-(m+m')\,l\}, \\ b &= \frac{1}{6} - \frac{2E\omega\kappa^2(m-m')}{pl^3} + \frac{6E\omega\kappa^2}{pl^4}\{2f-(m+m')\,l\} \end{aligned}\right\} \quad \text{......(ii).}$$

Let h be the distance of the 'extreme fibre' from the neutral axis and s the stretch in it, then $s/h = d^2y/dx^2$ and we find

$$s = \pm \frac{pl^2h}{2E\omega\kappa^2}\left\{\left(\frac{x}{l}\right)^2 - a\left(\frac{x}{l}\right) + b\right\} \text{..................(iii).}$$

Lamarle shows that s will take maximum values when

$$x = 0, \ = \tfrac{1}{2}al, \text{ and } = l.$$

These give, if $R = \frac{1}{2}pl^2/(E\omega\kappa^2)$ for the corresponding values of s:

$$\left.\begin{aligned} s_0/h &= \frac{R}{6} - \frac{m-m'}{l} + \frac{3}{l^2}\{2f-(m+m')\,l\}, \\ s_1/h &= \frac{R}{12} + \frac{m-m'}{l} + \frac{9}{Rl^4}\{2f-(m+m')\,l\}^2, \\ s_2/h &= \frac{R}{6} - \frac{m-m'}{l} - \frac{3}{l^2}\{2f-(m+m')\,l\} \end{aligned}\right\} \text{......... (iv).}$$

If the terminals of the rod had been simply built-in horizontally on the same level, we should have had $m = m' = f = 0$, and therefore the maximum stretch $= \frac{1}{6}Rh$; if the terminals had been simply supported we should have had $s_0 = s_2 = 0$, and therefore

$$2f = (m+m')\,l \text{ and } \tfrac{1}{6}R = (m-m')/l,$$

whence the maximum stretch would $= \frac{1}{4}Rh$.

[573.] Lamarle discusses the values of s_0, s_1, s_2 given by (iv) at considerable length and shows that their maximum will be least if:

$$2f = (m+m')\,l.$$

We have then so to choose $(m-m')/l$ that the greater of $s_0 \,(= s_2)$ and s_1 may be as small as possible. This gives us

$$(m-m')/l = R/24,$$

and

$$s_0 = s_1 = s_2 = Rh/8.$$

We are thus able to reduce the stretching effect of the load from $\frac{1}{6}Rh$ (or $\frac{1}{4}Rh$ as the case may be) to $\frac{1}{8}Rh$.

Various special cases are considered in which one or both terminals have given slopes, or in which there is a given difference of height. Lamarle shows that as a rule it is possible to reduce the greatest strain due to the load from 50 to 100 per cent. by properly building-in the ends (pp. 241—9).

He remarks:

Il y a lieu de faire observer que les quantités m, m' et f sont toujours très-petites relativement à l, et que souvent même elles sont de l'ordre des grandeurs dont on néglige de tenir compte dans la pratique. Sous ce rapport, l'influence considérable que peut exercer sur la résistance d'une pièce soumise à la flexion un changement très-minime apporté dans la disposition des supports mérite de fixer toute l'attention des constructeurs. Il est visible, en effet, qu'alors même qu'on voudrait s'en tenir aux conditions généralement adoptées, l'on devrait néanmoins procéder avec une extrême précision, et mettre le plus grand soin à éviter tout défaut de pose dans le sens où l'effet produit serait une diminution rapide de résistance (pp. 248—9).

Lamarle concludes this first part of his *Note* with an extension to the case of a beam passing over three points of support. He shows that if the middle support be lower than the terminal supports by

$$f = \frac{pl^4}{24E\omega\kappa^2}(8\sqrt{2} - 11),$$

the resistance of the beam will be increased by almost 50 per cent. By giving the terminals slopes determined by

$$m = -\frac{pl^3}{48E\omega\kappa^2},$$

and sinking the middle support by

$$f = \frac{pl^4}{96E\omega\kappa^2},$$

we increase the resistance of the beam by 100 per cent.

At the same time I must observe that it would be almost impossible in practice to insure that these slopes and deflections were accurately adjusted, and any slight sinking of the supports, due even to their elasticity, would upset the results entirely.

[574.] The *Deuxième Partie* of Lamarle's memoir is entitled: *Extension générale des résultats précédemment obtenus pour les cas de deux ou trois supports.* We have seen that the strength of the beam for a single span will be a maximum, if

$$2f = (m + m')\,l, \text{ and } (m - m')/l = R/24,$$

or from equations (ii) $\qquad a = 1, \quad b = \frac{1}{8}.$

Hence we find: $$\frac{d^2y}{dx^2} = R\left\{\left(\frac{x}{l}\right)^2 - \frac{x}{l} + \frac{1}{8}\right\} \quad \text{(vi)}.$$

This is true whatever be the value of m, provided we properly select m' and f; or, the above equations give m' and f as functions of m, and thus enable us to make the resistance of any particular span a maximum. We easily find that the points of inflexion, or those of zero bending moment, are given by

$$x = \frac{2 \pm \sqrt{2}}{4}\,l \quad \text{(vii)},$$

while the deflection

$$= mx - \frac{Rx^2}{48}\left\{1 - 4\left(\frac{l-x}{l}\right)^2\right\} \qquad \text{(viii)}.$$

Further

$$\left.\begin{aligned} m' &= m - \frac{Rl}{24}, \\ f &= ml - \frac{Rl^2}{48} \end{aligned}\right\} \qquad \text{(ix)}.$$

The equations (ix) must hold for each individual span, whence if m_n denotes the slope of the tangent at the end of the nth span and m_0 the given slope at the first terminal we have

$$\left.\begin{aligned} m_n &= m_0 - n\frac{Rl}{24}, \\ \text{and,} \quad f_n &= m_0 l - (2n-1)\frac{Rl^2}{48} \end{aligned}\right\} \qquad \text{(x)}.$$

Lamarle supposes all the spans equal and equally loaded, but the results may be easily extended to unequal spans. The total depression in the former case of the $(n+1)$th support below the first is

$$\begin{aligned} F_n &= f_1 + f_2 + \dots + f_n \\ &= nl\left(m_0 - n\frac{Rl}{48}\right). \end{aligned}$$

[575.] Lamarle deals with two special cases on his pp. 509—13. In Case (i) he supposes everything to be symmetrical about the middle of the beam and the terminals to be built-in at the proper slope. This is given by

$$m_0 = -m_n = n\frac{Rl}{48},$$

if there be n spans each of length l. Further we have

$$F_r = r(n-r)\frac{Rl^2}{48},$$

so that the proper depression of the rth support is determined.

In Case (ii) the terminals are not supposed to be built-in, but simply supported. We may then suppose the last spans in Case (i) to terminate at their points of inflexion. These are given by (vii), or the last spans will have lengths $l' = \frac{2+\sqrt{2}}{4}l$; whence, if the total length of the beam be $2L$, we have

$$(n-2)l + 2l' = 2L \qquad \text{(xi)},$$

and l', l and the corresponding differences in height of the points of support are easy to find.

[576.] On pp. 513—7, Lamarle works out the case of a uniformly loaded continuous beam of length $2L$ resting on $n+1$ points of support placed at equal distances, and finds for the maximum stretch s:

$$s/h = \frac{pL^2}{n^2 E\omega\kappa^2}\left\{1 - \frac{1}{\sqrt{3}}\frac{1-(\sqrt{3}-2)^n}{1+(\sqrt{3}-2)^n}\right\}.$$

The maximum stretch s' obtained for Case (ii) of the preceding article is (using Equation (xi)):

$$s'/h = \frac{pl^2}{16E\omega\kappa^2} = \frac{pL^2}{(2n-2+\sqrt{2})^2 E\omega\kappa^2},$$

whence
$$s/s' = \left(2 - \frac{2-\sqrt{2}}{n}\right)^2\left(1 - \frac{1}{\sqrt{3}}\frac{1-(\sqrt{3}-2)^n}{1+(\sqrt{3}-2)^n}\right) \quad \text{...... (xii).}$$

Lamarle has the following results:

$n=$	2	3	4	5	6	7	8	∞
$s/s'=$	1·4571	1·3028	1·4724	1·4927	1·5311	1·5517	1·5697	1·6906

Thus the increase of strength is a minimum for $n=3$ and then increases from 30 to nearly 70 per cent.

Supposing the beam $2L$ to have consisted of n separate simply supported spans we should have had for the maximum stretch s'':

$$s''/h = \frac{pL^2}{2n^2 E\omega\kappa^2},$$

and therefore[1]

$$s''/s' = \tfrac{1}{2}\left(2 - \frac{2-\sqrt{2}}{n}\right)^2 = 2 + \frac{\cdot 171573}{n^2} - \frac{1\cdot 171573}{n},$$

so that the advantage increases with n from about 46 to 100 per cent. Lamarle gives an interesting comparative table of results on p. 521.

If we wanted to realise the absolute maximum of resistance in the beam it would be necessary to fix the terminals at the slopes given in our Art. 575. This might be done by prolonging the beam over the terminal points of support up to the points of inflexion given by $x = \frac{2-\sqrt{2}}{4}l$ and then pivoting these new terminals. But it seems to me that this would often be practically difficult.

[577.] The memoir concludes with a brief indication of a similar theory for the case of an isolated load and for continuous beams of unequal span (pp. 524—5). On the whole we may say the memoir is very suggestive as showing what slight changes in the terminal slope and in the relative height of the supports will largely affect the resistance of a simple or continuous beam. It

[1] Lamarle has a misprint here (p. 519).

serves rather as a warning to constructors of the difficulties associated with the realisation of the theoretical stresses in structures of this kind than as a practical means of largely increasing their resistance.

[578.] L. F. Ménabréa: *Sopra una teoria analitica dalla quale si deducono le leggi generali di varii ordini di fenomeni che dipendono da equazioni differenziali lineali, fra i quali quelli delle vibrazioni e della propagazione del calore ne' corpi solidi. Annali di Scienze mathematiche e fisiche* (Tortolini), T. VI. Rome, 1855, pp. 363—370. This memoir is translated into French, pp. 170—180 of *Crelle's Journal für Mathematik,* Bd. 54, Berlin, 1857. It contains nothing on the vibrations of elastic solids that is of real importance.

[579.] O. Schlömilch: *Die gleichgespannte Kettenbrückenlinie. Zeitschrift für Mathematik und Physik,* Bd. I. Leipzig, 1856, pp. 51—55. This is an investigation of the proper area of the cross-section of the chains of a suspension bridge in order that the stress may be equal at each point. The paper contains references to earlier literature on the subject. At the conclusion of the paper the author remarks that an approach to such a suspension-chain exists in *Hungerford Bridge,* London;

> doch ist nichts über die ihr zu Grunde liegende Theorie bekannt geworden; wahrscheinlich haben auch die in der Praxis gewandten und kühnen, mit der Theorie aber meistens wenig vertrauten englischen Ingenieure überhaupt nach gar keinen Formeln construirt, sondern sich hier wie bei unzähligen anderen Gelegenheiten auf empirische Versuche und graphische Methoden verlassen.

[580.] G. Mainardi: *Note che risguardano alcuni argomenti della Meccanica razionale ed applicata. Memorie dell' I. R. Istituto Lombardo di Scienze,* T. 6, pp. 515—39. Milano, 1856.

Pp. 519—21 of this memoir are entitled: *Equilibrio di un filo elastico,* but the discussion seems to me obscure and does not appear to involve the proper number of elastic constants. It certainly adds nothing to the treatment of the problem by Saint-Venant and Kirchhoff: see our Arts. 1597*—1608*, 198 (*f*), and Chapter XII.

[581.] Von Autenheimer: *Zur Theorie der Torsion cylindrischer Wellen. Zeitschrift für Mathematik und Physik,* Bd. I. Leipzig, 1856, pp. 212—216.

A circular cylinder built in at one end is subjected to torsion by a couple at the other having for axis the axis of the cylinder. The author endeavours to measure the effect on the resistance of the longitudinal stretch of a fibre owing to the torsion. Let ϕ be the angle of torsion, l the length of the cylinder and a its radius, M the moment of the applied couple. He finds:

$$M = \mu \frac{\pi a^4}{2} \cdot \frac{\phi}{l} + \frac{\pi}{24} E a^6 \left(\frac{\phi}{l}\right)^3.$$

If s be the longitudinal squeeze of the cylinder:

$$s = \frac{a^2}{4} \left(\frac{\phi}{l}\right)^2,$$

and if

$$E = \frac{5\mu}{2}$$

we have

$$M = E \frac{\pi a^4}{5} \frac{\phi}{l} \cdot \left\{1 + \frac{5}{6} s\right\}.$$

So long as s lies within the elastic limit it will hardly exceed 1/1000, hence the effect on the couple of the stretch produced by torsion is negligible in practice. Even if we were to proceed up to $s = 1/50$ before rupture, the effect would only just become measurable experimentally.

The same matter has been dealt with by Wertheim, (Section II. of this Chapter), Saint-Venant (Art. 51), and Clerk-Maxwell (Art. 1549*).

[582.] Carl Holtzmann: *Ueber die Vertheilung des Drucks im Innern eines Körpers. Einladungs-Schrift der k. polytechnischen Schule in Stuttgart zu der Feier des Geburtsfestes seiner Majestät des Königs Wilhelm von Württemberg auf den* 27. *September*, 1856.

The earlier part of this paper reproduces the analysis of stress due to Cauchy and Lamé, leading up to their stress-ellipsoids and the shear-cone (see our Arts. 610* and 1059*). This occupies pp. 1—9. I do not think there is anything of novelty or importance in the treatment. The latter part of the paper applies the results so obtained to the discussion of stress in three special cases, namely those of:

(*a*) *A perfect fluid.* The fundamental equation of hydrostatics is deduced and a remark added that the disappearance of the shearing stress is not true for portions of the fluid where capillary action is called into play.

(*b*) *The stability of earth.* The earth is supposed to be bounded by two horizontal planes and a third vertical one, and the minimum and maximum pressures on the vertical plane are calculated, corresponding to the limits at which the earth will overcome the resistance of the plane, or the pressure on the plane overcome the resistance of the earth. A more general investigation has been undertaken by Lévy and Boussinesq: see our Art. 242. Rankine published important results in 1856—7 (see our Art. 453), but most probably Holtzmann had not seen them, and the special case worked out by him is simply dealt with and is of considerable interest.

(c) *A simple beam centrally loaded and subjected at the same time to continuous load on its upper surface.* Holtzmann supposes the beam of rectangular section and deals with the stress as uniplanar, in the manner of Jouravski, Bresse, Rankine, Kopytowski, Scheffler and Winkler: see our Arts. 183 (*a*), 468, 535, 556, 652 and 665. It is needless to repeat that the method is illegitimate and the results erroneous, except for the case of a section whose breadth is infinitely small as compared with the height (i.e. in practice the thin webs of girders).

[583.] H. Resal: *Mémoire sur le mouvement vibratoire des bielles. Annales des Mines*, Tome IX., pp. 233–79, Paris, 1856.

This paper contains an important application of the usual theory of the vibrations of bars (due to Bernoulli) to ascertain what influence the vibrations of a connecting rod have upon the forces which it exerts on the crank-pin and piston-head. The treatment is only approximate, terms of the third order in the ratio of the length of crank to that of connecting rod being neglected. But the results obtained are of very considerable interest, especially the analysis of the origin of the various types of longitudinal and transverse vibrations which occur. The danger of isochronism between a free period of vibration of the rod and the time of a complete revolution of the crank is brought out (p. 248): see our Art. 359 and ftn. p. 243. Resal considers at some length the effect on the magnitude of the vibrations of the connecting rod produced by putting a counterbalance upon it at or beyond the crank-pin. He shows that its influence is to produce a constant dilatation in the connecting rod and also to increase under ordinary conditions the amplitude of the transverse vibrations of the rod by one-third (p. 275). His analytic results are, however, too lengthy for citation here, even if their discussion did not carry us beyond the proper limits of our subject. They ought certainly to be consulted by those having to deal practically with the stresses in connecting rods.

[584.] H. Resal: *Recherches sur les tensions élastiques développées par le serrage des bandages des roues du matériel des chemins de fer. Annales des Mines*, Tome XVI., pp. 271–86, Paris, 1859.

This is an interesting paper although it involves several rather doubtful assumptions. It is well known that the true tire of a wheel is made slightly less in inner circumference than the outer circumference of the false tire (*faux bandage*) upon which it is placed when expanded by heat. On cooling the whole material of the wheel—spokes, false tire and true tire—is in a state of elastic strain, and Resal endeavours to ascertain on the Bernoulli-Eulerian theory of flexure the stresses in these various members. One assumption which his theory requires is that the linear dimensions of the cross-section of the tire must be small as compared with the length of tire between two spokes, and I do not think he has fully regarded this point, when he applies his theory to special cases of very close spokes. He also disregards the sliding effect produced by flexure and neglects the square of the ratio of the linear dimensions of the tire to the radius of the wheel.

[585.] Resal begins with a lemma of the following kind. Suppose a circular arc of radius ρ_0 to receive at the point defined by the radial angle θ the very small displacement towards the centre defined by $\rho_0 e$, then the change in curvature $1/\rho - 1/\rho_0$ at that point is measured by

$$1/\rho - 1/\rho_0 = \left(e + \frac{d^2e}{d\theta^2}\right) \Big/ \rho_0 \text{.......................(i).}$$

Let ρ_0 be the initial radius of the external circumference of the false tire, $\rho_0 (1 - \epsilon)$ the initial internal radius of the true tire, $\rho_0 (1 - e)$ the radius vector corresponding to the polar angle θ (measured from a spoke) of a point on the common circumference after strain, let β be the squeeze of this circumference at the same point, then ω_1 and ω being the cross-sections of the false and true tires, the squeeze in a 'fibre' distant y from this circumference in the false tire is easily found to be

$$\beta + \frac{y}{\rho} - \frac{y}{\rho_0} = \beta + \frac{y}{\rho_0}\left(e + \frac{d^2e}{d\theta^2}\right) \text{ .... (ii).}$$

Hence there is a total negative traction across the section of the false tire given by

$$E\omega_1 \left\{\beta + \frac{\bar{y}_1}{\rho_0}\left(e + \frac{d^2e}{d\theta^2}\right)\right\} \text{ .... (iii),}$$

and a total moment given by

$$E\omega_1 \left\{\beta \bar{y}_1 + \frac{\kappa_1^2}{\rho_0}\left(e + \frac{d^2e}{d\theta^2}\right)\right\} \text{.......................(iv),}$$

where $\bar{y}_1$ is the distance of the centroid of ω_1 from the common circumference and κ_1, the swing-radius of ω_1 about a line through that circumference.

For the true tire we easily deduce the expressions:

$$E\omega\left\{-\beta+\frac{\bar{y}}{\rho_0}\left(e+\frac{d^2e}{d\theta^2}\right)+\epsilon\right\} \text{ .... (v)},$$

$$E\omega\left\{-\beta\bar{y}+\frac{\kappa^2}{\rho_0}\left(e+\frac{d^2e}{d\theta^2}\right)+\epsilon\bar{y}\right\} \text{..................(vi)},$$

for the total positive traction and the moment with a similar notation. We suppose with Resal that the stretch-modulus for the material of both tires is the same. Hence subtracting (iii) from (v) and adding (iv) and (vi) we have:

For the total traction in the cross-section:

$$Q=E\Omega\left\{-\beta+\frac{\bar{Y}}{\rho_0}\left(e+\frac{d^2e}{d\theta^2}\right)\right\}+E\epsilon\omega \text{ .... (vii)}.$$

For the total moment:

$$M=E\Omega\left\{-\bar{Y}\beta+\frac{K^2}{\rho_0}\left(e+\frac{d^2e}{d\theta^2}\right)\right\}+E\epsilon\omega\bar{y} \text{ .... (viii)},$$

where $\Omega=\omega_1+\omega$, $\bar{Y}$ is the distance of the centroid of Ω from the common circumference, and K the swing-radius of Ω about an axis through the common circumference perpendicular to the plane of the wheel.

[586.] Now if we take the cross-section *cc* midway between two spokes which make an angle 2α with each other, the total stress over *cc* must consist of a couple, Em say, and a thrust at the common cir-

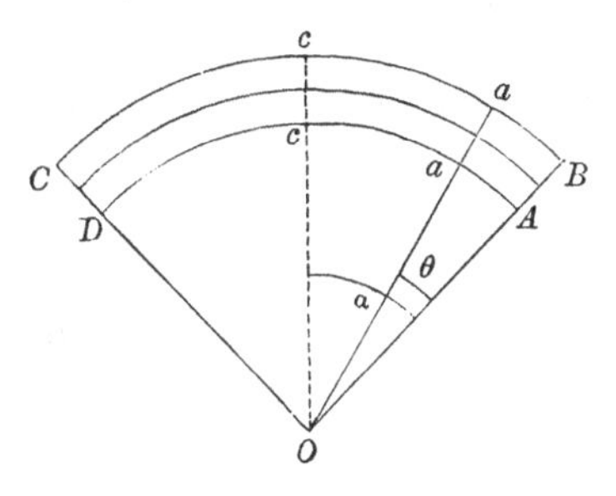

cumference perpendicular to the cross-section given by Ep, say. Hence for the cross-section *aa* we find at once from (vii) and (viii), since no forces act on the element *aacc* except at the cross-sections *cc* and *aa*:

$$-\beta\Omega+\frac{\bar{Y}\Omega}{\rho_0}\left(e+\frac{d^2e}{d\theta^2}\right)=p\cos(\alpha-\theta)-\epsilon\omega \text{ .... (ix)},$$

$$-\bar{Y}\beta\Omega+\frac{\Omega K^2}{\rho_0}\left(e+\frac{d^2e}{d\theta^2}\right)=-p\rho_0(1-e)\{\cos(\alpha-\theta)-\cos\alpha\}+m,$$

or,
$$-\bar{Y}\beta\Omega+\frac{\Omega K^2}{\rho_0}\left(e+\frac{d^2e}{d\theta^2}\right)=-p\rho_0\cos(\alpha-\theta)+m' \text{ ...... (x)},$$

when we neglect the products of small quantities (e.g. pe) and write m' for the terms on the right which do not involve θ. Eliminating successively β and e we find:

$$\frac{\Omega K_0^2}{\rho_0}\left(e+\frac{d^2e}{d\theta^2}\right) = -p\cos(\alpha-\theta)\{\rho_0+\bar{Y}\}+m'+\epsilon\omega\bar{Y}\ldots\ldots(\text{xi}),$$

$$\Omega K_0^2\beta = -p\cos(\alpha-\theta)\{K^2+\rho_0\bar{Y}\}+m'\bar{Y}+\epsilon\omega K^2\ldots\ldots(\text{xii}),$$

where K_0 is the swing-radius of Ω about a line in its plane through its centroid perpendicular to the plane of the tire.

The integral of equation (xi), remembering that $de/d\theta = 0$ for $\theta = \alpha$, is easily found to be

$$\frac{\Omega K_0^2}{\rho_0}e = -\frac{p}{2}(\rho_0+\bar{Y})\{(\alpha-\theta)\sin(\alpha-\theta)+m''\cos(\alpha-\theta)\}+m'+\epsilon\omega\bar{Y}\ldots\ldots(\text{xiii}),$$

where m'' is an undetermined constant. Equations (xii) and (xiii) contain the complete solution of the problem.

[587.] It remains to determine the constants p, m' and m''. It is easy to see that the total shear must vanish at the cross-section midway between two spokes, but that at a spoke cross-section it will not vanish but be equal to $p\sin\alpha$. Let σ be the cross-section of the spokes, l their length, then if their stretch-modulus be the same as for the tires, we have for their negative traction the expression

$$E\sigma\rho_0 e_0/l,$$

where e_0 is the value of e for $\theta = 0$.

Now consider the element of the wheel between two midway cross-sections, we have at once from statical considerations

$$2Ep\sin\alpha = E\sigma\rho_0 e_0/l,$$

or,

$$p = \tfrac{1}{2}\frac{\sigma\rho_0 e_0}{l\sin\alpha}\ldots\ldots\ldots\ldots\ldots\ldots\ldots\ldots(\text{xiv}).$$

The shearing force at AB (see figure on our p. 403)

$$= Ep\sin\alpha = \frac{E}{2}\frac{\sigma\rho_0 e_0}{l};$$

but this may be put $=\mu$ times the slide into the cross-section, or

$$\mu\Omega de/d\theta = E\sigma\rho_0 e_0/(2l)\ldots\ldots\ldots\ldots\ldots\ldots\ldots\ldots(\text{xv}).$$

Resal here *assumes* that the total shear is equal to the continued product of the slide-modulus into the total area of the cross-section into the complement of the angle the strained circumference common to the two tires makes with the radius at the spoke. Saint-Venant, however, finds values from about $\frac{2}{3}$ of this product for a rectangle to $\frac{5}{7}$ for a circle: see our Arts. 90 and 96.

Differentiating (xiii) and applying (xv) we obtain m'' in terms of e_0, thus since (xiv) gives p in terms of e_0 we have only to find e_0 and m'.

But putting $e = e_0$ and $\theta = 0$ in (xiii) we at once obtain m' in terms of e_0. Thus it remains to find e_0.

We have not yet made use of the condition that the angle cOA (see figure p. 403) retains a constant value or is equal to α. Now if ds be an element of the common circumference of the tires before and ds' after strain, we have obviously

$$\int_0^{\alpha\rho_0} ds = \rho_0 \alpha \quad \ldots\ldots\ldots\ldots\ldots\ldots\ldots\ldots \text{(xvi)}.$$

But $$(ds - ds')/ds = \beta \text{ and } ds' = \rho_0 (1 - e)\, d\theta,$$
hence $$ds = \rho_0 (1 + \beta - e)\, d\theta,$$
and substituting in (xvi) we have

$$\int_0^{\alpha} (\beta - e)\, d\theta = 0 \quad \ldots\ldots\ldots\ldots\ldots\ldots \text{(xvii)}.$$

Hence from (xii), (xiv) and (xvii) we can find e_0.

[588.] The calculations indicated in the previous article are carried out by Resal, who gives pp. 281–2 rather lengthy values for m' and e_0. He then returns to (ii) and to a similar expression for the squeeze in the true tire in order to find the maximum value of the traction or squeeze in the tire. Equating such value to the safe elastic limit, we find a maximum value for ϵ, or for $\rho_0\epsilon$ the difference in the radii of the two tires (pp. 282–3). As special problems Resal treats the case when the spokes are so close that $\sin\alpha$ may be replaced by α and further investigates a minimum safe value for ϵ in the case of a wheel turned by a crank having regard to the necessity of the moment of the friction between the two tires being greater than the moment of the force in the crank about the axis of the wheel. He does not, however, lay stress on this result (p. 286).

We have sufficiently indicated Resal's method of dealing with such problems to suggest to the student of this subject how he may complete or extend it.

[589.] H. Resal: *De l'influence de la suspension à lames sur le mouvement du pendule conique. Annales des Mines,* T. XVIII., pp. 1–16, Paris, 1860.

This is the application of the simple Bernoulli-Eulerian theory of flexure to the problem of the suspension by elastic laminae of a *balancier conique* due to Redier. The paper contains nothing further bearing on the theory of elasticity.

[590]. Mahistre: *Note sur les vitesses de rotation qu'on peut faire prendre à certaines roues, sans craindre leur rupture sous l'effort de la force centrifuge. Comptes rendus,* Tome XLIV., pp. 236–9, Paris, 1857.

This is only an extract from a longer memoir and the line of argument is scarcely intelligible from its brevity. As the problem has been satisfactorily dealt with (so far as that is possible on the Bernoulli-Eulerian theory) by Resal, we shall make no attempt to unravel Mahistre's obscure statements. We merely note that if S be the resistance to rupture of the metal in kilogs. per sq. metre, N the number of turns of the wheel per minute, R the mean radius of its rim, D its specific gravity, then Mahistre finds that to avoid rupture we must have a relation of the form:

$$N < \frac{30}{\pi R}\sqrt{\frac{gS}{D}}.$$

See our Art. 646.

[591.] The four memoirs by Poinsot on the impact of bodies published in *Tomes* 2 and 4 of the *Journal de Liouville*, 1857 and 1859, have nothing to do with elasticity, although their titles are cited by the writer of the article *Elasticitätstheorie des geraden Stosses* in the *Encyklopädie der Naturwissenschaften: Handbuch der Physik*, (see S. 296, Bd. I., of that work).

[592.] J. H. Koosen: *Entwickelung der Fundamentalgesetze über die Elasticität und das Gleichgewicht im Innern chemisch homogener Körper. Annalen der Physik*, Bd. CI., S. 401–52. Leipzig, 1857.

This is entitled: *Erste Abhandlung*, but I can find no trace of a *Zweite Abhandlung* having been published; perhaps its non-publication is hardly a loss. The author obtains the equations of elasticity for an isotropic medium practically in the same manner as Cauchy or Poisson, and he finds (S. 419) for the type of tractive stress:

$$\widehat{xx} = A\left(1 + \frac{du}{dx}\right) + (B - A)\left(3\,\frac{du}{dx} + \frac{dv}{dy} + \frac{dw}{dz}\right).$$

He apparently thinks there is something novel in this result, but the equation had been long previously obtained by Cauchy, who showed that A measures the initial stress: see our Art. 616* and our account of Saint-Venant, Art. 129. Koosen does introduce novelty, however, by retaining in general the coefficient A and supposing this *Molecularspannung* is somehow equilibrated by temperature exchanges between the elastic body and surrounding bodies (S. 425–6). I do not understand his reasoning on this point, nor in the following pages, and believe it to be incorrect. The equations involving an exponential of the time on S. 435 and the consequences drawn from them on the following pages are a mystery to me, and I should be inclined to describe the whole of this lengthy paper as no contribution to our subject, were I not obliged to confess that I have frequently been unable to follow its drift.

[593.] R. Hoppe: *Ueber Biegung prismatischer Stäbe. Annalen der Physik*, Bd. 102, S. 227–245, Leipzig, 1857. Reprinted in the *Zeitschrift des Vereins deutscher Ingenieure*, Bd. I., S. 308–13, Berlin, 1857.

This paper opens with the words:

Auf den bekannten Erfahrungssatz, nach welchem die zur Dehnung oder Zusammendrückung eines elastischen festen Körpers nach *einer* Dimension hin erforderliche Kraft den Volumincrementen proportional ist, lässt sich die Berechnung der Biegung eines prismatischen Stabes nur unter der Annahme gründen, dass sein Querschnitt weder in seinen Dimensionen, noch in seiner normalen Stellung zu allen Längenfasern eine Aenderung erleide. Die Bestimmung jeder ungleichmässigen Dehnung oder Compression nach *mehr* als einer Dimension, welche durch jene Annahme umgangen wird, erfordert die Zuziehung neuer empirischer Grundlagen oder Hypothesen; denn das unveränderte Volum selbst der kleinsten Theile begründet noch nicht das Gleichgewicht der darin befindlichen Spannungen (S. 227).

After reading this paragraph and remembering the researches of Saint-Venant and Kirchhoff (see our Chapters X. and XII.), it hardly seemed needful to study closely the present memoir. On examination, however, Hoppe's *Annahme* does not seem to have made his results any more incorrect than most investigations based on the Bernoulli-Eulerian hypothesis. His treatment, however, is somewhat obscure and does not appear to contribute anything of novelty or importance to the subject of flexure. It is based on the principle of virtual velocities and indicates the solution in elliptic integrals, but both these had been proposed and adopted previously: see the references under *Rods* in the index to our Vol. I.

[594.] J. Stefan: *Allgemeine Gleichungen für oscillatorische Bewegungen. Annalen der Physik*, Bd. CII., S. 365–87, Leipzig, 1857.

This paper deduces in the first place the general equations for the vibrations of an elastic medium when there are three rectangular planes of symmetry by Cauchy's method (S. 365–7): see our Art. 616*.

The author then goes on to investigate the like equations by Green's method, and afterwards considers the special cases of uniaxial symmetry and of isotropy. He deduces the equations which must be satisfied for the reflection and refraction of light at the common boundary of two such media. It seems to me, however, that in the media with biaxial and uniaxial symmetry he is tacitly supposing the crystalline axes to be parallel in the two media: see his equations S. 379–80 and 384. Thus he is really dealing with a very limited case of reflection and refraction at the common boundary. Stefan makes no attèmpt to solve his equations, and I do not think his paper can be considered as a valuable contribution to either optical or elastic theories.

[595.] E. Phillips: *Des parachocs et des heurtoirs de chemin de fer* *Comptes rendus*, Tome XLV., pp. 624–7, Paris, 1857.

The author commences by citing a formula from his memoir on springs (see our Art. 493 (*c*)) for the resilience of a spring of any form built-up of elastic laminae. The total elastic work to be obtained from a spring is $EVs^2/6$, where E is the stretch-modulus, V the volume, and s the safe or limiting stretch at the surface of all the component laminae.

Let w be the weight of a train in French tons, v its velocity in kilometres per hour, g gravitational acceleration, we must have:

$$\frac{1}{2}\frac{w}{g}v^2 = EVs^2/6,$$

if the spring be able to bring the train to rest without the spring being elastically damaged. Phillips takes 7·82 as the mean density of steel, 20,000,000 kilogs. per sq. mm. for E, and ·01 as the limit of s for very good steel. Thus he finds if W be the weight of the spring in kilogrammes:

$$W = \cdot 0952 \times w \times v^2 \quad \text{.....................(i).}$$

In this he neglects the friction of the laminae upon one another. He remarks, that if U' be the work due to this friction, it may be shown by the processes of his memoir of 1852 that:

$$U'/U < 2\phi\,(n-1)\,\frac{e}{L}.$$

where $U = EVs^2/6$, $\phi =$ coefficient of friction for steel on steel, $n =$ number of laminae, e their thickness and L the half length of the spring, so that U' will generally be small as compared with U.

Phillips next takes various values for w and v, for w from 90 to 600 tons and for v from 60 to 20 kilometres per hour, the values for W vary from 21 to 31 tons. Hence he concludes that it would be impossible to protect a train against collisions by causing it to carry at its ends buffers or springs of this enormous weight. On the other hand suitable buffers can be easily constructed to protect the masonry etc. at a terminus from the impact of a train with a small speed. In this case he takes v' to measure the velocity of the train in metres per second, he supposes that, as such springs are repeatedly loaded, s should not be taken greater than ·004 and he finds in French measure

$$W = 7{\cdot}7112 \times w \times v'^2.$$

For example if $v' = 1$ metre and $w = 30$ tons, W is the fairly reasonable weight of 230 kilogrammes, or about the weight of three ordinary carriage buffer-springs (70 to 80 kilogrammes). The memoir was referred to a commission then sitting to investigate the causes of accidents arising from the impact of railway wagons.

[596.] Deloy: *Extrait d'une Note relative à l'application de la théorie de M. Phillips à la construction d'un ressort de locomotive d'une nouvelle espèce.* *Comptes rendus*, Tome XLV., pp. 752–5, Paris, 1857.

This note gives details of a special kind of spring made by Gouin et Cie for the Lyons railway. Deloy calculated by Phillips' formulae (see our Arts. 489—90) the deflection of this spring under a load of 10,000 kilogs. and found it ·0478 metres, experiment gave it as ·048 metres. The experiment was repeated several times with the same result.

> Tous les ressorts nouveaux du chemin de fer de Lyon sont construits d'après la théorie de M. Phillips. J'ai commencé des essais pour déterminer la flexibilité de ces ressorts (p. 754).

The results of these experiments show such a noted agreement between experiment and oft abused theory that they deserve citing here. The deflections in metres were as follows:

	Experiment.	Theory.
Series of locomotive springs, 12 laminae, 3 matrix-laminae	·0377	·0357
,, ,, ,, 11 ,, 4 ,,	·067	·0646
,, tender ,, 9 ,, 4 ,,	·037	·03547
,, wagon ,, 7 ,, — —	·155	·15494

It cannot be denied that this is strong evidence in favour of the practical accuracy of Phillips' theory, more especially so when we consider the irregularities of material and manufacture in such technical products as railway springs.

[597.] E. Phillips: *Du travail des forces élastiques dans l'intérieur d'un corps solide, et particulièrement des ressorts: Comptes rendus*, T. XLVI., pp. 333–6 and *Supplément*, p. 440. Paris, 1858.

In this memoir Phillips remarks that it is generally impossible to apply Clapeyron's Theorem as suggested by its discoverer to springs (see our Arts. 608–9), because the value of the principal tractions cannot be found. He notes that he himself in an earlier memoir has applied the Bernoulli-Eulerian theory to springs and he cites his chief results: see our Arts. 483–508. The last page and the *Supplément* deal with the experimental stress which a steel bar may be subjected to without permanent extension; according to Phillips this stress is 40 to 50 kilogs. per sq. mm. It is difficult to understand whether Phillips means this as the safe load for bars liable to impact, or the real limit to a statically applied elastic stress.

[598.] We must now turn to a series of memoirs published in this decade and dealing with the problem of the reactions of bodies resting on several points of support. We note first:

Francesco Bertelli: *Ricerche sperimentali circa la pressione dei corpi solidi ne' casi in cui la misura di essa, secondo le analoghe teorie meccaniche si manifesta indeterminata e intorno alla relazione fra le pressioni e la elasticità de' corpi medesimi. Memoria Postuma. Mem. dell' Accad. delle Scienze di Bologna.* T. I., pp. 433–461, Bologna, 1850.

The memoir is divided into two parts, of which the first was read to the *Accademia* on February 16, 1843 and the second on March 28, 1844. It relates to the problem of the statically indeterminate reactions which arise when a body rests on more than two colinear or more than three non-colinear points of support. The problem occupied as large a share of attention in Italy in the first half of the present century, as that of solids of equal resistance in the second half of the seventeenth century, and the memoirs relating to it have almost as little permanent scientific value. Bertelli gives a very interesting account of the history of the problem (pp. 436–40 and 447–61)[1], and is apparently of the opinion that its solution cannot be reached without the aid of the theory of elasticity,—a view which had not met with general acceptance at the time when his memoir was read. He also describes a particular kind of dynamometer for measuring the indeterminate reactions (pp. 441–3). This he terms *il piesimetro.* Some experimental results obtained by means of such dynamometers are cited but no numerical details are given and they are too vague to be of service in testing for example the theory of continuous beams (pp. 443–6).

[599.] A. Dorna: *Memoria sulle pressioni sopportate dai punti d'appoggio di un sistema equilibrato ed in istato prossimo al moto. Memorie dell' Accad. delle Scienze di Torino,* Serie II., T. XVIII., pp. 281–318, Turin, 1857.

This is the first Italian memoir which attempts to deal with

[1] For the history of science the problem is of value as showing how power is frequently wasted in the byways of paradox. I give a list, which I have formed, of the principal authorities for those who may wish to pursue the subject further.

Euler: *De pressione ponderis in planum cui incumbit. Novi Commentarii Academiae Petropolitanae,* T. XVIII., 1774, pp. 289–329.
,, *Von dem Drucke eines mit einem Gewichte beschwerten Tisches auf eine Fläche* (see our Art. 95*), *Hindenburgs Archiv der reinen und angewandten Mathematik.* Bd. I., S. 74. Leipzig, 1795.
D'Alembert: *Opuscula,* T. VIII. *Mem.* 56 § II., 1780, p. 36.
Fontana, M.: *Dinamica, Parte* II.
Delanges: *Mem. della Società Italiana,* T. V., 1790, p. 107.
Paoli: *Ibid.* T. VI., 1792, p. 534.
Lorgna: *Ibid.* T. VII., 1794, p. 178.
Delanges: *Ibid.* T. VIII. *Parte* i., 1799, p. 60.
Malfatti: *Ibid.* T. VIII. *Parte* ii., 1798, p. 319.
Paoli: *Ibid.* T. IX., 1802, p. 92.
Navier: *Bulletin de la Soc. philomat.,* 1825, p. 35 (see our Art. 282*).
Anonym.: *Annales de mathém. par Gergonne,* T. XVII., 1826–7, p. 75.
Anonym.: *Bulletin des Sciences mathématiques,* T. VII., 1827, p. 4.
Vène: *Ibid.* T. IX., 1828, p. 7.
Poisson: *Mécanique,* Tome I., 1833, § 270.
Fusinieri: *Annali delle Scienze del Regno Lombardo-Veneto,* T. II., 1832, pp. 298–304 (see our Art. 396*).
Barilari: *Intorno un Problema del Dottor A. Fusinieri,* Pesano, 1833.
Pagani: *Mémoires de l'Acad. de Bruxelles,* T. VIII., 1834, pp. 1–14 (see our Art. 396*).
Saint-Venant: 1837–8: see our Art. 1572*.
,, 1843: see our Art. 1585*.
Bertelli: *Mem. dell' Accad. delle Scienze di Bologna,* T. I. 1843–4, p. 433.
Fagnoli: *Ibid.* T. VI., 1852, p. 109.

Of these writers only Navier, Poisson and Saint-Venant apply the theory of elasticity to the problem. Later researches of Dorna, Ménabréa and Clapeyron will be referred to in their proper places in this History as they start from elastic principles.

the problem of the body resting on more than three points of support from a rational standpoint,—that is to say, which makes direct appeal to the theory of elasticity. We have already referred to the earlier literature of this subject (see our Art. 598 and ftn.) and a memoir on very similar lines to this by Ménabréa will be considered later (see our Art. 604). Dorna's paper begins so well that we can only regret it does not end better. We say 'it begins well,' for it has not the flavour of mediaeval metaphysics traceable even so late as Fagnoli (see our Art. 509).

[600.] Dorna notes that if we give a virtual displacement to a system consisting of a rigid body resting on any number of points of support, then the sum of the virtual moments of these points of support must be zero independently of the virtual moments of the applied forces of the system. Hence if Q be a reaction and δq its virtual displacement, we must have:

$$\Sigma Q\delta q = 0 \qquad \text{(i)}.$$

To obtain δq Dorna now makes the following supposition; suppose that each point of support is connected with the rigid body by an elastic string of infinitely small length l and cross-section ω,—these being the same for all such strings,—and of stretch-modulus E *supposed to vary from string to string and to be that of the material of the supporting body in the neighbourhood of the supporting point* (p. 286), then we shall have

$$\delta q = \frac{l}{E\omega}\,\delta Q,$$

and consequently (i) will become:

$$\Sigma\,\frac{Q\delta Q}{E} = 0 \qquad \text{(ii)}.$$

Other relations between the δQ's will be given by the statical equations of equilibrium, whence either by eliminating the dependent δQ's or by the principle of indeterminate multipliers we have sufficient equations to find all the unknown reactions (pp. 291, 300 etc.).

[601.] Dorna's method is perfectly logical if we adopt his hypothesis namely (i) that the supports only are elastic, and the supported body rigid, (ii) that we may really introduce this string-link with stretch-modulus equal to that of the supporting material to explain what physically does happen at the point of support. Now the first hypothesis is just the reverse of what is usually assumed by practical engineers in calculating the reactions of continuous girders,—they suppose the supports rigid and the supported girder elastic[1]. Further the second hypothesis seems to me legitimate only in the case in which the support is a column of uniform cross-section with a reaction in the direction of its length, and even in this case the cross-sections of the column ought to be retained in Equation (ii) unless they happen to be all equal. To apply this hypothesis as Dorna does even to cases in which the reaction is perpendicular to the axis of support is to neglect entirely the distinction between the elastic coefficients of stretch and slide. Thus he deduces the extraordinary result:

la pressione, riferita all' unità di superficie, che una base piana di sostegno sopporta sotto l' azione di una forza diretta attraverso al suo centro di gravità, è la stessa, sia che questa operi a perpendicolo della base, sia che operi nella stessa base (p. 306).

[602.] Of the special applications which Dorna makes of his theory we may briefly note the following:

Problema II. A heavy rigid body rests on n colinear points of support, (pp. 290–2). This appears correct if the n points be supposed vertical columns of equal height and cross-section.

Problema III., (pp. 293–6) and *Problema* IV., (pp. 296–9). These are the general case of distribution of normal pressure over the cross-section of an elastic cylinder, and the special case when the cross-section is rectangular. The investigations are correct, but present no novelty except in the fact of their deduction from equation (ii) of our Art. 600. The results agree with those which flow from the theory of neutral axis and load-point and had long before been established by Bresse: see our Arts. 812* and 515–6, and compare Clifford's *Elements of Dynamic*, Book IV., pp. 14–28.

Problema V., (pp. 299–304). This supposes the general case in which any number of isolated points, or of continuous points composing a surface are connected by elastic string links with points on the surface of a rigid body supposed to be in contact with them. The analysis is not without interest, but I cannot consider that this problem corresponds to any physical reality, certainly not to a rigid

[1] This point has been dealt with by the Editor in a *Note on Clapeyron's Theorem: Messenger of Mathematics*, Vol. xx., pp. 129–35, Cambridge, 1890.

surface resting on any number of points or on an elastic surface as Dorna supposes.

Problema VI., (pp. 304–6), *Problema* VII., (pp. 306–7) and *Problema* VIII., (pp. 307–8) are absolutely inadmissible applications of *Problema* V.

Problema IX., (pp. 308–14). This is an attempt to generalise the theory of the neutral-axis and the load-point to pressure applied to a curved surface. The results obtained are all based on the hypothesis of *Problema* V., and are therefore physically inadmissible. Analytically they are not without interest as leading to theorems which are analogous to those which hold for the instantaneous axis of rotation of a rigid body and which were first discovered by Poinsot.

Problema X., (pp. 315–6) supposes a rigid body to rest on a portion of a spherical elastic surface. The results are inadmissible.

The memoir concludes with a *Nota* (pp. 316–8) containing a second demonstration of Equation (ii) of our Art. 600.

[603.] E. Clapeyron: *Calcul d'une poutre élastique reposant librement sur des appuis inégalement espacés: Comptes rendus*, Tome XLV., pp. 1076–1080, Paris, 1857.

This is only a *résumé* of a memoir, which I think was never published. It deals with the problem of a continuous beam and gives the equation of the three moments usually termed "Clapeyron's Theorem." Clapeyron states it only for the case of uniformly loaded spans of uniform cross-section. Let M_1, M_2, M_3 be three bending-moments at successive supports and l_{12}, l_{23} the intermediate spans, p_{12}, p_{23} their loads per foot-run,—then:

$$l_{12}M_1 + 2(l_{12} + l_{23}) M_2 + l_{23}M_3 = \tfrac{1}{4}(p_{12}l_{12}^3 + p_{23}l_{23}^3).$$

It will be seen that Clapeyron only deals with a very special case of his theorem, which has been much extended by later writers: see Heppel in our Art. 607 or Weyrauch: *Theorie der continuirlichen Träger*, S. 8–9.

Clapeyron mentions Navier as having said a few words on the problem in the *Bulletin de la Société Philomathique*, 1825; I suppose he refers to pp. 35–7. He cites Belanger as having studied in his course of lectures on construction at the *École des Ponts et Chaussées* the case of two spans, and other writers as having propounded the general equation, but left it complicated by the presence of the reactions. His own practical work on French railway bridges led him to investigate a formula free from the reactions.

He then applies his formula to the case of a bridge of seven equal spans. A remark on p. 1078 as to a defect in the design of the Britannia Bridge does not as a matter of fact apply as that bridge owing to its mode of construction is not a continuous beam in the theoretical sense: see our Art. 1489*.

[604.] L. F. Ménabréa: *Nouveau principe sur la distribution des tensions dans les systèmes élastiques: Comptes rendus*, T. XLVI., Paris, 1858, pp. 1056–1060.

Ménabréa here states a very important elastic principle, the application of which by Maxwell, Cotterill and others to framework and continuous beams has been of considerable service. I do not think the statement of the principle by Ménabréa sufficiently indicates that his proof only applies to what we now term a 'frame' or bit of 'framework', and that the links of such a frame must be supposed subjected to traction only and to be of uniform cross-section, which may vary, however, from link to link. A generalisation of the principle based upon Clapeyron's Theorem (see our Art. 608) is easily obtained and will be considered later.

Ménabréa states what he terms the *principe d'élasticité* in the following words:

Lorsqu'un système élastique se met en équilibre sous l'action de forces extérieures, le travail développé par l'effet des tensions ou des compressions des liens qui unissent les divers points du système est un minimum (p. 1056).

The proof given is essentially as follows: Let T be the traction in any element of the frame of length l and section ω. Then applying the principle of virtual work so that none of the points to which external force is applied have virtual displacements we must have:

$$\Sigma T\omega\delta x = 0,$$

where $\delta x =$ variation in the extension x of any link. But

$$T = Ex/l,$$

if E be the stretch-modulus of the link. Hence:

$$\delta T = \frac{E}{l}\,\delta x,$$

and

$$\Sigma \frac{l\omega}{E} T\delta T = 0 \quad \text{.........................(i)},$$

or, $$\delta\left(\Sigma\frac{l\omega T^2}{2E}\right)=0 \qquad\text{(ii)}.$$

But $l\omega T^2/2E$ is the work done by the traction T in the link l. Thus the principle is proved that the variation of the strain energy for the whole frame is zero. Ménabréa does not prove that this energy is a *minimum*. He terms (i) the *équation d'élasticité*.

[605.] Let there be n points united by m links, then there will be $3n$ equations of equilibrium for the n points; suppose in addition p equations of equilibrium between the external forces, then we shall have $3n-p$ equations between the m tractions, hence $m-3n+p$ tractions will be independent so far as the ordinary equations of statics go and require to be ascertained by (i). The method is indicated by Ménabréa in the following words:

Puisque pendant les variations infiniment petites des tensions qu'on a supposées, l'équilibre subsiste toujours, on pourra différentier, par rapport aux diverses valeurs de T, les $3n-p$ équations précédentes qui fournissent le moyen d'éliminer, de l'équation d'élasticité (i), un égal nombre de variations δT. On égalera à zéro les coefficients des diverses variations δT restantes dans l'équation (i). Ces coefficients seront des fonctions des forces extérieures et des tensions elles-mêmes; ainsi ces nouvelles équations unies à celles d'équilibre seront en nombre égal à celui des tensions à déterminer.

En général ces équations sont du premier degré. Dans bien des cas, l'emploi des coefficients indéterminés peut faciliter la solution du problème (p. 1058).

[606.] Ménabréa indicates how the following case should be dealt with, but I do not feel quite confident as to the exact form of elastic system he is dealing with, or as to the correctness of an assumption he makes. Suppose the system to be resting on a number of fixed points and P, Q, R to be the components of the reaction at such a point a, b, c parallel to the axes. Let X, Y, Z be types of components of applied force at x, y, z. The equations of statics give:

$$\left.\begin{array}{c}\Sigma X+\Sigma P=0;\ \Sigma Y+\Sigma Q=0;\ \Sigma Z+\Sigma R=0;\\ \Sigma(Xy-Yx)+\Sigma(Pb-Qa)=0;\ \Sigma(Zx-Xz)+\Sigma(Ra-Pc)=0;\\ \Sigma(Yz-Zy)+\Sigma(Qc-Rb)=0\end{array}\right\}\ldots\text{(iii)}.$$

Now P, Q, R are evidently components of the total traction ωT in the link to the point a, b, c, and therefore we should expect to have

$$\frac{l\omega T\delta T}{E}=\frac{l\,(P\delta P+Q\delta Q+R\delta R)}{E}$$

But Ménabréa writes:

Pour plus de généralité on peut supposer les *coefficients d'élasticité relatifs* des points fixes différents suivant les trois directions des axes; nous les représenterons par ϵ', ϵ'', ϵ'''; ainsi l'équation d'élasticité sera

$$\Sigma\left(\frac{1}{\epsilon'}P\delta P+\frac{1}{\epsilon''}Q\delta Q+\frac{1}{\epsilon'''}R\delta R\right)=0\ldots\ldots\ldots\ldots\text{(iv)}.\quad \text{(p. 1059)}.$$

I do not follow this at all. It would seem as if Ménabréa thought his theorem true for other strains than those produced by longitudinal traction in bars of uniform cross-section. This it certainly is *not*, in the form in which he has proved it. He appears further to put $\delta T=0$ for all links not going to fixed points, or, what is the same thing, to suppose the virtual displacements to be zero for such links. Taking the variation of (iii) we have:

$$\left.\begin{array}{c}\Sigma\delta P=0,\quad \Sigma\delta Q=0,\quad \Sigma\delta R=0,\\ \Sigma(b\delta P-a\delta Q)=0,\quad \Sigma(a\delta R-c\delta P)=0,\quad \Sigma(c\delta Q-b\delta R)=0\end{array}\right\}\ldots\ldots\text{(v)}.$$

Multiplying (v) by the indeterminate multipliers A, B, C, D, E, F respectively we have on adding to (iv):

$$\left.\begin{array}{l}P=-\epsilon'\,[A+Db-Ec]\\ Q=-\epsilon''\,[B+Fc-Da]\\ R=-\epsilon'''\,[C+Ea-Fb]\end{array}\right\}\ldots\ldots\ldots\ldots\ldots\ldots\text{(vi)}.$$

Substitute these values of P, Q, R in (iii), and we have six equations from which to find the multipliers and so can determine P, Q, R.

Ménabréa remarks that if we take $\epsilon'=\epsilon''=\epsilon'''$, and choose our origin and direction of axes so that

$$\Sigma\epsilon a=0,\quad \Sigma\epsilon b=0,\quad \Sigma\epsilon c=0,\quad \Sigma\epsilon bc=0,\quad \Sigma\epsilon ac=0,\quad \Sigma\epsilon ab=0,$$

we obtain the elegant forms for P, Q, R first given by Dorna in a memoir of 1857; see our Art. 599.

Here ϵ for any link equals the $E/(l\omega)$ of our notation.

For earlier researches in this same direction Ménabréa refers to Vène, Pagani and Mossotti besides Dorna. The memoirs of Vène and Pagani are those probably which we have cited in the footnote to our p. 411, while the reference to Mossotti is possibly to his *Meccanica razionale*. Ménabréa concludes by referring to a memoir he is about to publish, dealing more fully with the whole subject. I do not think he published this, or returned to the matter till a memoir of 1869.

[607.] J. M. Heppel: *On a method of computing the Strains and Deflections of Continuous Beams, under various Conditions of Load. Proceedings of the Institution of Civil Engineers.* Vol. XIX., pp. 625-643, London, 1859-60.

This paper deduces, apparently as a novelty, Clapeyron's theorem connecting the bending-moments at three successive points of support of a continuous beam, when the load system consists for each span of a uniformly distributed load and an isolated central load. The consideration of the latter load is the author's addition to Clapeyron's work. Let l_1, l_2 be the spans and M_1, M_2, M_3, the successive support bending moments, p_1l_1, p_2l_2 the total uniform loads and W_1, W_2 the isolated central loads, then:

$$8l_1M_1 + 16(l_1 + l_2)M_2 + 8l_2M_3 = 2p_1l_1^3 + 2p_2l_2^3 + 3W_1l_1^2 + 3W_2l_2^2 \ldots \text{(i)}.$$

Further the reaction R_{12} at the support between the spans l_1, l_2 is given by:

$$\left.\begin{aligned} R_{12} &= \frac{p_1l_1 + W_1}{2} + \frac{M_2 - M_1}{l_1} \\ &+ \frac{p_2l_2 + W_2}{2} + \frac{M_2 - M_3}{l_2} \end{aligned}\right\} \ldots\ldots\ldots\ldots\ldots\ldots\text{(ii)}.$$

The author also calculates the points of maximum-stress and of contraflexure (i.e. zero bending moment), and shows how the deflections may be obtained. I do not think there is any novelty in the methods used, but there are some interesting numerical applications to the Britannia Bridge, to a bridge on the Madras Railway and to a 'continuous rail of infinite length[1].'

[608.] E. Clapeyron: *Mémoire sur le travail des forces élastiques dans un corps solide élastique déformé par l'action de forces extérieures: Comptes rendus*, Tome XLVI, Paris, 1858, pp. 208–212.

This I presume to be only a *résumé* of the original memoir which so far as I can ascertain was never published.

Clapeyron had been led by a study of various kinds of springs to the conclusion that the resilience of an elastic body varies as its volume. He does not appear, however, to have known that Young and Tredgold had long previously reached this result. It led him to consider how the work of an elastic body could be expressed. In a memoir of 1833 Lamé and he had noted that on the uni-constant hypothesis if W be the work and E the stretch-modulus:

$$2W = \frac{1}{E}\iiint[A^2 + B^2 + C^2 - \tfrac{1}{2}(AB + BC + CA)]\,dxdydz,$$

where A, B, C are the principal tractions and the integration is

[1] A long series of memoirs on continuous beams will be found discussed in Section III. of this Chapter.

over the volume of the elastic solid. I hold that this result of the memoir of 1833 was due entirely to Clapeyron, for Lamé in his *Leçons*, of 1852, giving the formula in the form

$$2W = \frac{1}{E}\iiint [A^2 + B^2 + C^2 - 2\eta (AB + BC + CA)]\, dx dy dz,$$

due to bi-constant isotropy (η being the stretch-squeeze ratio), terms it *Clapeyron's Theorem*, and Clapeyron here speaks of it as he would do only if it were entirely due to himself.

[609.] Clapeyron proceeds after stating this formula in its modified form to suppose only one principal traction T, when we have:

$$2W = \frac{1}{E}\iiint T^2\, dx dy dz.$$

He then applies this to the calculation of W for various simple cases of rods under traction or flexure etc. and also for railway springs.

He remarks that if a framework be constructed in such a manner that the cross-sections of the various members are proportional to their total stresses, and these stresses are merely longitudinal tractions, then

$$2W = \frac{1}{E}\, VT^2,$$

where V is the volume of the whole framework. Hence if T be the safe tractional stress, and the load P be applied at one point with a resulting deflection f:

$$Pf = \frac{1}{E} VT^2.$$

Thus the same volume V of material distributed in different ways will give a maximum P for a minimum f; the resilience, however, will be quite independent of the particular distribution.

Un prisme posé de champ sur deux appuis porte plus que posé à plat dans le rapport de la hauteur à la largeur de la section; sa résistance à un choc est la même (pp. 210–11).

[610.] The remainder of the memoir treats of the question of uni-constancy. Dealing with one experiment of Coulomb's and eleven of Duleau's on torsion (see our Arts. 119*, 229*, and Vol. I., p. 873). Clapeyron finds that for iron $E = 5\mu/2$, or $\lambda = \mu$, very closely indeed. But from some experiments made in the workshops of the *Chemin de fer du Nord* on the compressibility of caoutchouc and on its stretch-modulus, Clapeyron concludes that for this substance

$$\lambda/\mu = 2201.$$

Thus while uni-constancy is very nearly true for the metals usual in construction, it appears to be quite impossible for caoutchouc. This is the well known argument from "squeezing india-rubber",—but it is one the validity of which is very doubtful: see our Arts. 924*, 1322*, 192 (*b*) and ftn. Vol. I., p. 504. We cannot accept it as a conclusive demonstration of bi-constant isotropy, until india-rubber has been demonstrated to satisfy all the other relations of a bi-constant isotropic *elastic* body[1]; this has not been done either by Clapeyron or by the several distinguished scientists who have used this argument. Other experiments on caoutchouc differ widely from Clapeyron's. See our Art. 1322*.

[611.] Clapeyron in the course of his discussion notes that the shear in a case of torsional stress gives rise to two principal tractions making angles of 45° with the direction of the shear, hence he states that the torsional resilience $= \frac{4}{5} T^2 V/E$. This is only true for the case of uni-constant isotropy. We see from our Arts. 493 (*c*) and 609 that in this case the resiliences of torsional, flexural and tensile springs of the same volume and material are as 24 : 5 : 15.

[612.] J. H. Rohrs: *On the Oscillations of a Suspension Chain. Transactions of the Cambridge Philosophical Society*, Vol. IX., pp. 379–98, Cambridge, 1856. This paper was read on December 8, 1851. It does not presuppose elasticity in the chain and so does not properly belong to our subject, but the general conclusions on p. 395 as to the vibrations of suspension bridges are of considerable interest.

[613.] P. van der Burg: *Ueber die Art Klangfiguren hervorzubringen und Bemerkungen über die longitudinalen Schwingungen. Annalen der Physik*, Bd. CIII. S. 620–4, Leipzig, 1858. This paper

[1] For example the slide modulus of india-rubber as determined by torsion and by pure slide experiments must be shown to have the same value as if it had been obtained by experiments on compressibility and traction. Roughly, from Clapeyron's experiments I find $\mu = 5$ kilogrammes per square centimetre.

contains miscellaneous information with regard to vibrating bars and plates. In particular the author recommends the following plan as leading to very correct Chladni-figures:

Man stellt nämlich einen Stab senkrecht auf eine Klangscheibe, fasst ihn in der Mitte mit der vollen linken Hand fest an, drückt ihn ziemlich stark auf die Scheibe, und streicht den oberen Theil von oben nach unten mit der vollen rechten Hand mittels eines Tuches, das mit pulverisirtem Harz bestreut ist; sobald ein reiner Ton entsteht, tritt sogleich die Figur sehr correct hervor (S. 621).

[614.] V. von Lang: *Zur Ermittelung der Constanten der transversalen Schwingungen elastischer Stäbe. Annalen der Physik*, Bd. CIII. S. 624–8, Leipzig, 1858. This does not seem any real contribution to the theory of elastic vibrations of rods. It proves an equation of the well-known form:

$$\int_0^l X_r X_s \, dx = 0,$$

where X_r and X_s are two solutions of Poisson's equation of the type:

$$d^4y/dx^4 + m^2y = 0,$$

by the lengthy process of substituting their values and actually integrating through the length l of the rod: see our Art. 468*.

[615.] Edward Sang: *Theory of the Free Vibrations of a Linear Series of Elastic Bodies. Edinburgh Royal Society Proceedings*, Vol. III., Part I., p. 358, Part VI., (*Alligated Vibrations*) pp. 507–8, Edinburgh, 1856–7. The first part is only referred to by title, the sixth part is accompanied by a short *résumé* of results, but this is not sufficient to indicate whether the original memoir is of real importance.

[616.] J. Stefan: *Ueber die Transversalschwingungen eines elastischen Stabes. Sitzungsberichte*, Bd. XXXII., S. 207–41, Wien, 1858. This paper proves 'by brute force' that the integral along the length of a rod of the product of two of the functions X_r and X_s which occur in Poisson's solution of the equation

$$\frac{d^2y}{dt^2} + a^2\frac{d^4y}{dx^4} = 0$$

is zero if r and s be different, and evaluates the integral when r and s are equal. The method adopted is longer than Poisson's original method, and I do not see that Stefan has really contributed anything to the previous discussion of the problem by Euler, Poisson, Seebeck and others.

He states that the method of integration by parts will not give the value of $\int X_r^2 dx$ as it leads in this case to an indeterminate form 0/0. This form, however, can be evaluated by the processes of the differential calculus and we can thus more briefly than by Stefan's laborious integrations deduce the value of $\int X_r^2 dx$. This was pointed out by V. von Lang in a paper entitled: *Einige Bemerkungen zu Herrn Dr T. Stefans Abhandlung: Ueber die Transversalschwingungen eines elastischen Stabes*, which appeared also in the *Sitzungsberichte*, Bd. XXXIV., S. 63–9, Wien, 1859.

[617.] J. Petzval: *Ueber die Schwingungen gespannter Saiten. Denkschriften der mathem. naturwiss. Classe der k. Akademie*, Bd. XVII., S. 91–136, Wien, 1859. An abstract of this memoir is given in the *Sitzungsberichte*, Bd. XXIX., S. 160–72, Wien, 1858.

This memoir commences by deducing the differential equations for the vibrations of an elastic string, when its mass per unit length is variable, its weight taken into account, and other variations not dealt with in ordinary treatments of the subject are considered (S. 91–6). The remainder of the memoir is devoted to the case in which two pieces of uniform string of different mass per unit length are united together to form a single piece. The author instead of considering the equality of the displacements and tensions at the joint treats this as a special case of varying mass. He obtains a solution involving a discontinuous function, and investigates at great length of analysis a problem which is easily dealt with by the ordinary equations for a vibrating string. I have not tested the results given for the notes, nodes etc., but these might be useful for purposes of comparison with the same quantities obtained by other processes. The author speaks of his problem as a *bisher nie in Betracht gezogenen Fall*, but this seems to me hardly probable although I am unable to give any reference to its earlier discussion. Possibly Duhamel has treated this case: see our Art. 897*.

[618.] E. Winkler: *Formänderung und Festigkeit gekrümmter*

Körper, insbesondere der Ringe. Der Civilingenieur, Bd. IV., S. 232–46, Freiberg, 1858.

This is an important memoir, both from the theoretical and practical standpoint, although many of its results require correction and modification. Some of these corrections have been made in *Kapitel* XL. (*Ringförmige Körper*) of the author's well-known treatise: *Die Lehre von der Elasticität und Festigkeit*, Prag, 1867, but this treatise does not cover anything like the same area as the memoir. I propose therefore to indicate the correct analysis and compare its results with those of Winkler.

The importance of the subject will be sufficiently grasped when I remind the reader that it is the only existing theory of the strength of the links of chains. To investigate the strength of such links by the complete theory of elasticity would involve even for the case of anchor rings an appalling investigation in toroidal and allied functions, while for the oval chain links with studs in ordinary use any successful attempt at a general investigation seems inconceivable. We shall have the less hesitation, however, in applying the Bernoulli-Eulerian theory, if we remember how close an approximation Saint-Venant's researches on flexure have shown it to be in the case of *straight* bars. At the same time, we are certainly going to put it to the very limit of its application, namely to *curved* bars in which the dimensions of the cross-section are not very small as compared with either the length or the radius of curvature of the central axis. It is non-fulfilment of the latter condition which renders Bresse's investigations for curved rods (see our Arts. 514 and 519) inapplicable without modification, and the former introduces, failing further experimental confirmation, an element of uncertainty into the results of an undoubtedly important theory.

[619.] Remembering that we need not assume adjacent cross-sections of our link to remain undistorted, if we only suppose them to be approximately equally distorted (see our Art. 84), we can easily investigate an expression for the stretch at any point by a method akin to that which results from the Bernoulli-Eulerian theory. We assume the central line of the link to lie in one plane and this plane to be that of the system of applied force and further to cut each cross-section of the link in a principal axis. These cross-sections will be supposed uniform, each of area ω and swing radius κ about a line (central axis) through the centroid perpendicular to the plane of the central line. (Central and neutral *axes* are straight lines lying in the plane of the cross-section; central and neutral *lines* the loci of points in which those axes meet the 'plane of the link'.).

We shall use the following notation:

s_v = stretch in a direction perpendicular to the cross-section at distance v from the central axis.

s_0 = stretch at points on the central axis.

v_0 = distance of the neutral axis from the central axis.

E = stretch-modulus of the material (not necessarily isotropic) in the direction of the central line.

$E\omega h^2$ = flexural rigidity of the link (no longer $E\omega\kappa^2$).

ρ_0 = radius of curvature of unstrained central line at any point.

x_0, y_0 = coordinates of a point on central line referred to the axes of symmetry of the link before strain.

x, y = the coordinates of the same point after strain.

$\Delta x, \Delta y = x - x_0, y - y_0$ respectively.

$d\sigma_0, d\sigma$ = elements of arc σ of central line before and after strain.

ϕ_0, ϕ = angles the tangent to the central line at any point makes with axis of x before and after strain, taken to increase with σ_0; $\Delta\phi = \phi - \phi_0$.

M = the bending moment at any cross-section, being the couple which must be applied (taken positive when it increases ϕ) for equilibrating the stresses if the material beyond the length σ of central line be cut away.

P = the *total* traction (i.e. negative thrust) at the same cross-section.

c_1, c_2 = the distances from the central axis to the 'extreme fibres', or what with an extension of terms we shall venture to call *intrados* and *extrados*. When we do not wish to particularise one or other of these, we shall simply use c for either.

Q = the *total* longitudinal pull on the link; this we shall suppose to be applied in the direction of the axis of y, which axis is taken to coincide with the greater axis of symmetry of the link, if there be one.

$\Delta\mathrm{b}, \Delta\mathrm{a}$ = increments of length of half the major and minor axes b and a (i.e. axes in directions of y and x respectively) of the link.

R = unknown reaction of the stud of the link supposed to coincide with the axis of x, if the link have one. Clearly

$$R = -\frac{E\omega}{n}\frac{\Delta\mathrm{a}}{\mathrm{a}} \quad \text{.........................} \quad \text{(i)},$$

where n is a quantity depending on the dimensions and materials of the stud and link. Winkler's result (39) S. 236 is really the same as this, although he puts it in a form apparently allowing for variation of the cross-section in the stud.

$d\omega$ = an element of the area of the cross-section.

$$\omega h^2 = \int \frac{\rho_0 v^2}{\rho_0 + v}\, d\omega.$$

Thus remembering the symmetry of the cross-section we have:

$$\int \frac{\rho_0\, d\omega}{\rho_0 + v} = \omega\left(1 + \frac{h^2}{\rho_0^2}\right) \quad\text{.........................} \quad \text{(ii)},$$

$$\int \frac{\rho_0\, v}{\rho_0 + v}\, d\omega = -\frac{\omega h^2}{\rho_0} \quad\text{............................} \quad \text{(iii)}.$$

Approximately:

$$\omega h^2 = \omega\kappa^2 + \frac{1}{\rho_0^2}\int v^4 d\omega + \frac{1}{\rho_0^4}\int v^6\, d\omega + \text{..................} \quad \text{(iv)}.$$

In some cases (e.g. Bresse's theory of arches) it is sufficiently approximate to put $h = \kappa$, retaining only the first term in (iv).

For a rectangle, I find if $2c$ be its height:

$$\left.\begin{aligned} h^2 &= \frac{\rho_0^3}{2c} \log \frac{\rho_0 + c}{\rho_0 - c} - \rho_0^2, \\ &= c^2\left(\frac{1}{3} + \frac{1}{5}\frac{c^2}{\rho_0^2} + \frac{1}{7}\frac{c^4}{\rho_0^4} + \ldots\ldots\right) \end{aligned}\right\} \text{...............} \quad \text{(v)},$$

which allows of easy calculation.

For a circle, if c be its radius:

$$\left.\begin{aligned} \omega h^2 &= \frac{\rho_0^4}{2}\int_0^{2\pi}\left(\frac{1}{\sin^2\theta}\log\frac{\rho_0^2}{\rho_0^2 - c^2\sin^2\theta} - \frac{c^2}{\rho_0^2}\right) d\theta, \\ \text{or,}\qquad h^2 &= \frac{c^2}{4}\left\{1 + \frac{1}{2}\frac{c^2}{\rho_0^2} + \frac{3\,.\,5}{6\,.\,8}\frac{c^4}{\rho_0^4} + \frac{3\,.\,5\,.\,7}{6\,.\,8\,.\,10}\frac{c^6}{\rho_0^6} + \ldots\right\} \end{aligned}\right\} \text{....} \quad \text{(vi)}.$$

The values of h^2 for some other sections may be easily found[1].

[620.] Let BOB', AOA' be the two principal axes of the curve formed by the central line $ABA'B'$, L a point on this central line, LT

[1] Its value for a trapezoidal section, symmetrical about the line joining the mid-points of the parallel sides is,—if d_1, d_2 be the lengths of those sides:

$$\rho_0^2\left(-1 + \frac{\rho_0}{\omega}\left[\left\{d_2 + \frac{d_1 - d_2}{c_1 + c_2}(c_1 + \rho_0)\right\}\log\frac{\rho_0 + c_1}{\rho_0 - c_2} - (d_1 - d_2)\right]\right).$$

Cf. Bach: *Elasticität u. Festigkeit*, S. 308–9. This is useful in the case of certain types of hooks.

the tangent there, $LTX = \phi_0$, $AL = \sigma_0$, then we readily find on the Bernoulli-Eulerian hypothesis:

$$s_v = \frac{s_0 + v\dfrac{d(\Delta\phi)}{d\sigma_0}}{1 + \dfrac{v}{\rho_0}} \text{ ....(vii).}$$

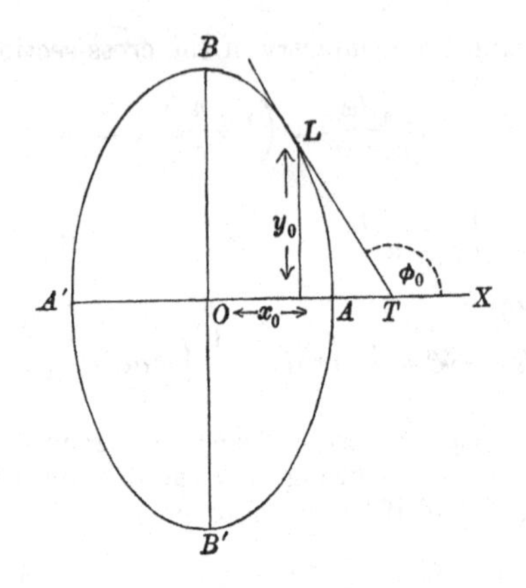

But

$$P = \int E s_v \, d\omega,$$

$$M = \int E s_v \, v d\omega.$$

Whence by (ii) and (iii)

$$P = E s_0 \omega \left(1 + \frac{h^2}{\rho_0^2}\right) - \frac{E\omega h^2}{\rho_0} \frac{d(\Delta\phi)}{d\sigma_0} \text{ ....(viii),}$$

$$M = -\frac{E\omega h^2 s_0}{\rho_0} + E\omega h^2 \frac{d(\Delta\phi)}{d\sigma_0} \text{ ....(ix).}$$

From (viii) and (ix) we find to determine s_0 and $\dfrac{d(\Delta\phi)}{d\sigma_0}$:

$$E\omega s_0 = P + \frac{M}{\rho_0} \text{ .... (x),}$$

$$E\omega h^2 \frac{d(\Delta\phi)}{d\sigma_0} = M + \frac{Ph^2}{\rho_0} + \frac{Mh^2}{\rho_0^2} \text{ ....(xi).}$$

We shall represent the right-hand sides of (x) and (xi) by p and m respectively. The usual formulae for arched ribs replace p and m by their first terms P and M: see our Art. 519.

Winkler in his memoir adds the term $P\kappa^2/\rho_0^2$ to (x) which I think is incorrect. He has the form (x) on S. 270 of his treatise.

Substituting in (vii) we find:

$$E\omega s_v = P + \frac{M}{\rho_0} + \frac{M}{h^2}\,\frac{\rho_0 v}{\rho_0 + v} \text{ ....(xii).}$$

Whence if T be the maximum traction in any section (T_1 the safe negative, T_2 the safe positive traction):

$$T\omega = P + \frac{M}{\rho_0} + \frac{M}{h^2}\,\frac{\rho_0 c}{\rho_0 + c} \text{ ....(xiii),}$$

and P and M must be given their values at the section of maximum stress while T is put equal to T_1 or T_2 to obtain the condition of safe loading.

Further from (xii) we find for the position of the neutral axis:

$$\frac{v_0}{\rho_0} = -\frac{P + \dfrac{M}{\rho_0}}{P + \dfrac{M}{\rho_0}\left(1 + \dfrac{\rho_0^2}{h^2}\right)} \text{ ....(xiv).}$$

For approximate values, if we neglect terms of the order $(v/\rho_0)^3$, we have:

$$s_v = \frac{Mv}{E\omega\kappa^2}\left(1 - \frac{v}{\rho_0} + \frac{\kappa^2}{v\rho_0}\right) + \frac{P}{E\omega} \text{ .... (xv),}$$

$$v_0 = -\kappa^2\left\{\frac{P}{M} + \frac{1}{\rho_0} - \frac{P\kappa^2}{M\rho_0^2}\left(2 + \frac{M}{P\rho_0} + \frac{P\rho_0}{M}\right)\right\} \text{........(xvi).}$$

Winkler in his memoir does not give (xii) to (xiv). He has (xv), but his approximation to s_v to the order $(v/\rho_0)^4$ seems to me wrong, while in his formula corresponding to (xvi) he has 1, where I have 2 in the second bracket. See his pages 234–6. Thus I think his final results cannot be depended upon.

[621.] From the consideration that $\cos\phi = dx/d\sigma$, and therefore $x = \int \cos\phi\,(1 + s_0)\,d\sigma_0$, we easily deduce:

$$\Delta x = -\frac{y_0}{E\omega}\int \frac{m}{h^2}\,d\sigma_0 + \frac{1}{E\omega}\int \frac{my_0}{h^2}\,d\sigma_0 + \frac{1}{E\omega}\int p\,dx_0 \text{....(xvii).}$$

Similarly from $\sin\phi = dy/d\sigma$, we find:

$$\Delta y = \frac{x_0}{E\omega}\int \frac{m}{h^2}\,d\sigma_0 - \frac{1}{E\omega}\int \frac{mx_0}{h^2}\,d\sigma_0 + \frac{1}{E\omega}\int p\,dy_0 \text{......(xviii).}$$

These equations agree with Winkler's (S. 234), except that he has the wrong values for m and p, which ought to have the values given in our (x) and (xi). They further agree with Bresse's approximate equations (see our Art. 519) if we put M and P for our m and p.

The above theory is so far perfectly general and not confined to the case of links. We now proceed to the case of a link symmetrical about two axes and with a stud.

[622.] Let $4\nu_0$ be the unstrained length of the perimeter of the central line. Then we find if χ be the angle the normal makes with the axis of x:

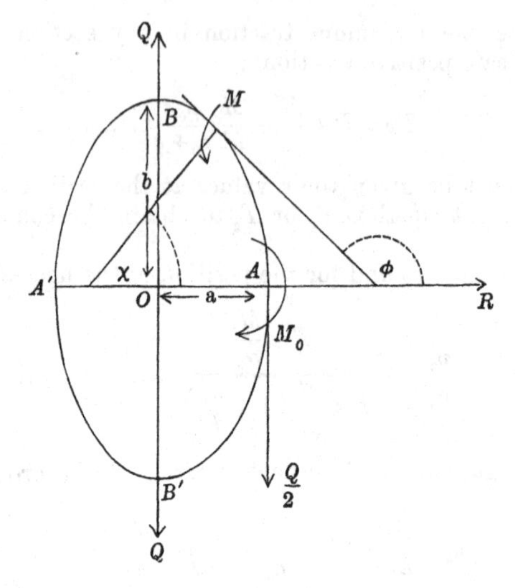

$$P = \tfrac{1}{2}(R \sin \chi + Q \cos \chi) \text{..................(xix).}$$

$$M = M_0 - \frac{R}{2} y + \frac{Q}{2}(\mathrm{a} - x) \text{..................(xx).}$$

Further, $\Delta\phi = 0$ at A and B, whence

$$\Delta\phi = \int_0^{\sigma_0} \frac{m}{E\omega h^2} d\sigma_0 \text{........................ (xxi),}$$

and

$$0 = \int_0^{\nu_0} \frac{m}{E\omega h^2} d\sigma_0 \text{........................(xxii).}$$

We also find from (xvii), (xviii) and (xxii):

$$\Delta \mathrm{a} = -\frac{1}{E\omega}\int_0^{\nu_0} \frac{my_0}{h^2} d\sigma_0 + \frac{1}{E\omega}\int_0^{\mathrm{a}} p dx_0 \text{......... (xxiii),}$$

$$\Delta \mathrm{b} = -\frac{1}{E\omega}\int_0^{\nu_0} \frac{mx_0}{h^2} d\sigma_0 + \frac{1}{E\omega}\int_0^{\mathrm{b}} p dy_0 \text{......... (xxiv).}$$

These values agree with those of Winkler's treatise but not with those of his memoir (S. 236).

[623.] Let us first apply these results to the case of a circular link of radius a. Here h^2 is constant and given, if the cross-section be as usual circular and of radius c, by (vi) with ρ_0 put equal to a.

Finding m from (xi), (xix) and (xx), and substituting in (xxii) we have:

$$\left(M_0 + \frac{Qa}{2}\right)\frac{\pi}{2}\left(1 + \frac{h^2}{a^2}\right) = \frac{Qa}{2} + \frac{Ra}{2} \text{ .... (xxv).}$$

Case (i). Suppose there to be no stud (*e.g.* an anchor ring). Then $R = 0$, and we find from (xix) and (xx),

$$\left.\begin{aligned} P &= \frac{Q}{2}\cos\chi, \\ M &= \frac{Qa}{2}\left(\frac{1}{\frac{\pi}{2}\left(1 + \frac{h^2}{a^2}\right)} - \cos\chi\right) \end{aligned}\right\} \text{ .... (xxvi),}$$

while
$$p = \frac{Qa^2}{\pi(a^2 + h^2)}, \quad m = Qa\left(\frac{1}{\pi} - \tfrac{1}{2}\cos\chi\right).$$

Further from (xxiii) and (xxiv) we have:

$$\left.\begin{aligned} \Delta a &= -\frac{Qa^3}{E\omega h^2}\left(\frac{1}{\pi} - \frac{1}{4}\right) + \frac{Qa^3}{E\omega\pi(a^2 + h^2)} \\ \Delta b &= \frac{Qa^3}{E\omega h^2}\left(\frac{\pi}{8} - \frac{1}{\pi}\right) + \frac{Qa^3}{E\omega\pi(a^2 + h^2)} \end{aligned}\right\} \text{ ...... (xxvii).}$$

Putting $h^2 = \kappa^2$ and neglecting the second terms as compared with the first, the results in (xxvii) agree with Saint-Venant's of 1837 (see our Art. 1575*). They differ by a factor $\frac{1}{2}$ in the second terms from those of Winkler's memoir even when h^2 is put equal to κ^2 in the first and neglected in the second terms. They agree except in the sign of the first term in the value of Δb with those of Winkler's treatise, S. 373. Winkler's results in the memoir for P and M agree to a first approximation with our (xxvi). See his S. 237–40.

For the position of the neutral axis we have from (xiv):

$$v_0 = -\frac{ah^2}{a^2 + h^2}\,\frac{1}{1 - \frac{\pi}{2}\cos\chi} \text{ .... (xxviii).}$$

This agrees with the result in the memoir, if h^2 be neglected in the denominator.

Lastly we find from (xiii) for the traction:

$$T\omega = \frac{Qa^2}{\pi(a^2 + h^2)} \pm \frac{ca}{a \pm c}\,\frac{Qa}{h^2}\left(\frac{a^2}{\pi(a^2 + h^2)} - \frac{1}{2}\cos\chi\right) \text{...(xxix),}$$

the upper sign referring to the extrados and the lower to the intrados. The result given in the memoir does not agree with this even to a first approximation.

[624.] Winkler traces in his Fig. 5, *Tafel* 33, for $a = 6c$ the form of the neutral line, and in Fig. 6 the tractions in extrados and intrados.

The latter are certainly incorrect. I have retraced both figures in the accompanying plate, where the stress is measured from the central axis along the radii in the scale: $T\omega/Q = \frac{1}{8}$ inch. The dotted lines are the curves obtained from the usual formula

$$T\omega = \pm \frac{Mc}{\kappa^2}.$$

It will be seen to give results often very divergent from those calculated from (xxix). The following are the numerical results for this case:

$$h^2 = \frac{c^2}{4} \times 1{\cdot}014{,}135; \qquad \frac{v_0}{c} = -\frac{\cdot 026{,}651}{\cdot 636{,}620 - \cos\chi};$$

$$v_0 = \infty, \text{ for } \chi = 50^\circ\ 27'\ 35'';$$

$$\frac{\Delta a}{a} = -\frac{Q}{E\omega} \times 9{\cdot}383{,}44;$$

$$\frac{\Delta b}{a} = \frac{Q}{E\omega} \times 10{\cdot}878{,}80;$$

$$\text{For extrados: } \frac{T\omega}{Q} = 6{\cdot}727{,}75 - 10{\cdot}142{,}35 \cos\chi;$$

$$\text{For intrados: } \frac{T\omega}{Q} = -8{\cdot}660{,}27 + 14{\cdot}199{,}29 \cos\chi;$$

$$\text{Old formula: } \frac{T\omega}{Q} = \pm(7{\cdot}586{,}75 - 12 \cos\chi).$$

For extrados $T = 0$ for $\chi = 48^\circ\ 27'$ (old formula $50^\circ\ 47'$);

Maximum positive traction $(\chi = 90^\circ) = \frac{Q}{\omega} \times 6{\cdot}727{,}75$,

Maximum negative traction $(\chi = 0) = -\frac{Q}{\omega} \times 3{\cdot}414{,}60$.

The old formula gives $\frac{Q}{\omega} \times 7{\cdot}586{,}75$ and $-\frac{Q}{\omega} \times 4{\cdot}413{,}25$ respectively.

For intrados $T = 0$ for $\chi = 52^\circ\ 25'$ (old formula $50^\circ\ 47'$);

Maximum positive traction $(\chi = 0) = \frac{Q}{\omega} \times 5{\cdot}539{,}02$;

Maximum negative traction $(\chi = 90^\circ) = -\frac{Q}{\omega} \times 8{\cdot}660{,}27$.

The old formula gives the same values of the traction for intrados as for extrados with the signs reversed.

[625.] It may be shown that the absolutely greatest traction is a negative one and occurs in the intrados at B and B'. For wrought iron, of which the links of chains are usually made, it would be sufficient to consider this traction[1], but there would have to be an investigation of the positive tractions in the case of cast iron.

[1] The 'fibrous' character of wrought-iron causes bars of this material to have a safe limit higher in tensile than in compressive stress, although for practical purposes they are frequently taken equal.

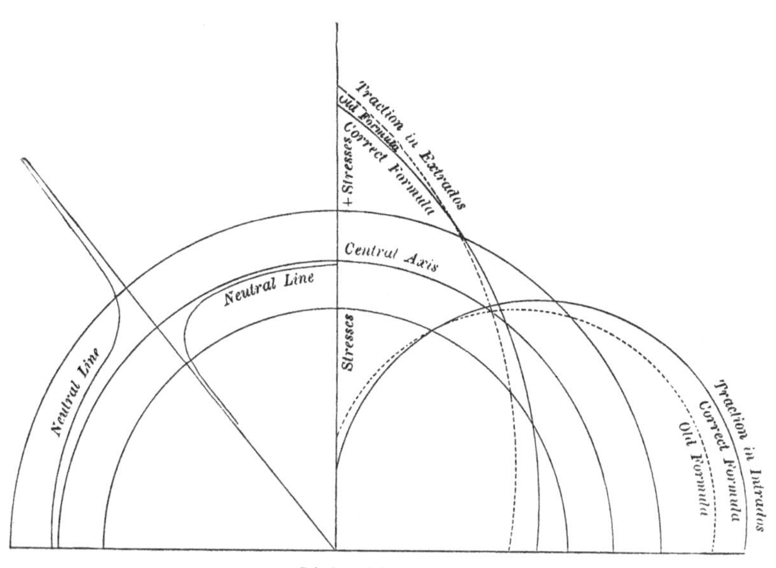

Link without stud.

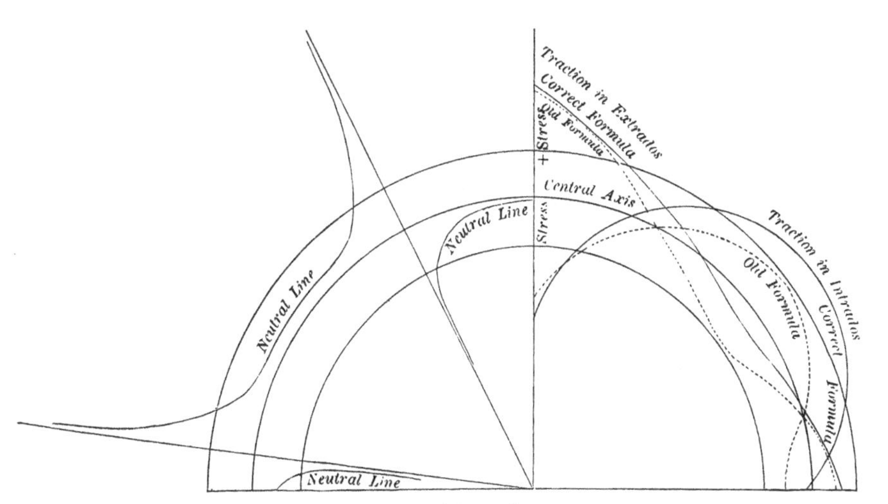

Link with stud.

To face Part I. *p.* 430.

The maximum negative traction in the intrados occurs at $\chi = 90°$, and equals $\dfrac{Qa^2}{\omega\pi(a^2+h^2)}\left\{\dfrac{ca^2}{(a-c)h^2} - 1\right\}$.

The maximum positive traction in the intrados occurs at $\chi = 0$, and equals $\dfrac{Qa^2}{\omega\pi(a^2+h^2)}\left(1 - \dfrac{ca^2}{(a-c)h^2}\right) + \dfrac{Qa^2c}{2\omega(a-c)h^2}$.

The maximum positive traction in the extrados occurs at $\chi = 90°$, and equals $\dfrac{Qa^2}{\omega\pi(a^2+h^2)}\left\{1 + \dfrac{ca^2}{(a+c)h^2}\right\}$.

The latter will be greater than the former if

$$4a^3 > \pi(a^2+h^2)(a+c),$$

which will generally be the case, e.g. if $a = 6c$.

In wrought iron our condition for safety is thus:

$$T_2 > \frac{Qa^2}{\pi^2c^2(a^2+h^2)}\left\{\frac{ca^2}{(a-c)h^2} - 1\right\} \quad \text{..............(xxx)};$$

or, to a first approximation, the diameter

$$2c > \sqrt[3]{\frac{32Qa}{\pi^2T_2}} \quad \text{.....................(xxxi)}.$$

This value of $2c$ may then be substituted in the small terms of (xxx), and a new approximation found. The result (xxxi) agrees with that given on S. 372 of the treatise, but the other results of this article are not given in it, and are erroneously given in the memoir. More exactly, neglecting only terms of the order $\left(\dfrac{c}{a}\right)^4$, I find that the cubic to determine the limiting value of c/a is:

$$\left(\frac{\pi^2T_2a^2}{4Q} - \frac{5}{16}\right)\left(\frac{c}{a}\right)^3 - \frac{1}{4}\left(\frac{c}{a}\right)^2 - \frac{3}{4}\left(\frac{c}{a}\right) - 1 = 0 \quad \text{......(xxxii)}.$$

This differs entirely from the cubic given in the memoir, and in the treatise (S. 372) Winkler has $\frac{3}{16}$ instead of $\frac{5}{16}$ for the second term of the first bracket.

[626.] *Case* (ii). Suppose the circular link has a stud.

Then we have from (xix), (xx) and (xxv),

$$P = \tfrac{1}{2}(R\sin\chi + Q\cos\chi) \quad \text{.................(xxxiii)},$$

$$M = \frac{1}{\pi}\,\frac{a(Q+R)}{1+\dfrac{h^2}{a^2}} - \frac{a}{2}(R\sin\chi + Q\cos\chi) \quad \text{........(xxxiv)};$$

whence
$$m = \frac{a}{\pi}(Q+R) - \frac{a}{2}(R\sin\chi + Q\cos\chi) \quad \text{........(xxxv)},$$

$$p = \frac{Q+R}{\pi}\,\frac{a^2}{a^2+h^2} \quad \text{..............................(xxxvi)}.$$

From (i) and (xxiii) we find:

$$\left.\begin{aligned} R &= \xi Q, \\ \text{where} \qquad \xi &= \frac{\dfrac{a^2}{h^2}\left(1 - \dfrac{\pi}{4}\right) - \dfrac{a^2}{a^2 + h^2}}{n\pi + \dfrac{a^2}{h^2}\left(\dfrac{\pi^2}{8} - 1\right) + \dfrac{a^2}{a^2 + h^2}} \end{aligned}\right\} \text{............(xxxvii).}$$

This value is not given in the treatise; it differs, even when we take only the first approximation, from the value given by Winkler in the memoir.

From (xxiv) we have:

$$\Delta \mathrm{b} = \frac{aQ}{E\omega}\left\{\frac{a^2}{h^2}\left(\frac{\xi}{4} + \frac{\pi}{8} - \frac{\xi + 1}{\pi}\right) + \frac{\xi + 1}{\pi}\,\frac{a^2}{a^2 + h^2}\right\} \text{....(xxxviii),}$$

while,

$$\Delta \mathrm{a} = -\frac{aQ}{E\omega}\, n\xi \text{ ....(xxxix).}$$

Let $\xi = \tan \epsilon$, then we easily find for the position of the neutral line from (xiv):

$$\frac{v_0}{a} = -\frac{\dfrac{2\sqrt{2}\sin(45^\circ + \epsilon)}{\pi}\,\dfrac{h^2}{a^2 + h^2}}{\dfrac{2\sqrt{2}\sin(45^\circ + \epsilon)}{\pi} - \cos(\chi - \epsilon)} \text{ ....(xl).}$$

Finally for the tractions in extrados and intrados from (xiii) we have:

$$\frac{T\omega}{Q} = \frac{\xi + 1}{\pi}\,\frac{a^2}{a^2 + h^2}\left\{1 \pm \frac{a^2 c}{h^2(a \pm c)}\right\} \mp \frac{a^2}{2h^2}\,\frac{c}{a \pm c}\,\frac{\cos(\chi - \epsilon)}{\cos \epsilon} \text{ ... (xli),}$$

where the upper sign refers to the extrados.

[627.] Now ξ is positive, hence ϵ will be found to be an angle in the first quadrant; $\cos(\chi - \epsilon)$ cannot thus be negative and we shall get the maximum positive traction in the extrados by making $\cos(\chi - \epsilon)$ as small, or $\chi - \epsilon$ as large as possible, irrespective of sign. Thus we must put $\chi = \pi/2$ or 0 according as ϵ is $<$ or $> 45^\circ$, or $\xi <$ or > 1; the former generally holds. Hence the maximum positive traction in the extrados

$$= \frac{Q}{\omega}\left\{\frac{\xi + 1}{\pi}\,\frac{a^2}{a^2 + h^2}\left(1 + \frac{a^2 c}{h^2(a + c)}\right) - \frac{a^2}{2h^2}\,\frac{c}{a + c}\,\xi\right\} \text{.........}(\alpha).$$

The maximum negative traction in the extrados will be at $\chi = \epsilon$ and so equals:

$$\frac{Q}{\omega}\left\{\frac{\xi + 1}{\pi}\,\frac{a^2}{a^2 + h^2}\left(1 + \frac{a^2 c}{h^2(a + c)}\right) - \frac{a^2}{2h^2}\,\frac{c}{a + c}\,\frac{1}{\cos \epsilon}\right\} \text{........}(\beta).$$

For the intrados the maximum positive traction will be obtained by putting $\chi = \epsilon$, and so equals:

$$\frac{Q}{\omega}\left\{\frac{\xi + 1}{\pi}\,\frac{a^2}{a^2 + h^2}\left(1 - \frac{a^2 c}{h^2(a - c)}\right) + \frac{a^2}{2h^2}\,\frac{c}{a - c}\,\frac{1}{\cos \epsilon}\right\} \text{.........}(\gamma).$$

For the maximum negative traction in the intrados we must have $\chi - \epsilon$ as great as possible, or as a rule we put $\chi = \pi/2$. Thus it equals:

$$\frac{Q}{\omega}\left\{\frac{\xi+1}{\pi}\frac{a^2}{a^2+h^2}\left(1-\frac{a^2c}{h^2(a-c)}\right)+\frac{a^2}{2h^2}\frac{c}{a-c}\,\xi\right\} \quad\text{.........}(\delta).$$

These values (α)—(δ) must be calculated for any given link of definite material. (δ) has in general, regardless of sign, the greatest value. Hence, if the links be made of wrought iron for which the safe tensile and compressive stresses may be taken as equal (see our Art. 625), we have, if T_2 be the safe maximum compressive stress:

$$Q < \frac{\omega T_2}{\dfrac{\xi+1}{\pi}\dfrac{a^2}{a^2+h^2}\left(1-\dfrac{a^2c}{h^2(a-c)}\right)+\dfrac{a^2}{2h^2}\dfrac{c\xi}{a-c}} \quad\text{.........(xlii).}$$

This equation also gives us the proper ratio of c to a when the value of Q is given.

Results (xxxviii) to (xlii) differ very considerably from Winkler's. He makes the maximum stress to be tensile and not compressive.

[628.] Let us suppose the link of our Art. 624 to have a cast iron stud placed in it, and let us take its modulus to be one-half that of the wrought iron link and its mean cross-section to be two-thirds that of the link[1], then:

$$n = E\omega/(E'\omega') = 3,$$

and we find:

$$\xi = {\cdot}676{,}098.$$

For a special elliptic link I find in Art. 640, $\xi = {\cdot}359{,}813$. Winkler finds in his treatise (§ 372) for an oval link $\xi = {\cdot}5612$, but I have not verified his arithmetic. Thus it appears that in the stud of a circular link there may be nearly double the stress that there is in that of an elliptic link.

For the stretches in the two axes we have from (xxxviii) and (xxxiv),

$$\frac{\Delta \mathrm{a}}{a} = -\frac{Q}{E\omega}\times 2{\cdot}028{,}295,$$

$$\frac{\Delta \mathrm{b}}{a} = \frac{Q}{E\omega}\times 4{\cdot}534{,}677.$$

The first is less than a fourth, the second less than a half of the values for the same link without stud. The total extension of a chain made of links having studs would only be about $\frac{5}{12}$ of the extension of a chain of the same length under the same load having the same links without studs. We may note that in general:

$$\xi = -\frac{\Delta \mathrm{a}_0}{a}\frac{E\omega}{Q}\Big/\left(n+\frac{\Delta \mathrm{b}_0}{a}\frac{E\omega}{Q}\right), \quad\text{and}\quad \frac{\Delta \mathrm{b}}{a} = \frac{\Delta \mathrm{b}_0}{a} + \xi\,\frac{\Delta \mathrm{a}_0}{a},$$

[1] These agree pretty closely with the numbers chosen by Winkler in his treatise, § 372, for an oval ring with stud.

where $\frac{\Delta b_0}{a}$ and $\frac{\Delta a_0}{a}$ are the stretches in the semi-axes of an equal link without stud. These simplify the calculations for a link with stud.

For the neutral line from (xl):

$$\frac{v_0}{c} = -\frac{\cdot 037{,}091}{\cdot 883{,}962 - \cos(\chi - \epsilon)}, \text{ where } \epsilon = 34^\circ\ 3'\ 45'',$$

and it passes to infinity when

$$\chi = 61^\circ\ 56'\ 20'' \text{ and } 6^\circ\ 11'\ 14''.$$

I have traced the neutral line in the lower figure of the plate, p. 430.

Finally for the tractions in extrados and intrados (T and T' say) we have:

$$T = \frac{Q}{\varphi}\{11\cdot 276{,}31 - 12\cdot 242{,}90 \cos(\chi - \epsilon)\},$$

$$T' = \frac{Q}{\omega}\{-14\cdot 515{,}44 + 17\cdot 140{,}06 \cos(\chi - \epsilon)\}.$$

The maximum value of T is positive and occurs at $\chi = 90^\circ$, its value being $\frac{Q}{\omega} \times 4\cdot 419{,}13$. The maximum numerical value of T' is negative, and occurs also when $\chi = 90^\circ$, its value being $-\frac{Q}{\omega} \times 4\cdot 915{,}34$. In the case of wrought iron the latter gives the limit to strength. Thus we see that the circular link with a stud of the above character in it is about 1·76 times as strong as the link without stud. Winkler in his memoir makes it 2·5 times as strong, but his analysis leading to a tensile limit is, I think, incorrect. In the treatise the only case of a link with a stud which he works out is an oval link. Here he finds his maximum stress compressive and the ratio of strengths with and without stud = 2 088. I have not verified his arithmetic, but the results of the treatise seem more probable than those of the memoir.

The traction in the extrados vanishes for

$$\chi = 11^\circ\ 8'\ 33'' \text{ and } 56^\circ\ 58'\ 57'',$$

that in the intrados for

$$\chi = 1^\circ\ 56'\ 8'' \text{ and } 66^\circ\ 11'\ 22''$$

The curves of stress in extrados and intrados will be found traced on the right-hand side of the lower figure of the plate, p. 430. These curves are very interesting especially when compared with the curves in the upper figure, as they show the influence of the stud. The dotted curves give the values of the tractions calculated from the formula, $T\omega = \pm Mc/\kappa^2$, where M is given its value from (xxxiv) after h^2 has been put equal to κ^2. We find:

$$T = \pm \frac{Q}{\omega}\{12\cdot 716{,}77 - 14\cdot 485{,}38 \cos(\chi - \epsilon)\},$$

which vanishes for

$$\chi = 5° 27' 10'' \text{ and } 62° 40' 20''.$$

We see that the old formula gives results diverging considerably from the true ones.

[629.] The diagrams on the plate, p. 430, referred to in Arts. 624 and 628, indicate a useful rule for welding anchor rings and others of circular form. The weld ought to be subjected to the least positive traction; hence the proper place to weld them does not seem to be at the section to which load is applied, but in the case of a ring without a stud about 40° from this section, in the case of a ring with a stud about 30° from the same section. As in the former case the ring can generally slip round so that the load may be applied at every section, we ought to provide for the welded section being able to sustain easily a traction equal to the greatest traction, which occurs in this case when the welded joint is the loaded section.

[630.] The next portion of Winkler's memoir is entitled: *Ring dessen Axe aus zwei geraden und zwei halbkreisförmigen Theilen besteht* (S. 240–2). The analysis of this as that of the previous cases is incorrect; it is not reproduced in the treatise. There is also a difficulty about this case which does not seem to have been noticed by Winkler and which also reappears in the case of the oval link formed of four circular arcs which he discusses in the treatise[1]. The difficulty arises from the discontinuity in the tractions at the sections for which there is an abrupt change of curvature; thus, while to satisfy the statical conditions we make a continuous change of bending moment and thrust at these sections, there is an abrupt change of traction owing to the application of the Bernoulli-Eulerian hypothesis. The exact distribution of the stress over such sections seems on that hypothesis to be arbitrary, but it probably may be safely taken equal to the mean of the tractions on either side. I do not think this peculiarity invalidates the solution for sections at small distances from those of discontinuity. An interesting but I expect difficult problem would be to analyse the nature of the stress at such a section by the general theory of elasticity.

[1] I have not verified Winkler's analysis for this oval link, which replaces the link with straight sides and the elliptic link of the memoir. It is worked out for special numerical cases with and without a stud, but no attempt is made in the treatise to draw stress curves as in the memoir.

[631.] I shall give my own analysis of the link with semi-circular ends and flat sides and compare the results with Winkler's. Let a be

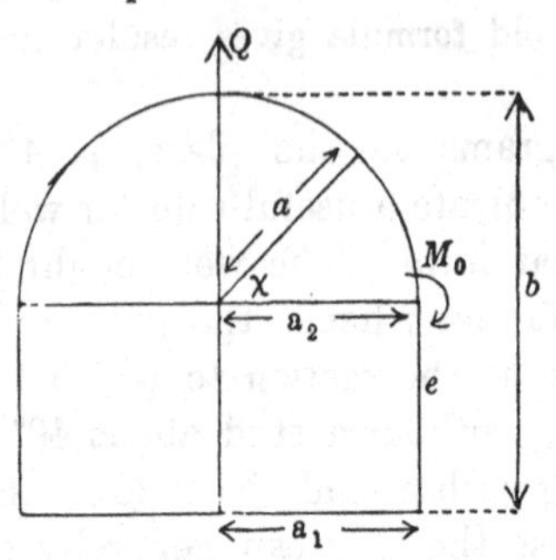

the radius of the semicircular ends, $2e$ the length of each of the straight parts, $b = a + e$; Δa_1 = change in semi-diameter of the link, between the mid-points of the straight parts; Δa_2 = change in semi-diameter of base of semi-circular part; M_0 the bending-moment at the joint of semi-circular and straight parts, and let the rest of the notation be as before except that subscript $_1$ refers to the straight and $_2$ to the circular parts.

We easily find:

$$M_1 = M_0, \quad P_1 = \tfrac{1}{2}Q, \quad m_1 = M_0, \quad p_1 = \tfrac{1}{2}Q,$$

since $\rho_0 = \infty$ for the straight parts. Further

$$h_1^2 = \kappa^2,$$

$$M_2 = M_0 + \frac{Q}{2} a (1 - \cos\chi), \quad P_2 = \frac{Q}{2} \cos\chi \text{ ....(xliii)},$$

whence

$$m_2 = \left(M_0 + \frac{Qa}{2}\right)\left(1 + \frac{h_2^2}{a^2}\right) - \frac{Qa}{2}\cos\chi, \quad p_2 = \frac{M_0}{a} + \frac{Q}{2} \text{ .... (xliv)}.$$

Let $h_2^2 = q\kappa^2 = qh_1^2$, where q is given by Equation (vi) of our Art. 619. Then from Equation (xxii) we have:

$$\left.\begin{aligned} M_0 + \frac{Qa}{2} &= \zeta\frac{Qa}{2} \\ \text{where} \quad \zeta &= \frac{qe + a}{qe + \dfrac{a\pi}{2}\left(1 + \dfrac{h_2^2}{a^2}\right)} \end{aligned}\right\} \text{ ....(xlv)}$$

$$\left.\begin{aligned} \text{whence} \quad m_2 &= \frac{Qa}{2}\left\{\zeta\left(1 + \frac{h_2^2}{a^2}\right) - \cos\chi\right\}, \\ p_2 &= \zeta\frac{Q}{2}, \\ M_2 = \frac{Qa}{2}(\zeta - \cos\chi), &\quad M_1 = \frac{Qa}{2}(\zeta - 1) \end{aligned}\right\} \text{ ....(xlvi)}$$

For the tractions we have from equation (xiii),

$$\left.\begin{aligned} T\omega &= \frac{Q}{2}\left\{\zeta \pm \frac{a^2}{h_2^2}\frac{c}{a \pm c}(\zeta - \cos\chi)\right\} \text{ for the curved parts} \\ T\omega &= \frac{Q}{2}\left\{1 \pm \frac{ac}{\kappa^2}(\zeta - 1)\right\} \text{ for the straight parts} \end{aligned}\right\} \text{(xlvii).}$$

It may be shown that, since $\zeta < 1$, the greatest positive traction occurs as a rule at $\chi = \pi/2$, but that the negative traction at the same section is greater. Hence for wrought iron this negative traction becomes the measure of safe loading. If T_2 be the limit to safe compressive stress, we must have:

$$Q < \frac{2\omega T_2}{\zeta\left\{\dfrac{a^2}{h_2^2}\dfrac{c}{a-c} - 1\right\}} \text{ .... (xlviii).}$$

For a circular link without stud we have from (xxx):

$$Q < \frac{2\omega T_2}{\dfrac{2a^2}{\pi(a^2+h^2)}\left\{\dfrac{a^2}{h^2}\dfrac{c}{a-c} - 1\right\}}$$

The latter, if we take $h^2 = h_2^2$, will therefore give a greater permissible value than the former for Q, if, $\dfrac{2a^2}{\pi(a^2+h^2)}$ be $< \zeta$, which is easily seen to be always true whatever e may be Hence we do not gain increased strength when we elongate a given circular link by inserting straight pieces at the sides. In fact the longer the straight pieces the weaker the link, till the weakness reaches a maximum with $\zeta = 1$, for $e = \infty$.

[632.] For the neutral line I find:

$$\left.\begin{aligned} \frac{v_0}{a} &= -\frac{h_2^2}{a^2}\frac{\zeta}{\zeta\left(1 + \dfrac{h_2^2}{a^2}\right) - \cos\chi} \text{ for the curved parts,} \\ \frac{v_0}{a} &= -\frac{\kappa^2}{a^2(\zeta - 1)} \text{ for the straight parts} \end{aligned}\right\} \text{...(xlix).}$$

Further for the change in the semi-axes we have by (xxiii) to (xxiv):

$$E\omega\Delta a_1 = -\int_0^e \frac{m_1 y_0}{h_1^2}dy_0 - \int_0^{\pi/2}\frac{m_2(e + a\sin\chi)\,ad\chi}{h_2^2} + \int_0^e p_2 dx_0,$$

whence

$$E\omega h_2^2\Delta a_1 = \frac{Qa}{2}\left\{\tfrac{1}{2}qe^2 + ae + \tfrac{1}{2}a^2 - \zeta\left(ae\frac{\pi}{2} + a^2 + q\kappa^2\frac{e\pi}{2a} + \tfrac{1}{2}qe^2\right)\right\} \text{...(l).}$$

This result would agree with Winkler's were we to put $q = 1$, or $h_2^2 = \kappa^2$, throughout.

We easily find:

$$\Delta a_1 - \Delta a_2 = \frac{M_0 e^2}{2E\omega\kappa^2},$$

or,
$$E\omega h_2^2 \Delta a_2 = E\omega h_2^2 \Delta a_1 + \frac{Qa}{4}(1-\zeta)\, qe^2 \text{ .... (li).}$$

For Δb we have:

$$E\omega h_2^2 \Delta b = \frac{Qa}{2}\left\{q\kappa^2 \frac{e}{a} + qae + \frac{\pi}{4}a^2 - \zeta\,(qae + a^2)\right\} \text{ ..... (lii).}$$

To a first approximation (i.e. if $q = 1$) (lii) agrees with Winkler's result, but my value for Δa_2 appears to be quite different from his.

Winkler traces the neutral line and stress-curves for the particular numerical case of $a = \frac{5}{2}c$ and $b = 4c$, or the length of the straight piece $\frac{3}{4}$ of the diameter of the cross-section of the link. Equation (xlix) shows us that the neutral axis for the curved part is similar to that for a circular link without stud, while for the straight piece, it is a straight line parallel to the straight piece and outside the link, since ζ is < 1. The stress curves are thus similar to those of the upper figure on the plate, p. 430, for the curved parts, and are straight lines for the straight pieces. I have not redrawn Winkler's curves, which are wrong owing to his erroneous formulae. They present, however, no novelty beyond those we have already dealt with.

[633.] The concluding pages of the memoir (S. 242–6) are entitled: *Ring mit elliptischer Axis*, and deal with elliptic links with and without studs. Not only is Winkler's analysis incorrect, but even as an approximation the terms he neglects are of equal importance with those he retains. He expands also certain expressions in terms of the eccentricity in very slowly converging series, which would be better replaced by elliptic integrals, whose values could be found in Legendre's tables. The case of an elliptic link is not dealt with in the treatise.

The following considerations by which Winkler selects a numerical case will, perhaps, show the difficulty I feel in accepting his analysis. He argues that to prevent the jamming of two links we must have for elliptic links (axes $2a$, $2b$) of circular cross-section (diameter $2c$) $c = < \frac{a^2}{b} - c$, whence $a/b > = 2c/a$. Redtenbacher, for a link without stud, says b should equal $3{\cdot}6c$, or we must have $a/b > = {\cdot}745$. He further takes for a link with stud $b = 4c$ (whence a/b should be $> = {\cdot}71$). In both cases, however, he puts $a/b = {\cdot}69$ which allows jamming. Winkler takes $a/b = {\cdot}71$ and $b = 4c$, whence he finds $a = 2{\cdot}82c$, and the eccentricity

$e = \cdot 709{,}22$[1] Thus e is not a small quantity and his series converge very slowly. Further his least radius of curvature $= a^2/b = 2c$ nearly. But he puts throughout $h^2 = \kappa^2$, or by our Equation (vi) he neglects terms of the order $\frac{1}{2}\frac{c^2}{\rho_0^{\ 2}}$ or 1/8 of those he retains; thus his expansions of the elliptic integrals to high powers of e are futile, for his results on other grounds are not necessarily correct to the first place of decimals. To retain the term in $(c/\rho_0)^4$ in h^2 leads to enormous complexity of calculation, but I propose to retain the term $(c/\rho_0)^2$ neglected by Winkler so that even in the very eccentric link chosen by him, we may hope to get within two per cent. of the true result, while for values of a^2/b large as compared with c we shall have all the accuracy requisite in practice. In what follows I indicate only the general outlines of my analysis.

[634.] Let χ be the angle the normal at any point of the elliptic central axis makes with the minor axis a, and let the radius to the corresponding point of the auxiliary circle make an angle ψ with a; let e equal the eccentricity $= \sqrt{1 - a^2/b^2}$

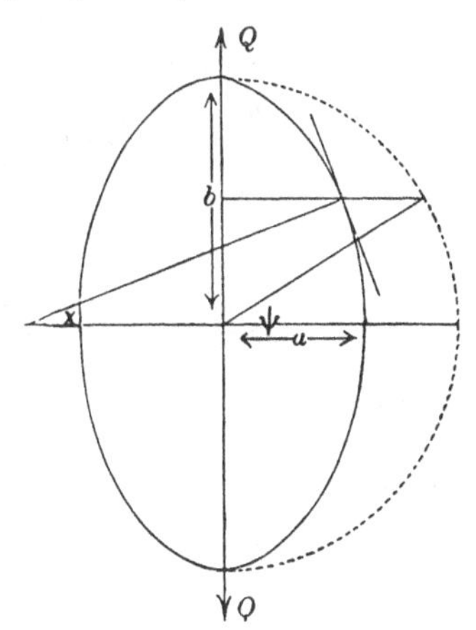

Then we easily find, with the notation of the previous articles:

$$\left.\begin{aligned} \tan\chi &= \frac{a}{b}\tan\psi, & \rho_0 &= \frac{b^2}{a}(1 - e^2\sin^2\psi)^{\frac{3}{2}}, \\ d\chi &= \frac{a}{b}\frac{d\psi}{1 - e^2\sin^2\psi}, & d\sigma_0 &= b\sqrt{1 - e^2\sin^2\psi}\,d\psi \end{aligned}\right\} \quad \text{......} \quad \text{(liii).}$$

[1] His values for a and e do not seem to agree with those he has chosen for a/b and b/c!

If there be a stud with resistance $R=\xi Q$, we have from (xix) and (xx):

$$P=\frac{Q}{2}(\xi\sin\chi+\cos\chi)=\frac{Q}{2}\frac{\xi\sqrt{1-e^2}\sin\psi+\cos\psi}{\sqrt{1-e^2\sin^2\psi}}\;\ldots\ldots\text{(liv)},$$

$$\left.\begin{aligned}M&=M_0-\frac{R}{2}y+\frac{Q}{2}(a-x)=\frac{Q}{2}(\eta a-\xi b\sin\psi-a\cos\psi),\\ \text{where}\quad M_0&+\frac{Qa}{2}=\eta\frac{Qa}{2}\end{aligned}\right\}\ldots\text{(lv)}.$$

Further from (vi) since $\kappa^2=c^2/4$:

$$1/h^2=1/\kappa^2\left(1-\tfrac{1}{2}\frac{c^2}{\rho_0{}^2}-\tfrac{1}{16}\frac{c^4}{\rho_0{}^4}-\right),$$

so that neglecting quantities of the order $(c/\rho_0{}^4)$ we have:

$$1/h^2=1/\kappa^2\left\{1-\tfrac{1}{2}\frac{c^2a^2}{b^4}\frac{1}{(1-e^2\sin^2\psi)^3}\right\}\;\ldots\ldots\ldots\ldots\text{(lvi)}$$

Further $m=M+\dfrac{Ph^2}{\rho_0}+\dfrac{Mh^2}{\rho_0{}^2}$ and therefore:

$$\frac{m}{h^2}=\frac{Q}{2}\left\{\frac{\eta a-\xi b\sin\psi-a\cos\psi}{\kappa^2}-\frac{a^2}{b^4}\frac{\eta a-\xi b\sin\psi-a\cos\psi}{(1-e^2\sin^2\psi)^3}+\frac{a}{b^3}\frac{\xi a\sin\psi+b\cos\psi}{(1-e^2\sin^2\psi)^2}\right\}\ldots\ldots\text{(lvii)}.$$

Whence from (xxii) with the notation of the footnote below[1] we have

$$\left.\begin{aligned}\eta\left(\frac{b^2}{\kappa^2}\gamma_1-\frac{a^2}{b^2}\gamma_{10}\right)=\xi&\left\{\frac{b^3}{\kappa^2 a}\gamma_3-\frac{a}{b}(\gamma_{11}+\gamma_8)\right\}\\ &+\frac{b^2}{\kappa^2}\gamma_4-\frac{a^2}{b^2}\gamma_{12}-\gamma_9\end{aligned}\right\}\ldots\ldots\text{(lviii)}.$$

[1] The following integrals ($\Delta\psi=\sqrt{1-e^2\sin^2\psi}$) will be of value in this discussion; the γ''s are their values for the special case of our Arts. 636–40 where $e^2=1/2$:

$$\gamma_1=\int_0^{\pi/2}\Delta\psi d\psi;\quad \gamma_1'=1{\cdot}350{,}644.$$

$$\gamma_2=\int_0^{\pi/2}\frac{d\psi}{\Delta\psi};\quad \gamma_2'=1{\cdot}854{,}075.$$

(These are the ordinary complete elliptic integrals, their usual symbols E and F being discarded to avoid confusion with the elastic moduli.)

$$\gamma_3=\int_0^{\pi/2}\sin\psi\,\Delta\psi d\psi=\frac{1}{2}+\frac{1-e^2}{4e}\log\frac{1+e}{1-e};\quad \gamma_3'={\cdot}811{,}613.$$

$$\gamma_4=\int_0^{\pi/2}\cos\psi\,\Delta\psi d\psi=\frac{\sqrt{1-e^2}}{2}+\frac{1}{2e}\sin^{-1}e;\quad \gamma_4'={\cdot}908{,}914.$$

$$\gamma_5=\int_0^{\pi/2}\frac{\sin^2\psi}{\Delta\psi}d\psi=\frac{\gamma_2-\gamma_1}{e^2};\quad \gamma_5'=1{\cdot}006{,}862.$$

$$\gamma_6=\int_0^{\pi/2}\frac{\cos^2\psi}{\Delta\psi}d\psi=\frac{\gamma_1}{e^2}-\left(\frac{1}{e^2}-1\right)\gamma_2;\quad \gamma_6'={\cdot}847{,}213.$$

For a link without stud we have only to put $\xi = 0$, and we find

$$\eta = \frac{\frac{b^2}{\kappa^2}\gamma_4 - \frac{a^2}{b^2}\gamma_{12} - \gamma_9}{\frac{b^2}{\kappa^2}\gamma_1 - \frac{a^2}{b^2}\gamma_{10}} \quad \text{.....................(lix).}$$

Winkler retains only the first terms in the numerator and denominator, thus he should have $\eta = \gamma_4/\gamma_1$.

$$\gamma_7 = \int_0^{\pi/2} \frac{d\psi}{(\Delta\psi)^3} = \frac{\gamma_1}{1-e^2}; \ \gamma_7' = 2\cdot701{,}288.$$

$$\gamma_8 = \int_0^{\pi/2} \frac{\sin\psi}{(\Delta\psi)^3} d\psi = \frac{1}{1-e^2}; \ \gamma_8' = 2.$$

$$\gamma_9 = \int_0^{\pi/2} \frac{\cos\psi}{(\Delta\psi)^3} d\psi = \frac{1}{\sqrt{1-e^2}}; \ \gamma_9' = 1\cdot414{,}214.$$

$$\gamma_{10} = \int_0^{\pi/2} \frac{d\psi}{(\Delta\psi)^5} = \frac{1}{3(1-e^2)}\left(\frac{4-2e^2}{1-e^2}\gamma_1 - \gamma_2\right); \ \gamma_{10}' = 4\cdot166{,}526.$$

$$\gamma_{11} = \int_0^{\pi/2} \frac{\sin\psi}{(\Delta\psi)^5} d\psi = \frac{1}{3}\frac{3-e^2}{(1-e^2)^2}; \ \gamma_{11}' = 3\cdot333{,}333.$$

$$\gamma_{12} = \int_0^{\pi/2} \frac{\cos\psi}{(\Delta\psi)^5} d\psi = \frac{1}{3}\frac{3-2e^2}{(1-e^2)^{\frac{3}{2}}}; \ \gamma_{12}' = 1\cdot885{,}618.$$

$$\gamma_{13} = \int_0^{\pi/2} \frac{\sin^2\psi}{(\Delta\psi)^3} d\psi = \frac{1}{e^2}\left(\frac{\gamma_1}{1-e^2} - \gamma_2\right); \ \gamma_{13}' = 1\cdot694{,}426.$$

$$\gamma_{14} = \int_0^{\pi/2} \frac{\cos^2\psi}{(\Delta\psi)^3} d\psi = \frac{\gamma_2 - \gamma_1}{e^2}; \ \gamma_{14}' = 1\cdot006{,}862.$$

$$\gamma_{15} = \int_0^{\pi/2} \frac{\sin\psi\cos\psi}{(\Delta\psi)^3} d\psi = \frac{1}{e^2}\left(\frac{1}{\sqrt{1-e^2}} - 1\right); \ \gamma_{15}' = \cdot828{,}427.$$

$$\gamma_{16} = \int_0^{\pi/2} \frac{\sin^2\psi}{(\Delta\psi)^5} d\psi = \frac{1}{3e^2(1-e^2)}\left(\frac{1+e^2}{1-e^2}\gamma_1 - \gamma_2\right); \ \gamma_{16}' = 2\cdot930{,}476.$$

$$\gamma_{17} = \int_0^{\pi/2} \frac{\cos^2\psi}{(\Delta\psi)^5} d\psi = \frac{1}{3e^2}\left(\gamma_2 - \frac{1-2e^2}{1-e^2}\gamma_1\right); \ \gamma_{17}' = 1\cdot236{,}050.$$

$$\gamma_{18} = \int_0^{\pi/2} \frac{\cos\psi\sin\psi}{(\Delta\psi)^5} d\psi = \frac{1}{3e^2}\left\{\frac{1}{(1-e^2)^{\frac{3}{2}}} - 1\right\}; \ \gamma_{18}' = 1\cdot218{,}951.$$

$$\gamma_{19} = \int_0^{\pi/2} \frac{\cos\psi\sin\psi}{\Delta\psi} d\psi = \frac{1}{e^2}(1 - \sqrt{1-e^2}); \ \gamma_{19}' = \cdot585{,}786.$$

$$\gamma_{20} = \int_0^{\pi/2} \sin^2\psi\Delta\psi d\psi = \frac{1}{3e^2}\{(1-e^2)\gamma_2 - (1-2e^2)\gamma_1\}; \ \gamma_{20}' = \cdot618{,}025.$$

$$\gamma_{21} = \int_0^{\pi/2} \cos^2\psi\Delta\psi d\psi = \frac{1}{3e^2}\{(1+e^2)\gamma_1 - (1-e^2)\gamma_2\}; \ \gamma_{21}' = \cdot732{,}619.$$

$$\gamma_{22} = \int_0^{\pi/2} \cos\psi\sin\psi\Delta\psi d\psi = \frac{1}{3e^2}\{1 - (1-e^2)^{\frac{3}{2}}\}; \ \gamma_{22}' = \cdot430{,}964.$$

[635.] The second relation between ξ and η must now be found from (xxiii). After some reductions we have since:

$$E\omega\Delta a = -nRa = -n\xi Qa$$

$$2n\xi = \eta\left(\frac{b^2}{\kappa^2}\gamma_3 - \frac{a^2}{b^2}(\gamma_{11}+\gamma_8)\right) + \xi\left(-\frac{b^3}{a\kappa^2}\gamma_{20} + \frac{a}{b}(\gamma_{16}+2\gamma_{13}-\gamma_5)\right.$$
$$\left. -\frac{b^2}{\kappa^2}\gamma_{22} + \frac{a^2}{b^2}(\gamma_{18}+\gamma_{15}) + \gamma_{15} - \gamma_{19}\right) \quad \text{..............(lx).}$$

Equations (lviii) and (lx) enable us to completely solve the problem. If there be no stud we put $\xi = 0$ on the right and replace $n\xi$ by $-\frac{E\omega}{Q}\frac{\Delta a}{a}$ on the left of (lx), using (lix) with it. If there be a stud (lviii) and (lx) give the values of ξ and η.

The following is the value of Δb

$$\frac{2E\omega}{Q}\frac{\Delta b}{b} = \eta\left\{-\frac{a^2}{\kappa^2}\gamma_4 + \frac{a^4}{b^4}\gamma_{12} + \frac{a^2}{b^2}\gamma_9\right\} + \xi\left\{\frac{ab}{\kappa^2}\gamma_{22} - \frac{a^3}{b^3}(\gamma_{18}+\gamma_{15}) - \frac{a}{b}(\gamma_{15}-\gamma_{19})\right.$$
$$\left. + \frac{a^2}{\kappa^2}\gamma_{21} - \frac{a^4}{b^4}\gamma_{17} - 2\frac{a^2}{b^2}\gamma_{14} + \gamma_6\right\} \quad \text{.........(lxi).}$$

Further the neutral line is given by

$$v_0 = -\frac{\frac{b^2}{a}\left\{\frac{\eta a}{b} - e^2\left(\xi\sin^3\psi - \frac{b}{a}\cos^3\psi\right)\right\}(1-e^2\sin^2\psi)^{\frac{3}{2}}}{\frac{\eta a}{b} - e^2\left(\xi\sin^3\psi - \frac{b}{a}\cos^3\psi\right) + \left(\frac{\eta a}{b} - \xi\sin\psi - \frac{a}{b}\cos\psi\right)\left(\frac{b^4}{a^2\kappa^2}(1-e^2\sin^2\psi)^3 - 2\right)},$$
$$\text{..................(lxii),}$$

and the traction in extrados and intrados by

$$T\omega = P \pm \frac{M}{c}\left(4 \mp \frac{3c}{\rho_0} + \frac{2c^2}{\rho_0^2} \mp \frac{2c^3}{\rho_0^3}\right) \quad \text{......... (lxiii)}[1],$$

P, M, and ρ_0 being substituted from (liii)–(lv).

The values of these quantities might be traced for a link either with or without stud as in the case of the circular link, but the discussion must be omitted here, and we confine ourselves to the consideration of a numerical example.

[1] The full discussion of these tractions would be complex, but the maximum maximorum after the investigations of Arts. 625, 627–8, and 631 may be assumed to exist at the loaded sections, $\psi = \pm\pi/2$. We have then, if cb/a^2 be small:

for the extrados:

$$\frac{T\omega}{Q} = \tfrac{1}{2}\xi + \frac{\eta a - \xi b}{2c}\left(4 - \frac{3cb}{a^2} + \frac{2c^2b^2}{a^4} - 2\frac{c^3b^3}{a^6}\right),$$

for the intrados:

$$\frac{T\omega}{Q} = \tfrac{1}{2}\xi - \frac{\eta a - \xi b}{2c}\left(4 + \frac{3cb}{a^2} + \frac{2c^2b^2}{a^4} + \frac{2c^3b^3}{a^6}\right).$$

These do not appear to agree with the results on S. 245 of Winkler's memoir even when we neglect c^2/a^2.

[636.] As a numerical example we cannot take exactly Winkler's case because his numbers do not appear to be consistent. Suppose we put $c = a^2/b - c$ and $b = 4c$, then $a^2/b^2 = \cdot 5$, or $e^2 = 1/2$, whence $a/b = e = \cdot 707,107$ and $a = 2\cdot 828,427c$.

[637.] Applying these values and those of the footnote on pp. 440–1 to equations (lviii) and (lx), we find:

$$84\cdot 357,945\,\eta = 69\cdot 687,551\,\xi + 55\cdot 813,461 \quad\text{..............(lxiv)},$$

$$49\cdot 276,540\,\eta = (2n + 52\cdot 180,746)\,\xi + 26\cdot 315,392, \quad\text{......(lxv)},$$

or, if the link be of wrought-iron and the stud of cast-iron of the same relative dimensions as in our Art. 628, $n = 3$ and:

$$49\cdot 276,540\,\eta = 58\cdot 180,746\,\xi + 26\cdot 315,392 \quad\text{.........(lxvi)}.$$

[638.] First suppose $\xi = 0$ in (lxiv), then we have for η

$$\eta = \cdot 661,627 \quad\text{........ ....(lxvii)},$$

while Winkler has for the corresponding quantity ·670,32. I believe that ·662 is very near the correct value.

Putting $n\xi = -\dfrac{E\omega}{Q}\dfrac{\Delta a}{a}$ and $\xi = 0$ in (lxv) I find:

$$\frac{\Delta a}{a} = -3\cdot 143,638\,\frac{Q}{E\omega} \quad\text{.....................(lxviii)}$$

Winkler's result after some reductions yields with considerable divergence from mine:

$$\frac{\Delta a}{a} = -3\cdot 568,917\,\frac{Q}{E\omega}.$$

From (lxi),
$$\frac{\Delta b}{b} = 2\cdot 255,656\,\frac{Q}{E\omega} \quad\text{.............. ....(lxix)},$$

while Winkler has $\dfrac{\Delta b}{b} = 2\cdot 134,83\,\dfrac{Q}{E\omega}$. For the oval link in the treatise he finds $\dfrac{\Delta b}{b} = 2\cdot 252,5\,\dfrac{Q}{E\omega}$.

Lastly for the maximum positive and negative tractions we have:

$$T\omega = 2\cdot 641\,Q \quad\text{and}\quad T\omega = -6\cdot 050\,Q$$

of which the latter is the greater and may therefore be taken to measure the strength of a wrought-iron link of these dimensions.

Winkler's numbers give $T\omega = -5\cdot 204\,Q$, a result which appears much too small. The fact is that for our present case neither (lxiii) nor the series in the footnote on p. 442 are sufficiently approximate. Supposing the value of η to be correct, I have obtained the value above by using (xiii). For an oval ring with $a = 2\cdot 5c$, $b = 4c$, made up of four circular arcs Winkler in his treatise (p. 375) finds $T\omega = -6\cdot 3735\,Q$, which tends to confirm the result we have found for the elliptic link.

[639.] Let us compare the strengths, weights and longitudinal extensions of three links, one elliptic, one circular and the third with circular ends and flat sides. Suppose them all to have a longitudinal semi-axis $b = 4c$, and for the ellipse suppose $a/b = 1/\sqrt{2}$ as above; for the flat-sided link suppose the curvature of its circular ends equal to that of the ellipse at the ends of its longer axis; this involves in the notation of our Art. 631, $a = e = \frac{1}{2}b$.

If T', T'', T''' be the maximum compressive stresses, w', w'', w''' the weights of the links, $\Delta b'$, $\Delta b''$, $\Delta b'''$ the semi-extensions we find from Equations (xxvii), (xxix), (xlviii), (lii) and Art. 638:

$$T' : T'' : T''' :: 6\cdot050 : 6\cdot159^1 : 4\cdot904$$
$$w' : w'' : w''' :: 5\cdot4026 : 6\cdot2832 : 5\cdot1416$$
$$\Delta b' : \Delta b'' : \Delta b''' :: 2\cdot2557 : 4\cdot9242 : 1\cdot3085.$$

Whence, generalising from the results of these particular links, it would appear that elliptic and circular links *of the same length* are not very different in strength, that the elliptic link stretches about only one half of what the circular one does and weighs less; but that the link with flat sides and circular ends is, if *of the same length*, stronger than either of the others, less heavy and stretches considerably less than the elliptic one. Thus such a link is distinctly the best of the three forms considered, and in fact is frequently adopted in practice.

[640.] Suppose the link to have a cast-iron stud. Then we find the following values of η and ξ from equations (lxiv) and (lxvi).

$$\eta = \cdot958{,}866, \quad \xi = \cdot359{,}813,$$

values certainly not to be trusted beyond the third decimal place. Winkler (S. 244–6) finds the very different values:

$$\eta = 1\cdot211{,}500, \quad \xi = \cdot631{,}804.$$

We have also from equations (lx) and (lxi):

$$\frac{\Delta a}{a} = -1\cdot079{,}440\frac{Q}{E\omega}, \qquad \frac{\Delta b}{b} = 1\cdot455{,}831\frac{Q}{E\omega}$$

Winkler has for the numerical coefficients for the case of the ellipse in the memoir ·055,353 and ·866,752 and for the oval link in the treatise 1·808,305 and ·743,774. I think these coefficients in both the memoir and treatise are incorrect. For the oval link Winkler has Δb actually $< \Delta a$ (S. 376) and this seems extremely improbable for a

[1] This value differs in the first place of decimals from that given by Winkler in his treatise, S. 372, but his approximations are very rough.

link with a stud. We notice that the effect of studs of the character considered above on elliptic links is to reduce the stretch of the chain to less than two-thirds of the stretch in a chain made of links without studs.

Finally for the maximum compressive stress I find from (xiii):

$$T\omega = -3\cdot9353\,Q.$$

Winkler in his memoir has (S. 246):

$$T\omega = -2\cdot092{,}872\,Q,$$

a value, I think, much too small. For the oval link of his treatise he finds (S. 376):

$$T\omega = -3\cdot082{,}472\,Q,$$

which differs more widely than I should have anticipated from my result for the elliptic link. It will be noticed that the strength of the elliptic link with stud is more than 1·5 times as great as that of the link without stud.

[641.] In the above articles I have corrected and developed Winkler's theory as the best yet available for stress and strain in the links of chains. The calculations have been laborious and I cannot hope they are even now absolutely free from error, still I believe that I have avoided some of the slips of Winkler. The theory can only be approximate at best, and the six places of decimals to which some of the results are calculated must not be supposed to suggest any real accuracy beyond the first two or three figures, unless the dimensions of the cross-section are small as compared with the radius of curvature of the link. This remark applies particularly to Winkler's numerical example of an elliptic link, which with certain modifications we have followed. More accurate results would have been obtained by taking the eccentricity still $= 1/\sqrt{2}$, but b equal to 6 or even 10 times c. The formulae we have given for the ellipse may be readily applied to centrally loaded elliptic arches as well as to complete elliptic springs.

The *absolute strength* of chains will be found in reality to be greater than would be given by the above formulae for the maximum compressive stress. Such formulae ought only to be applied to obtain the *fail limit*: see our Arts. 5 (*e*) and 169 (*g*). Before rupture is reached set has changed the shape of the link, and the links press upon and hold each other, till in some cases the absolute strength of a chain appears to be close upon the absolute shearing or even tensile strength of the material: see Section III. of this Chapter.

[642.] E. Winkler: *Festigkeit der Röhren, Dampfkessel und Schwungringe. Der Civilingenieur*, Bd. VI., S. 325–62 and S. 427–62. Freiberg, 1860. This is a lengthy analytical memoir, which so far as its methods are concerned is more likely to be intelligible to the mathematician than to the practical engineer. It commences with a brief reference to Scheffler's memoir on tubes, remarking that his hypotheses, that there is no longitudinal expansion resulting from lateral pressure, and that the maximum traction is the measure of strength, are both alike unacceptable: see our Art. 654.

[643.] The first section of the paper entitled: *Allgemeines* (S. 326–38) contains a general discussion of the resolution of stress and strain, remarks on the value of the stretch-squeeze ratio (η)—which Winkler proposes to take either $\frac{1}{4}$ or $\frac{1}{3}$ according to the material,—the consideration of a stretch limit of strength, and finally expressions for the stresses in terms of the shifts in the case of cylindrical coordinates and bi-constant isotropy.

[644.] Section II. (S. 338–47) contains the theory of right cylindrical tubes with open ends; there is nothing of real importance in the section which had not already been given by Lamé, or which we have not reproduced in a more general and accurate form from Saint-Venant: see our Arts. 1012*, 1087*–8* and 120.

Section III. (S. 348–62) deals with the same form of tubes with closed ends either hemi-spherical or plane. The treatment of spherical shells presents no novelty and Winkler seems to have missed Lamé's method of fitting the cylindrical and spherical parts of a boiler by a proper choice of thicknesses: see our Arts. 1038* and 125.

The treatment of the flat ends, or of circular plates (S. 355–62) under a uniform surface pressure, is based upon the assumption that lines in the plate perpendicular to the mid-plane before strain, remain perpendicular to that plane after strain.

This problem had already been fully worked out for a thin plate by Poisson (see our Arts. 495* and 502*), and another problem very like it for plate of moderate thickness (2ϵ) has been considered in our Arts. 329–30. Winkler assumes that even with surface pressure (cf. our Art. 325) we may neglect the traction perpendicular to the mid-plane of our plate, or put in the notation of our Art. 329, $\widehat{zz}=0$; this leads to the equation:

$$\frac{d\widehat{rz}}{dr}+\frac{\widehat{rz}}{r}=0,$$

whence as in Art. 330 we ought to have $\widehat{rz}$ of the form:

$$\widehat{rz}=\text{const.}\times\frac{\epsilon^2-z^2}{r}.$$

Winkler by not very intelligible reasoning deduces:

$$\widehat{rz}=\text{const.}\times r\,(\epsilon^2-z^2),$$

or, a value which does not satisfy the body-stress equations. It follows that his values for $\widehat{rr}$, $\widehat{\phi\phi}$ and for u, w are all wrong. Thus neither his results nor the inferences he draws from them as to the thickness for the plane ends of cylindrical boilers need further consideration.

[645.] The fourth section of the memoir (S. 427–48) is entitled: *Einfluss der Endflächen, des Gewichts der Röhre und des ungleichen Wasserdruckes.* This investigation seems to me absolutely unreliable and quite as nugatory as that of Scheffler. In the first place (with the notation of our footnote p. 79) Winkler neglects the traction $\widehat{\phi\phi}$, i.e. the traction perpendicular to a meridian plane of the cylinder (mean radius a), in the next place he assumes w to be of the form $f_1(z)+f_2(z)\,r$, then he takes u to be independent of r, which he says is legitimate if 2ϵ, the thickness of the cylindrical wall, be very small. Lastly he neglects a term $16\eta\,\frac{\epsilon^2}{a^4}\,\frac{d^2u}{dz^2}$ as compared with d^4u/dz on the ground that ϵ^2/a^2 is very small (Equation 119, S. 430). On his own showing the ratio of d^4u/dz^4 to d^2u/dz^2 is of the order $1/l^2$ where l is the length of the cylinder; hence for a cylinder in which l/a is great his results will not be correct, and were it only for this assumption, i.e. they would not be true for flues.

In the part of the memoir which deals with the influence of the weight of the cylindrical tube, Winkler supposes a ring cut out of the cylinder by two planes perpendicular to its axis *at unit distance* and calculates the effect of the weight of this ring in deforming itself after the manner of his memoir of 1858 (see our Art. 622). But I have elsewhere given reasons (see our Arts. 1547*, 537) for questioning such a method of treatment. We might just as fitly apply it to solve the problem of the cylindrical shell subjected to external and internal pressures. All Winkler really works out in these pages is the effect of weight in distorting a thin circular belt of unit breadth placed in a vertical plane. In doing this he neglects quantities of the order (thickness/diameter)^2.

If a be the radius of the circular ring, 2ϵ its thickness supposed of rectangular cross-section ($2\epsilon\times 1$), ρ its density, Winkler finds for the maximum bending moment M', which is found at the lowest point, or point of support:

$$M' = -3a^2\epsilon g\rho.$$

Further he has:

$$\text{Maximum compressive stress} \quad = \frac{3M'}{2\epsilon^2} = -\frac{9a^2 g\rho}{2\epsilon},$$

$$\text{Extension of horizontal diameter} = 3(4-\pi)\frac{a^4 g\rho}{2E\epsilon^2},$$

$$\text{Compression of vertical diameter} = 3(8-\pi^2)\frac{a^4 g\rho}{4E\epsilon^2}.$$

These results are correct for a slender belt resting on its lowest point and subjected only to the action of its own weight, i.e. when terms, whose ratio to those retained equals ϵ^2/a^2, are neglected[1]. They have no legitimate application to the case of a heavy cylindrical shell. (S. 442.)

Winkler's further investigation by a similar process of the strain in a cylinder which is only partially filled with water and is thus subjected to different internal pressures in its lower and upper portions seems to me equally questionable. (S. 442–46.)

Allowing for the weight of the cylinder and of the water in it, Winkler finally gives for the thickness 2ϵ of a cylindrical boiler of internal radius r_1, σ being the density of water (S. 447):

$$2\epsilon = r_1\left[\frac{29}{48}\frac{P}{T} + \frac{12}{n}\frac{r_1 g\rho}{T} + \sqrt{\left(\frac{29}{48}\frac{P}{T} + \frac{6}{n}\frac{r_1 g\rho}{T}\right)^2 + \frac{15 g\sigma r_1}{4nT}}\right] + \tau.$$

Here τ is a constant added to allow for wear and tear (see Section III. of this Chapter), and n is a factor (which Winkler puts $= 3$) taken to *reduce* the values calculated for the effects of the weights of the cylinder and of the water inside it. P is the steam pressure and T the safe tensile stress of the material. The author says (S. 446):

Diese Werthe sind allerdings zu gross, da die Deformirung des Kessels durch die Einmauerung, durch die Boden, sowie durch etwaige Bänder, welche man um die Kessel legt und welche zur Erhöhung der Sicherheit sehr zu empfehlen sind, geschwächt wird.

A theory which gives such *large* values that they have to be corrected by arbitrary factors can hardly be considered satisfactory. I give the result for what it may be worth, but express no confidence whatever even in its approximate accuracy. Winkler reduces it to numbers and compares it with a formula which he says is usual in Prussia (Brix's formula with an exponential: see Section III. of this Chapter). The two frequently give very divergent results.

[1] I find for such a ring of any cross-section ω in the notation of this work:

$$\text{Maximum bending moment} \quad = -\frac{3\omega\rho g a^2}{2},$$

$$\text{Stretch of horizontal diameter} = \frac{\rho g a}{E}\frac{a^2}{\kappa^2}\left(1-\frac{\pi}{4}\right),$$

$$\text{Squeeze of vertical diameter} \quad = \frac{\rho g a}{E}\frac{a^2}{\kappa^2}\left(2-\frac{\pi^2}{4}\right),$$

which agree with Winkler's results for the special case.

[646.] Section V. of the memoir is entitled: *Schwungringe* (S. 448–62). Winkler first gives a theory of fly wheels in which the influence of the spokes is neglected. He further supposes the traction perpendicular to the meridian plane uniform across the cross-section whence he easily finds for its value

$$\widehat{\phi\phi} = \rho a^2 \left(1 + \frac{\kappa^2}{a^2}\right) \omega^2,$$

ω being the spin of the wheel, ρ the density of its material, and κ the swing-radius of the cross-section about an axis in its plane through its centroid and a the distance of that centroid from the axis of the wheel. Winkler (S. 450) goes even so far as to apply this formula to mill- and grind-stones Compare our Art. 590.

In a *genauere Theorie* (S. 451–4) Winkler puts $\widehat{zz} = 0$ and $\widehat{zr} = 0$. This appears to be really identical with Maxwell's theory (see our Arts. 1550*–51*). Winkler finds, if T be the safe limit of tractive stress that:

$$\omega < 2\sqrt{\frac{T}{\rho\left[(1-\eta)\,r_1^2 + (3+\eta)\,r_2^2\right]}},$$

r_1 and r_2 being the inner and outer radii of the rim. This theory for $\eta = 1/4$ gives a result for an entire disc almost in agreement with that given by the first theory (S. 454).

[647.] Finally (S. 454–62) Winkler attempts to take into account the influence of the spokes. He practically follows the lines of Resal's investigation (see our Art. 584), except that he treats first the case when the portion of the rim of the wheel between two spokes can be considered as pivoted at the spokes. Winkler's results are complex and not put into a form which permits of easy citing. I have not verified them. He takes the values of s_0 and $d\,(\Delta\phi)/d\sigma_0$ given in his memoir of 1858 (see our Art. 620, Equations (x) and (xi)), thus his results if his analysis be correct, might be more exact than Resal's for the case when Resal's ϵ^2 (see our Art. 585) is not negligible. It usually will be negligible in practice.

The memoir is rather cumbersome and while containing some interesting points is spoiled by a number of assumptions for which no strong reasons are given, if indeed they exist.

[648.] Hermann Scheffler: *Theorie der Festigkeit gegen das Zerknicken nebst Untersuchungen über die verschiedenen inneren Spannungen gebogener Körper und über andere Probleme der Biegungstheorie mit praktischen Anwendungen.* Braunschweig, 1858, S. 1–138.

The author of this book—a practical architect—had already published a volume entitled: *Theorie der Gewölbe, Futtermauern und eisernen Brücken* (see Section III. of our present Chapter), of which he writes that the object was identical with that of the present work—namely to bring more closely together the scientific and practical sides of the subject. The present volume deals with the buckling load and strength of struts, the slide of beams under flexure, and the calculation of the stresses in continuous beams—all problems which have much exercised the practical engineer. Two of these problems had received fairly complete theoretical solutions before 1858, but, as so often occurs, the mathematical investigation failed to reach the hands of the technologists.

[649.] S. 1–58 of the work are devoted to the discussion of the strut problem. We have already seen what erroneous results Euler's theory for the buckling load of struts gives when the length is not very much greater than the diameter of the cross-section. This fact caused Hodgkinson to entirely discard that theory in favour of an empirical formula, and led Lamarle to limit the theory to such struts as had not passed the elastic limit before the buckling load was reached: see our Arts. 958*–961*, 1258*. Lamarle's limitation is quite unrecognised by Scheffler. He starts from Euler's formula for the buckling load P of a column (see Arts. 67* and 74*), or

$$P = E\omega\kappa^2\pi^2/l^2$$

and shews that this does not agree with experiment. He modifies the theory as given by Euler and Lagrange by placing, as a result of the compression, the 'neutral axis' in an eccentric position, he thus obtains for a doubly-pivoted strut the formula

$$P = E\omega \frac{1}{1 + \dfrac{l^2}{\pi^2\kappa^2}}$$

See *Corrigenda* to our Vol. I. p. 2.

But this formula does not give absolute values agreeing with experiment, so that Scheffler after citing one or two other semi empirical formulae by various authors, proceeds to propound a modified theory of his own. Briefly the modification consists in the hypothesis that the longitudinal load on the terminal sections of the column or strut is not exactly *central*. By this means he endeavours to explain the discrepancy between theory and experiment. In doing this he adopts a true eccentric position for the neutral axis, but assumes in comparing his theory with Hodgkinson's experiments that the proportionality of stress and strain holds up to rupture. We will examine one of Scheffler's results somewhat at length:

Let the longitudinal load P applied at a distance b from the axis of the strut, produce a deflection at the mid-point $= a - b$. Let l be the length of the strut, which will be supposed doubly-pivoted, and $\omega\kappa^2$ the moment of inertia of the cross-section about a line through its centroid perpendicular to the load plane. Then if we take as axes the direction of the vertical load (x) and the perpendicular upon this from the mid point of the central line of the strut (y), we easily deduce for the equation to the distorted central line:

$$y = b \sec \frac{l}{2\beta^{\frac{1}{2}}} \cos \frac{x}{\beta^{\frac{1}{2}}} \quad \text{..................(i)},$$

where $$\beta = \left(\frac{E\omega}{P} - 1\right)\kappa^2$$

The curvature at the mid-section of the central line

$$= 1/R_0 = -\left(\frac{d^2y}{dx^2}\right)_0 = \frac{b}{\beta} \sec \frac{l}{2\beta^{\frac{1}{2}}} \quad \text{..................(ii)},$$

and the deflection $(a - b)$ is obtained from

$$a = b \sec \frac{l}{2\beta^{\frac{1}{2}}} \quad \text{..................(iii)}.$$

Equation (iii) gives a relation between the load and the deflection which, introducing the value of β, leads to:

$$P = \frac{E\omega}{1 + \dfrac{l^2}{4\kappa^2\left(\cos^{-1}\dfrac{b}{a}\right)^2}} \quad \text{..................(iv)}.$$

If $b = 0$, this coincides with the value given for the buckling load on the hypothesis of the eccentric neutral axis.

To find the elastic strength of the strut, if the elastic limit be reached first in compressive stress, say at the value C, we have to equate C to the maximum compressive stress which arises in the extreme 'fibre' at the mid cross-section. Let the distance of this fibre from the central axis be h, then the stress

$$= \frac{P}{\omega} + E\frac{h}{R_0} = C,$$

or, $$h = R_0\left(\frac{C}{E} - \frac{P}{E\omega}\right) = \frac{\beta}{b}\left(\frac{C}{E} - \frac{P}{E\omega}\right) \cos \frac{l}{2\beta^{\frac{1}{2}}}.$$

Substituting for β and remembering that unity in all practical cases may be neglected as compared with E/C or $E\omega/P$ we find:

$$\frac{hb}{\kappa^2} = \left(\frac{C\omega}{P} - 1\right) \cos \frac{l}{2\beta^{\frac{1}{2}}}.$$

Let us put $P/\omega = p$, and $p_0\omega$ equal the value of the buckling load as given by Euler's theory, then we have

$$\frac{hb}{\kappa^2} = \left(\frac{C}{p} - 1\right) \cos\left(\frac{\pi}{2}\sqrt{\frac{p}{p_0}}\right) \quad \text{..................(v)}.$$

This equation (v) agrees with Scheffler's equation (53) S. 25, and gives the limiting safe load p per unit section for any doubly-pivoted strut. At the same time it must be noted that b is a perfectly arbitrary constant. Till some hypothesis is made with regard to it (v) only shews us that non-central application of the thrust does influence the value of p, but does not indicate the amount.

Scheffler supposes the terminals hemispherical. In this case, if ϕ be the angle between the central line and the direction of the load:

$$\tan\phi = -\frac{dy}{dx} = \frac{b}{\beta^{\frac{1}{2}}}\tan\frac{l}{2\beta^{\frac{1}{2}}}.$$

Further $$b = h\sin\phi,$$

whence we find $$b = h\sqrt{1-\frac{\beta}{h^2}\cot^2\frac{l}{2\beta^{\frac{1}{2}}}} \quad\text{..................(vi).}$$

Scheffler takes $b = \frac{h}{n}\sin\frac{l}{2\beta^{\frac{1}{2}}}$ and says n is a constant depending only on the material and form of the end (S. 26). It is quite true that (vi) substituted in (v) leads to a very complicated expression for p, but why this is capable of being "replaced practically" by the simpler formula Scheffler takes, I fail to understand. I am compelled to look upon his 'coefficient of correction' n as a function of the load p. Substituting his value of b in (v) we have:

$$\frac{h^2}{n\kappa^2} = \left(\frac{C}{p}-1\right)\cot\left(\frac{\pi}{2}\sqrt{\frac{p}{p_0}}\right) \quad\text{................(vii).}$$

From this formula he calculates the value of p, putting for:

Cylindrical	columns	of cast	iron	$n = 6$,
"	"	" wrought	"	$n = 24$,
Square	"	" oak		$n = 6$,
"	"	" deal		$n = 3$.

The results obtained are compared, *not* with the numbers of Hodgkinson's actual experiments but with the results calculated from Hodgkinson's empirical formulae. There is a general agreement, but it does not seem to me sufficient to overcome the difficulties I feel with regard to the value chosen for b: see S. 29–39 of the book.

[650.] Scheffler makes the eccentricity of the load a function of the material, which it must be confessed is difficult to understand. Further the eccentricity is *not small* as compared with the linear dimensions of the cross-section. Supposing it were and the ends truly hemispherical, then we should have: $\tan\phi = \phi = \sin\phi = b/h$ and therefore,

$$\beta^{\frac{1}{2}}/h = \tan l/(2\beta^{\frac{1}{2}}),$$

or, $$\cot\left(\frac{\pi}{2}\sqrt{\frac{p}{p_0}}\right) = \frac{2h}{l}\left(\frac{\pi}{2}\sqrt{\frac{p}{p_0}}\right).$$

Hence, if $d = 2h$ be the diameter and this be small as compared with l, we have after some approximations:

$$p = p_0\left(1-\frac{2d}{l}\right) \quad\text{........................(viii).}$$

This, however, gives as a rule far too large results, i.e. values of p which far exceed those given by Hodgkinson's experiments for *rupture*. Thus we cannot suppose the load applied close to the centre of the terminal cross-section, if the eccentricity is to account for the observed differences.

We may, however, obtain what is, perhaps, theoretically a better formula than Scheffler's in the following manner. We do not know what function b is of the deflection, but as we attempt to centre the load, when the deflection is zero, we will assume the eccentricity b of the loading to be proportional to the deflection $(a-b)$, and thus:

$$b = b_0\left(1 - \frac{b}{a}\right).$$

For a very long strut b is insensible as compared with the deflection $(a - b)$ and therefore as compared with a. Hence b_0 is the value of b for a very long strut. The terminal section of such a strut, in whatever manner the load be distributed over it, cannot have any 'fibres' in tension, hence the limiting position of the load point must correspond to the neutral axis just touching the section. This would be the farthest distance of the load point from the centre, and would I think be not an unreasonable condition of things to assume as existing in a long strut just before the limiting stress is reached. In this case $h \times b_0 = \kappa^2$, and therefore $b = (\kappa^2/h)(1 - b/a)$. Using (iii) and (v) we have:

$$1 - \cos\left(\frac{\pi}{2}\sqrt{\frac{p}{p_0}}\right) = \left(\frac{C}{p} - 1\right)\cos\left(\frac{\pi}{2}\sqrt{\frac{p}{p_0}}\right),$$

whence $$p/C = \cos\left(\frac{\pi}{2}\sqrt{\frac{p}{p_0}}\right) \quad\ldots\ldots\ldots\ldots\ldots\ldots \text{(ix)}.$$

For a very short strut p_0 is immensely greater than p, or we have as we should expect $p = C$.

For a short strut in which p/p_0 is small we may expand the cosine and we have after some reductions:

$$p = \frac{p_0 C}{p_0 + \frac{\pi^2}{8} C} \quad\ldots\ldots\ldots\ldots\ldots\ldots \text{(x)}.$$

This agrees very fairly with the Gordon-Rankine formula: see our Art. 469. For example that formula in our present notation gives

$$p = \frac{p_0 C}{p_0 + \frac{E}{nC}\pi^2 C},$$

where n is a certain constant empirically selected. For cast-iron we have $E = 16{,}000{,}000$, $C = 80{,}000$ and $n = 1{,}600$ (according to Rankine), hence it follows $\frac{E}{nC} = \frac{1}{8}$. For wrought-iron we have $\frac{1}{10 \cdot 8}$ instead of $1/8$, and for timber (taking $E = 2{,}000{,}000$, say) about the same.

When p/p_0 is not small we must use (ix) as it stands, unless $p = p_0$ nearly. In this case p_0 must be small, i.e. l very large. Hence p/C is small, and putting $\sqrt{p/p_0} = 1 - x$ we find

$$\sin \frac{\pi x}{2} = p/C, \text{ or } x = \frac{2}{\pi} p/C.$$

Thus we deduce

$$p = \frac{p_0 C}{C + \frac{4}{\pi} p_0} \text{..........................(xi)},$$

which gives the correction on $p = p_0$ for a large value of l. Corresponding formulae for the cases of doubly-built-in and built in pivoted struts are easily deduced.

[651.] On S. 43–58 Scheffler deals with a variety of cases in which the terminal loads on the strut are inclined to its central line as well as eccentric. His results are all fairly easy deductions from the ordinary theory, but some of them—e.g. those for rods under the action of three forces (S. 48–49)—are very interesting and would probably give accurate forms for metal ribbons under such loading. S. 58–73 deal with braced girders with parallel straight booms. The calculation of the stresses in the bracing bars would as a rule be now dealt with graphically. It is difficult to understand how any of the bracing bars in Fig. 28 can be in tension, yet I imagine the alternate ones ought to be. The whole investigation does not seem in the light of recent work to have any importance. Scheffler points out that for bracing bars it is usual to take the length not more than 24 times the least diameter of the cross-section, but that for this ratio the buckling strength of wrought-iron struts is $\frac{2}{3}$ the compressive strength and therefore very nearly equal to the tensile strength. Hence for practical purposes the tensile strength can always be taken to determine the dimensions of a bar. As in most practical cases bracing bars are subject to alternate stress, this, if correct, would give the convenient rule that the dimensions are to be determined from the maximum load without regard to its sign.

[652.] Scheffler next endeavours to introduce the conception of slide into the theory of beams under flexure. This is done very much on the lines of Jouravski and Bresse: see our references Art. 582 (c). If x be the direction of the central axis of the beam, y perpendicular to the plane of flexure and in that plane, this theory fails because it deals only with the shear and omits to consider the shear $\widehat{yz}$; it likewise omits all consideration of $\widehat{xy}$. As Saint-Venant's great memoir of 1854 had solved the problem, there is no need to enter into Scheffler's struggles of four years later date. It is characteristic of his method that the equality $\widehat{zx} = \widehat{xz}$ is announced as "die bemerkenswerthe Thatsache dass in jedem Punkte des Balkens *die horizontalen und vertikalen Abscherungskräfte pro Flächeneinheit einander gleich sind*" (S. 79), and the discovery of this remarkable fact is attributed to Laissle and Schübler!

The discussion of the distribution of stress in a beam under flexure (S. 82–92) is historically interesting as one of the early attempts in this direction, and is quite as accurate as those which are still to be found in several English textbooks.

On S. 112–138 Scheffler returns to the influence of slide in beams under flexure. His results here seem to be entirely erroneous. Thus in the notation of our Art. 83, he finds that for a rectangular cross-section we must have:

$$\mu \frac{d^2u}{dz^2} + E\frac{d^2u}{dx^2} = 0 \text{ ....(S. 120)},$$

which equation leads to an absurdity when we combined it with the result of substituting the value of F from (18') in the first equation of (19') of the same article. I have not thought it worth while to follow out the whole of Scheffler's analysis. His first assumption that u is independent of y is at least one fruitful source of error.

[653.] On S. 95–109 we have a method described for dealing with the problem of continuous beams, or beams passing over several points of support and having only transverse loading in a vertical plane. The method depends upon certain fairly obvious relations between the position of the points of support, the points of inflexion, and the points of maximum or minimum curvature on the central axis. Scheffler obtains an easy geometrical construction for the points of maximum and minimum curvature in the case of a *uniformly loaded beam* (S. 103), which might find its way into practical textbooks. For any general system of loading the graphical methods of Mohr, Culmann and Ritter or the application of Clapeyron's Theorem (see our Art. 603) are, I think, superior to what is here suggested. Three pages (109–111) on the bending of a beam into a given shape present no novelty and seem of no practical interest.

A criticism of Scheffler's work by Grashof will be found in the *Zeitschrift des Vereins deutscher Ingenieure. Dritter Jahrgang*, 1859, S. 338–43. Grashof rejects Scheffler's theory of buckling and his treatment of braced bars (S. 58–73). On the other hand he praises certain of the later portions of the work.

[654.] H. Scheffler: *Die Elastizitätsverhältnisse der Röhren, welche einem hydrostatischen Drucke ausgesetzt sind, insbesondere die Bestimmung der Wanddicke desselben. Eine für das Ingenieurwesen wichtige Erweiterung der Biegungstheorie.* Wiesbaden, 1859,

S. 1–67. This is a reprint from the *Organ für die Fortschritte des Eisenbahnwesens* for 1859.

The memoir deals with a very important problem in hydraulics and gunnery, namely the strength of tubes subjected to internal pressure and the effect obtained by strengthening them by belts or bands of very inelastic metal. If the author's analysis could be trusted such belts while reducing the stress at certain points increase it at others. Accordingly, as he takes a stress limit instead of a stretch limit for safety, he concludes that such bands have in reality no strengthening effect. Whether they have or not is certainly not determined by the present memoir for the analysis is vitiated by errors of a most vital kind, so that I do not see any reason for supposing the results to be even approximately true.

[655.] The author begins by referring to the paper by Blakely (see Section III. of this Chapter). He then shews why certain empirical formulae proposed by Barlow and Brix for the strength of an endless tube subjected to external and internal fluid pressure are erroneous. He next proceeds to deduce the formula of Lamé, which is curiously enough given quite correctly although the method of deducing it is entirely erroneous. With the notation of the footnote on our p. 79, he is really assuming:

$$\widehat{\phi\phi} = E\frac{u}{r} \text{ and } \widehat{rr} = E\frac{du}{dr},$$

or, in other words he puts the tractions equal to the stretches multiplied by the stretch-modulus although he is not dealing with a rod under pure traction. This error he repeats, when he considers a tube surrounded by rigid belts. Compared with this it is a small matter that he considers it justifiable to neglect the shear $\widehat{zr}$. The algebra is prodigious, but the results so pretty, that we might well wish them to be true, but the writer is hopelessly at sea in his physical conceptions of elasticity. His hypotheses lead in fact to

$$\widehat{\phi\phi} - \widehat{rr} = E\left(\frac{u}{r} - \frac{du}{dr}\right) = 2\mu\left(\frac{u}{r} - \frac{du}{dr}\right),$$

since he supposes symmetry round the axis of the cylindrical tube. Hence we are compelled to suppose $E = 2\mu$, or du/dr a constant, both incompatible with other results of the investigation. The real solution for the case Scheffler proposes requires the two types of Bessel's functions of zero order, and then the conditions at a belt will tax the powers of a very first-rate analyst[1].

So far as the results of the earlier portions of the memoir are concerned, Saint-Venant has completed the subject in his paper of 1860:

[1] The indestructibility of error is suggested by the fact that Virgile makes precisely the same mistakes five years later. See *Comptes rendus*, T. LX. 1865, p. 960.

see our Arts. 120–2. The latter part of the memoir involves problems hitherto still unsolved, and of first-class importance for the theory of ordnance.

[656.] Eduard Zetzsche: *Zur Bestimmung des Querschnitts eines Körpers dessen absolute Festigkeit in Anspruch genommen wird. Schlömilchs Zeitschrift für Mathematik u. Physik*, Bd. IV., S. 341–52. Leipzig, 1859. The author applies the theory of a uniform vertical prism under terminal traction to the case where there is not only terminal traction but also weight as a body-force. He then investigates the proper form for a solid of equal resist ance subject to terminal traction and gravitational body force. He does not notice that his method is one only of approximation, for in both his cases the cross-sections no longer remain plane, and in the first the sides of the prism no longer remain vertical: see our Arts. 1070* and 74.

We indicate all the contents of this article which are of any value in the following remarks,—supposing that the stretch is uniform across each cross-section,—which is obviously not the case.

Let ω_0 = terminal cross-section to which a traction P is applied; let ω = section at distance x from this terminal, $g\rho$ = weight of unit volume of the material, and s_x = stretch across any cross-section; then by resolving vertically we have:

$$\int_0^x g\rho\omega dx \pm P\omega_0 = \omega \times Es_x.$$

For equal resistance Es_x must be a constant $= T$, the limit of safe elastic traction, therefore:

$$\int_0^x g\rho\omega dx \pm P\omega_0 = \omega T \quad\text{(i)}.$$

Or, differentiating, $\quad g\rho\omega = Td\omega/dx,$

$$\omega = Ce^{\beta x}, \text{ where } \beta = g\rho/T.$$

But $\quad x = 0, \quad \omega = \omega_0,$

hence $\quad \omega = \omega_0 e^{\beta x} \quad\text{(ii)}.$

Now if $x = 0$ in equation (i), $\omega = \omega_0$ and therefore we must have $P = T$, this is only possible if we take the positive sign, which is obviously a condition for the material being entirely in a condition of limiting traction. Thus equation (ii) gives us the area of the cross-sections, and if we know their form we can determine the curve which by its revolution generates the form of the column of 'equal resistance.

[657.] Gustav Zehfuss: *Ueber die Festigkeit einer am Rande aufgelötheten kreisförmigen Platte. Schlömilchs Zeitschrift für Mathematik u. Physik*, Bd. V., S. 14–24. Leipzig, 1860.

This paper involves, S. 16–21, an investigation of the equation for the elastic equilibrium of a plate on the hypotheses proposed by Kirchhoff in 1852, i.e the equation is not deduced from the general principles of elasticity. I do not think anything in this investigation calls for special notice, or is of any particular value. There is a statement at the commencement of the article which is not absolutely true, namely: that, when a body is strained beyond the elastic limits, its stretch-modulus varies with the strain. The stretch-modulus of a body may remain sensibly constant and practically equal to its original value nearly up to rupture: see pp. 441, 887, 889 of our Vol. I.

[658.] On S. 14–15 we have results of the following kind. If $K' = \lambda$, $K = \lambda + 2\mu$ of our notation, then

$$K = \frac{\epsilon^2 + \epsilon'^2}{\epsilon^3 - 2\epsilon'^3 - 3\epsilon\epsilon'^2}, \quad K' = \frac{\epsilon\epsilon' + \epsilon'^2}{\epsilon^3 - 2\epsilon'^3 - 3\epsilon\epsilon'^2},$$

where ϵ is the stretch produced by unit traction ($= 1/E$ of our notation), and ϵ' = the corresponding transverse squeeze ($= \eta/E$). These results had already been given by Cauchy in a somewhat different form.

[659.] S. 22–23 give the solution of the differential equation for a plate, supposing it to be uniformly loaded with a total load $\pi l^2 Q$, to be of thickness h, of radius l and to be built in at its edge. The result for the deflection z at distance r from the centre is (see our Art. 398),

$$z = \frac{3}{16} \frac{Q}{h^3} \frac{K}{K^2 - K'^2} (l^2 - r^2)^2$$

This gives for the stretches at the surface (see our Art. 398):

$$s = \frac{h}{2} \frac{d^2 z}{dr^2}.$$

Whence if s_0 be the safe stretch limit we have, to determine the proper thickness for a given load Q from:

$$h = l \sqrt{\frac{3}{4} \frac{Q}{s_0} \frac{K}{K^2 - K'^2}}.$$

This seems to me the proper condition of safety, but my numbers do not agree with those of Zehfuss. I do not think the remarks of his concluding paragraph are correct.

[660.] E. O. Winkler: *Die inneren Spannungen deformirter, insbesondere auf relative Festigkeit in Anspruch genommener Körper. Erbkams Zeitschrift für Bauwesen*, Jahrgang X., S. 93–108, 221–36, 365–80, Berlin, 1860.

The first part of this memoir only reproduces general results such as the body-stress equations and the analysis of traction and shear given long before by Cauchy and Lame. The republication at this date may have been serviceable in Germany considering the ignorance of the general theory of elasticity manifested by Scheffler and by Laissle and Schübler, but it has not historical importance.

[661.] The second part of the memoir is entitled: *Theorie der relativen Festigkeit.* Winkler takes as his elastic body that which would be generated by a plane closed figure whose centroid described a plane curve (*central line*) so that the plane of the figure was perpendicular to the plane of the curve, the form of the figure changing during the motion in any arbitrary manner. At any given section he takes for axis of x the tangent to the plane curve, which lies in the plane of xy and he supposes this plane to contain the direction of gravity and that of the load system.

He then says that the ordinary theory of flexure has neglected $\widehat{yy}$ and $\widehat{zz}$ or found erroneous values for them, and cites in this respect the researches of Poncelet, Scheffler, Laissle and Schübler. He further states that Scheffler, and Laissle and Schübler have attempted to take into account $\widehat{yx}$, but neglected $\widehat{zy}$ and $\widehat{xz}$. He declares that in general all these stresses differ from zero, but remarks that $\widehat{yz}$ will usually be quite negligible and proceeds to neglect it. He thus reduces his body-stress equations to the form:

$$\begin{aligned}
\frac{d\widehat{xx}}{dx} + \frac{d\widehat{xy}}{dy} + \frac{d\widehat{xz}}{dz} + X_0 &= 0,\\
\frac{d\widehat{xy}}{dx} + \frac{d\widehat{yy}}{dy} \qquad\quad + Y_0 &= 0,\\
\frac{d\widehat{xz}}{dx} \qquad\quad + \frac{d\widehat{zz}}{dz} \qquad\quad &= 0.
\end{aligned}$$

He then writes down the body surface equations on the assumption that there is a uniform surface pressure p (S. 223). The equations thus obtained he cannot solve, and so he takes refuge in hypotheses almost as incorrect as those of the writers he has previously cited. He first assumes $\widehat{xx}$ to have the same value for all points on a line in the cross-section perpendicular to the load plane (or parallel to the axis of z). He further takes $\widehat{xy}$ or the shear in the cross-section parallel to the load plane uniform along the same line, although the breadth of the cross-section changes continuously with the height (i.e. with y):

> wie z. B. beim rechteckigen und kreisförmigen Querschnitt. Bei dem ersteren unterliegt diese und die vorige Annahme überhaupt keinem Zweifel (S. 223).

It is perhaps needless to remind the reader that Saint-Venant five years before the publication of this paper had shown that these hypo theses which 'admit no doubt' are absolutely untenable for the cross-sections in question Further for a thin rib Winkler takes $\widehat{xz}$ constant for the whole length of the rib—'In den übrigen Theilen kann natürlich die vorige Annahme beibehalten werden' (S. 223).

[662.] It is needless to follow Winkler's analysis further. It seems to me that the modifications he introduces into the Bernoulli-Eulerian theory do not tend to correct it in the case in which the cross-sections are incapable of treatment by Saint-Venant's method (**T** and **I** cross-sections etc), while when the cross-sections fall under the cases treated by Saint Venant the true theory is not a bit more complex than Winkler's lengthy process (see our Arts. 87–98). As for the case in which central line is not a straight line and the cross-section varies, I doubt whether he has found even an approach to an approximate solution.

[663.] The second part of the memoir discusses principal tractions and applies them to the theory of rupture. The work is inferior to what had been done several times previously and takes a tractive and not a stretch limit of strength (S. 229–30). Winkler applies this discussion of traction to several examples (S. 230–3) and concludes this part by the consideration of the effect of a rapidly moving load on the deflection of a girder or beam. Here he has to return to the Bernoulli-Eulerian theory for a solution. He considers first an isolated load.

His reasoning in this case is the following. Suppose M the mass of the moving load, ρ the radius of curvature of the central line of the beam immediately under the load supposed at the centre, v the velocity of the load, l the length and m the mass of the beam. Then there is a centrifugal force Mv^2/ρ acting downward and consequently the terminal reaction R is given by:

$$R = \tfrac{1}{2}\left(Mg + mg + \frac{Mv^2}{\rho}\right) \quad \ldots\ldots\ldots\ldots\ldots \ \ldots \ \text{(i)}.$$

But the bending moment at the centre or

$$\begin{aligned} E\omega\kappa^2/\rho &= \tfrac{1}{2}Rl - \tfrac{1}{8}mgl \\ &= \tfrac{1}{8}mgl + \tfrac{1}{4}Mgl + \tfrac{1}{4}\frac{Mlv^2}{E\omega\kappa^2}\,\frac{E\omega\kappa^2}{\rho}, \end{aligned}$$

whence

$$E\omega\kappa^2/\rho = (\tfrac{1}{8}mgl + \tfrac{1}{4}Mgl)\left(1 + \tfrac{1}{4}\frac{Mlv^2}{E\omega\kappa^2}\right), \text{ approximately}\ldots\ldots\text{(ii)},$$

since $\dfrac{Mlv^2}{E\omega\kappa^2}$ is very small.

R can then be found from (i), and Winkler gives for the approximate central deflection[1]

$$f = \tfrac{5}{384}\frac{mgl^3}{E\omega\kappa^2} + \tfrac{1}{48}\frac{Mgl^3}{E\omega\kappa^2}\left\{1 + \tfrac{1}{4}\frac{v^2}{E\omega\kappa^2}(M + \tfrac{1}{2}m)\,l\right\}.$$

[1] Winkler has $\frac{5}{584}$ and $\frac{1}{84}$ for the numerical coefficients, but I presume these to be misprints for $\frac{5}{384}$ and $\frac{1}{48}$.

These results shew that Winkler was quite unaware of the labours of Stokes and Phillips (see our Arts. 1276*–91* and 372–7, 552–4), to say nothing of Homersham Cox, who had proceeded on these very lines, with the like inexact results: see our Art. 1433*. He concludes:

Wirkliche numerische Berechnungen zeigen, dass selbst bei bedeutenden Lasten und sehr grossen Geschwindigkeiten die Vermehrung der Beanspruchung nur äusserst gering ist (S. 234).

This is hardly however experimentally confirmed: see our Arts. 1418*, 1420* and 1375*.

[664.] The last section of the second part (S. 234–6) is entitled: *Einfluss eines bewegten Zuges.* It deals with what we have termed Bresse's problem (see our Arts. 382 and 540), and presents no novelty. Winkler's results agree with those previously obtained by Bresse, but he does not refer to him. Some numerical calculations are given to show that the increment of bending moment and deflection due to the velocity of the load are very small.

[665.] In the third and last part of his memoir, Winkler applies the formulae of his second part to various special cases. Thus he finds (S. 365) for a cantilever of rectangular cross-section ($h \times b$) under bending moment M and total shear Q that:

$$\widehat{xx} = -\frac{12My}{bh^3}, \quad \widehat{xz} = 0, \quad \widehat{xy} = \tfrac{3}{2}Q\,\frac{h^2 - 4y^3}{bh^3}$$

Comparing these with Saint-Venant's results (Art. 95) we see that they are incorrect.

I have again no confidence in the results Winkler gives for beams with varying cross-sections or with I sections. Thus I think the paper failed in achieving the purpose proposed by its author.

[666.] In conjunction with Winkler's attempt to solve an already solved problem I may briefly refer to the following somewhat later memoir in this place:

George Biddell Airy: *On the Strains in the Interior of Beams. Phil. Trans.* 1863, pp. 49–80. This memoir was received on November 6 and read December 11, 1862. By 'strains' the late Astronomer Royal here understands what we now term stresses. Having regard to the full and able treatment of the flexure of rectangular beams by Saint-Venant in his memoir on flexure of 1854 (see our Art. 69) it seems unnecessary to analyse this paper at any length. It may suffice to remark in this place that a solid rectangular beam cannot be considered as built-up of a number of parallel plates, still less can the stresses be expanded in integer powers of x and y (Cartesian coordinates in the cross-section) in the manner adopted by Airy in § 14. The tables and diagrams of the memoir cannot be considered of value, but fortunately the plaster models and tabulated numbers of Saint-Venant effectually accomplish the objects Airy had in view when writing the paper.

667. C. Neumann: *Zur Theorie der Elasticität. Journal für Mathematik*, Vol. 57, S. 281–318, Berlin, 1860.

The object of this memoir is not to add anything to the theory of elasticity, but to obtain the fundamental equations of elasticity in a new way. The memoir consists of two parts: in the first the ordinary equations referred to rectangular axes are obtained; in the second these are transformed so as to give the equations referred to a system of triple orthogonal surfaces, which were first investigated by Lamé.

The first paragraph of the memoir explains its object:—

Es existiren bekanntlich zwei Methoden, um die für das Gleichgewicht und die Bewegung eines elastischen Körpers geltenden Differential-Gleichungen abzuleiten, von denen die eine von *Navier*, die andere von *Poisson* herrührt. Die erste geht von der Berechnung der *Kraft* aus, welche ein einzelnes Molecül des Körpers von allen übrigen Molecülen empfängt, die zweite von der Berechnung des *Druckes*, welchen ein Flächen-Element im Innern des Körpers erleidet. Im vorliegenden Aufsatze gebe ich eine dritte Methode zur Ableitung dieser Gleichungen; ich bestimme zuerst das *Potential* der auf ein einzelnes Molecül von allen übrigen Molecülen ausgeübten Wirkung; erhalte daraus für das Potential aller, im ganzen Körper stattfindenden, Molecular-Wirkungen zusammengenommen ein dreifaches, über den Raum des Körpers ausgedehntes, Integral; und gelange dann durch Variation dieses Integrals—in ähnlicher Weise also wie *Gauss* in der Theorie der Capillarität—zu den Bedingungs-Gleichungen, welche erfüllt sein müssen, wenn sich der elastische Körper unter der Einwirkung äusserer Kräfte im Gleichgewicht befinden soll.

The memoir is a fine piece of mathematical analysis[1]

[668.] Neumann supposes his material homogeneous and isotropic. Further he assumes uni-constant isotropy or he uses only one *elastic* constant in his results (Poisson's $k =$ our λ). He

[1] The following misprints may be noted:

On S. 285 observe that Neumann assumes the result in Moigno's *Statique*, p. 703. S. 294, at the top, for the first K read H. S. 297, equation (26): for $2u_1 + \Theta$ read $2u_1 + \bar{\Theta}$. S. 315, in (55): for $\frac{1}{a}, \frac{1}{b}, \frac{1}{c}$, read in each case $\frac{1}{abc}$.

starts indeed from the assumption that intermolecular force is a function only of the individual molecular distance; thus he neglects *aspect* and *modified action*. The second constant K which appears in his results is not an elastic constant, but an initial stress equivalent to the $\widehat{xx}_0$ of our Art. 616* (see second set of formulae on our p. 329). The following remark shows how it arises and why its value is taken to be the same in all directions:

Während der primitiven Lage sollen die Molecüle gleichförmig und ohne Bevorzugung irgend welcher Richtungen durch den Raum hin vertheilt gewesen sein. Ob damals Gleichgewicht herrschte, oder ob es äusserer Kräfte bedurft hätte, um die Molecüle während jener Lage festzuhalten, mag dahin gestellt bleiben (S. 282).

[669.] Neumann's work, as an investigation on the grounds of uni-constant isotropy, is extremely good,—only alas! such an investigation has not much practical value now that more and more bodies are observed to be aeolotropic. Perhaps the part which will best repay study is the method by which he surmounts the difficulties attaching to the expression of the surface-forces in terms of the strains, when we cannot sum over the whole of a sphere of molecular action. These difficulties had been noticed by Jellett: see our Arts. 1532*–3*, but Neumann, I think, surmounts them and shows that surface-forces can be really expressed in terms of elastic constants having the same values as at points of the body remote from the surface (S. 289–92).

[670.] It will not be without interest to compare Neumann's and Sir W. Thomson's methods of reaching the general equations of elasticity.

Let $2mF$ be the potential of the molecular forces on the molecule m, or the total influence of all the other molecules on m[1]. Then Neumann's results on S. 292–3 are, I believe, really perfectly general and have no relation to any particular law of molecular force, or to any magnitude of strain. We may state them in the language and notation of the present work as follows. Let u_x, u_y, u_z, v_x, v_y, v_z, w_x, w_y, w_z represent the nine first fluxions of the shifts, u, v, w, ρ_0 the initial density, $1/\nabla$ the determinant

$$\begin{vmatrix} 1+u_x & u_y & u_z \\ v_x & 1+v_y & v_z \\ w_x & w_y & 1+w_z \end{vmatrix},$$

[1] The total potential energy of the system $=\frac{1}{2}\Sigma 2mF=\Sigma mF$, or $F=w$, the work of the elastic strain per unit mass of the body at m.

let Q_x be the x component of the force necessary to hold the molecule m in equilibrium when it lies within the volume of the body, and P_x the corresponding component when the molecule is near the surface of the body, then if n be the direction of the normal to the surface measured inwards and $\overline{nx}$, $\overline{ny}$, $\overline{nz}$ the angles it makes with the axes, we have:

$$\left.\begin{aligned} Q_x &= \frac{d}{dx}\left(\frac{dF}{du_x}\right) + \frac{d}{dy}\left(\frac{dF}{du_y}\right) + \frac{d}{dz}\left(\frac{dF}{du_z}\right), \\ P_x &= \rho_0 \left(A_x \cos \overline{nx} + A_y \cos \overline{ny} + A_z \cos \overline{nz}\right) \end{aligned}\right\} \text{.............} \quad \text{(i)},$$

where

$$\left.\begin{aligned} A_x &= \left(\frac{dF}{du_x}(1+u_x) + \frac{dF}{du_y} u_y + \frac{dF}{du_z} u_z\right)\nabla, \\ A_y &= \left(\frac{dF}{du_x} v_x + \frac{dF}{du_y}(1+v_y) + \frac{dF}{du_z} v_z\right)\nabla, \\ A_z &= \left(\frac{dF}{du_x} w_x + \frac{dF}{du_y} w_y + \frac{dF}{du_z}(1+w_z)\right)\nabla \end{aligned}\right\} \text{.....} \quad \text{(ii)}.$$

Similar values hold for Q_y, Q_z and for P_y, P_z in terms of the corresponding quantities B_x, B_y, B_z, C_x, C_y, C_z, obtained from (ii) by cyclical interchanges. These results are deduced by assuming F a function of the first nine shift fluxions and applying the method of virtual moments.

It must be noted that Q_x is the *force per unit of mass* of the material at the point $x+u$, $y+v$, $z+w$, while P_x is the force per unit area of the surface of the material. In the course of his work Neumann shews that if ξ, η, ζ be the displaced coordinates of the point x, y, z $(= x+u,\ y+v,\ z+w)$ then:

$$Q_x = \frac{1}{\nabla}\left(\frac{dA_x}{d\xi} + \frac{dA_y}{d\eta} + \frac{dA_z}{d\zeta}\right) \text{..................} \quad \text{(iii)}.$$

Thus it would appear that A_x, A_y, A_z are what are generally termed the stresses across the elementary face perpendicular to the axis of x. Neumann does not consider under what conditions we shall have relations of the type $A_y = B_x$.

[671.] Sir W. Thomson (*Phil. Trans.* 1863, p. 610, or Thomson and Tait's *Natural Philosophy*, Second Edition, Part II. p. 462) takes the work w a function of the six quantities $2\epsilon_x$, $2\epsilon_y$, $2\epsilon_z$, $2\eta_{yz}$, $2\eta_{zx}$, $2\eta_{xy}$ of our Art. 1619* which he represents by $A-1$, $B-1$, $C-1$, a, b, c respectively. He deduces the general equations for the equilibrium of a body under no body-forces and finds they are of the type

$$\frac{dA'_x}{dx} + \frac{dA'_y}{dy} + \frac{dA'_z}{dz} = 0 \text{ ....} \quad \text{(iv)},$$

where

$$A'_x = 2\frac{dw}{dA}(u_x+1) + \frac{dw}{dc} u_y + \frac{dw}{db} u_z$$

$$= \frac{dF}{d\epsilon_x}(1+u_x) + \frac{dF}{d\eta_{xy}} u_y + \frac{dF}{d\eta_{xz}} u_z \text{ in our notation,}$$

$$= \frac{dF}{d\epsilon_x}\frac{d\epsilon_x}{du_x} + \frac{dF}{d\eta_{xy}}\frac{d\eta_{xy}}{du_x} + \frac{dF}{d\eta_{xz}}\frac{d\eta_{xz}}{du_x}$$
$$= \frac{dF}{du_x}.$$

Thus Thomson's equation given as (iv) above becomes:

$$\frac{d}{dx}\left(\frac{dF}{du_x}\right) + \frac{d}{dy}\left(\frac{dF}{du_y}\right) + \frac{d}{dz}\left(\frac{dF}{du_z}\right) = 0,$$

and is only a special case of Neumann's (i) cited in our previous article.

It seems more symmetrical and concise to write the quantities A'_x, A'_y,..., B'_x, B'_y,..., as dF/du_x, dF/du_y,..., dF/dv_x, dF/dv_y, ..., as Neumann has done. We must be careful to note that these expressions (A'_x...) are not the stresses, except for very small strains when

$$A_x = \frac{dF}{du_x} = A'_x.$$

Generally
$$\left.\begin{aligned} A_x &= (1 + u_x) A'_x + u_y A'_y + u_z A'_z, \\ A_y &= v_x A'_x + (1 + v_y) A'_y + v_z A'_z, \end{aligned}\right\} \quad \ldots\ldots\ldots\ldots\ldots \text{(v)},$$

whence we can at once express the stresses in Thomson's notation.

I believe that Neumann was the first to give these generalised equations and the generalised expressions for stress.

[672.] Supposing the strain to be small and in particular uni-constant isotropy to hold, Neumann shows (p. 285) that we may express F by

$$\left.\begin{aligned} 2F &= H + 2K\theta + (K + 3k)\theta^2 + (K + k)T + (2K + 4k)V, \\ \text{where}\quad \theta &= u_x + v_y + w_z, \\ T &= (v_z - w_y)^2 + (w_x - u_z)^2 + (u_y - v_x)^2 \\ &= 4\tau^2 \text{ if } \tau \text{ be the resultant twist,} \\ V &= (v_z w_y - v_y w_z) + (w_x u_z - w_z u_x) + (u_y v_x - u_x v_y) \end{aligned}\right\} \ldots \text{(vi)},$$

and H, K, k are constants depending on the molecular summations. The value of K is physically explained at once as the value of the stress A (or B_y or C_z) when the strains are all zero.

Writing: $\phi = F + KV$, and neglecting squares of small quantities, we may put as types:

$$Q_x = \frac{d}{dx}\left(\frac{d\phi}{du_x}\right) + \frac{d}{dy}\left(\frac{d\phi}{du_y}\right) + \frac{d}{dz}\left(\frac{d\phi}{du_z}\right),$$

$$P_x = \rho_0\left\{\frac{d\phi}{du_x}\cos\overline{nx} + \frac{d\phi}{du_y}\cos\overline{ny} + \frac{d\phi}{du_z}\cos\overline{nz}\right\} \ldots\ldots \text{(vii)},$$

whence the ordinary uni-constant equations of elasticity can be at once deduced.

[673.] The *Zweiter Abschnitt* of Neumann's memoir is occupied by a transformation of the equations and results given above to an-orthogonal, and ultimately as a limitation to orthogonal curvilinear coordinates. The deduction of the equations in curvilinear coordinates is hardly likely to be a short process. Neumann's possesses an elegance which can hardly be postulated of Lamé's original investigation, but at the same time the latter part of it requires considerable modification, if it is to be adapted to bi-constant isotropy. We have already referred to Bonnet's investigation of the uni-constant curvilinear equations (see our Art. 1241*), and we shall have occasion to refer to others, *e.g.* that of Borchardt in Crelle's *Journal der Mathematik*, Bd. 76, S. 45–58, 1873.

[674.] E. Phillips: *Mémoire sur le spiral réglant des chronomètres et des montres. Journal de Mathématiques*, Deuxième Série, T. v., pp. 313–366. Paris, 1860. This memoir[1] was presented to the Academy and was favourably reported on by Lamé, Mathieu and Delaunay on May 28, 1860.

Phillips introduces his memoir with the following remarks:

> Quelque important que soit le régulateur dont il s'agit, sa théorie n'avait pas encore été établie, la forme essentiellement complexe de ce ressort introduisant dans l'application de la théorie de l'élasticité des équations différentielles tellement compliquées, qu'il serait absolument impossible de les intégrer. J'ai pourtant été assez heureux, par des combinaisons particulières, pour vaincre ces difficultés dans tout ce qui touche au problème, et c'est cette théorie qui fait l'objet de ce Mémoire (p. 314).

Phillips, as in his memoir on railway-springs, adopts the Bernoulli-Eulerian theory of flexure; that is to say he puts the bending moment equal to the product of the flexural rigidity ($E\omega\kappa^2$) and the change in curvature. He thus supposes the flexure to take place without slide. Of this assumption he writes:

> Je me hâte d'observer que, dans une Note placée à la fin du Mémoire que j'ai présenté à l'Académie des Sciences, je démontre que, dans le problème actuel, ce principe est une conséquence rigoureuse de la théorie mathématique de l'élasticité (p. 315).

The note referred to is printed in an extended version of the memoir published in the *Annales des mines*: see our Arts. 677–8.

[1] As an earlier research in this direction I may refer to G. Atwood: *Investigations, founded on the Theory of Motion, for determining the Times of Vibration of Watch Balances. Phil. Trans.*, 1794, p. 119.

Let G be the couple, X, Y the components of force applied to one end of the spiral spring taken as origin of coordinates. Let $1/\rho - 1/\rho_0$ be the change in curvature due to strain at the point x, y of the spring. Then we easily see that we must have on the Bernoulli-Eulerian hypothesis:

$$E\omega\kappa^2\left(\frac{1}{\rho}-\frac{1}{\rho_0}\right)=G+Yx-Xy \quad\text{.................(i).}$$

Suppose l the length of the spiral, then integrating equation (i) along the length we easily find

$$E\omega\kappa^2(\phi-\phi_0)=Gl+l(Y\bar{x}-X\bar{y}) \quad\text{.................(ii),}$$

where $\phi-\phi_0$ is the angle between the new and old positions of the tangent at the force-end of the spring, and $\bar{x}$, $\bar{y}$ are the coordinates of the centroid of the spiral. If the force-end of the spiral be fixed at a constant angle to the balance of the watch attached to the spring, $\phi-\phi_0$ is the angle through which the balance has turned. Hence if we can put $Y\bar{x}-X\bar{y}=0$, we have the couple $G=E\omega\kappa^2(\phi-\phi_0)/l$, or it is proportional to the angle through which the balance has turned. Isochronism thus follows.

Phillips investigates at some length the conditions under which we may put $Y\bar{x}-X\bar{y}=0$, for example it would obviously be satisfied if the spiral so moved that its centroid remained at the fixed end of the spring. He also deals with a number of problems bearing on watch and chronometer springs which have, however, more interest for the historian of mechanics than for the historian of elasticity.

[675.] On pp. 352—4 an expression is deduced for the strain-energy of the spiral or the work required to displace its normal at the 'balance' end through any given angle. If s, s_0 be the stretches in the spiral, then the work needful to carry it from the one state of strain to the other is $EV(s^2-s_0^2)/6$, where V is the volume. This is an illustration of Young's theorem in resilience, see p. 875 of Vol. I. and our Arts. 1384*, 493 and 609.

[676.] E. Phillips: *Mémoire sur le spiral réglant des chronomètres et des montres. Annales des mines*, Tome XX., pp. 1–107. Paris, 1861. This is the completer form of the memoir recommended by a Commission of the Academy for publication in the *Recueil des savants étrangers.* We have already referred to the portion published in Liouville's *Journal*: see our Art. 674, and touched on those parts more closely associated with the theory of elasticity. There is a good deal of additional matter here of a very interesting kind, thus the influence on isochronism of temperature and of friction in the balance are taken into account, and a considerable number of curves which are theoretically suitable forms for the terminal of the spiral are given. By aid of these the centre of gravity of the spiral may be retained in the axis of the balance, one of the conditions for its efficient working. *Chapitre* II. entitled: *Des expériences faites à l'appui de la théorie précédente* (pp. 76–95) is remarkably interesting; it gives a very considerable number of experiments on the isochronism of spirals with or without terminals curved to the theoretical forms. The memoir is an excellent example of a high standard of theoretical knowledge applied to an important practical problem.

[677.] To the memoir as it stands in the *Annales* is attached a *Note* entitled: *Pour faire voir que, dans les circonstances que présente le problème actuel, les principes sur lesquels est fondée sa solution et qui rentrent dans la théorie de l'axe neutre, sont non-seulement parfaitement d'accord avec l'expérience, mais avec la théorie mathématique de l'élasticité* (pp. 95–107). Phillips remarks in a footnote that his demonstration is an extension of that which Saint-Venant has applied to the strain of a straight rod bent by a couple: see the *Leçons de Navier*, p. 34 and our Art. 170. It somewhat resembles the general treatment of the rod problem due to Kirchhoff: see our Chapter XII.

The assumptions made by Phillips in his theory of the spiral spring are that, when a couple is applied to its terminal the strain is such, that: 1° all the points primitively in a cross-section remain in a cross-section and that the strained cross-section remains perpendicular to the central line, 2° the central line remains unstretched. It is obvious that these are the ordinary assumptions of the Bernoulli-Eulerian theory extended to rods with an initially curved central axis. The problem is how far are they true for a spiral acted upon by a couple. Let us assume them to be true and investigate the resulting shifts and consequent stresses. In addition to 1° and 2° above Phillips makes the further assumptions involved in the following remarks:

J'appelle ligne neutre le lieu géométrique des centres de gravité de toutes les sections transversales, ce lieu étant une courbe quelconque, mais que je suppose plane, en négligeant, pour le spiral cylindrique, la très-faible inclinaison des spires. J'admets que toutes les sections transversales sont égales et qu'elles sont partagées symétriquement par un plan, que j'appellerai plan horizontal, passant par la ligne neutre.

J'imagine que sans changer la longueur de la ligne neutre, et tout en satisfaisant aux conditions de position et d'inclinaison assignées à ses deux extrémités, on déforme celle-ci dans son plan, d'après la loi $1/\rho - 1/\rho_0 =$ constante, ρ_0 et ρ étant les rayons de courbure en un quelconque de ses points: le premier avant la déformation et le second après. On a vu précédemment qu'il est possible de satisfaire géométriquement à cette condition en donnant aux courbes extrêmes une forme déterminée (pp. 95–6).

[678.] At any point O of the central line let its tangent be taken as axis of x, its normal as axis of z and the axis of y perpendicular to the horizontal plane containing these and defined above. Let u, v, w be the shifts parallel to these axes of a point P on a cross-section, infinitely near to that through O, and u_0, v_0, w_0 the shifts of the centroid of this latter cross-section. Then Phillips shows by easy geometrical analysis that we must have:

$$u = xz\,(1/\rho - 1/\rho_0), \quad v = v_0, \quad w = w_0 - \tfrac{1}{2}x^2\,(1/\rho - 1/\rho_0).$$

To determine v_0 and w_0 he assumes that the three stresses $\widehat{yy}$, $\widehat{zz}$ and $\widehat{yz}$ are zero as in Saint-Venant's theory of flexure: see our Art. 77. This leads him to the values

$$v_0 = -\eta yz\,(1/\rho - 1/\rho_0), \quad w_0 = \tfrac{1}{2}\eta\,(y^2 - z^2)\,(1/\rho - 1/\rho_0).$$

The values of u, v, w are now completely known. They will be found 1° to satisfy the body shift equations, 2° to give $\widehat{xz}$ and $\widehat{yz}$ zero values, and make

$$\widehat{xx} = Ez\,(1/\rho - 1/\rho_0).$$

Hence obviously if the neutral-axis goes through the centroid of each cross-section, there will be a zero total traction and the total system of stress over a cross-section will be represented by the couple, i.e. the bending moment:

$$E\omega\kappa^2\,(1/\rho - 1/\rho_0),$$

which is constant since $(1/\rho - 1/\rho_0)$ is assumed constant. Further the surface stress-equations are satisfied at every point. Thus if the force given by $Ez\,(1/\rho - 1/\rho_0)\,d\omega$ be applied to each element $d\omega$ of the cross-sections which bound a small portion of the spiral, this portion will be in elastic equilibrium, but since the cross-sections remain *plane* any number of such portions can be put together, and it is only necessary to apply such forces to the terminal cross-sections of any length of spiral. Thus by the principle of the elastic equipollence of statically equivalent loads (see our Arts. 8–9, 21 and 100) we see that Phillips' solution on the basis of the Bernoulli-Eulerian theory is really rigid on the complete mathematical theory, provided the terminals of his spiral are acted upon by couples of the magnitude

$$E\omega\kappa^2\,(1/\rho - 1/\rho_0). \qquad \text{(pp. 106–7).}$$

[679.] It will be noted that the above investigation is in no wise dependent on the central line being initially a spiral. It would seem that the above values of the shifts would apply to *a rod with its central line in the form of any plane curve whatever,* when its terminals were acted upon by equal and opposite couples. Their effect is to produce a constant change of curvature, and the Bernoulli-Eulerian theory is rigidly true.

[680.] Another interesting memoir by Phillips may be not unfitly considered here, although it belongs to a somewhat later date. It is entitled: *Solution de divers problèmes de Mécanique, dans lesquels les conditions imposées aux extrémités des corps, au lieu d'être invariables, sont des fonctions données du temps, et où l'on tient compte de l'inertie de toutes les parties du système, Journal de Mathématiques,* Tome IX., pp. 25–83. Paris, 1864. This interesting paper unfolds a valuable method for the treatment of various mechanical problems involving the longitudinal and transverse vibrations of rods. The author deals with the solution of problems in which the *shift at one end of the rod is a given function of the time*, or in which the rod itself, subject to a given system of load, is moving in space. The value of such solutions lies in their application to the stresses in various moving portions of machines. The mode of solution adopted is the determination of the special arbitrary functions involved in the general solution $u = F(x + at) + f(x - at)$.

[681.] The memoir is divided into two chapters: the first deals with the following problems:

Problem (i). *To determine the relative shifts of the parts of a rod, one end of which is subjected to a given motion, and the other is free when each point of the rod moves parallel to its axis* (pp. 25–38). Phillips treats in detail the cases when the motion imposed on the end is uniformly accelerated ($u_0 = \frac{1}{2}ft^2$), p. 29, and when it is harmonic ($u_0 = \alpha - \alpha \cos at$), p. 35.

Problem (ii). *To find the stresses in* AB *a connecting rod,* AC *and* BD *being two parallel revolving cranks of equal length* r *and having a spin* ω. *The section, length and weight of the connecting rod are given and the constant resistance* Q *is supposed to be applied at* B *tangentially to the circumference of* BD. (pp. 38–45.)

Problem (iii). *To find the stresses in a rod one end of which is subjected to a harmonic motion, while the other is attached to a piston under the action of steam.* (pp. 45–55.)

Phillips to simplify matters replaces the compound harmonic

action of the steam, by a single harmonic term of the same period but of different phase from the harmonic motion of the other terminal (difference equal to $\pi/4$). He supposes this to represent simply and approximately the mean action of the steam.

Problem (iv). *A crank* OM *turns uniformly round* O *which is fixed. A force, for example that of steam, acts upon the extremity* M *in a constant direction* MC. *The law of this force being given by a harmonic term of the same period as the rotation, to determine the strain in the crank* (pp. 55–61).

Problem (v). *One end of a cord being fixed and the other caused to vibrate transversally with harmonic motion, it is required to find the transverse vibrations of the string* (pp. 61–65).

This is the case for example of a string one end of which is fixed to a massive tuning-fork set vibrating harmonically.

The following two problems (vi) and (vii) treat the same string when *both* ends are caused to vibrate in a certain manner,—not however the most general possible.

[682.] In the second chapter Phillips adopts the solution in Fourier's series of the partial differential equation for the longitudinal vibrations of rods, but he first breaks up his shift u into two components

$$u = u_1 + U$$

and chooses u_1 in such a manner that it causes the terms resulting from the special terminal conditions, which are functions of the time, to disappear from his equations; U will then be found by the ordinary methods for evaluating the coefficients of Fourier's series.

Thus in his Problems (i) to (iii) (pp. 71–79) Phillips verifies results of his first chapter. In his last Problem (iv), however, he passes to somewhat different considerations. He makes use of Poisson's solution for the transverse vibrations of a rod to solve the following problem:

The two terminals of a connecting rod receive the same harmonic motion perpendicular to its length. It is required to find the strain (pp. 80–3).

Analytically this amounts to solving the equation:

$$\frac{d^2u}{dt^2} = -k^2 \frac{d^4u}{dx^4} \quad \ldots\ldots\ldots\ldots\ldots\ldots\ldots\ldots\ldots\text{(i)},$$

subject to the conditions:

$$u = r \sin \omega t, \quad d^2u/dx^2 = 0 \text{ when } \left\{ \begin{smallmatrix} x=0 \\ x=l \end{smallmatrix} \right\} \text{ ....(ii)}.$$

In addition Phillips supposes that initially, or for $t = 0$,

$$u = 0, \quad du/dt = \omega r \text{ for all points from } x = 0 \text{ to } x = z \text{...(iii)}.$$

The form of the special integral, which removes the time terms from the terminal conditions, is easily found to be

$$u_1 = \{A_1 \sin(\sqrt{\omega/k}\, x) + B_1 \cos(\sqrt{\omega/k}\, x)$$
$$+ C_1 \sinh(\sqrt{\omega/k}\, x) + D_1 \cosh(\sqrt{\omega/k}\, x)\} \sin \omega t.$$

Thus if, $u = u_1 + U$ we easily find U by Poisson's process (see our Art. 468*), while A_1, B_1, C_1, D_1 are determined from the four equations (ii). Equations (iii) give the constants of Poisson's solution.

SECTION II.

Physical Memoirs including those of Kupffer, Wertheim and others.

GROUP A.

Memoirs on the correlation of Elasticity to the other physical properties of bodies.

[683.] A. J. Ångström: *Om de monoklinoedriska kristallernas molekulära konstanter. Kongl. Vetenskaps-Akademiens Handlingar för år* 1850, *Sednare Afdelningen*, Vol. 38, pp. 425–61, Stockholm, 1851. This memoir[1] was presented on March 7, 1851. It is an important contribution to a subject still very obscure, notwithstanding the investigations of Plücker, Sénarmont, Wiedemann and Ångström: see the references in our Chapter XII., Section I. Its topic is the exact nature of the relation between the various axes of a crystal—the axes of figure, of elasticity, of electrical conductivity, the thermal, the optic, and the magnetic axes. French and German translations of parts of Ångström's paper

[1] There is an earlier memoir by Ångström in the Upsala memoirs of the previous year, which I have not examined. It belongs to the theory of light, and endeavours to show that the optical properties of gypsum and of crystals of the monoclinohedric system can only be explained by supposing the elasticity of the ether has relation to a system of oblique axes.

will be found in the *Annales de Chimie*, T. 38, pp. 119-127, 1853, (by Verdet) and *Poggendorffs Annalen der Physik*, Bd. 86, pp. 206-237, 1852.

[684.] Neumann's identification of the principal crystalline axes had been, at least for certain types of crystals, discovered in later researches to be inaccurate: see our Arts. 788*-793* and Chapter XII., Section I. Ångström s investigations with regard to gypsum are some of the most important in this direction.

Detta studium bör dessutom för de klinoedriska kristallerna blifva så mycket mera fruktbärande, som hos dessa ett *nytt bestämnings-element* framträder, nemligen den *olika riktningen af de principala elasticitetsaxlarne* (p. 427).

[685.] The first section of the memoir is occupied with the determination of the optic axes of gypsum (pp. 428-38). Ångström shows like Neumann in his later work, that the optical axes of elasticity are not fixed but vary with the temperature and the colour of the light: see our Chapter XII., Section I.

The second section of the memoir (pp. 438-49) is entitled: *Klangfigurer hos gipsen* and investigates the axes of acoustic symmetry by means of Chladni's figures. The theory of the nodal lines for a substance of the elastic complexity of gypsum has not I think been worked out, Ångström takes rather arbitrary curves to represent the lines, although very possibly they give close enough approximations to the acoustic axes.

The third section of the memoir (pp. 449-51) is entitled: *Ledningsförmåga för värmet.* This confirms in part Sénarmont's results, but the author believes that the isothermals change with change of temperature.

Ehuru försöken icke äga all den noggranhet man kunde önska, tror sig dock författaren kunna sluta, att isothermerna i det symmetriska planet hos gipsen verkligen förändra läge med temperaturen, och att denna förändring sker åt samma led och är tillika af ungefärligen samma storlek som de optiska elasticitets-axlarnes vridning vid en lika temperaturförändring (p. 451).

The fourth section entitled: *Gipsens utvidgning genom värme* (pp. 451-3) cites Neumann's results and determines the absolute extension of gypsum. It shows that there is a direction in which gypsum *apparently* shrinks with increasing temperature (p. 453)

The fifth section entitled: *Gipsens hårdhet* (pp. 453-5) cites

Frankenheim s results: see our Art. 825*. Ångström's results in part confirm, in part modify Frankenheim's. Those of Franz he rejects as unsatisfactory: see our Art. 839.

The sixth section is entitled: *Gipsens förhållande till elektricitet och magnetism* (pp. 455–6) and cites the results of Plücker, Wiedemann etc.: see our Chapter XII., Section I. Ångström's researches confirm Wiedemann's for electricity, but he could not confirm Plücker's for magnetism.

In the seventh section (pp. 457–8) we have a *résumé* of the results[1] for axes of all kinds:

Sammanföras de resultater, vi i det föregående erhållit, bekommer man följande öfversigt af de olika axelsystemernas läge i det symmetriska planet, hvarvid α betecknar lutningen emellan den *fibrösa* genomgången och den axeln, som faller inom de båda genomgångarnes spetsiga vinkel:

	α
Optiska axlarnes medellinie	14°
Minsta utvidgningen för värme	12°
Största hårdheten omkring	14°
Magnetisk attraktion omkring	14°
Största ledningsförmågan för värmet	50°
Största elasticitets axeln i akustiskt hänseende	53°
Minsta ledningsförmågan för elektricitet	62°

It will thus be seen that these axes group themselves in two distinct sets which probably connotes some inter-relation of the corresponding physical quantities. Ångström makes some not very conclusive remarks on the reason for these groupings (pp. 457–8).

[686.] Section VIII. is devoted to felspar (pp. 458–60). On p. 460 a system of results for this crystal is given, partly based on the experiments of Brewster, Sénarmont and Plücker, partly on Ångström's own experiments. We have for the angle α between the given directions and the base of the fundamental prism of felspar:

	α
Optiska polarisationsaxeln	4°, 1
Diamagnetiska axeln	4°, 1
Hårdheten	4°, 1 ?

[1] I have purposely refrained from translating the Swedish as there seems to me a certain amount of vagueness in the expressions used by Ångström.

Största ledningsförmågan för värmet .... 60°
Akustiska axeln .... 63° + } .
Minsta ledningsförmågan för elektricitet .... 63°

Thus the like two groups recur.

Further discussion of results is given in Section IX. (pp. 460–1), while Section X. (p. 461) sums up as follows:

Slutligen och såsom hufvudresultat af det föregående anser sig författaren hafva på experimentel väg bevisat *oriktigheten af det vanliga antagandet, att kristaller hafva 3ne rätvinkliga elasticitetsaxlar*, sa vidt nemligen satsen gäller de monoklinoedriska kristallerna; och att tvertom ej blott *kristallernas form utan äfven deras optiska, thermiska och akustiska fenomener ovillkorligen häntyda på tillvaron af snedvinkliga elasticitetsaxlar, konjugataxlar.*

[687.] The theoretical relation of three rectangular and unequal axes of elasticity supposing them to exist to the various physical vectors the position of which is given by Angström seems in the present state of our knowledge of the correlation of the various branches of physics somewhat obscure. The planes of cleavage at any rate would probably take up a variety of positions relative to the three axes of elasticity depending on the exact relative magnitude of the constants of cohesion, and we should hardly expect them to make any definite angle (such as 45° or 90°) with these axes. How far Ångström's opinion that it is impossible to admit three rectangular axes of elasticity in crystals of the monoclinohedric system is correct must be left to the decision of those who have a wider knowledge of the properties and structure of crystals than the present writer.

[688.] James Prescott Joule. The first contribution of this physicist to our subject, namely the memoir of 1846: *On the Effects of Magnetism upon the Dimensions of Iron and Steel Bars*, has already been briefly referred to: see our Art. 1333* and its footnote. This memoir was published in the *Philosophical Magazine*, Vol. XXX. pp. 76–87, 225–241, and is reprinted in the *Scientific Papers*, Vol. I., pp. 235–264. Joule commenced to experiment in 1841–2 (see *Sturgeon's Annals of Electricity*, Vol. 8, p. 219), so that he was really the first investigator in this field : see our Art. 1333*. He obtained the following results:

(i) Magnetisation [? below a certain critical value] increases the length of a bar, but

(ii) It does not perceptibly increase its bulk owing to a lateral contraction.

(iii) The elongation is [? perhaps approximately below the critical value] in the duplicate ratio of the magnetic intensity of the bar.

The bars for which Joule deduced these results were of annealed and unannealed iron and of steel.

(iv) When iron wires are submitted to longitudinal tension and then magnetised, the increase of tension diminishes the elongation due to magnetism and with more than a certain tension increase of magnetisation produces a shortening effect.

(v) When iron bars are subjected to pressure the amount of the pressure does not seem to sensibly affect the magnitude of the elongation due to a given magnetic intensity.

(vi) The shortening effect when a wire is under tension is very nearly proportional to the product of the magnetic intensity in the wire into the current traversing the coil. [Hardly warranted by Joule's own experiments and scarcely confirmed by later investigators.]

In a particular experiment with iron wire one foot long and a quarter of an inch in diameter the tension at which magnetisation would produce no elongation for the electric currents employed in the experiments was conjectured to be about 600 lbs. By this I take Joule to mean the *total* tension: see his p. 232. *Scientific Papers*, Vol. I., p. 254. With regard to the apparently diverse results (iii) and (vi), Joule remarks:

The law of the square of the magnetism will still indeed hold good where the iron is sufficiently below the point of saturation, on account of the magnetism being in that case nearly proportional to the intensity of the current. For the same reason, on examination of the previous tables, it will be found that the elongation is, below the point of saturation, very nearly proportional to the magnetism multiplied by the current. The necessity of changing the law arises from the fact that the elongation ceases to increase after the iron is fully saturated; whereas the shortening effect still continues to be augmented with the increase of the intensity of the current (pp. 232–3. *Scientific Papers*, p. 255).

(vii) Shortening effects in the case of iron wire are proportional *caeteris paribus* to the square root of the tension. [Scarcely proven.]

In the case of hardened steel wire, however, the shortening effects were found not to increase sensibly with increase of tension.

(viii) No magnetic influence on strain could be found in the case of copper wires.

These results of Joule's have been considerably modified (as indicated above) by more recent researches and new light has been thrown on the whole subject by Villari, Shelford Bidwell, Ewing and others in memoirs to be discussed later.

[689.] The next paper of Joule's touching on our subject is entitled: *On the Thermo-electricity of Ferruginous Metals, and on the Thermal Effects of stretching Solid Bodies. Proceedings of the Royal Society*, Vol. VIII., pp. 355–6, 1857; *Scientific Papers*, pp. 405–7. This paper records that experiments on the stretching of metals showed a decrease of temperature in the metal when the load was applied and an increase when it was removed. The experiments were on iron wire, cast iron, copper and lead. Joule writes:

The thermal effects were in all these cases found to be almost identical with those deduced from Professor Thomson's[1] theoretical investigation, the particular formula applicable to the case in question being $H = \frac{t}{J} \times Pe$, where H is the heat absorbed in a wire one foot long, t the absolute temperature, J the mechanical equivalent of the thermal unit, P the weight applied, and e the coefficient of expansion per 1° (p. 355).

The same results occurred with gutta percha, but they were exactly reversed in the case of vulcanised india-rubber, which was heated by loading and cooled by unloading. Sir William Thomson suggested that loaded vulcanised india rubber would be found to be shortened when heated, a result Joule found in accordance with experiment as well as theory.

[690.] *On the Thermal Effects of the Longitudinal Compression of Solids. Proc. Royal Soc.*, Vol. VIII., pp. 564–6, 1857; *Scientific Papers*, Vol. I. pp. 407–8. In this paper Joule continues his experimental verifications of Thomson's thermo-elastic theory. He finds that for metal pillars and cylinders of vulcanised india-rubber heat is evolved by compression and absorbed on removing compressive force. His investigations lead him to determine how far the "force of elasticity in metals is impaired by heat," or what may be the effect of tensile stress on expansion by heat. He makes experiments on a helical spiral of steel wire and on one of copper wire, and he supposes such spirals, like J. Thomson (see our Art. 1382*–3*) to resist extension only by torsion. He thus finds that for the steel wire the force of torsion is decreased ·00041 by each degree of temperature' (C°), while the number for copper wire is ·00047. Kupffer found for steel wire ·000471 and for copper ·000691: see our Art. 754, where however, these results are given for a degree R°.

[1] Now Sir William Thomson.

[691.] *On some Thermo-dynamic Properties of Solids. Phil. Trans.*, 1859, Vol. CXLIX, pp. 91–131; *Scientific Papers*, pp. 413–73. This contains a detailed account of experiments similar to those referred to in the two previous memoirs (see our Arts. 689–90) bearing on the thermo-elastic relations of metals and india-rubber. Joule had found that a helical spring showed no sensible thermal changes when compressed, and he attributed this to the equal and opposite thermal effects produced in its compressed and extended portions. At the suggestion of Sir William Thomson, he undertook to investigate independently the "heat developed by longitudinal compression and that absorbed on the application of tensile force."

[692.] The portion of the memoir which really concerns us begins with § 18 and is entitled: *Experiments on the Thermal Effects of Tension on Solids.* Joule made careful experiments to measure the thermal increase H in degrees centigrade due to the stress, and he compared his experimental results with the formula of Sir W. Thomson[1]

$$H = \frac{t}{J}\frac{pe}{sw},$$

where t = temperature Centigrade from absolute zero,

J = mechanical equivalent of the thermal unit in foot-pounds,

p = total load in lbs. (negative of course for a tension),

e = longitudinal expansion per degree Centigrade.

s = specific heat, and w = mass in lbs. of a foot length of the bar.

As a measure of the coincidence of experiment and theory I think it well to cite the following results, noting that Joule took some of his constants from Dulong and Petit, Lavoisier and Laplace, etc., others he ascertained experimentally for his own specimens.

Values of H in degrees Centigrade.

	By Experiment.	Calculated.
Iron bar	− ·115	− ·11017
	− ·124	− ·1099
	− ·1007	− ·1069
Hard Steel	− ·162	− ·125
Cast Iron	− ·1605	− ·112
	− ·1481	− ·115
Copper	− ·174	− ·154
Lead	− ·0531	− ·0403
	− ·0758	− ·0550
Gutta-percha	− ·0284	− ·0306
	− ·0524	− ·0656

[1] See Sir William Thomson's *Collected Papers*, Vol. I., p. 203, and Vol. III., p. 66.

[693.] Joule next turns to the curious thermo-elastic phenomena presented by india-rubber §§ 33–58 (*Scientific Papers*, pp. 429–440), which he discusses at considerable length. He refers to the discoveries of Gough: see our footnote, Vol. I., p. 386. Joule confirms Gough's conclusions, which might be deduced from Thomson's formula by supposing e negative. These conclusions are in accordance with those previously noted by Joule: see our Art. 690. Besides Gough's conclusions Joule deduces from his experiments the following results:

(*a*) India-rubber softened by warmth, may be exposed to 0^0 Fahr. for an hour or more without losing its pliability, but a few days rest at a temperature considerably above the freezing-point will cause it to become rigid.

(*b*) A large amount of elastic after-strain exists in india-rubber.

(*c*) Moderate stretching weights produce little heat or even a slight cooling effect, but after a certain weight is reached there is a rapid increase of heating effect.

(*d*) When by keeping india-rubber at rest at a low temperature for some time it has become rigid, it ceases to be heated when stretched by a weight, and, on the contrary, a cooling effect takes place as in the metals and gutta-percha.

(*e*) For *vulcanised* india-rubber results similar to (*d*) hold, but that the specific gravity is increased by stretching it, as Gough supposed (Vol. I. p. 386 ftn. (4)), appears to be exactly contrary to Joule's experience, § 45 (*Scientific Papers*, p. 434).

(*f*) The slight cooling effect referred to in (*d*) produced by weak tensile forces disappears for vulcanised india-rubber when the temperature of the thong is a few degrees higher than $7^0{\cdot}8$ C.

(*g*) The effect of heat on a thong of india-rubber under tension predicted by Thomson was experimentally measured §§ 50–58 (*Scientific Papers*, pp. 433–40) and the numbers (p. 438) agreed with theory perhaps as closely as could be expected with a material of this kind.

(*h*) A rise of temperature removes from vulcanised india-rubber set produced by earlier experiments at high tensions.

(*i*) For vulcanised india-rubber H (see our Art. 692) $= +\cdot 137$ by experiment (corrected 'for elongation of rubber by use' to $+\cdot 155$), and $= +\cdot 114$ by calculation.

[694.] Joule next turns his attention to *wood* which presents some remarkable thermo-elastic properties and leads him to rather inconsistent results.

The discrepancies arose apparently from considerable elastic after-strain and from the effects of moisture on the wood in altering its elastic condition. Thus different hygrometric conditions could cause the wood under tension either to expand or contract, as the case might be, when its temperature was raised, and here again there were great differences according as the wood was strained with or across the grain. Joule's general results are stated in § 75 (*Scientific Papers*, Vol. I., p. 450). Removal of set by heating and also elastic after-strain were observed in *whalebone* as well as wood §§ 84–5 (*Scientific Papers*, Vol. I., pp. 454–6).

[695.] The next portion of Joule's memoir (§§ 94–122, *Scientific Papers*, Vol. I., pp. 459–71) is devoted to the *Thermal Effects of Longitudinal Compression on Solids.* Here again Joule compares the heat evolved in degrees Centigrade with that calculated by Thomson's formula, and he finds the following *mean* results, where we omit those for vulcanised india-rubber and wood, the apparent agreement in the case of wood cut across the grain disappearing if the individual results are analysed:

Values of H in degrees Centigrade.

Thermal Effect from	Experiment.	Theory.
Wrought Iron	·115	·108
Cast Iron	·118	·107
Copper	·146	·124
Lead	·087	·079
Glass	·017	·011

Evidently the experimental results are somewhat in *excess* of the theoretical. Joule attributes this discrepancy to "experimental error or to the incorrectness of the various coefficients which make up the theoretical results" (§ 120). He also considers that elastic after-strain would introduce a small error into the thermo-elastic formula, but that this is too small to be capable of measurement in his experiments (§ 121).

[696.] The last paragraphs of the memoir (§§ 122–6, *Scientific Papers*, Vol. I., pp. 471–3) contain an experimental verification of a principle of Thomson's namely that:

if a spring be such that a slight elevation of temperature weakens it, and the full strength is recovered again with the primitive temperature, work done against that spring by bending or working in whatever way must cause a cooling effect.

Now Joule had found a diminution in the slide-modulus of steel of ·00041 per degree Centigrade (§ 120 of the memoir, or see our Art. 690), hence the effect of compressing or stretching a helical steel spring ought to be to cool it slightly. By taking the mean of a hundred experiments on compression and then of a hundred on extension Joule was able to measure this slight cooling effect. He found it ·00306, theoretically it should have been ·00403. The three-thousandth part of one degree Centigrade as measured by Joule is rather a small quantity to draw definite conclusions from, but he found in these numbers sufficient evidence of the truth of the theory and concludes his memoir with the words:

Thus even in the above delicate case is the formula of Professor Thomson completely verified. The mathematical investigation of the thermo-elastic qualities of metals has enabled my illustrious friend to predict with certainty a whole class of highly interesting phenomena. To him especially do we owe the important advance which has been recently made to a new era in the history of science, when the famous philosophical system of Bacon will be to a great extent superseded, and when, instead of arriving at discovery by induction from experiment, we shall obtain our largest accessions of new facts by reasoning deductively from fundamental principles (§ 126; *Scientific Papers*, Vol. I. p. 472).

[697.] One or two other memoirs of Joule's may be just referred to here, although falling into a later period.

(*a*) *On a Method of Testing the Strength of Steam Boilers. Memoirs of the Literary and Philosophical Society of Manchester*, 3rd Series, Vol. I. pp. 175 and 233, 1861. This contains nothing with regard to the strength of the materials of the boilers tested.

(*b*) *On a new Magnetic Dip Circle.* A memoir of this title was published in the *Manchester Proceedings*, Vol. VIII. 1869, p. 171, and when it was republished in the *Scientific Papers*, Vol. I. p. 575, Joule added (pp. 579–583) some account of experiments on the strength of silk and spider filaments made in 1870. These experiments show the large influence of elastic after-strain, and further prove that silk and spider filaments, like caoutchouc, when under tension, become shorter if the temperature be raised. The effect of moisture tends to obscure both after-strain and temperature effects. Numerous experimental measurements are given.

(*c*) *On the Alleged Action of Cold in rendering Iron and Steel brittle. Manchester Proceedings*, Vol. X. pp. 91–4, 1871; *Scientific Papers*, Vol. I. pp. 607–610. *Further Observations on the Strength of Garden Nails* will be found on pp. 127–8 and 131–2 of the same volume of the *Proceedings*, or on pp. 610–13 of the *Scientific Papers*.

Joule brings evidence against the hypothesis that cold renders iron and steel brittle from: (i) Experiments on iron and steel wires, part of which were in contact with a freezing mixture and part at about 50° F. The wires broke *outside* the freezing mixture. These were pure tractive experiments. (ii) Flexure experiments on steel darning needles. The average strength of the metal at 12° F. was found to be slightly greater than at 55° F. (iii) Impact experiments on warm and cold cast-iron garden nails, broken by the blunt edge of a steel chisel falling upon the middle of the nail terminally supported. These experiments were not in favour of the hypothesis that frost makes cast-iron brittle.

[698.] C. Matteucci: *Sur la rotation de la lumière polarisée, sur l'influence du magnétisme et sur les phénomènes diamagnétiques en général. Annales de Chimie,* T. XXVIII., pp. 493–9, Paris, 1850.

A heavy glass prism (presented to the author by Faraday) placed in a strong electro-magnetic field rotated the plane of polarised light. Matteucci records somewhat vaguely in this note the effect produced on this power of rotation by compressing the glass. He finds that:

(i) Before compression the rotations to the right or to the left according to the sense of the current are the same for the same intensity of current. After compression the one rotation is much greater than the other; the one being twice to thrice the other.

(ii) The greater rotation is always that which is produced by the passage of the current which acts in the same sense as the compression[1].

(iii) The maximum rotation of the compressed glass is sometimes greater, sometimes less than that which occurs when the glass is not compressed; when the rotation produced by the compression is very much greater than that which the electro-magnet produces in the uncompressed glass, then the maximum rotation due to the action of the electro-magnet on the compressed glass is equal to or greater than that which the electro-magnet produces in the uncompressed glass; the contrary occurs when the rotation produced by the electro-magnet in the uncompressed glass is equal to or greater than that produced by the compression; in this case the maximum rotation is equal to or less than that produced upon the uncompressed glass.

(iv) Other glasses such as flint and crown exhibit the like phenomena. But the electro-magnetic field produces no sensible rotatory action on pieces of crown glass subjected only to slight compression.

Matteucci holds that this last result will explain to some extent why crystals do not exhibit rotatory power in the magnetic field. The electro-magnet further produced no rotatory power in compressed laminae of quartz and annealed glass: see Wertheim's results cited in our Arts. 786 and 797 (*d*).

(v) When the compression was removed the glass resumed its previous magnetic rotatory power.

(vi) There was a sensible although hardly measurable interval between the instant of closing the circuit and the instant at which the

[1] The 'rotation' due to the compression alone seems to have been measured by the angle through which the Nicol's prism of a bi-quartz analyser had to be turned in order that the two halves of the image should have the same colour.

maximum rotation was attained. This interval was greater when the glass was compressed than when it was uncompressed.

[699.] Matteucci further records some experiments in which he placed vibrating square plates of glass, brass and iron between the poles of an electro-magnet. He found that the Chladni-figures or system of nodal lines were the same whether the current was passing or not, whence he argues that very different groups of atoms (*groupes d'atomes*) must be affected by the action of magnetism and by the influence of acoustic vibrations (p. 499).

[700.] C. Matteucci. Some account of a memoir by this physicist entitled: *Sur l'influence de la chaleur, de la compression,......sur les phénomènes diamagnétiques* will be found on pp. 740–4 of the *Comptes rendus*, T. XXXVI., Paris, 1853. The only part of the account which concerns us is entitled: *Compression du bismuth* and occurs on p. 742. Matteucci writes:

J'ai trouvé qu'une aiguille prismatique de bismuth, comprimée dans le sens de son axe, se dirige toujours équatorialement, quelle que soit la face qui est suspendue horizontalement; son pouvoir diamagnétique est considérablement augmenté par la compression. Si l'aiguille de bismuth a été comprimée perpendiculairement à son axe, elle se dirige dans la ligne des pôles quand les faces comprimées sont verticales, et dans la ligne équatoriale si les faces comprimées sont horizontales. Cette propriété persiste après avoir chauffé l'aiguille de bismuth jusqu'à une température peu inférieure à la fusion du métal.

[701.] C. Matteucci: *Sui fenomeni elettro-magnetici sviluppati dalla torsione. Il nuovo Cimento*, Tomo VII. pp. 66–97, Pisa, 1858. *Annales de Chimie*, T. LIII. pp. 385–417, Paris, 1858; *Comptes rendus*, T. XLVI. pp. 1021–4, Paris, 1858.

Parte I. of this memoir is entitled: *Di un nuovo caso d'induzione elettro-magnetica* (pp. 67–81). Matteucci begins by describing his apparatus and mode of experimenting. Briefly an iron rod supported perpendicular to the magnetic meridian was placed in circuit with a galvanometer and to this rod any amount of torsion in either sense could be given. Round the rod was placed a coil of three or four turns of wire, through which a current could be sent in either direction and this served to magnetise the rod.

When a current from two to four Grove elements was sent round the coil magnetising a bar of half-hard iron (*ferro semiduro*), then at the moment when it was started a small deflection of

$\frac{1}{4}$ to $\frac{1}{2}$ a scale division was exhibited by the galvanometer needle of the secondary circuit. But the result was quite different when a sudden torsion was given to the bar:

Perchè non vi sia difficoltà alcuna a concepire il risultato dell' esperienza principale, supporremo che per l' azione della spirale magnetizzante si formi un polo sud (o attratto dal polo nord della terra) in quella estremità della verga che è volta verso l' est, e un polo nord all' estremità opposta, che è quella fissata nel centro della ruota. Supporremo finalmente che l' osservatore che deve torcere la verga magnetizzata guardi la ruota. Nel momento in cui è applicata alla verga una certa torsione elastica che può essere di 5 fino a 20, o 25 gradi secondo la qualità del ferro, in modo che lo zero della ruota giri alla sinistra dell' osservatore, l' ago del galvanometro è spinto a 10 o 20 o 30 gradi o più, indicando una corrente diretta nella verga dall' estremità sud all' estremità nord. L' ago torna subito allo zero o al suo punto d' equilibrio e se allora si fa cessare bruscamente la torsione, l' ago indica una nuova corrente in senso contrario della prima. Ripetendo la stessa torsione in senso contrario, cioè verso la destra dell' osservatore, si ha di nuovo una corrente della stessa intensità di quella ottenuta colla torsione a sinistra, ma diretta in senso contrario cioè dal nord al sud nella verga. Anche in questo caso la detorsione sviluppa una corrente che è in senso contrario della corrente prodotta dalla torsione corrispondente (pp. 68–9).

Reversing the magnetising current, we have secondary currents in the reversed sense. The phenomena repeat themselves so long as the rod is subjected only to elastic torsion. Like all induced currents the secondary currents vary in intensity with the time in which a given torsion is produced.

Matteucci develops in this earlier part of his memoir (p. 70) the theory (rejected by Wiedemann) that the iron bar may be looked upon as a bundle of conducting fibres which are converted into spirals by the torsion (see our Art. 713), and he supposes that the magnetising coil has greater induction on this bundle of spirals than on the bundle of fibres parallel to its axis.

Matteucci further notices that when he first twists the bar and then closes the primary circuit he likewise obtains an induced current, and that its magnitude is more constant for the same torsion than that obtained by reversing the order of proceeding. Opening the primary circuit[1] when the bar is twisted gives a less induced current, however, than the process of magnetising, and Matteucci attributes this to the residual magnetism (p. 71). After the primary circuit has been opened and closed several times, the bar reaches a definite permanent magnetisation and the induced currents at closing and opening become the same (p. 75). Experiments were also made on steel rods with a greater or less degree of hardness; the phenomena were the same in

[1] Termed by Matteucci the *demagnetisation*.

general character as for the iron rods, but the induced currents were much less in magnitude.

[702.] On pp. 76–7 we have various experimental results connecting the induced current with the length, diameter and angle of torsion of the rod; on the latter page are also some statements with regard to the influence of torsional set or 'tort.' Matteucci found that for rods of hard and half-hard iron of ·4 m. in length and of diameters of 4 mm. and upwards the current was proportional to the angle of torsion, and he further concluded that set had not the power of developing a current, see his p. 77, to be compared with Wiedemann's results in our Arts. 713–4. Matteucci found that the induced currents due to twisting did not increase in proportion to the strength of the primary current[1], but began to diminish after this reached a certain intensity (cf. Wiedemann in our Art. 714). Further conclusions as to the difference in magnitude of the currents induced, according as untwisting was followed by the opening of the primary circuit or the reverse order was adopted, are given on pp. 79–81, but the results are not stated with the clearness and precision of Wiedemann's: see our Art. 714. Indeed the memoir suffers from the want of a general statement of results, and the *leggi determinate* to which Matteucci refers on p. 81, have to be drawn from a rather confused mass of experimental statements.

[703.] *Parte II.* of the memoir is entitled: *Delle variazioni nello stato magnetico di una verga di ferro prodotte dalla torsione* (pp. 82—8).

Matteucci opens this *Parte* with an historical *résumé* of his own[2] and Wertheim's earlier investigations (see our Arts. 812 and 811, 813 *et seq.*). Wertheim had not obtained for rods of cast steel any diminution of the magnetisation by elastic torsion (see our Art. 814 (ix)), but Matteucci asserts (p. 83) that he has found small variations of the magnetisation with torsion in a variety of cast-steel bars. He sums up the conclusions to be drawn from the scarcely sufficient experiments recorded in this part as follows:

1°. La torsione elastica di una verga magnetizzata a saturazione determina una diminuzione nella sua forza magnetica, la quale persiste per tutto il tempo in cui la torsione dura; colla detorsione la forza magnetica è ristabilita come prima.

2°. Dalle relazioni che esistono fra le variazioni determinate dalla torsione e detorsione nella forza magnetica di una verga e le correnti indotte nella spirale esterna, è dimostrato che quelle variazioni sono la cagione delle correnti stesse (pp. 87–8).

See Wiedemann's results cited in our Art. 714.

[1] Matteucci says 'magnetisation of the rod', which he erroneously takes to be proportional to the strength of the primary current.

[2] *Comptes rendus*, T. XXIV. p. 307.

[704.] *Parte III.* (pp. 88–97) of the memoir is entitled: *Spiegazione dei fenomeni descritti.* This portion of the memoir, after a remark that the phenomena of induction described in the preceding parts can only be produced in iron and some other magnetic bodies, proceeds to develop a second 'bundle of fibres' theory, namely: that each fibre is a separate iron rod and that these rods after being converted into magnets are then twisted by the torsion round the current in the direction of the axis of the bundle. This theory is supported by rather vague reasoning which does not seem to meet the objections which Wiedemann has raised against it: see our Art. 713.

On the whole Matteucci, while doubtless the first to discover many points relating to the influence of stress upon magnetism, had not that power of marshalling his experimental facts and clearly stating the conclusions to be drawn from them which is characteristic of both his French and German rivals in the same field. The memoirs of Wertheim and Wiedemann are models of physical research, but we must confess to finding Matteucci's letter-press, never broken by a symbol or a formula and only occasionally relieved by a thin scattering of experimental numbers, wearisome reading.

[705.] In a footnote on pp. 95–7 of the memoir Matteucci records some earlier results as to the effect of stretching three magnetised iron wires of 1·5 mm. diameter. A wire was placed along the common axis of two spiral coils, one in circuit with a galvanometer, and the other used for magnetising; on the stretching or unstretching of the wire when magnetised an induced current was observed in the galvanometer. Matteucci measured by means of a certain astatic system, described in the second part of the memoir, the changes in the magnetisation of the wire due to the stretch, and he found induced currents corresponding to the changes in magnetisation. After demagnetisation (? opening the magnetising circuit) the currents obtained by stretching or unstretching the iron wire were much stronger than when the magnetising current was flowing. These phenomena were most marked in annealed iron wire, but the induced currents were in the *opposite sense* to those in the case of hard iron wire. Thus stretching appeared to diminish the magnetisation of annealed and increase that of hard iron wire. If the current in the magnetising coil were however broken, a stretch indicated an increase of magnetisation for annealed in the same way as for hard iron wire.

When an iron wire magnetised by a surrounding coil was put in circuit with a galvanometer, no trace of a current along the wire was observed at the moment when it was stretched or unstretched.

These results become more intelligible in the light of the later researches of Villari, Ewing and others.

[706.] G. Wiedemann: *Ueber die Torsion, die Biegung und den Magnetismus. Verhandlungen der naturforschenden Gesellschaft in Basel,* Vol. II., Basel, 1860, S. 168–247.

This important paper was reproduced in a rather fragmentary manner in various volumes of *Poggendorff's Annalen*.

The following scheme shows the corresponding pages and will enable the reader to whom only the *Annalen* are accessible to identify our quotations:

Verhandlungen.	*Poggendorff.*			
S. 169–172	= Bd. CVI.,	1859.	S. 161–164,	(α).
S. 172–184	= Bd. CVI.,	1859.	S. 174–183,	(α).
S. 184–193	= Bd. CVII.,	1859.	S. 439–448,	(β).
S. 193–196 and S. 201–7	= Bd. C.,	1857.	S. 235–244,	(γ).
S. 197–201	= Bd. CVI.,	1859.	S. 170–174,	(α).
S. 207–223	= Bd. CIII.,	1858.	S. 563–577,	(δ).
S. 223–227	= Bd. CVI.,	1859.	S. 164–168,	(α).
S. 227–247	= Bd. CVI.,	1859.	S. 183–201,	(α).

We shall cite the pages of the *Annalen* by the Greek letters.

[707.] Wiedemann commences his memoir with the following account of its object:

Eine Reihe von Beobachtungen hatte mich vermuthen lassen, dass die durch mechanische Mittel hervorgebrachten Aenderungen der Gestalt der Körper nach ganz ähnlichen Gesetzen von den dieselben bedingenden Kräften abhängen, wie die Magnetisirung der magnetischen Metalle von den dieselbe bewirkenden magnetisirenden Kräften. Ich habe deshalb die Gesetze der Torsion und Biegung der Körper einerseits ebenso wie die der Magnetisirung des Eisens und Stahles anderseits in dieser Beziehung einer neuen Untersuchung unterworfen, deren Resultate ich im Folgenden mitzutheilen mir erlaube (S. 169).

[708.] The first section of the memoir is entitled *Torsion*, and occupies S. 169–84. The section opens with an account of the apparatus employed (S. 169–72; α, S. 161–4), and then Wiedemann continues:

Drähte von verschiedenem Stoffe wurden mit Hülfe dieses Apparates durch aufsteigende Gewichte L tordirt, welche stets so lange wirkten, bis der Draht eine constante Torsion angenommen hatte. Die dieser Belastung L entsprechende temporäre Torsion L des Drahtes wurde an der Kreistheilung abgelesen. Nach dem Heben der drehenden Gewichte wurde wiederum einige Zeit gewartet, bis die zurückbleibende permanente Torsion T_0 vermittelst der Spiegelablesung bestimmt wurde. Nach der Torsion des Drahtes wurde er allmählig durch entgegengesetzt drehende Gewichte...detordirt, wieder tordirt u. s. f. Dabei wurde sorgfältigst jede Erschütterung des Apparates vermieden (S. 172–3; α, S. 174).

This passage enables us without describing Wiedemann's apparatus to grasp his method of procedure. The intervals which elapsed between the experiments served to remove as far as possible elastic after-strain. Wiedemann's experiments were upon annealed iron and brass wires; the numerical results of his experiments are given on S. 174–7 (α, S. 175–7) and they are in part represented graphically in Fig. 3 of his Plate I. His general conclusions for torsion including certain temperature effects, which are based on less careful experimental methods (*mit manchen Fehlerquellen behafteten Versuche,* S. 183), are given on S. 178–84 (α, S. 177–83). I do not cite them here, as we shall return to a general statement of conclusions for torsion and magnetism when considering the sixth section of the memoir.

[709.] The second section of the memoir is entitled: *Biegung,* and occupies S. 184–93 (β, S. 439–48). It shows that results similar to those holding for torsion hold for flexure also. Wiedemann's apparatus is described on S. 184–5 (β, S. 439–40). His experiments were made on annealed brass rods built-in at one end and bent in a horizontal plane. The numerical details are given on S. 187–9 (β, S. 442–4) and the general conclusions on S. 189–91 (β, S. 444–6). These are of such great interest and anticipate so much of Bauschinger's later work that we cite them here:

(i) If a rod previously unbent be bent by a series of increasing loads, the elastic flexures which the rod exhibits while subjected to these loads increase more rapidly than the loads.

(ii) After removal of the loads the rod exhibits flexural sets or *bents*[1]; these begin with the smallest loads and increase in a far more rapid ratio than the corresponding loads.

(iii) If a bent rod have its bent removed (*entbogen*) by the application of reversed loads, then the bent decreases somewhat more slowly than the loads increase. To produce complete unbending a considerably smaller load is necessary than that which produced bending.

(iv) If the rod after the first bending and unbending is repeatedly bent and unbent, then the bents do not increase so much more rapidly than the loads as is the case in the first bending; on the contrary they become more and more nearly proportional to the loads, being greater for small loads than in the first case. The bent due to the maximum load decreases gradually to a definite limit after repeated loadings. On the other hand the load necessary for unbending the rod increases with

[1] Isaac Walton uses this word of a fishing rod, Wilkins of a bow and Richard Hooker for the set of 'an obstinate heart.'

repeated loading and unloading—the load which removed the first bent now leaving a residual bent.

Thus in one set of experiments with Wiedemann's units a bending load of 240 produced a bent of 89, and an unbending load of 211 left a bent of only 1, but after repeated operations the same bending and unbending loads produced a bent of only 44·8 and left a bent of 24·4.

(v) If a rod has been so often bent and unbent that the same bending load always produces the same bent, then when the rod is left to rest for awhile, it returns a little towards its primitive condition.

This result was only based on one experiment. Indeed in these experiments on flexure upon only one occasion was 15 hours left between two series, the other series being carried on continuously and therefore their results were probably somewhat affected by after-strain: see S. 188 (β, S. 443) of the memoir.

(vi) It is obvious that if a definite load $-L$ deprives a rod of bent, neither this load nor any less load repeated in the same direction as $-L$ will give the rod a bent in the direction opposite to that of the first bent. But the load $+L$ on the contrary will produce a greater or less bent of the rod.

(vii) If a rod, which possesses a bent B (which may be $=0$) be brought to another bent B' by a load L, and then by a load $-L'$ opposed to L be brought to a bent B'' which lies between B and B', then to bring the rod again to the bent B' the load L will be again needful.

(viii) If a rod be shaken while subjected to a bending load, this increases the elastic flexure; if it be shaken after the removal of the load, this decreases the bent. If a rod be bent and then deprived of bent by reversed load, shaking produces anew a bent in the sense of the initial bent.

These results are at least qualitatively the same as for torsion. The temperature effect is not so great in the case of flexure as in that of torsion: see our Art. 754.

[710.] The practical value of these results has only been fully brought out by the more elaborate experiments of Bauschinger on larger masses of material, but Wiedemann certainly draws from his more limited range of experiments conclusions which to some extent anticipate those of the careful Munich technical elastician: see also our Arts. 749 and 767. These are given on S. 191–3 of the memoir (β, S. 447–8) for both torsion and flexure.

Wiedemann remarks that it depends merely on the sensibility of our apparatus whether we are able or not to measure the set of the smallest loads (see our Art. 1296*):

Demnach ist der Begriff der sogenannten Elasticitätsgränze, wie man ihn gewöhnlich fasst, durchaus ein nur für die Praxis willkührlich eingeführter, insofern man dieselbe da ansetzt, wo eben für bestimmte Beobachtungsmethoden die permanenten Gestaltveränderungen der Körper sichtbar werden (S. 192; β, S. 447).

Thus if a body be bent to set and afterwards deprived of bent, or torted[1] to set and afterwards deprived of tort, and this process be repeated, then on the bending or torting by any less load there will always be a sensible set, which becomes more nearly proportional to the load as the process is more often repeated. There is here then no limit of elasticity. Thus although we return to a state of no torsion or flexure, that is, of no *apparent* strain, the elastic condition of the material has quite changed in character. What we term the 'state of ease' has, for one sense of loading, been reduced to a vanishingly small range. On the other hand if a set has been produced by a load L, no load less than L in the same sense will produce any set. Thus, as we have frequently noted, L marks the elastic limit or state of ease. To obtain a state of ease, which starts from the position of no apparent strain and embraces a load L_1, we must proceed as follows:

First apply a load L_2 in the opposite sense to L_1 and then a load L_3 in the same sense as L_1 which just undoes the bent or tort produced by L_2, thus L_1 by (vii) will not produce any set provided we have taken L_2 so great that L_3 is greater than L_1.

The suggestiveness of these results will be still more apparent as we come in the course of our *History* to further experimental investigations bearing on the state of ease.

[711.] Section III. of Wiedemann's memoir is entitled: *Magnetisirung von Eisen und Stahl* (S. 193–210; γ, S. 235–244 and δ, S. 563–6). This deals with the problems of temporary and residual magnetism and the effect of temperature on magnetism. The results obtained are very similar to those obtained for elastic strain and set in the previous sections of the memoir, but to discuss them here would lead us beyond our limits: see our Art. 714.

[712.] Section IV. is entitled: *Einfluss der Torsion auf den Magnetismus der Stahlstäbe* (S. 210–16; δ, S. 566–71). This problem had already been considered by Wertheim (see our Arts. 811–18), and Wiedemann commences by quoting Wertheim's results as to the magnetic equilibrium produced by repeated torsions in iron and steel bars: see our Art. 814. Wiedemann confirms and extends Wertheim's conclusions, measuring the changes in magnetisation by direct magnetometric means and not as Wertheim by induced currents in a coil surrounding the rod.

The only result of this section which is not cited in the general results of the sixth section is iv. (S. 216; but v. in δ, S. 571), and this accordingly may be noted here:

[1] I use the noun *tort* for torsional set and the verbs *to tort* and *to detort* for the processes of twisting and untwisting when it seems advisable to emphasise the tort part of the strain produced.

If by torsion more magnetism be withdrawn from a steel bar than could be withdrawn by repeated changes of temperature within definite limits (in the experiments 0° to 100° C.), then any loss of magnetism produced by a rise of temperature within those limits is restored when the bar is again cooled to the previous temperature.

[713.] Section v. of the memoir is entitled: *Einfluss der Magnetisirung auf die Torsion der Eisen- und Stahldrähte* (S. 217–27, δ, S. 571–7, and α, S. 164–8). The influence of magnetism in reducing torsional set or tort is here noted and measured. Iron wires which have no tort do not appear to be twisted by magnetism. As most of the results of this section are restated in the following section, we shall not specially cite them now; they deserve, however, careful attention from those interested in the mutual relations of magnetism and set[1]. Wiedemann gives cogent reasons for rejecting Matteucci's hypothesis that an iron wire may be looked upon as a bundle of parallel fibres, which are converted by torsion into spirals and which magnetisation by producing mutual repulsion again straightens: see our Art. 704. He also rejects the hypothesis that the phenomena observed can be due to the heat produced in the wire by magnetisation (S. 222–3; δ, S. 576–7).

At the conclusion of the section the author promises in a future paper to deal with the influence of bending on magnetism, but at the same time he notes the great difficulties which stand in the way of experimental investigation (S. 227).

[714.] The sixth and final section of the memoir is entitled: *Vergleichung der Resultate und Versuch einer Theorie* (S. 227–47; α, S. 183–201). We first find a comparison of the properties of magnetism and torsion which, although pressed rather far, contains a good deal of matter novel at the time. I reproduce it here:

Torsion.	Magnetism.
1. The temporary torsions of a wire twisted for the first time by increasing loads increase more rapidly than these loads.	1. The temporary magnetisations of a bar magnetised for the first time by increasing galvanic currents increase more rapidly than the intensity of those currents.
2. The torsional sets or torts of the wire increase still more rapidly.	2. The permanent magnetisations of the bar increase still more rapidly.

[1] On S. 227 Wiedemann gives in grammes weight a measure of the detorting force of magnetism in a special case (α, S. 167–8).

3. To completely detort the wire a much less load is required than to tort it.

3. To completely demagnetise the bar a much weaker current is required than to magnetise it.

4. By repeated tortings and detortings the torts of the wire approach nearer and nearer proportionality with the corresponding loads. The torts are greater than at the first torting.

4. By repeated magnetisations and demagnetisations of a bar, the permanent magnetisations approach nearer and nearer proportionality with the intensity of the magnetising currents. The magnetisations are greater than at the first magnetising.

5. By repeated application of the same torting and detorting loads L and $-L'$ the maximum of tort reached by the torting sinks and the minimum reached by the detorting rises to a certain definite limit.

5. By repeated application of the same magnetising and demagnetising currents J and $-J'$ the maximum of magnetisation reached by the magnetising sinks and the minimum reached by the demagnetising rises to a certain definite limit.

6. The wire, if torted beyond the limits of the repeated tortings and detortings, conducts itself as if torted for the first time.

6. The bar, if magnetised beyond the limits of the repeated magnetisations and demagnetisations, conducts itself as if magnetised for the first time.

7. A torted wire, detorted by the load $-L$, cannot by repetition of this load be torted in a sense opposite to that of the initial torting. The load $+L$ torts it, however, in the first sense.

7. A magnetised bar, which is demagnetised by a current of intensity $-J$ cannot by repetition of this current be magnetised in a sense opposite to that of the initial magnetisation. The current $+J$ magnetises it, however, in the first sense.

8. If a wire having the tort A be brought by the load b to the torsion B, and afterwards be brought to any other torsion C, which lies between A and B, then to obtain the torsion B again we have only to apply the same load b. Here A can be zero, and B greater or less than A.

8. If a bar having permanent magnetism A be brought by the current b to the magnetisation B and afterwards be brought to any other magnetisation C, which lies between A and B, then to obtain the magnetisation B again we have only to apply the same current b. Here A can be zero, and B greater or less than A[1].

---

[1] Important qualifications of the above statements as to magnetisation, especially of 1–4, will be found on S. 192–200 (a, S. 172–3). Wiedemann apparently omits them in this *résumé* as he wishes only to emphasise the correspondences between torsion and magnetisation. These statements are thus very far from representing accurately the complete results of his purely magnetic experiments.

9. Shaking (*Erschütterung*) during the application of a twisting load increases the torsion of a wire.

9. Shaking during the application of a magnetising current increases the magnetisation of a bar.

10. The tort of a wire after release of the load is lessened by shaking.

10. The residual magnetisation in a bar after cessation of the current is lessened by shaking.

11. A torted and then partially detorted bar loses part of its tort by shaking or gains tort afresh according to the magnitude of the detorting.

11. A magnetised and then partially demagnetised bar loses still more of its magnetisation by shaking or gains magnetism afresh according to the magnitude of the demagnetisation.

---

12. Tort in an iron wire decreases owing to its magnetisation, but in a ratio decreasing with increasing magnetisation.

12. Residual magnetisation in a steel bar decreases owing to torsion, but in a ratio decreasing with increasing torsion.

13. Repeated magnetisations in the same sense scarcely continue to decrease sensibly the tort of a wire. A magnetisation in the opposite sense, however, produces afresh a large decrease of the tort.

13. Repeated torsions in the same sense scarcely continue to decrease sensibly the residual magnetisation of a steel bar. A torsion in the opposite sense, however, produces afresh a large decrease of the magnetisation.

14. If a wire by repeated magnetisation in opposite senses is so far detorted as is possible by the given range of magnetisation, then by magnetisation in one sense the wire shows a maximum and by magnetisation in the opposite sense a minimum of tort.

14. If a bar by repeated torsion and detorsion is so far demagnetised as is possible by the given range of torsion, then by torsion in one sense it shows a maximum, by torsion in the opposite sense a minimum of residual magnetisation.

15. A torted wire which has been slightly detorted loses by magnetisation much less of its tort than one which has only been torted. If the wire be further detorted, it exhibits at first by slight magnetisation an increase of tort, this by increasing magnetisation rises to a maximum and then decreases. The more the wire has been detorted the greater

15. A magnetised bar which has been slightly demagnetised loses by torsion much less of its magnetism than one which has only been magnetised. If the bar be further demagnetised, it exhibits at first by slight torsion an increase of its magnetisation, this by increasing torsion rises to a maximum and then decreases. The more the bar has been demagnet-

must be the magnetisation in order to reach this maximum. If the wire has been very much detorted, then its tort increases even on the application of very great magnetisation.

ised the greater must be the torsion in order to reach this maximum. If the bar has been very much demagnetised, then its magnetisation increases even on the application of very great torsions.

16. If a wire be magnetised while subject to the twisting load, then its torsion increases for slight and decreases again for greater magnetisations.

16. If a steel bar be twisted while under the influence of a magnetising current, its magnetisation increases for slight but decreases again for greater torsions.

---

17. A wire torted at the ordinary temperature loses tort by heating and regains a part of its loss on cooling. The changes increase with increasing tort.

After repeated changes of temperature the wire reaches a stable condition in which a definite tort corresponds to each temperature, decreasing as the temperature rises.

17. A bar magnetised at the ordinary temperature loses residual magnetisation by heating and regains a part of its loss on cooling. The changes are proportional to the magnetisation. After repeated changes of temperature the bar reaches a stable condition in which a definite residual magnetisation corresponds to each temperature, decreasing as the temperature rises.

18. A wire torted at the ordinary temperature and then partly detorted, loses by heating so much the less of its tort the more it has been detorted. Its tort on cooling is less than before if the detorting has been slight, it is greater if the detorting has been large.

18. A bar magnetised at the ordinary temperature, and then partly demagnetised loses by heating so much the less of its residual magnetisation the more it has been demagnetised. Its magnetisation on cooling is less than before if the demagnetisation has been slight, it is greater if the demagnetisation has been large.

19. A wire torted at a higher temperature loses tort on cooling. On a second warming it loses still further and only by the second cooling regains a part of its loss. If the wire is shaken before the first cooling, it gains at once in tort.

19. A bar magnetised at a higher temperature loses residual magnetisation on cooling. On a second warming it loses still further and only by the second cooling regains a part of its loss. If the bar is shaken before the first cooling, it gains at once in magnetisation.

A conception of the advance made by Wiedemann may be formed by comparing the above statements with those of Matteucci and Wertheim, the most important previous investigators in this field: see our Arts. 701–5 and 811–8 especially comparing 12 and 13 above with (ii), (vi) and (vii) of Arts. 813–4.

It will be seen that the laws of torsional set (*tort*)—which is what Wiedemann refers to when he speaks generally of a wire being "torted" in the above analysis—are similar to those of flexural set (*bent*), and their investigation constitutes a wide field for research which is only in the present decade being thoroughly explored.

[715.] On the basis of these analogies Wiedemann attempts a *mechanical* as distinguished from a *hydromechanical* or *aetherial* explanation of magnetisation (S. 233–47; *a*, S. 189–201). Like W. Weber, he supposes the ultimate magnetic element to be a polar molecule, and the axes of these molecules to be initially turned in all conceivable directions. He then attempts by general descriptive reasoning to account for the above relations and analogies between magnetism and strain. As a type of the general reasoning I quote the following paragraph:

Erschütterungen setzen die Theilchen der Körper in Bewegung, die Reibung der Ruhe zwischen ihnen wird gewissermassen in eine Reibung der Bewegung verwandelt. Daher werden in allen Fällen die Theilchen mehr den gerade auf sie wirkenden Kräften folgen können, und es müssen Erschütterungen eine Zunahme der temporären, eine Abnahme der permanenten Torsionen und Magnetisirungen bewirken (S. 239; *a*, S. 193).

The perusal of this type of descriptive (as distinguished from quantitative) reasoning leaves the mind almost as unsatisfied after as before, and Wiedemann himself freely acknowledges that his theoretical considerations do not fully explain all the observed phenomena (S. 247, *a*, S. 201). They do not, however, reduce in the least the value of the experimental part of this important memoir.

[716.] Resal: *Recherches sur les effets mécaniques produits dans les corps par la chaleur. Comptes rendus*, T. LI., pp. 449–50, Paris, 1860.

This is an abstract from a memoir presented to the Academy. The author supposes a body which is submitted to a uniform surface pressure to be heated. The heat expended is then divisible into two portions, one of which does work against the uniform pressure, the other does internal work (*travail que l'on peut considérer comme le résultat du développement ou de l'introduction dans le système matériel de nouvelles forces moléculaires essentiellement répulsives*, p. 450). The object of the memoir is the discovery of an expression for the latter work in the case of homogeneous bodies.

Resal gives the following expression for it in solids:

$$T = \frac{wE\alpha^2}{3\pi c},$$

where α = coefficient of linear dilatation,

E = stretch-modulus,

w = specific weight,

c = specific heat.

No proof is given of this formula, nor do I understand how it is deduced.

[717.] Hermann Vogel: *Ueber die Abhängigkeit des Elasticitätsmoduls vom Atomgewicht. Annalen der Physik*, Bd. CXI., S. 229–239, Leipzig, 1860.

This is an endeavour to find a relation between the stretch-modulus and the coefficient of thermal expansion, but neither the theoretical reasoning nor the numerical results are satisfactory.

Let a be the coefficient of linear thermal expansion, c the specific heat and w the specific weight of a prismatic metal rod of unit length, unit cross-section and unit (? absolute) temperature. The quantity of heat of a volume of water equal to that of the rod being unity, then the amount of heat in the metal rod equals cw. This amount of heat produces an extension in length equal to a, and therefore unit quantity of heat produces an extension equal to $a/(cw)$.

Vogel then continues:

Derselbe Stab erleidet durch eine, in der Richtung der Länge wirkende, dehnende, der Gewichtseinheit gleiche Kraft eine Ausdehnung, die man den *Dehnungsquotienten* nennt.

Ist nun die Arbeit, welche die Wärmeeinheit zu leisten vermag, eine constante Grösse, so werden die Ausdehnungen, welche verschiedene Metalle durch die Wärmeeinheit erfahren, in demselben Verhältnisse zu einander stehen, wie ihre Dehnungsquotienten (S. 230).

Vogel denotes by *Dehnungsquotient* the reciprocal of our stretch-modulus; and the first paragraph is intelligible, but I do not understand the second, for the amount of heat communicated not only dilates the body but also raises its temperature, and even if there were no heat expended in raising the temperature, the extensions which different metals receive from unit quantity of heat ought to be as the reciprocals of their dilatation-moduli rather than as those of their stretch-moduli. These two sets of ratios will not necessarily be equal unless we presuppose uni-constant isotropy.

[718.] It is possible of course if the amount of heat used in raising temperature be proportional to the total amount of heat applied to a body that we may have, on the uni-constant hypothesis, a relation of the form:

$$E : E' :: \frac{cw}{a} : \frac{c'w'}{a'},$$

or,

$$\frac{Ea}{cw} = \text{a constant} \quad\text{.......................(i).}$$

Hence it is worth while noting what numerical results Vogel gives. He finds for the metals the mean value of $Ea/(cw) = 2{\cdot}44$, exactly agreeing with its value for silver. The minimum is 1·85 for lead and the maximum 3·18 for zinc. Below zinc stands iron with 2·79, and above lead are platinum with 2·01 and gold with 2·10. Thus the presumed constant has a rather wide range, which may be due to error in the theory, to the fact that the quantities were not determined from the same specimens of metal, or to the need of replacing the stretch-modulus by the dilatation-modulus.

[719.] According to Dulong and Petit and Regnault, if A be the atomic weight,

$$Ac = \text{a constant}.$$

Hence, it must follow that

$$\frac{EaA}{w} = \text{a constant} \quad\text{.......................(ii).}$$

Or, the product of the stretch-modulus, the coefficient of thermal expansion, the atomic weight and the reciprocal of the specific weight is a constant.

The exactitude of (ii) seems even less than that of (i). The constant is 6·03 for lead, rising to 10·22 for zinc, the mean value being 7·716, which is not very different from that for tin (7·69). Vogel remarks of these results:

In Anbetracht des Umstandes, dass alle in der Formel EaA/w enthaltenen Werthe, A ausgenommen, innerhalb gewisser Gränzen schwanken und noch dazu von verschiedenen Beobachtern an verschiedenen Metallstücken bestimmt worden sind, ist eine solche Uebereinstimmung immerhin merkwürdig genug (S. 233).

[720.] Vogel then draws attention to a result of Masson's referred to in our Art. 1184*, 7°, namely that the product of the reciprocal of the stretch-modulus (*coefficient d'élasticité*) and the atomic weight or a multiple of the atomic weight is a constant. This would only be true according to Vogel's theory on the assumption that the ratios of the values of aE^2/w for the metals were as whole numbers. As a matter of fact they are nearly as 30 : 15 : 5 : 3 for iron, copper, silver and tin (S. 235). Vogel draws as easy corollaries from his formula the following statements:

(i) If the values of a/w are in a very simple ratio to each other, then the product of the stretch-modulus and the atomic weight or a multiple of it is constant.

For example, in the case of copper and silver, the values of a/w are very nearly equal, and we have $EA = 397391$ for copper, and $= 409482$ for silver.

(ii) If for different metals Ea is constant, then their specific volumes (or the values of A/w) are equal.

For example, in the case of iron and copper, $A/w = 3{\cdot}6$, while the values of Ea are ·2458 and ·2157 respectively.

Natürlich kann hier nicht von absoluter, sondern nur von annähernder Uebereinstimmung der Werthe von A/w und von Ea die Rede seyn (S. 236).

[721.] Vogel in conclusion refers to Wertheim's result (see our Art. 1299*) that $E\left(\frac{A}{w}\right)^{\frac{7}{3}}$ is approximately constant for metals. Vogel combines this with (ii) and finds: $a \propto \left(\frac{A}{w}\right)^{\frac{7}{3}}$. He shows that for silver, iron and cadmium there is some approach to this law (S. 238).

He does not refer to Person's results, which are in some respects akin to his own: see our Art. 1388*.

While Vogel's theory is wanting in accuracy, and he himself admits that his formulae must not be pressed too far, still the numerical results of his paper are sufficient to show that careful experimental investigation in this field might lead to the discovery of results of great value, and for this reason the paper has been more fully referred to here than at first sight it appears to deserve.

GROUP B.

Kupffer's Memoirs with Zöppritz's theoretical Discussion of Kupffer's Results.

[722.] In 1849 the Russian government established a Central Physical Observatory in St Petersburg and appointed A. T. Kupffer as Director. According to the rules the Director had to furnish a yearly report on the experiments conducted in the Observatory as well as on other matters to the Minister of Finances. Thus arose the *Compte rendu annuel* of the *Observatoire physique central.* In these *Comptes rendus* for the years 1850 to 1861 will be found accounts of the researches in elasticity carried on by Kupffer. In 1860 he published the first volume of a great work entitled: *Recherches expérimentales sur l'élasticité des métaux faites à l'observatoire physique central de Russie,* Tome I., St

Petersburg, 1860. This first volume is devoted to the experimental study of flexure and the transverse vibrations of elastic laminae with a view to the discovery of the elastic properties of metals. A second volume was to be devoted especially to metals prepared in Russia and a third to torsion and torsional oscillations. Further Kupffer promised to consider the resistance of metals strained beyond their elastic limit and also up to rupture. Only the first volume of this important work was ever published; experiments partly covering the ground of this volume, and partly that of the proposed succeeding volumes, will be found in the above mentioned *Comptes rendus* up to 1861: see also our Art. 1389* After this date they ceased and Kupffer died in 1865. Separate memoirs by Kupffer belonging to the period 1850–60 are also considered in our Arts. 745–57. His researches are among the most elaborate and careful that have ever been made on the elasticity of metals. We shall commence our consideration of them by noting points in the *Comptes rendus* not embraced in the volume of 1860.

[723.] *Compte rendu annuel*. Année 1850 (St Petersburg, 1851). Pp. 1–11 are occupied with a description of the apparatus recently erected and of the experiments made on the elasticity of metals at the new observatory. The torsional experiments referred to are chiefly those of the memoir of 1848: see our Art. 1389*. The experiments on flexure are the earliest of the series described in the work of 1860, namely: the determination of the stretch-modulus by the transverse oscillations of a clamped-free rod. One or two points may be noted:

(*a*) Kupffer as a rule uses in his experiments the symbol δ (sometimes δ') for the extension of a rod of unit length and unit radius (circular cross-section) under the traction of unit force. On p. 9 of the *Compte rendu* for 1850 he gives a formula: 'où δ désigne le coefficient d'élasticité du métal.' On p. 19 of the *Recherches expérimentales* he writes: 'on désigne par $1/\delta'$ ce que l'on appelle ordinairement le *coefficient d'élasticité*.' Here δ or $\delta' = 1/(\pi E)$, where E is the stretch-modulus. Elsewhere in the *Recherches* he uses δ for $\pi\delta'$ and terms it the *dilatation élastique* (pp. xv and xxxi). On p. 299 of the *Recherches* he says let $\beta =$ *l'accroissement du coefficient de dilatation élastique*, and then uses a formula involving β and δ in such fashion that he evidently means δ to be the *coefficient de dilatation élastique*. His experiments really go to show that the stretch-modulus (and presumably the slide-modulus) *decreases* with increase of the temperature, or that δ increases with increase of temperature. If this be so, he must in the paragraph of the memoir of 1848 cited in our Art. 1395* mean by *coefficient* *d'élasticité* the quantity δ, although both in that memoir and in the *Recherches* he defines this coefficient as either $1/\delta$ or $1/\delta'$. This really follows from the results in our Arts. 1392* and 1396*. Hence in the remarks following the citation in our Art. 1395* the words 'slide-modulus increases' and 'is probably increased' should be replaced by 'slide-modulus decreases' and 'is probably decreased' respectively. This confusion in terms is not confined to the *coefficient d'élasticité;* it is occasionally difficult to understand what Kupffer means by *la force élastique du métal,* a term which he freely uses in summing up his results.

In the present notice of his experiments (p. 4) he refers in the following words to Wertheim's results on the relation of temperature to the elastic moduli (see our Arts. 1298* and 1301*, 5°):

Ces mêmes expériences [i.e. those on torsion of 1848] m'ont fait voir que les changements de température exercent une influence sensible sur la force élastique des fils métalliques, qui augmente, lorsque la température diminue, et réciproquement. Les expériences de M. Wertheim......avaient déjà signalé cette influence pour de grands intervalles de température; mes expériences étaient assez rigoureuses pour la préciser pour les différences de température de 10 à 15° R. M. Wertheim est arrivé à des résultats fort différents des miens, et la loi qu'il a trouvée n'est pas aussi simple que celle que je viens d'énoncer; mais comme nos valeurs ont été obtenues par des méthodes d'observation très différentes, elles ne sont pas exactement comparables. Cette question a encore besoin d'être traitée à fond, et le sera assurément, puisque la Société Royale des Sciences de Göttingue en a fait une question de prix pour l'année 1852.

[724.] (*b*) A second point worth noting is a suggestion, made I believe for the first time, to determine the mechanical equivalent of heat from the force necessary to produce a given stretch. It is contained in the following words:

Nous avons vu dans ce qui précède qu'on peut déterminer, avec une très grande précision, la dilatation qu'un fil éprouve par l'action d'un poids; évaluer ensuite la dilatation de ce même fil par la chaleur, n'est-ce pas évaluer en poids la force mécanique de la chaleur? (p. 5).

The reasoning, however, by which Kupffer deduces the mechanical equivalent of heat seems to me very doubtful, and the agreement of his value for it with Joule's must I think be looked upon as a happy coincidence.

The same numerical results as are here given are repeated in a paper in the *Bulletin,* but the reasoning there is somewhat different: see our Art. 745.

In the first place Kupffer makes an appeal to the theorem, due to Poisson, that the same traction applied to the terminal sections of a bar produces double the stretch that it would do if applied all over the surface. This is easily proved on the uni-constant hypothesis, but I fail to see that it is properly applicable to the present problem, where it would seem we ought to deal with equal quantities of work spent in these two forms of strain rather than with equal tractions. Kupffer then continues:

Un cylindre, dont la longueur et le rayon sont égaux à l'unité, est allongé de cette même unité (c'est-à-dire d'un pouce), par un poids $p=1/\delta$, où δ désigne l'allongement que ce même cylindre éprouve par la traction de l'unité de poids (c'est-à-dire d'une livre); on peut donc évaluer la force élastique du cylindre, en disant qu'elle élève le poids p à la hauteur d'un pouce. En échauffant ce même cylindre de 0° à 80° R., il s'allonge de la quantité a; d'après l'hypothèse que nous avons adoptée plus haut, il s'allongerait de la quantité $2a$, si l'effet de la chaleur n'avait lieu que dans une seule direction comme la traction; la quantité de chaleur, qui produit cet allongement est égal à wmd/d', où w est la quantité de chaleur, qu'il faut pour élever de 0° à 80° R. la température d'un cylindre d'eau, dont la hauteur et le rayon sont égaux à l'unité, m la chaleur spécifique et d la densité du corps élastique, et où d'...est la densité de l'eau*: nous aurons donc l'expression $wmd/(2ad')$ pour la quantité de chaleur, qui produirait un allongement d'un pouce; ou, comme les causes doivent être égales, lorsque les effets sont égaux, nous aurons évidemment $p=wmd/(2ad')$. Mais nous avons aussi $p=1/\delta$ (p. 6)......

* J'appelle densité le poids de l'unité de volume ou d'un pouce cube....

Hence Kupffer reaches as his final equation:

$$1/\delta = wmd/(2ad');$$

and by substituting the numerical values of the quantities involved, he finds a magnitude for w agreeing closely with Joule's.

[725.] But Kupffer obtains this result by a compensation of errors. In the first place the elastic work corresponding to p and unit extension ought to be $\frac{1}{2}p$ and not p. And further it is not evident that 'the effects are equal' (*les effets sont égaux*), for in the case of a pure elastic strain we have the body at temperature 0° say, but in the application of heat we have the same strain *together* with the body at a temperature of 80° R. Suppose H the quantity of heat given to the body and let it be held at the strain produced by this amount of heat and cooled down to temperature 0°, and in doing so let H' be the amount of heat communicated to the refrigerator, and h the amount of heat the body would give off in being strained at constant temperature zero up to the same expansion, then the heat equivalent to the mechanical strain would seem to be $H-H'+h$ and not H as Kupffer assumes. There is, I think, no reason for assuming $H'-h$ indefinitely small as compared with H, indeed Kupffer's result seems to indicate (since he has dropped the $\frac{1}{2}$) that $H'-h=\frac{1}{2}H$ approximately in his case, otherwise his errors would not compensate each other as they appear to do[1].

[1] Kupffer's results are quoted without any apparent questioning in some modern works, e.g. G. Helm, *Die Lehre von der Energie*, S. 91, just as they were cited in the *Philosophical Magazine*, *Poggendorffs Annalen*, and other journals without demur in 1852. In the *Fortschritte der Physik* for 1852, S. 373–7, Helmholtz remarks that the argument of the famous St Petersburg physicist is too brief to be open to intelligent criticism, and he shows that Kupffer's formula is not identical with any of the known equations of Thermodynamics. He does not, however, distinctly state that it cannot be true. Compare Vogel's paper discussed in our Arts. 717–21.

[726.] (*c*) The last point to be noted in the present paper is the experimental discovery of after-strain in metals. Both Seebeck and Clausius had suggested its existence (see our Arts. 1402* and 474), but no physicist had distinctly seen and measured its effect, so far as I am aware, before Kupffer.

The following sentences give his conclusions:

(i) La flexion qu'une verge encastrée par une extrémité et libre de l'autre éprouve par une charge quelconque, suspendue à son extrémité libre, augmente avec le temps, et ne s'arrête qu'après un temps plus ou moins long, quelquefois après plusieurs jours seulement.

(ii) Lorsqu'une verge est restée fléchie pendant quelque temps, ce n'est qu'après un intervalle de temps plus ou moins long, qu'elle revient exactement à sa première position.

(iii) Une verge fléchie par un poids, pendant un instant seulement, revient tout de suite et exactement à sa première position, aussitôt que le poids a été ôté, mais cela n'a lieu que jusqu'à une certaine limite; lorsque le poids dépasse cette limite, la verge ne revient plus tout de suite à sa première position; elle n'y revient qu'après longtemps ou pas du tout (p. 11).

The last statement shows the possibility of *set* combined with after-strain arising from instantaneous loading.

[727.] *Compte rendu annuel.* Année 1851 (St Petersburg, 1852). Pp. 1–11 give an account of experiments to determine the elastic constants of iron and brass by different methods. Kupffer finds that for brass pure traction and flexure experiments give practically the same value for the stretch-modulus, but that this value differs considerably from the value deduced on the uni-constant hypothesis from the slide-modulus as determined by the method of torsional vibrations. Nor is the ratio of the slide- to the stretch-modulus the same for brass and for iron wire. This would be an argument against uni-constancy, if we could assume Kupffer's wires to have been isotropic (pp. 1–5). Kupffer next refers to the various effects which strain, annealing etc. have on the stretch-modulus, as obtained by the method of transverse vibrations of a bar (pp. 5–7), and then he deals with the influence of the resistance of the air on torsional vibrations (pp. 7–10). These matters will be more fully dealt with in our discussion of Kupffer's great work of 1860.

[728.] *Compte rendu annuel.* Année 1852 (St Petersburg, 1853). Pp. 1–19 furnish a further account of flexure experiments to determine the stretch modulus. The experiments were made partly by oscillatory, partly by statical methods. $1/\delta' = 1/(\pi\delta)$ is defined as *le coefficient d'élasticité du métal*[1] (p. 6). With regard to experiments by these

[1] δ and δ' exactly change meanings in the *Comptes rendus* and the *Recherches*! compare pp. 19 and 133 of the latter work with p. 9 of the *Compte rendu* for 1850 or p. 6 of that for 1852; i.e. $\delta = \pi\delta' = 1/E$ in the *Recherches*, but $\delta' = \pi\delta = 1/E$ in the memoir of 1848 and the *Comptes rendus*.

different methods Kupffer finds (pp. 13 and 19) in Russian measure the following results for $1/E$:

Material	Statical Method	Oscillatory Method
Soft Steel Lamina, No. 5	$10^{-13} \times 296,020$	$10^{-13} \times 297,952$
Soft Cast Steel, No. 6	$10^{-13} \times 301,055$	$10^{-13} \times 300,623$
Platinum Lamina	$10^{-13} \times 358,600$	$10^{-13} \times 358,438$
Brass Rod	$10^{-13} \times 592,913$	—
Iron Rod, No. 3	$10^{-13} \times 329,270$	$10^{-13} \times 322,363$ (*Recherches*, p. 283)

So far as this table goes we see that with the exception of the first steel lamina, δ' is greater and therefore the stretch-modulus E is less when calculated by the statical method. The differences are small except in the case of iron. The exception in the case of one kind of steel agrees with the difference found by Kennedy and Wüllner (see ftn. p. 702 of our Vol. I., where obviously the conclusion should say 'in favour of' and not 'opposed to' Wertheim's result). The numbers given for the steel lamina No. 6 and for platinum in the *Recherches* vary a good deal (see pp. 219–229, and 264–5) and therefore are perhaps not very exact. If E_s be the static and E_k the kinetic stretch-modulus we have:

For platinum, $E_k/E_s = 1{\cdot}00045$, For iron, $E_k/E_s = 1{\cdot}0214$.

These ratios are less than those obtained by Weber and Wertheim (see our Arts. 705* and 1403*, Weber's numbers give for iron 1·072, for platinum 1·21 and for copper 1·09). Results like these differ, however, so enormously from those calculated by Sir W. Thomson (Article: *Elasticity* in the *Encyclopaedia Britannica*, § 76, *Thermodynamic Table II.*, or *Mathematical and Physical Papers*, Vol. III., p. 71) and also among themselves, that but little faith can be placed in them. In fact they depend in Kupffer's case on the *last three* figures of his values, but an examination of the individual experiments shows that these three figures vary very greatly from one experiment to another. Kupffer gives no value of δ' for the brass rod in the *Compte rendu*, but the values he gives for such rods in the *Recherches* (pp. 111–21) make $E_k/E_s < 1$. Hence Kupffer's results will not allow us to assert that the kinetic method always gives larger stretch-moduli than the static, and that the differences are too great to be explained solely by thermal action. Even if we could allow for influence of after-strain (see our Arts. 1402* and 474) the extreme difficulty of determining the modulus accurately to six places of figures would render any evaluation of the ratio of specific heats by Kupffer's process impossible.

Kupffer himself attributes the difference between the values of the modulus as obtained by the two methods to the fact that in the case of transverse oscillations the rods oscillate elliptically and never in a plane, and he holds that this tends to diminish slightly the duration of the oscillations (ftn. p. 19). He does not demonstrate this, and I do not see why it should be true: see as to other difficulties our Art. 821 and footnote.

[729.] *Compte rendu annuel.* Année 1853 (St Petersburg, 1854). The continuation of experiments on the determination of the stretch-modulus by static and kinetic methods is described on pp. 1–7. Kupffer notes that the static flexural method gives results more in accordance with themselves than the kinetic (on voit encore ici, que les valeurs de δ', obtenues par la flexion, sont d'une exactitude bien supérieure à celle, qu'on peut obtenir par des oscillations transversales, p. 4). A series of experiments on the static flexure of cast-iron is described on pp. 6–7. Kupffer remarks that flexural set always occurs with this kind of iron, and that when this is subtracted from the total flexure due to the load the deflections are still not proportional to the loads, but increase more rapidly than the loads. Thus for two bars (i) and (ii) of specific gravities 7·124 and 7·130 respectively we have

$$\text{(i)}\quad\begin{cases}\delta' = 10^{-13}\times 622{,}724 \text{ for a total load of } 1 \text{ lb.}\\ \quad = \ldots\ \times 636{,}762 \ \ldots\ldots\ldots\ldots\ldots\ldots\ 1\cdot125 \text{ lbs.}\\ \quad = \ldots\ \times 653{,}590 \ \ldots\ldots\ldots\ldots\ldots\ldots\ 1\cdot375 \text{ lbs.}\end{cases}$$

For this bar $\delta' = 10^{-13} \times 559{,}288$ from transverse oscillations.

$$\text{(ii)}\quad\begin{cases}\delta' = 10^{-13}\times 589{,}100 \text{ for a total load of } 1 \text{ lb.}\\ \quad = \ldots\ \times 601{,}650 \ \ldots\ldots\ldots\ldots\ldots\ldots\ 2 \text{ lbs.}\\ \quad = \ldots\ \times 620{,}860 \ \ldots\ldots\ldots\ldots\ldots\ldots\ 3 \text{ lbs.}\\ \quad = \ldots\ \times 636{,}980 \ \ldots\ldots\ldots\ldots\ldots\ldots\ 4 \text{ lbs.}\end{cases}$$

For this bar $\delta' = 10^{-13} \times 564{,}137$ from transverse oscillations[1]. These and similar results are in general conformity with Hodgkinson's experiments: see our Arts. 969* and 1411*, and conclusively show the want of exact meaning in the term stretch-modulus for the case of cast-iron.

[730.] In this year Kupffer also began a series of experiments on the dilatation by heat of the same metal bars as he had been experimenting on elastically. The observations were made by taking each bar as a pendulum, the bob being so attached to the bar that the distance of its centre of gravity from the axis of oscillation depended only on the length of the bar. The results of experiments on two brass bars only are given. These bars were taken from the same casting but one had been vigorously hammered. The coefficients of linear expansion were measured for an increment of 1° between 25° R. and 30° R. We have:

	Coefficient.
Cast brass	·000,025,727.
Hammered brass	·000,024,980.

[1] This number is incorrectly given in the *Compte rendu*: see the *Recherches*, p. 87.

Thus the ratio = 1·030 : 1 about. The ratio of the specific gravities was 1 : 1·035, or the coefficients of expansion were nearly inversely proportional to the specific gravities.

[731.] The *Compte rendu* for this year also contains a scheme for an extensive series of experiments on the entire elastic life of materials prepared in Russia. This scheme is perhaps the most complete ever drawn up for a detailed investigation of the cohesive and elastic properties of metals. The commission proposed in it would have achieved on a more catholic and more scientific (physical as distinct from empirico-technical) basis for many metals what the English commission did for iron only: see our Art. 1406*. Such experiments as Kupffer made in this direction would have occupied the second volume of his *Recherches;* what they were we can only gather from subsequent numbers of the *Comptes rendus.* The programme is drawn up with a view to the industrial use of metals, and I only regret that our space does not permit of its reproduction here. Elastic properties, as well as those of set and rupture, are taken into full consideration; further the influence of the various processes of manufacture, of working, of temperature-effect, of impulsive and long continued stress on one and all of these properties are dealt with. As a scheme for further physico-technical researches in elasticity, or for a treatise on the subject, Kupffer's programme would still, with a few modifications in the light of more recent discoveries, be of very great value. It occupies pp. 11–14 of the *Compte rendu annuel.*

[732.] *Compte rendu annuel.* Année 1854 (St Petersburg, 1855). The account of elastical researches occupies pp. 1–28. It commences with some further remarks on flexural measurements chiefly directed to investigate the effect of 'working' on the metals. Kupffer concludes that "l'élasticité des métaux est considérablement augmentée par le travail qu'ils subissent dans le laminage, l'écrouissage et en passant par la filière" (p. 3). By an augmentation of the elasticity is to be understood a smaller value of δ' or a greater value of the stretch-modulus.

[733.] The major portion of this report is occupied with experiments on torsion (pp. 4–28). These were made with an apparatus similar to, but far more exact than that used for the experiments described in the memoir of 1848: see our Art. 1389*. An account of this apparatus will be found in the *Compte rendu* for 1850 and it is repeated here (pp. 4–5). The apparatus involves an oscillatory method of experiment, but one used by Kupffer with extreme accuracy and careful determination of all the possible sources of disturbance. The real slide-modulus μ is to be obtained from Kupffer's δ by the relation $1/\delta = \frac{5}{2}\pi\mu$: see our Art. 1390*. Kupffer's μ (p. 6) is *not* our slide-modulus, but $= \frac{1}{2}\pi\mu$, i.e. it is the moment of the force necessary to turn through unit angle a cylinder of unit radius and unit length.

[734.] Kupffer confirms his former result (see our Art. 1391*) as to the law connecting the duration of the oscillations with the amplitudes. He finds that a/P_s (in the notation of our article referred to), now written ψ, depends largely on the nature of the material and the working it has been subjected to. This quantity ψ is termed by Kupffer the *coefficient of fluidity*. Kupffer's 'fluidity' of metals is a property corresponding to Sir W. Thomson's 'viscosity' (see our Chapter devoted to that physicist), and as it appears to be the first real consideration of the matter, I quote p. 15 of Kupffer's remarks:

ψ a une valeur constante pour chaque fil, mais varie considérablement d'un fil à l'autre, comme le prouvent non seulement les expériences que je viens d'exposer et qui se rapportent au fer et à l'acier, mais aussi toutes les observations qui vont suivre.

Les observations précédentes donnent

pour le fil de fer $\psi = \cdot 000616$,

pour le fil d'acier $\psi = \cdot 00003736$.

C'est-à-dire la valeur de ψ est 17 fois plus grande pour le fer, que pour l'acier.

De là il suit que l'accroissement, que la durée des oscillations éprouve lorsque les amplitudes augmentent, ne peut être un effet de la résistance de l'air, ni une conséquence de la loi générale de l'élasticité, quelle qu'elle soit d'ailleurs (que l'élasticité soit proportionnelle aux accroissements de la distance entre les molécules, ou qu'elle suive une autre loi relativement à ces distances); cela doit être une propriété inhérente aux corps élastiques, qui varie d'un métal à l'autre, qui varie même pour le même métal, selon le travail qu'il a subi.

J'ai fait voir, par des expériences rapportées dans mon Compte rendu de l'année 1851, que l'amplitude des oscillations diminue aussi bien dans le vide, que dans l'air, cette diminution ne peut donc pas être non plus un effet de la résistance de l'air, cette résistance la fait seulement diminuer plus rapidement. La position d'équilibre, à laquelle il faut rapporter toutes les forces, qui font osciller un fil métallique, se déplace continuellement et toujours dans le sens des oscillations; de sorte que cette position d'équilibre oscille avec le fil même autour d'une position moyenne, qui est celle du fil complètement revenu au repos. Il paraît que les molécules des corps solides possèdent la propriété non seulement de s'écarter les unes des autres, en produisant une résistance proportionnelle aux écarts, mais aussi de glisser les unes sur les autres, sans produire aucun effort. Cette propriété est possédée à un haut degré par les fluides; je la nommerais donc volontiers la fluidité des corps solides; le coefficient ψ pourrait être appelé coefficient de fluidité; la malléabilité des métaux paraît en dépendre, et peut être aussi leur dureté; des expériences ultérieures nous apprendront jusqu'où va cette analogie.

Le coefficient de fluidité peut varier beaucoup dans le même métal, deux autres fils de fer de ·04801 et de ·08099 de rayon ont donné $\psi = \cdot 000393$ et $\psi = \cdot 000494$. Pour un fil de cuivre jaune de ·09518 de rayon, il a été trouvé égal à ·000284, pour un autre, dont le rayon était égal à ·0807 on a eu $\psi = \cdot 000930$. Mais il varie surtout d'un métal à l'autre: on a:

pour le platine $\psi = \cdot 0001376$,

pour l'argent $\psi = \cdot 0003650$,

pour l'or $\psi = \cdot 000300$.

Here we have a very clear description of the action of viscosity in metals, a property which has much exercised physicists upon its frequent rediscoveries since Kupffer's investigations of 1848–1854.

[735.] But Kupffer's torsion experiments led him to consider several other points connected with torsional vibrations which have been largely dealt with in recent years. Thus:

(i) On pp. 16–23 he shows how the resistance of the air may be taken into account and eliminated.

(ii) On p. 23 he refers to the reduction of the observations to a constant temperature: see our Arts. 1392* and 1396*.

(iii) On pp. 23–28 he discusses what effect the traction of a wire has on its torsional resistance. This is important as it is necessary to allow for the weight of the vibrator.

Kupffer had in the *Compte rendu* for 1851 given the following result, where M and M' are respectively the torsional rigidities of the wire[1] without and with a traction which produces a stretch s in the wire:

$$M' = M(1 - 3s).$$

Kupffer remarks that Neumann of Königsberg (the great *Franz*) had sent him the result

$$M' = M(1 - \epsilon s),$$

where ϵ can vary between the limits 1 and 3, as the result of a mathematical investigation in which it is not assumed that the elastic coefficients are altered or the proportionality of stress and strain abrogated. The investigation is not given, but it is easy to replace it. Let η be the stretch-squeeze ratio, and let the wire be of length l and radius r, then we have for the torsional rigidity without traction

$$M = \mu \frac{\pi r^4}{2l},$$

and with traction

$$M' = \frac{\mu \pi r^4 (1 - \eta s)^4}{2l(1 + s)} = M\{1 - (1 + 4\eta) s\}.$$

Now η can take all values from 0 to $\frac{1}{2}$ for bi-constant isotropy: see our Art. 169 (*d*). Hence Neumann's statement follows. That Kupffer's experiments gave $1 + 4\eta = 3$ or $\eta = \frac{1}{2}$ for his wires, brass and steel, I attribute, not to the fact that those wires had bi-constant isotropy approaching its limit, but to their being really aeolotropic.

[736.] A result also due to Franz Neumann and recorded by Kupffer in a note on p. 24 deserves notice. He says that Neumann had shown by fixing small mirrors to a rectangular bar under flexure

[1] *Torsional rigidity* of a wire is a convenient term for the torsional moment per unit angle of torsion.

that a cross-section perpendicular to the axis is no longer a rectangle but a *trapezium*[1]. This had been previously shown by E. Clark for set (see our Art. 1485*), and it is a physical confirmation of Saint-Venant's theory so well exhibited in his plaster-models of flexure: see our Arts. 92 and 95, and also the *Leçons de Navier*, p. 34.

Kupffer further notices that Neumann had experimentally demonstrated that the volume of a wire increases under traction up to the elastic limit, but that if it is stretched beyond this limit, the volume remains constant, i.e. that set is unaccompanied by change in volume. According to Kupffer, Neumann had also shown experimentally that the value of the stretch-squeeze modulus is not constant (e.g. $\frac{1}{4}$ according to Poisson, or $\frac{1}{3}$ according to Wertheim) but varies with the nature of the metal. Kupffer does not state what was the method used in Neumann's experiments (*expériences également ingénieuses et précises*).

[737.] *Compte rendu annuel.* Année 1855 (St Petersburg, 1856). This report deals with the influence of heat on the elasticity of metals. This as we have seen (Art. 723) was the subject of a prize offered by the Royal Society of Göttingen. It was awarded to Kupffer in November, 1855.

He divides his researches under two heads:

(i) Influence of an increase of temperature on elasticity, lasting only while this temperature is maintained.

(ii) Changes produced by an increase of temperature on elasticity after the thermal influence has ceased. Of these he writes:

On verra dans le cours de ces recherches, que ces deux actions de la chaleur sur les corps élastiques sont très différentes, elles peuvent même être opposées; lorsque la température d'un corps élastique augmente, son élasticité diminue toujours; mais lorsque l'action de la chaleur cesse et lorsque le corps élastique est revenu à sa température initiale, son élasticité ne revient pas toujours à la même valeur, mais elle a souvent changé considérablement; tantôt on la trouve augmentée, tantôt on la trouve diminuée (p. 2).

Kupffer points out that the elasticity of metals can be easily in-

[1] Turning to our Art. 95 we obtain for the tangent of the angle ψ through which a small mirror would be turned if fixed at the middle of a vertical side of a cantilever at a distance ζ from the loaded end

$$\tan\psi = \eta_1 \frac{P\zeta b}{E\omega\kappa^2} \quad \text{(i)}.$$

If a small mirror were fixed to the middle of the top of the beam at the same distance, it would be turned through an angle ψ_1, given by

$$\tan\psi_1 = \frac{P\zeta c}{E\omega\kappa^2}, \text{ approximately} \quad \text{(ii)}.$$

Hence $\tan\psi = \eta_1 \frac{b}{c} \tan\psi_1$, and we have what appears to be a practical optical method of determining the stretch-squeeze ratio η_1. It might also be found by substituting directly the value of E in (i).

vestigated in three different ways and the effects of heat on all these ought to be considered. These are:

Statical Traction—Longitudinal Vibrations,
Statical Flexure—Transverse Vibrations,
Statical Torsion—Torsional Vibrations.

[738.] He points out how the investigations in these directions are affected by secondary elastic properties, more particularly by elastic after-strain. He now attributes to this property the augmentation of the duration of the oscillations, which he had found in torsional oscillations to vary as the square root of the amplitude: see our Art. 1391*. In other words he supposes elastic after-strain to be the origin of the property he has termed *fluidity*, or of our more modern *viscosity*. Sir W. Thomson seems to think also that the viscosity may be due wholly or partially to elastic after-strain: see our ftn. p. 390, Vol. I.

[739.] Returning to the formula of our Arts. 1391* and 734, or

$$P_0 = P_s(1 - \psi\sqrt{s}),$$

we note that Kupffer now states that he has found more accurately how ψ varies with the size of his wire. If r be the radius of the wire, l its length and ν a constant coefficient which depends on the elastic properties of the material, then:

$$P_s = P_0\left(1 + \nu r\sqrt{\frac{s}{l}}\right).$$

Kupffer terms ν the "true coefficient of fluidity or ductility." We may perhaps term it the "after-strain (or viscosity) coefficient for torsional vibrations": see our Art. 751 (*d*).

[740.] The rest of the memoir is occupied with details taken from the great memoir on thermo-elasticity: see our Arts. 748–57. If the temperature be raised from t to t' and the stretch and slide-moduli change from E, μ to E', μ' respectively, then Kupffer gives the values of β_f and β_T for various metals, where:

$$E' = E\{1 - \beta_f(t' - t)\},$$
$$\mu' = \mu\{1 - \beta_T(t' - t)\}.$$

These values are determined by transverse and torsional vibrations[1].

[1] Kupffer neither here nor in his memoir clearly states whether he has attempted to eliminate the effect of heat in lengthening his wire, and so affecting the torsional vibrations. If he has not, then, by our Art. 735 (iii), the torsional moment is altered, and thus the slide-modulus will appear to be altered. The alteration would be given by a formula of the form $\mu' = \mu(1 - \epsilon s)$, where s is the thermal stretch $= \alpha(t' - t)$, α being the coefficient of linear thermal dilatation. Now for brass Kupffer has found (see our Art. 730) $\alpha = \cdot000{,}025{,}727$ and $\epsilon = 3$ nearly, hence $\mu' = \mu\{1 - \cdot000{,}077{,}181(t' - t)\}$, but β_T for the like brass $= \cdot000{,}6982$. Thus the purely lengthening effect of change of temperature on the wire would only account for about 1/9 of the change in μ.

It should be noted that β_τ here is twice the β of our Art. 1396*. The results are considered at length in our Arts. 752–4. We merely note now that the values of β_τ are given for higher ranges of temperature than in the memoir of 1848: see our Arts. 1392* and 1396*. The effect also on ν of changes in temperature are noted as in the memoir above referred to.

[741.] *Compte rendu annuel.* Année 1856 (St Petersburg, 1857). Pp. 57–66 give an account of the elastical researches carried out during the year in the *Observatoire physique central.* One or two points may be noted:

(*a*) Three laminae were formed from the same piece of cast brass, the first remained as originally cast, the second was vigorously rolled (*fortement laminé*) and the third vigorously hammered (*fortement martelé*). It was found that their *stretch-moduli were nearly in the ratio of the squares of their densities.* The same result was very nearly true for specimens of English and Swedish wrought-iron (compare Art. 759 (*e*)).

On voit par ce qui précède, combien l'influence du martelage et du laminage sur l'élasticité des métaux est grande (p. 58).

The result is important if only approximate.

(*b*) Kupffer regards (pp. 59–62) from a very insufficient theoretical standpoint the effect of a stretch produced by heat or load on the value of the elastic constants as obtained by experiment. He seems to have considerable difficulties with Neumann's formula (see our Art. 735 (iii)), largely due, I think, to his assumption that wires possess isotropy. He wants (p. 62) to reject the formula

$$M' = M(1 - \epsilon s)$$

as an explanation of the effect of traction on torsion when he finds values of ϵ greater than 3, although this would in fact not necessarily indicate anything more than aeolotropy: see our Art. 308 (*b*).

He gives the results of some experiments on the value of ϵ when successive set-stretches are given to a wire under torsion; ϵ begins by being as great as 6 and diminishes to about 3·4 as the sets are continued.

(*c*) The report concludes with the results of a number of Kupffer's experiments giving the elastic moduli in kilogramme-millimetre units: see our Art. 772.

[742.] *Compte rendu annuel.* Année 1857 (St Petersburg, 1858). This contains:

(*a*) Values of the stretch-moduli for various kinds of Russian steel and comparison with the values for English steel (pp. 55–6).

(*b*) Proposals to measure the value of gravitation at different points of the earth by the difference in the periods of transverse vibration of an elastic rod clamped vertically and with a weight attached to its upper or free extremity (pp. 60–1).

[743.] *Compte rendu annuel.* Année 1858 (St Petersburg, 1860). A few results for the stretch-modulus of copper, steel, aluminium and tin are given in French measure (p. 51).

Compte rendu annuel. Année 1859 (St Petersburg, 1861). This contains nothing concerning elasticity but a notice of the completion of the printing of the first volume of the *Recherches* (p. 41).

Compte rendu annuel. Année 1861 (St Petersburg, 1862). On pp. 45–48 numerical values are given of the inverse of the stretch-moduli and of the specific gravities of various metals, principally different kinds of Russian and Austrian iron and steel.

The *Comptes rendus* for the years 1862–4 give promises of further experiments on elasticity,—promises destined never to be fulfilled.

[744.] We now turn to the memoirs Kupffer published during this decade and note first two shorter ones which are printed in the *Bulletin*. We shall then pass to the long memoir on thermo-elasticity and conclude with an analysis of the *Recherches*.

[745.] A. F. Kupffer: *Bemerkungen über das mechanische Aequivalent der Wärme. Bulletin de la Classe physico-mathématique de l'Académie Impériale des Sciences*, T. x., cols. 193–7. St Petersburg, 1852. A reprint of this paper will be found in the *Annalen der Physik*, Bd. 86, S. 310–14, 1853. Suppose a cylinder of unit length and unit radius to receive extension δ under unit tractive load, and further when it is raised from freezing to boiling point of water let its extension be a. Then if m be the specific heat of the metal and S its specific gravity, it will take mS times the heat to raise the metal cylinder from 0° to 100° that it takes to raise a cylinder of water of the same radius and height through the same range of temperature.

Let c be the latter quantity in mechanical units, then we have cmS for the work done. Kupffer now continues:

Da nun die Ausdehnungen, die ein Drath erleidet, den angewandten Kräften proportional sind, so sieht man gleich, dass die Werthe von a und δ uns eine Vergleichung der ausdehnenden Kraft der Wärme mit der dehnenden Kraft eines Gewichts darbieten, oder mit andern Worten, dass jene Werthe uns ein Mittel an die Hand geben, das mechanische Aequivalent der Wärme zu bestimmen. Man muss hier nicht vergessen, dass die Wärme gleichmässig nach allen Seiten wirkt, wie ein Druck: nun hat aber Poisson gezeigt, dass ein Gewicht welches einen Drath um δ ausdehnt, als nach allen Seiten gleichmässiger Druck angewandt, eine lineäre Ausdehnung von $\frac{1}{2}\delta$ hervorbringen würde. Wir haben also $2a/\delta$ als das Verhältniss der mechanischen Wirkung der bezeichneten Wärmemenge zur mechanischen Wirkung eines Pfundes anzusehen. Um dieses Verhältniss in Zahlen auszudrücken, darf man nur für irgend eine Substanz die elastische Constante, den specifischen Wärmestoff und das specifische Gewicht, so wie auch ihre Ausdehnung durch die Wärme kennen (Col. 194).

Kupffer then gives the equation:

$$cmS = 2a/\delta,$$

and calculates c in Russian units for the results he has found for iron, brass, platinum and silver wires. The mean value of these results he reduces to English and French units and finds

$$J = 9921 \text{ inch-pounds for } 1^\circ \text{ F.},$$
$$= 453 \text{ kilogrammètres for } 1^\circ \text{ C.}$$

[746.] I do not follow Kupffer's reasoning. Putting aside the fact that he assumes the wires to possess uni-constant isotropy, he seems to me on this occasion to equate a quantity of heat or energy to a *force*. I have already alluded to the difficulties I feel with regard to Kupffer's method of treating this problem in Art. 725, and his argument here seems to me, although somewhat different, no clearer than that in the *Compte rendu annuel*.

[747.] A. F. Kupffer: *Untersuchungen über die Flexion elastischer Metallstäbe. Bulletin de la Classe physico-mathématique de l'Académie Impériale des Sciences,* T. XII., cols. 161–7. St Petersburg, 1854.

This contains matter which reappears in Kupffer's great work, notably the erroneous formulae for flexion: see our Arts. 760–2.

Ibid. T. XIV., cols. 273–84, and cols. 289–99. *Einfluss der Temperatur auf die Elasticität der festen Körper.* This contains matter which reappears in the memoir of 1852–7 (see our Arts. 748–57) and partially in the *Compte rendu annuel* (see our Art. 740) and the *Recherches* (see our Arts. 770–1), so that we need not discuss it further here.

[748.] A. T. Kupffer: *Ueber den Einfluss der Wärme auf die elastische Kraft der festen Körper und ins besondere der Metalle: Mémoires de l'Académie...de St Pétersbourg, Sixième Série, Sciences mathématiques, physiques et naturelles,* T. VIII., *Première Partie: Sciences mathématiques et physiques,* T. VI., pp. 397–494 (separate pagination 1–98), St Petersburg, 1857. This memoir, written in German, received the prize of the Royal Society of Göttingen in 1855: see our Art. 723. It was apparently read before the St Petersburg Academy on December 3, 1852.

It commences with a *Vorwort* describing its scope, of which the first paragraph may be cited here:

Die nachstehende Abhandlung ist aus einer grösseren Arbeit über Elasticität entnommen, die noch nicht beendigt ist, und die zu ihrer Zeit wird bekannt gemacht werden. Ich habe einstweilen in der Einleitung einige allgemeine und noch nicht bekannte Thatsachen aus der grösseren Schrift mittheilen zu müssen geglaubt, um den Leser zu zeigen, wie man die Elasticitätscoefficienten derselben Metalle sehr genau bestimmen könne, und bestimmt hat, für welche in dieser Schrift der Einfluss der Temperatur auf diese Coefficienten bestimmt worden ist. Indem ich durch Versuche erwies, dass der Einfluss der Temperatur bei Torsionsschwingungen ein anderer sein kann als bei Transversalschwingungen, war es interessant nachzuweisen, dass auch der Elasticitätscoefficient für die Torsionsschwingungen ein anderer ist, als für die Transversalschwingungen. Diese Mittheilungen führten zur Erwähnung des Coefficienten ν, den ich den Flüssigkeitscoefficienten genannt habe, und von dem meines Wissens vor mir noch nicht die Rede war, oder dessen Werth wenigstens vor mir noch nicht genau bestimmt worden ist (S. 399).

Thus Kupffer's discovery of viscosity and after-strain in metals dates at least from 1852. The *coefficient of fluidity* certainly appeared implicitly in the memoir of 1848 (see our Art. 1391*), but I do not think Kupffer had at that date clearly separated its effect from that due to the resistance of the air.

[749.] The *Vorwort* goes on to state that all the experiments on temperature have been made by vibrational as distinguished from statical methods; in this case by means of transverse and torsional oscillations.

Ich habe auch Versuche über den Einfluss der Temperatur auf das statische Moment der Elasticität gemacht, aber sie sind vollständig misslungen: bei fortdauernder Erwärmung war die bleibende Aenderung des Flexions- oder des Torsionswinkels so stark, dass die vorübergehende, mit der Erhöhung der Temperatur eintretende, und mit deren

Verminderung sich wieder vermindernde, ganz darin verschwand; die elastische Nachwirkung brachte noch mehr Verwirrung in die Resultate (S. 399).

The full complexity of elastic problems was fully appreciated by Kupffer and he foreshadows in the following words the direction of much of the research taken later by Bauschinger:

Ich sah daraus, dass um die Einwirkungen der Wärme auf das statische Moment der Elasticität zu finden, man vor allen Dingen ein Mittel haben müsste, die Einwirkung derselben Wärme auf die Verrückung der Gränzen der Elasticität und auf die Nachwirkung von ihrer Einwirkung auf die Elasticität selbst zu trennen; um ein solches Mittel zu finden, werden noch viele Arbeiten über die Gränze der Elasticität und über die Nachwirkung erforderlich sein, so dass die Lösung dieses Problems mir noch sehr ins Unbestimmte hinaus gerückt zu sein scheint. Man hat aber erst angefangen die Gesetze der Elasticität in ihrem ganzen Umfange zu studiren; bei jedem Schritte stösst man in diesen Untersuchungen auf neue Eigenschaften der elastischen Körper; je weiter man vorgeht, desto mehr Verwickelung. Bei solchen Umständen ist wohl in diesem Augenblick keine völlig abgeschlossene Arbeit über irgend eine Eigenschaft der elastischen Körper möglich (S. 400).

Notwithstanding our great increase in knowledge, the same words may almost be used of the science of elasticity to-day. The fact is that to grasp thoroughly the bearing and mutual relations of the secondary elastic properties we must know what is the real kinship between the various branches of physics when viewed from the standpoint of the molecule—and this is very far from being understood even forty years after Kupffer wrote.

[750.] The next portion of the memoir, termed *Einleitung*, occupies S. 401–427. It contains details of the methods of experiment and of the formulae adopted[1]. Several points here deserve notice:

(*a*) On S. 404–7 we have the details of the first scientific experiments on the elastic after-strain of *metals*[2], the existence of

[1] Kupffer, S. 402, defines the stress that can be called into play in a body by external pressure its 'elasticity.' This is another instance of his tendency to rather vague definition to which I have previously referred: see our Arts. 723 (*a*), 728 and footnote.

[2] Between Weber and Kupffer a few experiments on after-strain were made by R. Kohlrausch, and are referred to by him in an article on an electrometer in *Poggendorffs Annalen*, Bd. 72, 1847: see S. 393–6. His remarks amount to little more than the assertions that he has confirmed Weber's discovery of after-strain in silk threads, and finds that it is manifested also in the torsion of glass threads. He makes, further, some not very conclusive statements (S. 396–8) on the influence which rise of temperature has upon the torsional elasticity of silk threads, and upon the effect which boiling them in soapy water has on their elastic after-strain.

which had been doubted even by Wertheim and Saint-Venant: see our Arts. 819 (noting Art. 803) and 197. The experiments were made on the flexure of a cylindrical bar of steel and the continual decrease of the deflections for a period of several days after the removal of the load was clearly marked. The influence of elastic after-strain on the reduction of the amplitude and period of torsional vibrations *in vacuo* is also referred to on S. 407–8.

Die allmählige Abnahme der Schwingungsweiten (selbst im luftleeren Raum) lässt sich sehr gut durch die Nachwirkung erklären, weshalb auch schon Weber vorausgesehen hat, dass die Schwingungsweiten elastischer Körper in luftleerem Raum allmählich abnehmen würden, wie ich später durch Versuche bewiesen habe. Die Nachwirkung bringt hier dieselbe Wirkung hervor, wie die Friction beim Widerstande der Luft, und besteht wohl auch in Nichts anderem, als in einem mit Friction verbundenem Glitschen der Theile über einander: nur ist nicht zu übersehen, dass die Friction der Theilchen unter einander nicht zu erklären im Stande ist, warum der Stab oder der Draht, nach Aufhebung der ablenkenden Kraft, wieder zu seinem ursprünglichen Gleichgewichtszustande zurückkehrt; diese Erscheinung setzt offenbar eine gewisse Kraft voraus, welche jeden festen Körper, selbst wenn er durch Aenderung seiner Form in andere Gleichgewichtsbedingungen versetzt worden ist, dennoch immer wieder in längerer oder kürzerer Zeit zu seiner ursprünglichen Form (oder zu seiner ursprünglichen Gleichgewichtsbedingung) zurückführt, wenn die Abweichung von der ursprünglichen Gleichgewichtslage nicht gar zu gross gewesen ist (S. 407–8).

This passage seems to me to mark off the real distinction between after-strain and any frictional action between the parts of a body, and I think destroys the force of the comparison of a solid body's elastic after-strain with fluid action. It is a strong reason for not allowing elastic after-strain to be masked under the term 'viscosity': see the footnote p. 390 of our Vol. I.

(*b*) Kupffer shows that elastic after-strain is not proportional to the load and that accordingly the vibrations are not truly isochronous (see his S. 407–8). He further adds that working, temperature etc. have all great influence on the elastic after-strain as well as on the elastic fore-strain (S. 409).

[751.] (*c*) The next portion of the *Einleitung* is termed: *Transversalschwingungen elastischer Stäbe*, and occupies S. 409–419. This contains the formula for transverse vibrations, which I shall have occasion again to refer to when dealing with the *Recherches*. It must be looked upon, I suppose, as an empirical formula, to be justified by its agreement with the data of Kupffer's experiments, but I cannot see that theory at all justifies its form: see our Arts. 763–6. I shall return to this point more fully later. A series of experiments intended to show the good results obtainable from this theoretically questionable formula are given in this part of the *Einleitung*.

(*d*) The remainder of the *Einleitung* (S. 419–27) is occupied with the formulae for torsional vibrations. The method is that due to Gauss and presents some variations on that of the memoir of 1848, notably the equation $P_s = P_0\left(1 + \nu r\sqrt{\frac{s}{l}}\right)$ is used for the reduction of the periods: see our Art. 739.

Some interesting experiments as to the exactness of this formula are given on S. 423–426. Kupffer finds that for

$$\text{copper}\begin{cases}\text{unannealed}, & \nu = \cdot 04302 \text{ (to } \cdot 04828),\\ \text{annealed}, & \nu = \cdot 2365 \text{ (to } \cdot 2450),\end{cases}$$
$$\text{steel} \qquad \nu = \cdot 007122.$$

He shows that the coefficient ν of elastic after-strain is capable of immensely modifying the value of the elastic-modulus as determined by the method of torsional vibrations (S. 427). It should, however, be noted that the discrepancy he finds between the values of $\delta = 1/(\pi E)$ as found by transversal and torsional vibrations for copper wire need not be solely due to the influence of elastic after-strain. Kupffer's δ as obtained from torsional vibrations is $= 2/(5\mu\pi)$ and from transverse vibrations $= 1/(\pi E)$, but any want of *uni-constant isotropy* in the copper wire would not allow of our assuming $E = 5\mu/2$ or these values of δ to be equal. On the other hand the fact that steel wire with a very small ν (see above) gave for δ almost the same values when determined by torsional and by transverse vibrations may only point to a nearer approach to uni-constant isotropy in that material.

[752.] The next portion of the memoir is entitled: *Einfluss der Temperatur auf die elastische Kraft der festen Körper*. Kupffer divides the effects of heat into two main groups:

(i) Change of elasticity during the time the temperature is raised, the elasticity returning to its old state when the temperature is lowered to its first value.

(ii) Change of elasticity remaining after the heating has ceased, and the old temperature has been restored.

The first series of investigations as to (i) was upon the transverse vibrations of a rod clamped at one end so as to be vertical, the free end being loaded with weights of different magnitudes. If E_t be the stretch-modulus at temperature t, we have according to Kupffer

$$E_{t'} = E_t\{1 - \beta_f(t' - t)\},$$

where $t' > t$ and β_f is a constant. Kohlrausch takes the effect of temperature to be represented by an expression involving also the square of $(t'-t)$ so that the factor is then of the form $\{1-\beta(t'-t)-\gamma(t'-t)^2\}$. In most cases γ is very small, but if $t'-t$ be large the term in $(t'-t)^2$ might be sensible. Kupffer's first series of experiments were only made for the difference between external and internal winter temperatures—amounting to from 13 to 25 degrees Réaumur. The values of β_f were obtained from what I have spoken of as the questionable formula for transverse vibrations (see our Art. 751 (c)), but as the stretch-modulus probably appears as a factor of the correct formula—at least to a close degree of approximation,—serious error would hardly be introduced by the use of the formula.

Kupffer neglects the effect of heat in expanding the rods, remarking that the changes of temperature only altered their dimensions insensibly: see, however, the ftn. on our p. 509. At the same time he notes that the least change in the distance from the point of clamping to the centre of gravity of the vibrating load would have made an important alteration in the period of oscillation (S. 430 and ftn.). He does not seem to have noted that the dimensions of the rod would also have been slightly different in the positions when the weight and the clamped end were respectively uppermost. Both these causes might somewhat effect the values of β_f he gives for the different metals. They are reproduced in the Table I. below from his S. 451. S. 431–51 are occupied with numerical details of the observations.

I.

Values of β_f for one degree Réaumur found from changes of temperature lying between $-15°$ R. and $15°$ R., the changes being not much more than 20°,

$$E_{t'}=E_t\{1-\beta_f(t'-t)\},\quad t'>t.$$

Metal.	β_f	β'_f
Silver	·000563 (mean)	——
Brass (hammered)	·000471 (mean)	·000478 (mean)
,, (cast)	·000533 (mean)	·000533 (mean)
Brass (rolled[1]) (1st kind)	·000536	——
,, ,, (2nd kind)	·000476	·000500
Platinum	·000201 (mean)	——
Plate Glass	·000125 (mean)	——
Cast Iron (soft)	·001795 (mean)	·00188 (β_f=·001618 for same specimen)
Steel (soft, rolled)	·000348 (mean)	——
English Forged Steel (1st kind)	·000320	——
,, ,, ,, (2nd kind)	·000256	——
Steel (soft, cast)	·000242 (mean)	——
Swedish Wrought Iron	·000456	·000381
Rolled Iron Bar (1st kind)	·000442	——
,, ,, ,, (2nd kind)	·000463	·000488
English Wrought Iron	·000376	——
Rolled Iron Plate (across fibre)[2]	·000353	——
,, ,, ,, (along fibre)	·000425	——
Copper	·000560	·000598 (but elasticity had been permanently altered)
Zinc (rolled)	·000644	——
Lead (rolled)	·003035	——
Gold	·000394	——

[1] *Tafelmessing*. [2] i.e. bar cut out of plate perpendicular to direction of rolling.

[753.] In the third column of the above Table I. I have placed the mean β'_f of some of the results of Kupffer's experiments included in the following section of his memoir (S. 551–63) entitled: *Einfluss der Temperatur auf die Elasticität der Metalle bei höhern Temperaturen.* (i) *Bei Transversalschwingungen.* The change in temperature here was a rise from about 14° R. to 79° R. In all cases except those of Swedish wrought-iron where there appears to be a reduction, and of cast-brass where there is no sensible change β'_f appears to be $> \beta_f$; in the former case the experiments do not seem to have been made on the same specimens, so that not much stress can be laid on the result. We see therefore that the introduction of Kohlrausch's term $\gamma (t' - t)^2$ with a *positive* value of γ would be in accordance with Kupffer's results.

[754.] Our author next ((ii) *Bei Torsionsschwingungen*) determines the effect of a like *large* change in temperature on the slide-modulus. Assuming in his memoir uni-constant isotropy Kupffer speaks of this effect as an alteration in the stretch-modulus. Without this assumption, however, we may gather the following results from S. 464–8 of his work:

II.

Values of β'_τ for one degree Réaumur found for changes of temperatures between 15° R. and 79° R., the changes being about 65°, where $\mu_{t'} = \mu_t \{(1 - \beta_\tau (t' - t)\}$, $t' > t$.

Metal.	β'_τ
Copper	·0008634
Best Viennese Pianoforte Wire (Steel)	·0005885
Very soft Brass Wire	·0006982
Very hard Brass Wire { unannealed	·0004258
Very hard Brass Wire { annealed	·0004861

Thus, so far as we can compare the materials of these wires with those of the bars in the previous article, we see that β'_τ for copper and steel is greater than β'_f, or that the slide-modulus for these metals diminishes with the rise of temperature more rapidly than the stretch-modulus. Kupffer's result for copper differs widely from that of Kohlrausch, but supposing Kohlrausch's brass wire to have been of the sort that Kupffer terms very hard, then they agree fairly closely for this metal.

[755.] The next section of the memoir is entitled: *Beobachtungen über den Einfluss vorübergehender Temperaturerhöhungen auf die Elasticität der Metallstäbe.* It occupies S. 469–492.

Da die Wärme den Agregatzustand des gehämmerten, oder gewalzten, oder gehärteten Metalls bleibend ändert, so ist zu vermuthen, dass der Elasticitätscoefficient sich ebenfalls durch vorübergehende Temperaturänderung bleibend ändert (S. 469).

The experiments were made by means of the transverse vibrations of rods exactly as in the method referred to in our Art. 752. The change in temperature was produced by heating the rods with a Berzelius' spirit lamp, sometimes to incandescence. The stretch-modulus was determined before and after this thermal process.

[756.] Kupffer concludes his memoir by an investigation of the effect of heat on elastic after-strain (*Einfluss der Temperatur auf die elastische Nachwirkung* S. 492–4). We have already seen that Kupffer attributes to elastic after-strain a considerable portion of the reduction of the amplitudes of torsional oscillations. Hence if the wire be subjected to any thermal process the effect of this on its after-strain property will be shown by the difference, if any, in the number of oscillations made between the same amplitudes before and after the thermal process—the resistance of the air being the same in both cases. Thus the change in the after-strain coefficient, if not its absolute value in either case, could be ascertained without the need of experimenting *in vacuo*. Details of the experiments on the various metals in the case of both elasticity (see the previous article) and after-strain are given in the memoir; we summarise them in the following Table:

III.

Temperature Effect on Metals.

Metal.	Cyclic Effect of rise of Temperature.		Permanent Effect after a heating nearly or quite to incandescence.	
	Coefficient of after-strain ν.	Stretch- & Slide-Moduli E & μ.	Coefficient of after-strain ν.	Stretch-Modulus E.
Silver	increases	diminish	increased (if almost to incandescence)	diminished (if almost to incandescence)
			increased (if to softening)	increased (if to softening)
Brass	do.	do.	diminished (even if to incandescence)	increased with slight heating and then diminished
Copper	do.	do.	increased (if to incandescence)	diminished (if to incandescence)
			diminished (if not to incandescence)	increased (if not to incandescence)
Zinc	do.	do.	slightly diminished (if heated till covered with a coating of oxide)	much increased (if heated till covered with a coating of oxide)
Platinum	do.	do.	diminished (if heated short of or up to incandescence)	increased (if heated short of or up to incandescence)
Cast Iron	do. (considerably, even at temperature so low as that of boiling water)	do.	No experiments	No experiments
Steel	do.	do.	diminished (if tempered very hard after annealing)	increased (if tempered very hard after annealing)
			increased (if soft, and heated up to incandescence)	increased (if soft, and heated up to incandescence)
Wrought Iron	do.	do.	No experiments	No experiments
Gold	do.	do.	much increased	diminished

The general law thus seems to be that processes which increase the coefficient of after-strain or 'viscosity' decrease the elastic-constants and *vice-versâ*, but there are exceptions to this rule.

Kupffer speaks simply of the 'elasticity' as being increased or diminished. I have put stretch- or slide-modulus according as his method was that of transverse or torsional vibrations, in order that there may be no assumption that even in questions of thermal influence these necessarily exhibit the same tendencies.

[757.] In conclusion we may remark that this memoir of Kupffer's is of very considerable value although we cannot feel thoroughly satisfied with his use of the experimental method of transverse vibrations, and could have wished a more complete investigation of β_f and β_T for a greater variety of temperatures. Still to have demonstrated the existence of after-strain in metals and indicated its changes with temperature is no small service, while the absolute measurements of the thermal coefficients are at least valuable for comparison.

[758.] A. T. Kupffer: *Recherches expérimentales sur l'élasticité des métaux faites à l'observatoire physique central de Russie.* Tome I. folio, (all published), pp. i–xxxii and 1–430, with nine plates. St Petersburg, 1860.

This work contains some of the most carefully made experiments on the stretch-moduli of different metals and the effect of temperature upon them, which we have to record in this period. The experiments seem to have been conducted with extreme accuracy;—unfortunately the formulae used by Kupffer do not appear equally accurate, and it may be questioned whether very useful labour might not still be spent in revising Kupffer's numbers with the aid of a more accurate elastic theory.

The preface to the work explains its scope and the contents of the projected remaining volumes: see our Art. 722. It also states the relation between Russian, English and French measures[1]. It occupies pp. i–ix.

[759.] The *Introduction* occupies pp. xi–xxx. One or two brief remarks may be made here.

[1] A Russian foot = an English foot; a Russian inch = an English inch = 2·540 centimetres. A Russian pound = ·9 English pounds nearly = 409·512 grammes (or 1 kilogramme = 2·442 Russian pounds).

For comparison of specific gravities we may note: a cubic inch of water at the normal temperature $13\frac{1}{3}^0$ R. (=62^0 F.) and *in vacuo* weighs very nearly ·04 Russian pounds.

(*a*) On p. xii Kupffer remarks that the formula of Euler for the transverse vibrations of a rod clamped at one end and loaded at the other does not give accurate results. He seems to think the formula theoretically correct, but this is not the fact. It is only an approximation which neglects the inertia of the rod.

(*b*) The author insists upon the importance of a national institution for experiments on the resistance of materials. This importance is no less to-day than in 1860,—greater also in a manufacturing country like England than in Russia.

Je crois, qu'un établissement spécial, consacré à des expériences sur la résistance des matériaux entre et hors des limites de l'élasticité, où l'on pourra mettre à l'épreuve les productions métalliques de toutes les usines du pays, avant et à mesure qu'elles sont livrées au commerce, ne présenterait pas seulement des données certaines pour la rectification des devis de construction, mais contribuerait aussi puissamment au perfectionnement des méthodes de fabrication, puisque chaque fabricant désirera que ses productions fussent notées le plus haut possible. Rien n'entrave les perfectionnements dans la fabrication des métaux, comme l'incertitude où le gouvernement ou le public se trouvent relativement à leur qualité, et si, à cause de cette incertitude, ils sont toujours taxés de la même manière, qu'ils soient bons ou mauvais. L'élévation des prix, que la confiance publique accorde à certaines usines anciennes et connues, n'a pas d'autre source que l'épreuve du temps, qui pourrait être considérablement abrégée par des expériences préliminaires (p. xiii).

(*c*) The doubtful formulae for flexure and transverse vibrations to which I have referred in Arts. 747 and 751, and to which I shall return in Arts. 760–6, are given on pp. xv–xvii and pp. xx–xxiv.

(*d*) On p. xviii Kupffer cites experiments confirming Hodgkinson's result—namely that the stretch-modulus of cast-iron decreases rapidly with the load. Of this he remarks:

La rapidité, avec laquelle la dilatation élastique de la fonte augmente avec la charge, me semble prouver, que nous n'avons pas ici affaire à une autre loi des dilatations et des compressions, mais à une autre propriété des corps élastiques que quelques métaux seulement possèdent et qui cache la véritable loi (p. xix).

He promises to return to this matter in a later volume, but we have no later trace of it, I think, in his published work. (See our Arts. 729 and 767, however.)

(*e*) Remarks on the relation of the stretch-modulus to the density are given on p. xxvii. In the case of brass Kupffer shows that, after working different specimens of the same piece, the moduli were as the *cubes* of the densities (compare Art. 741 (*a*)). Our hope, however, of finding any general law connecting modulus and density is even to-day very small. He further notes the effect of working in producing a difference in the stretch-modulus for different directions (p. xxviii).

[760.] The first portion of the text of Kupffer's work deals with the preliminary experiments and the theory of the statical deflection of bars. It occupies pp. 1–44. He remarks (p. 2) that he had noticed the fact of the distortion of the contour of the cross-sections by flexure. This had already been observed experimentally by Clark for set (see our Art. 1485*) and theoretically by Saint-Venant for elastic strain (see our Art. 170). Thus a rectangular contour becomes a trapezium with slightly curved sides : see our Art. 736.

Kupffer then turns to the formula for flexure which he states as follows for the case of a horizontal cantilever :

$$\left.\begin{aligned} 1/E &= \tfrac{1}{4}\frac{dab^3}{l^2L\,(\frac{3}{8}p+p')}, \text{ for a rectangular section,} \\ 1/E &= \tfrac{3}{4}\frac{d\pi r^4}{l^2L\,(\frac{3}{8}p+p')}, \text{ for a circular section,} \end{aligned}\right\} \quad \ldots\ldots\text{(i)}.$$

$$d = \tfrac{2}{3}L\tan\phi, \text{ for both, } \ldots\ldots\ldots\ldots\ldots\ldots\ldots\ldots\text{(ii)},$$

where : d is the total deflection of the free end,

ϕ is the angle the tangent at the free end makes with the horizontal,

l is the length of the bar,

L is the horizontal distance of the free end from the built-in end after flexure,

p' is the load at the free end,

p is the weight of the bar,

a is the horizontal, b the vertical side of the rectangular cross-section, r the radius of the circular cross-section.

(See pp. xvi, xvii, 11, 19, 45, 50 etc.)

The angle ϕ can be measured by the angle between the reflected and incident rays of light on a small mirror attached to the free end of the bar, and thus the stretch-modulus E can be determined. This Kupffer did with very great caution and accuracy.

The formulae above occur frequently in his works on elasticity, and we have now to ask how far they are as accurate as his measurements really require.

[761.] Neglecting slide we have on the Bernoulli-Eulerian hypothesis

$$E\omega\kappa^2/\rho = p'(L-x) + \frac{p}{l}\int_x^L \frac{ds}{dx'}(x'-x)\,dx' \quad \ldots\ldots\ldots\ldots\text{(iii)},$$

where x is the horizontal distance of the element ds of the central axis of the rod from the built-in end, ρ the radius of curvature at ds and $E\omega\kappa^2$ the flexual rigidity of the bar: see our Art. 79.

First suppose the deflections so small that we may neglect $(dy/dx)^2$ or put $1/\rho = d^2y/dx^2$ and $ds = dx$. Then we easily find

$$E\omega\kappa^2 \tan\phi = \frac{l^2}{2}\left(p' + \frac{p}{3}\right),$$

and

$$E\omega\kappa^2 d = \frac{l^3}{3}\left(p' + \frac{3p}{8}\right).$$

Hence

$$d = \frac{2l}{3}\tan\phi \cdot \frac{1}{1 - \dfrac{p}{24p' + 9p}} \quad \text{.........(iv).}$$

Thus we see that Kupffer's formula (ii) cited above is not true even for small flexures unless $\dfrac{p}{24p' + 9p}$ be very small.

In several of the experiments p' and p were about the same order of magnitude, so that the values of d derived from this formula would have an error of 2 or 3 per cent. Further it is to be noted that Kupffer replaces l by L, and that many of his rods were so flexible that L could differ from l very considerably without the elastic limit being passed. Suppose then the difference between l and L to be so considerable that we must take it into account. Then we ought to solve the equation (iii) above to at least a second approximation, but this leads to very complex results. To test the accuracy of Kupffer's formulae however, it is sufficient to take the case when $p = 0$. We then find, if χL^2 or $p'L^2/(E\omega\kappa^2)$ be a small quantity, that to a second approximation:

$$\left.\begin{aligned} d/L &= \tfrac{1}{3}\chi L^2 + \tfrac{1}{35}(\chi L^2)^3 \\ l/L &= 1 + \tfrac{1}{15}(\chi L^2)^2 \\ \tan\phi &= \tfrac{1}{2}\chi L^2 + \tfrac{1}{16}(\chi L^2)^3 \end{aligned}\right\} \quad \text{.....................(v).}$$

Hence:

$$d = \tfrac{2}{3}L\tan\phi\,\{1 - \tfrac{11}{280}(\chi L^2)^2\} \quad \text{..... ....(vi).}$$

We thus see that Kupffer's formula neglects the term in $(\chi L^2)^2$, but this is just the order of the difference between l and L. Thus his results would have been as satisfactory, if he had always taken l for L. But in some of his observations the difference between l and L is so considerable[1] that he does not feel able to neglect it; *in these cases therefore his numerical results are still liable to the same order of error as if he had replaced L by l in his formulae.*

Further the deflection due to slide is of the order $\dfrac{\kappa^2}{l^2}$, and if we include terms of the order $\left(\dfrac{p'l^2}{E\omega\kappa^2}\right)^2$, we cannot neglect slide unless $(\kappa/l)^3$ is small as compared with $p'/E\omega$. For these reasons I do not think Kupffer's values of E are necessarily so accurate as his attempted distinction between l and L would lead us to believe.

[1] For example, $l = 28{\cdot}003$, $L = 27{\cdot}685$ (pp. 17–18); $l = 13{\cdot}9607$, $L = 13{\cdot}7985$ (pp. 39–40).

[762.] A value to a second degree of approximation of the stretch-modulus obtained from flexure is given by Saint-Venant in his *Leçons de Navier*, p. 84, footnote. He supposes however, that p is distributed uniformly along L and not along l; further in his equation, p. 82 (at the top of the footnote), I do not see why the second term on the right, which he admits is only approximate, is really admissible to the degree of approximation required. It is equivalent in our case to replacing the integral on the right of equation (iii) by $\frac{1}{2}(L-x)^2/\cos\phi$. I have not succeeded in solving equation (iii) to a second approximation.

If $p=0$, we easily find from (v)

$$\chi L^2 = 2\tan\phi - \tan^3\phi,$$

but

$$L^2 = l^2\{1 - \tfrac{2}{15}(\chi L^2)^2\} = l^2\{1 - \tfrac{8}{15}\tan^2\phi\}.$$

Hence

$$\chi = \frac{1}{l^2}\, 2\tan\phi\,\{1 + \tfrac{1}{30}\tan^2\phi\},$$

or

$$\frac{1}{E} = \frac{\kappa^2}{l^2}\frac{\omega}{p'}\, 2\tan\phi\,\{1 + \tfrac{1}{30}\tan^2\phi\}.$$

For a circular section:

$$\frac{1}{E\pi} = \frac{r^4}{2l^2 p'}\tan\phi\,\{1 + \tfrac{1}{30}\tan^2\phi\}.$$

Kupffer uses:

$$\frac{1}{E\pi} = \frac{r^4}{2lLp'}\tan\phi,$$

which really

$$= \frac{r^4}{2l^2p'}\tan\phi\,\{1 + \tfrac{8}{30}\tan^2\phi\}.$$

He further replaces $\tan\phi$ by $\phi\tan 1'$. In most cases I do not think that the term with $\tan^2\phi$ really affects his results, but the distinction between L and l ought not then to have been preserved.

Kupffer's own deduction of the result $d = \frac{2}{3}L\tan\phi$ is absolutely erroneous; he assumes the form of the elastic line to be a semi-cubical parabola, which is of course quite inadmissible (p. 11).

[763.] We next turn to the formula which Kupffer has adopted for the transverse vibrations of a loaded elastic rod clamped at one end. This formula has been largely used in his researches. It will be found discussed in his volume on pp. xix–xxv and pp. 126–135. Kupffer's experiments were made in the following manner. A bar, of which the weight of the vibrating part was p, was loaded with a weight p' at one end; the other end was then firmly clamped and the bar set vibrating about a vertical position, first with the weight p'

uppermost, and secondly with the clamped end uppermost, t_1 and t, the periods of the complete oscillations in these two positions, were then observed.

Kupffer gives the following empirical formula for E, the stretch-modulus, in the case of a rectangular bar of cross-section $a \times b$, oscillating parallel to the side b:

$$E = \frac{9}{2} \frac{Ig}{ab^3} \frac{t_1^2 + t^2}{t_1^2 - t^2} \sqrt{\frac{\lambda}{\sigma}} \quad \text{(i)},$$

where:

I = total moment of inertia of bar and load about the clamped end.

λ = length of a simple pendulum having the same period as the pendulum which would be formed by the bar of weight p and the load p' supposed rigid and capable of freely oscillating about an axis through the clamped end.

$\sigma = \frac{1}{2} \frac{t_1^2 t^2}{t_1^2 - t^2} \frac{g}{\pi^2}$, or, according to Kupffer, σ is the length of a simple pendulum, which would have the same period as the bar "si le barreau n'avait point d'élasticité, et si la pesanteur agissait seule" (p. 133). By this Kupffer means that the 'elasticity of the bar is supposed zero,' but I do not grasp the exact bearing of this. Practically he calculates the value of σ from t_1 and t as given above.

[764.] Let us examine a little more closely into the formula for a vibrating rod. The complete period T of the fundamental note of a 'clamped-free' bar is given by

$$T^2 = \frac{4\pi^2 l^3 p}{ab^3 Eg} \frac{12}{m^4},$$

where m is the least root of

$$\cos m \cosh m + 1 = 0.$$

See our Art. 49* or the footnote Vol. I. p. 50.

Seebeck, in the memoir referred to in our Art. 471 (see his p. 140), finds $m = 1{\cdot}875104$, hence we have

$$m^4 = 12{\cdot}3624.$$

Thus if found from the fundamental note

$$E = \frac{4\pi^2 l^3 p}{ab^3 T^2 g} \times {\cdot}970688 \quad \text{(ii)}.$$

Now Kupffer says (p. xx. and p. 134) that, if a free-clamped bar oscillates very rapidly, so that the influence of gravity is inappreciable whatever be the position of the bar, then Euler has given the following formula for its stretch-modulus:

$$E = \frac{4\pi^2 l^3 p}{ab^3 T^2 g} \quad \text{(iii)}.$$

Formula (ii) would, of course, be a close approximation if a horizontal bar were oscillating in a horizontal plane and its weight p produced no vertical deflection worth taking into account. For example a short bar in which a was considerably greater than b.

Kupffer cites (iii) indeed as due to Euler but I do not know to what memoir of Euler's he is referring. He supposes it to hold in cases in which (ii) is really true. It is therefore not surprising that he found formula (iii) in small agreement with observation.

[765.] But the point to be noticed is that even with an unloaded bar, Kupffer found the action of gravity was sensible and different according as the bar was placed vertically with the free end upwards or downwards, in other words gravity produced different effects upon the periods in the two cases. Thus for an unloaded brass bar $t_1/2 = \cdot 31625$ seconds and $t/2 = \cdot 28200$ seconds (p. 135).

We cannot therefore apply formula (ii) still less (iii) to this case. We are bound to take gravity into account. Indeed Kupffer's bars must have been so flexible that they vibrated with something of a pendulous nature about the clamped end. He measured the transverse vibrations not by the note but by the eye:

Une lame, qui est assez longue et assez mince, pour que ses oscillations transversales soient appréciables à la vue, oscille plus lentement, lorsque son extrémité libre est en haut que lorsqu'elle est en bas, la formule d'Euler n'est donc plus applicable directement (p. xx).

Kupffer modifies the formula (iii) and deduces from it (p. xxi):

$$E = \frac{9}{2}\frac{Ig}{ab^3}\frac{t_1^2 + t^2}{t_1^2 - t^2} \quad\text{.......................}\quad \text{(iv)},$$

but I am unable to accept his reasoning. This formula still not giving results in accordance with experiment, he proceeded to further modify it and found that it would give values agreeing among themselves and with those obtained from flexure experiments if the right hand side were multiplied by $\sqrt{\lambda/\sigma}$. I cannot find any formula in the least agreeing with Kupffer's by attempting to solve the problem by the assumption of a form for the normal-function. Indeed, as Lord Rayleigh has pointed out for a similar case—where, however, gravity and the inertia of the bar are neglected—there seems to be not one but two principal periods: see *The Theory of Sound*, Vol. I. § 183. This formula must therefore be treated as a purely empirical formula, and I find it accordingly difficult to draw any comparison between the values of the stretch-moduli as found from the statical and from the vibrational methods. On the other hand for comparative values of E, as for the same bar affected only by temperature, possibly (i) may give good results. Kupffer in this case works with the formula:

$$\frac{E}{E'} = \frac{t_1^2 + t^2}{t_1^2 - t^2}\cdot\frac{t_1'^2 - t'^2}{t_1'^2 + t'^2} \quad\text{.......................}\quad \text{(v)},$$

supposing the part $\frac{9}{2}\frac{Ig}{ab^3}\sqrt{\frac{\lambda}{\sigma}}$ of (i) to be but slightly influenced by temperature.

[766.] The problem of a vertical rod clamped at one end and loaded with a weight at the other was first attempted by Ménabréa in the memoir referred to in our Art. 551 (g). He did not, however, form and solve the equation for the notes, and the equations which give rise to the transcendental equation for the notes do not look promising (especially when as in Kupffer's case the weight of the rod and the load are not very different). One thing appears clear, I think, from Ménabréa's results, that there would be two different series of notes, nor does there seem any reason why both of these should not have been present in Kupffer's experiments, nor for the terms involving the fundamental note of one series being negligible as compared with those involving that of the other. Kupffer observed the time of, say, a thousand transits of a mark on his rod across the mid-thread of his telescope; dividing the time by the number (1000) of vibrations, he considered the result to be the time t_1 or t (as the case might be) of an oscillation of the rod, and substituted in the formula given above. There thus seems to me considerable doubt as to what period t_1 or t really denotes, and till the theory of this vibrating motion is fully worked out, it does not seem possible to derive all the profit from Kupffer's experiments that their accurate methods of observation would justify. We shall see later (Arts. 774 *et seq.*) that Zöppritz also has not surmounted the difficulties which arise in dealing analytically with this case.

[767.] The details of the first series of experiments on the statical flexure of bars of rectangular cross-section are given on pp. 51–109. Such bars Kupffer terms laminae (*lames*), while those of circular cross-section, details of experiments on which he gives on pp. 109–125, he terms rods (*verges*). The former set of experiments contains most interesting evidence as to after-strain in cast-iron: see pp. 83–4, 88–9, and to its imperfect elasticity (see our Art. 729 and Vol. I., p. 891, Note D), that is, the apparent decrease of its stretch-modulus with increase of the load (p. 87). This appears still more markedly if we can accept the value of the stretch-modulus given by Kupffer from transverse vibrations as that for vanishingly small loads.

The values Kupffer gives for $\delta\,(=1/E)$ as obtained by statical and vibratory methods do not show that E is invariably greater or less when measured by the one or by the other method, but this does not seem to me very conclusive as those values are obtained from formulae which I cannot recognise as sufficiently exact for this purpose.

Kupffer notes on p. 125 that the original limits of elasticity (state of ease) can be altered by repeated alternating stress (extension of state of ease: see our Vol. I. pp. 886–8 and Arts. 709, 749). The flexure experiments were made on various kinds of steel, cast and wrought iron, brass and platinum.

[768.] On pp. 135–268 we have details of experiments on the transverse vibrations of bars of rectangular cross-section, and on pp. 268–94 on those of bars of circular cross-section. The value of E is found to depend to some extent on the length of the vibrating portion of the bar. This divergence Kupffer thinks is due rather to the variation of E along the bar than to the effect of the resistance of the air acting on different lengths. It seems to me it may also be partially due to defects in Kupffer's empirical formula: see his pp. 153, 172 etc.

The experiments cover most of the principal metals: brass, steel, iron, silver, gold, platinum, zinc bars and brass, copper and steel wires.

[769.] Pp. 294–7 are entitled: *Oscillations transversales des lames horizontales, dont une extrémité est encastrée,* and would, if more extensive, have been most valuable for comparison and investigation of Kupffer's empirical formula for the vertical vibrating rod. We have here a case to which existing theory ought directly to apply. Unluckily Kupffer only gives the details of a few experiments on a steel bar, and substitutes the results in an empirical formula instead of the theoretical one.

He adopts the following formulae:

For the transverse vibrations of a bar in a horizontal plane,

(i) loaded at the free end:

$$E = \frac{9}{2} \frac{4\pi^2}{ab^3} \frac{Il'}{T_1^2} \sqrt{\frac{\sigma}{\lambda}} \quad \text{.......................(vi)},$$

(ii) without load:

$$E = \frac{4\pi^2}{gab^3} \frac{pl^3}{T_1^2} \sqrt{\frac{\sigma}{\lambda}} \quad \text{.......................(vii)},$$

where: T_1 = duration of the oscillations (= twice Kupffer's T_1),

p = weight of vibrating part of rod,

l' = distance of centroid of load from clamped end,

and the remainder of the notation is the same as in our Art. 763. If we supposed p/p' very small, we might obtain a formula suitable for case (i) by supposing the bar to take at each instant the statical form corresponding to the actual deflection at l'. We then find:

$$E = 4\frac{4\pi^2 Il'}{ab^3 T_1^2} \quad \text{......................(viii).}$$

On the other hand Equation (vii) ought to be the same as Equation (ii) of our Art. 764. Thus we ought to find at least approximately that with load $\sqrt{\sigma/\lambda} = \cdot 8889$, and without load $\sqrt{\sigma/\lambda} = \cdot 9707$. From the values given on p. 296 I find with load $\sqrt{\sigma/\lambda} = \cdot 9410$ (in a case, however, for which p/p' is not very small), and without load $\sqrt{\sigma/\lambda} = \cdot 9908$. I do not clearly understand what the physical meanings of σ and λ are supposed to be (neither in the latter case equals $\frac{2}{3}l$), and the above results show that their values when substituted do not give any close relation between Kupffer's empirical formulae and our (viii) or (ii) above. We cannot delay longer over the matter now, but there seems to be sufficient ground for suggesting that Kupffer's experiments and formulae require cautious dealing with. See Zöppritz's investigations referred to in our Arts. 774–84.

[770.] The remainder of Kupffer's work is devoted to the influence of heat on the elasticity of metals. A great part of this is reproduced from the memoir of 1852: see our Art. 748, and thus does not require further discussion here. As the measurements are chiefly based on Equation (v) of our Art 765, by putting $E_{t'}/E_t = 1 - \beta_f(\tau' - \tau)$ where $\tau' - \tau$ is the rise of temperature and β_f the thermal constant required, I do not think there is the same difficulty about the trustworthiness of the results as in the case of absolute measurements.

P. 299 gives the formula; pp. 300–2 describe the apparatus and method of experimenting; pp. 302–341 give the details of the experiments on various metals the results of which have already been tabulated in our Art. 752. These pages indicate how the value of β_f differs for ordinary and for high temperatures, and according as the metal has been cast, hammered or rolled.

[771.] Pp. 341–373 deal with the influence of a past change in temperature on the elasticity. These experiments have already been considered in our discussion of the memoir of 1852: see our Art. 755. The remarks in the memoir on after-strain are omitted in the *Recherches* as they would have fallen under the head of *Torsion*, the topic of the projected third volume of the work.

Pp. 377–425 are entitled: *Additions*, and are occupied with the details of experiments on the determination of E for steel and copper bars by the method of transverse vibrations. The experiments on steel were made with a view of determining $\sqrt{\lambda/\sigma}$, especially when the bar being vertical the load at the free end was such that it buckled in the position of equilibrium.

On p. 425 Kupffer finds that for a soft and for a rolled copper bar without load $\sqrt{\lambda/\sigma}=1\cdot02200$ and $1\cdot02625$ respectively. These give $\sqrt{\sigma/\lambda}=\cdot978,47$ and $\cdot974,42$, the theoretical value, if the bar were horizontal, being $\cdot970697$ (see Art. 764 above). Hence I am inclined to think that if Kupffer had placed his bars horizontally and allowed them to oscillate horizontally without any load at the free end, he would have obtained better results and these in good accordance with a well established theoretical formula.

[772.] Kupffer prefixes to his work (pp. xxxi–ii) a table of the quantity $\delta\,(=1/E)$, or in his units the number of millimetres which a bar of one metre length and one square millimetre cross-section would be extended by a load of one kilogramme. The density of each material is also tabulated, but there is no obvious relation between the density of a substance and its δ. Indeed it is not always the denser specimen of a metal which has the least δ, although this is generally true.

As Kupffer's work is not accessible to all and his numbers are not to be found cited in the ordinary text-books of elasticity, I give in Table IV. on p. 531 certain of his results for metal bars in which I have taken mean values for the different specimens whenever the number is followed by (m), and have added some of the results for brass, iron, steel and copper *wires*.

The numbers in brackets with the letter Z. attached to them I have calculated from Zöppritz's discussion of Kupffer's experiments. Zöppritz obtains his results from a *more accurate* theory, but they are deduced from a very small range of Kupffer's measurements and so are more liable to the influence of error in the individual experiment or to fault in the individual specimen.

It should be noted that Wertheim's value of the stretch-modulus for gold is considerably larger (by $\frac{1}{8}$) than Kupffer's and consequently his value of δ smaller. Wertheim finds $\delta=\cdot115,674$ to $\cdot112,971$. But the effect of annealing was to send δ up to $\cdot179,051$, so that the form of treatment or working appears to alter the modulus of gold very greatly. See the memoir referred to in our Art. 1292*.

[773.] A useful *résumé* of the various memoirs on elasticity by both Kupffer and Wertheim will be found in the *Bibliothèque universelle de Genève: Archives des sciences physiques et naturelles*, T. 25, pp. 40–58, Genève, 1854. Tables of the numerical results of both investigators are likewise given.

IV.

Metal.	$\delta=1/E$. (For units see Art. 772)	Density.
Brass, Cast	·112,365 (m.) (·109,806, Z. (m.))	8·2648 (m.)
,, Rolled	·090,016 (m.) (·093,572, Z.)	8·5047 (m.)
,, Hammered	·088,400 (m.) (·088,285, Z.)	8·5538 (m.)
Iron, Wrought (English)	·048,892 (m.) (·049,848, Z.)	7·6957 (m.)
,, ,, (Swedish)	·046,929 (m). (·047,553, Z.)	7·8114 (m.)
,, Rolled in bars	·049,937 (m.)	7·6449 (m.)
,, Plate (in direction of fibre)	·056,732 (·051,951, Z.)	7·6763
,, ,, (perpendicular to direction of fibre)	·052,225 (·047,366, Z.)	7·6775
Steel, Soft, Rolled	·046,938 (·047,558, Z.)	7·835
,, ,, Cast	·047,114 (m.) (·047,906, Z.)	7·838 (m.)
,, Wrought (English)	·047,432 (m.)	7·8335 (m.)
,, Remscheid	·047,069 (m.) (·047,393, Z.)	7·8321 (m.)
Tin (English)	·19,673	7·263
Aluminium	·13,940	2·739
Copper, Vigorously rolled	·078,213 (·079,076, Z.)	8·907
,, Soft ('passé au rouge')	·077,093 (·077,961, Z.)	8·930
Iron, Cast	·088,490 (m.) (·088,755, Z.)	7·1272 (m.)
Platinum	·056,467	21·122
Silver	·126,632 (·128,650, Z.)	10·494
Gold	·132,832 (·134,916, Z.)	19·264
Zinc, Rolled (Belgian)	·099,815 (·103,456, Z.)	7·1517
Wires: Brass (diameter 4 mm.)	·090,602 (m.) (·096,386, Z.)	8·4150 (m.)
Wires: Iron (,, ,,)	·051,384	7·5326
Wires: ,, (,, 5·5 mm.)	·050,782	7·6620
Wires: Steel (,, 3·5 mm.)	·048,525 (·051,878, Z.)	7·7572
Wires: Copper (,, 4 mm.)	·065,017	8·9241
Wires: ,, (,, 5·5 mm.)	·066,937	8·9427

It deserves to be noted how little the working alters the stretch-modulus of steel. Zöppritz's results are chiefly based on isolated experiments and specimens.

[774.] Three memoirs of K. Zöppritz bearing on Kupffer's investigations will be best considered in this place although they belong to a somewhat later date. The first is a *Habilitations-*

Dissertation, published in Tübingen, 1865, and entitled: *Theorie der Querschwingungen eines elastischen, am Ende belasteten Stabes.* Its purpose is explained in the following words of the *Einleitung*:

Die vorliegende Arbeit wurde veranlasst durch das Erscheinen von Kupffers *Recherches expérimentales*........; und mir von meinem hochverehrten Lehrer, Herrn Professor Neumann in Königsberg empfohlen mit dem Wunsche, dass es mir gelingen möge, durch eine strenge, auf die Principien der Elasticitätslehre basirte Theorie diese Versuche in der Ausdehnung für die Wissenschaft zu verwerthen, wie es dem für ihre Anstellung gebrauchten Aufwand an Zeit, Mitteln und Mühe entsprechend sei (S. 1).

After reference to the labours of Euler, D. Bernoulli, Poisson and Seebeck, the *Einleitung* draws attention to an erroneous theory of the vibrations of a loaded, weightless rod given by Lippich (*Poggendorffs Annalen*, Bd. 117, S. 161, 1862). It then points out the difficulty of solving generally the differential equation for the vibrations of a heavy loaded rod, especially for the case when the weights of the rod and load are, as in Kupffer's investigations, not very different, and finally gives a *résumé* of the contents of the paper.

[775.] The first section (S. 3–8) is entitled: *Ableitung der Differentialgleichungen für die Bewegung eines schweren, am Ende belasteten Stabes.* As in Kupffer's experiments the rod is supposed vertical, clamped at one end, and loaded at the other or free end.

Zöppritz adopts the method of Lagrange (*Statique, Sect. V. Art.* 42) and deduces by a not very luminous or satisfactory process the general equation. We can obtain Zoppritz's form at once, if ω be the cross-section of the rod, by writing equations (i) of our Art. 780 in the form:

$$E\omega\kappa^2\frac{d^4y}{dx^4} - \frac{d}{dx}\left(\frac{dy}{dx}\int_x^l \Delta\omega X dx\right) - \Delta\omega Y = 0.$$

Here,

$$\int_x^l \Delta\omega X dx = \pm\left(m\frac{l-x}{l} + M\right)g,$$

$$\Delta\omega Y \equiv -\Delta\omega\frac{d^2y}{dt^2},$$

where m $(=\Delta\omega l)$ and M are the masses of the rod and load respectively. Hence

$$E\omega\kappa^2\frac{d^4y}{dx^4} \mp (m+M)g\frac{d^2y}{dx^2} \pm \frac{mg}{l}\frac{d}{dx}\left(x\frac{dy}{dx}\right) + \Delta\omega\frac{d^2y}{dt^2} = 0.$$

Zöppritz writes $\dfrac{E\kappa^2}{\Delta} = b^2, \quad \dfrac{Mg}{\omega\Delta} = 2a^2,$
whence we find:

$$b^2 \frac{d^4y}{dx^4} \mp (2a^2 + gl)\frac{d^2y}{dx^2} \pm g \frac{d}{dx}\left(x\frac{dy}{dx}\right) + \frac{d^2y}{dt^2} = 0 \quad \text{.........(i):}$$

The upper sign corresponds to the free end downwards, the lower to it upwards[1].

For terminal conditions we have:

$$\left.\begin{array}{ll} \text{at } x = 0, \quad y = 0, \quad dy/dx = 0, & \\ \text{at } x = l, & \left\{\begin{array}{l} d^2y/dx^2 = 0, \\ M\dfrac{d^2y}{dt^2} = \mp Mg\dfrac{dy}{dx} + E\omega\kappa^2\dfrac{d^3y}{dx^3}, \\ \text{i.e. } \dfrac{2a^2}{g}\dfrac{d^2y}{dt^2} = \mp 2a^2\dfrac{dy}{dx} + b^2\dfrac{d^3y}{dx^3} \end{array}\right. \end{array}\right\} \quad \text{.............(ii).}$$

[776.] If we neglect the terms involving g explicitly in equation (i), equations (ii) remaining the same, we have the case of a loaded weightless rod. This case is discussed in the second section of the memoir (S. 9–16), while the case of a heavy unloaded rod is discussed in the second memoir (see our Arts. 780–1).

Zöppritz, taking the *upper sign*, solves (i) for the former case by assuming

$$y = \Sigma X_n (C_n \cos m_n t + D_n \sin m_n t),$$

where X_n is of the form

$$C_1 \cosh \alpha x + C_2 \sinh \alpha x + C_3 \cos \alpha' x + C_4 \sin \alpha' x,$$

and

$$\left.\begin{array}{l} \alpha = \pm \dfrac{1}{b}\sqrt{a^2 + \sqrt{a^4 + b^2 m_n^2}}, \\ \alpha' = \pm \dfrac{1}{b}\sqrt{-a^2 + \sqrt{a^4 + b^2 m_n^2}} \end{array}\right\} \quad \text{.................(iii).}$$

To determine m_n he obtains the transcendental equation (S. 12):

$$0 = \frac{gb^2 m_n^2}{a^2} + \cosh \alpha l \left\{\frac{g}{a^2}(2a^4 + b^2 m_n^2)\cos \alpha' l - 2\gamma a b m_n \sin \alpha' l\right\}$$
$$+ \sinh \alpha l \{2\gamma \alpha' m_n b \cos \alpha' l + g b m_n \sin \alpha' l\} \quad \text{......(iv),}$$

where $\gamma = \sqrt{a^4 + b^2 m_n^2}$.

The *least* root of this equation would give the periodic time observed by Kupffer, for rods whose weight is insensible as compared with the load. The above case corresponds to the clamped end uppermost and

[1] It will be noted that Zöppritz neglects the influence of the rotatory inertia of the rod which would introduce the term $-\kappa^2 \dfrac{d^4y}{dt^2\,dx^2}$ into (i).

the load undermost. For load uppermost we must change the signs of g and a^2 in (iii) and (iv). Zöppritz makes no attempt to solve (iv) for any of the large range of numerical examples given in Kupffer's *Recherches.*

[777.] The second section concludes by demonstrating that if

$$y = X_n \{C_n \cos m_n t + D_n \sin m_n t\}$$

be a solution of the general equation (i) of a loaded heavy rod, then

$$\int_0^l X_n X_{n'} dx + \frac{2a^2}{g} (X_n X_{n'})_{x=l} = 0,$$

if n be not equal to n'.

Zöppritz indicates how this enables us to determine the arbitrary constants of the solution in terms of the initial conditions.

[778.] The third and final section of the memoir is entitled: *Angenäherte Anwendung auf den schweren Stab* (S. 17–24). There are some interesting points in it, but the reasoning seems occasionally questionable.

Suppose a rod of weight w to have the weight W attached to one end; further suppose the rod to remain straight and rigid while the effect of its flexural rigidity is replaced by a restorative couple $\epsilon l\phi$, when the rod is inclined at an angle ϕ to the vertical position, l being the length of the rod. Then the equation of small oscillations would be

$$\frac{Wl^2 + \frac{1}{3}wl^2}{g} \frac{d^2\phi}{dt^2} = -\epsilon l\phi \mp (Wl + \tfrac{1}{2}wl)\phi,$$

the negative sign corresponding to the fixed end uppermost. Hence if $T = 2\pi/m$ be the periodic time we have:

$$m^2 = \frac{4\pi^2}{T^2} = \frac{g}{l} \frac{\epsilon \pm (W + \frac{1}{2}w)}{W + \frac{1}{3}w},$$

$$= g \frac{\epsilon l \pm (S + S')}{J + J'} \quad \text{...................(v)},$$

where S and S' are the first and J and J' the second moments of the *weights* of load and rod about the point of support[1].

To compare this very questionable result, which suggested Kupffer's formula, with what really takes place, Zöppritz returns to equation (i) and assumes the principal vibration to be of the form

$$y = X \cos (mt + \alpha),$$

[1] Zöppritz has dropped an l as factor of his ϵ in either his equation (62) or (65), which equation is not clear from his definition of ϵ.

so that by substitution we have, taking the upper sign,

$$b^2 \frac{d^4X}{dx^4} - (2a^2 + gl)\frac{d^2X}{dx^2} + g\frac{d}{dx}\left(x\frac{dX}{dx}\right) - m^2X = 0.$$

Integrate from 0 to x, and we find:

$$b^2 \frac{d^3X}{dx^3} - \frac{dX}{dx}\{2a^2 + g(l-x)\} - m^2\int_0^x X dx = C.$$

To determine the constant C put $x=l$, and use equation (ii); we thus have

$$C = -m^2\left(\frac{2a^2}{g}X_l + \int_0^l X dx\right).$$

Substitute this value of C and integrate again between the limits 0 and l; then remembering (ii) and integrating where necessary by parts, we have:

$$b^2\left(\frac{d^2X}{dx^2}\right)_0 + X_l 2a^2 + g\int_0^l X dx = m^2\left\{\int_0^l xX dx + \frac{2a^2}{g}lX_l\right\}.$$

Whence substituting the values of a^2 and b^2 and putting

$$1/R_0 = \left(\frac{d^2X}{dx^2}\right)_0$$

we have:

$$m^2 = g\frac{\left(\frac{E\omega\kappa^2 l}{f_l R_0} + Wl + w\int_0^l \frac{f}{f_l}dx\right)}{Wl^2 + w\int_0^l \frac{xf}{f_l}dx} \qquad \text{(vi)},$$

where $1/R_0$ is the maximum curvature at the clamped end, f_l the maximum deflection at the free end, and f the deflection at the distance x when the rod takes the maximum deflection at the free end.

Comparing (v) and (vi) we see that they will be identical, if we take:

$$\epsilon = \frac{E\omega\kappa^2}{f_l R_0}, \quad S' = w\int_0^l \frac{f}{f_l}dx, \quad J' = w\int_0^l \frac{xf}{f_l}dx.$$

Obviously f can never be greater than xf_l/l or the maximum values of

$$\int_0^l \frac{f}{f_l}dx \text{ and } \int_0^l \frac{xf}{f_l}dx$$

are $l/2$ and $l^2/3$ respectively.

I cannot, however, agree with Zöppritz's arguments on S. 20, that to a close approximation we may suppose these quantities equal to their maximum values. I think a far closer approximation would be obtained by giving them their values for the statical relationship of f to f_l. I have not worked out the ratio of f to f_l for the general statical case, but it can be found in terms of Bessel's functions in the manner indicated in the footnote to Vol. I. p. 46. If we suppose W only to act,—certainly a more reasonable hypothesis than supposing the bar to remain straight,—we find:

$$\int_0^l \frac{f}{f_l} dx = \cdot 3634 l \text{ and } \int_0^l \frac{xf}{f_l} dx = \cdot 2687 l^2$$

instead of $\cdot 5l$ and $\cdot \dot{3} l^2$ which Zöppritz adopts.

An approximation to the value of R_0 might also be made from statical considerations, but while results for dynamical deflections based on such considerations are generally fairly approximate, those for curvature are often very erroneous: see our Art. 371 (iii).

Zöppritz himself suggests (S. 21) putting for R_0 the value calculated in the second part of his paper, but this would involve the solution of (iv) for every experiment and an appalling amount of labour for the reduction of any series of observations. For a rod with the free end uppermost we must change R_0 to R'_0, f to f', and alter the signs of W and w in the numerator of (vi).

[779.] For the case of an unloaded rod, if $2\pi/m$ and $2\pi/m'$ be the periods when the free end is undermost and uppermost respectively:

$$m^2 = g \frac{\dfrac{E\omega\kappa^2 l}{R_0} + w \displaystyle\int_0^l f dx}{w \displaystyle\int_0^l xf dx},$$

$$m'^2 = g \frac{\dfrac{E\omega\kappa^2 l}{R'_0} - w \displaystyle\int_0^l f' dx}{w \displaystyle\int_0^l xf' dx}.$$

Suppose $2\pi/m_0$ the period when the rod is weightless, then since this must be the same whichever way the rod is placed we have

$$m_0^2 = \frac{gE\omega\kappa^2 l}{R_0} \Big/ \left(w \int_0^l xf dx \right) = \frac{gE\omega\kappa^2 l}{R_0'} \Big/ \left(w \int_0^l xf' dx \right),$$

or,

$$m^2 = m_0^2 + g \frac{\displaystyle\int_0^l f dx}{\displaystyle\int_0^l xf dx},$$

$$m'^2 = m_0^2 - g \frac{\displaystyle\int_0^l f' dx}{\displaystyle\int_0^l xf' dx}.$$

Zöppritz, who seems to me to have treated R_0 and R'_0 as *absolute constants independent of w*, then puts (S. 22)

$$m^2 + m'^2 = 2m_0^2 \qquad \text{(vii)}.$$

In this he supposes the second members on the right hand of the expressions for m^2 and m'^2 to be equal, i.e. *he does not distinguish between f and f', but assumes without comment that they can be taken equal.*

For the gravest note of a weightless free-clamped rod we have

$$E = \frac{wl^3}{\omega\kappa^2 g} \frac{m_0^2}{(1{\cdot}875{,}104)^4},$$

(see our Art. 764 or Lord Rayleigh's *Theory of Sound*, Vol. I. pp. 207 and 224).

Whence for a rod of rectangular section $a \times b$,

$$E = \frac{wl^3}{1{\cdot}0302 ab^3 g} m_0^2 \quad\text{.....................(viii).}$$

Equations (vii) and (viii), if T and T' be the times of half oscillations (i.e. $T = \pi/m$, $T' = \pi/m'$) give:

$$E = \frac{w}{g} \frac{l^3}{ab^3} \frac{\pi^2}{1{\cdot}0302} \frac{1}{2}\left(\frac{1}{T^2} + \frac{1}{T'^2}\right) \quad\text{.....................(ix).}$$

This agrees with the result of Zöppritz's third memoir (see our Art. 783, Equation (xi)), but in the present paper (S. 20) he has the number 1·019 instead of 1·0302 in the denominator.

He concludes by calculating E for one series of experiments made by Kupffer on an iron bar. Owing to the numerical error just referred to, the results cannot be very accurate.

The exactness with which this approximate theory gives the fairly accurate formula (ix) is, considering the assumptions, somewhat surprising and suggests the use of like methods in similar cases.

[780.] The second of Zöppritz's memoirs is entitled: *Theorie der Querschwingungen schwerer Stäbe,* and occupies S. 139–56 of Bd. 128 of *Poggendorffs Annalen*, Leipzig, 1866. Zöppritz first forms the equations for the equilibrium of a thin prismatic rod of uniform cross-section and density built-in at one end and acted upon by any body-forces in such wise that the flexure takes place in one plane.

The equation for the deflection y at distance x from the built-in end is easily found to be

$$\frac{E\kappa^2}{\Delta}\frac{d^4y}{dx^4} - \frac{d^2y}{dx^2}\int_x^l X dx + X\frac{dy}{dx} - Y = 0 \quad\text{................(i),}$$

where X and Y are the body-forces per unit mass in the plane of flexure parallel to the axis and perpendicular to it respectively, while l is the length and Δ the density of the rod. Further,

$$\left.\begin{array}{lll}\text{when } x = 0, & y = 0, & dy/dx = 0;\\ \text{when } x = l, & d^2y/dx^2 = 0, & d^3y/dx^3 = 0.\end{array}\right\} \quad\text{..............(ii).}$$

Zöppritz now takes the special case of $X = \pm g$, $Y = -d^2y/dt^2$, or that of a vertical rod vibrating transversely in a vertical plane under its own weight. X will be $+g$ or $-g$ according as the fixed or free end of the rod is uppermost. The equation (i) now becomes:

$$\frac{E\kappa^2}{\Delta}\frac{d^4y}{dx^4}+\frac{d^2y}{dt^2}\mp g\frac{d}{dx}\left\{(l-x)\frac{dy}{dx}\right\}=0\ldots\ldots\ldots\ldots\text{(iii)}.$$

The solution of (iii), subject to (ii), will therefore correspond to Kupffer's experimental determination of the period of oscillation of a heavy vertical rod built-in at one end.

[781.] Zöppritz writes (iii) in the form:

$$\frac{d^4y}{dx^4}+\frac{1}{b^2}\frac{d^2y}{dt^2}=p\frac{d}{dx}\left\{\left(1-\frac{x}{l}\right)\frac{dy}{dx}\right\}\quad\ldots\ldots\ldots\ldots\text{(iv)},$$

where $p=\pm\Delta lg/E\kappa^2$ and $b^2=E\kappa^2/\Delta$.

Now Zöppritz remarks that p, or as I think he should say pl^2, is a very small quantity owing to the magnitude of E as compared with Δlg, and hence it will generally be legitimate to neglect its square (i.e. if l^2/κ^2 is small as compared with $E/\Delta lg$). He assumes y to be of the form:

$$y=\Sigma\,(u_m+pv_m)\cos(mt+\alpha)\ldots\ldots\ldots\ldots\ldots\ldots\text{(v)},$$

where $\dfrac{d^4u_m}{dx^4}-\dfrac{m^2}{b^2}u_m=0$, or u_m is the type of term given by the Euler-Poisson solution for the vibrations of a weightless rod. Neglecting p^2, v_m will be given by the equation (S. 146):

$$\frac{d^4v_m}{dx^4}-\frac{m^2}{b^2}v_m=\frac{d}{dx}\left\{\left(1-\frac{x}{l}\right)\frac{du}{dx}\right\}\ldots\ldots\ldots\ldots\ldots\ldots\text{(vi)}.$$

Equation (vi) is then solved subject to the conditions (ii). This is followed by a rather long algebraic investigation of the equation for the notes. If $\lambda=l\sqrt{m/b}$, Zöppritz finds (S. 153):

$$0=1+\cosh\lambda\cos\lambda+\frac{pl^2}{8\lambda^3}[\lambda^2(\cosh\lambda\sin\lambda+\sinh\lambda\cos\lambda)$$

$$-2\lambda\sinh\lambda\sin\lambda+4(\cosh\lambda\sin\lambda-\sinh\lambda\cos\lambda)]\ldots\ldots\text{(vii)},$$

a result which I have not verified. If we put $p=0$, the equation reduces to Euler's form

$$1+\cosh\lambda_0\cos\lambda_0=0,\qquad\text{(see our Art. 49*)}$$

the root of which corresponding to the fundamental tone is $\lambda_0=1{\cdot}875104$.

Assuming $\lambda=\lambda_0+p\chi$, χ is found to be given by

$$\chi=\frac{l^2}{\lambda_0^3}\times{\cdot}39342,$$

or

$$\lambda=\lambda_0\pm{\cdot}39342\,\frac{gl^3}{b^2\lambda_0^3}\quad\ldots\ldots\ldots\ldots\ldots\ldots\text{(viii)}.$$

See S. 148–56 of the memoir[1].

[1] In the equations of lines 9 and 10 on S. 156 read l^3 for l^2.

[782.] The third memoir of Zöppritz is entitled: *Berechnung von Kupffers Beobachtungen über die Elasticität schwerer Metallstäbe,* and is published in *Poggendorffs Annalen,* Bd. 129, S. 219–237. Leipzig, 1866. It directly applies the result (viii) of our preceding article to Kupffer's numbers. This result, however, only covers a very limited range of Kupffer's work, for his most important experiments were made with heavy rods having weights attached to their free extremity. Zöppritz does indeed indicate how this latter problem might be treated and describes the stages of the theory so far as he has worked it out (S. 221):

Diese Arbeit ist indessen eine wegen der algebraischen Rechnung äusserst mühselige und zeitraubende und die Endform der Gleichung so complicirt, dass ich mich bis jetzt noch nicht zu einer Berechnung der Kupffer'schen Beobachtungen danach habe entschliessen können und somit der grösste Theil dieses ausgezeichneten Materials noch unvollkommen benutzt liegen bleibt[1].

Zöppritz also concludes, although on different grounds (S. 220), that Kupffer's formula which we have criticised in our Arts. 763–5 is inadmissible.

[783.] Let T be the time of a *half*-oscillation, then $T = \pi/m$, whence, since $\lambda = l\sqrt{m/b}$, we have, neglecting $p^2\chi^2$, from (viii) of Art. 781,

$$\frac{l^4}{b^2}\frac{\pi^2}{T^2} = \lambda_0^4\left\{1 \pm 4\nu\,\frac{gl^3}{b^2\lambda_0^4}\right\} \quad\text{(ix)},$$

where $\nu = \cdot 39342$.

Whence to the same degree of approximation:

$$b^2 = \frac{E\kappa^2}{\Delta} = \frac{l^4}{\lambda_0^4}\left\{\frac{\pi^2}{T^2} \mp 4\nu\,\frac{g}{l}\right\},$$

or,

$$E = \frac{w}{g}\frac{l^3}{\omega\kappa^2}\frac{1}{\lambda_0^4}\left\{\frac{\pi^2}{T^2} \mp 4\nu\,\frac{g}{l}\right\} \quad\text{(x)},$$

where w is the weight of the rod.

Thus (x) gives the stretch-modulus when the period T of a half-oscillation is known. If T be the half period when the free end is downwards, T' when the free end is upwards, we have,

$$E = \frac{w}{g}\frac{l^3}{\omega\kappa^2}\frac{\pi^2}{2\lambda_0^4}\left\{\frac{1}{T^2} + \frac{1}{T'^2}\right\} \quad\text{(xi)}.$$

[1] Elsewhere he refers to the invaluable material which, owing to the want of a strict theory, has not yet been used in a manner corresponding to what the excellence of the instruments and methods employed would warrant (S. 219).

By subtracting the two values of E we have

$$\pi^2\left(\frac{1}{T^2}-\frac{1}{T'^2}\right)=8\nu\frac{g}{l}.$$

Let σ be the length of the simple pendulum equivalent to the rod when oscillating about an extremity supposed pivotted; then $\sigma=\frac{2}{3}l$, or

$$\sigma=\frac{16}{3}\nu\frac{T^2T'^2}{T'^2-T^2}\frac{g}{\pi^2} \quad \text{.....................(xii).}$$

Remembering that in the notation of our Art. 763, $t_1=2T'$ and $t=2T$, we see that, there being no terminal load, Kupffer puts for *his* σ

$$\sigma=\frac{2T^2T'^2}{T'^2-T^2}\frac{g}{\pi^2},$$

or he omits the factor $\frac{8}{3}\nu=1\cdot04912$ in his estimation of the values of σ.

For the case of a rod of rectangular section $(a\times b)$ we have $\omega\kappa^2=ab^3/12$, and with the rest of the notation as in our equation (i), Art. 763, we find from (xi) by the aid of (xii):

$$E=\frac{9}{2}\frac{Ig}{ab^3}\frac{t_1^2+t^2}{t_1^2-t^2}\times\frac{32\nu}{\lambda_0^4} \quad \text{.....................(xiii).}$$

The factor $32\nu/\lambda_0^4=\dfrac{1\cdot04912}{1\cdot03020}=1\cdot01837$, and instead of this Kupffer has $\sqrt{\lambda/\sigma}$ where λ is the symbol defined in our Art. 763. Kupffer obtains values of $\sqrt{\lambda/\sigma}$ such as 1·02016, 1·02581, 1·02399 etc. (see his pp. 136, 148, 156 etc.) when the bars are unloaded. Thus his E will be slightly too large and his δ too small. Kupffer works, I think, for the unloaded beams from a formula like equation (i) of our Art. 763 and not from one of the type of (ix) in our Art. 779 with the number 1·0302 replaced by $\sqrt{\lambda/\sigma}$ as Zöppritz (S. 237) seems to imagine. Hence the amount of his error is measured by the ratio of $\sqrt{\lambda/\sigma}$ to 1·01837 and not to 1·0302 as Zöppritz states. Further Kupffer endeavours to allow for the effect of certain parts of his apparatus in his value of $\sqrt{\lambda/\sigma}$ (see his value of λ, p. 134, which contains i'), which Zöppritz's theory of course does not include. Thus the error in his formula is not so large as might be imagined, and his results agree very closely with those calculated by Zöppritz (S. 223—33) for the case of *unloaded* bars. In many cases the difference is within the limits of experimental error. But this is not the case when we compare Kupffer's values of the stretch-modulus for rods carrying a load with the results calculated by Zöppritz for unloaded rods; i.e. it is, as we have indicated in Art. 771, in the case of the loaded rod that Kupffer's formula is inadmissible. Zöppritz sums up his results as follows:

Kupffers Werthe für δ sind fast durchweg kleiner als die meinigen. Bei den Messingstäben 1, 2, 4, 5, 8 betragen die Abweichungen nur $\frac{1}{140}$ bis $\frac{1}{70}$ des ganzen Werthes und steigen bei No. 6 auf $\frac{1}{15}$ negativ. Bei Stahl ist die Abweichung nur höchst unbedeutend negativ, ebenso bei den meisten Eisenstäben. Bei den Eisenblechstreifen 1 und 2 sind aber Kupffer's Werthe um ein volles Zehntel zu gross. Bedeutendere Abweichungen zeigen auch Gold $\frac{1}{65}$, Zink $\frac{1}{30}$, und Remscheid-Stahl No. 17 ebenfalls $\frac{1}{30}$. Die Werthe, welche Kupffer aus den Versuchen *ohne* angehängtes Gewicht berechnet hat, kommen den meinigen im Allgemeinen viel näher, manche fallen innerhalb der Fehlergränzen mit ihnen zusammen; leider aber hat Kupffer gerade diese Beobachtungen verworfen, weil ihm die Resultate zu schlecht mit den übrigen, bei angehängtem Gewicht angestellten, übereinstimmten (S. 234).

[784.] Zöppritz notes four misprints of Kupffer's on the latter's pp. 270—2 and corrects them (S. 229). Further on S. 233 he gives the value in French measure of those few of Kupffer's experimental results which allow at present of calculation by accurate theory. He compares them with the numbers obtained by Wertheim and other earlier experimentalists. These results on S. 233 may be said to represent all of Kupffer's work which has probably a numerical exactness equivalent to the excellency of his experimental methods. We have tabulated them in a somewhat different form alongside Kupffer's results on our p. 531. They are the numbers in brackets with the letter Z attached.

Zöppritz's three memoirs certainly throw a great deal of light on the degree of accuracy in the results obtained by Kupffer from empirical formulae of a doubtful character. They form also an interesting chapter in the theory of vibrating rods.

Group C.

Wertheim's Later Memoirs[1].

[785.] Wertheim and Breguet: *Expériences sur la vitesse du son dans le fer* (Extrait). *Comptes rendus*, T. 32, pp. 293–4. Paris, 1851. This paper gives some account of experiments by the authors on the velocity of sound in the iron wire used for telegraphing between Paris and Versailles. Biot had found the velocity of sound in cast iron to be 10·5 times that in air, although the theoretical value of the velocity deduced from the stretch-modulus was 12·2 times that of air. After describing their method of making the experiments, the authors continue:

Ces expériences ont donné en moyenne une vitesse de 3485 mètres par seconde, tandis que 2 mètres du même fil de fer tendus sur le sonomètre longitudinal rendent un son de 2317 vibrations, d'où l'on déduit une vitesse de 4634 mètres.

La vitesse linéaire, d'après l'expérience directe dans le fer, est donc

[1] For Wertheim's earlier researches see our Arts. 1292*–1351*.

de beaucoup inférieure, et à la vitesse théorique, et à celle que l'on déduit du procédé de Chladni ; la différence est dans le même sens et plus grande encore que celle qui résulte de l'expérience de M. Biot sur la fonte.

[786.] G. Wertheim: *Mémoire sur la polarisation chromatique produite par le verre comprimé. Comptes rendus,* T. 32, pp. 289–292. Paris, 1851. This is an abstract of a memoir which contained the details of certain experiments made with the aid of apparatus described in a sealed packet; this packet had been opened and read by the President on February 3 of the same year (see *Comptes rendus,* T. 32, pp. 144–5. Paris, 1851).

The object of the investigations described in the memoir was to ascertain whether the very different doubly-refracting powers of glass or crystals of different materials are really inherent in their substance, or are due to different states of initial stress (*tensions moléculaires*) in different crystals.

En d'autres termes : si, dans différents corps homogènes et dans les mêmes directions, on pouvait produire des compressions et des dilatations égales, ces corps acquerraient-ils le même pouvoir biréfringent ou auraient-ils des pouvoirs divers ? (p. 290.)

Wertheim experimented on crown, plate and flint glass, all substances with different specific gravities and very different stretch-moduli. He loaded these with different weights till he produced the same difference of phase between the two rays, which he rather unfortunately terms the 'ordinary and extraordinary' rays. Suppose this to be obtained in any case by a vertical squeeze s, then the horizontal stretch $=\eta s$, where η is the stretch-squeeze ratio. Then Wertheim found that the ratio $(1+\eta s)/(1-s)$ was the same for the various kinds of glass. He terms this the *rapport des deux densités linéaires*, and he puts $\eta=1/3$ on his own hypothesis: see our Art. 1319*. I do not know quite why he should thus define it, but it is interesting to know that it is the same for all kinds of glass. Since η is taken by Wertheim a constant, it is the same thing as saying that it requires the same squeeze to produce the same doubly-refracting power in all kinds of glass, which is one way in which Wertheim himself states his result. Since $s=T/E$, the measurement of the load T required to produce a given doubly-refracting power gives a means of ascertaining the stretch-modulus of the glass.

Wertheim points out in the conclusion of the paper that the magnetic rotation of the plane of polarisation is the more feeble the greater the mechanical strain: see our Arts. 698, (iv) and 797, (*d*).

[787.] *Rapport sur divers Mémoires de M. Wertheim,* by Regnault, Duhamel, Despretz and Cauchy (rapporteur). *Comptes rendus*, T. 32, pp. 326–330. Paris, 1851. The Commission appears to have dealt with those memoirs of Wertheim which supported his hypothesis that $\eta = 1/3$; see our Arts. 1319*–26* and 1339*–51*. They sum up briefly the experimental arguments he has given for $\eta = 1/3$, but remark that this value is impossible on the ordinary (i.e. rari-constant) theory of elasticity. But that theory is based on the supposition that each molecule may be reduced to a point:

Si l'on suppose, au contraire, chaque molécule composée de plusieurs atomes, alors, suivant la remarque faite par l'un de nous, dès l'année 1839 (compare our Art. 681*), les coefficients compris dans les équations des mouvements vibratoires cesseront d'être des quantités constantes, et deviendront, par exemple, si le corps est un cristal, des fonctions périodiques des coordonnées. Or, en développant ces fonctions et les inconnues elles-mêmes, suivant les puissances ascendantes et descendantes des fonctions les plus simples de cette espèce, représentées par des exponentielles trigonométriques convenablement choisies, on obtiendra des équations nouvelles desquelles on déduira, par élimination, celles qui détermineront les valeurs moyennes des inconnues. D'ailleurs les équations définitives, trouvées de cette manière, seront encore des équations linéaires et à coefficients constants, qui ne pourront devenir isotropes et homogènes, sans reprendre la forme obtenue dans la première hypothèse. Mais le rapport entre les deux coefficients $(\lambda + \mu, \mu)$ que renfermeront alors les équations dont il s'agit ne deviendra pas nécessairement égal à 2, quand les pressions intérieures s'évanouiront; et l'on verra par suite disparaître l'objection proposée (p. 329).

Cauchy thus sees clearly what Wertheim never appears to have done, namely: that the latter's theory is incompatible with the uni-constancy of the early elasticians. Cauchy attempts a reconciliation by means of his suggestion of 1839. The objections to this hypothesis have been considered by Saint-Venant: see our Art. 192, (*d*).

The Commissioners speak highly of Wertheim's memoirs and recommend their insertion in the *Recueil des savants étrangers.*

[788.] G. Wertheim: *Note sur la double réfraction artificiellement produite dans des cristaux du système régulier. Comptes rendus*, T. 33, pp. 576–9. Paris, 1851.

Wertheim holds that the optic axes of these crystals under pressure do not coincide with their elastic axes, but make with them an angle which changes when the same force is applied in different directions to the body. His reasoning is founded on the statement that it is impossible to suppose the 'coefficient of elasticity' to be different in different directions in these crystals. It would contradict Neumann's statements, which, however, had been already corrected by Neumann himself as well as by Ångström: see our Arts. 788*–93* and 683–7. Wertheim's arguments are based on experiments on alum, which he says ought to have its elasticity equal in all directions.

[789.] G. Wertheim: *Deuxième Note sur la double réfraction artificiellement produite dans des cristaux du système régulier. Comptes rendus*, T. 35, pp. 276–8. Paris, 1852. This *Note* gives a series of results similar to those referred to in the previous article. We may note the following points:

(*a*) Crystals of cubic form act under external force like homogeneous bodies, the same force in *any direction* perpendicular to two faces of the crystal produces the same difference of phase between the 'ordinary and extraordinary' rays.

(*b*) For rock-salt and fluor-spar the difference of phase in the two rays is the same when the compression is the same as Wertheim found for the different kinds of glass (see our Art. 786), i.e. they have the same specific doubly-refracting power as glass.

(*c*) Alum which crystallises, Wertheim says, in *cubo-octaèdre* does not act like a body optically homogeneous, "although its elasticity is equal in all directions." Under pressure the elastic and optic axes do not coincide.

This is a restatement of the conclusion in our Art. 788. Various results as to the effect of pressure on other forms of crystals of the regular system are given. Wertheim sums up these results with the following statement, which seems somewhat doubtful in so far as it definitely asserts that the elasticity of a body is independent of the various changes of form (? sets) which the body has previously undergone:

Tous ces phénomènes: l'inégale compressibilité optique, aussi bien que la rotation de l'ellipsoïde optique, paraissent avoir leur origine dans les effets permanents produits par les tensions ou pressions qui ont lieu pendant l'acte de la cristallisation; on sait que l'élasticité mécanique ou moléculaire est indépendante des changements de forme que le corps a subis antérieurement; mais l'élasticité optique en conserve pour ainsi dire l'empreinte (p. 278).

[790.] G. Wertheim: *Note sur des courants d'induction produits par la torsion du fer. Comptes rendus,* T. 35, pp. 702–4. Paris, 1852. The contents of this *Note* are stated among many others in our discussion of the second part of the great memoir on *Torsion:* see our Art. 813.

[791.] G. Wertheim: *Mémoire et thèse sur la relation entre la composition chimique et l'élasticité des minéraux. Conclusions. Cosmos* T. IV., pp. 518–20. Paris, 1854.

This paper gives the results of a memoir by Wertheim which I do not think was ever published.

Wertheim considers the elasticity of a metal to be independent of its manufacture and to depend only on its density, chemical constitution, and crystalline form. The coefficient of elasticity increases with the density, but more rapidly; a slight chemical change has a great influence on the elasticity. On passing from the amorphic to the crystalline stage a body changes its density, and it has yet to be determined how far crystallisation *directly* affects elasticity (the densities of graphite and diamond[1] are as 1 : 2, their elasticities, if graphite is like other carbons, are as 1 : 20). A body which can crystallise in two different forms with the same density (e.g. pyrites) and composition can have different elasticities in the two forms. When bodies enter into chemical combination each retains its own elasticity, which is not destroyed by the action of the chemical forces, but only modified by the elasticities of the other bodies in the combination (Wertheim appears to have arranged bodies in tables according to their chief constituent, e.g. iron, nickel or manganese, but I do not know that these tables were ever published). The following conclusion seems of sufficient importance to be cited at length:

D'après ce qui précède, on pourrait être tenté de calculer l'élasticité d'un corps composé en prenant la moyenne entre les élasticités des corps composants, et en attribuant aux corps gazeux ou liquides une élasticité hypothétique qu'ils auraient à l'état solide, par un procédé analogue à celui dont on s'est servi pour le calcul des densités et des points de fusion. En effet, les sulfures et les arséniures se prêtent assez bien à ce mode de calcul, mais il est complétement en défaut pour les oxydes ; l'oxyde magnétique est doué d'une élasticité inférieure à celle du fer métallique, tandis que les sesqui-oxydes de fer ont une élasticité supérieure à celle-ci. Il faudrait donc, d'après le premier, attribuer à l'oxygène solide une élasticité inférieure à celle du fer ; et d'après le second, lui en attribuer une supérieure (p. 519).

When two bodies of "analogous composition" or "which belong to the same type are compared, the elasticity is always the greatest for that in which the molecules are closest (*les plus rapprochées*), but this relation does not hold for bodies of entirely different composition. For these on the contrary the elasticity and molecular distance diminish

[1] The *Encyklopädie der Naturwissenschaft, Handbuch der Physik*, Bd. I., S. 155, gives the ratio of the *mean* density of graphite to the density of diamond as 2·25 : 3·52, considerably less than 1 : 2.

as they become more complex" (p. 520). The memoir concludes by suggesting the need for new hypotheses as to the grouping of molecules and as to molecular weight; such hypotheses, however, could only be verified by a wider range of experiments than, Wertheim states, he had at that time undertaken. He promises to complete his researches in this direction.

[792.] G. Wertheim: *Mémoire sur la double réfraction temporairement produite dans les corps isotropes, et sur la relation entre l'élasticité mécanique et l'élasticité optique. Annales de chimie et de physique*, T. XL., pp. 156–221. Paris, 1854. This memoir is translated into English in the *Philosophical Magazine*, Vol. VIII., pp. 241–61 and 342–57. London, 1854. It is an attempt to investigate the relation between stretch and traction and the question of the equality of the stretch- and squeeze-moduli by means of photo-elasticity. There are references in the memoir to the researches of Brewster, Neumann and Maxwell: see our Vol. I. p. 640 Arts. 1185*, and 1556*. I do not think, however, that Wertheim has sufficiently expressed his indebtedness to these authors.

[793.] The memoir commences with twelve pages entitled *Historique* (pp. 156–168). Here it is pointed out that the theory of elasticity assumes: (i) the proportionality of stress and strain; and (ii) the equality of the stretch- and squeeze-moduli. There are, however, various experimental investigations, which throw doubt on the truth of these assumptions, notably those of Hodgkinson (see our Arts. 234*, 1411*–12*). Wertheim remarks on the great difficulty of making direct experiments on compression. For sensible squeezes we require a long bar of material, and this will certainly buckle unless supported at the sides. But if supported at the sides the disturbing action of friction is introduced. This disturbing action Vicat had met with in his experiments on the compressibility of lead (see his memoir of 1833 referred to in our Art. 724*). Wertheim also attributes the large values of the squeeze-modulus for wrought-iron obtained by Pictet (see his memoir of 1816 referred to in our Art. 876*) to the same cause. The difficulties attending experiments on compression have hindered the undertaking of any important series of *direct* experiments except those of Hodgkinson. These Wertheim discusses at very considerable length on pp. 159–166. He takes Hodgkinson's results and removing the sets, he calculates from the purely elastic stretches and squeezes the values of the stretch- and squeeze-moduli for forged and cast-iron. He obtains the following results:

(i) Removing the set, the proportionality of stretch and traction for wrought-iron holds almost up to rupture.

(ii) Removing the set, the stretch increases more rapidly than the traction for cast-iron. (Wertheim holds that this result may be due to defects almost unavoidable in the method of experiment.)

(iii) Removing the set, then for eight series of experiments on cast-iron the squeezes diminish with the pressures in three series, in one series they increase, and in four there is a sensible proportionality. Two of the last series of experiments are for a mixture of cast-irons (Leeswood and Glengarnock)[1]. Wertheim considers that Hodgkinson's experiments are very far from giving any conclusive answer as to the legitimacy of the assumptions made in the usual theory. He proposes therefore to investigate them afresh by the aid of photo-elastic measurements. For the theory of photo-elasticity he claims (p. 168) some precedence for Fresnel over Neumann. He refers to a memoir of Fresnel's written in 1819 and only published in 1846, five years after Neumann's (see *Annales de Chimie...*, 3e série, T. XVII.). Fresnel's paper in nowise detracts from the transcendent merits of Neumann's great memoir. That memoir was based upon Brewster's experimental researches, and the discovery of double refraction by pressure is the real contribution to be attributed to Fresnel. The statement of the fundamental equations of photo-elasticity and their application to the wide range of phenomena observed by Brewster is undoubtedly due to Neumann.

[794.] Pp. 169–185 describe Wertheim's apparatus, which is constructed with his usual ingenuity; not the least valuable part is the differential arrangement described on pp. 181–2 for use when the loads are small. It lies beyond the scope of our *History* to do more than refer to accounts of physical apparatus. Suffice it to say that Wertheim shows that the difference of the equivalent air-paths of the two rays is approximately proportional to the loads, and by means of a very complete table on p. 180 he is enabled to measure the loads by a scale

[1] These results assume of course that we may suppose the set not to affect the stretch- and squeeze-moduli; they are not based on experiments in which the bars have been previously reduced to a state of ease embracing the maximum load. Supposing squeeze set to be produced by *lateral* stretch, we should not expect the squeeze-modulus to be so sensibly affected by set as the stretch-modulus until the pressure was 4 to 6 times as great as the traction (i.e. granted $\eta = \frac{1}{4}$ to $\frac{1}{6}$).

Thus the compression load of about 350 cwt. which limits the compression experiments ought not to be compared, so far as equality of the stretch- and squeeze-moduli is concerned, with a load of more than about 70 cwts. in the traction experiments. It will then be found that the difference between the stretch- and squeeze-moduli is not great.

of colours of the two images. He thus carries out exactly what Brewster had proposed in the *Chromatic Teinometer:* see our Art. 698* and ftn., Vol. I. p. 640. The reader will easily understand how this colour scale of stress can be applied to the problems stated in the previous article.

[795.] Taking Neumann's theory (see our Art. 1191*, Eqn. (iv)) we see that air thickness answering to the colour measured in Newton's scale is proportional to the stretch (or squeeze) in the case of a prism under pure positive (or negative) traction. According to Wertheim's experiments it is proportional to the load; we have thus a method of ascertaining whether stress is here proportional to strain, and if so whether the constant of proportionality is the same for both positive and negative stress. Wertheim's own theory and his comparison of its results with those deduced from Neumann's seem to me somewhat obscure. Thus he says:

"Let O be the velocity of light in the air; O_o and O_e the ordinary and extraordinary velocities in the substance[1], which possesses for the time double refraction" (p. 199).

Then if h, l, b be the height, length and breadth of the prism subjected to a *total* traction P in the sense h, these dimensions become

$$h(1+s), \quad l(1-\tfrac{1}{3}s), \quad b(1-\tfrac{1}{3}s),$$

where s is the stretch produced by P. Wertheim puts this down without stating that he is assuming that $\lambda = 2\mu$ and consequently $\eta = 1/3$, his own particular theory of the inter-constant relation for isotropy. He then continues: "The two rays have to traverse the distance $l(1-\frac{1}{3}s)$; and consequently the difference of their equivalent air-paths d, after they leave the prism is proportional to $l(1-\frac{1}{3}s)(O/O_o - O/O_e)$ and to the dilatation s, we have then:

$$d = sl(1-\tfrac{1}{3}s)(O/O_o - O/O_e)\text{"} \quad \text{.................(i)}.$$

Assuming s^2 negligible, and stress proportional to strain, or,

$$s = P/(Elb),$$

we find:

$$d = \frac{P}{Eb}(O/O_o - O/O_e) \quad \text{..................(ii)}.$$

Now: "O/O_o and O/O_e are the two indices of refraction I_o and I_e; and for $d = \pm 1$ and $b = 1$, we have $P = C$, and accordingly,

$$I_e = I_o \mp E/C \quad \text{...........................(iii)},$$

where it is necessary to take the negative sign for positive traction and the positive sign for negative traction."

I do not understand Equation (i). I should have thought that if

[1] Wertheim following the analogy of natural double refraction speaks of 'ordinary and extraordinary' rays. Homogeneous plane polarised light, being incident normal to the face $h \times b$ of the prism, will be decomposed into two rays, one with vibrations parallel to h, the other with vibrations parallel to b. These rays travel with velocities differing from each other and from the velocity of light in the unstrained prism. Neither ray has thus a special claim to be termed 'ordinary.'

O^o and O_e were the ray velocities the value of d must, for small s, be $l\,(O/O_o - O/O_e)$, which agrees with Neumann's value in our Art. 1191*, if we remember that his δ is not the length of the equivalent air-path, but the thickness of air for the corresponding colour of the Newtonian scale. Further, since Wertheim makes $O/O_o - O/O_e$ or $I_o - I_e$, a constant in Equation (iii) depending on the elastic nature of his material, he can hardly mean O_o and O_e to be the *velocities* of the two rays. They are really the velocities divided by the stretch. Comparing Wertheim's Equation (i) above with Neumann's (iv) of our Art. 1191*, we see that:

$$\frac{p-q}{V^2}(1+\eta) = \frac{1}{O_o} - \frac{1}{O_e},$$

or, that the constant $I_o - I_e$, which Wertheim terms "the true measure of the double refraction,"

$$= \frac{p-q}{V}(1+\eta) \times r$$

in terms of the photo-elastic constants p and q of Neumann, where r is the refractive index and V the velocity of light in the unstrained material. Taking $r = O/V$ and assuming it $= 1{\cdot}543$ for plate glass, Neumann's values for p/V and q/V (see Art. 1193*) give $I_o - I_e = {\cdot}158$ according to my calculations and with uniconstant isotropy. Wertheim gives ·157 for its theoretical value for $\eta = \frac{1}{4}$, and, ·168 for $\eta = 1/3$. His experimental determination gives ·191. The difference is considerable, but in both Neumann's and Wertheim's results all the elastic and optic constants were not determined for the same kind, still less for the same piece of glass.

It may be noted that Wertheim terms the constant C above (Equation (iii)) the "*coefficient d'élasticité optique.*" Thus C is proportional to the inverse of what Maxwell terms the "optical effect": see our Arts. 1543*, 1544*, and 1556*. Wertheim's name seems well chosen, as we have from Equation (ii), $d = P/(Cb)$, an equation analogous to $s = P/(Ebl)$; thus C is the load corresponding in a prism of unit breadth to unit difference of equivalent air-paths (p. 196).

[796.] Wertheim's first three experimental results verify the relation $d = P/(Cb)$, where C depends on the material and not on the size of the prism taken (Experimental Laws, $1^\circ - 3^\circ$, pp. 189–90). A fifth law stated on p. 197 is that the difference (d) of the equivalent air-paths of the two rays is independent of the wave length; thus the dispersion accompanying the double refraction is insensible.

Wertheim's fourth experimental law is the answer to the problems he stated at the commencement of his memoir. According to Neumann's theory d is proportional to the stretch s, and by Wertheim's experiments d is given as a function of P. Hence if d be plotted up to P, the curve ought to be a straight line if P be proportional to s, or if stress be proportional to strain. Further this line ought to pass through the origin without change of slope if the stretch- and squeeze-moduli are equal. The following is Wertheim's conclusion:

Double refraction or the difference of the equivalent air-paths of the two rays is proportional to the mechanical stretch or squeeze, but these are not rigorously proportional to the tractions. In taking the tractions for abscissae and the stretches and squeezes for ordinates, a curve is obtained for the pressures concave to the axis of abscissae, and another for the tensions convex to the same axis; these curves straighten themselves as the stresses increase till they coincide with one and the same straight line which corresponds to the elastic modulus usually adopted for both stretch and squeeze (p. 191).

What Wertheim's experiments go to show is a want of proportionality between d and P. He gives (pp. 192–3) reasons for supposing that s and d are proportional, hence it follows that s and P are not. It must be remembered that Wertheim has been plotting up his curves starting with initially very small loadings quite within the elastic limit, and that it is not till these small loadings are passed that the exact proportionality of stress and strain appears to commence. In other words the slopes of the tangents to the stress-strain curves at the origin are not what we are to understand by the statical moduli. Are we to take then the slope of one or other tangent at the origin as the modulus obtained by vibrational methods, or are we to suppose no proportionality between very small stresses and strains? This latter view is opposed to the isochronism of sound vibrations. Of course in experiments involving delicate measurements of this kind, it is always possible to raise a suspicion as to the accuracy of the results. Here is what Wertheim himself says of his results:

Les résultats que nous venons d'obtenir ne sont d'aucune importance pour la pratique des constructions ; ces différences sont trop petites pour être prises en considération lorsqu'il s'agit de l'emploi des matériaux, et nos expériences prouvent que l'on peut continuer en toute sûreté de se servir d'un même coefficient d'élasticité pour calculer les effets des tractions et des compressions. Mais ces résultats acquièrent une grande importance lorsqu'on les considère au point de vue de la théorie des forces moléculaires ou de celle des oscillations mécaniques et des vibrations sonores ; je crois qu'ils fourniront la solution d'un certain nombre de questions qui sont restées en suspens, et sur lesquelles je me propose de revenir dans une autre occasion (pp. 196–7).

[797.] Some other points in the memoir deserve notice :

(*a*) The "double-refractive power" (= in Wertheim's notation $I_o - I_e$) depends in some undiscovered way on the other elastic and optical properties of the bodies. According to Wertheim it is not the same for all substances (he finds on p. 202 values from ·2182 for crown glass to ·0875 for 'Flint Faraday' and ·0641 for 'alum inactif'); neither does it stand in any simple relation to the density, nor is it a function of the refractive index only.

We note that Neumann's value of the "double-refractive power" contains $(1+\eta)$ and r^2 as factors, if it be written in the form

$$\frac{p-q}{O}(1+\eta)\,r^2,$$

and hence throws us back on the determination of p and q as functions of the elastic and optic constants (pp. 204–5).

(*b*) Wertheim holds that there is no relation between the two kinds—natural and artificial—of double refraction. To convince oneself of this, he says, it is only requisite to consider the forces which it is necessary to apply to an isotropic body in order to produce for equal thickness the same double refraction which arises from the passage of a ray across a plate of doubly refracting crystal cut parallel to the axis. For example he takes Iceland spar and ordinary crown glass, for which he says "the differences of the two indices of refraction are the same." Now I have already referred to his obscurity about the quantities O_o and O_e and indicated that his I_o and I_e are not the true refractive indices. It seems to me that $s(I_o - I_e)$ is the real difference of the refractive indices for the strained material. Does he then mean that this or that the "double-refractive power" $I_o - I_e$ for crown glass is equal to the difference of the indices for Iceland spar? If he does mean, as he says, the difference of the indices, then the force required to make the crown glass refract as the Iceland spar is $P = Es$, or P would be the pressure required to produce the necessary strain in the glass for the given difference of indices. On the other hand if he means that the "double-refractive power" of crown glass is equal to the difference of the indices of Iceland spar, then we have $P = E$, as he says, or a pressure is required a thousand times greater than would crush the glass (p. 204).

This apparent confusion leads me to doubt the accuracy of the values given for I_e in the table p. 202. E/C is presumably found from the experiments and equals $I_o - I_e$ in Wertheim's notation, but why is I_o = to the refractive index for the isotropic material? Since it is $s(I_o - I_e)$ which is the difference of the refractive indices of the 'ordinary and extraordinary' rays, this appears a perfectly arbitrary assumption.

(*c*) In a footnote on p. 206 Wertheim objects to Maxwell's having referred (in the memoir of 1850) to his hypothesis that $\lambda = 2\mu$, while citing *only* his experiments on caoutchouc as evidence for it, and neglecting all the other experimental evidence in favour of it. He also not unreasonably objects to Maxwell's taking cork and jelly as types of isotropic homogeneous bodies, but Maxwell has not been the only offender in this respect.

(*d*) Wertheim considers what rotatory effect magnetic force has on the plane of polarisation when an isotropic body in the magnetic field is subjected to positive or negative traction. His experiments were on bodies which possess to a high degree the magnetic rotatory power when isotropic ('tels que les flints'). The result was always the same, a small stress rendered this power relatively feeble. The exact load at which it disappeared was not capable of accurate measurement, but, the light being homogeneous, all rotation had disappeared when the difference of the equivalent air-paths of the two rays had become equal to half the wave length of the light. Wertheim considers that for all natural or artificial doubly-refracting media the magnetic rotatory power is in inverse ratio to the doubly-refracting power,—when the one is most energetic the other is feeblest. Thus it is to be noted that purely mechanical forces can apparently annul the action of magnetism on the optical medium[1] (pp. 207–9).

(*e*) On pp. 209–216 Wertheim describes what he terms the *Dynamomètre Chromatique*. This is merely a variation of Brewster's *Chromatic Teinometer*: see our Art. 698*, Vol. I. p. 640, ftn. Brewster's Teinometer is based on flexural stress, Wertheim's on traction, but the idea is exactly the same in both. Wertheim, it is true, makes considerable practical application of his instrument, and describes accurately its structure and use, but he ought to have acknowledged the source from which he had taken his idea, as he elsewhere refers to the very paper of Brewster's in which an account of the *Teinometer* is given.

Wertheim, assuming the accuracy of his Teinometer, shows what very large errors may arise in the manometric measurements of pressure in a large hydraulic press (p. 215).

(*f*) He applies his Teinometer to ascertain the squeeze-modulus of diamond. Turning to Equation (ii) of Art. 795, we have

$$d = \frac{P}{Eb}(I_o - I_e).$$

Now d, b, and P can be measured, hence if we knew $I_o - I_e$ we should have E. Here I fail to follow Wertheim, he assumes $I_o - I_e$, instead of $s(I_o - I_e)$, as the difference of the refractive indices again. He puts $I_o = 2{\cdot}470$, Brewster's value for unstrained diamond, and assumes for I_e a value equal to that of fluor-spar. Thus he obtains $E = 10{,}865$ kilos per sq. mm., or about the value for annealed copper—"et nullement en rapport avec sa grande dureté" (p. 217).

Pp. 217–21 contain a *résumé* of the results of this memoir,—a memoir which is undoubtedly of value, but which requires somewhat cautious and critical reading.

[1] Wertheim (p. 208) refers to experiments of Bertin and Matteucci in the same direction, but without giving the *loci* of their memoirs: see our Arts. 698, (iv) and 786.

[798.] G. Wertheim: *Mémoire sur la torsion. Comptes rendus*, T. 40, pp. 411–414, and *Mémoire sur les effets magnétiques de la torsion*, in the same volume, pp. 1234–7. Paris, 1855. These contain extracts from the great memoir on Torsion: see our Arts. 799 *et seq.*

[799.] G. Wertheim: *Mémoire sur la Torsion. Annales de chimie et de physique*, T. 50. Première Partie (*Sur les effets mécaniques de la torsion*), pp. 195–321. Seconde Partie (*Sur les effets magnétiques de la torsion*), pp. 385–431. Paris, 1857. This paper was presented to the Academy on February 19, 1855.

Wertheim exhibits here as in other work all his merits and demerits,—excellency and width of experimental investigation, ignorance or misapplication of theory, which leads him to misinterpret the results of some even of his own experiments.

[800.] The memoir, after a brief statement of the problem of torsion, opens with an account of the history of that subject (*Historique*, pp. 196–202). Here reference is made to Coulomb (Art. 119*) and Biot (Art. 183*) for the theory of torsion; to Poisson, Cauchy, Lamé and Clapeyron for the general equations of elasticity; to Neumann, Stokes and Maxwell as arriving at the same results by different processes but as adding nothing essential; to Heim[1] and Segnitz (Art. 481) as determining the shortening of a prism by torsion; to Savart (Art. 333*), Duleau (Art. 229*), Bevan (Art. 378*), Giulio (Art. 1218*) and Kupffer (Art. 1389*) as experimentally verifying the laws of torsion, or as determining by its means the elastic constants; reference is made also to Saint-Venant, whose work Wertheim seems totally to have misunderstood, probably to a great extent through insufficient analytical knowledge. The footnote on p. 199 is neither just to the results which Saint-Venant had published before 1857, nor does it apparently grasp his position. It is one thing to agree that a certain coefficient of correction is necessary, it is another to accept Wertheim's erroneous theory of torsion and his purely empirical relation between the coefficients of bi-constant isotropy ($\lambda = 2\mu$) in order to deduce that coefficient. We cannot enter at length into

[1] Heim in the work referred to in our Art. 906* deals (S. 237–47) by a cumbersome analysis with the stretch in the 'fibres' of a prism due to torsion. What is material on this point has been said in our Art. 51.

the details of this controversy which we have had occasion several times to refer to (see our Arts. 1339*–43*, 1628*–30*, and 191–2). It must suffice to state here that we hold Wertheim to have been in the wrong throughout, and occasionally, we fear, influenced by the dread that Saint-Venant's brilliant theoretical achievements would throw into the shade his own very valuable experimental researches. The lesson to be learnt from the controversy is the ever-recurring one, namely, the need that physicists should have a sound mathematical training, or, failing this, leave the theoretical interpretation of their results to the mathematician.

[801.] After a description (pp. 202–205) of the apparatus adopted for the experiments, Wertheim states the problems he proposes to deal with. They are the following:

(i) Whatever may be the magnitude of the *elastic* strain, are the angles of torsion still rigidly proportional to the moments of the torsional couples and to the lengths of the prisms to which torsion is applied?

(ii) What is the relation between torsional elastic strain and torsional set (*tort*)?

(iii) Is torsional elastic strain accompanied by change of volume, and if it be, what is the relation of that change to the torsional couple and to the shape of the prism?

(iv) How far is the accordance with experiment of the formulae of torsion modified by the aeolotropy of the material or by the shape of the prism?

(v) How far do the results of torsional experiments confirm Wertheim's theory that $\lambda = 2\mu$, or tend to demonstrate uni-constant isotropy, $\lambda = \mu$?

Wertheim's experiments were made on 65 prisms, partly on circular, partly on square, rectangular and elliptic bases, some being solid and some hollow. The materials were steel, iron, brass, glass, and in a few cases oak and deal.

In the experiments on torsion the terminal cross-sections of the prism were fixed at a constant distance from each other, *i.e.* the length of the prism could not change with the torsion (p. 202).

[802.] Wertheim takes the opportunity afforded by the hollow prisms to investigate in Regnault's manner the value of the stretch-slide ratio η. The hollow prism blocked at the ends is filled with fluid communicating with a capillary tube passing through one of the terminal blocks. The prism is then stretched with a given stretch s_1. If the prism were isotropic the change in unit volume of the hollow ought to be $s_1(1-2\eta)$. This change in volume can be measured by the amount the fluid has advanced or receded in the capillary tube. There are considerable difficulties in making the experiment,—e.g. Wertheim found that the results depended to some extent on the diameter of his capillary tube (see pp. 209–10 and the numerical tables). The results are given in a table on p. 212. Wertheim puts $\eta = 1/3$ and thus takes the change in unit volume to be 1/3 of the stretch. The calculated results do not agree very closely with the observed, being greater for cylinders of brass and less for those of iron. Rectangular prisms of brass give fairly good results. At the same time $\eta = 1/3$ gives generally better results than could be obtained from $\eta = 1/4$. As Wertheim, however, admits that despite the annealing his prisms were *not isotropic*, there is no real reason why η should be equal to 1/4. This want of isotropy Wertheim considers beyond the reach of the then existing theory, and an attempt he makes to deal with it on the basis of Cauchy's equations is not successful. The difficulty was fully overcome in Saint-Venant's paper of 1860: *Sur les divers genres d'homogénéité*: see our Arts. 114–125. While probably the aeolotropy accounts for the variety of the results, Wertheim also notices that the fluid itself may affect chemically the material of the tube or the cement which fastens its terminals (p. 216).

Pp. 216–221 are entitled: *Sur les effets optiques produits par la torsion*, and mainly describe the difficulty of making the necessary experiments. Wertheim concludes from experiments made on glass only that:

> Ces expériences prouvent qu'il s'agit seulement d'une double réfraction ordinaire qui devient positive ou négative selon que la torsion a lieu vers la droite ou vers la gauche; on ne peut rien en conclure en ce qui concerne un corps parfaitement homogène, et elles ne peuvent servir ni à confirmer ni à infirmer les prévisions de l'analyse de M. Neumann. (See our Art. 1195*.)

[803.] Pp. 221–225 are entitled: *Sur quelques faits généraux et indépendants de la forme de la section transversale.* Wertheim commences by dividing the angle of torsion into two parts, ψ_1 the elastic part and ψ_2 the set part. He recognises the after-strain discovered by Weber (*l'effet secondaire découvert par M. Weber, et qui est insensible dans l'allongement des métaux*) to be sensible in torsion experiments on metals, but he disregards it because

> ses effets se confondent avec ceux des oscillations tournantes que la barre exécute autour de chaque position d'équilibre avant de revenir au repos (p. 221).

As after-strain can be observed for more than twenty minutes in steel wires, I am somewhat doubtful as to the exact meaning of the sentence cited.

The following general conclusions are drawn by Wertheim:

(*a*) There is no point at which set can be said to commence (thus Wertheim had not reduced his prisms to a state of ease).

(*b*) The set-angle bears no obvious relation to the elastic angle. It is not proportional to the length of the prism nor to the load couple, although of course it varies with these. It begins to increase at first very gradually, then more rapidly, and finally just before the bar breaks (ou se mette à *filer*) becomes incapable of determination. (This seems to point to the early stages of set being merely due to the 'working' of the individual specimen.)

(*c*) The angles which measure the elastic strain are not rigorously proportional to the load-couples applied.

Wertheim attributes this result to two causes: the first that stretch is not proportional to traction as shown by his paper of 1854 (see our Art. 796), whence, as he holds torsion involves a longitudinal traction, torsional stress ceases to be proportional to strain; and the second that as the torsion increases the cross-sections contract, and so the 'moment of resistance' of the prism decreases. (Both these causes seem to me quite insignificant except for very large strains, which of course do not fall within the ordinary theory of elasticity. The effect of the traction on the torsional couple is given in our Art. 735, (iii)).

(*d*) The angles of torsion are not rigorously proportional to the lengths of the prisms.

(*e*) The interior cavity of a hollow prism, whatever be its form, is diminished under the influence of torsion. This diminution is proportional to the length of the prism and to the square of the angle of torsion per unit length of prism.

For the case of hollow circular cylinders Wertheim gives the following formula for the diminution δV of the cavity V:

$$\delta V/V = -\tau^2 a_1^2,$$

where τ is the angle of torsion per unit length of the prism and a_1 the inner radius of the hollow cylinder (p. 226). This gives results fairly in accordance with his experiments. He propounds a partial theory on pp. 229–235, which I am not able to accept. It contains the conclusions cited below, which I think are erroneous:

Il se présente d'abord la question de savoir si, ainsi que nous le supposons, la diminution de volume que nous venons de trouver représente réellement celle qu'aurait éprouvée un cylindre solide de même matière que la paroi du tube, de dimensions telles, qu'il remplît toute la cavité intérieure de celui-ci, et qui aurait été soumis à la même torsion temporaire? L'affirmative ne me semble pas douteuse (p. 229).

We can test the result in the manner of our Art. 51. With the notation of that article the longitudinal squeeze of a solid cylinder of radius a is $\frac{\tau^2 a^2}{4}$ if the cylinder be allowed to shorten. The cylinder, however, in Wertheim's experiments was maintained at length l. Hence there would be a longitudinal tension corresponding to a stretch of $\frac{\tau^2 a^2}{4}$, or if η be the stretch-squeeze ratio its radius a would become $a\left(1 - \eta\frac{\tau^2 a^2}{4}\right)$ and therefore

$$V + \delta V = \pi a^2 l \left(1 - 2\eta \frac{\tau^2 a^2}{4}\right),$$

or
$$\frac{\delta V}{V} = - 2\eta \frac{\tau^2 a^2}{4} \quad \text{.......................................(i)},$$

$$= - \frac{\tau^2 a^2}{8} \text{ for uni-constant isotropy,}$$

$$= - \frac{\tau^2 a^2}{6} \text{ on Wertheim's hypothesis } (\lambda = 2\mu).$$

On the other hand the longitudinal squeeze is not equal to $\frac{\tau^2 a_1^2}{4}$ for a hollow cylinder of inner radius a_1. To ascertain its value ϵ ($\equiv$ the η of the notation of our Art. 51) we must put instead of the equation at the middle of our page 42:

$$\int_{a_1}^{a_2} 2\pi r dr E \left(\frac{\tau^2 r^2}{2} - \epsilon\right) \sin P'PN = 0.$$

This gives us
$$\frac{\tau^2 (a_2^4 - a_1^4)}{8} = \epsilon \frac{a_2^2 - a_1^2}{2},$$

or,
$$\epsilon = \frac{\tau^2 (a_2^2 + a_1^2)}{4},$$

that is the *double* of its previous value if a_2 be nearly equal to a_1. Hence by the same reasoning as before the internal radius a_1 becomes

$$a_1 \left(1 - \eta\tau^2 \frac{a_2^2 + a_1^2}{4}\right).$$

Hence the new volume of the cavity

$$= V + \delta V = \pi a_1^2 l \left(1 - 2\eta\tau^2 \frac{a_2^2 + a_1^2}{4}\right),$$

or,
$$\frac{\delta V}{V} = - 2\eta\tau^2 \frac{a_2^2 + a_1^2}{4} \quad \text{.................................(ii)},$$

$$= - \frac{\tau^2}{8} (a_2^2 + a_1^2) \text{ for uni-constant isotropy,}$$

$$= - \frac{\tau^2}{6} (a_2^2 + a_1^2) \text{ on Wertheim's hypothesis.}$$

Thus (ii) gives for the hollow cylinder a value at least double that for a solid cylinder of the radius of the hollow. Our theoretical investigation, however, gives a value for $\delta V/V$ for these cylinders only about a *third to a fourth* of that given by the formula which Wertheim holds established by experiment. I am unable to explain this discrepancy between the above theory and experiment. Possibly it arises from the difficulty in Wertheim's apparatus of the terminal sections contracting and hence in some way there may result a tendency to an inward buckling of the sides of the cylinder.

[804.] On pp. 235–7 we have a *résumé* of the results for hollow and solid circular cylinders. In comparing the experimental results with theory Wertheim takes $\eta = 1/3$, and it appears to agree better than $\eta = 1/4$, but if we adopt bi-constant isotropy then there is no particular reason for taking $\eta = 1/4$, and if we suppose rari-constant aeolotropy the same remark holds. In both cases we should choose a value of η best fitting with the experiments and varying from one material to another. Hence it is impossible to agree with Wertheim's statement:

Quant à l'exactitude de la constante que j'ai introduite dans la formule, nul doute ne peut subsister à cet égard ; tous les angles calculés seraient avec l'ancienne formule de $\frac{1}{18}$ plus petits, et il en résulterait entre le calcul et l'expérience un désaccord constant et de beaucoup supérieur à la limite des erreurs, désaccord qui, dans les torsions considérables, atteindrait souvent l'importance de plusieurs degrés (p. 237).

This statement is a fair enough argument against the uni-constancy of the material of Wertheim's prisms, but is of no value in favour of a general law that $\eta = 1/3$. He appears to consider that there is really a theoretical reason for this particular value, so that it has more claim on our attention than $\eta = \cdot 3$ say, while in fact it has a purely empirical basis: see our Arts. 1324*–6*.

[805.] Wertheim next passes to the torsion of prisms on elliptic bases. Here his method is very singular. He writes:

On obtient la formule pour la torsion de ces cylindres, en substituant dans la formule que M. Cauchy a trouvée pour les prismes rectangulaires, à la place du moment d'inertie du rectangle par rapport à l'axe, le moment polaire de l'ellipse (p. 238).

Wertheim has a footnote to 'l'ellipse'—"Voyez les ouvrages de MM. Persy, Poncelet, Moseley et Weissbach."

Now Cauchy's formula for rectangular prisms is quite wrong (see our Arts. 661*, 684*, 25 and 29), and if it were correct it could not be applied in the manner suggested to elliptic prisms. But the wrong formula for rectangular prisms, erroneously assumed to hold for elliptic prisms, does give the true result for the latter as Saint-Venant had shown ten years before this memoir (see our Art. 1627*). Wertheim in this manner reaches Saint-Venant's formula for prisms of elliptic cross-section (see our Art. 18) without referring to its discoverer. The footnote can hardly serve to do more than mystify the reader. In our Art. 1623* it has been pointed out that Saint-Venant in 1847 separated the *gauchissement* into two elements, a distinction which he afterwards dropped. This distinction, however, made no change in the facts or formulae he deduced for torsion. Now Wertheim (taking however $\eta = 1/3$) finds a close agreement between what is really Saint-Venant's formula and the results of his own experiments. He writes:

Les résultats moyens des expériences s'accordent donc avec le calcul d'une manière satisfaisante même pour les cylindres 12 et 14 qui ont pour bases des ellipses d'une forte excentricité; dans la théorie de M. de Saint-Venant cet accord prouverait que le premier gauchissement qui seul existerait dans des cylindres elliptiques, et dont l'influence sur le moment de résistance à la torsion n'a pas été déterminée par ce géomètre, serait complétement négligeable sous ce rapport (p. 239).

The only intelligible reading of this passage would seem to be that Wertheim had a different theory from Saint-Venant for elliptic prisms and that the theory of the latter was demonstrated by the experiments to be erroneous. These conclusions would be the exact opposite of the truth. Saint-Venant's reply to "cette observation de l'honorable et consciencieux expérimentateur" is polite but complete (see the *Leçons de Navier*, pp. 629–31, and our Art. 191).

[806.] The next section of Wertheim's memoir is entitled: *Sur la torsion des prismes homogènes à base rectangulaire*, and occupies pp. 239–53.

Before entering on the matter of this section we must remind the reader that Saint-Venant had in 1847 given the true theory for rectangular prisms (see our Art. 1626*) and shown wherein Cauchy's theory was erroneous. Further that in 1854 Cauchy had acknowledged the justice of Saint-Venant's criticism (see our Art. 684*). Wertheim's memoir was read in 1855 but not published till two years later, after, indeed, the appearance of Saint-Venant's great memoir on *Torsion*, which was printed in a volume of the *Mémoires des savants étrangers* dated 1855. Hence it seems unaccountable that Wertheim should without comment adopt Cauchy's formula as the theoretical view of the subject, and apply to it a numerical coefficient of correction which depends in an unknown manner on the ratio of the sides of the rectangular base.

Wertheim commences by comparing the experiments of Duleau and Savart on rectangular prisms with Cauchy's formula and deducing a coefficient of correction. This is close to its value as given by Saint-Venant's theory: see our Arts. 31, 34 and 191.

The next point dealt with is the diminution of volume of the interior of a hollow rectangular prism or tube under torsion. Wertheim gives the formula,

$$\delta V/V = -\frac{(a_1^2 + b_1^2)^4}{16\,(a_1 b_1)^2}\,\tau^2,$$

where $2a_1$, $2b_1$ are the sides of the hollow and τ the angle of torsion per unit length of the prism. But he remarks:

il n'est pas impossible que la théorie après de nouveaux progrès conduise à une formule différente de celle-ci, et qui ne s'accorde pas moins bien avec les expériences (p. 243).

This seems possible as there is a mistake somewhere in this

empirical formula, for the left-hand side is a numerical quantity, but the right-hand side is an *area*. Possibly Wertheim intended to write in the denominator $16(a_1b_1)^3$.

If we go back to Art. 51 and *assume* the shortening of the 'fibre' at any point of the hollow rectangle to be still $\frac{1}{2}\tau^2 r^2 - \epsilon$, where r is the distance of the fibre from the axis of the prism, we easily find $\epsilon = \frac{1}{2}\tau^2\kappa^2$, where κ^2 is the swing-radius of the section about the axis of the prism. In the case of a hollow rectangle with lengths of inner sides $2a_1$, $2b_1$, of outer sides $2a_2$, $2b_2$, and uniform thickness we easily find:

$$\kappa^2 = \tfrac{1}{3}(a_2^2 + b_2^2 + 2a_1b_1).$$

Whence with the same reasoning as before:

$$\delta V/V = -2\eta \frac{a_2^2 + b_2^2 + 2a_1b_1}{3} \frac{\tau^2}{2}.$$

If the prism be very thin we have:

$$\delta V/V = -\tfrac{1}{12}(a_1 + b_1)^2 \tau^2 \text{ for uni-constant isotropy,}$$

$$= -\tfrac{1}{9}(a_1 + b_1)^2 \tau^2 \text{ on Wertheim's hypothesis.}$$

For the case of a square section $(a_1 = b_1)$ these give only about one *third to a half* of Wertheim's results, thus differing almost as much as in the case of a hollow circular cylinder (see our Art. 803, (*e*)).

If we were to multiply the above results by $(1 + 2\eta)/2\eta$, or by 3 or 2·5 according to the hypothesis adopted, they would then agree fairly well with Wertheim's experiments. But this amounts to supposing that Wertheim's terminal conditions were of such a nature that there was for a *thin* prism a reduction of sectional dimensions given by $\frac{1}{2}(1 + 2\eta)$. Thus according to Wertheim if the torsional couple be so large that it would produce, when the ends of the prism were not fixed, a sensible longitudinal squeeze, then, if the ends be fixed, there will be a diminution of the linear dimensions of any internal cavity of about $(1 + 2\eta) \times$ this squeeze.

[807.] So far as Wertheim's own experiments on solid prisms of rectangular cross-section go, the 'coefficient of correction' was very nearly that required by Saint-Venant's theory,—the errors were such as were not unlikely to occur in material which was hardly isotropic and in torsions carried in many cases beyond the limit of linear elasticity. See Saint-Venant's *Leçons de Navier*, pp. 622–629.

The formula for the torsion of hollow rectangular tubes given on p. 250 is of course wrong: see our Art. 49.

[808.] *Sur la torsion des corps non-homogènes* is the title of the section of the memoir which occupies pp. 253–258. Wertheim supposed that for sheet-iron and wood there are three rectangular axes of elasticity, but his struggles to reach a theoretical formula for this case were not successful. The true formulae for prisms of elliptic and rectangular cross-sections are given in our Arts. 46–7. Wertheim obtains by a series of inadmissible hypotheses a formula for a rectangular prism which corresponds to some extent with Saint-Venant's for an elliptic prism. He applies it, not very satisfactorily, to his experiments on wood.

One point in this section deserves to be noticed, namely that for hollow cylinders of sheet-iron there was an *increase* instead of a decrease of internal capacity produced by torsion (p. 254). This cannot be explained by the formula (ii) of our Art. 803 (e) unless we put η negative, which is, however, impossible. Wertheim's own formula is equally inapplicable.

[809.] The following section (pp. 258–269) deals with the torsional vibrations of homogeneous bodies. So far as the theory of this section goes it is partly erroneous (e.g. for rectangular prisms) and partly hypothetical. It was a retrograde step to publish it after Saint-Venant's memoir of 1849: see our Arts. 1628–30*. Saint-Venant shows in the *Leçons de Navier* (see pp. 635–645, especially p. 643) that Wertheim's experimental observations are in complete accord with the formula

$$\frac{n}{n'} = \sqrt{\frac{E}{\mu}\,\frac{\omega\kappa^2}{\nu}}$$

given in our Art. 1630*. Wertheim introduces as before a 'coefficient of correction' for the bars of rectangular section.

In a footnote (pp. 264–6) he corrects a slip of Cauchy's in his *Exercices*, T. IV., p. 62. This slip is also noted by Saint-Venant in a footnote to p. 641 of his edition of the *Leçons de Navier*. It is not of importance, however, as the corrected formula is itself wrong.

There is only one remark in this section which it seems interesting to quote. Possibly the influence which produced the effect observed was after-strain:

Je profiterai de cette occasion pour faire remarquer que cette dépendance mutuelle entre l'intensité du son et son élévation n'a pas seulement lieu pour les vibrations tournantes; c'est au contraire un fait général dont on a pu faire abstraction pour faciliter les calculs, mais dont il faudra tenir compte actuellement. Les sons des corps solides montent en s'éteignant, tandis que ceux des liquides et des gaz baissent à mesure qu'ils s'affaiblissent. En ce qui concerne les corps solides, ces inégalités proviennent évidemment de ce que l'allongement qu'ils éprouvent par l'effet d'une faible traction n'est ni rigoureusement égal à la compression produite par cette même force lorsqu'elle agit comme pression, ni rigoureusement proportionnel à cette force (on this point Wertheim refers to the memoir discussed in our Art. 792). Maintenant, lorsque l'on se sert de vibrations longitudinales pour déterminer le coefficient d'élasticité, on trouve nécessairement une valeur plus ou moins élevée selon que l'on considère comme le vrai son fondamental de la barre ou du fil un son plus ou moins faible. Ordinairement on n'emploie à cet effet que les sons les plus faibles, parce que ce sont en même temps les plus purs, et qu'on les reproduit plus facilement avec le sonomètre, la sirène, ou avec l'instrument quelconque qui sert à la détermination du nombre de vibrations. On comprend donc qu'en opérant ainsi, on obtiendra toujours un coefficient d'élasticité trop élevé, et que cette différence ne disparaîtrait que si l'on pouvait, pour ces déterminations, se servir de sons tellement intenses, que leurs amplitudes fussent égales aux allongements et aux compressions considérables, que l'on emploie pour la détermination directe de ce même coefficient (pp. 259–60).

The inequality of the elastic constants as found by statical and vibrational methods has, indeed, been disputed: see our Arts. 767 and 824. But if the pitch of the fundamental note really depends as Wertheim asserts on the intensity of the disturbance, it must necessarily follow. If this assertion were true then the argument of Stokes in favour of the linearity of the stress-strain relations from the tautochronism of sound vibrations falls to the ground: see our Arts. 928* and 299. The matter would be clearer if the effects of after-strain which Wertheim holds "se confondent avec ceux des oscillations tournantes" could be eliminated in all cases of vibrations.

[810.] We now pass to the section of the memoir entitled: *Sur la rupture des corps homogènes produite par la torsion* (pp. 269–80). Wertheim distinguishes two kinds of rupture, which he considers characteristic respectively of hard and soft bodies (des corps *roides* et des corps *mous*). In the first class rupture occurs by slide, in the second by stretch of the fibres converted into helices. As hard bodies he takes glass, tempered steel and sealing wax; as soft certain sorts of iron (fer doux), cast-steel and brass, the second metal forming the transition from one class to the other. The distinction does not seem to me very real or necessary. I imagine that sealing wax might be made to show a very great change in form before rupture if a small twisting force were applied to it for a very long time. Our figure reproduces the

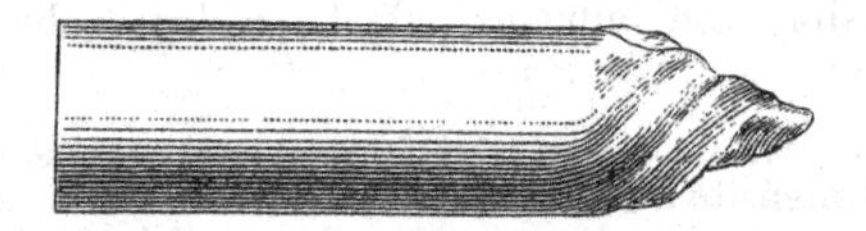

rupture-surface according to Wertheim for cylinders of sealing wax or of glass of small diameter. As he remarks, this surface certainly merits, were it feasible, analytical treatment.

To the *hard* bodies Wertheim applies a theory of strength deduced from the hypothesis that elasticity lasts up to rupture. But when applying this theory he always supposes rupture to occur in the 'outermost fibre.' The theory of elasticity, however, only makes the strain greatest in this fibre in the case of a right circular cylinder. In addition Wertheim puts $\mu = \frac{3}{8}E$, a result which flows from his hypothesis that $\lambda = 2\mu$. He gives, even admitting this assumption, a totally wrong expression for the strength of a rectangular prism (p. 275).

For *soft* bodies Wertheim believes rupture to take place by the stretch of the extreme fibres when they become helical. He appeals for the case of a right circular cylinder to Weisbach's *Mechanik*, and gives for the stretch in a surface fibre $\frac{\tau^2 a^2}{2}$. With the notation of our Art. 51, it is $\frac{\tau^2 a^2}{2} - \eta = \frac{\tau^2 a^2}{4}$, and this must be less than T/E, where T is the rupture traction. This gives for the safe angle of torsion

$$\tau a < 2\sqrt{T/E}.$$

Now if S be the shear at which rupture would take place by pure slide we have $\tau a < S/\mu$. Hence in order that rupture should take place by longitudinal stretch we must have

$$2\sqrt{T/E} < S/\mu,$$

or

$$\frac{S}{\mu} > 4 \frac{T}{S} \frac{\mu}{E}.$$

Now according to Wertheim's hypothesis $E = \frac{8}{3}\mu$, and as a rule T and S are not very different (see our Vol. I. p. 877); hence we must have S/μ something like 3/2, which seems quite absurd as S is at most 1/500 part of μ. Wertheim's whole treatment, however, of rupture is very unsatisfactory. He applies the proportionality of stress and strain, which does not extend beyond the fail-limit (see our Arts. 5 (*e*) and 169 (*g*)), and although he uses this elastic theory he places his fail-points in the surface fibres of his prism farthest from instead of nearest to the axis. The only grain of satisfaction to be found in these pages is the confirmation of Saint-Venant's theory to be found in the following words:

> Le fer fibreux se rompt par l'allongement des fibres extrêmes ; longtemps avant la rupture on y remarque souvent des fentes profondes et parallèles à l'axe, surtout vers le milieu des petites faces des prismes (p. 278).

See our Arts. 23, 30, etc.

Wertheim states that for practical construction it is worth noting that a torsional set in soft iron increases the resistance to torsional elastic strain (p. 280).

The *Première Partie* of the memoir concludes with a summary of results on pp. 280–6 and with tables of the experimental measurements—possibly the most valuable portion of the whole paper—on pp. 288–321.

[811.] The second part of the memoir is entitled: *Sur les effets magnétiques de la torsion* (pp. 385–431).

We have already referred to other experiments of Wertheim

himself and to those of de la Rive, Joule, Matteucci, Wiedemann etc. on the relations between magnetism and stress. Wertheim in this memoir proposes to deal with the influence of shearing or rather torsional stress on the magnetic properties of a body. He commences with some account of the early researches on the influence of mechanical action on magnetism. Gilbert was apparently the first to observe that the regular or irregular vibrations of a bar of iron affect its state of magnetisation or the rate at which it develops magnetisation. Gay-Lussac was among the first to analyse these effects, while Réaumur[1] offered an explanation of them which would hardly be considered satisfactory to-day. Scoresby[2] added to previous knowledge by showing that the same mechanical actions which cause an iron bar to acquire magnetism when parallel to the direction of magnetic force, produce a loss when the bar is perpendicular thereto; further that repeated blows cause a highly tempered and strongly magnetised bar of steel to lose a large part of its magnetisation whatever may be its position relative to the magnetic poles of the earth. Baden-

[1] *Mémoires de l'Académie Royale des Sciences*, Paris, 1723 (Edition Amsterdam, 1730). *Expériences qui montrent avec quelle facilité le fer & l'acier s'aimantent, même sans toucher l'aimant*, pp. 116–149. See also the *Histoire*, pp. 7–8. Réaumur notes the effect of hammering in magnetising a bar: Après le premier coup de marteau, cette vertu est encore faible; on l'augmente si on applique une seconde fois la pointe de l'outil sur un morceau de Fer, & qu'on frappe sur l'autre bout une seconde fois. Cette opération simple, répétée un nombre de fois, ajoutera toujours à la nouvelle force attractive; mais il y a un terme par de-là lequel on répéteroit inutilement l'opération, la vertu de l'outil n'y gagneroit plus rien (p. 119).

This is probably the first scientific notice of the effect of impulsive stress on magnetism.

[2] William Scoresby: *Transactions of the Royal Society of Edinburgh*, Vol. IX. pp. 243–58, 1823, gives an account of the influence of impulsive stress (hammering) on the production of magnetism in iron and steel bars. A *résumé* of his results is given in the *Edinburgh Philosophical Journal*, Vol IV., 1821, pp. 361–2. We extract the following:

4. A bar of soft iron, held in any position, except in the plane of the magnetic equator, may be rendered magnetical by a blow with a hammer or other hard substance; in such cases, the magnetism of position seems fixed in it, so as to give it a permanent polarity.

5. An iron bar with permanent polarity, when placed anywhere in the plane of the magnetic equator, may be deprived of its magnetism by a blow.

6. Iron is rendered magnetical if scoured or filed, *bent or twisted*, when in the position of the magnetic axis, or near this position; the upper end becoming a south pole and the lower end a north pole; but the magnetism is destroyed by the same means, if the bar be held in the plane of the magnetic equator.

9. A bar-magnet, if hammered when in a vertical position, or in the position of the magnetic axis, has its power increased, if the south pole be upward, and loses some of its magnetism if the north end be upward.

10. A bar of soft steel, without magnetic virtue, has its magnetism of position fixed in it, by hammering it when in a vertical position; and loses its magnetism by being struck when in the plane of the magnetic equator.

Powell[1] following Scoresby was apparently the first to deal with the effect of torsion on the magnetisation of an iron bar placed in the magnetic meridian but inclined at different angles. He also showed that a straight bar magnetised and then bent loses a great part of its magnetisation. De Haldat[2] observed that sound vibrations have less effect than irregular impulses in destroying magnetisation, but according to Wertheim his views on the influence of torsion are incorrect. A memoir by E. Becquerel of which the title, Wertheim says[3], will be found in the *Comptes rendus*, February 9, 1845, had not yet been published when Wertheim wrote, but Wertheim was able to state briefly one of Becquerel's conclusions:

Un fil de fer doux est chargé d'un poids à son extrémité inférieure, et une partie de ce fil vertical est placée au centre d'une spirale dont le circuit comprend un galvanomètre; on observe un courant de même sens pour toutes les torsions, que celles-ci aient été effectuées dextrorsum ou sinistrorsum, et un courant de sens opposé pour toutes les détorsions quel que soit leur sens (p. 387).

[812.] Finally Wertheim cites a note of Matteucci to Arago[4]. This ought to have been noticed in connection with our Arts. 1333*–36*; It gives an earlier date to several of Matteucci's results published in the memoir of 1858: see our Art. 701. In it Matteucci arrives at the following conclusions:

(i) A bar of soft iron or steel being magnetised by the passage round it of a spiral current, the first torsions of the bar increase the strength of the magnetisation.

(ii) This effect is independent of the sense in which the torsion is applied, i.e. whether it is in the same or the opposite direction to the current.

(iii) When the current has ceased the same torsions tend to decrease the magnetism, and this whether they are applied immediately after the cessation of the current or several days after.

(iv) If the same mechanical stresses be applied at short intervals successively they cease to have the same magnitude of effect.

Le magnétisme acquis par les mêmes actions de torsion, données successivement soit dans un sens, soit dans le sens opposé, soit alternativement, va toujours en diminuant; si l'on continue toujours, on voit apparaître les signes du magnétisme qui se détruit qui sont remplacés par des signes du magnétisme qui s'accroît, et tous ces faits oscillent dans les mêmes limites (p. 388).

[1] Thomson's *Annals of Philosophy*, Vol. III., 1822, pp. 92–5.

[2] *Annales de Chimie*, T. XLII., 1829, pp. 39–43.

[3] I cannot find even the title of this memoir in the *Comptes rendus* for 1845. The memoir in T. XX. (pp. 1708–11), contains nothing material to the present point and was read on June 9.

[4] *Comptes rendus*, T. XXIV., 1847, p. 301.

(v) When the current has ceased the same repeated actions rapidly destroy the magnetism.

[813.] Wertheim in a note communicated to the Academy in 1852 and printed in the *Comptes rendus*, T. xxxv. p. 702, had announced results not in perfect accord with Matteucci's and he repeats the contents of this note in the present memoir. They are as follows:

(i) In so far as a bar of iron has not attained a state of magnetic equilibrium torsions and detorsions[1] act upon it as all other mechanical disturbances, i.e. they tend to facilitate its magnetisation when under the influence of a current or terrestrial magnetism, and they tend to facilitate its demagnetisation when it is under no such influence.

(ii) In both cases as soon as magnetic equilibrium is established, whether the bar be or be not under the influence of magnetic induction, all elastic torsion, whatever be its sense, produces partial demagnetisation, while elastic detorsion restores the primitive magnetisation.

(iii) When an iron bar or a bundle of iron wires under the action of a current or terrestrial magnetism receives a large torsional set, then all elastic torsion or detorsion which is applied to it in the sense of the torsional set produces a partial magnetisation, and all elastic torsion or detorsion in the opposite sense produces a demagnetisation (p. 389): see our Art. 815, (xiv) and (xv).

[814.] Wertheim in the memoir under consideration discusses the experiments which confirm the results of the previous article. He gives in addition certain amplifications and corrections of them. Among the latter we may note:

(iv) The purely mechanical actions of torsion and detorsion are in themselves insufficient to magnetise iron (p. 401). This result, as Wertheim remarks, is initially probable.

(v) The torsion of a bar under magnetic influence enables it to take a much greater permanent magnetisation than it would otherwise be capable of (p. 401).

(vi) When the bar has taken all the temporary and permanent magnetisation of which it is capable under the action of the given external magnetising force, then torsion diminishes, and the corresponding detorsion restores its magnetisation (p. 401).

This is only an ampler statement of (ii).

(vii) When the external magnetising influence is removed torsion and detorsion (as other mechanical disturbances) rapidly destroy the temporary magnetism, but they continue indefinitely to exercise in-

[1] *détorsion*, by which I presume Wertheim means a release from a state of torsion, not a *negative* torsion, but his language is obscure.

fluence on the permanent magnetism, i.e. the latter is diminished by the torsions and restored by the detorsions (p. 401).

This is an amplification of (ii) and it is important to notice that Wertheim now makes a distinction between a *temporary* and a *permanent* magnetisation.

(viii) The effect of torsion is generally greater than the opposite effect of detorsion (p. 401).

This may possibly have been only apparent, i.e. due to Wertheim's mode of experimenting.

(ix) Whatever may be the magnetic state of the bar, provided it be one of equilibrium, the effects of the torsions are proportional to the angles of torsion, but the magnitude of these effects appears to depend more on the magnitude of the permanent than on that of the temporary magnetisation (pp. 401–2).

Wertheim follows up these results (pp. 402–4) by some remarks on the different effects produced by torsion on different materials, e.g. soft iron, hard iron and untempered cast steel (see our Art. 703).

(x) The effects of torsion diminish with the elapse of time as the iron loses a part of its magnetisation. There appeared however to be a limit to this diminution as iron bars of any quality gave perceptible magnetic results when twisted six months after their magnetisation (p. 407).

(xi) Je dois faire remarquer ici une anomalie que j'ai observée plusieurs fois et qui me semble tout à fait inexplicable : elle consiste en ce que les fers durs donnent souvent, immédiatement après l'interruption du courant, des déviations plus fortes qu'ils n'en avaient fourni tant que le courant passait ; dans ces cas la diminution ne se fait sentir qu'après quelque temps (p. 407).

The 'deviations' referred to are those of a galvanometer, connected with a coil round the bar, and were caused by the induced currents whereby Wertheim measured the changes in magnetisation of the bar. The further current, which he himself mentions in (xi.), is that which produced the magnetising force on the bar.

(xii) Wertheim was unable to obtain any sensible results in the case of torsion applied to diamagnetic bodies (p. 407).

[815.] The next points to which Wertheim turns are of considerable interest. Suppose the torsional set to be zero or negligible, then suppose any elastic torsional strain given to the bar and let it be magnetised in the strained state. Will the magnetisation be a maximum in this state, in the state of zero strain, or in any other state? Wertheim found that:

(xiii) The maximum of magnetisation always coincides with the position of zero strain (p. 409).

He next turned to the problem of torsional set. Set he found exercised no influence, if it preceded magnetisation. But supposing the set was applied during the time the bar was under the influence of magnetising force, what would be the position of maximum magnetisation; would it coincide with the position in which the bar would have no elastic torsional strain? The angle between the positions of zero torsional couple and maximum magnetisation is termed the *angle de rotation du maximum.* Wertheim found that for harder sorts of iron (*fer dur, ou même demi-dur*) very large torsional sets were not necessary in order that this angle of rotation should be sensible; on the other hand it was very difficult to obtain sensible measurements when soft iron bars and not wires were used. A table of numerical results is given on p. 413. We may note:

(xiv) The angle of rotation is less than the elastic limit to torsional strain measured from the new position of zero elastic strain, and is in the direction of the torsional set.

(xv) Torsional strains when less than this angle of rotation produce increasing magnetisation, when greater than this angle decreasing magnetisation, which becomes less than the magnetisation at zero strain for double the angle of rotation (pp. 411–12).

It will be noted that (xiv) and (xv) sensibly modify (iii) of Wertheim's *Note* of 1852: see our Art. 813. The latter statement is only true provided the torsions do not exceed double the angle of rotation.

[816.] Wertheim now turns to the last of his experimental investigations. A bar having been given an elastic torsional strain while under the influence of the magnetising force, what will be the character of its magnetism when the magnetising force is suddenly removed? He found that:

(xvi) For all qualities of iron the effect of removing the magnetising force (stopping the current in the coil) while there is an elastic torsional strain is to rotate the position of maximum magnetisation in the direction of the temporary strain, but the angle of rotation is always less than the angle of this torsional strain (p. 414).

The phenomena of (xiv), (xv) are especially marked in hard iron, those of (xvi) in soft iron. Some additional information will be found on pp. 414 and 419, while pp. 415–8 are occupied with tables of the experimental results.

[817.] On pp. 419–428 Wertheim discusses how far the phenomena he has described can be accounted for by any known theory of magnetism. His results, as might be supposed, are negative. Thirty years later we have hardly reached a really valid theory of the relation between strain and magnetism, although we see more exactly their physical relations. The two-fluid theory, the *force coercitive* of Coulomb, or even the hypothesis of Matteucci—that the magnetic effect of strain is a secondary effect of its change of volume—give Wertheim no aid. It is curious that Wertheim takes refuge in a wave theory of the ether. We may not be able to follow his somewhat vague reasoning, but it is not without interest to note that he holds that magnetisation as a polarisation—or a bringing into concordance—of pre-existing discordant ether-vibrations surrounding the molecules, is a hypothesis far better fitted than those before cited to account for magnetic phenomena.

The memoir concludes with some rather general remarks on the effect of earth-strain (produced by other celestial bodies, change of temperature, earthquakes etc.) on terrestrial magnetism, and on possible methods of correcting compass-deviations in iron ships.

[818.] Historically the importance of this memoir of Wertheim's seems considerable. He noticed a number of novel phenomena, although he did not see the necessary limitations to some of his statements—in particular he did not discover the existence of a 'critical twist,' except in so far as this is implied by (xiv)-(xvi) for the cases of *previous* torsional set under magnetisation or of elastic torsional strain with sudden cessation of the magnetising force. Wertheim's results must therefore be read with due regard to more recent researches: see the references to *Magnetisation under Stress* in the Index to this Volume, also Wiedemann, *Lehre von der Elektricität*, III. S. 692, and J. J. Thomson, *Applications of Dynamics to Physics*, pp. 59–62.

[819.] Wertheim: *Mémoire sur la compressibilité cubique de quelques corps solides et homogènes. Comptes rendus*, T. LI. pp. 969–974. Paris, 1860. (Translated in the *Philosophical Magazine*, Vol. XXI., pp. 447–451. London, 1861.)

Wertheim refers to his memoir of 1848 (see our Art. 1319*) and to the value 1/3 which he there proposes for the stretch-squeeze ratio η, and which he holds has been confirmed by subsequent experiments. He remarks that several distinguished mathematicians, without doubting the accuracy of his experiments, have yet endeavoured to bring them into unison with the results of uni-constant isotropy by the aid of *hypothèses très diverses, mais malheureusement aussi très arbitraires* (p. 970). Wertheim refers in the first place to Clausius: see our Art. 1400*. Clausius did not deny the homogeneity of Wertheim's materials, but, as we have noticed, supposed like Seebeck (Art. 474) that elastic after-strain had affected his results. Wertheim rejoins that no one has yet observed after-strain in metals or glass. This statement was absolutely incorrect even in 1860: see our Arts. 726, 748 and 756.

Wertheim next remarks that he does not assert that $\eta = 1/3$ holds for all metals, but only for those upon which he has experimented. This, I think, is not in complete accordance with his earlier statements, but it allows him to maintain that he is not in disagreement with Lamé and Maxwell, nor even with Clapeyron's results for vulcanised caoutchouc: see our Arts. 1163*-4*, 1537* (and footnote) and 610.

He next proceeds to criticise Saint-Venant's hypothesis of aeolotropy, or rather of varied distributions of elastic homogeneity (see our Arts. 114, *et seq.*), in language which suffices to prove that he has not understood it.

Finally Kirchhoff's memoir of 1859 (see Section II. of our Chapter XII.) with its direct determination of η for brass and tempered steel is discussed. Wertheim holds that Kirchhoff's apparatus and his mode of experimenting were likely to produce error (*sont autant de circonstances fâcheuses:* p. 973).

He takes comfort in the fact that the mean of the values given by Kirchhoff for η (·294 for tempered steel and ·387 for brass) is not very far from 1/3. He will not affirm that $\eta = 1/3$ for steel, but he holds that Kirchhoff's experiments do not demonstrate its improbability. Putting aside Clapeyron's experiments on caoutchouc, Wertheim sees no fact that has yet been deduced to show that η varies from body to body. He promises to present shortly a memoir to the Academy on this subject.

[820.] The last mentioned memoir (*Expériences sur la Flexion?*) has never, so far as I know, been published. Scarcely a month (January 19, 1861) after the presentation of the memoir discussed in the last article Wertheim in a fit of melancholy committed suicide by throwing himself from the tower of the Cathedral at Tours. A bibliography of Wertheim's papers and some criticism of his methods by Verdet will be found in *L'Institut*, T. XXIX., pp. 197–201, 205–9 and 213–6. Paris, 1861.

GROUP D.

Memoirs on the Vibrations of Elastic Bodies[1].

[821.] A. Baudrimont: *Recherches expérimentales sur l'élasticité des corps hétérophones. Annales de chimie et de physique*, T. XXXII., pp. 288–304. Paris, 1851.

[1] See also Arts 433–9, 471–4, 510, 534, 539–41, 546–8, 550–9, 583, 612–7, 680–2, 722–86 *passim*, and 809 of this Chapter.

Baudrimont uses the word *isophone* to denote a body, the elasticity of which is the same in all directions, or an isotropic body; *hétérophone* he applies to aeolotropic bodies, but especially to bodies having axes or planes of elastic symmetry. The object of this paper is to present a preliminary investigation of the notes given by rods vibrating laterally. Baudrimont's ultimate object, however, is to calculate the stretch-moduli in different directions of an aeolotropic material by means of the notes given by the lateral vibrations of rods so cut from the material that their axes are in the given directions.

It is first needful to ascertain how far, what Baudrimont terms Euler's formula, is accurate for such rods. This formula gives for the frequency f:

$$f = n^2 \frac{\kappa}{l^2} \sqrt{\frac{E}{\rho}} \quad \text{.............................(i)},$$

where n is a mere numeric depending on the graveness of the note, E is the stretch-modulus in the direction of the length l of the rod, ρ the density of the material and κ the swing-radius of the cross-section about an axis through its centroid perpendicular to the plane of vibration (see Lord Rayleigh's *Theory of Sound*, Vol. I. § 171). Thus for the gravest note of a given material the frequency varies directly as the swing-radius and inversely as the square of the length.

Equation (i) is obtained theoretically: (*a*) by supposing the cross-sections to remain plane after bending and perpendicular to the axis of the rod, (*b*) by assuming the rod not to diverge much from absolute straightness, and (*c*) by concentrating the inertia of each cross-section at its centroid.

Baudrimont by a series of experiments on ice, metal, quartz, and wooden bars, believes that he has demonstrated that the laws which hold for isotropic and aeolotropic bodies are the same, but that the frequency of the notes is not inversely as the square of the length of the rod.

Lord Rayleigh has given a correction for the rotatory inertia of the cross-section (*Theory of Sound*, Vol. I. § 186), but assuming Baudrimont's experimental results to be true[1], this correction is very far from accounting for the divergence between Euler's formula and physical fact, even when the ratio of length to diameter is as great as 30, 40, or even 50. The correction is in the right direction but not nearly large enough. It is obvious that we must for sound vibrations accept the assumption (*b*). Hence if we are to trust Baudrimont's results, the formula obtained from the Bernoulli-Eulerian theory for the notes of rods is very inaccurate so long as the ratio of length to diameter of a

[1] I suspect some large source of error, which might possibly have arisen in clamping the rods. See the remarks on the difficulty of determining the stretch-modulus by lateral vibration, in a memoir by Wertheim in the *Annales de chimie et de physique*, T. XL., p. 201. Paris, 1854.

rod is less than 30 to 40. The complete theory, which ought to be deduced from the general equations of elasticity, would like Saint-Venant's theory of the flexure of beams take account of slide; it would be interesting to ascertain the order of the modification such a theory would introduce into the expression for the frequency: see for the case of *longitudinal* vibrations Chree, *Quarterly Journal of Mathematics*, Vol. XXI., p. 287. If we accept Baudrimont's results it is obvious that the stretch-modulus as calculated from Euler's formula for lateral vibrations must diverge very considerably from that obtained by pure tractional loading, except when the length of the rod is immensely greater than its diameter. Such rods it would be difficult to procure in many of the aeolotropic bodies (crystalline materials for example) whose elasticity Baudrimont proposes to investigate by the method of transverse vibrations.

[822.] Montigny: *Procédé pour rendre perceptibles et pour compter les vibrations d'une tige élastique. Bulletins de l'Académie Royale...de Belgique*, T. XIX. 1re Partie, pp. 227–50. Bruxelles, 1852.

This is an extension of a method suggested by Antoine (*Annales de chimie et de physique*, T. XXVII. pp. 191–209. Paris, 1849) of rendering sonorous vibrations visible by combining a motion of translation with that of vibration. Montigny used the following arrangement to render visible the vibrations of a rod:

Si l'extrémité de la tige autour de laquelle les vibrations doivent s'effectuer, est fixée perpendiculairement à un axe de rotation, si, lors de sa révolution rapide, l'extrémité libre éprouve un choc contre un obstacle fixe, les vibrations transversales de la tige, excitées de cette manière dans le plan de sa révolution, la rendent visible sur toute sa longueur dans des positions rayonnant du centre, et qui se trouvent également espacées (p. 228).

After some general reasoning as to what it is the eye really sees in this combination of motions Montigny concludes that:

Il résulte de là que l'œil ne perçoit la tige qu'à chaque vibration double, et que nous devrons prendre pour le nombre des vibrations simples, effectuées dans un temps donné, le double des images de la tige perçues pendant le même intervalle de temps (p. 229).

The rotation round the axis is so arranged that after a complete revolution the rod returns to visibility at the same position as before; this can always be obtained by quickening or slackening its spin. Hence if t be the time of a revolution and n the number of images of the rod seen, the number of vibrations of the rod will be $2n$ and their period $t/2n$. The positions of visibility arise where the velocity due to the spin is almost equal and opposite to the velocity of an element of the rod due to lateral vibration. It seems to me that Montigny is using the word vibration for what we in England term the *half* oscillation and that with our terminology we should have the period of oscillation equal to t/n.

Montigny applies his method to verifying some of the well-known theoretical laws of the vibration of rods. His experiments confirming theory are thus opposed to the results obtained by Baudrimont and cited in our Art. 821. This difference between his own and Baudrimont's results our author discusses at some length (pp. 241–7), and the inference certainly is that there was some large source of error in Baudrimont's experiments. The memoir concludes by noting how the new method may be rendered available for technical purposes, for example, in finding the stretch-moduli in the case of iron and wood[1].

[823.] A. Masson : *Sur la corrélation des propriétés physiques des corps. Annales de chimie et de physique,* T. LIII., pp. 257–93. Paris, 1858. This memoir was presented to the Academy, March 2, 1857. It is only the first chapter of the *Première Partie* (*Vitesse du son dans les corps*) with which we are concerned. Masson after some slight discussion of the relation between the stretch-modulus, the coefficient of thermal expansion, the specific heat of a material and the mechanical equivalent of heat,—which is based upon Kupffer's erroneous hypothesis (see our Arts. 724–5 and 745),—proceeds to describe the experiments by which he has measured the velocity of sound in metals (pp. 260–4).

As he had previously found that the velocity of sound deduced from the longitudinal vibrations of a metal rod increased with the

[1] While referring to memoirs dealing with methods of rendering vibrations visible I may note the following paper which escaped record in its proper place in our first volume :

E. F. August : *Ueber einige isochrone Schwingungen elastischer Federn. Zwei Abhandlungen physicalischen und mathematischen Inhalts.* Berlin, 1829. This was published in the *Program des Cölnischen Real-Gymnasii.* It contains some account of simple school experiments for proving Taylor's laws for vibrating strings (here represented apparently by fine brass wire spirals) with no more complex apparatus than the stand of an Atwood's machine. The effect of isochronous vibrations is rendered visible by the oscillating of the machine violently for one length only of the spring under a given load. The paper concludes (S. 4–10) with a rather clumsy demonstration of the formula for the period of vibration of a weight suspended by such a spring and with some experimental confirmation of its accuracy.

length of the rod, the diameter remaining constant, he replaced the rod by a wire of very small diameter: see our Art. 821. He took wires as a rule of 1·5 metres length and of diameters form ·1 to ·9 mm. The wires were placed horizontally and kept stretched, one end being passed over a pulley and attached to a weight. The vibrations were measured by the aid of a sonometer, and in all the experiments the periods of a great number of harmonics as well as that of the fundamental vibration were measured on each wire.

[824.] The densities for a number of metals are tabulated on p. 263, and the corresponding velocities of sound are given on p. 264. These velocities were found for gold, brass, copper, silver, platinum, iron, zinc, lead, tin, aluminium, cadmium, palladium, steel, cobalt and nickel. Direct experiments were also made on some of these metals to find their stretch-moduli. Masson gives the following among other results on p. 264:

Stretch-modulus in kilogrammes per sq. millimetre.

	From Sound Experiments.	From Traction Experiments.
Gold	8247	6794
Brass	9783	9446
Silver	7421	7080
Platinum	16932	15924
Iron	19993	18571

These are in general agreement with Wertheim's results except in the case of iron: see our Art. 1297* and compare with Art. 728.

[825.] (*a*) A. Terquem: *Note sur les vibrations longitudinales des verges prismatiques. Comptes rendus*, T. XLVI. pp. 775–8. Paris, 1858.

(*b*) Same author and title. *Comptes rendus*, T. XLVI. pp. 975–8. Paris, 1858.

(*c*) *Idem: Étude des vibrations longitudinales des verges prismatiques libres aux deux extrémités. Annales de chimie et de physique*, T. LVII. pp. 129–190. Paris, 1859.

(*d*) J. Lissajous: *Note sur les vibrations transversales des lames élastiques. Comptes rendus*, T. XLVI. pp. 846–8. Paris, 1858.

(*e*) J. Bourget et Félix Bernard: *Sur les vibrations des membranes carrées. Premier Mémoire. Annales de chimie et de physique*, T. LX. pp. 449–479. Paris, 1860.

These five memoirs deal with the nodes of vibrating bars and the nodal lines of square membranes, and so belong more particularly to the theory of sound[1]. They contain, however, references to the elastical researches of Wertheim, Savart, Germain, Poisson, Lamé etc., while (*a*) and (*b*) ought to have had a reference to Seebeck: see our Arts. 471–5. The papers (*a*), (*b*) and (*c*) have special application to the case in which a rod is able to vibrate longitudinally and transversely with the same tones and to the

[1] I may take this opportunity of referring to a memoir on the nodal lines of plates which escaped my attention in the first volume.

It is by Giovanni Paradisi and entitled: *Ricerche sopra la vibrazione delle lamine elastiche. Mem. dell' Accad. delle Scienze di Bologna*, T. I. P. 2, pp. 393–431. Bologna, 1806.

The memoir is among the earliest which followed the publication of Chladni's researches. The author made experiments on plates of rectangular (including square) and equilateral form, the material of the plates being glass, brass, silver, tin, wood (walnut and maple) and bone. The material was observed to influence the note but not the nodal lines. The author found that the nodal lines (*le curve polvifere*) and the centres of vibration (*centri di vibrazione*) were such that the point of support of the plate might be anywhere on a nodal line and the point of disturbance (*punto del suono, il centro primario*) at any other of the centres of vibration (*centri secondari di vibrazione*) without any change in the system of nodal lines. By the centres of vibration 'primary and secondary' Paradisi appears to denote the points of maximum vibration corresponding to the loops in the vibrations of a rod or string. Paradisi asserts that with the same point of support and the same centre of disturbance plates can be made to vibrate with one, two or more different tones, according to the manner in which the vibrations are excited and that each such tone has a different system of nodal lines (pp. 416–9);

> Dallo stesso triangolo sospeso nel centro, e suonato alla metà della base in *S*, secondo che si preme più o meno l' arco, ricaviamo un tuono diverso; talvolta un tuono acuto, talvolta un medio, e talvolta un grave. Questi tre tuoni i quali sono i soli che possono ricavarsi dal punto *S* dispongono la polvere in tre diverse maniere (p. 417).

He supposes that the nodal lines must be due to one or other of two causes; (1) that they are the locus of points at which the plate is at rest, (2) that they are the locus of points at which, although the points themselves are in motion, the forces on the grains of powder are in equilibrium (p. 397). He chooses the latter alternative, notwithstanding his experimental demonstration that the nodal system remains unchanged if the points of the vice which supports the plate be moved along a nodal line. His arguments in favour of this alternative are far from convincing and his comparison of the nodal lines and centres of disturbance in plates with wave motion in strings and water is unsatisfactory (pp. 399–401, 404–5). His diagrams showing the manner in which the lines of powder are gradually formed in experiment are however interesting.

Finally his attempt to form on his hypothesis a differential equation connecting the nodal lines with the centres of vibration may be dismissed as absolutely fruitless. It is based upon the assumption that the *unknown* force on any grain of powder upon a nodal line is along the tangent to that nodal line, which would cause the powder to move along the nodal line and not remain at rest there (pp. 429–31).

resulting nodes. Lissajous confirms Terquem's conclusions by a very different experimental method. The excitement of transversal tones by longitudinal vibrations had been already noted by Savart[1]: see our Art. 350*. Bourget and Bernard give interesting figures of the nodal lines of square membranes.

The title of another acoustic paper by Terquem bearing on a kindred subject but belonging to the next decade may be cited here:

(*f*) *Note sur la co-existence des vibrations transversales et tournantes dans les verges rectangulaires. Comptes rendus*, T. LV. pp. 283–4. Paris, 1862.

[826.] J. Lissajous: *Mémoire sur l'étude optique des mouvements vibratoires. Annales de chimie et de physique*, T. LI. pp. 147–231. Paris, 1857. This classical memoir deserves at least a reference here. By means of the image of a bright point reflected from a small mirror attached to a vibrating elastic body, the image being given a translatory or oscillatory motion perpendicular to the direction of the vibration produced in it by the vibrating body, we obtain an optical representation of the vibrations of the body. Lissajous shows how vibrations may be analysed, and vibrations in the same or perpendicular directions optically compounded. His methods are as important for the investigation of the vibrations of large masses of elastic material as for the ordinary purposes of acoustics.

[827.] F. P. Le Roux: *Sur les phénomènes de chaleur qui accompagnent, dans certaines circonstances, le mouvement vibratoire des corps. Comptes rendus*, T. L. pp. 656–7. Paris, 1860.

This note draws attention to the fact that if a vibrating rod of wood, ivory, steel etc. be clamped at a point which is not a node of the free vibrations, this point rapidly rises in temperature. Various experiments are described by which this rise in temperature can be easily rendered sensible.

Le Roux concludes that when any vibratory motion is damped, the kinetic energy of the vibrations will be converted into heat in the neighbourhood of the parts damped.

[1] The subject is briefly referred to by Lord Rayleigh: *Theory of Sound*, Vol. I. § 158.

GROUP E.

Elastic After-Strain in Organic Tissues.

[828.] E. Weber: *Muskelbewegung*, S. 1–122, Zweite Abtheilung, Bd. III. of R. Wagner's *Handwörterbuch der Physiologie*. Braunschweig, 1846. This article contains a considerable number of experiments on the elasticity of muscle (S. 70–99) and some attempt to explain their elastic action (S. 100–117). The treatment, however, is rather physiological than physical. The general results of the writer as to the elasticity of muscle are given in S. 121–2, and we cite the following:

27. Die Thätigkeit des Muskels besteht nämlich nicht nur in einer Aenderung seiner (natürlichen) Form, die sich verkürzt, sondern auch in einer Aenderung seiner Elasticität, die sich vermindert.

28. Weil die Elasticität des Muskels sich beim Uebergange zur Thätigkeit beträchtlich vermindert, übt ein Muskel durch seine Verkürzung eine weit geringere Kraft aus, als er ausüben würde, wenn seine Elasticität unverändert dieselbe wie im unthätigen Zustande bliebe.

29. Die Elasticität des thätigen Muskels ist sehr veränderlich; sie vermindert sich bei Fortsetzung der Thätigkeit immer weiter. Diese fortschreitende Abnahme der Elasticität bei fortgesetzter Thätigkeit ist die Ursache der Erscheinungen der Ermüdung und der grossen Kraftlosigkeit, welche die Muskeln während derselben zeigen.

Weber also points out that the elasticity is more imperfect in dead than living muscle, and that there is a great difference in the general elastic properties of the two conditions.

[829.] W. Wundt: *Ueber die Elasticität feuchter organischer Gewebe. Archiv für Anatomie, Physiologie und wissenschaftliche Medicin.* Jahrgang 1857, S. 298–308. Berlin, 1857.

After referring to the experiments of Wertheim and E. Weber (see our Arts. 1315* and 828) Wundt remarks that these experiments leave us without any simple conception of the stretch-modulus in the case of moist organic tissues, and that we are thrown back on an empirical stress-strain diagram. At the same time he takes exception to Wertheim's experimental methods, chiefly on the grounds that they were made too long after the death of the tissue-bearing individual, and that sufficient regard was not paid to the time-element, which is so important a factor in the elastic after-strain of such tissues.

Wundt's objects in this paper are to measure: (i) the ultimate extensions by given loads, and (ii) the temporary extensions in given intervals of time.

[830.] On S. 301–3 Wundt describes his apparatus, especially his means for keeping the tissue moist. As to his results he concludes that *the ultimate extensions are proportional to the loads*, but he comes to no definite conclusions as to after-strain (*die vorliegende Untersuchung hat zu keinem für die Kenntniss der elastischen Nachwirkung bemerkenswerthen Resultat geführt*, S. 303).

The following diagram clearly indicates the results of experiments on a frog's muscle of 2·79 mm. length; AB is the stretch-traction curve for ultimate extensions, where two abscissa-divisions represent 1 gramme and

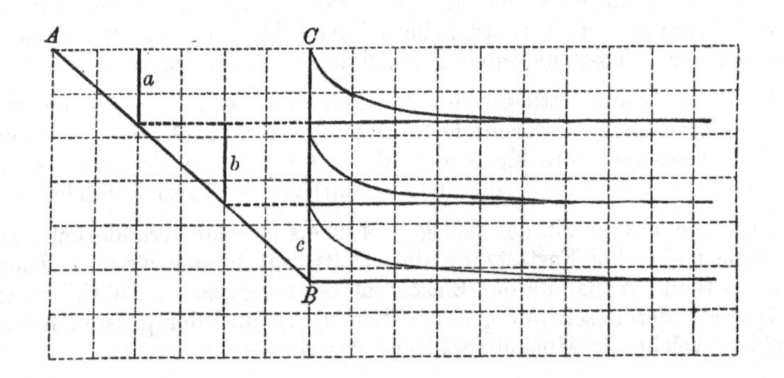

the ultimate extension for 1 gramme = ·272 mm. The three heavy line ordinates a, b, c, respectively ·272 mm., ·254 mm. and ·242 mm. long, are projected on CB and, one abscissa-division measuring 10 minutes, the after-strain curves for these three loadings are given to the right of CB, so that after each increase of load we see the extension gradually increasing up to linear elasticity. Wundt points out that the line AB is within the limits of experimental error straight, and his after-strain curves are distinctly of interest. He concludes also that the limits within which this proportionality of traction and final stretch holds are wider the fresher is the tissue and the less it has been previously loaded. The following are the stretch-moduli in grammes per sq. millimetre, the loads being from 1 to 10 grammes and the temperature 10° to 15° C.:

Artery 72·6; muscle 273·4; nerve 1090·5; sinew 1669·3.

The experiments were made on artery (calf), muscle (ox and frog), nerve (calf), tendon (calf), but Wundt only gives details of some few of them (S. 307–8).

[831.] A. W. Volkmann: *Ueber die Elasticität der organischen Gewebe. Archiv für Anatomie, Physiologie u. s. w., herausgegeben von C. B. Reichert und E. du Bois-Reymond*, Bd. I. S. 293–313. Leipzig, 1859.

W. Wundt: *Ueber die Elasticität der organischen Gewebe. Zeitschrift für rationelle Medicin von Henle und Pfeufer*, Bd. VIII., S. 267–279. Leipzig, 1860.

These papers relate to a controversy of considerable interest upon the exact form taken by the stress-strain relation for organic tissues. We have already referred in our first volume to Wertheim's researches (see our Art. 1315*), but his chief results may be cited here in order that the reader may understand the point in dispute between Volkmann and Wundt. They are as follows

(i) Wertheim recognised after-strain to exist in human tissues. He found it to vary with their dryness but to be only a very small proportion of the total strain when the latter was measured in the first few minutes after loading.

(ii) He represented the immediate stress-strain (stretch-traction) relation by an equation of the form:

$$s_x^2 = a\widehat{xx}^2 + b\widehat{xx}.$$

If a be positive as Wertheim found it, the stretch-traction relation is thus hyperbolic[1]. Set was excluded from the measurement of strain (see our Arts. 1315*–18*).

These experiments of Wertheim are in agreement with those of E. Weber, who also found that the stretch-traction relation for muscles was not linear: see our Art. 828.

[832.] As we have seen W. Wundt published in 1857 a paper (see our Art. 829), in which he asserted that if regard were only paid to elastic after strain, it would be found that the stress was proportional to the strain for organic tissues.

It is at this point that Volkmann took up the matter, and made an attempt to measure elastic fore-strain by itself (see Vol. I. p. 882). He adopted an ingenious method of tracing by a *Kymographion* the longitudinal vibrations of a muscle or artery suspended vertically and suddenly loaded but without any impact. The load then oscillated about the mean position which was that of statical equilibrium. This mean position altered with the time the weight was left oscillating owing to elastic after-strain. The mean positions are those of maximum speed in the oscillating weight, and they correspond to the points of inflexion on the diagram of the oscillations which is drawn on the revolving cylinder of the kymographion. The first point of inflexion ought to give the elastic fore-strain. Unluckily Volkmann found that

[1] The hyperbolic form of this curve is really confirmed by the researches of C. S. Roy: see the *Journal of Physiology*, Vol. III. pp. 125–59, corrected Vol. IX. pp. 227–8. Cambridge, 1880 and 1888.

the points of inflexion were not easily determined with great accuracy, as slight errors in the motion of the cylinder or of the tracing pen influenced their position. Under these circumstances he gave up the idea of measuring the first mean position, and contented himself in each series of experiments by measuring all the strains at the *same* small interval of time after the instant of loading.

In eight series of experiments he compares his observed stretches with those given by Wertheim's relation in (ii) above. The result is a very close agreement. Volkmann finds for silk thread, for human hair, for an artery, for a nerve (*nervus vagus* of man) that a is positive; on the other hand for muscle it is negative, or the stretch-traction relation is *elliptic*. Permanent set appears to have been sensible only in the final experiments of any series. In the last series of experiments (S. 307) Volkmann subtracted the set before applying Wertheim's formula and again found it to hold. He thus considers that formula to be proved for elastic fore-strain, i.e. for primary strain within the elastic limits.

I may note that Volkmann seems to think this stretch-traction relation something peculiar to organic bodies, distinguishing them from inorganic bodies. But as we have seen (see our Vol. I. p. 891) that the stretch-strain relation within the elastic limits for certain metals is not linear—whatever else it may be,—it is not necessary on this account to suppose that an absolute distinction must exist between the elasticity of organic and of inorganic substances.

[833.] The remainder of the paper is a criticism of Wundt's experiments, chiefly based on the ground that the time-element had not been taken into account, and that accordingly the strain measured by him was neither fore- nor after-strain. Further Volkmann holds that Wundt's experiments cover such a small range of loads, that for that range the stress-strain curve might approximately be taken as straight. An attempt to show that some of Wundt's experiments contradict his own hypothesis is, I think, fairly met in Wundt's reply.

[834.] In Wundt's reply, the title of which I have given in Art. 831, he does not I think do justice to the care with which the experiments of Wertheim and Volkmann appear to have been conducted. Against E. Weber's and Wertheim's results Wundt cites their want of caution in drying the tissues, in noting the influence of set, the effect of physical change (as *rigor mortis*), and above all the existence of after-strain, which he asserts was left out of consideration. Now it seems to me that Wertheim does reckon with all these factors and especially refers to the latter: see our Art. 1317*. Wundt suggests also that the weights applied were such as to change the elastic modulus of the body, i.e. its elastic constitution.

He defends his own limited range of loads on the ground that only for such loads as he applied do set and elastic after-strain cease to be so considerable that elastic fore-strain can be easily measured. (We may note here that Wundt seems to consider that want of proportionality between stress and strain would certainly mark organic substances sharply off from inorganic substances!) He objects to Volkmann's method of getting rid of after-strain by making his measurements of stretch at a constant interval after loading. He remarks that the empirical formula given by W. Weber for after-strain involves a constant itself depending on the load (the c of our Art. 714*). He complains also that Volkmann gives no evidence that he has proved the unaltered condition of the elasticity of his material after each experiment by allowing it to return to its original condition as to load. This, he holds, is especially necessary to free successive experiments from the after-strain of previous ones.

I do not think these objections of Wundt have really great force, because Volkmann's observations were made while the elastic after-strain was an exceeding small quantity, and because his notice of the existence of set shows that he must have examined whether his tissues returned to their original lengths. A further and supposed conclusive argument of Wundt's against Volkmann, namely, that the elliptic nature of the stress-strain relation would prove that by increase of load the muscle would ultimately be shortened instead of extended, is simply absurd. What the formula really denotes is the elliptic form of the relation *within the limits of elasticity*. Had Wundt examined the values of the constants given by Volkmann he would have found that the extension would have become enormous—far beyond the limits of rupture—before the stretch began to decrease with the load.

[835.] On S. 274—6 Wundt compares the accuracy for tissues of Hooke's law as deduced from his own experiments, with its accuracy as deduced from Wertheim's. But it is no argument to assert that because the former experiments give results less divergent from Hooke's law than the latter do, therefore this law must hold for the latter as well as the former. The fact is that Hooke's law may hold for neither within the range of loads applied.

On S. 277 we are treated to a proof—by means of Taylor's Theorem (!!)—that stress *must* be proportional to strain for all bodies whatever within certain limits of strain: see our Arts. 928* and 299.

Having deduced from Taylor's theorem that the stress-strain relation to a second approximation *must* be of the form:

$$\widehat{xx} = As_x + Bs_x^2,$$

Wundt, by squaring and some absolutely illegitimate process of neglecting the cube in preference to the fourth power, deduces that

$$\widehat{xx}^2 = A^2(s_x^2) + B^2(s_x^2)^2.$$

This he naïvely remarks is an equation of the hyperbolic form of Wertheim and Volkmann, whose results he then attributes to the fact that the strains considered by them exceed those for which the first term of Taylor's series suffices. It is perhaps needless to remark that Wundt, if a good physiologist, is but a poor mathematician and physicist.

Group F.

Hardness and Elasticity.

[836.] We now reach a number of memoirs dealing with the hardness of materials, a subject intimately related to their elasticity and strength. It will perhaps not be out of place here to refer briefly to the earlier history of the subject. I owe my references chiefly to M. F. Hugueny's *Recherches expérimentales sur la Dureté des Corps*[1] and to the memoir of Grailich and Pekárek[2], but I have in every case consulted the original authorities for myself, and I have often amplified the notices of these writers when the original papers were not accessible to them or had escaped them.

(*a*) Apparently the first writer to make any reference to the scientific measure of hardness and the variation of hardness with *direction* is Huyghens. In his *Traité de la Lumière* (Leyden, 1690) after suggesting a grouping of flat spheroidal molecules as suited to explain the optical phenomena of Iceland spar he continues (pp. 95–6):

Tout cecy prouve donc que la composition du cristal est telle que nous avons dit. A quoy j ajoute encore cette expérience; que si on passe un cousteau en raclant sur quelqu'une de ces surfaces naturelles, & que ce soit en descendant de l'angle obtus équilateral, c'est-à-dire de la pointe de la piramide, on le trouve fort dur; mais en raclant du sens contraire on l'entame aisément. Ce qui s'ensuit manifestement de la situation des petits spheroides; sur lesquels, dans la premiere manière, le cousteau glisse; mais dans l'autre il les prend par dessous, à peu près comme les écailles d'un poisson.

(*b*) Musschenbroek concludes his *Physicae Experimentales et Geometricae Dissertationes* (Leyden, 1729) with a chapter entitled: *Tentamen de corporum Duritiâ* (pp. 668–672), that portion of his work (*Introductio ad Cohaerentiam corporum frmorum*) to which we have referred in our Art. 28* δ. His method, of which he speaks very diffidently, was to count the number of the blows of a mass swung like the bob of a pendulum which are required in order to drive a chisel through a slab of definite thickness of the given material. He supposes that the number of blows divided by the specific gravity of the material may be taken as a measure of its hardness. He tested in this way the hardness of a great number of specimens of wood and of some of the

[1] This work will be dealt with under the year 1865.
[2] See our Arts. 842–4.

more usual metals. He gives the following ascending order of hardness for metals: lead, tin, copper, Dutch silver of small value, gold, brass, Swedish iron. Obviously Musschenbroek's definition of hardness would involve absolute strength rather than set.

(*c*) Torbern Bergman writes in 1780 that testing gems by their hardness is usual. The following passage is taken from his *Opuscula Physica et Chemica*, Vol. II., p. 104, Upsaliae, 1780 (*De Terra Gemmarum*, pp. 72–117). *English Translation*, Vol. II. *Physical and Chemical Essays.* London, 1784 (Of the Earth of Gems, pp. 107–8).

The species of gems is used to be determined by the hardness; and by that quality particularly, together with the clearness, has their goodness been estimated. The spinellus is particularly worthy of observation, which is not only powdered by the sapphire, but even by the topaz; as also the crysolith, which is broken down by the mountain crystal[1], the hardness of which seems rather to be owing to the degree of exsiccation than the proportion of ingredients. The analysis of spinellus, of crysolith, and other varieties, will sometimes illustrate the true connection; otherwise, after the diamond, the first degree of hardness belongs to the ruby, the second to sapphire, third to topaz, next to which comes the genuine hyacinth, and fourth the emerald.

(*d*) A. G. Werner in 1774 in his treatise on mineralogy gave a first scale of hardness. This was somewhat extended by R. J. Haüy in his *Traité de Minéralogie*, Paris, 1801.

In Tome I. (p. 221) Haüy defines hardness in a vague way, and gives (pp. 268–71) in four groups the substances (i) which scratch quartz, (ii) which scratch glass, (iii) which scratch calcspar and (iv) which do not scratch the latter substance. In these lists he confines himself to substances usually termed *stones*. On p. 348 of Tome III. Haüy gives the following list of the usual metals in order of hardness: lead, tin, gold, silver, copper, platinum, iron or steel. Perhaps the only importance of Haüy's work for the theory of hardness lies in the fact that he appears to have first suggested the 'mutual scratchability' of substances as a measure of their relative hardness.

Ultimately Mohs' modification of Haüy's scale was adopted by mineralogists. In his *Grundriss der Mineralogie*, 1822, Bd. I. S. 374 he gives the following order: (i) Talc; (ii) Gypsum; (iii) Calcspar; (iv) Fluorspar; (v) Apatite; (vi) Adularia; (vii) Rock Crystal (Rhombohedric Quartz); (viii) Prismatic Topaz; (ix) Sapphire; (x) Diamond. In this scale each member was able to scratch all preceding members. Mohs gave numbers to these classes and placed other bodies with decimal places between these numbers by testing the relative hardness of two nearly equally hard bodies by their resistance to a file and the comparative noise. In 1836 Breithaupt attempted to introduce a scale of hardness of 12 classes, but it does not appear to have met with any wide acceptance.

[1] The Latin version has *crystallo montana*; I suppose *rock crystal*. See also p. 113 of the *Opuscula* for a further remark on the hardness of diamond.

(*e*) The conception of relative hardness based upon the power of one body to scratch a second is evidently very unscientific. Huyghens had shown a century earlier that the hardness of a body varies with direction, and its power to scratch varies also with the nature of the edge and face. The latter fact was well brought out by a memoir of Wollaston entitled: *On the Cutting Diamond*, *Phil. Trans.* 1816, pp. 265–9. This memoir draws a distinction between *cutting* and *scratching*, which has been unfortunately lost sight of by later writers on hardness. Wollaston shows that the diamond irregularly tears the surface unless its natural edge, which is the intersection of two curved surfaces and thus a curved line, be so held that a tangent to it lies in the plane face of the material to be cut and is also the direction of motion of the diamond. The curved surfaces must also be held as nearly as possible equally inclined to the plane face. By paying attention to similar principles Wollaston succeeded in getting sapphire, ruby and rock crystal to cut glass for a short time with a clean fissure. It required a fissure of only $\frac{1}{200}$ of an inch deep to produce a perfect fracture.

Further evidence in the same direction is given by C. Babbage in his work *On the Economy of Machinery and Manufactures* (London, 1832). After some remarks (p. 9) to the same effect as Wollaston's on the proper position for working the diamond he continues:

> An experienced workman, on whose judgment I can rely, informed me that he had seen a diamond ground with diamond powder on a cast-iron mill for three hours without its being at all worn, but that, changing its direction with reference to the grinding surface, the same edge was ground down (p. 10).

(*f*) L. Pansner in a pamphlet published in St Petersburg in 1813 seems to have been the first to adopt the plan of testing minerals, not by scratching them upon each other but by means of a series of diamond and metal points. Later in a memoir entitled: *Systematische Anordnung der Mineralien in Klassen nach ihrer Härte, und Ordnungen nach ihrer specifischen Schwere*, published in both Russian and German in the *Mémoires de la Société Impériale des Naturalistes*, T. v., pp. 179–243, Moscow, 1817, we find him classifying minerals as follows: (*a*) *Adamanti-Charattomena* (scratchable by a diamond, but not by a steel graver); (*b*) *Chalybi-Charattomena* (by a steel but not by a copper graver); (*c*) *Chalco-Charattomena* (by a copper but not by a lead graver); (*d*) *Molybdo-Charattomena* (by a lead graver); (*e*) *Acharattomena* (those whose hardness cannot be tested by scratching). These classes formed by relative hardness are again subdivided according as the specific gravity of the mineral is less than 1, less than 2 etc., into (0) *Natantia*, (1) *Hydrobarea*, (2) *Di-hydrobarea*, (3) *Tri-hydrobarea* etc., etc. Pp. 183–202 (erroneously paged 173) are occupied with a table of several hundred minerals thus classified, with the specific gravities to four places of decimals. The remainder of the memoir does not relate to hardness but to a classification of inorganic substances by other physical characteristics.

Pansner was followed by Krutsch who also states that he had used metal needles to scratch bodies in his *Mineralogischer Fingerzeig*, Dresden, 1820, but I have been unable to find a copy of this work.

(*g*) The first experimentalist to obtain results of value from this method was Frankenheim in his *De crystallorum cohaesione: Dissertatio Inauguralis*, Bratislaviae, 1829. Of this work I have been unable to procure a copy. Its contents are, however, embodied and extended in the same author's later book *Die Lehre von der Cohäsion*, Breslau, 1835: see our Arts. 821* and 825*. Frankenheim's results were obtained by scratching with the metal needle held in the hand and judging relative hardness by the pressure and pull necessary to produce a scratch. He applied this method to test the relative hardness of crystalline surfaces in different directions. It cannot be said that such a method is capable of really great scientific accuracy, but we shall have occasion later to compare some of Frankenheim's results with those of other experimentalists.

(*h*) About 1822 Barnes of Cornwall had noted that a circular plate of soft iron if revolving with very great rapidity is capable of cutting the hardest steel springs and files. His experiments were repeated by Perkins in London, and accounts of them were published in most of the physical and technical journals. Further experiments were made by Darier and Colladon in 1824 (*Bibliothèque universelle des Sciences et Arts*, T. xxv. pp. 283–89, Geneva, 1824, or Schweigger's *Jahrbuch der Chemie u. Physik*, Bd. XIII. S. 340–6. Halle, 1825,) and these physicists showed that when the iron disc moved with a circumferential speed of less than 34 ft. per second it was easily torn by hardened steel, but that with a speed of 35 ft. 1 in. per second the iron began to affect the steel, till at 70 ft. per second only small fragments of iron were thrown off, although the steel was violently attacked (pp. 265–6). They further showed that the effect was not produced by the softening of the steel during the process, nor, at any rate initially, by the particles of steel which cling later to the iron disc and increase the effect. They attributed the result to the influence in some way of the impact, and supposed it to depend, not on the cohesion of the iron, but on each particle of the iron acting for itself. Chalcedony was slightly attacked and quartz was torn by the rotating iron disc (p. 287). A disc made of copper mixed with one-fifth tin produced no effect on steel, and a copper disc was itself attacked by steel even at a circumferential speed of 200 ft. per second (p. 288).

In Silliman's *American Journal of Science*, Vol. x. p. 127, and p. 397, 1826, will be found further facts with regard to the above phenomenon in letters to Silliman from T. Kendall and I. Doolittle (see also Schweigger's *Jahrbuch der Chemie und Physik*, Bd. XVII. S. 77–81, 1826). The former points out that when the iron cuts the steel, it is the latter which gets hot, but that when it fails to do so, the iron takes even a blue colour from the heat. He considers that the steel is in the process heated up to that particular heat ('black heat') at which it is easily fragile, which it is not at a less or greater heat.

He thus holds that the results are associated with a particular temperature, and attributes to the inability of copper to produce this temperature in steel its failure when rapidly rotated to cut the latter metal. I. Doolittle notes that although he could cut steel with a rotating iron plate, he found it quite impossible to cut perfectly gray and soft cast-iron.

I have cited the above results to show that in measuring relative hardness the velocity with which the graver moves is in itself of importance; hence the report referred to by Darier and Colladon that the Chinese cut diamonds with iron may not after all be so entirely mythical.

(*i*) Seebeck in a school program of 1833 (*Ueber Härteprüfung an Krystallen,—Prüfungsprogramme des Berliner Real-Gymnasiums*), of which I have been unable to procure a copy, invented a more scientific instrument for measuring hardness. He placed a loaded needle or scriber on the crystal and measured the hardness in any direction by the least weight which would just cause the needle to scratch the crystal when the latter was drawn under the point by the hand. Seebeck writes of this method:

Bei der hier angeordneten Bestimmungsmethode ist es nur der Druck der Spitze gegen die Fläche welcher gemessen wird; etwas anders würde es sein, wenn bei constantem Drucke die zum Verschieben nöthige Kraft gemessen würde; auch hier würde man wohl, wenigstens bei einem ziemlich starken Drucke zwischen den verschiedenen Richtungen des Krystalls, Unterschiede finden, andere zwar als die vorigen, aber mit ihnen zusammenhängende. Bei der Prüfung mit der Hand (Frankenheim) werden sich beide Wirkungen durch das Gefühl ziemlich vermischen, wenn man auch vorzüglich auf den gegen die Fläche ausgeübten Druck achtet.

Franz, as we shall see later (Art. 837), experimented much in Seebeck's manner except that he used a conical point, instead of a needle, and did not draw the crystal by hand.

The 'Sklerometer' of Grailich and Pekárek (see our Art. 842) is a more complete form of Seebeck's instrument.

Seebeck did not make very much use of his machine, but he confirmed with it Huyghens' statement (see our Art. 836 (*a*)) and made some experiments on calcspar, gypsum and rock salt. In the case of the first substance Seebeck's results differ from those of Frankenheim (*Die Lehre von der Cohäsion*, S. 335) and Franz: see our Art. 839.

(*j*) It will be seen that Seebeck did a good deal to advance the scientific conception of hardness, and to produce an instrument which would measure those differences in the hardness of crystals which had been first noted by Frankenheim, namely: (i) hardness in different senses of the same direction, (ii) hardness in different directions of the same face, (iii) hardness in different faces of the same crystal.

But none of the various ideas of hardness held by these writers clearly distinguish: (i) between set and rupture, (ii) between shearing, tensile and compressive actions. Yet it seems very clear that relative hardness may be different according as the instrument applied merely produces set or actually tears the material: that is according to the manner in which it produces an effect, whether for example by indent or by scratch. The reader will find it well to bear in mind this obscurity in reading our *résumés* of later memoirs on the subject.

(*k*) To the researches of Ångström on the hardness of gypsum and felspar and to those of Wade on the hardness of metals, we have referred in other parts of this chapter see our Arts. 685 and 1040–3.

The following ten articles are devoted to some account of the researches on hardness due to the decade 1850–60.

[837.] R. Franz: *Ueber die Härte der Mineralien und ein neues Verfahren dieselbe zu messen. Poggendorff's Annalen*, Bd. 80, S. 37–55. Leipzig, 1850.

Franz defines the hardness of a mineral as follows:

Mir scheint nämlich die Härte eines Minerals diejenige Kraft desselben zu seyn, welche seine Theilchen zusammenhaltend, dem Körper, der diese zusammenhangenden Theilchen trennen will, Widerstand leistet. [So far this might stand as a definition of cohesion.] Sie ist also diejenige Kraft des Minerals, welche das Eindringen eines Körpers in das Mineral verhindert, und zugleich der Fortbewegung einer in die Oberfläche eingedrückten Spitze sich entgegenstellt. Das Maass dieser Widerstandskraft ist nun aber offenbar der Druck, welcher angewandt werden muss, um den Körper zum Eindringen in das Mineral zu bringen (S. 37).

It seems to me that this manner of determining hardness may really measure two different kinds of resistance, viz. the resistance to entry and the resistance to tearing after entry. Franz assumed that in measuring these resistances he was measuring one and the same property—hardness[1]

[838.] Seebeck in 1833 had already drawn attention to these different methods of measuring hardness, viz. by (i) the least load on a scriber drawn normal to the surface of a mineral which will produce a scratch, (ii) the least load parallel to the surface which will draw a scriber which has already entered the mineral across the surface. Franz's apparatus differed little from Seebeck's, except that he has two separate instruments for measuring the resistances (i) and (ii), and in (i) the mineral mounted in a car is drawn

[1] A criticism of Franz's methods on rather different grounds will be found on pp. 39 and 48 of Hugueny's *Recherches expérimentales sur la Dureté des Corps*, Paris, 1865.

on its bed by a winch and not pulled with the hand: see our Art. 836 (*i*). Franz used the first method to determine considerable differences of hardness and the second for slight differences, such as occur for example in the relative hardness in different directions of the same crystalline surface. He used for scribers a steel cone of 54° vertical angle and a diamond crystal. The steel cone was sharpened daily before experimenting till it was just sharp enough to scratch under a given load a standard bit of gypsum.

[839.] We may note the following results:

Talc. No variation of the hardness with different directions was found (S. 40).

Gypsum. The hardness in the plane of most perfect cleavage was investigated. The maximum hardness was found in a direction making an angle of about 20° with the shorter diagonal of the rhombus and approximating to the second direction of cleavage; the minimum hardness in a direction about perpendicular to this (S. 41–3). Ångström rejects Franz's results as untrustworthy. See our Art. 685.

Iceland Spar. On the rhombohedric surface the greatest hardness was found in the shorter diagonal when the scriber moved in the direction from the obtuse to the acute angle of the rhombohedron. The minimum hardness was in the same direction but in the reversed sense. In the direction of the greater diagonal both senses gave the same value. Frankenheim, according to Franz, found the greatest hardness in the greater diagonal, the least in the same direction as Franz. Franz demonstrated Frankenheim's supposed error by causing the scriber to describe circles on the face; when it went round clockwise the deepest furrow was made exactly at the points where there was scarcely a trace of a furrow when it went round counter-clockwise and *vice versâ*. He refers for Frankenheim's error to S. 337 of the work discussed in our Arts. 821* and 825* On reference to this page, it will be found that Frankenheim says nothing about the directions of least and greatest hardness in the rhombohedric surface, but that on S. 335, where he does, he writes:

> Am grössten ist die Härte parallel der kurzen Diagonale, wenn man nach der scharfen Ecke zieht......Die Härte auf der langen Diagonale steht zwischen der Harte auf beiden Richtungen der kurzen Diagonale, allein dem Maximum näher, als dem Minimum.

In a footnote Frankenheim even corrects an error of Seebeck's who found the minimum hardness in the direction of the longer diagonal. Franz's 'correction' of Frankenheim's 'error' is thus rather gratuitous. I think he could not have carefully read Frankenheim's work. What the latter indeed says about the hardness in the longer diagonal being nearer the maximum than the minimum is confirmed neither by Franz nor by

Grailich and Pekárek (see our Art. 844). Franz returns to the real error of Seebeck and the imaginary error of Frankenheim on S. 53–5. In fig. 4 on Plate II. he gives a curve of hardness for the rhombohedric surface of Iceland spar. I believe he was the first to make use of these curves of hardness, in which *radii-vectores* measure the hardness in a given direction; the credit of them has been recently attributed to Exner.

Fluor-Spar. The least differences in hardness are found in the octahedric surface. In the cubic surface, the greatest hardness is in the diagonal, the least in lines parallel to the sides of the cube (S. 45).

[840.] Various rather scanty results are given for the hardness in certain planes and in a few directions of *Apatite*, *Felspar*, *Quartz*, *Topaz*, *Sapphire* and *Syenite* (in this case somewhat more complete, with a curve of hardness, Fig. 5 on Plate II.). From all these results Franz draws the following conclusions (S. 49–51):

(i) The directions of the greatest and least hardness in the same crystalline surface are intimately associated with the directions of cleavage.

(ii) The direction which is the softest in planes which cut the planes of cleavage is that which is perpendicular to the direction of cleavage, the hardest direction in the crystal is that which is parallel to the planes of cleavage.

(iii) If the surface of the crystal is cut by two directions of cleavage, then in this surface the hardest direction approaches the direction of easiest cleavage.

(iv) Of the different surfaces of the same crystal, that one is the hardest which is intersected by the plane of most perfect cleavage.

(v) If the direction of an easy cleavage is not perpendicular to the surface of investigation, then the hardness is greatest when the acute angle between the surface and the plane of easy cleavage points in the direction opposed to that of the motion of the scriber; it is least in the opposite direction. (Compare with these the almost identical results of Frankenheim on S. 337–8 of his work above cited.) On S. 51–3 Franz gives details of the mean hardness of a considerable number of minerals.

[841.] A. Kenngott: *Ueber ein bestimmtes Verhältniss zwischen dem Atomgewichte, der Härte und dem specifischen Gewichte isomorpher Minerale. Jahrbuch der k. k. geologischen Reichsanstalt.* Jahrgang iii., Vierteljahr[1] IV., S. 104–116. Wien, 1852.

This memoir does not state particulars as to the manner in

[1] Each Vierteljahr has a separate pagination.

which the hardness of the various substances discussed has been determined. The author supposes his atoms to be liquid and spherical; he states that they can or must be treated as liquid if they are to group themselves into molecules and as such into crystals (S. 104). As to the numerical results given in the memoir, I am unable to express any opinion as to their value, but the conclusions which the author draws from his chemical data appear to be summed up in the following paragraph which occurs on S. 114–5, and which I content myself with citing:

Wenn die hier vorgeführten Beispiele zeigen, dass bei isomorphen Species, welche homolog zusammengesetzt sind, ein bestimmtes Verhältniss zwischen dem Atomgewicht, dem Atom- oder Molecül-Volumen, dem specifischen Gewichte und der Härte vorhanden ist, so dass mit dem relativen specifischen Gewichte in geradem, oder dem Atomvolumen in umgekehrtem Verhältnisse die Härte steigt und fällt, und bei gleichen gleich ist, während die Krystallgestalten wegen der übereinstimmenden Gruppirung übereinstimmend sind, weil die gleichgeordneten Atome der einen die Masse in einem dichteren Zustande enthalten als die Atome der anderen, so zeigen sie auch gleichzeitig, dass auf diese Differenzen der Härte und des relativen specifischen Gewichtes die Stellung in der elektrochemischen Reihe oder das elektrochemische Verhältniss der verbundenen Atome ohne Einfluss ist. Aus diesem Grunde habe ich die Atome in der elektrochemischen Reihenfolge vorangestellt, darunter die Verbindungen der ersten Ordnung und in denselben die höheren, wo es dergleichen gibt, und es wird daraus ersichtlich, dass nicht durch den stärkeren elektrochemischen Gegensatz die grössere Härte und das grössere relative specifische Gewicht hervorgerufen sind.

[842.] J. Grailich und F. Pekárek: *Das Sklerometer, ein Apparat zur genaueren Messung der Härte der Krystalle. Sitzungsberichte der k. Akademie der Wissenschaften.* Bd. XIII., *Math. Naturwiss. Classe,* S. 410–36. Wien, 1854.

This memoir opens with an interesting historical account of the various modes of classifying or measuring hardness (S. 410–21). The authors note how unscientific was the earlier use of the word 'hard' by palaeontologists and mineralogists, and then record the researches of some of the writers to whom we have referred in our Art. 836.

[843.] Grailich and Pekárek describe on S. 421–3 the principles of their own *sklerometer* (σκληρὸς = hard). It is essentially based on Seebeck's ideas. They use, however, a conical scriber of 20° to 30° vertical angle, pull the sleigh containing the crystal by a weight, and have accurate means of levelling and rotating into any azimuth the polished surface of the crystal to be scratched. The description of their apparatus occupies S. 423–6, and it is easily grasped in principle from the plate which accompanies the memoir.

They used it in three different ways (S. 426–32). First they counted the number of times the crystal must be drawn in any direction under the scriber in order to make a visible scratch, this involved a constant minimum load on the scriber. Secondly they put a constant maximum load on the scriber and determined the force necessary to draw the crystal in any given direction. Obviously the load on the scriber must be sufficient to produce a scratch even in the hardest direction. Or thirdly they measured the least load on the scriber which would produce a scratch when the crystal was drawn in a given direction. This method they found to be the most exact, and their experiments on Iceland spar were made in this manner.

[844.] The memoir concludes with the details of these experiments on Iceland spar (pp. 432–6). The authors found that the hardness depended not only on direction but also on *sense*. The accompanying figure taken from their memoir gives their general conclusions, where the numbers are the loads on the scriber in centigrammes which just sufficed to produce a scratch.

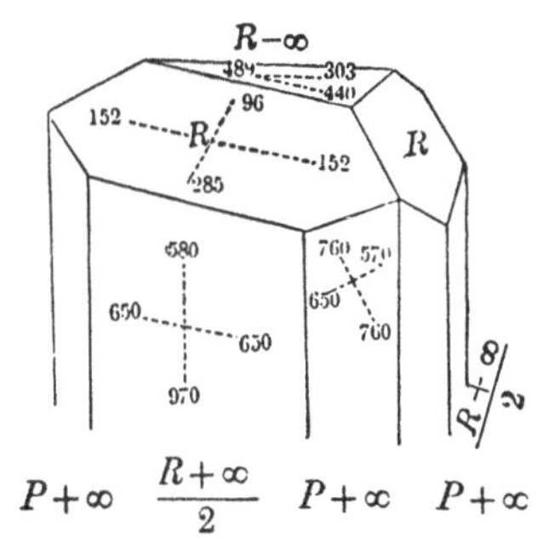

Sklerometric properties of rhombohedric carbonate of lime.
Hardest surface: $R+\infty$.
Softest surface: R.
Hardest direction: 970 centigrammes.
Softest direction: 96 centigrammes.

[845.] F. C. Calvert and R. Johnson: *On the Hardness of Metals and Alloys. Manchester Memoirs,* Vol. xv. pp. 113–121. Manchester, 1860. The hardness of the metals was tested by the weights which would drive a steel point 3·5 mm. into a disc of the metal in half-an-hour. It is worth noting that in the tables given of the relative hardness of metals and alloys, *cast-iron* stands at the top[1] I do not understand the reference to the "half-hour" in the method of experimenting, nor how the load could be so regulated as to drive the steel point into the disc just 3·5 mm. in half-an-hour.

The method differs from that of the continental experimentalists and approaches more nearly that of Wade (see our Art. 1040), but it is open to the same objection as the methods of Seebeck and Franz, i.e. that a steel point driven 3·5 mm deep would sometimes produce set and sometimes rupture.

[846.] Clarinval: *Expériences sur les marteaux pilons à cames et ressorts et sur la dureté des corps. Annales des mines.* T. XVII., pp. 87–106 Paris, 1860. This paper gives an account of a *marteau pilon* invented in 1848 by Schmerber and its application in ascertaining the relative hardness of various substances. Clarinval finds that, the hardness of lead being taken as unity, tin has a hardness of about 4, and very hard iron heated to the temperature usual in forging from 1·4 to 2 5, the increase depending on the cooling of the metal during the series of experiments (pp. 98–102). He compares these results with those obtained by F. C. Calvert and R. Johnson in the *Manchester Memoirs* of 1848 (see our Art. 845), who with unity for lead give 1·7 for tin. Clarinval attributes this divergence to want of chemical purity in his own specimen. At the same time he remarks that he much prefers his own method of experimenting for practical purposes (p. 106).

GROUP G.

Memoirs on Elasticity, Cohesion, Cleavage etc.

[847.] *Volpicelli: Cosmos,* T. I. pp. 214–15. Paris, 1852. We find here a note attributing to this writer *une méthode qui nous semble nouvelle, pour la détermination des coefficients d'élasticité.* The coefficients in question are the so-called coefficients of restitution in the

[1] Staffordshire cold blast cast-iron being taken as 1000, we have: steel 958 (?), wrought-iron 948, platinum 375, copper 301, aluminium 271, silver 208, zinc 183, gold 167, cadmium 108, bismuth 52, tin 27 and lead 16. Probably these are not very trustworthy results as absolute numbers.

theory of impact of elastic bodies, or the kinetic coefficients of elasticity in Newton's theory. The 'new' method is the one used by Newton and described by him in the *Principia*: see footnote to p. 26 of our Vol. I.

[848.] J. T. Silbermann: *Mémoire sur la mesure de la variation de longueur des lames ou règles soumises à l'action de leur propre poids; pour servir de correctif aux mesures linéaires. Comptes rendus*, T. XXXVIII., pp. 825–8. Paris, 1854. This memoir remarks on the effect of the weight of a standard scale of length in elongating it when it is supported vertically by one terminal and not placed horizontally: see also our Art. 1247*. It gives the detail of some experiments to ascertain this elongation for certain bars used as scales of measurement.

[849.] Ch. Brame: *Sur la structure des corps solides. Comptes rendus*, T. XXXV., pp. 666–9. Paris, 1852. This letter to M. Babinet discusses the cleavages of various substances and is not directly concerned with our subject.

[850.] A. Laugel: *Du Clivage des roches. Comptes rendus*, T. XL., pp. 182–5. Paris, 1855, with a *Supplément* on pp. 978–80.

This memoir, of which only a *résumé* is given, was an attempt to extend the methods of Lamé and Resal: see our Arts. 561–70. The author apparently starts from Lamé's ellipsoid of elasticity and supposes that at each point of the earth's surface one principal plane of the ellipsoid will be horizontal. He then states a number of propositions which he says he has demonstrated with regard to the planes of cleavage. It is not evident how he has obtained them from the ellipsoid of elasticity or how, if found, they would necessarily be true, for I see no reason for associating cleavage with a stress rather than a strain surface see our Arts. 1367* and 567 (*b*). Numerical measurements of the inclinations of the planes of cleavage in various localities are given and are compared with what are termed calculated values,' but the method by which the latter have been obtained is not explained. The *Supplément* contains further results professing also to be based on the ellipsoid of elasticity bearing on rupture and the general elevation of mountain chains by eruption, but it is difficult to understand, from the vague description given of the memoir, whether the statements made have any real basis in the theory of elasticity.

[851.] P. Boileau: *Note sur l'élasticité du caoutchouc vulcanisé. Comptes rendus*, T. XLII., pp. 933–7. Paris, 1856.

The author made experiments on springs' composed of alternate plates of iron and annular discs of vulcanised caoutchouc. He found that the squeeze of such a 'spring' was very far from being proportional to the load. The increments of squeeze for increments of charge amounting to ·2 kilog. per sq. centimetre are tabulated, and it will be found that they reach a maximum for a load of about 4·7 kilog. per square centimetre, and then decrease, far less rapidly, however, than they have increased. The writer neglects apparently the squeeze of the iron plates as compared with that of the caoutchouc. There are various other irregularities in the way in which the increments of squeeze alter. Thus after a load of 11·5 kilog. they become very small indeed, and after set has begun they appear to have alternate periods of slow and rapid alteration up to rupture. The author attributes these complicated phenomena to the peculiar molecular structure of caoutchouc and to its thermal characteristics. He notices also elastic after-strain in the springs. Finally he proposes 14 kilog. per sq. centimetre for static and 10 for impulsive or repeated loading as the proper limit for vulcanised caoutchouc of good quality.

[852.] There is a paper on the strength of ice in the *Moniteur Industriel*, No. 2417, Paris, 1860, but I have been unable to find a copy of this periodical.

[853.] W Fairbairn and Thomas Tate: *On the Resistance of Glass Globes and Cylinders to Collapse from external pressure and on the Tensile and Compressive Strength of various kinds of Glass. Phil. Trans.*, pp. 213–247. London, 1859. The paper was received May 3 and read May 12, 1859.

[854.] This is a memoir which in some senses is characteristically British. Its authors display little theoretical knowledge and small acquaintance with the works of previous writers or investigators, but at the same time they present us with a number of useful experimental results, which would have been of very much greater value had the researches been directed by any regard to theory. There is no reference to the experiments of Oersted, Colladon and Sturm, Regnault or Wertheim, nor to the theories of Poisson or Lamé: see our Arts. 686*–91*, 1310–11* 1227*, 1357* and 535*, 1358* It is true that the results of the mathematical theory of elasticity will only apply approximately, if they apply at all, to absolute strength; still a comparison of Saint-Venant's results (see our Art. 119) with those of this memoir would be of value, even if we did not adopt an empirical stress-strain relation at rupture such as that suggested in our Art. 178. The words 'hard,' 'rigid,' 'homogeneous' are used in a rather vague manner in the memoir and without precise definition.

[855.] I cite the following experimental results (pp. 216 and 221):

Glass	Tenacity per sq. inch in lbs.	Crushing load per sq. inch in lbs.
Flint	2286—2540[1]	27,582
Green	2896	31,876
Crown	2546	31,003

Not much weight can be laid on the very few tenacity experiments. The glass was in cylindrical bars and the flint-glass annealed; the increase of tenacity, however, with the diminution of section points to a skin change of elasticity or to a cylindrical distribution of homogeneity. In the case of crushing, fracture occurred frequently with a load of only 2/3 that of the crushing load. We have in the fracture surfaces a strong argument in favour of rupture by transverse stretch: see our Arts. 169 (*c*) and 321 (*b*). The authors write:

The specimens were crushed almost to powder from the violence of the concussion, when they gave way; it however appeared that the fractures occurred in vertical planes, splitting up the specimen in all directions. This characteristic mode of disintegration has been noticed before, especially with vitrified brick and indurated limestone. The experiments following on cubes of glass which were exposed to view during the crushing process, illustrated this subject further: cracks were noticed to form some time before the specimen finally gave way; then these rapidly increased in number, splitting the glass into innumerable irregular prisms of the same height as the cube; finally these bent or broke, and the pressure, no longer bedded on a firm surface, destroyed the specimen (p. 221).

[856.] For cut glass cubes the following results were obtained:

Glass	Crushing load per sq. inch.
Flint	13,130
Green	20,206
Crown	21,867

In these cut glass cubes (sides about 1″) the skin-effect was probably quite lost or reduced to a minimum, so that we find the crushing load of such cubes is to that of the glass cylinders on the average only as 1 : 1·6. *Supposing we might assume* the fracture load to have been 2/3 of the crushing load, we should have for green glass rupture occurring by transverse stretch under a load of 13,471 lbs., but rupture by longi-

[1] Area of section in the first case ·255 and in the second ·196 sq. inches.

tudinal stretch takes place at 2896 (or probably a little less, if skin-effect were allowed for)—the ratio of these numbers is about 4 6 : 1, or somewhat different from the 4 : 1 of the theoretical limiting loads for *elastic* stretch and squeeze.

In some *General Observations* on these crushing experiments (pp. 223–4) the authors refer to Coulomb's theory of compressive strength; they are apparently unaware of its erroneous character: see our Arts. 120* and 169 (*c*).

[857.] Section III. of the memoir (pp. 224–231) is devoted to the 'resistance of glass globes and cylinders to internal pressure.' There were only 17 experiments, 14 on spherical, 1 on cylindrical and 2 on ellipsoidal vessels. Section IV. (pp. 231–240) deals with external pressure. Here 11 experiments were made on glass spheres and 12 on glass cylinders. It is obvious that such a very narrow range of experiments (4 vessels of the last set were not ruptured) cannot be considered as a very satisfactory basis for the purely empirical formulae given in Section V. as a deduction from the experimental results (pp. 241–247).

[858.] These empirical formulae are the following:

External Pressure.

P = external pressure at rupture in lbs. per sq. inch.

d = diameter of the sphere or cylinder, l = length of the cylinder.

τ = thickness of the glass.

p = pressure P reduced to unity of thickness taken to be $\tau = 01$ inch.

Then if C, C', a, a', β, β' be constants the authors *assume* we can represent P by:

$$\left.\begin{aligned} P &= \frac{C\tau^{a}}{d^{\beta}} \text{ for spheres,} \\ P &= \frac{C'\tau^{a}}{d^{\beta'}\, l^{a'}} \text{ for cylinders,} \end{aligned}\right\} a \text{ being the } \textit{same for both.}$$

They conclude that

(*a*) For spheres:

$$P = 28{,}300{,}000 \times \tau^{1\cdot 4}/d^{3\cdot 4}.$$

This formula, however, gives calculated results varying in some cases from the experimental by ± 1/4 of their value, and does not therefore seem to me worthy of much credit (p. 243).

(*b*) For cylinders:

$$P = 740,000 \times \tau^{1\cdot 4}/dl.$$

This formula gives values of P differing in some cases by $+ 1/3$ to $- 1/7$ from their experimental values and is not deserving of more confidence than the previous one.

[859.] The authors next deal with internal pressure, and adopt, as *experimentally* proved for vessels of glass, the formula

$$P = \frac{\omega T}{A},$$

where P = bursting pressure, T = the tenacity of the material, ω = the area of a longitudinal cross-section of the material, that is, the area of the rupture-surface, and A = the area bounded by a longitudinal section of the vessel.

From the experiments in Section III. the authors find in lbs. per sq. inch:

$$T = 4200, \text{ for flint glass,}$$

$$= 4800, \text{ for green glass,}$$

$$= 6000, \text{ for crown glass.}$$

Thus the mean tenacity = 5000 or nearly *twice its value* as given by direct tractive experiment: see our Art. 855. The authors remark:

The tenacity of glass in the form of thin plates is about twice that of glass in the form of bars (p. 246)...This difference is no doubt mainly due to the fact that thin plates of this material generally possess a higher tenacity than stout bars, which, under the most favourable circumstances, may be but imperfectly annealed (pp. 216–7).

[860.] The memoir in its concluding paragraph assumes that the *mean* ratio of the tensile and compressive strengths of glass is equal to the mean ratio of the tensile and crushing strengths or as 1 : 11·8 nearly. It seems to me that we ought to take fracture rather than crushing to powder as the limit of compressive strength, in which case the results for the flexural strength of glass bars deduced on p. 247 from formulae and not from experiment would be much modified.

It should be noted that Saint Venant's theory for cylinders and spheres of thickness small as compared with the radius does not lead to formulae for safe loading of the type given in the preceding articles for bursting pressures: see our Arts. 120 and 124.

GROUP H.

Minor Notices chiefly of Memoirs on Molecular Structure.

[861.] J. Szabó: *Einfluss der mechanischen Kraft auf den Molecular-Zustand der Körper. Haidingers Berichte über die Mittheil ungen von Freunden der Naturwissenschaften in Wien.* Bd. VII., 1849, S. 164–73. Wien, 1851. This paper brings a good deal of rather discursive evidence to show that bodies of the same chemical constitution can exist in more than one physical condition, and that the application of such mechanical processes as scratching, vibrating, changing the temperature etc., suffices to throw the body from one condition into the other. The author cites for example black and red cinnabar and wrought iron in the fibrous and crystalline conditions. The paper is not of any permanent value, and is a collection of old rather than of novel facts.

[862.] O. L. Erdmann: *Ueber eine merkwürdige Structurveränderung bleihaltigen Zinnes. Berichte über die Verhandl. der k. sächsischen Gesellschaft der Wissenschaften. Mathematisch-physische Classe,* Jahr gang 1851, S. 5–8. Leipzig, 1851.

At the repair of an organ said to date from the 17th century in the *Schlosskirche* at Zeitz, the pipes were found to be strangely crystallised in certain places, *die ohne Ordnung, jedoch ziemlich gleichmässig vertheilt standen und von verschiedener Grösse, von der eines Silbergroschens bis zu der eines Thalers waren.* The crystallised parts were extremely brittle, the rest of the metal ductile. Analysis showed the constitution of both parts to be chemically the same, so that the difference was in mechanical structure. Erdmann attributed this change of structure to the vibrations which the pipes had undergone, but hazarded no conjecture as to the manner in which the crystallisation was distributed.

Jedenfalls dürfte aber die mitgetheilte Beobachtung nicht ohne Interesse in Bezug auf das von einigen Technikern noch immer bezweifelte Krystallinischwerden von eisernen Achsen, Radreifen u. s. w. sein, wenn dieselben, wie beim Eisenbahnbetriebe, fortwährenden Erschütterungen ausgesetzt sind (S. 8).

[863.] D'Estocquois: *Note sur l'attraction moléculaire. Comptes-rendus,* T. 34, p. 475. Paris, 1852. This note merely refers to a paper which the author had submitted to the Academy. He states that he has proved that, if the molecules of a liquid all attract or all repel each other according to some inverse power of the distance, then they cannot retain the liquid condition "à moins que cette puissance ne soit le carré." No further reference is given to the mode in which this singular result has been deduced, beyond the statement that it depends on the equation of continuity.

[864.] Sir David Brewster: *On the Production of Crystalline Structure in Crystallised Powders by Compression and Traction. Edinburgh Royal Society, Proceedings,* Vol. III., 1853, pp. 178–180. Edinburgh, 1857. Evidence is given in this paper of the effect of compression on powders and of traction on 'soft-solids' in producing doubly refracting properties.

[865.] I have given a reference to several memoirs by Séguin in Art. 1371*. The molecular theory expounded in them formed the subject of a quarto volume of 55 pages and two plates published in 1855 at Paris. It is entitled: *Considérations sur les causes de la Cohésion, envisagées comme une des conséquences de l'attraction Newtonienne et résultats qui s'en déduisent pour expliquer les phénomènes de la Nature.*

The author in his preface speaks somewhat sorrowfully of the neglect which his memoirs read before the *Institut* have met with, and also somewhat slightingly of the advantages of mathematical analysis. His present work, he tells us, aims at providing a basis for the discussion *in the future* of molecular action:

Tout le monde sait, que chaque question scientifique a son heure et son moment, qu'il ne dépend pas de la volonté d'un seul homme de faire avancer ou retarder. Cette heure et ce moment viendront, je l'espère, et alors ma cosmogonie se trouvera forcément à l'ordre du jour.

Séguin's cosmogony is based on the hypothesis that the ultimate elements of bodies, here termed molecules, are of infinitely small volume and infinitely great density. This idea he appears to have gained from a conversation with Herschel in 1823 (p. 2, ftn. 3). Séguin supposes the density to increase inversely as the diminishing radii of the molecules which are taken to be spherical. By arranging these molecules in files and supposing them to obey the Newtonian law of gravitation, he endeavours to explain some of the features of cohesion, i.e. to obtain from the Newtonian law a sufficiently great cohesive force. The whole of his calculations are of a most crude, insufficient and often obscure kind. I must confess that I am in many places unable to follow his reasoning. The density of the molecule has to be immensely greater than the density of the earth (pp. 8–9); this might be intelligible, but as he puts the molecules of a bar of iron in contact, it seems to me that he makes a bar of iron of a different order of density to the earth. Perhaps this difficulty may be got over by a right interpretation of the following words:

Si l'on considère la vaste échelle sur laquelle Dieu a tout créé, tout fait, tout ordonné ! et le témoignage de nos sens, tout comme notre raison, doivent être, en pareille matière, complétement éliminés comme tendant à rétrécir et restreindre nos idées dans la sphère de nos conceptions qui sont si éloignées de l'intelligence des œuvres du Créateur (p. 20).

Further Séguin tells us that in the beginning matter created by

God consisted of infinitely small, infinitely dense molecules uniformly distributed through space. Then:

> au *fiat lux* la matière reçut de Dieu la faculté de s'attirer en raison directe des masses et inverse du carré des distances, et je considère que cette attribution que la matière inerte a recue de Dieu, constitue pour elle une espèce de vie matérielle (p. 41).

The *fiat lux* of the Jewish cosmogonist has received many interpretations, but scarcely any so grotesque as this of the French physicist and member of the *Institut*!

The reader will probably agree with the view expressed in our first volume (Arts. 163*-72*, 752*-8*) that the Newtonian law is insufficient to account for the phenomena of cohesion. What might be said for Herschel's idea, does not, however, seem to me to have been said in an intelligible fashion by Séguin[1]. I feel, indeed, reluctantly compelled to class him with Eisenbach and Père Mazière. From the *Polytechnische Bibliothek* 1887, No. 9, S. 133, I see that a reprint of Séguin's work has just appeared in Lyons. I venture to doubt whether 'son heure et son moment' has even yet arrived.

[866.] R. P. Bancalari: *Sur les forces moléculaires. Cosmos*, VIII., pp. 501-3. Paris, 1856. Bancalari appears to have published a memoir in the preceding year in which he is said to have established the remarkable proposition that: *the resultant of the molecular forces in a body is directly proportional to the increments or decrements of intermolecular distance and inversely proportional to the cubes of the same distances.* The methods by which the law of gravitation and Hooke's law are deduced from this proposition seem to me very unsatisfactory, and have not encouraged me to examine the original memoir for more particulars than *Cosmos* provides.

[867.] J. Zaborowski: *De triplici in materia cohaerendi statu. Disquisitio physica.* Posaniae, 1856. This is a quite worthless metaphysical dissertation which asserts that cohesion, treated as either negative or positive, is really adhesion and depends on the absolute continuity of matter. The author appears quite ignorant of the enormous advances which had been made in physical science between the time of Bacon and the middle of the nineteenth century, and the sole interest of his pages lies in their demonstration of the possibility of atavism in science.

[1] The theory of Herschel has been dealt with by Sir William Thomson in a paper published in the *Proceedings of the Royal Society of Edinburgh*, Vol. IV., pp. 604-6, 1862, and reprinted in the *Popular Lectures and Addresses*, Vol. I., pp. 59-63. London, 1889. Sir William, of course, is suggestive and clear, but his conclusion that:

> It is satisfactory to find that, so far as cohesion is concerned, no other force than that of gravitation need be assumed,

seems to me far too optimistic.

[868.] C. S. Cornelius: *Ueber die Bildung der Materie aus ihren einfachen Elementen. Oder: Das Problem der Materie nach ihren chemischen und physikalischen Beziehungen mit Rücksicht auf die sogenannten Imponderabilien.* Leipzig, 1856.

This tract of xi + 64 pages professes to explain chemical, cohesive and gravitational forces by a new atomic theory. The method of procedure, although making frequent appeals to physical and chemical facts, is so metaphysical that I have not been able to perform the *gewisse Denkoperationen, die ihren Grund mehr oder weniger im Thatsächlichen haben,* which would have allowed me to reach the principles on which the author bases the sensible properties of matter. I am the more disappointed in this as the author assures us that his investigation is *in ihrer Art vollständig,* and it appears not only to explain gravitation and elasticity but to remove in general any difficulty about the mutual action between body and soul. It would appear that the author arrives on S. 18 at precisely Boscovich's definition of an atom, although he associates it with the names of Ampère, Cauchy, Séguin, Moigno and Faraday, together with a metaphysician or two. After this I can only follow an occasional passage here and there. It seems that a true element of matter must be *ein völlig intensives Eins,* but a contradiction arises from the fact that *ein sich selbst gleiches substantielles Eins* cannot influence its kith and kin *Da jedes dem anderen hinsichtlich der Qualität völlig gleich ist, so kann keinem etwas von dem anderen widerfahren* (S. 20). However by a *dauernder Act innerer Thätigkeit* an element can produce motion in the unlike. Hence arises a vibratory motion of a sphere of ether all round an atom. At this point we are rather abruptly introduced to mass and pressure, shown how action at a distance takes place, and given a demonstration of the law of gravitation. Strangely enough an atom treated as a pulsating ether-squirt *does* go a considerable way to explain chemical and cohesive forces. Perhaps some scientist who is capable of performing the required *Denkoperationen, die ihren Grund mehr oder weniger im Thatsächlichen haben* will be able to say whether the author has any inkling of this. If so metaphysicians have a royal road to truth quite out of the ken of the ordinary scientist.

[869.] Vogel: *Zur Theorie der Glasthranen: Erdmanns Journal für praktische Chemie,* Bd. 77, S. 481-2. Leipzig, 1859. The writer of this note placed 'Prince Rupert's drops (*larmes bataviques*) in hydrofluoric acid so that the outer coat including the major part of the tail of the drop was dissolved away in 48 hours. The drop did not break up, and no effect was produced by breaking away the fragment left of the tail. A slight blow of the hammer, however, caused the drop to burst. The author concluded that the outer surface of the drop was not that which preserved the inner material in a state of great strain, or its removal would have brought about the bursting of the drop.

[870.] A. Bouché: *Recherches sur l'attraction moléculaire. Mémoires de la Société Académique de Maine et Loire,* T. VI., pp. 229-333.

Angers, 1859. T. VIII., pp. 133–144. Angers, 1860. T. X., pp. 181–249. Angers, 1861.

This is an elaborate attempt to explain the phenomena of gravitation, cohesion and chemical affinity by means of the law of intermolecular attractive force R,

$$R = \frac{f}{d^2}\left(1 - \frac{a}{d}\right) \text{ .... (i)},$$

where d is the distance of two molecules, a is a very small constant distance and f another constant.

Bouché obtains this law by simply combining Newton's law of gravitation with Mariotte's law that the pressure of a gas varies as its density, while the density of a gaseous mass must vary inversely as the cube of the intermolecular distance He proposes in the first paper to apply this law to distances less than interplanetary and greater than gaseous intermolecular distances.

[871.] Bouché works out at very considerable length the results which flow from accepting this law in the cases of planetary action, of the pressure of gases &c., but there is nothing very conclusive in these results, or that could not in general terms have been almost foreseen from the nature of the formula itself. The second part of the memoir consists of rather indefinite philosophical reasoning. In the third paper (p. 223) in *Tome* X., Bouché makes a a function of the temperature and obtains an expression for the pressure p of the form:

$$p = f\left\{\frac{A}{d^2} - \frac{Ba(1 + K\theta)}{d^3}\right\},$$

where A, B, K are additional constants and θ is the temperature. Of this formula he now writes:

Nous regarderons cette formule comme vraie dans toute l'étendue des intervalles planétaire et gazeux, et pour les valeurs de θ aussi grandes qu'on veut (p. 223).

There is again much indefinite discussion, and we conclude the memoir with the feeling of having made no real progress in understanding how far such a law as (i) will carry us in explaining intermolecular action. The same form of force has been discussed by Saint-Venant and Berthot: see our Art. 408.

[872.] J. G. Macvicar, D.D.: *An adaptation of the Philosophy of Newton, Leibnitz and Boscovich to the Atomic Theory. Proceedings of the Philosophical Society of Glasgow*, Vol. IV., pp. 32–80. Glasgow, 1860. This paper deals with atomic and molecular phenomena from the metaphysical standpoint. The remarks on elasticity (pp. 55–6) are unintelligible to me, and some critics might term them nonsense.

SECTION III.

Technical Researches.

GROUP A.

Treatises and Text-books dealing with the Strength of Materials from the Technical Standpoint.

[873.] THE decade with which we are dealing is marked by the publication of many works treating of the strength of materials and the theory of structures. I cannot hope to have formed even an approximately complete list of works of this character, but it is probable that those I have considered in the following articles are very fair representatives of their class and suffice to indicate the progress of technical research and applied elasticity.

[874.] A work by G. F. Warr entitled: *Dynamics, Equilibrium of Structures and the Strength of Materials* was published in London in 1851. There is an interesting chapter, now of course quite out of date, on bridge-structure (pp. 117–232), and one on the strength of materials (pp. 232–282), which contributes, however, nothing of value to the history of our subject.

[875.] C. L. Moll and F. Reuleaux: *Die Festigkeit der Materialien, namentlich des Guss- und Schmiedeisens.* (*Besonderer Abdruck aus der Constructionslehre für den Maschinenbau*), Braunschweig, 1853, 72 pages. This work is a synopsis of formulae rather than a treatise. It emphasises, however, an important principle, which has too often been forgotten by technical writers, namely that the rupture strength of a material is not a true guide to its use in construction. The authors adopt what they term a *Coefficient der stabilen Festigkeit* as a measure of the stress permissible in a material. This coefficient is based upon the *elastic limit*, but we are not told how the elastic limit is to be determined, while we know that within a certain range, it can safely be extended without injuring the material. For cast-iron they take the elastic limit in compression double its magnitude in extension (7·5 kilogs. per sq. millimetre), and they suggest that upon this result the best practical section for a cast iron beam ought to be based, and not upon Hodgkinson's results as to rupture strength: see our Arts. 243*–4*, 176 and 951.

We may note that the authors appear to have had no conception of shear or slide. They take (S. 65) the *Drehungsmodel* (*sic!*) always $\frac{2}{5}$ of the stretch-modulus without any mention of aeolotropy or multi-constancy. Further their views on torsion and the resulting formulae are completely erroneous (S. 12–13 and the footnote (!)); and, notwithstanding their assumption of uni-constancy, they treat all elastic bodies as built-up of 'fibres' (S. 2). Lastly they give copious values for the moments of inertia of various cross-sections. I have not tested all these, but some of them are certainly wrong and others inconsistent with the results given by later writers (e.g. XVIII. S. 23).

[876.] In the year 1853 was published the fifth volume of Morin's *Leçons de mécanique pratique*, the first volume of which had appeared in 1846. This fifth volume forms the first edition of the well-known *Résistance des matériaux*, a work which in several editions extending over a long course of years has had great influence as a book of reference for students of technical elasticity[1].

A note giving an account of this work by Morin himself, will be found in the *Comptes rendus*, Tome XXXVI., pp. 284–7. Paris, 1853. Morin states that his work is not intended as a complete treatise on the strength of materials; it is only the text of lectures delivered by him in the *Conservatoire des Arts et Métiers* during the years 1851–2. His object in the work has been to remove doubts which have arisen with regard to the ordinary theory of elasticity owing to its extension to problems lying outside its proper limits. Those limits, however, contain, he maintains, really all that is needful for most practical constructions: for it is not the *absolute* but the *elastic* strength of a material which ought to determine the proportions of any piece of it.

A second edition of Morin's *Résistance des matériaux* appeared in 1856, and a note by Morin on its presentation to the *Académie* will be found in the *Comptes rendus*, Tome XLIII., pp. 939–41. Paris, 1856. It is therein remarked that the additions made to the volume tend further to demonstrate the applicability of the ordinary theory to small strains. Thus by very careful measurements on the flexure of wooden, wrought-iron and cast-iron beams, Morin states, that he has demonstrated that the resistances to stretch and squeeze are "within the elastic limit" equal, *i.e.* that the stretch- and squeeze-moduli are initially equal: see, however, our Arts. 1411* and 793. The difficulty here is to grasp the exact meaning of the term "elastic limit." Morin uses in one place (p. 940) the phrase "premières flexions et celles que l'on peut sans danger admettre dans les constructions," but this seems equally vague.

[1] The third edition in two volumes appeared in 1862.

[877.] We cannot analyse all the separate editions of Morin's work and must content ourselves therefore with some remarks on the first edition and a notice of additions in the last edition at a later stage of our history. A German translation of the first edition under the title: *Die Widerstandsfähigkeit der Baumaterialien* will be found on S. 196–264, Jahrgang XVIII., and S. 194–343, Jahrgang XIX., of *Försters Allgemeine Bauzeitung*, Wien, 1853 and 1854. The first part of this concludes with a bibliography of earlier works on elasticity and the strength of materials, having special reference to technical researches; most of the works referred to will be found quite sufficiently dealt with in our first volume.

[878.] The *Première Partie* of Morin's work is entitled: *Extension*, and occupies pp. 1–60. This section is very characteristic of his methods. While G. H. Love (see our Arts. 894–905) exaggerates the discrepancies between theory and practice and would reduce elasticity to an empirical science, Morin on the other hand seems to me to disguise the real difficulties which occur, and so to some extent his book tends to check that development of theory which invariably follows when any discordance with experience is clearly recognised. He endeavours to reconcile the insufficient theory of Navier and Poncelet with the experimental conclusions of Hodgkinson, Fairbairn and others.

Thus he assumes: (i) that for every given material the limits of elasticity are absolute and not relative to the working; (ii) that perfect elasticity necessarily ceases with the proportionality of stress and strain (pp. 2–3); (iii) that the limit of safe stress for practical purposes is this elastic limit (pp. 3 and 7). On p. 48 we have a table of absolute limits of elasticity and the corresponding safe charges for a great variety of materials. Now to-day we are certain that the limit of elasticity is relative to the working and previous loading of the individual specimen, and further does not necessarily connote proportionality of stress and strain (see our Vol. I. Note D, p. 891 and Art. 796). Hence it is difficult to consider Morin's treatment of safe-loading as satisfactory. Indeed he himself remarks that further experiments on the elastic limit are needed and proposes to fall back on 1/10 of the rupture stress for wood, stones and cements, and 1/6 of the rupture stress for metals as the safe permanent stress. He gives tables of stresses thus calculated on pp. 54–6.

[879.] We may briefly note one or two other points in this first part.

(*a*) On pp. 5–17 the results of experiments by Bornet, Ardant and Hodgkinson for wrought- and cast-iron are given and stress-strain (stretch-traction) curves are plotted out (Plate I.) These are very valuable and suggestive for the comparison of various types of hard and soft iron, and their relative technical advantages. Compare our Arts. 817*, 983*–4* and 1408* *et seq.*

(*b*) On pp. 17–28 various elementary theoretical and empirical formulae are given for cylindrical and spherical shells subjected to internal pressure. These are applied to numerical examples in the case of boilers and hydraulic presses. Notably it is shown that the presses used to raise the tubes of the Britannia bridge were dangerously weak (pp. 25–7).

(*c*) We may note here the formula adopted at that date by the French Government for the thickness τ of boilers of plate iron of internal diameter d, subjected to N atmospheres of internal pressure:

$$\tau = \cdot 0018 Nd + \cdot 003,$$

d and τ being measured in metres.

Here ·003 is a constant introduced to allow for the wear of the material, and the safe tractive stress for plate iron is taken to be 3,000,000 kilogs. per sq. metre. As the rupture traction of plate iron equals about 30,000,000 kilogs. per sq. metre, and according to Morin we ought to take 1/6 of this for safe loading the formula leaves a considerable margin of safety. We refer to this formula here as it recurs in many French and even in German books of this period: see, for example, our Art. 1126.

(*d*) On pp. 28–31 the experiments of Fairbairn and Clarke on plate iron are considered. Morin seems to hold that Fairbairn's results are really correct for the better kind of plates, but this should be compared with our Arts. 1497* and 902. He then passes to Fairbairn's experiments on rivets and cites his result that absolute tractive and shearing strengths are practically equal: see our Arts. 1480* (ii) and 1499*–1500*. He compares it with that of Gouin et Cie[1], who found for iron rivets the tractive and shearing strengths about 4000 and 3200 kilogs. per sq. centimetre respectively, or very nearly in the ratio of 5/4, which is what the theory of uniconstant isotropy would give for the ratio of the corresponding *fail* limits in traction and shear: see our Arts. 5 (*e*), 185, and Vol. I., p. 877.

[1] The details of Gouin et Cie's experiments were given in the *Mémoires...de la Société des ingénieurs civils*, Année 1852, pp. 155–7, Paris, 1852: see our Art. 1108.

(*e*) After a *résumé* of experiments on wood (see *Wood*, Index, Vol. I.) Morin gives some details of experiments on the strength of iron cables by which it would appear that the French navy at that date had cables considerably stronger than those of the English navy (pp. 42–7). I may note one point which seems to me suggestive. In the experiments of Captain Brown the absolute strength of the iron employed to make the links of the chain cables was about 40 kilogs. per sq. mm., but the strength of the chain cable was only 34 kilogs. per sq. mm. or the ratio of the two $= \frac{5}{4}(1 - \frac{1}{17})$, or nearly 5/4. But this by the preceding paragraph is the ratio of the shearing to the tractive strength in wrought-iron. Hence it appears to me that Brown's cables were possibly destroyed by shearing and not tensile stress: see our Art. 641.

(*f*) On p. 49 Morin, reasoning, however, only from Wertheim's experiments on *wires*, states that annealing does not effect the elasticity of iron and steel, but does that of copper, gold, platinum and silver. He suggests, however, that this would not hold true for larger masses of iron, as for example axles, kept at a moderately high temperature for a long period. He believes that such masses would change from a soft and fibrous to a crystalline condition, and he cites an experiment of his own, where moderate and continuous annealing during five months and twelve days produced this effect. Hence he concludes that it is not advisable to anneal axles and other large pieces of metal: see our Arts. 1295*, 1463*–4*, 891 (*d*), and 1070.

(*g*) On pp. 57–60 Poncelet's results for elastic and absolute resilience are reproduced (see our Arts. 981*, 988*–92*). These resiliences are the areas of the corresponding parts of the stress-strain curves. It does not, however, seem to me true that because the rupture resilience of soft iron is greater than that of hard iron, the former ought to be employed for bodies like iron-cables, etc. subjected to impulses. Repeated impulses with less resilience than the elastic resilience of a hard iron bar, but greater resilience than that of a soft iron bar, would leave the former undamaged but wear the latter out. It is only when the resilience of the impulse is likely to be greater than the elastic resilience of both hard and soft iron, that it is advantageous to use the latter. See on this point Cavalli's remarks cited in our Arts. 1085–9.

Further I must note that Morin's method of equating the kinetic energy of a falling body to the total resilience of a bar does not seem to me to give a true limit to the height from which the body may fall on the bar without destroying it. It has first to be shown that the kinetic energy of the falling body will not be absorbed by one element of the bar, but be distributed throughout its volume. This Morin has not attempted to do. The complete solution of any problem of resilience is one of great complexity: see our Arts. 362–71, 401–7 and 410–14.

[880.] The *Deuxième Partie* of the work is entitled: *Résistance des corps solides à la compression* and occupies pp. 61–123. We may note briefly one or two points:

(*a*) Pp. 61–76 deal with the resistance of wood. Morin compares the results obtained experimentally by Rondelet and Hodgkinson Rondelet in his *Traité de l'art de bâtir* (see our Art. 696*) does not seem to have distinguished between rupture by pure compression and rupture by buckling. He gives a table of the following kind for wooden columns:

Ratio of height to least dimension of cross-section	1	12	24	36	48	60	72
Crushing Strengths	1	5/6	1/2	1/3	1/6	1/12	1/24

This is a purely empirical table and it would not be necessary to refer to it here, had not a recent writer apparently adopted these numbers for the crushing strengths of apparently all kinds of material[1] Obviously a different kind of strain appears the moment the block becomes long enough to buckle, and these numbers are at best only approximately true for the particular type of wood upon which Rondelet was experimenting.

Hodgkinson on the other hand found for short blocks (height double the diameter) that the crushing load P was proportional to the area of cross-section, while for wooden struts (see our Art. 965*) of length l and rectangular cross-section $a \times b$ $(b<a)$ he adopted a formula of the type $P = Kab^3/l^2$, where K is a constant depending on the material. Morin adopts Hodgkinson's results in preference to Rondelet's and gives the following values for K, when a and b are measured in centimetres, and l in decimetres:

Strong oak: 2565.
Weak oak: 1800.
Red and strong white deal and resinous pine: 2142.
Weak white deal and yellow pine: 1600.

For safe loading 1/10 of P may be taken (pp. 68 and 73).

Obviously $100\,K$ = crushing strength of a cube of the material of one centimetre side, or K equals crushing strength in kilogrammes per sq. decimetre of such cubical blocks. The numbers we have cited can only be treated as roughly approximate, for the crushing strength varies greatly with the degree of moisture, age, etc. of the wood: see our Arts. 1312*–4*.

[1] F. Auerbach in the *Handbuch der Physik* (*Dritte Abtheilung* of the *Encyklopädie der Naturwissenschaften*), Bd. I., S. 312.

(*b*) Pp. 76–89 give details of the experimental results of Rondelet, Clark and Vicat on the crushing strengths of stone, mortar, cement and brick: see our Arts. 696*, 1478* and 724*–30*. Morin after citing at length Vicat's results on the compression of cylinders and spheres between parallel tangent planes, and the pyramidal or conical surface of rupture remarks:

Les nombreuses observations que nous avons recueillies à Metz, M. Piobert et moi, sur la rupture des projectiles brisés par le choc, ont montré que dans ce cas la rupture se fait d'une manière analogue, avec cette différence que le point choqué est le plus ordinairement le sommet déprimé d'une pyramide à cinq faces quand la vitesse du choc n'est pas très-considérable, et qu'aux grandes vitesses cette pyramide se change en un cône à génératrice curviligne, qui est presque toujours multiple ou formé de plusieurs autres cônes conaxiques, et dont l'axe diminue de longueur à mesure que la vitesse du choc augmente (p. 81).

Supposing we assume that the rupture surfaces are practically in close agreement with the surfaces of maximum elastic stretch. I think an explanation of these extremely interesting conical and pyramidal rupture surfaces, to which I have frequently had to refer (see our Arts. 730*, 949*, 1414* and 1446*), might be deduced by Hertz's method of investigation (*Crelle's Journal*, Bd. 92, 1882, S. 156–71).

(*c*) Pp. 90–100 deal with Hodgkinson's experiments on cast-iron: see our Arts. 1410*–5*. Morin adopts a *mean* value for the squeeze-modulus, which does not seem to me justified by Hodgkinson's results: see our Art. 1411*. Pp. 100–105 deal with wrought-iron and a comparison of its action under compression with that of cast-iron. Morin gives graphical representations of Hodgkinson's results. Then follows a discussion of Hodgkinson's experiments on cast-iron pillars (see our Arts. 954*–65*), which are represented by numerical tables (pp. 108–9), more easy to work from than Hodgkinson's formulae, and also graphically by curves on Plates II. figs. 4 and 5, III. figs. 1 and 2. Pp. 115–23 deal by approximate methods with the compression in arched ribs. Here a circular rib is treated as a parabolic arch and supposed to be loaded uniformly per foot-run of the horizontal chord, although Morin (p. 116) speaks of the load as being often in great part due to the weight of the arch. Thus Morin really only deals with a part of the complete expression as worked out by Bresse see our Art. 525. A table on pp. 118–9 gives the compressive stress in a number of existing arches and viaducts in France on these assumptions.

[881.] The *Troisième Partie* on *Flexion* occupies pp. 124–431, or embraces the bulk of the volume.

(*a*) After some general remarks on the experiments of Duhamel du Monceaux, Dupin, Duleau, Hodgkinson, etc., which he holds tend to confirm the customary axioms of the Bernoulli Eulerian theory, Morin proceeds with a slight historical preface (pp. 140–4) to develope that theory in the usual manner. The usual problems are solved and the fail-limit or safe load for a beam is deduced from the formula for the bending moment (see our Art. 173).

$$M_0 = \frac{T_0 \omega \kappa^2}{h},$$

where T_0 is taken as a quantity to be determined by flexure experiments and not to be assumed from pure tensile results. Morin gives a table of its values for different materials on p. 169, but he does not note that it is part of the 'paradox in the theory of beams' that it varies from one form of section to another: see our Arts. 173 and 930. Besides the fact that in a great variety of special cases the deflection of simple beams is worked out, there is nothing calling for special notice in the whole of these pages (pp. 138–232).

(*b*) Morin next proceeds to apply these theoretical results to the experimental determinations of the elasticity and strength of wood made by Barlow (see our Art. 188*), and of cast-iron made by several English Engineers for the *Report of the Iron-Commissioners* (see our Art. 1406*). In the latter case Morin concludes that the flexure of cast-iron for all practical purposes obeys closely enough the laws of the Bernoulli-Eulerian theory (p. 268), but I hardly think the facts warrant this conclusion. He next deals with rolled and plate iron girders, considering especially a great variety of T and double-T beams (pp. 269–90). Next we have a long account of the theory and construction of tubular bridges in plate-iron, with details of Fairbairn's experiments (pp. 290–322): see our Arts. 1465* *et seq.* Then follow descriptions of a girder in plate-iron designed by Brunel, of the experiments of James and Galton on travelling loads, and remarks on the alteration of structure in axles by Marcoux[1] and Arnoux[2]: see our Arts. 1417* and 1463*–4*

(*c*) The remaining portion of the *Troisième Partie*, namely pp. 354–431, is devoted to a discussion of roof trusses (*charpentes*) in wood and iron. The theory employed is analytical, and appeals only to the elementary principles of statics combined with the Bernoulli-Eulerian theory of beams. Morin cites Ardant's results (p. 361 etc.: see *Addenda* to our Vol. I., pp. 5–10), and gives formulae and tables which might possibly be still of service in the design of roof-trusses. The only articles which call for special notices are §§ 324–6, which deal with the first experiments ever made, I believe, to test the stresses calculated for the members of a frame. In these experiments Morin was assisted by Tresca and Kaulek. The tie rods to be tested

[1] Marcoux considered that axles were weakened by prolonged vibration, but that they did not change their structure from fibrous to crystalline (p. 350).

[2] Arnoux considered that axles were weakened by prolonged service and that there was a structural change (p. 353).

were replaced by chains containing dynamometers, and the chains were carefully screwed up to the original length of the replaced tie rod. The stresses measured by the dynamometers agreed with the results of calculation to a degree sufficiently accurate for practical purposes,—in all cases but one with less than 6 p.c. difference and often with considerably less (pp. 396–400).

[882.] The *Quatrième Partie* is entitled *Torsion* and occupies pp. 432–53. The statement on pp. 432–3, that the absolute displacements are proportional to the distances of the displaced elements from the axis of the prism under torsion, is only true for prisms of circular cross-section, and the application of Coulomb's theory to bars of rectangular cross-section (p. 438) is of course incorrect Some experiments on the resistance of cast-iron shafting to torsion made at Mulhausen and some others made by Carillion in Paris are cited on pp. 444–51, but the theory given of rupture by torsion (p. 448) seems to me obscure if not erroneous, and this portion of the work is not satisfactory.

Considering the date at which the work was published, it was extremely good of its kind, although Love's book is in many points of more practical service. It has in later editions progressed with the advance of technical elasticity and we shall have occasion to refer to it again.

[883.] H. Tellkampf: *Die Theorie der Hängebrücken mit besonderer Rücksicht auf deren Anwendung*. Hannover, 1856. This is a useful *résumé* in 120 pages of the theory of suspension bridges from the practical side. Attention may be drawn to the *Sechstes Kapitel* entitled: *Oscillationen der Hängebrücken,* S. 99–114, which developes various, not absolutely rigid, theories of impact. We may note especially § 49 (S. 107–12), which applies a theory of impact similar to that of Hodgkinson, Cox and Saint-Venant to the case of a weight falling on the centre of a suspension chain. This may be taken as an example of a theory, which if somewhat hypothetical still gives results probably accurate enough in practice: see our Arts. 943*, 1434*, and compare with Arts. 366–71.

[884.] We are justified in asserting that the period with which we are dealing in this chapter marks a great improvement in the type of text-books for practical technologists and students. Notably in this respect we owe much to J. Weisbach, Morin and J. M. Rankine. The first edition of Weisbach's *Ingenieur-Mechanik* appeared in 1846, the second in 1850, the third and fourth in 1856 and

1863 respectively. Polish, Swedish, Russian, English and American translations have appeared. E. B. Coxe's translation of the fourth edition (Trübner, 1877) is the form most accessible to English readers. Section IV. and the Appendix deal with a number of elastic problems and profess in the fourth edition to incorporate the then recent book of Lamé, Rankine and Bresse. An interesting sign of the progress of our science is the continual remodelling of these portions of the book in successive editions. While the work shows greater advance over the earlier text-books on the strength of materials and while some points of it might even be of service to-day, it must nevertheless be read with caution. In the English edition of 1877 we still find an unsatisfactory and even erroneous treatment of flexure, torsion and of the theory of struts, while the theory of combined stress exhibits the same errors as Weisbach's earlier memoir on that subject: see our Art. 1377*–8* Further, contrary to Weisbach's opinion, Kupffer's experiments show that the stretch-modulus can be found with some degree of accuracy from transverse vibrations by means of the formula cited in § 5 of the *Appendix*. The book so far as our subject is concerned is entirely replaced on the theoretical side by Grashof's text-book; neither of them can, however, be considered satisfactory from the physico-technical side. An account of Weisbach's labours is given by Rühlmann: *Vorträge über Geschichte der technischen Mechanik*, 1885, S. 415—24.

Other German text-books of this period, to which I have found frequent reference in memoirs dealing with the strength of materials, are discussed in the three following articles.

[885.] G. Rebhann: *Theorie der Holz- und Eisen-Constructionen mit besonderer Rücksicht auf das Bauwesen.* Wien, 1856. This is a text-book of technical elasticity and bridge-construction containing xiv + 602 pages. It was probably a serviceable students' work at the time it was written but it embraces nothing, I think, of permanent or historical importance.

[886.] H. Scheffler: *Theorie der Gewölbe, Futtermauern und eisernen Brücken.* Braunschweig, 1857 (454 pages and XVIII. plates). So far as this work deals with masonry structures it may be considered to lie entirely outside our field as it does not appeal to any *elastic* principles, i.e. any relation between stress and strain. S. 375–454 deal with the *Theorie der eisernen Brücken.* They are, however, only concerned with straight girders, and with a few cases of continuous beams. They present no particular grace of method and no originality of result, at least, so far as a cursory examination of this and a more thorough acquaintance with other writings of the same engineer allow me to judge.

[887.] Fr. Laissle and Ad. Schübler: *Der Bau der Brückenträger mit wissenschaftlicher Begründung der gegebenen Regeln und mit besonderer Rücksicht auf die neuesten Ausführungen.* Stuttgart, 1857.

I have been unable to find a copy of a second edition of this work, but there appears to have been an edition or issue of a somewhat similar work by these authors published in 1869 and 1870, of which I have seen a French translation entitled: *Calcul et Construction des ponts métalliques*, Bruxelles, (1871?). The later work is very much more extensive than that of 1857, involving about 600 pages in two volumes with numerous plates, while the former has only 156 pages and four plates.

[888.] The authors in their preface state that the ordinary theory of beams due to Navier has not proved itself incorrect but rather incomplete, and that their object is to supplement rather than replace it—*wir haben hiebei streng den Gang der Wissenschaft beibehalten.* They refer especially to the work of Schwedler (see our Arts. 1004–5) as having been of special service to them. They also mention the works of Rebhann and Scheffler (see our Arts. 885–6) as having appeared while their work was in course of preparation.

[889.] There is little deserving of note in the present day in the book. The statement on S. 7 is of course erroneous; the moment of the tractions in the longitudinal fibres of a beam about the neutral axis, as well as the total shear in a cross-section, were certainly not first introduced by Schwedler in 1851, although he may have been the first to use the symbols $\Sigma(Xy)$ and $\Sigma(Y)$ for them. Nor was Schwedler, I think, the first to show that the slope of the bending moment curve is the total shear $\left\{\text{i.e. in symbols } \frac{d\Sigma(Xy)}{dx} = \Sigma(Y)\right\}$, or that points of zero total shear are points of maximum bending moment; but our authors seem to think so on S. 9.

The discussions on shearing stress and the resolution of stresses on S. 18–25 are all old work, and the former only a rough approximation at best. The treatment of the buckling load of struts on S. 25–7 follows Schwarz's work and is as obscure as the original: see our Art. 956. The general discussions on simple and continuous beams, and on plate and lattice girders are reproductions of the results published in various articles in the *Civilingenieur*, *Erbkams Zeitschrift* and the journals of the German and the Austrian *Ingenieur-Vereine* to which we have drawn elsewhere sufficient attention. The book denotes progress in Germany in the theory of bridges, but it is in no way superior to the works of Bresse, Morin or Love, published about the same time in France. It was favourably reviewed by Grashof in the *Zeitschrift des Vereins deutscher Ingenieure*, 1858, S. 312–21.

[890.] L. Molinos and C. Pronnier: *Traité théorique et pratique de la construction des ponts métalliques*, Paris, 1857. The text is in quarto and contains viii + 340 pages; there is also an atlas of plates in folio. This is an extremely well got-up work dealing with the practical side of bridge-structure. The early chapters, containing experimental details and theoretical investigations of flexure and bending moment in simple and continuous beams, are chiefly based on the researches of Hodgkinson, Fairbairn, the Iron-Commission, Clapeyron and Belanger. The practical labours of Stephenson, Brunel and Cowper, as well as the numerous English and French researches on riveting receive ample attention. Indeed the book presents the best historical picture of the state of bridge construction in 1857, both from the mechanical and theoretical sides, that I have come across. A number of cases of continuous beams will be found worked out with practical applications on pp. 253–78, and the comparative criticism of the various types of metal bridges with which the work closes might possibly be still of service. The book is certainly pleasant reading after the laboriously written and poorly printed treatises to which we have referred in the immediately preceding articles.

[891.] *The Useful Metals and their Alloys: Orr's Circle of the Industrial Arts*, London, 1857. This book is the joint production of J. Scoffern, W. Fairbairn, W. Truran and others. Chapters XII.–XXIII. deal with iron and steel and structures made from them, and present a fairly complete picture of the current knowledge with regard to them at that time. We may note a few points of the work:

(*a*) Chapter XII. entitled: *The strength and other properties of Cast Iron* (pp. 210–19) gives details of the various influences which alter the tenacity of cast-iron. Thus the tenacities of cast-iron prepared with cold and hot blast respectively are nearly as 1 : ·8; remeltings increasing the density will increase the tensile strength 2 to 3 times; maintaining the iron in fusion, which has much the same influence, will also nearly double the tensile strength (p. 215); casting 'under a head,' rapidity of cooling, etc. which increase the density, produce increase of strength.

(*b*) Chapter XIII. (pp. 220–251) and Chapter XIV. (pp. 252–69) giving accounts of the preparation of wrought-iron and of 'Recently patented refining processes' (notably the Bessemer, 1856) can be still read with interest, especially by those who wish to understand how it is possible for the processes of working to produce such totally different physical characteristics as occur in the various types of iron.

(*c*) Chapter XV. is entitled: *Metals which alloy with iron* (pp.

270–309); it is devoted rather to their chemical constitution than to their elastic properties.

(*d*) Chapter XVI. is entitled: *On wrought iron in large masses* (pp. 310–33), and is principally occupied with the effect of various methods of working on the tenacity and ductility of wrought-iron intended for ordnance: see our Arts. 879 (*f*), and 1065–7. Chapter XIX. deals with the subject of cast-iron for ordnance (pp. 385–397).

(*e*) Chapters XVII. (*Steel manufacture*) and XVIII. (*Application of steel...*), pp. 334–84, discuss the processes of making steel (including the then recently introduced patent processes of Heath, Bessemer, Uchatius, etc.) and especially treat of the varying physical characteristics due to difference of chemical constitution or to working.

(*f*) Chapters XXI.–XXIII. deal with the application of cast- and wrought-iron to various types of structures. Pp. 410–33 form a practical treatise on the strength of various types of beams; pp. 433–41 deal with iron floors and roofs with considerable detail as to strength and cost; pp. 442–66 treat of girders and bridges for railways, etc. with a *résumé* of the experiments of Fairbairn, Hodgkinson and others on tubular bridges as well as details of strength; pp. 467–78 are occupied with the application of iron to shipbuilding, and give a *résumé* of Fairbairn's experiments on rivets and plates.

It will be seen from this brief account of the contents that the book is calculated to give the reader a very fair knowledge of the condition of applied elasticity in 1857. Novelty in results is of course not to be expected in a work of this kind.

[892.] A work by E. Roffiaen entitled: *Traité théorique et pratique sur la résistance des matériaux*, 1858, might possibly contain some contribution to our subject from the technical side, but I have been unable to find a copy: see however our Art. 925.

[893.] J. B. Belanger: *Théorie de la Résistance et de la Flexion plane des Solides.* Paris, 1858. The first edition of this book, a reprint of lectures at the *École centrale des Arts*, contains 104 pages. The second issued in 1862 and somewhat augmented and modified contains xii + 148 pages. My references will be to the pages of the second edition as the more accessible.

Chapters I. and II. of the book deal with pure traction and torsion, the latter by the old erroneous theory, and offer nothing of note. Chapters III. to VI. are occupied with the discussion of flexure on the Bernoulli-Eulerian hypothesis. In the last of these Chapters various cases of continuous beams are worked out with some detail, and on p. 67 Clapeyron's Theorem of the three moments is given for the case of uniform loading, the two spans having unequal flexural rigidity and the supports not being on the same level. In the notation of our Art. 607 the theorem then takes the form

$$4\frac{l_1}{\epsilon_1}M_1+8\left(\frac{l_1}{\epsilon_1}+\frac{l_2}{\epsilon_2}\right)M_2+4\frac{l_2}{\epsilon_2}M_3=\frac{p_1l_1^3}{\epsilon_1}+\frac{p_2l_2^3}{\epsilon_2}-24\left(\frac{y_1-y_2}{l_1}+\frac{y_3-y_2}{l_2}\right),$$

where $\epsilon_1=E_1\omega_1\kappa_1^2$ and $\epsilon_2=E_2\omega_2\kappa_2^2$ are the flexural rigidities and y_1, y_2, y_3 the heights of the three points of support corresponding to the two spans.

Both simple and continuous beams are dealt with analytically, and their treatment presents no novelty of method. Very simple geometrical proofs based on Mohr's theorem might be given for most of the results stated in these pages.

Chapter VII. modifies the previously given theory of flexure by introducing Jouravski's treatment of slide (see our Art. 183 (*a*)). Chapter VIII. deals on the old lines with solids of equal resistance and Chapter IX. discusses struts without throwing any new light on that difficult subject. Chapter X. treats some simple cases of beams braced by tie-rods; Chapter XI. contains a very insufficient treatment of arched ribs, while the last Chapter XII. after an elementary treatment of the problem of the indefinitely thin right cylindrical shell, practically reproduces Bresse's treatment of a slightly elliptic flue: see our Art. 537.

It is somewhat remarkable that a book certainly not standing at the level of then existing knowledge should have reached a second edition; still more noteworthy that reference to it should be met with at the present day.

[894.] G. H. Love: *Des diverses Résistances et autres Propriétés de la Fonte, du Fer et de l'Acier et de l'emploi de ces métaux dans les constructions*, Paris, 1859. This work contains xxxi + 357 pages. It contributed largely in its day to a knowledge of the physical properties of cast-iron and steel, and forms the opposite pole in technical literature to Morin's book published in 1853. Morin attempted to show that the Bernoulli-Eulerian theory suffices in practical elasticity, Love tried to discredit it altogether. The mean of these views is probably nearer the truth; there are many phenomena of great practical importance, which cannot be explained by existing mathematical theories, while on the other hand these theories *used with proper limitations* are capable of being made of great service in directions not hitherto considered. To those who wish to ascertain the exact results of the technical experiments conducted in the preceding decade, Love's book will still be of great service and suggestiveness.

[895.] After a copious *Table des Matières* (pp. v–xxiii), the book opens with an *Introduction* (pp. xxv–xxxi) in which the author refers to a first publication of his work in 1852[1], and explains why at that time he depended upon English experiments for so much of his data. In the present volume he takes due account of recent French work. He pays, however, a high compliment to Hodgkinson on pp. xxvi–xxvii:

Aussi longtemps que l'emploi du fer fut restreint aux anciens usages, ou que l'industrie n'éprouva qu'un mouvement graduel et modéré, personne ne songea à constater l'insuffisance des anciennes données pratiques et à vérifier le plus ou moins d'exactitude des formules fournies par la Théorie. Mais à peine l'industrie des chemins de fer prenait-elle naissance en donnant une grande extension à l'emploi de la fonte, que M. Hodgkinson commença ses essais sur cet utile métal. Expérimentateur consciencieux, il rejeta toute idée préconçue, comme celle de la *limite d'élasticité*, de nature à limiter le cadre de ses expériences, et, par suite, à donner des notions incomplètes ou inexactes sur les propriétés de la fonte. Il pensa, sans doute, dans sa probité scientifique, qu'il n'avait pas le droit de présenter des expériences tronquées comme celles que nous avaient léguées la plupart de ses prédécesseurs, pour venir en aide à la Théorie. Son but, plus sage, plus utile, était de faire connaître les propriétés du métal aussi complétement que possible; ce à quoi il ne pouvait arriver évidemment qu'en poussant, dans tous les cas, ses essais jusqu'à la rupture, au lieu de les arrêter, comme les autres expérimentateurs, en des points variant avec l'imagination ou la fantaisie particulière de chacun. Le résultat le plus saillant de ces essais faits sur une très-grande échelle et jusques en ces derniers temps, fut un démenti donné *à la limite de l'élasticité*. M. Hodgkinson démontra, en effet, qu'il n'existait, pour *la fonte*, aucun point fixe où l'élasticité commençait à s'altérer; que cette altération se produisait sous les plus petites charges, pour le fer comme pour la fonte.

This paragraph expresses concisely Love's view and the nature of his attack on the theorists. If there be no limit of elasticity, there can be no truth in the ordinary theory, he argues, and thus elasticity becomes a purely empirical science.

[896.] *Livre Premier* of the work is entitled: *Du fer, de la fonte, et de l'acier soumis à des efforts de traction*, and its first chapter is devoted to the extension of these metals (pp. 1–67). In this chapter (pp. 2–3) Love states the general conclusions of the old theory, perhaps

[1] I suppose this to refer to the memoir: *Résistance du fer et de la fonte basée principalement sur les recherches expérimentales les plus récentes faites en Angleterre*, or possibly to a reprint of it. It was published in the *Mémoires...de la Société des Ingénieurs civils*, Année 1851, pp. 163–272. Paris, 1851.

a little too unfavourably, and then formulates the following propositions in opposition to them, propositions which give the key-note to his book (pp. 3–5):

(i) La proportionalité entre l'allongement et la charge n'existe pas *pour la fonte* d'une manière absolue, et pour *le fer doux* cette loi ne peut s'affirmer en général que pour les charges comprises entre zéro et la moitié de celle qui produirait la rupture instantanée.

Does Love here mean (*a*) that cast-iron cannot be reduced to a state of ease, or (*b*) that if it can be, there is not direct proportionality of stress and strain? Would a cast-iron tuning fork give no note?

(ii) Un allongement permanent se manifeste sous les plus petites charges, et le point où les allongements croissent beaucoup plus vite que ces charges est très variable, même dans les fers de même provenance. Par conséquent la limite d'élasticité, en tant qu'elle existe, n'a pas le caractère défini qu'on lui a attribué, et perd forcément toute importance aux yeux du praticien.

The state of ease would here again be an important factor.

(iii) Sous la même charge la fonte s'allonge beaucoup plus que le fer.

This is stated because certain engineers had held the reverse to be true[1]; Love's statement would certainly follow for the state of ease from the greater value of the stretch-modulus of wrought-iron.

(iv) Les écarts considérables de résistance observés sur les échantillons de fer ou de fonte de même calibre, mais de provenances diverses, ne permettent en aucune façon de compter sur une *moyenne de résistance*. Il en résulte que lorsqu'on ne connaît pas la résistance particulière du métal dont on dispose, la prudence conseille d'adopter le taux minimum de résistance fourni par l'observation.

This is only an argument in favour of establishing testing laboratories independent of the manufacturers, possibly as government institutions.

(v) Le fer et la fonte, soustraits aux chocs ou aux vibrations, supportent indéfiniment les charges les plus voisines de celles capables de produire *la rupture instantanée*.

That this is highly questionable follows from the experiments of Wöhler and others: see our Arts. 991, 992 and 997, etc. Most ordnance makers and users would certainly be glad if it were true

(vi) Les formules tirées de la théorie en vigueur ne peuvent être appliquées avec quelque sécurité qu'après avoir subi des transformations importantes.

The legitimate application depends entirely on the limits within which the formulae are applied and on various modifications which may be made in the definitions of the quantities involved.

[1] Love writes: L'opinion contraire s'est généralement accréditée chez les praticiens (ftn. p. 2). He does not, however, cite examples.

[897.] The major part of this first chapter is occupied with details of the experiments of Hodgkinson (see our Arts. 969*, 1411*–12*, 1449*, etc.), of Bornet (see our Art. 817*), of Vicat (see our Arts. 721*–36*), of Leblanc (see our Art. 936*), and of the more recent French experimenters on steel, Gouin and Lavalley, Jackson, Petin and Gaudet, and Tenbrinck, whose results appear to be published in Love's work for the first time.

Love adopts Hodgkinson's formula for cast-iron: see our Art. 1411*. He admits a proportionality of stress and strain for a first stage of the elastic life of wrought-iron and steel, and he gives a formula for iron wire or cable (pp. 61–3) which is based upon the fact that such a wire only becomes straight under a definite load, the wire or cable itself always being manufactured under an initial traction. This practically consists in adding to the stretch-modulus the constant traction under which the cable was manufactured (see, however, our Art. 241). It seems to me that this traction would form an indefinitely small part of the stretch-modulus (i.e. 300 to 1158240 in the example on p. 62 !) and might well be neglected. The real point, I think, to be noted is that no stretch-traction relation would hold till the applied traction reached the constant traction under which the cable had been manufactured.

[898.] Noting the discordance between various observers' results on extension Love remarks :

> que, dans l'état actuel des choses, ce que l'on possède sur l'allongement des métaux usuels laisse énormément à désirer et que des renseignements plus précis seront difficiles à obtenir. Tandis, qu'au contraire, les faits de rupture présentent une constance sur laquelle on peut se reposer avec sécurité ; qu'ils ne peuvent, dans leur interprétation, laisser de prise à l'invention ou à l'imagination comme les allongements. Car il est évident que si deux expérimentateurs peuvent différer sur la question de savoir si, à un moment donné, une barre a atteint, sous une certaine charge, son degré définitif d'allongement, il est impossible qu'ils ne tombent pas d'accord immédiatement sur un fait aussi tranché, aussi brutal que celui de rupture. D'ailleurs les expériences sur la rupture étant les plus simples et plus faciles, tout fait une loi de fixer cette phase de la résistance des solides, comme le point de départ, la seule base de toute formule pratique de résistance (pp. 58–59).

To the last sentence we can only put a very large query, but the first sentences express a very real and oft neglected experimental difficulty.

[899.] Chapter II. (pp. 68–93) of Love's work is devoted to the absolute strength of cast-iron. The author commences by citing Tredgold's extraordinary statements on the absolute strength of cast-iron (*Practical Essay on the Strength of Cast-iron...*, p. 252, Edn. 4) due to the 'paradox in the theory' (see our Arts. 999* and 178), and then proceeds to analyse the early experiments of Minard and Desormes and of Hodgkinson (see our Arts. 940*, 966* and 1408*). These are followed by details of experiments on French cast-iron, in most cases here published for the first time. Love repudiates any mean value for the absolute strength of cast-iron (p. 74), and considers that it must be determined *de novo* for each particular sort.

[900.] Chapter III. (pp. 94–135) deals with various cast-iron structures, as tubes, cylinders, hydraulic presses etc., in which Love supposes the principal stress to be tractive. Love gives an interesting *résumé* of the various empirical formulae for cast-iron pipes. He objects to such formulae on the ground that they do not allow sufficient play for the variation in strength of the metal employed, but concludes by adding a new formula of his own. Let τ be the thickness in centimetres of the pipe, N the number of atmospheres of internal pressure, T the absolute strength in kilogs. per sq. centimetre, D the diameter in centimetres; then Love puts (p. 102):

$$\tau = \frac{6ND}{2T} + \cdot 7.$$

Morin in his *Résistance des matériaux* puts:

$$\tau = \cdot 85 + \cdot 00238ND.$$

It might seem that Love's formula must be better than Morin's which takes no account of possible differences in the value of T, but as Love determines his constant term ($\cdot 7$) for a particular kind of iron from the Fourchambault foundry the advantage is not so obvious. His formula gives far less thicknesses in all cases than any of the other formulae then in use, and thus certainly does not err on the side of safety: see the Table of comparative results p. 103. The reason of this divergence is that Love takes for his formula a less factor of safety (about 6), and allows less ($\cdot 7$ instead of $\cdot 85$ or even 1) for the wear and tear of the surfaces of the pipe.

Pp. 113–7 of this section of the work are devoted to tubes as used for the foundations (piers) of bridges.

For the cylinders of steam-engines Love retains the above formula, increasing, however, the constant term $\cdot 7$ to $1 \cdot 5$ centimetres, as he considers there is greater wear. He compares results calculated from this formula with those given by other formulae (pp. 117–20).

[901.] The remainder of the chapter is devoted to the discussion of hydraulic presses.

Love adopts again the same formula as for pipes, only, having regard to the thicknesses with which we have to deal in such cases, he now neglects the constant $\cdot 7$. He cites also formulae of Barlow and Redtenbacher (pp. 121–2). It is strange that these formulae, based on no theory whatever, should have retained their places in the text-books so long after Lamé's investigations (see our Arts. 1013* and 1038*). Love discusses at some length the hydraulic presses used for raising the tubes of the Britannia Bridge (see our Art. 1474*) and the dimensions and presumed strength of various other presses in practical use (pp. 123–37).

[902.] Chapter IV. (pp. 138–87) is entitled: *Resistance finale à la rupture par traction du fer et de l'acier*. It cites the experiments on bars of iron of Rondelet, Duleau, Martin, Brunel, Tenbrinck, etc. (see our Arts. 696*, 226*, 817*), and gives in a fairly concise form their results as to absolute strength, final stretch, stricture, temperature of the section of rupture and the nature of the rupture-surface (pp. 138–50). The experiments of Séguin, Leblanc and Dufour on iron wire (see our Arts. 984*, 936* and 692*) are then discussed (pp. 150–9). Love shows that if the absolute strength be plotted up to the area of the cross-section we obtain a curve with several maxima of strength, which maxima themselves appear to lie on a regular curve. Such a curve would probably depend very much on the preparation of the wire, and Love himself is compelled to conclude that the tenacity of each special make of wire ought to be independently determined (p. 157).

He then turns to iron plate and cites the experiments of Navier (see our Art. 275*), Clark and of Lavalley, those of the latter being here published for the first time. Finally we have a brief reference to Fairbairn's results (see our Art. 1497*). Love considers that these only show that iron plate can be prepared by special processes to be equally strong in and across the direction of the rolling[1], but they do not invalidate the conclusion of other experimenters that the absolute strength and the ultimate extension are considerably less perpendicular than parallel to the 'fibres.' After some few pages on the absolute strength of various special kinds of iron Love discusses the resistance of steel to traction (pp. 176–87). He publishes for the first time experimental results due to Tenbrinck and Lavalley. The discussion is solely of practical value and has special reference to the kind of steel produced at that date.

[903.] Chapter V. (pp. 188–212) is entitled: *De la résistance à la rupture par traction de la tôle assemblée par des rivets et accessoirement de la résistance des rivets au cisaillement.* Love cites the experiments made for the tubular bridges (see our Arts. 1480*–2*) on the proportion of riveting strength due to shearing strength and friction respectively, and considers the amount of confirmation Clark's results receive from experiments made for MM. Gouin et Cie. Both sets of experiments go to show that the additional strength due to the friction produced on the cooling of the rivet is from 1200 to 1300 kilog. per sq. centimetre of the section of the rivet (p. 191). On the other hand while Clark found the absolute shearing strength of rivet-iron only 2/3 the absolute tractive strength, Lavalley determined it at 3/4.

[1] Il suffirait, paraît-il, de croiser les mises du paquet au lieu de les placer dans le même sens (p. 171).

Love adopts Clark's value; if we suppose uniconstant linear elasticity to hold up to rupture we should have the ratio equal to 4/5 : see our Vol. I., p. 877. The remainder of the chapter forms an interesting practical discussion on the various modes of riveting and the resulting theoretical and experimental strengths.

[904.] Chapter VI. (pp. 213–338) is entitled: *Application du fer et de l'acier sous leurs diverses formes aux appareils et constructions usités dans l'industrie.* This chapter consists entirely of practical applications, and the small amount of theory applied is often of a rather dubious character (e.g. pp. 214, 217 etc.). The topics dealt with are: riveted boilers (pp. 213–25), water pipes of plate iron (pp. 225–32), water reservoirs of plate iron (pp. 232–42), iron chain-cables (invented by Captain Brown[1] and first used by him on board the *Penelope*, 1811), the best form of link for chains and the few details known of their strength (pp. 242–75), and lastly the cables of iron wire and bar-iron for suspension bridges with a lengthy discussion of the various applications of such bridges, the strength of their various parts, their advantages and dangers (pp. 275–338)[2].

[905.] The final chapter (pp. 339–57) of Love's work is entitled: *De certaines résistances du fer et de la fonte se rapprochant plus particulièrement de la résistance à la rupture par traction.* This is devoted to such subjects as the strength of screws under a traction which does not turn them (pp. 340–3), so that rupture is produced by shearing off the thread, on punching (pp. 343–5), on the resistance of iron and steel to torsion (pp. 345–51), and on the strength of railway axles and their journals (pp. 351–7) Several of these matters are treated with greater detail and more exact theory in other works of this period: see our Arts. 966–7, 1043, 1049, 957–9 and 988–1003.

The work concludes with an appendix giving sheets prepared with blank columns for various details on the local preparation and strength of the different kinds of metals: these were to be filled in by experimenters and returned to the author.

In conclusion we may remark that the book was distinctly the best practical treatise on the strength of iron and steel produced in the years 1850–60, and that even to the present day it may be consulted on some points with advantage.

[906.] W. Fairbairn: *Useful Information for Engineers.* The first edition of the *First Series* appeared in 1855 and a fifth edition of this series in 1874, the first edition of the *Second Series* in 1860,

[1] For the history of chain cables: see *Transactions of the Institution of Naval Architects*, Vol. I., pp. 160—70. London, 1860.

[2] The first suspension bridge was built in America by James Finley at Jacob's Creek in the year 1796; the first in Great Britain was due to Samuel Brown and crossed the Tweed at Berwick, being built in the year 1819; and the first in France was due to Séguin aîné and dates from 1821.

and a second edition in 1867, the first edition of the *Third Series* in 1866. Our references will be in each case to the pages of the last edition mentioned. The work consists of reprints of Fairbairn's original researches, and of more popular articles and lectures by him. It has played a considerable part in developing a more rational scientific education for engineers.

[907.] In the *First Series*, *Lecture* II. deals in a popular manner with the strength of boilers (pp. 28–53) and *Lectures* VI. (pp. 127–153) and VII. (pp. 154–7) have further details of the strength of the materials used in boiler construction. *Lecture* X. is a popular account of the strength of the material used in iron-ship building. In the *Appendix* is a reprint of Fairbairn's Royal Society paper on the strength of wrought-iron plates: see our Arts. 1495*–1503*.

[908.] The only other part of this *Series* which needs notice is the second portion of the *Appendix* entitled: *Experimental Researches to determine the Strength of Locomotive Boilers, and the causes which lead to Explosion* (pp. 321–40). This paper originally appeared in *The Civil Engineer and Architect's Journal*, Vol. 17, 1854, pp. 219–223. See also the *Mechanic's Magazine*, Vol. 60, pp. 393–5. A series of experiments was first made on the absolute strength of the fire box and exterior shell of a locomotive boiler (pp. 325–8). This was followed by an attempt to find relations between the temperature of the steam, the time and the pressure in a boiler when the safety valve is screwed down and the fire kept going. It is shown that under these circumstances a boiler will burst in from about 20 to 40 minutes (pp. 328–331). Fairbairn next deals with the strength of the flat surfaces or sides of a fire box and with the strength of the stays (pp. 331–7). Two experiments were made in which two pair of parallel plates one of copper (·5″ thick) and the other of iron (·375″ thick) were stayed together with one stay to the 25□″ and one stay to the 16□″ respectively. Fairbairn says that the weakest part of the box was not in the copper but in the iron plates which gave way by stripping or tearing asunder the threads or screws in the part of the iron plate at the end of a stay. In the first experiment, however, the head of one of the stays was drawn through the copper plate. The pressures at which the fire boxes gave way were respectively 815 and 1625 lbs. per square inch and thus immensely greater than what could be borne by any other part of the boiler. It is not easy to see theoretically why the strengths should be nearly as 1 : 2 in the two cases of one stay to the 25 and one stay to the 16 sq. inches respectively.

The paper concludes with *Experiments to determine the Ultimate Strength of Iron and Copper Stays generally used in uniting the Flat Surfaces of Locomotive Boilers* (pp. 338–40). Here iron and copper stays were screwed and riveted into iron and copper plates. Fairbairn concludes that:

the iron stay and copper plate (not riveted) have little more than one-half the strength of those where both are of iron; that iron stays, screwed and riveted into iron plates, are to iron stays screwed and riveted into copper plates as 1000 : 856; and that copper stays, screwed and riveted into copper plates of the same dimensions, have only about one half the strength of those where both the stays and plates are of iron (p. 340).

Hence so far as regards strength iron is much superior to copper as a stay, but its inferior conducting powers and probably inferior durability have still to be taken into account.

[909.] In the *Second Series* we may note as popular lectures involving only elementary theorems in the strength of materials: *Lectures* V. and VI. (pp. 100–37) on the *Strength of Iron Ships* see also *Transactions of the Institution of Naval Architects*, Vol. I., pp. 71–97, London, 1860. This is a subject on which Fairbairn had, as one of the earliest constructors of iron-vessels, a great right to be heard and these lectures are thus of considerable interest from the standpoint of the history of technical elasticity. *Lecture* VII. (pp. 138–56): *On Wrought Iron Tubular Cranes* with experiments on their deflection and set, and a theory of their strength by Tate, is also of interest: see our Art. 960. *Lecture* VIII. (pp. 157–73) returns to the old subject of boiler strength, appealing, however, to the then recently published memoir on the strength of flues: see our Art. 980.

In the second part of this volume entitled: *Experimental Researches* we have reprints of the memoirs on cylindrical vessels of wrought-iron (see our Art. 980), on glass globes and cylinders (see our Arts. 853–6), on the tensile strength of wrought-iron at various temperatures (see our Art. 1115)[1], and on the resistance to compression of various kinds of stone (see our Art. 1182). On pp. 328-9 will be found some experiments on Irish Basalt or Whinstone to be added to the results of this memoir. The specimens of this stone "fractured by vertical fissures splitting up into thin prisms, wedge-shaped usually at one end."

[910.] The *Third Series* contains the following papers dealing more or less closely with our subject: *Lecture* VI. (pp. 98–124) entitled: *Iron and its Appliances*, which returns again to the strength of boilers; a paper *on the Construction of Iron Roofs* (pp. 204–43), this gives details of the trusses of large iron roofs and the calculation of the stresses in their members; a paper *On the mechanical properties of the Atlantic Cable* (pp. 276–89), this is a reprint from the *Report of the British Association*, 1864, pp. 408–15, and gives details of the absolute strength, stretches and ultimate elongations of a great variety of cables as well as of their several parts, central core, covering wires and gutta percha sheath; finally a reprint (pp. 290–316) of Fairbairn's Royal Society memoir of 1864 (*Phil. Trans.* pp. 311–25) on the effect of impact and repeated loading on wrought-iron girders. The experi-

[1] See also *The Artizan*, 1856, pp. 227-8, and *Dinglers Polytechnisches Journal*, Bd. 150, 1858, S. 105-8 and S. 288-95.

ments embraced in this paper had formed the subject of a communication to the British Association in 1860 (see our Art. 1035) and various accounts of them had appeared in the technical journals[1]. We defer our full analysis of the memoir and criticism of the methods of experiment until we come to deal with the technical memoirs of the decade 1860–70.

[911.] The titles of two other books by Fairbairn may be just noted here:

(*a*) *On the Application of Cast and Wrought Iron to Building Purposes.* London, 1854. The third edition has added to it a section on *Wrought Iron Bridges.* A fourth edition appeared in 1870.

(*b*) *Treatise on Iron Ship Building, its History and Progress, as comprised in a Series of Experimental Researches on the Laws of Strain; the Strengths, Forms and other Conditions of the Material, etc.* London, 1865.

[912.] Another technical text-book, the contents and method of which are much like those of this decade is A. Ritter's *Lehrbuch der technischen Mechanik* The first edition was published in 1865, and the third edition which I have used appeared at Hannover in 1874. The *Fünfter Abschnitt* entitled: *Statik elastischer Körper* (S. 479–563), and the *Sechster Abschnitt* (S. 564–616): *Dynamik elastischer Körper*, belong to our subject. The work in its third edition is still a fairly useful text-book for the engineering student. The part on elasticity and the strength of materials contains one or two points, to which I may refer as interesting, and one or two grievous errors, against which the student should be warned.

[913.] Let us assume that it is legitimate to apply the Bernoulli-Eulerian theory of beams to a cantilever, which has a constant thickness (h) in the vertical plane of flexure, but in a horizontal plane perpendicular to the plane of flexure, is in the form of an isosceles triangle of base b and height l. Then, if $1/\rho$ be the curvature at distance x from the free end of the cantilever under load P and $E\omega\kappa^2$ be the flexural rigidity, we have:

$$\frac{E\omega\kappa^2}{\rho} = Eh\frac{xb}{l}\frac{h^2}{12\rho} = Px,$$

or

$$\rho = \frac{Eh^3b}{12lP} \quad \text{.............................(i).}$$

[1] E.g. *The Artizan*, 1860, pp. 219–21, and 1861, pp. 228–31.

Thus the cantilever has uniform curvature. Further if f be the terminal deflection, then f equals $l^2/(2\rho)$ very nearly, or

$$f = \frac{6l^3P}{Ebh^3} \quad \text{..............................(ii).}$$

If S be the maximum traction to which the material ought to be subjected :

$$S = \frac{Eh}{2\rho} = \frac{6lP}{h^2b},$$

which gives for the minimum requisite breadth b at the built-in end :

$$b = \frac{6lP}{h^2S} \quad \text{..............................(iii).}$$

Such formulae can at most be supposed to hold only when the triangular cantilever changes its cross section very gradually, i.e. when b/l is very small.

Ritter now builds up a spring formed of several laminae by cutting

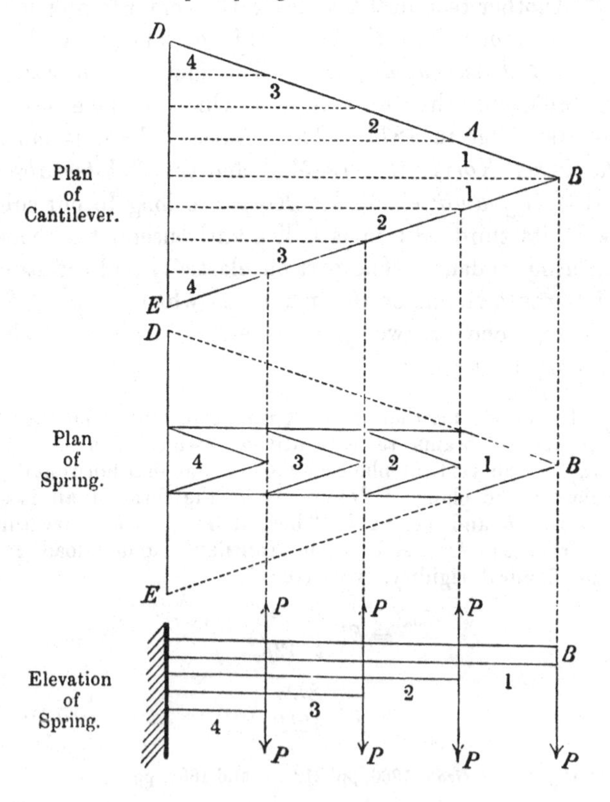

up the triangular cantilever and replacing the several parts as indicated in the accompanying figures. He thus gets the half of a laminated spring of the ordinary form, and one which has besides a very easy theory. He supposes the pointed end of each lamina to press on the lamina above with a force equal to P, the load at the end of the spring. The result is that the triangular part of each lamina supports as a cantilever a load P at its apex, which causes it to take the curvature given in (i), while the rectangular part of each lamina is acted upon by a couple $Pl/4$ which will also be found to give it the curvature determined by (i). Thus each lamina is bent in exactly the same way as if it were a part of the triangular cantilever discussed above, while the spring itself is a solid of equal resistance, whose deflection is given by (ii) and whose proper breadth can be determined by (iii). For this special case the result appears to agree with that of Phillips: see Eqn. (xxv) of our Art. 496. Ritter does not, however, demonstrate clearly how and why the action of the apex of one lamina on the lamina above must equal P.

[914.] On S. 521–6, we have another added to the already numerous methods of calculating the maximum safe loading for a strut. Suppose the strut bent to a central deflection f, its ends being pivoted. Then if $\omega\kappa^2$ be as usual the moment of inertia of the cross-section and h the diameter of the section in the plane of flexure, it is easy to see that the maximum compressive stress T due to a longitudinal load P is (see our Art. 832*):

$$T = \frac{P}{\omega}\left(1 + \frac{fh}{2\kappa^2}\right).$$

Now, Ritter argues that f cannot be as great as it would be in the case of a circular flexure, when

$$f = l^2/(8\rho), \text{ nearly,}$$

if l be the length of the strut and $1/\rho$ its uniform curvature. But if δ be the stretch (or squeeze) due *solely to the bending* at the central section

$$\delta = h/(2\rho),$$

and thus the maximum of $fh = l^2\delta/4$, whence we find on this hypothesis:

$$P = \frac{\omega T}{1 + \frac{\delta}{8}\frac{l^2}{k^2}}$$

If T be the maximum safe compressive stress; this will give a minimum limit for the safe maximum load P, supposing the safety to be rendered doubtful by compression before extension. So far there is no ground for much criticism. But what is δ to be taken as? Ritter says it is to be put equal to *das Verkürzungsverhältniss, welches der Elasticitäts-Grenze entspricht* (S. 524). This would be at least something greater than its real value which is solely due to the bending and therefore the tendency would be to err on the side of safety. Ritter would thus make $\delta = T/E$, where T is the maximum compressive stress, although he does not express himself in this manner. We thus obtain finally:

$$P < \frac{\omega T}{1 + \frac{1}{8}\frac{T}{E}\frac{l^2}{\kappa^2}} \text{ ....}(a).$$

This result agrees with that which I have obtained by a very different method in Eqn. (x). Art. 650, if it be remembered that C and p_0 of that article $= T$ and $\pi^2 E\kappa^2/l^2$ of this respectively.

Hence it seems extremely probable that (a) may be considered as a fairly efficient measure of the safe load for struts, although the process by which Ritter deduces it is extremely questionable.

The application on S. 528 to the case of an eccentric load upon a strut seems to me quite illegitimate.

[915.] (a) On S. 528–33 we have a wholly inadmissible theory of shear, which leads to the slide-modulus being always one-half of the stretch-modulus. This is applied to deduce an erroneous theory of torsion on S. 533–6.

(b) The following sections deal very fully with stresses in a great variety of roof trusses and bridge frames. These stresses are deduced by taking a section cutting three bars only, and by equating the moment of the stress in one bar about the intersection of the other two to the bending moment of the girder at the section. This valuable method, especially useful in testing graphical work, is now generally termed *Ritter's Method* (S. 537–55). The following pages (S. 555–63) deal with frame arches having a pin-joint, and therefore zero bending moment, at the crown. These important frames have been largely used in German engineering practice. For a still more complete discussion of the application of *Ritter's Method* to the stresses in various types of frames, we must refer the reader to his *Elementare Theorie und Berechnung eiserner Dach und Brücken-Constructionen*, Hannover, 1862 (Second edition, 1873[1]).

[916.] (a) In the *Sechster Abschnitt*, Ritter turns in the first place to problems of resilience. Thus he calculates in the usual elementary manner (S. 567) that the total longitudinal resilience of a bar $= \frac{1}{2}$ (volume) $\times\, T^2/E$, where T is the maximum traction allowable. But like most elementary writers he equates this result to the kinetic energy of the impulse-giving body, quite forgetting that it does not follow that such a body will communicate its kinetic energy to the *whole* bar uniformly and not expend it in producing strain in one part only. That it is not distributed *statically* in the case of either transverse

[1] An English translation by H. R. Sankey appeared in 1879 (London, Spon).

or longitudinal impact is shown in our Arts. 361–71 and 401–7, and the same remark holds good for torsional impulse.

(*b*) On S. 567–8 there is a paragraph entitled: *Einfluss der Fehlstellen*, in which the author remarks:

Es ergiebt sich also das bemerkenswerthe Resultat: dass die Widerstandsfähigkeit eines Körpers gegen mechanische Arbeit, oder gegen lebendige Kraft bewegter Massen, durch *Verminderung* der Materialmenge unter Umstände vergrössert werden kann.

This paradox is easily accounted for by a flaw in Ritter's argument, which it is surprising should have survived a third edition.

(*c*) The erroneous theory of torsion and a consequently wrong expression for torsional resilience reappear on S. 571–3. The theory of the torsional pendulum on S. 573–5 is also incorrect.

(*d*) *Capitel* XXV. (S. 575–608) involves little more than the usual theory of impact of particles and of uniplanar bodies. The method by which the problem on S. 589–91 is treated seems to me very doubtful indeed. Ritter endeavours to ascertain what the velocity of a cylindrical shot must be in order that the shot may penetrate an iron plate of a given thickness, and he obtains a solution by equating the kinetic energy of the shot to the product of the maximum total shearing resistance of the hole punched in the plate into the semi-length of the shot. It seems to me that a better result would have been obtained by equating the kinetic energy of the shot to the work of punching, or to

$$\frac{S^2}{2\mu}\tau\pi d,$$

where S is the absolute shearing strength, d the diameter of the shot, τ the thickness of the plate and μ the slide-modulus. This of course supposes elasticity to hold up to rupture, which is an approximation for some sorts of steel only. Ritter supposes the work done by the total shear ($S\tau\pi d$) in *flattening the shot* to be the total shear into the semi-length of the shot, but I do not understand this, nor his method of deducing Equation (823).

(*e*) *Capitel* XXVI. (S. 608–16) deals in an elementary fashion, but not always correctly, with the stresses produced in elastic bodies owing to the relative accelerations of their parts. For example, if a thin ring of radius a and uniform density ρ be rotating with spin α about an axis perpendicular to its plane it is easy to show that the maximum stress $=\rho\,(\alpha a)^2$ (S. 613), but when Ritter attempts to apply a similar theory to a rotating circular disc he goes hopelessly wrong, for he assumes the stress *uniform* across the whole length of a diameter. There are other parts of this chapter which seem to me very doubtful, but it is impossible to devote further space to their discussion.

[917.] A third edition of Rühlmann's *Grundzüge der Mechanik* was published at Leipzig in 1860, but I have not examined this work, as the number of technical text-books is too great to be examined individually. So far as my experience goes they rarely contain any novelty in the domain of elasticity—beyond an occasional theoretical heresy.

GROUP B.

Memoirs dealing with the Application of the Theory of Elasticity to special Technical Problems.

[918.] F. Fink: *Versuche über die Tragfähigkeit gespannter und ungespannter Hölzer. Polytechnisches Centralblatt,* Jahrgang 1851, Cols. 1485–8, Leipzig. (Extracted from the *Gewerbeblatt f. d. Grossh. Hessen,* 1851, S. 237.) This paper does not appear to contain more than the statement of the fact that a wooden bar subjected to transverse load supports more when its ends are built-in than when they are simply supported. In the actual case the ends were not built-in but subjected to tractive load. Fink does not work out the theory of this case, but it obviously approximates roughly to built-in ends; the breaking loads were, however, in general *more than double* their values for simply supported ends.

[919.] A. Wöhler: *Notiz über die Berechnung der Durchbiegung elastischer Körper. Erbkams Zeitschrift für Bauwesen,* Jahrgang III., S. 433–6, Berlin, 1853. In this paper Wöhler enquires what the unstrained form of the central line of a cantilever must be in order that when it is loaded at its free end with a weight P the central line may become straight.

For a cantilever of uniform cross-section it seems to me we have the unstrained form of the central line given by:

$$y = \frac{P}{6E\omega\kappa^2}(x^3 - 3l^2x + 2l^3),$$

where y is measured vertically upwards from the strained position of the central line and the origin is at the free end.

If the section be not uniform, we have on the Bernoulli-Eulerian hypothesis an equation of the form

$$\frac{d^2y}{dx^2} = \frac{Px}{E\omega\kappa^2}$$

to integrate where $\omega\kappa^2$ is a function of x.

For 'beams of equal resistance' we must have

$$Px/(\omega\kappa^2) = T_0/z,$$

where T_0 is the uniform traction in the fibre at maximum distance z from the neutral axis. Hence

$$\frac{d^2y}{dx^2} = \frac{T_0}{E}\frac{1}{z},$$

or if $T_0/E = n$,

$$d^2y/dx^2 = n/z,$$

which we can integrate as soon as z is given. None of these results agree with Wöhler's, who obtains differential equations of the first order only, and integrates them for a number of special cases, which the reader can easily construct for himself. (e.g. $y^{m+1} = px^m$, etc.)

[920.] C. R. Bornemann: *Ueber relative Festigkeit: Polytechnisches Centralblatt*, Jahrgang 1853, Cols. 1297–1308. (Extracted from *Der Civilingenieur*, Bd. I., S. 18.) The author notes the discrepancies which occur in the text-book formulae for the strength of beams (in great part due to the 'paradox': see our Art. 930), and proposes to use a stretch-limit instead of a stress-limit in the 'extreme fibre'. He is, I think, right in preferring a strain- to a stress-limit, but otherwise the paper only contributes what must, I think, be regarded as empirical formulae for relative strength.

[921.] C. R. Bornemann: *Graphische Tabelle über die relative Festigkeit. Der Civilingenieur*, Neue Folge, Bd. I., S. 18–25. Freiberg, 1854.

Bornemann by means of logarithmic scales, or what Lalanne would term an *abaque* (see *Annales des ponts et chaussées*, T. XI., 1846, pp. 1–69), represents by a system of parallel straight lines the families of curves given by the formula:

$$\text{Bending Moment} = T \times \frac{\omega\kappa^2}{h},$$

for the flexure of beams[1]. Here T is the safe limit to tractive

[1] The idea of applying the method of logarithmic coordinates to the graphical representation of the strength of materials, as well as the method itself is due to Lalanne. See his: *Mémoire sur les tables graphiques et sur la géométrie anamorphique. Annales des ponts et chaussées*, T. XI., pp. 1–69, 1846; also his book *Méthodes graphiques pour l'expression des lois empiriques ou mathématiques à trois variables*, Paris, 1878. Louvel gave *abac* diagrams for the resistance of iron bars in double-T in the *Portefeuille des conducteurs des ponts et chaussées*, 4e Série, Nos. 6 and 7. Phillips' formulae for springs (see our Arts. 483-508) have been reduced to

stress, $\omega\kappa^2$ is the moment of inertia of the cross-section about the "neutral axis" and h is the distance of the "extreme fibre" from that axis. For example for a beam of rectangular cross-section ($b \times 2h$) we have

$$\text{Bending Moment} = \frac{2T}{3} \times b \times h^2.$$

There would thus be three entries into the graphic table, i.e. bending moment, breadth and height of beam, supposing T a definite constant. Bornemann after some rather lengthy discussion chooses values of the constant T for wrought-iron, cast-iron and wood, and his table contains the system of lines for rectangular beams, with some few lines at a different slope for beams of circular, hollow circular, square and I sections. The method is of real worth for considering the relative strength of beams of diverse cross-sections. But I think Bornemann's example of the method is of small value, since T varies greatly with the different kinds of iron and wood and further varies with the shape of the cross-section. The first entry ought not to be M the bending moment, but M/T, in which case the same set of lines answer for all materials. The method has been further discussed by Vogler: *Anleitung zum Entwerfen graphischer Tafeln*, S. 37, Berlin, 1877. I have for some time used a table constructed like Bornemann's, but with M/T as the variable, for calculating beams with either a uniform load or an isolated central load. The body of Bornemann's paper is taken up by details as to the best practical values for the constants in formulae for absolute and relative strength, and to this part of it we have briefly referred in the preceding article.

[922.] O. Ortmann: *Zur Theorie der Widerstandsfähigkeit der Baumaterialien. Försters Allgemeine Bauzeitung*, Jahrgang xx., S. 243–62. Wien, 1855.

Ortmann in the Jahrgang VIII. (1843) of this journal (S. 408–40, under the title *Theorie des Widerstandes fester elastischer Körper*) had

graphical *abac* representation by Lévy-Lambert in the *Annales des ponts et chaussées*, T. xx., 1880, 2e Semestre, pp. 59–65, while Chéry has given similar diagrams for the strength of beams of wood, iron, etc. of the principal forms in use in the *Pratique de la résistance des matériaux dans les constructions*, Paris, 1877, p. 16, Plates 7 to 24. An interesting discussion of the method is given by P. Terrier in his French translation of Favaro's *Calcul graphique*, Paris, 1885, pp. 208–224, with very copious references. See also Lalanne's *Description et usage de l'abaque*, Paris, 1845 and 1851, English translation, London, 1846.

drawn attention to the fact that in many cases of flexure there is a normal thrust on the cross-sections of a beam and therefore that in such cases the 'neutral axis' does not pass through the centroids of the cross-sections. He says in the later memoir that Morin and Rebhann (see our Arts. 876 and 885) have neglected this fact[1]. There does not, however, seem any real novelty in Ortmann's own investigations. He attempts to take into account the effect of impulsive stress, but certainly does not get further than, if as far as, Poncelet had done many years previously (see our Art. 988*). Some of the results obtained seem also very questionable both in hypothesis and in method of analysis, and the paper does not seem to require from us more than a note of caution.

[923.] With: *Ueber den Widerstand der Baumaterialien. Organ für die Fortschritte des Eisenbahnwesens*, Bd. 8, S. 200. Wiesbaden, 1853. This paper contains a general *résumé* of the theory of the strength of materials combined with a statement of supposed experimental facts. It appears to be based chiefly on Morin's work: see our Art. 876, and contains assertions with regard to the proportionality of stretch and traction in cast-iron which are certainly incorrect. It has no present value.

[924.] A long paper: *Ueber zusammengesetzte Festigkeit* was read by Grashof to the *Berliner Bezirksverein* of German engineers on March 7, 1858. It is printed *in extenso* on S. 183–225 of the *Zeitschrift des Vereins deutscher Ingenieure*, Jahrgang III., 1859. It follows very much the lines of Weisbach (see our Art. 1377*) and contains nothing of intrinsic importance. We shall refer to Grashof's methods later when dealing with his well known treatise on elasticity.

[925.] F. Roffiaen: *Widerstandsfähigkeit der Baumaterialien. Praktische Beispiele für die Stärkebestimmungen der verschiedenen Verbandstücke der Holz- und Eisenkonstrukzionen nebst Bemerkungen über Bauten, die aus diesen Materialien ausgeführt sind. Försters Allgemeine Bauzeitung*, Jahrgang XXIV., S. 257–320. Wien, 1859. I presume this is a translation of the work referred to in my Art. 892, but which was inacessible to me in the original French. There is, however, no statement that it is a translation.

The first part applies the ordinary theory of elasticity to frames built-up of straight wooden beams, the second to curved wooden arches. In both cases the author appeals to Ardant, see the *Addenda* to our Vol. I., pp. 4–10. The third part deals with iron-structures, giving numerical examples for cases of wrought, cast and plate iron girders,

[1] This is hardly correct: see p. 151 of the first edition (1853) of Morin's work.

and later the author passes to structures combining wood and iron (S. 301–4). There does not seem to me any special novelty in these portions of the work, nor any special permanent technical value in the numerical examples worked out. In an *Anhang* (S. 304–20) the author develops a new theory of flexure and applies it to several examples. This theory as detailed on S. 304–5 is very obscure, but it seems to amount to taking the stretch- and squeeze-moduli different, and on this supposition calculating the position of the 'neutral axis' which no longer pass through the centroid of the cross-section. The ultimate resistances to tension and compression are also taken to be in the ratio of these moduli. The application by Roffiaen, however, of the theory seems perfectly arbitrary. Thus his (v) and (xi) S. 305 are quite erroneous, and I do not grasp the meaning of (vi) S. 304, nor its application in paragraph 2 of S. 306. Indeed, what is new in the investigation seems to me wrong. The hypothesis itself is at least as old as 1822: see our Arts. 234*–40*.

[926.] Klose: *Ueber die Festigkeit und die zweckmässigste Form eiserner Träger, Hannoverische Bauzeitung (Architecten- u. Ingenieur-Verein)*, Hannover, 1854, S. 523.

The author at rather needless length of analysis based on Navier's theory of ribs (see our Art. 257*), shows that a cantilever in the form of a circular arc if loaded at the free end perpendicular to the tangent at the built-in end is no stronger than a straight cantilever of the same cross-section and of length equal to half the chord of double the arc of the circular cantilever. This result is only true when the radius of the cantilever is great as compared with the linear dimensions of the cross-section (see our Arts. 519 and 621). Klose tested this theoretical result for four rods six feet long and with cross-sections squares of one inch, supported at their ends and centrally loaded. The details of the experiments show a remarkable agreement between the deflections of the rods for the same weights whether they were of straight or circular central line and this agreement lasted up to rupture, which also occurred at the same load for both types. The circular rib therefore has no advantage as a cantilever over the straight beam, and further has the disadvantage of considerable lateral yielding, the values of which are tabulated in the experiments.

[927.] The second part of the paper deals with the comparative strength of cast-iron beams of two special cross-sections. The first is a ⊤ in which the web is trapezoidal, and the second a ⊥ akin to Hodgkinson's strongest section (see our Art 244*). The beams were 14′ between the points of support, but not, as I think they ought to have been, of the same height, the former being $8\frac{1}{2}''$ and the latter 10″; the areas of the cross-sections were practically equal. Klose found the ⊥ much the stronger section. But it is remarkable that he found the theoretical value of the stress in the 'extreme fibre' at rupture was the same for both cross-sections. He does not state whether this stress was also equal to the absolute tensile strength of his cast-iron. He applies without hesitation the Bernoulli-Eulerian theory of flexure to find the conditions of rupture.

[928.] A. Junge: *Ueber die Tragkraft gesprengter Balken. Polytechnisches Centralblatt*, 1855, Cols. 844–54. (Extracted from *Der Civilingenieur*, Neue Folge, Bd. 2, S. 79. Freiberg, 1855.)

Der gesprengte Balken besteht aus zwei über einander liegenden Theilen, welche an ihren Enden fest verbunden sind, übrigens aber durch dazwischen gestellte Spreizen auseinander gehalten werden (Col. 844).

The present paper investigates whether such a girder, which has initial strain, is theoretically stronger than one in which the two booms are united so as to form a simple girder. Junge supposes the two booms exactly equal. He shows by means of tables for wood and wrought-iron that the limits within which the split beam is stronger are rather narrow:

Die Gefahr die günstigste Spannung zu überschreiten liegt also sehr nahe (Col. 854).

Several such girders had had to be removed after a short period of use, and they appear now to have gone out of fashion. They were introduced by Laves of Hannover in a work entitled: *Ueber die Anwendung und den Nutzen eines neuen Constructions-Systems nebst erläuternder Beschreibung desselben*, 1839.

I have not verified Junge's analysis; he assumes that the curvature of the initially strained booms is circular (Col. 851).

[929.] Baumgarten: *Note sur la valeur du coefficient d'élasticité de la fonte à l'appui du rapport de MM. Collet-Meygret et Desplaces sur le viaduc de Tarascon. Annales des ponts et chaussées*, 1er Semestre, 1855, pp. 225–233. Paris, 1855. This paper gives some results on the flexure of beams of considerable size, the cross-sections being ⊤ and ⊥ the beams had a varying cross-section from middle to end, and might be conceived of as parabolic or as 'solids of equal resistance.' Baumgarten supposes a beam of length $2l$ divided into $2m$ parts of equal length, and the cross-section in each of these parts to be uniform for the part, then he gives the following formula for the deflection f under a central load $2P$:

$$f = \frac{l^3P}{3Em^3}\left\{\frac{3(m-1)m+1}{\omega_1\kappa_1^2} + \frac{3(m-2)(m-1)+1}{\omega_2\kappa_2^2} + \frac{3(m-3)(m-2)+1}{\omega_3\kappa_3^2} + \dots\dots + \frac{3(m-n)(m-n+1)+1}{\omega_n\kappa_n^2} + \dots\dots + \frac{3.1.2+1}{\omega_{m-1}\kappa^2_{m-1}} + \frac{1}{\omega_m\kappa_m^2}\right\},$$

where $\omega_n \kappa_n^2$ is the moment of inertia of the cross-section of the nth part about its central axis.

By means of this formula he finds different values of E for the two pieces on which he has experimented, and both values differ largely from that usually adopted for cast-iron.

An attempt to explain the 'beam paradox' (pp. 231–3), with which Baumgarten concludes his memoir, falls into the old fallacy of supposing the proportionality of stress and strain to last up to rupture: see our Arts. 173, 507, 930–8 and 1053.

[930.] William Henry Barlow. *On the existence of an element of Strength in Beams subjected to Transverse Strain, arising from the Lateral Action of the fibres or particles on each other, and named by the author the "Resistance of Flexure." Phil. Trans.* 1855, pp. 225–242. This paper was received on February 23 and read March 29, 1855.

It deals with the "old beam paradox :" see our Arts 173, 507 and 542. Barlow expresses it thus: "the strength of a bar of cast iron subjected to transverse strain cannot be reconciled with the results obtained from experiments on direct tension, if the neutral axis is in the centre of the bar" (p. 225). By a series of experiments (pp. 225–8) Barlow shows (as many previous experimenters: see our Arts. 998*, 1463*(*e*) and 876) that the neutral line within the limits of experimental error coincides with the central line. He then endeavours to explain by 'lateral adhesion,' i.e. shearing stress, the increased absolute strength. We must here note one or two points: (i) the formula adopted by Barlow from the Bernoulli-Eulerian theory of beams supposes the material to remain elastic up to rupture,—this is certainly not true; (ii) it assumes the elasticity also linear, this again is hardly true for cast-iron; (iii) even if with Saint-Venant we introduce the proper slide terms into the formula they would make no sensible difference except for very short beams; (iv) to account for the 'paradox' we must suppose stress-strain relations other than linear to hold in the neighbourhood of rupture, i.e. such relations as those suggested in our Arts. 1411* and 178. It is to such relations, giving us results *depending not only on the moment of inertia of the cross-section, but more generally on its shape*, that we must look for light on the so-called 'paradox.'

Barlow gives (p. 231) an empirical formula for the breaking strength of 'doubly ribbed open beams'. The cross-sections were not of sufficient variety to give any evidence that the quantity he terms 'resistance to flexure' is really independent of the shape of the cross-section.

[931.] W. H. Barlow. *On an Element of Strength in Beams subjected to Transverse Strain, named by the author "The Resistance of Flexure." Second Paper. Phil. Trans.*, 1857, pp. 463–88. This paper was received March 12 and read March 26, 1857. It is a continuation of the subject dealt with in the memoir referred to in the preceding article.

Let T be the breaking tensile stress, W the breaking load, $E\omega\kappa^2$ the flexural rigidity of the beam, h the distance from the neutral axis, of the 'fibre' furthest removed, l the length of the beam, and $1/\rho$ the curvature at the mid-point. The beam is supposed weightless, doubly supported and centrally loaded. Then we have, assuming the correctness of the Bernoulli-Eulerian theory up to rupture:

$$T = Eh/\rho, \qquad E\omega\kappa^2/\rho = \tfrac{1}{4}Wl,$$

or

$$W = 4T\omega\kappa^2/hl \text{.........................(i).}$$

Now in his first memoir Barlow assumes that T consists of two parts, the first T_1 due to tensile strength and the second T_2 to what he terms 'resistance to flexure'—produced by 'lateral adhesion.' The former part he holds to be constant, the latter to vary when the beam is hollow as the 'depth of metal into the deflection.' This value of T_2 seems curious, but as an empirical expression we may perhaps allow it to stand, however little it may have to do with 'lateral adhesion.'

[932.] In the second memoir, however, Barlow goes much further, he expresses his traction $\widehat{zz}$ at distance x from the neutral axis by the formula (p. 472)

$$\widehat{zz} = \frac{Ex}{\rho} + T_2 = \frac{T_1 x}{h} + T_2 \text{ .... (ii).}$$

He then proceeds to take the moment of these tractions at the middle of the beam round the neutral axis, and equates them to the bending moment, or we have:

$$\frac{T_1 \omega\kappa^2}{h} + T_2 \omega\overline{x} = \tfrac{1}{4}Wl.$$

At first sight it would appear that the second term ought to vanish, but Barlow evidently intends that T_2 shall change sign on crossing the neutral axis so that if ω_1, ω_2 be the areas of cross-section above and below the neutral axis and $\bar{x}_1$, $\bar{x}_2$ the *numerical* distances of their centroids from this axis we have:

$$T_1 \frac{\omega\kappa^2}{h} + T_2 (\omega_1 \bar{x}_1 + \omega_2 \bar{x}_2) = \tfrac{1}{4} Wl \quad \text{.................(iii)},$$

or for a section symmetrical about the neutral axis:

$$T_1 \frac{\omega\kappa^2}{h} + T_2 \omega \bar{x}_1 = \tfrac{1}{4} Wl.$$

For a rectangle ($2b \times 2a$), $\quad (\tfrac{1}{3} T_1 + \tfrac{1}{2} T_2)\, 4ab^2 = \tfrac{1}{4} Wl$,

for a circle (b) $\quad \left(\frac{\pi}{4} T_1 + \tfrac{4}{3} T_2\right) b^3 = \tfrac{1}{4} Wl.$

From these and similar results Barlow seeks to find the values of T_1 and T_2 and to ascertain whether they are constant.

[933.] We must now inquire whether there is any ground for the formula (ii).

Since the total longitudinal load is zero, we must have:

$$\frac{T_1 \omega \bar{x}}{h} + T_2 (\omega_1 - \omega_2) = 0.$$

Hence we see that unless $\omega_1 = \omega_2$ the neutral line will not coincide with the central line. W. H. Barlow himself has only dealt with symmetrical sections, but Peter Barlow in an appendix to the paper, pp. 483–8, treats of the non-coincidence of the neutral and central lines in the case of ⊥ sections. There is no experimental investigation of whether this non-coincidence, a clear result of the theory used, is real or not.

Barlow (p. 472) defines T_2 as "the resistance of flexure acting as a force evenly spread over the surface of the section." He has previously, by a reasoning which I fail to follow, deduced that this 'resistance of flexure' is due to lateral cohesion (p. 472). Now whatever be the experimental value of a formula such as (ii) I think we may safely say: (*a*) that the quantity T_2 can have nothing whatever to do with shearing strength, (*b*) that the formula gives a discontinuous change of tractive stress at the neutral line, i.e. suddenly from T_2 to $-T_2$, (*c*) that the constancy of the value T_2 as a term in the traction, all over the surface and whether the traction is negative or positive, is exceedingly improbable.

[934.] But we may still inquire whether there may not be an approximation to the truth in Barlow's formula. The first term $T_1 x/\rho$ might be a portion of the traction due to elasticity, the second T_2, a constant, a portion due to plasticity. Now before rupture it does not seem improbable that a part of the beam may be plastic and another elastic, although it is perhaps hardly likely that these two stages occur at the same time in all fibres. The outer fibres will be enervated and their tractions more nearly constant, the inner especially in the neighbourhood of the neutral axis will still remain elastic. Thus as a *mean result for the whole cross-section* Barlow's expression for the traction does not appear so unreasonable as when we associate it, as its author has done, with any idea of 'lateral cohesion.'

[935.] The chief value of the memoir however lies in the tabulation of the results of experiments on the absolute strength of beams of diverse section under flexure. We reproduce some of these results as they cannot fail to be of value to any one investigating a theory of rupture by flexure. So far as the cast-iron beams are concerned, it may be questioned whether an allowance ought not to be made for the 'defect in Hooke's law,' even if we use Barlow's plastico-elastic formula. This allowance will probably account for most of the difference: see our Art. 1053.

[936.] Cast Iron Beams. (Barlow's 'open girders' are omitted.)

Form	l in inches	ω in sq. inches	Dimensions of cross-section etc. in inches	T from formula (i) in lbs.	T_2 from formula (iii) in lbs.	T_1 in lbs.
Rectangle	60	2	2 × 1 height, 2	41,709	15,654	17,971
Square	60	1	side vertical	45,630	17,892	19,399
Square	60	1	diagonal vertical	53,966	17,523	19,213
Circle	60	1	—	51,396	19,158	20,236
I Section	48	2·60	{equal flanges 2 × ·5 and web 1 × ·5 about}	37,508	—	20,942
H Section	48	2·59	{equal flanges 2×·51 and web ·98 × ·51 about}	43,358	—	18,460
Square	60	4	large 2 × 2	39,094	—	16,644
Circle	60	5	large $2\frac{1}{2}$ diam.	39,560	—	15,902
Circle	60	3·79	large $2\frac{1}{4}$ diam.	44,957	—	17,778
Square	60	4	{large, diagonal vertical}	47,746	—	16,878

These numbers bring out sufficiently the following points:

(i) That T [as calculated from formula (i)] varies with the form of the section. Had we included the 'open girders,' we should have seen that its value varies from 25,000 to 54,000 lbs. about. The average tensile strength of the metal is however only 18,750 lbs.

(ii) The values of T_2 and T_1 are by no means constant, but the values of T_1 are much more nearly that of the tensile strength of the material.

(iii) The large bars are relatively weaker than the small: see our Arts. 952*, 1484* and 169 (*e*) and (*f*). Barlow takes as mean values:

$$T_2 = 16{,}573 \text{ lbs.} \quad \text{and} \quad T_2/T_1 = \cdot 847.$$

The mean value of T_2 is computed from those given in the above table and from the mean results of six series of experiments on 'open girders'

[937.] Barlow then proceeds to investigate how far other experiments give confirmatory results. He considers:

(*a*) Hodgkinson's Experiments: *Iron Commissioners' Report.* (See our Art. 1413*.)

Mean ratio of T_2 to $T_1 = \cdot 853$. This *average* compares well with Barlow's ·847, but the values of T_2/T_1 range from ·516 to 1·185, or the ratio must be held to vary with each quality of metal. No information as to variety of cross-section is given.

(*b*) Wade's Experiments. *American Report on Canon Metal.* (See our Art. 1043.)

Here the cross-sections of the cast-iron bars were square and circular, and the breaking load under flexure and the tensile strength are given. From the former Barlow calculates by his formula the value of T_1, it agrees with Wade's determination of the tensile strength pretty closely. I suppose, although Barlow does not state it distinctly, that he has taken $T_2/T_1 = \cdot 9$.

The following remarks of Barlow following on these experiments may be cited (pp. 479–80):

> If the metal were homogeneous and the elasticity perfect, it is probable that the resistance of flexure would be precisely equal to the tensile resistance, instead of bearing the ratio of nine-tenths as found by experiment. It is evident, however, that it varies in different qualities of metal, and that the tensile resistance does not bear a constant ratio to the transverse strength.

Barlow, after showing from Wade's experiments that a decrease in the absolute tensile strength may be accompanied by an increase in the absolute flexural strength, continues:

> It is easy to conceive also, that the resistance to flexure might be supposed to maintain nearly the same proportion to the tensile resistance in bodies similarly constituted, as for example crystalline substances, yet great variation may be expected to occur between crystalline and malleable and fibrous substances.

I see no theoretical reason why it should be probable that $T_1 = T_2$ for homogeneous and perfectly elastic bodies.

(c) Peter Barlow's Experiments on Wrought-Iron (pp. 480–3).

Here we find W. H. Barlow no longer treating of absolute strength but of stresses "just sufficient to overcome the elasticity"—which seems to me a very different matter. He remarks that he has done this because the material yields by bending and not by fracture. The mean result is very nearly $T_1/T_2 = 2$. I see, however, no reason for applying formula (ii) to this case, and W. H. Barlow himself admits that the experiments are insufficient. On p. 481 he gives experiments confirming the coincidence of neutral and central lines in the case of wrought-iron.

(d) Further results of Hodgkinson's Experiments on Cast-Iron taken from the *Manchester Memoirs*, Vol. v.: see our Art. 237*. These are given in the form of an appendix by Peter Barlow (pp. 483—8); the experiments in question are those on the cross-section of greatest absolute strength. Peter Barlow takes $T_1 = T_2$ owing to the difficulty of finding a mean value for T_1. This leads to a series of values for T_1 varying from 14,000 to 16,000 lbs., values not varying more among each other than those for the tensile strengths of 50 square cast-iron bars given on p. 9 of the *Iron Commissioners' Report*: see our Art. 1408*. For *large* iron castings owing to their relative weakness, Barlow remarks, T_1 ought to be taken much less, probably not more than 10,000 lbs.

[938.] We may conclude then that:

(i) There is no theoretical basis of sufficient validity for Barlow's formula, also that the term containing T_2 cannot arise from "lateral adhesion," and that the name "resistance of flexure" is thoroughly bad; but,

(ii) there is sufficient evidence to show that for a considerable range of *cast-iron* beams of varied cross-section the formula gives results for the absolute flexural strength accurate enough in practice.

[939.] Jouravski: *Sur la résistance d'un corps prismatique et d'une pièce composée en bois ou en tôle de fer à une force perpendiculaire à leur longueur. Annales des ponts et chaussées, Mémoires,* 1856, 2e Semestre, pp. 328–51. Paris, 1856. This is an extract from a work in three volumes 4to. on bridges built on Howe's or the American system.

The memoir is an attempt to improve the old Bernoulli-Eulerian theory of flexure by introducing the consideration of the lateral adhesion of the fibres. Jouravski remarks that given a rectangular beam its strength will be diminished by dividing it into two equal beams by a horizontal plane; hence there is an element of strength in the beam due to the lateral adhesion of the fibres, or in our own words there is shearing-stress along the neutral axis of a beam which in itself forms an element of strength. Jouravski's investigation, suggestive as it is, is not, however, correct, for he really supposes this shearing stress to be uniform all along the neutral axis of a cross-section. This of course, as Saint-Venant has shown, is not the case, and the full treatment of the problem has been given by him in his memoir on flexure: see our Art. 69. At the same time Saint-Venant has praised Jouravski's idea and in his *Leçons de Navier* (see our Art. 183 (*a*) and his p. 390) adapted it to the case of a rectangular beam, the dimensions of which in the plane of flexure are much greater than those perpendicular to that plane. The memoir applies this incomplete theory of the slide-element in flexure to various numerical cases to which, I think, no importance can be attached. See my remarks on the similar investigations of Winkler and Airy in Arts. 661-6.

[940.] J. Dupuit: *Note sur la poussée des pièces droites employées dans les constructions. Comptes rendus*, T. XLV., pp. 881-2. Paris, 1857. The brief extract given here of the memoir does not enable us to judge of its contents. The author apparently finds fault with the ordinary theory of beams, because it does not take account of the fact that the 'fibres' cannot slip over the points of support and states that this produces a great side thrust on the points of support. Further the reactions themselves modify the points of support and therefore their resistance. There is no hint in the paper of how the author proposed to allow for these sources of error and I do not think the paper was ever published in full.

[941.] Fabré: *Sur la résistance des corps fibreux. Comptes rendus*, T. XLVI., p. 624. Paris, 1858. A memoir under this title was presented to the Academy in 1858. The author had concluded from very fine measurements that the ordinary theory of beams is incorrect, the *central line* being always compressed or elongated. I cannot find that the memoir was ever published.

[942.] G. Rebhann: *Relative Widerstandsfähigkeit eines an beiden Enden festgehaltenen prismatischen Trägers. Försters Allgemeine Bauzeitung.* Jahrgang XVIII., S. 130-7. Wien, 1853. This is an application of the Bernoulli-Eulerian theory of beams to ascertain the increased strength obtained by building-in the terminals of a beam. There is no novelty to record: see our Arts. 571-3 and 943—5.

[943.] F. Grashof: *Ueber ein im Princip einfaches Verfahren, die Tragfähigkeit eines auf relative Festigkeit in Anspruch genommenen prismatischen Balkens wesentlich zu vergrössern. Zeitschrift des Vereins deutscher Ingenieure,* Jahrgang I., pp. 75–80. Berlin, 1857. Grashof, having seen Lamarle's results (see our Art. 571) without his analysis, gives an investigation of some special cases where the strength of simple beams would be increased by fixing their terminals at definite slopes.

[944.] F. Grashof: *Ueber die relative Festigkeit mit Rücksicht auf deren möglichste Vergrösserung durch angemessene Unterstützung und Einmauerung der Träger bei constantem Querschnitt derselben. Zeitschrift des Vereins deutscher Ingenieure,* Jahrgang II., S. 22–31, and Jahrgang III., S. 23–8, 45–8, 155–160. Berlin, 1858 and 1859. This memoir starts with a criticism of Scheffler's work referred to in the next article. It then proceeds to a consideration of the following problem: Suppose a simple beam uniformly loaded and having a concentrated load at any one point of it, at what angles must the ends be built-in in order to ensure the maximum of strength? It will be observed that this is little more than the simpler case of Lamarle's memoir (see our Arts. 571–3). Grashof however works out a very great number of special cases, as when the isolated load is at the centre, the terminals are built-in horizontally, etc., etc. There seems to be no novelty of method or result in the paper, and its technical importance is minimised by the difficulty we have already referred to of practically building-in the terminals of a beam at a required angle: see our Art. 573.

In the last part of the memoir Grashof deals with continuous beams passing over points of support at different heights, but as he takes here only the case of continuous loading, his analysis is not more general than Lamarle's and is in no way superior to it.

[945.] H. Scheffler: *Ueber die Versuchung der Tragfähigkeit der Brückenträger durch angemessene Bestimmung der Höhe und Entfernung der Stützpunkte. Organ für die Fortschritte des Eisenbahnwesens,* Bd. XII. S. 97. Wiesbaden, 1857. Further: *Ueber die Tragfähigkeit der Balken mit eingemauerten Enden,* Ibid. Bd. XIII. S. 51, 1858.

The first of these papers only deals with a very special case of what Lamarle and afterwards Grashof have treated generally, and in addition the analysis is very cumbersome. The second paper is controversial, a poor reply to Grashof's perfectly legitimate criticism.

[946.] H. *Festigkeits und Biegungsverhältnisse eines über mehrere Stützpunkte fortlaufenden Trägers.. Der Civilingenieur,* Neue Folge, Bd. IV., S. 62–73. Freiberg, 1858.

A continuous beam is supported on $(n+1)$ points of support not on the same level and forming n equal spans. The beam weighs p lbs. per foot-run, and a live load of p' lbs. per foot-run together with an isolated load Q cross the beam and occupy successively each span. The author finds the deflections and bending moments with great length of analysis. He considers in particular the case of five spans, and applies his results to various special numerical cases.

[947.] H. *Continuirliche Brückenträger*. *Der Civilingenieur*, Neue Folge, Bd. IV., S. 142–6. Freiberg, 1858. This memoir deals fully with the case of a beam of uniform cross-section resting on three points of support, when the mid-point of support is lower than the terminals and the live load covers one or both spans. Many numerical details are given for this case for variations of the ratio of live to dead loads, etc.

[948.] H. *Continuirliche Brückenträger*. *Der Civilingenieur*, Neue Folge, Bd. VI., S. 129–202. Freiberg, 1860. This paper is a continuation of the paper referred to under the same title in our Art. 947. The writer now supposes three unequal spans, the mid-span being longer than the two equal external spans and the mid-points of support lower than the terminals. There is a uniform dead load, and a uniform live load which may cover: (i) the two outside spans, (ii) the mid-span, (iii) all the spans, (iv) one outside span, (v) the mid-span and one outside span. All these separate cases are worked out with great detail and then more than 30 pages of numerical values are given for the reactions and maximum moments in these five different cases of loading for various ratios: (a) of the lengths of the spans, (b) of the live to the dead load, and for various amounts (c) by which the middle points of support may be supposed to be sunk below the terminals. Supposing these portentous tables to be correct we have here the most complete treatment the three-span continuous beam has ever received or is likely to receive.

[949.] E. Winkler: *Beiträge zur Theorie der continuirlichen Brückenträger*. *Der Civilingenieur*, Neue Folge, Bd. VIII., S. 135–182. Freiberg, 1862. This paper may be noted here as it belongs essentially to the same group as those referred to above. It opens with an investigation of results similar to those of Clapeyron, Heppel, etc., and then proceeds to discuss the effect of loading only the rth out of n spans. Winkler presents his results in the form of continued fractions. He then takes the case of equal spans and calculates reactions, deflections, etc. (S. 147–60). Next he passes to the investigation of the most dangerous load system, and then turns to tables of numerical detail for various types of loading. Then he deals by aid of copious tables (S. 171–182) with the case of a continuous beam of four spans. Thus we may say that before 1862, the complete analytical theory of continuous beams of any number of spans and complete numerical details of beams up to four spans had been published[1].

[950.] S. Hughes: *An Inquiry into the Strength of Beams and Girders of all Descriptions, from the most simple and elementary Forms,*

[1] For further investigations with regard to continuous beams belonging to this decade see our Arts. 535, 571–7, 598–607, 889, 890, and 893.

up to the Complex Arrangements which obtain in Girder Bridges of Wrought- and Cast-Iron. The Artizan: Vol. xv., pp. 145–50, 170–3, 194–8, 217–20, 244–47, (255), 267–71, Vol. xvi., pp. 3–6, 27–31, (44–5), 79–84, 129–133, 158–160, London, 1857 and 1858. These papers contain a practical treatise on bridge making, involving very little theory, but citing the experimental results of Hodgkinson, Fairbairn and others, and applying them to the details of various girders, beams and bridges. They do not appear, however, to demand any close analysis in our History at the present day.

[951.] W. R. R. *Notiz über die besten Querschnittsform eiserner Träger. Zeitschrift des Vereins deutscher Ingenieure,* Jahrgang II., S. 310. Berlin, 1858. The writer of this note criticises Hodgkinson's and Davis' beams of strongest section (see our Arts. 244* and 1023), saying that in practice we do not want to have the beam strongest at rupture, but strongest at receiving set. He supposes that the elastic limit of cast-iron in tension is half that in compression and that both elastic limits are equal for wrought-iron. How he reaches these numbers I do not know. A footnote by Grashof questions their accuracy, but does not seem to notice the relative nature of the elastic limit in general. Compare our Art. 875.

[952.] Albaret: *Nouvelles Annales de la Construction,* Juin, 1859. I have only seen an extract of this paper entitled: *Festigkeit von Metallträgern* in the *Hannöverische Bauzeitung,* 1860, S. 523. The author discusses by a method, which is not very intelligible in the extract, the form of the girder which containing a minimum of material can yet safely carry a given load. The paper does not appear to be of any importance.

[953.] Callcott Reilly: *On the Longitudinal Stress of the Wrought-Iron Plate Girder.* This was a paper read before the British Association in 1860. It is published in the *Civil Engineer and Architect's Journal,* Vol. XXIII., pp. 261–4 and 294–5; also in *The Artizan,* Vol. XVIII., pp. 209–12 and 220–1. London, 1860. The author supposes the limits of safe tensile and compressive stress to be different, taking for wrought-iron 5 and $3\frac{3}{4}$ (to 4) tons per square inch respectively (pp. 262 and 294).

He assumes that these safe limits are reached in compression and tension under the same bending moment. He further supposes the stress to be proportional to the strain and the stretch- and squeeze-moduli equal (p. 262). This fixes the neutral axis, and therefore, if there be no thrust, the position of the centroid of the cross-section. For a given bending-moment and I section of given total height and given thickness of web, we then have enough equations to determine the areas of the flanges, if their breadths be given: or for *thin* flanges, the areas even without the breadths. Reilly concludes his paper with a calculation of the girders for a bridge actually built from designs based on his theory.

[954.] L. Mitgau: *Ueber die relative Tragfähigkeit guss- und schmiede-eiserner Träger mit Rücksicht auf deren Verwendung zu baulichen Zwecken. Zeitschrift für Bauhandwerker*, Jahrgang 1860, S. 121–7, Braunschweig. This paper is of no importance for us, the greater portion of it being occupied with the calculation of the moments of inertia of T and I sections.

[955.] Blacher: *Application du calcul des ressorts. Mémoires et Compte-rendu des travaux de la Société des Ingénieurs civils.* Année 1850, pp. 143–52. Paris, 1850.

This memoir may be considered as quite replaced by that of Phillips (see our Art. 483), for it only treats a very special case of the latter memoir. The formulae for springs cited on pp. 147–9 are due to Clapeyron who had given them to Shintz (*sic*, Schinz?) from whom Blacher obtained them. The assumption from which the memoir starts ("Soit une série de lames d'égale épaisseur et d'égale largeur superposées de manière que l'une repose sur les extrémités de l'autre *par l'intermédiaire de petits tasseaux*" p. 144) is by no means so general or satisfactory as that of Phillips.

[956.] Schwarz: *Von der rückwirkenden Festigkeit der Körper. Erbkams Zeitschrift für Bauwesen*, Jahrgang IV., S. 518–30. Berlin, 1854.

This memoir after some general remarks on cohesion and elasticity proceeds to deduce a formula for the buckling load P of doubly pivoted struts in the following manner. If $E\omega\kappa^2$ be the flexural rigidity of the strut supposed of length l, then as in Euler's theory (see our Art. 67*):

$$P = \frac{\pi^2 E\omega\kappa^2}{l^2} \quad\text{.........................(i).}$$

If there were compression without buckling, we should have as the value for P

$$P = \omega C \quad\text{..............................(ii),}$$

where C is the safe compressive stress.

Now if δ be the limiting elastic squeeze and C' the corresponding compressive stress $C' = E\delta$,

$$\text{or}\qquad C' = \frac{l^2 P\delta}{\pi^2\omega\kappa^2} \quad\text{...................... (iii).}$$

Now Schwarz argues that the strut has to withstand buckling and compression at the same time, hence we must have

$$C' + C \text{ or } \frac{P}{\omega}\left(1 + \frac{l^2\delta}{\pi^2\kappa^2}\right)$$

less than the greatest safe compressive stress, say C_0. Thus he finds

$$P < C_0\omega \Big/ \left(1 + \frac{l^2\delta}{\pi^2\kappa^2}\right).$$

This formula is in fact akin to the empirical formula, of Gordon, Rankine and Scheffler, but the above process by which Schwarz deduces it remains a mystery to me.

[957.] A. C. Benoit-Duportail: *Calcul des essieux pour les chemins de fer*. *Le Technologiste*, 1856, pp. 315–25. Paris, 1856. Translated in the *Polytechnisches Centralblatt*, 1856, Cols. 705–14. This paper appears to be theoretically correct and is of considerable technical value.

Suppose $2P$ to be the load on the axle, b the distance from the mid-point of the wheel to the mid-point of the journal on which the load rests, then the bending moment throughout the axle is uniform and equal to Pb. If r be the radius of the axle and T the safe tractive strength of its material, then on the Bernoulli-Eulerian theory

$$Pb = \frac{T}{r} \times \pi r^2 \times \frac{r^2}{4} = \frac{T\pi r^3}{4} \quad \text{..................... (i).}$$

This formula holds whether the journals are placed inside or outside the wheels, the flexure in the two cases being, however, in opposite senses.

According to Benoit-Duportail the value of b lies between ·2 and ·3, metres, and T varies from 600 to 400 kilogs. per sq. centimetre. He thus finds for r in centimetres:

$$\begin{aligned} r &= \cdot 35 \sqrt[3]{P}, \text{ for } b = \cdot 2 \text{ m. and } T = 600 \text{ kilogs.} \\ &= \cdot 40 \sqrt[3]{P}, \quad \ldots = \cdot 25 \text{ m. and } T = 500 \ldots\ldots \\ &= \cdot 457 \sqrt[3]{P}, \quad \ldots = \cdot 3 \text{ m. and } T = 400 \ldots\ldots \\ &= \cdot 40 \sqrt[3]{P}, \quad \ldots = \cdot 2 \text{ m. and } T = 400 \ldots\ldots \end{aligned}$$

The values of r are then tabulated for various loads P (pp. 316–7).

[958.] The flexure is next calculated, as before on the assumption that the axle has a uniform cross-section. Since the flexure is circular this is easily done and for such axles as are in common use the amount is found to be 1·75 mm., which throws the top of the wheel from 3 to 4 millimetres out of the perpendicular (p. 317).

Il est évident que lorsque l'essieu tourne, la flexion change de position et que l'essieu prend un mouvement de flexion oscillatoire, analogue à celui qu'on opère pour faciliter la rupture des barres de fer ou des morceaux de bois, qui tend à altérer la qualité du fer. Mais il est à remarquer que lorsque les trains sont en marche, comme il faut un certain temps pour que la flexion se produise, l'amplitude des oscillations est d'autant moindre que la vitesse est plus grande et les effets d'altération qui se produisent sont beaucoup moins destructifs qu'on ne pourrait le craindre au premier abord (p. 317).

This seems to me to disregard the possibility of an accumulation of stress: see our Arts. 970 and 992.

[959.] The journal has to be treated somewhat differently from the body of the axle. In the first place it may be considered as a cantilever of length l and thus we have for its radius:

$$Pl = \frac{T\pi r^3}{4} \quad \text{.............................. (ii).}$$

This gives the radius required at the wheel-end or place of maximum stress, the journal being supposed *outside* the wheel.

But there is another point to be considered, namely, the friction of the load, which produces heat. It is found that for an axle that has been some time in use, the frictional surface is about one-third of the circumference. Hence the area of friction $= \frac{2}{3}\pi rl$. Or, if p be the mean pressure, we have, according to our author, $P = p \times 2rl$ about, that is

$$l = \tfrac{1}{2}P/(pr) \quad \text{............................ (iii).}$$

This reasoning is not, I think, satisfactory; p is normal to the journal surface at each point and we ought rather to have

$$P = p \times 2rl\sqrt{3}/2 = p \times 1{\cdot}7rl \text{ about.}$$

From (ii) and (iii) we find for the dimensions of the journal (p. 318):

$$\left.\begin{aligned} r &= \sqrt{P\sqrt{\frac{2}{pT\pi}}}, \\ l &= \frac{1}{2p}\sqrt{P\sqrt{\frac{pT\pi}{2}}} \end{aligned}\right\} \quad \text{.................. (iv).}$$

Benoit-Duportail states that $p = 25$ kilogs. per sq. centimetre is found by experience to be about the maximum mean pressure which will not overheat the journal. Putting $T = 600$ kilogs., he then finds:

$$r = {\cdot}08\sqrt{P} \text{ in centimetres, and } l = 3{\cdot}125r \text{..............(v).}$$

The value of l as given by this formula would then be retained for the journal. But the value of r would only be valid after there had been a certain wearing away due to use. The value of r therefore must be increased by from 1/10 to 1/12 of its value (p. 319).

If we had supposed, however, the load distributed uniformly over the length of the journal and not all acting at one end as in the extreme case, we should have had instead of (ii) the equation

$$\tfrac{1}{2}Pl_1 = \frac{T\pi r_1^3}{4} \quad \text{........................ (vi),}$$

which with (iii) leads us when $T = 600$ and $p = 25$ kilogs. to:

$$r_1 = \cdot 068 \sqrt{P} \text{ and } l_1 = 4{\cdot}3 r_1 \text{ .... (vii),}$$

instead of (v).

On the *Chemin de fer du Nord* for certain wagons P equals 3250 kilogs. This gives, correcting Benoit-Duportail's arithmetic:

$$r_1 = 3{\cdot}9 \text{ centimetres, and } l_1 = 16{\cdot}7 \text{ centimetres.}$$

Adding to r_1 a tenth of its value we have

$$r_1 = 4{\cdot}29 \text{ and } l_1 = 16{\cdot}7,$$

the values actually taken being $r_1 = 4$ and $l = 17$. Thus the theory leads to results in fair agreement with practice.

For engines and locomotives whose springs are much stiffer, T ought not to be taken so large, but $= 400$ kilogs., say, and p ought also to be reduced. We then have from (iii) and (iv) (p. 319):

for $p = 20$ kilogs., $d = 2r = \cdot 16 \sqrt{P}$, $l = 2d$,
$= 15$ kilogs., $d = 2r = \cdot 171 \sqrt{P}$, $l = 2{\cdot}3d$,
$= 10$ kilogs., $d = 2r = \cdot 189 \sqrt{P}$, $l = 2{\cdot}8d$.

Benoit-Duportail gives a table of the values of d and l for loads ($2P$) from 500 to 12000 kilogs. (pp. 320). Then follows with tables an investigation similar to the above for the case when the journals are inside the wheels (pp. 322–3). The memoir concludes with a discussion of the effect of wedging (*calage*) the wheels on to the axle (pp. 324–5).

[960.] W. Fairbairn: *Tubular Wrought-Iron Cranes. Institution of Mechanical Engineers. Proceedings*, 1857, pp. 87–98. This paper contains details as to the strength and deflection of these cranes: see our Art. 909.

[961.] Callon: *Rapport à la commission centrale des machines à vapeur sur la réponse des diverses commissions de surveillance des bateaux à vapeur aux questions posées par la circulaire ministérielle du* 15 *juillet* 1853. *Annales des ponts et chaussées. Mémoires* 1856, 2e semestre, pp. 71–102. Paris, 1856. I only refer to this memoir because in a *Note*, pp. 90–102, it seems to me to give a very doubtful and, I think, erroneous theory of the safe limit for the thickness of the walls of cylindrical boilers. I do not see how the theory of beams can be applied to this case, and if it be applicable I do not understand how $(dy/dx)^2$ could be neglected as it is on p. 92: see my remarks in Art. 537 on a like treament of the problem due to Bresse.

[962.] F. Gray: *Tredgold's Formula for the Thickness of Cast-Iron Cylinders. The Artizan*, Vol. XVII. pp. 289–90, London, 1859. Tredgold *On the Steam-Engine*, §§ 518–20, gives a formula for the proper thickness of cast-iron cylinders and pipes subjected to strain arising from unequal expansion. Thus he supposes one-half of a cylinder to be heated 300° F. higher than the other half. Tredgold's formula and his method are both very questionable. Gray points out a slip in his algebraical work and gives a corrected formula based on the same hypothesis. The whole procedure seems to me so questionable that I place no faith in Gray's corrected version of Tredgold's formula. Winkler has attempted a similar problem, in a manner which I think inadmissible also, but certainly better than Tredgold's: see our Art. 645. The problem is of some importance and ought, I think, to admit of a solution by accurate analysis.

[963.] Mahistre: *Mémoire sur les limites des vitesses qu'on peut imprimer aux trains des chemins de fer, sans avoir à craindre la rupture des rails. Comptes rendus*, T. XLIV., pp. 610–13. Paris, 1857.

This paper after referring to the Portsmouth experiments on the flexure of railway rails under a travelling load (see our Art. 1417*), proceeds to develop a formula for the maximum load which can cross with given velocity a doubly built-in rail without destroying its elastic efficiency. Mahistre treats the railway rails as built-in at the sleepers, and finds by a process which does not seem to me free from doubtful hypotheses the following formula for the maximum load which can travel with velocity V along the rail:

$$P = \frac{\left(E - 2T_0 \frac{h}{b}\right) \omega\kappa^2}{l\left(\frac{V^2}{g} + \frac{1}{2}\frac{E}{T_0} b - h\right)},$$

where $2l$ is the length of the rail, b its vertical diameter, $E\omega\kappa^2$ its flexural rigidity, h the height of the centre of gravity of the travelling load above the rail, T_0 the limit to elastic tractive stress, E the stretch-modulus of the material of the rail, and $4P$ the part of the weight of the locomotive which rests on the most charged pair of wheels.

If $2h/b$ be small compared with E/T, as I think it would generally be, this formula reduces to

$$P = E\omega\kappa^2 / . \, l\left(\frac{V^2}{g} + \frac{E}{2T_0} b\right)$$

This may be compared with a formula for P which I have deduced from Saint-Venant's result (xiii^b) for a *doubly supported* beam given in our Art. 375. Putting the Q of that article $= 2P$, I find *approximately* for P:

$$P = E\omega\kappa^2 / . \, l\left(\frac{4}{3}\frac{V^2}{g} + \frac{E}{2T_0} b\right)$$

Thus the difference occurs in the factor $4/3$ of the term V^2/g.

Mahistre neglects the inertia of the rail; he assumes it at each instant under the transit of the load to take the statical form which would be produced by the force $2P + \frac{2P}{g}\frac{V^2}{r}$, where r is the radius of curvature of the curve described by the centre of gravity of the load; and he further appears to assume that the centres of curvature of the path of the load and of the curve of deflection of the rail at the point of contact with the wheel must coincide at each instant.

I do not think either the first equation of his (1) or his equation (3) holds for a doubly built-in rail. They apply rather to *one simply supported*, and this on the assumption that the cross-section has its centroid in the mid-point of the vertical axis of symmetry. Further I do not follow the argument by which it is shown that the envelope of the successive curves of deflection is a circle; nor if it be a circle do I understand why equation (4) for the deflection and curvature of the strained form *at any instant* must hold. My surprise is rather that the author comes so close to the right result than that he differs from the formula of Phillips and Saint-Venant.

[964.] Robert Mallet: *On the increased Deflection of Girders or Bridges exposed to the Transverse Strain of a rapidly passing Load.* This paper was read at the Institution of Civil Engineers of Ireland and will be found printed in *The Civil Engineer and Architect's Journal*, Vol. XXIII. pp. 109–110. London, 1860. Mallet refers to the labours of Willis and to the experiments of James and Galton, and then adopts Morin's formula for the deflection, which is really due to Cox: see our Arts. 1417*–24*, 1433* and 881 (*b*). It supposes the beam to take its greatest deflection when the travelling load is at the centre and the deflection then to be that which would be due to a central load equal to the weight of the travelling load, together with a load equal to the instantaneous 'centrifugal force' of the travelling load. If the load statically placed at the centre of the beam would produce a bending moment M_S there, then I find on this theory that the bending moment M_D, when it is travelling with velocity V, is

$$M_D = \frac{M_S}{1 - \dfrac{M_S V^2}{E\omega\kappa^2 g}},$$

where $E\omega\kappa^2$ is as usual the flexural rigidity of the beam. Hence if M_0 be the safe bending moment, we must have

$$M_S < \frac{M_0}{1 + \dfrac{M_0 V^2}{E\omega\kappa^2 g}}.$$

But if w be the weight of the travelling load and $2l$ the length of the beam, $M_S = \frac{1}{2} wl$, or,

$$w < \frac{2M_0/l}{1 + \dfrac{M_0 V^2}{E\omega\kappa^2 g}}.$$

If $M_0 = T_0 \omega \kappa^2 / h$, we have

$$w < \frac{2T_0 \omega \kappa^2 / (lh)}{1 + \frac{T_0}{E} \frac{V^2}{gh}}.$$

Mallet does not reduce Morin's result to this simple form, but says rather vaguely that it accords with James and Galton's experiments. The above result will be found to agree with that of Mahistre's cited in the previous article, if we remember that $w = 2P$, $h = \frac{1}{2}b$ of the latter's notation: see also our Art. 663. The conclusions, however, of Mahistre and Mallet are erroneous, being founded on a method condemned at a much earlier date by Stokes: see our Art. 1433*. Mallet in a footnote notices Phillips' memoir (see our Art. 552), and commends it strongly to the reader, but does not seem to have noticed that its results contradict those he cites. He considers Willis and Stokes' work as excellent, but "past the usual range of practical men." There is nothing of further importance in the paper.

[965.] Lemoyne: *Note sur l'évaluation du poids équivalent à un cahot en ce qui concerne la résistance d'une poutre de pont; ou plus généralement: Détermination de la charge tranquille équivalente, quant à la flexion d'une pièce élastique reposant, par ses extrémités, sur deux appuis de niveau, au choc d'une masse déterminée tombant d'une hauteur connue sur le milieu de cette pièce. Annales des ponts et chaussées. Mémoires*, 1859, 1er semestre, pp. 326–33.

The method of this paper is not very satisfactory, and the whole matter has since been thoroughly discussed by Saint-Venant (see our Arts. 362–71 and 413).

Lemoyne argues as follows: Let the weight Q dropped from the height h produce the same deflection f as the weight P statically placed upon a bar, then to get a superior limit for the value of P we may suppose the work done in bending the bar in the two cases equal or:

$$Q(h + f) = Pf,$$

but $f = l^3 P / (48 E \omega \kappa^2)$, where l is the length of the bar and $E\omega\kappa^2$ its flexural rigidity, hence we find:

$$P = \frac{Q}{2}\left(1 + \sqrt{1 + \frac{192 E \omega \kappa^2 h}{l^3 Q}}\right) \quad \text{.................} \quad (a).$$

Il est à remarquer de plus, que la relation (a) fournirait seulement, dans la pratique, une limite supérieure du poids à déterminer; car cette expression n'est vraie, théoriquement, que pour le cas impossible de deux corps parfaitement élastiques, ce cas étant le seul où il n'y ait pas une perte de puissance vive par le fait même du choc (p. 329).

To obtain an inferior limit Lemoyne determines the kinetic energy absorbed in the elastic deformation. Let V' be the joint velocity after impact and W the weight of the bar, then if both W and Q were free we should have by the principle of momentum

$$\frac{Q}{g}\sqrt{2gh} = \frac{W+Q}{g}V'.$$

Thus, $$\frac{1}{2}\frac{W+Q}{g}V'^2 = \frac{Q^2h}{W+Q}.$$

Putting: $$Pf = \frac{Q^2h}{W+Q} + Qf,$$

we find for an inferior limit of P (p. 330):

$$P = \frac{Q}{2}\left(1 + \sqrt{1 + \frac{192E\omega\kappa^2 h}{l^3(W+Q)}}\right).$$

Lemoyne's reasoning seems to me very doubtful; in particular with his definitions of P and f I think $\frac{1}{2}Pf$ and not Pf is the work corresponding to P. He gives an example of these limits, and he finds that for a certain case P must lie between 13,260 and 3,210 kilogs., when $Q = 3{,}000$ kilogs. and the drop is ·08 metres. A further assumption leads him to limits of 11,000 and 5,142 kilogs. In neither case do the limits seem to me sufficiently close to be of much practical value.

[966.] F. Grashof: *Ueber die Berechnung der Festigkeit der Schraubengewinde. Zeitschrift des Vereins deutscher Ingenieure.* Jahrgang IV., S. 289–92. Berlin, 1860.

The author commences with a short historical account of the discussions which had taken place in Germany on the strength of screws. At a meeting in March, 1860, of the Hannoverian *Architekten- u. Ingenieur-Verein* Wittstein had given a formula for the strength of screws based on the assumption that the thread is a beam under flexure (!). Let a be the width and c the depth of the thread, b the circumference of the screw multiplied by the number of turns of the screw in the matrix, then Wittstein supposed that the tractive strength P of the screw is given by

$$P = mbc^2/a \qquad\qquad \text{(i)},$$

where m is a constant to be deduced by experiments on the rupture of screws and not from pure traction experiments on its material. This formula led to a discussion in which Rühlmann, Karmarsch, Kirchweger and others took part. Karmarsch, from experiments made by himself twenty years previously, concluded that the resistance of wooden screws was due to the shearing and not to the bending strength of the thread. Thus if S_0 be the absolute shearing strength of the material, we ought to have instead of (i)

$$P = S_0 bc \qquad\qquad \text{(ii)}.$$

Rühlmann (*Zeitschrift des Architekten- u. Ingenieur-Vereins f. d. Königreich Hannover*, Bd. VI. Heft 2 u. 3, where the details of the discussions will be found,—a periodical unfortunately inaccessible to me) considers that formulae (i) or (ii) will be true according as the thread does not or does accurately fit the matrix. He then proposes to use a formula given by Navier, in § 154 of his *Leçons*, for combined flexure and shear. This formula is erroneous, so that no weight can be laid on Rühlmann's results. In the course of his paper he refers to certain experiments on wrought-iron made in 1834, the details of which are published on S. 228 of the *Mittheilungen des Gewerbe-Vereins f. d. Königreich Hannover*, 1835, showing that the absolute shearing strength of wrought-iron is from 68 to 80 per cent. of its absolute tractive strength,—a result not very far from the $\frac{4}{5}$ obtained by extending the results of uniconstant isotropic elasticity to the phenomena of cohesion: see our Arts. 879 (*d*) and 903.

[967.] Such was the state of the problem when Grashof took it up. He considers (ii) to be the correct formula for tightly fitting screws, and that it is impossible to apply (i) to the case of a beam the height of whose cross-section is of the same dimensions as its length. It is necessary, he holds, in the case of a metal screw, which does not fit so closely as a wooden one, to take into account both the flexure and shear of the thread. He supposes the pressure P to be distributed on a cylinder round the spindle of the screw, the radius of which is slightly greater than the mean between the radii of the spindle and the thread. I hardly see that he justifies this assumption (S. 290). He then, after demonstrating at some length the error of Navier's formula,—a fact long before known from Saint-Venant's researches—proceeds to apply Saint-Venant's formula for combined flexure and shear to the case of the thread of a screw. He attributes this formula to Poncelet and says he first found a rational treatment of combined flexure and shear in Laissle and Schübler's work (see our Art. 889)! He then gives a numerical table showing the influence of shear and flexure respectively on short beams. This table is similar to one which had been previously given by Saint-Venant and which we have already cited in a later form (see our Art. 321 (*d*)). Grashof concludes from his formula,—into which, however, he has not introduced the differences between the stretch and slide moduli, and between the absolute tractive and shearing strengths,—that if a screw thread is not to give way by flexure we must have

$$P < \tfrac{8}{13}\, bc\, T_0 \text{ ....(iii)},$$

where T_0 is the absolute tractive strength, and if it is not to give way by shear we must have:

$$P < \tfrac{4}{5}\, bc\, T_0 \text{ ....(iv)}.$$

Thus a slightly loose screw would give way sooner than a tight one. Better results than these would be obtained on the same *hypothesis*—i.e. *that of the thread as a beam*,—from the conclusions of our Art. 321 (*d*).

I question, however, whether this hypothesis in the least approximates to the facts of the case. Were the thread cut through in several places parallel to the axis of the screw, would not its strength be weakened? Further, if it is to be treated as a beam, surely in practice its cross-section is not uniform and equal to bc? Lastly if we may suppose these assumptions to make no difference, would not better results be obtained by treating the thread as a very narrow *plate* built-in at one edge and loaded near the parallel edge? In this case we should not obtain a formula like (iii) in which T_0 is the absolute tractive strength. We should have to deal with a plate incapable of contracting in its own plane and the results would be again different.

GROUP C.

Experimental Researches on Shaped Material and Structures.

[968.] The remarks we have made on the papers of the two great engineering Institutions for the earlier period see our (Art. 1464*) hold in a slightly modified degree for much of the technical literature of the period 1850–60. The scientist stands aghast at the great mechanical results which have been obtained often by a defective, sometimes by a false theory. Perhaps it is only a consciousness of the large 'factor of safety' used which makes a railway journey endurable for a scientist after a perusal of some of the technical papers published in this decade!

[969.] The following papers in the *Proceedings of the Institution of Mechanical Engineers* may be just noted:

(*a*) 1850–1, January, pp. 19–31. *On Railway Carriage and Waggon Springs* by W. A. Adams, with an additional paper by the same author, April, pp. 14–26. These papers are interesting as giving an account of the various buffing and bearing springs then in use. There are details of experiments on the deflection, set and absolute strength of some of these springs. Several of them might be made the subject of interesting theoretical investigations.

(*b*) 1853, pp. 45–56. *On Improved India-Rubber Springs for Railway Engines, Carriages, etc.* by W. G. Craig. Similar remarks apply to this paper.

(*c*) 1857, pp. 219–226. *Description of a New convex-plate laminated Spring* by J. Wilson. The flat plates of the ordinary spring are replaced in this new spring by "grooved or trough plates." Details of deflection and set in springs will be found on p. 223.

(*d*) 1858, pp. 160–5. *On a new construction of Railway Springs*, by T. Hunt. This gives details of deflection and set.

[970.] *On the Fatigue and consequent Fracture of Metals.* F. Braithwaite. *Institution of Civil Engineers. Minutes of Proceedings*, Vol. XIII. 1853–4, pp. 463–7 (Discussion pp. 467–75). The word 'fatigue' is attributed by Braithwaite to Field (p. 473) and is used by him to denote a progressive destructive action arising from repeated loading. The paper is in very general language and the only evidence brought forward is drawn from numerous "unaccountable" accidents which the author attributes to a wearing-out of material due to repeated stress. Fairbairn in the discussion supported the old view that the variations in stress produce a change in the molecular structure, thus "wrought-iron assuming a crystalline instead of a fibrous arrangement" (p. 469): see our Arts. 1463*–1464* and 881 (*b*). Sewell held "that fractures were frequently owing to the arrest of the longitudinal wave of vibration by a transverse check." He believed that this would account for the action of fatigue at shoulders and angles (p. 471). This is really a true view although obscurely expressed, the wave of stress is reflected by such 'checks,' and the stress tends to double if not further multiply itself at such points[1]. This accumulation of vibratory stress owing to reflection can be easily demonstrated theoretically in the case of longitudinal and torsional vibrations, and I believe is the real reason why a vibrating body *appears* to give way under a stress less than the statical rupture stress. Thus 'fatigue' would only express the constantly increasing set due to an accumulated stress which exceeded the elastic limits. The vague way in which the latter term is used by Hawksley in the discussion is characteristic; he does not seem to have in the least grasped that the vibratory strain under a certain loading may be twice, or even more times, as great as the strain produced by the same load under statical conditions. It is accordingly the maximum vibratory strain and not the statical strain which must be less than the elastic limit, but the vibratory strain can only be deduced from the load by theoretical calculations, which are occasionally of a rather complex character.

[971.] C. R. Bornemann: *Festigkeitsversuche mit dreieckigen Stäben, Der Civilingenieur*, Neue Folge, Bd. I., S. 186–195. Freiberg, 1854. This is an attempt by means of experiments on wooden and cast-iron bars of triangular cross-section to ascertain whether the stretch- and squeeze-moduli of such materials are equal. The bars were of equilateral cross-section, and in the case of wood were of deal with the fibres apparently *in the plane of the cross-section and parallel to a median line.* The experiments with the wooden bars were made with the cross-section in three different positions relative to the plane of the load supposed vertical, i.e. (i) with the vertex upwards, (ii) with it downwards, in both cases one side being horizontal, and (iii) with a side vertical. Experiments were then made in which elastic flexure, set and ultimately flexural strength were measured. Similar bars of cast-iron with their

[1] The papers of Thorneycroft and McConnell referred to in our Art. 1464* both draw attention to the fact that axles almost invariably break at the wheel. McConnell attributes this position of the fracture to "the sudden stoppage or reaction of the vibratory wave at that place."

cross-sections in the same three positions relative to the load plane were experimented on. The details of both sets of experiments will be found on S. 187–92.

Bornemann finds that the relative strength of both wooden and cast-iron beams, calculated by extending the Bernoulli-Eulerian theory to rupture, would depend on the position of the cross-section relative to the plane of the load. Further that the elastic flexures in these positions require us to suppose that for cast-iron at least the stretch- and squeeze-moduli are unequal. He works out a theory of flexure on this hypothesis on S. 192–3, which is very similar to that of Hodgkinson: see our Arts. 234* and 925. His general conclusions stated on S. 195 are as follows. His experiments

(1) Die bei früheren Festigkeitsversuchen gemachten Beobachtungen bestätigen, nämlich die Proportionalität der Zunahme der Einbiegungen mit den Gewichtslagern, innerhalb gewisser Grenzen; dann die stärkere Zunahme der Einbiegungen bei höheren Belastungen, das Auftreten permanenter Einbiegungen bereits bei sehr geringen Belastungen (z. B. für Holz bei 1/57 der Bruchlast, für Gusseisen bei weniger als 1/20 derselben) oder bei sehr unbedeutenden Ausdehnungen der extremiten Fasern (für Holz bei einer Ausdehnung = ·00032, und für Gusseisen bei der Ausdehnung = ·00086), die stärkere Zunahme der permanenten Einbiegungen in der Nähe der Bruchbelastung, das plötzliche Eintreten des Bruches bei gusseisernen Barren und unter Bildung eigenthümlicher keilförmiger Hervorragungen an der Seite der comprimirten Fasern, endlich die Abnahme des Elasticitätsmodulus mit wachsenden Einbiegungen;

(2) Scheint sich aus diesen Versuchen ein Unterschied zwischen dem Elasticitätsmodulus der comprimirten und demjenigen der ausgedehnten Fasern herauszustellen, welcher aber für Holz wo die entsprechenden Werthe sich wie 1·054 : 1 verhielten nur sehr unbedeutend sein kann, für Gusseisen dagegen, wo die Elasticitätsmodeln sich wie 1·4939 : 1 verhielten, nicht übersehen werden könnte, sondern die Annahme einer andern Lage der neutralen Axe und die Einführung anderer Biegungsmanente nöthig machen würde. Als Elasticitätsmodulus der ausgedehnten Fasern ergab sich für Gusseisen im Mittel 9562500000 Kilogr. [per sq. metre], für Holz 1531955000 Kilogr. [per sq. metre].

(3) Gleichzeitig ergiebt sich aber auch eine Veränderlichkeit der neutralen Axe mit steigenden Belastungen, indem sie sich immer mehr dem Schwerpunkte zu nähern scheint.

(4) Die gewöhnliche Berechnungsweise der Festigkeit wird durch die Versuche nicht bestätigt, es scheint vielmehr, als ob zwischen den Festigkeitsmodeln der dem Druck und der dem Zuge ausgesetzten Fasern dieselbe Ungleichheit bestünde, welche zwischen den betreffenden Elasticitätsmodeln gefunden wurde.

The fourth result is of course the old 'beam paradox': see our Arts. 173, 178 and 930–8. The possibility of obtaining satisfactory results by making the stretch- and squeeze-moduli for cast-iron different, is, I think, doubtful, since the chief peculiarity of cast-iron is the non-linearity of its stress-strain curve from the very outset: see our Arts. 1411* and 793.

[972.] Eaton Hodgkinson: *Experimental Researches on the Strength of Pillars of Cast Iron from various Parts of the Kingdom. Phil. Trans.* 1857, pp. 851–899 with three plates. London, 1858.

This memoir may be treated as a supplement to that of 1840 considered in our Arts. 954*–965*. From the theoretical point of view it does not contribute much additional information, and from the practical it must be looked upon chiefly as modifying the constants obtained in the earlier investigations.

[973.] The pillars in the present researches were upwards of 10 ft. long and from 2·5 to 4 inches in diameter, partly solid and partly hollow. The material was cast-iron and the iron was of a variety of well-known qualities. A description of the testing machine will be found on pp. 851–2 (with plate xxxi); the experiments, made at University College, London, were at the joint cost of the Royal Society and Mr Robert Stephenson. Hodgkinson being unable to determine any point which could be described as that of incipient flexure, confined his attention to rupture loads.

The rupture load of a solid pillar of diameter d and length l, according to Hodgkinson's investigations of 1840 varied as $d^{3\cdot55}/l^{1\cdot7}$. In the present experiments he gives for Low Moor Iron the rupture load w in tons of a hollow pillar of internal and external diameters, d_1 and d inches, l feet long (p. 862):

$$w = 42\cdot347 \frac{d^{3\cdot5} - d_1^{3\cdot5}}{l^{1\cdot63}},$$

while in the researches of 1840 he gave (see our Art. 961*):

$$w = 46\cdot65 \frac{d^{3\cdot55} - d_1^{3\cdot55}}{l^{1\cdot7}}.$$

It is obvious that there is a very considerable difference between these results, and we are compelled to put even less faith in the formula than we might hope to do, when we notice that the powers of the diameters vary from 3·213 to 3·679, and the powers of the lengths from 1·5232 to 1·6724 in the different sets of experiments; still other values of the powers would have arisen, if the results of the experiments of 1840 had been taken into account. Further these empirical formulae are, we are told, to be limited to pillars of cast-iron, whose lengths are at least 30 times their diameter (pp. 864–6). It may be questioned whether the Gordon-Rankine formula or some of its numerous German equivalents would not give equally good, perhaps better, results: see our Arts. 469, 650 and 956.

For the different kinds of cast-iron Hodgkinson gives on p. 872 a summary of the values of m, where in tons $w = m \,.\, d^{3\cdot5}/l^{1\cdot63}$ for solid pillars. m, for d in inches and l in feet, varies from 33·6 up to 49·94. This summary may have value for practical purposes, but we can only afford space to refer to it here.

[974.] We may note one or two points of the memoir which have possibly a theoretical bearing:

(*a*) The relative strength of pillars with flat or bedded ends to those with rounded ends was found to be as 3·107 to 1 (p. 855 and compare p. 854, § 5). The theoretical incipient flexure loads are as 4 : 1. Probably a certain amount of strength was gained by the flattening of the rounded ends under the pressure. With one end flat and one rounded the ratio was as 2 : 1, which agrees with that given by incipient flexure very fairly: see our Art. 959* and *Corrigenda* to Vol. I. p. 2.

(*b*) Hodgkinson gives a theory of these ratios (pp. 855–58), but it is not very novel or sufficient. His remarks on the points of rupture (p. 858) seem to suggest that rupture ultimately takes place at the points of maximum elastic stress as given by Euler's theory. These are the points referred to in our Art. 959*.

(*c*) On pp. 861–2 the great loss of strength due to removing the external crust is referred to. Hodgkinson thus notes "that to ornament a pillar it would not be prudent to plane it". Further: "In experiments upon hollow pillars it is frequently found that the metal on one side is much thinner than on the other, but this does not produce so great a diminution in the strength as might be expected, for the thinner part of a casting is much harder than the thicker, and this usually becomes the compressed side" (p. 862).

The considerable differences between the crushing strength of iron at the core and towards the periphery of the casting are again referred to on pp. 866–870. Thus if R, R' be the resistances to crushing per square inch at the periphery and the core respectively, then $\dfrac{R-R'}{R}$ = the defect of resistance at the core, the resistance towards the periphery being taken to be unity. Hence if d be the diameter of the weaker core, d_1 of the pillar, Hodgkinson supposes (see our Art. 169, (*e*)):

$$w = \frac{\beta}{l^{1\cdot63}}\left(d_1^{3\cdot5} - \frac{R-R'}{R}\, d^{3\cdot5}\right),$$

where β is a constant. For Low Moor iron $\dfrac{R-R'}{R} = \dfrac{1}{4}$ nearly, and from $d = \cdot25 d_1$ to $d = \cdot8 d_1$, we see that w will vary from

$$\cdot998\,\frac{\beta d_1^{3\cdot5}}{l^{1\cdot63}} \text{ to } \cdot8855\,\frac{\beta d_1^{3\cdot5}}{l^{1\cdot63}}.$$

Thus the weakness of the core may have considerable effect. The reasoning by which Hodgkinson reaches the above formula is not very satisfactory, but it probably roughly expresses the effect of the variation of the strength of the material across the section. The point has been considered in other applications by Saint-Venant and Bresse: see our Arts. 169 (*e*)–(*f*) and 515.

Experiments showing the decrease of strength from the periphery to the core of castings are given on pp. 889–92, and might be useful for comparison with further theoretical investigation on this subject, although in many cases the evidence only proves irregularity in the casting; in one case even the greatest strength was near the centre.

(*d*) In the *Appendix* to the memoir, pp. 893–9, we have some experiments on six cast-iron columns of circular, square and triangular cross-sections. From the few results obtained it would appear that for the same quality of metal, the same weight and length, the circular, square and equilateral triangular cross-sections give loads varying as 55299, 51537, and 61056 respectively, or the *triangular* is distinctly the strongest and the square the weakest. In these cases the ends were *flat;* Hodgkinson seems to hold that this would not be true if the ends were rounded, but the experiment on a cruciform pillar, made in 1840, on which he bases his conclusion does not seem very satisfactory. The ratio of the corresponding *buckling* loads is on Euler's theory $9 : 3\pi : 3{\cdot}464\pi$, which makes the load for the triangle the greatest, and with roughly about the same ratio to that for the square as Hodgkinson gives for the rupture loads. But this theory applied to rupture makes the square stronger than the circle, which is the reverse of Hodgkinson's experience.

The rupture surfaces of the pillars experimented on are figured. The details of some experiments on wrought-iron columns and timber balks referred to on pp. 852–3 were not, so far as I know, ever published.

[975.] This is the last memoir of Hodgkinson's that we have to note. He died on June 28, 1861, after some years of bad health. An account of his life will be found in the *Mémoires et compte rendu des travaux de la Société des Ingénieurs civils*, Année 1861, pp. 505–10. Paris, 1861. Of this society he was an honorary member. The account is a translation by Love of a notice of the veteran technical elastician which appeared in the *Manchester Courier*, but on what date I do not know.

[976.] D. Treadwell: *On the Strength of Cast-Iron Pillars. Proceedings of the American Academy*, Vol. IV., pp. 366–73. Boston, 1860.

This paper is a mere *résumé* and criticism of earlier work. It refers to the labours of Euler, Rennie, Tredgold and Hodgkinson (see our Arts. 65*, 74*, 185*, 196*, 833*, and 954*–65*). It points out that Hodgkinson's formulae ought not to be trusted beyond the limits of his experiments, careful as the latter were. Treadwell further remarks that no practical directions founded upon Hodgkinson's experiments have been given in any engineering work[1], and that American architects are governed by Tredgold's formula, which leads to different and in many cases quite incongruous results. At Treadwell's suggestion a committee was appointed to draw up rules which should be consistent with the laws of strength for small as well as for large pillars. I am unaware whether this committee ever published any report.

[977.] B. B. Stoney: *On the Strength of Long Pillars. Proceedings of the Royal Irish Academy*, Vol. VIII., pp. 191–4. Dublin, 1864. This gives a very insufficient and unsatisfactory method of deducing a formula for the 'deflecting weight' of long struts. It practically only reaches Euler's result in an incomplete form: see our Art. 74*.

[978.] G. H. Love: A memoir by this writer may be noted although it belongs to a year outside the present decade. It is entitled: *Sur la loi de Résistance des piliers d'acier déduite de l'expérience pour servir au calcul des tiges de piston, bielles, etc. Mémoires et compte rendu des travaux de la Société des Ingénieurs civils*, Année 1861, pp. 119–66. Paris, 1861.

The memoir commences by noting the want of experiments on the strength of steel columns, and proposes to rectify this by experiments on three columns of steel with rounded ends of one centimetre diameter and of lengths 10, 20 and 30 centimetres respectively. From results obtained for these *three columns only*, and by a process which is not very intelligible to me, Love obtains the following empirical formula for the total crushing load P of a steel column:

$$P = C\omega\sqrt{\frac{1{\cdot}32}{l/d - 5{\cdot}50}} \quad\text{......................(i)},$$

where C = the crushing strength of a small block of steel,

ω = the cross-section of the column, supposed circular,

l = its length and d its diameter.

Presumably this formula only holds when $P < C\omega$ or when $l/d > 6{\cdot}82$, and this only for steel like that of the experiments having an absolute strength not less than 7600 kilogs. per sq. centimetre.

For columns with flat ends Love gives the formula:

$$P = C\omega\sqrt{\frac{28}{l/d + 35}} \quad\text{...................... (ii)}.$$

[1] This seems to me incorrect, as Hodgkinson's formulae got at once into the text-books—and have unfortunately remained there till to-day.

For columns of square cross-section Love multiplies the results (i) and (ii) by 1·53 as a factor in analogy with Hodgkinson's results for cast-iron[1].

These formulae may be compared with the empirical expressions for P in the case of columns of wrought-iron of circular cross-section which Love gives on p. 136. They are the following:

For flat ends: $$P = \frac{C\omega}{1\cdot 55 + \cdot 0005\,(l/d)^2},$$

For rounded ends: $$P = \frac{C\omega}{\cdot 0012\,(35 + l/d)^2}.$$

Such formulae may hold fairly closely for a limited range of experiments, but there ought to be great caution in applying them beyond that range, as their extreme diversity of form shows that they have no theoretical justification. I draw attention to them here only because such formulae are still frequently quoted in practical treatises on bridge-design often without the necessary note of caution; for example in Résal's *Ponts Métalliques*, Paris, 1885, p. 12. Love's insistance (p. 142 footnote) on the generality of his formulae does not seem to me warranted.

[979]. The remainder of the memoir consists of practical applications of these formulae and of a criticism of Euler's expression for the buckling load of struts (see our Arts. 74* and 649), which had been dogmatically applied to rupture.

Love concludes with throwing down the gauntlet to the theoretical elasticians in the following words:

Au reste, je reviendrai sur cette question dans un écrit auquel je mets en ce moment la dernière main, et qui traite de l'influence de *la méthode spéculative en général, et de la spéculation mathématique en particulier sur le progrès des sciences d'application;* et j'espère montrer que cette méthode, qui, en France, trône encore dans toutes les sciences et qui empêche l'avènement de la méthode baconienne, a été et continue d'être le plus grand obstacle aux progrès des sciences et de la société (p. 163).

[980.] William Fairbairn: *On the Resistance of Tubes to Collapse. Phil. Trans.*, 1858, pp. 389–413, with two plates. London, 1859. This memoir was read on May 20, 1858.

The experiments recorded in the memoir were undertaken at the joint request of the Royal Society and the British Association. The author thus describes their aim:

[1] I find by Art. 974 (*d*), that, if $w = \beta d^{3\cdot 5}/l^{1\cdot 63}$ and $w' = \beta' d'^{3\cdot 5}/l^{1\cdot 63}$ for columns of circular and square cross-section respectively, then $\beta' = 1\cdot 42\beta$ and not $1\cdot 53\beta$.

Their object is to determine the laws which govern the strength of cylindrical vessels exposed to a uniform external force, and their immediate practical application in proportioning more accurately the flues of boilers for raising steam, which have hitherto been constructed on merely empirical data (p. 389).

After referring to the great increase in the number of boiler-explosions owing to the rise of working pressure from 10 to 50 and even 150 lbs. per sq. inch, Fairbairn goes on to remark that it is impossible to treat flues, the ends of which are supported by rigid rings or securely fastened frames, as cylindrical tubes of indefinite length, or as tubes whose strength is unaffected by their length. He states that practical engineers have supposed boiler-flues to be equally strong at all parts of their length notwithstanding their built-in terminals, but that this is very far indeed, from the fact. Thus flues 35 feet long were found to be distorted under considerably less force than those 25 feet long (p. 390).

Pp. 390–2 describe the apparatus by which a large external pressure was applied to a tube. The air inside the tube was maintained at the atmospheric pressure by means of a small connecting pipe which also allowed the air to rush out on the collapse of the tube.

[981.] The tubes to be experimented on were composed of single thin plates bent to the required form on a mandril and then riveted. They were also brazed to prevent leakage into the interior. The tubes were closed by cast-iron terminals, to which they were securely riveted and brazed. They were supported at one end by a rod from the cast-iron terminal of the tube to the lower cover of the hydraulic cylinder, and the other terminal was united by a pipe to the upper cover of the cylinder. Fairbairn seems to think that this rod and pipe screwed tightly up to the covers hindered the buckling action of the pressure on the cast-iron terminals of the tube, but the great increase of strength in one experiment in which an iron rod was placed inside the tube between the cast-iron terminals so as to prevent their approach appears to me[1] to indicate that the pressure on those terminals may in some cases have acted as a buckling force, and the collapse may not have been entirely due to the lateral pressure on the tube (cf. Fairbairn's

[1] Fairbairn regards the increase as due to a tin ring left in by mistake.

experiment M. and second ftn. p. 393). This buckling action of the pressure would not occur in ordinary boiler flues.

[982.] The first series of experiments were on tubes of 4, 6, 8, 10 and 12 inch diameters, and from 19 to 60 inches in length (pp. 392–6). The general conclusion is that the 'pressure of collapse' varies inversely as the length. The tubes appear in these cases to have been lap-jointed (?) and made of plates of ·043 inches thickness. The forms of the collapsed tubes together with their cross-sections at positions of greatest collapse are depicted. The latter are generally star-shaped and of surprising regularity (up to even five angles).

The two tubes treated next were made of equal shape and size, but the one with a butt-joint and the other with a lap-joint. The one with a lap-joint showed more than $\frac{1}{3}$ less strength than that with a butt-joint, proving how much a slight deviation from the true circular form reduces the strength (pp. 396–7), and therefore how important it is to adhere to that form.

Fairbairn, as I have remarked, considered that his arrangement maintained a constant distance between the cast-iron ends of his tubes. He now gives some experiments in which the ends were left free to approach; in these no internal rod was placed inside the tube, nor were its ends connected with the covers of the enclosing cylinder. In these cases the pressure of collapse did not vary so exactly with the inverse of the length as in the previous results (pp. 397–8).

The experiments we have referred to up to this point were on tubes made of thin wrought-iron plates. The next three were on steel and iron tubes of somewhat different forms, and in each case with an internal longitudinal stay between the ends (pp. 399–400). These do not appear to be very conclusive. They were followed by two on elliptic tubes, which showed a great weakness as compared with circular tubes of like construction and size. Thus the strength was found to be less by one-half when a tube of circular section 60″ in length, 12″ diameter, and ·043″ thickness of plate was compared with one of the same length and thickness, but of elliptic cross-section $14'' \times 10\frac{1}{4}''$. The experiments were, however, too few to be really of theoretical value.

[983.] Fairbairn next turned to experiments on the strength of tubes subjected to internal pressure. The results are not very satisfactory for the data were too few. He concluded, however, that the strength was in this case, for lengths greater than one to two feet, nearly independent of the length for wrought-iron tubes; the difficulty arising from the fact that the tubes invariably gave way at the riveted joint was not overcome. The conclusion as to the bursting pressure being independent of the length was confirmed by experiments on leaden pipes (pp. 401–3).

[984.] Pp. 403–10 are entitled: *Generalisation of the Results of the Experiments.* Fairbairn states that T. Tate and W. Unwin assisted him in this matter. He assumes the following purely empirical formula for tubes collapsing under external pressure:

$$p = C\tau^n / (ld),$$

where p = pressure in lbs. per sq. inch at collapse, τ = thickness of plate of tube in inches, d its diameter in inches (whether internal or external not stated, but the difference is a small percentage), and l its length in feet, C and n being constants to be determined from the experimental data (p. 404).

For sheet-iron tubes Fairbairn gives as the mean of his experiments:

$$C = 806{,}300 \text{ and } n = 2{\cdot}19.$$

Approximately therefore we may take in practice:

$$p = 806{,}300\, \tau^2/(ld).$$

Fairbairn considers that l ought to be limited in the more exact formula to values between 1·5 and 10 feet.

For very thin tubes of 12″ diameter, the divergence, however, is considerable, and Fairbairn accordingly gives the following formula as a closer approximation to the results of his experiments (p. 408):

$$p = 806{,}300\, \frac{\tau^{2{\cdot}19}}{ld} - {\cdot}002\, \frac{d}{\tau}.$$

Here the second term on the right is negligible for all but very thin tubes.

It may well be doubted whether the experiments made by Fairbairn really permit of the generalisations involved in these formulae, and I feel inclined to lay still less stress on the formulae suggested for elliptic tubes, for riveted tubes subjected to internal pressure and for lead pipes given on pp. 409–10. These are all based on the result of only two or three experiments, which cannot be considered as sufficing in such difficult and delicate matters.

[985.] On pp. 410–13 Fairbairn states the practical conclusions as to boiler construction which may be drawn from his experiments. He points out that boiler flues are generally dangerously weak as compared with the outer shell of the boiler. Both have to resist the same pressure, but the rupture pressure of the former is given by

$$p = 806{,}300\, \tau^{2{\cdot}19}/(ld),$$

and that of the latter by $p' = 60{,}000\, \tau/d$. Hence we have for tubes of the same thickness and diameter

$$p' \Big/ p = \frac{1}{13{\cdot}44}\, \frac{l}{\tau^{1{\cdot}19}}.$$

So that the maximum internal pressure p' is greater than the maximum external pressure p, whenever (l in feet and τ in inches)

$$l > 13{\cdot}44\, \tau^{1{\cdot}19}.$$

In order to equalise the strengths of the shell and flues, Fairbairn suggests: (1) that butt-joints with longitudinal covering plates should be used, and (ii) that strong angle-iron ribs in the form of rings should be placed round the flues, two such ribs would increase the strength nearly three times by practically reducing the length to $\frac{1}{3}$ of itself. This result was confirmed by Fairbairn in an experiment: see Experiment F. p. 392.

[986.] F. Grashof: *W. Fairbairns Versuche über den Widerstand von Röhren gegen Zusammendrückung. Zeitschrift des Vereins deutscher Ingenieure*, Jahrgang iii., S. 234–43. Berlin, 1859. Grashof commences with a *résumé* of Fairbairn's experiments, and then attempts to draw a more accurate empirical formula from them than had been given by Fairbairn himself. If l be the length of the cylinder, d its diameter, τ its thickness, *all* now in inches, and p the collapsing pressure in lbs. per sq. inch, then Fairbairn found (see our Art. 984):

$$p = 9{,}675{,}600\tau^{2\cdot19}/(ld).$$

Grashof by a more careful selection of experiments and using the method of least squares concludes that for the whole range of Fairbairn's tubes

$$p = 24{,}469{,}500\,\tau^{2\cdot315}/(ld^{1\cdot278}) \ldots\ldots\ldots\ldots\ldots\ldots\text{(i)}.$$

He then considers in like manner an empirical formula for the tubes which had a thickness of $\frac{1}{8}''$ or more, which is the thickness usual in practice; he finds

$$p = 1{,}035{,}000\,\tau^{2\cdot081}/(l^{\cdot564}d^{\cdot889}) \ldots\ldots\ldots\ldots\ldots\ldots\text{(ii)},$$

a formula which is totally different in character from (i), p now varying nearly as $l^{-\frac{1}{2}}$ and not as l^{-1}. It seems to me to be very difficult to attach any weight, even practically, to formulae of this character. Grashof concludes his memoir by an attempt at the theoretical investigation of a formula for a tube of slightly elliptic cross-section. His method is very similar to that of Bresse (see our Art. 537), and seems to me to confuse in like manner the plate and bar elastic moduli. If C_0 be the highest safe compressive stress in the material, ϵ the ellipticity, and d the diameter of a circular tube having the same circumference as the ellipse, Grashof finds for the limiting pressure:

$$p = \frac{2C_0\,\tau/d}{1 + \dfrac{3\epsilon}{2}\dfrac{d}{\tau}}.$$

This may be deduced at once from our Art. 537 (*b*).

For ϵ very small we get the ordinary formula for the limiting pressure in a tube of circular section. For ϵ so large that the first term of the denominator may be neglected as compared with the second,

$$p = \frac{4C_0}{3\epsilon}\frac{\tau^2}{d^2}.$$

Grashof now supposes this formula to apply to all circular flues, faults of construction really causing them to be slightly elliptic. As there is no obvious way of determining ϵ for such flues in general, this formula really leads nowhere. To a "*freilich* sehr gewagte Betrachtung" given in a footnote, we do not suppose Grashof intends to give any weight.

According to the author the Prussian Government had adopted for circular flues a formula of the type:

$$p = \text{Constant} \times \tau^3/d^3,$$

but its theoretical or empirical origin appears to be unknown.

[987.] G. H. Love: *Sur la résistance des conduits intérieurs à fumée dans les chaudières à vapeur. Mémoires et compte rendu des travaux de la Société des Ingénieurs civils*, Année 1859, pp. 471–500. Paris, 1859.

In this memoir Love gives a *résumé* of the experiments of Fairbairn on the collapse of tubes: see our Arts. 980–5. On pp. 471–9 Fairbairn's experimental details are reproduced. On pp. 480–8 his conclusions are discussed and criticised. Love in particular rejects Fairbairn's idea that the manner in which the ends of the tube are fixed really causes the variation of the resistance with the change of length: see pp. 393–5 of Fairbairn's memoir. He believes the effect noted by Fairbairn to be solely due to the closing discs, which do not permit of the walls of the tube in their neighbourhood collapsing so easily laterally: see our Art. 981. His remedy, however, would agree with Fairbairn's, namely, riveting rings of ⊥-shaped iron round the tube at suitable intervals.

Love rejects Fairbairn's empirical formula, which he remarks does not give results sufficiently accurate even for practical purposes, and after considerable discussion (pp. 488–95) suggests the following empirical formula for the collapsing external pressure:

$$p = \frac{C\tau^2 + l\tau(5\tau - 1{\cdot}75)}{{\cdot}078ld},$$

where p is the pressure in kilogrammes per square centimetre,

τ is the thickness of the tube, l its length and d its diameter in centimetres,

C is the crushing strength in kilogrammes per sq. centimetre of the material of the walls of the tube.

This formula gives closer values than Fairbairn's for the rupture pressures, but it does not seem to me very satisfactory, especially as the pressures calculated from it are occasionally greater than those obtained experimentally. The remainder of the memoir (pp. 496–500) deals with the practical application of the above formula to boiler tubes.

[988.] J. E. McConnell: *On Hollow Railway Axles. Proceedings of the Institution of Mechanical Engineers*, 1853, pp. 87–100. This paper contains some interesting experiments on the comparative strength of solid and hollow axles, together with other experiments on axle journals. The writer finds that the hollow axle has nearly double the strength of what he terms the corresponding solid axle.'

[989.] W. Bender: *Mittheilungen über Versuche mit MacConnell'schen Hohlaxen. Polytechnisches Centralblatt*, Jahrgang 1856, Cols. 713–721 (Extract from the *Zeitschrift des österr. Ingenieur-Vereins*, 1856, Jahrgang viii.). This paper gives details of some not very conclusive experiments on the relative resistances to impact by a falling mass, to blows from a hammer and to torsion of hollow and solid railway wagon axles.

[990.] Kaumann: *Versuche über die Durchbiegung und die Elasticitätsgrenze für Axen der Eisenbahnfahrzeuge. Erbkams Zeitschrift für Bauwesen*, Jahrgang v., S. 412. Berlin, 1855. *Polytechnisches Centralblatt*, Jahrgang 1855, Cols. 1107–1110. The axles were clamped at any chosen cross-section and loaded as cantilevers. The flexures and the elastic limits were then noted. The paper contains nothing of permanent value.

[991.] A. von Burg: *Ueber die von dem Civil-Ingenieur Hrn. Kohn angestellten Versuche, um den Einfluss oft wiederholter Torsionen auf den Molekularzustand des Schmiedeisens auszumitteln. Wiener Sitzungsberichte.* Bd. VI., S. 149–52. Wien, 1851. This paper contains a very brief and insufficient account of Kohn's experiments on repeated small torsions. Another account is given in the *Zeitschrift des österr. Ingenieur-Vereins*, Jahrgang iii., 1851, S. 35. But the most satisfactory description of the experiments and apparatus will be found in the memoir discussed in the following article.

[992.] A. Schrötter: *Ist die krystallinische Textur des Eisens von Einfluss auf sein Vermögen magnetisch zu werden? Wiener Sitzungsberichte.* Bd. XXIII., S. 472–81. Wien, 1857.

This paper gives a very good account of the manner in which Kohn's experiments were made. In the first series a rotating wheel had three teeth, each producing a small torsion in the bar or spindle under test and then suddenly releasing it from all load. In the second series the wheel was replaced by an eccentric, and thus the torsion was gradually imposed and removed. More than 32,000 torsions were given in the course of an hour, that is to say as many as nine a second. It seems to me, then, very probable that there may have been an 'accumulation of strain,' i.e. it does not follow that because each individual total torsion gave a *mean* torsion which was below the elastic limit, that the bar or spindle was never subjected at any one point to strain beyond the elastic limit[1]. Waves of torsional strain would move

[1] I have calculated the value of this accumulation of strain, which can easily amount to two or three times that due to the individual total torsions supposing them to be applied statically. My results were communicated to the *Cambridge Philosophical Society* in 1888, but have not hitherto been published.

backwards and forwards along the bar and would hardly become insignificant in 1/9 of a second. Seven bars were first experimented on and then broken at different stages by a hydraulic press after the loading had been repeated from 32,000 up to over 128,000,000 times. Each bar was bent into a right-angular form ABC; A was built-in, B was embraced by a socket which allowed free rotation of the bar, AB was the vertical part of the bar receiving torsion by means of the horizontal bar BC, which was acted upon at C by the toothed wheel or eccentric. The cross-sections after rupture were examined with a lens. The seven bars gave the following results with the toothed wheel:

(i) After 32,400 torsions no change was observable in either AB or BC.

(ii) After 129,600 torsions no change was visible to the naked eye, but in AB the lens showed the fibrous structure already broken and as *ein Aggregat von feinen Nadeln.*

(iii). After 388,800 torsions the change in AB was visible to the naked eye and the rupture surface was *grobkörnig.*

(iv) After 3,888,000 torsions, the whole length of AB was affected, especially at the middle section, which we are told is the place of greatest torsion. But why it should be so, is not shown, and it does not seem theoretically probable.

(v) After 23,328,000 torsions the rupture surfaces in AB were *sehr grobkörnig* but showed still no *Blättchen.*

In all these cases the horizontal arm BC had shown no change in its rupture surfaces owing to the flexural vibrations it was subjected to.

(vi) After 78,732,000 torsions the bar AB showed, especially when ruptured at its central cross-section, a remarkable change, the rupture-surfaces were similar to those of cast zinc, and at the same time the horizontal arm BC began to alter its structure.

(vii) After 128,309,000 repeated torsions (occupying thirteen months) the bar AB showed no further change at its centre, only sections nearer to the ends began also to be *grossblättriger.* The horizontal arm BC also advanced in its structural changes.

Similar results were obtained with the eccentric, only the number of torsions had to be greater to produce the same changes.

[993.] Schrötter concludes from these results:

(i) That repeated torsions can change the structure of a bar from fibrous to crystalline and then to *blättrig*, and that the absolute strength decreases during these changes.

(ii) That the number of torsions required depends upon their magnitude. (He deduces this from the fact that the changes occur first at the centre of the bar AB, according to him the place of maximum torsion, but if we do not accept this hypothesis, the statement is still doubtless true.)

(iii) Impacts increase the effect of torsion, or without torsion produce ultimately the same structural changes.

(iv) The changes were due to mechanical action and not to the influence of variations in temperature.

The general conclusion then to be drawn from these results is the gradual destruction of wrought-iron by change of structure due to *rapidly* repeated loading or repeated vibratory impacts.

[994.] Schrötter applied to the *k. k. Handelsministerium* with a view to the institution of experiments to ascertain whether the magnetic properties of a bar of iron were changed by several million repeated torsions. If they were so, a ready means would have been found for testing the loss of strength due to such loading. These experiments were ultimately undertaken by Militzer on five bars which had been subjected respectively to no strain, and in round numbers to 4, 23, 29 and 79 million repeated torsions, and an account of these experiments is given on S. 477–80. The conclusions to be drawn from these experiments—supposing we adopt the theory that repeated loading changes iron from fibrous to crystalline are:—

That this important molecular change corresponds to no marked alteration in the capacity of the bar either: (i) to be magnetised by an electric current, or, (ii) to retain magnetism on the cessation of the current.

Militzer's field appears to have been a high one. A *few* torsions certainly change magnetic properties in a weak field and this without appreciable change of the mechanical properties: see our Arts. 714, (12)–(14), and 811–4.

[995.] *Ueber Gussstahl-Achsen. Dinglers Polytechnisches Journal*, Bd. 146, S. 65–8. Stuttgart, 1857. This paper gives the details of experiments made at the *Gussstahlfabrik des Bochumer Vereins für Bergbau und Gussstahlfabrication* on cast-steel railway axles. A weight of 1403 pounds was dropped from heights of from 3 to 36 feet upon the axles supported at points 3 feet apart. The flexural sets were noted, and after each few blows the top and bottom fibres of the axle *were reversed.* With this reversal 5 or 6 blows from 36 feet destroyed the axles tested.

[996.] H. Résal: *Note sur les formules à employer dans les épreuves des essieux de l'artillerie. Annales des mines*, Tome XIII., pp. 497–503. Paris, 1858.

The axles were tested by dropping a given weight upon them while they were simply supported at their ends. Résal gives a theory of this sort of impact, but as his theory depends solely on the principle of work and on Cox's hypothesis that the beam retains under central impact the statical form of the elastic line, it is not very satisfactory. The whole matter has been more thoroughly investigated by Saint-Venant: see our Arts. 362–71 and 410.

[997.] Wöhler: *Bericht über die Versuche, welche auf der Königl. Niederschlesisch-Märkischen Eisenbahn mit Apparaten zum Messen der Biegung und Verdrehung von Eisenbahnwagen-Achsen während der Fahrt, angestellt wurden. Erbkams Zeitschrift für Bauwesen*, Jahrgang viii., S. 642–52. Berlin, 1858.

This paper describes the first investigations of Wöhler on the repeated loading of railway axles, and not only is of historical interest as leading up to his later more important researches, but contains in itself much that is of considerable value. It should be read in conjunction with the memoir on the theory of axles referred to in our Arts. 957–9.

[998.] Wöhler designed two pieces of apparatus to measure respectively the flexure and torsion of the axles of railway wagons performing their usual service. The first apparatus, designed to measure flexure, recorded automatically the maximum approach and separation of two opposite points on the wheels, and by halving this it is obviously easy, supposing the flexure of the wheels insensible, to calculate the maximum flexure of the axle. The amount through which the point on each wheel was shifted was measured by a separate instrument, which recorded the shift by a needle scratching a slip of zinc. Wöhler remarks:

Beide Apparate zum Messen der Biegung sind so construirt, dass 1 Zoll Zeiger-Ausschlag während der Fahrt einer Bewegung *ac* am Umfange des Rades von $\frac{3}{16}$ Zoll oder einer Abweichung *am* von der normalen Lage von $\frac{3}{32}$ Zoll entspricht.

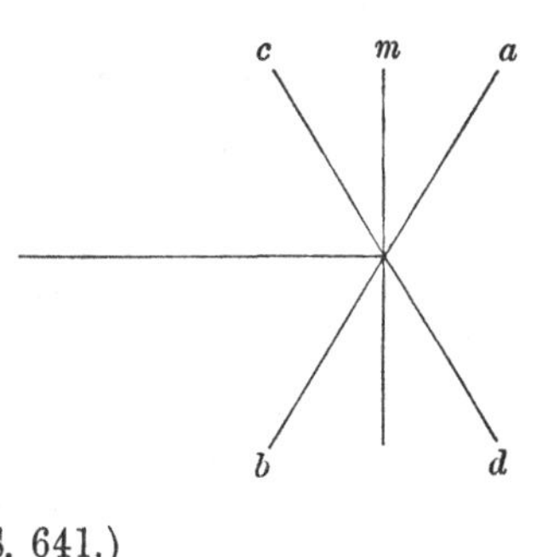

Die Seitenkraft, welche am Umfange des Rades angebracht werden muss, um eine gleiche Biegung der Achse, also einen einseitigen Zeiger-Ausschlag von $\frac{1}{2}$ Zoll hervorzubringen, ist für die Achsen von $3\frac{3}{4}$ Zoll Durchmesser in der Nabe, mit Rädern von $36\frac{1}{4}$ Zoll Durchmesser, $= 23\frac{1}{2}$ Centner[1], und für die Achsen von 5 Zoll Durchmesser in der Nabe, mit Rädern von $36\frac{3}{4}$ Zoll Durchmesser, $= 70\frac{1}{2}$ Centner. (S. 641.)

[1] The *Centner* = one hundredweight, or presumably 100 Prussian lbs. = 103 English lbs.

This *Seitenkraft* was ascertained by connecting opposite points on the rims of the wheels by a chain containing a dynamometer and then pulling them together. This does not seem to me an entirely satisfactory process. In the first place the axle is really bent by a couple on either journal. This couple changes, owing to unevenness in the permanent way, jolting, etc. during the journey. The ordinary load produces a certain flexure in the axle when the wagon is at rest; the same load applied, as it is, in *alternate* directions during the rapid rotation of the axle, we should expect to produce at least *twice* the flexure of the statical couple even if the way were perfectly level. Wöhler does not distinguish between the flexure produced by the ordinary load applied in alternate directions and the flexure which may arise by way of variation due to unevenness in the way, etc. Further, his dynamometer did not act as a couple applied to the journals would have done, but it would include in the shifts measured by his apparatus the flexure of the wheels themselves. Thus while it would measure the load corresponding to flexure produced by side pressure on the flanges of the wheels, it does not seem to me to have suitably recorded the flexure of the axle due to statical dead load or to variations in the journal couples due to motion. Indeed Wöhler appears to neglect the flexure due to statical dead-load, otherwise he ought surely to have stated whether the dynamometer was applied while the dead-load was on the journals and in what plane it was applied, as it might tend to increase or decrease the dead-load flexure according to the position of that plane. If the flexure due to dead-load was negligible, as compared with the vibratory flexure, then this ought certainly to have been stated. Was there no motion of the recording needle on the zinc when the axle was *slowly* turned round?

[999.] The apparatus for torsion measured only what might be termed the *integral* torsion of the axle, i.e. the angle one wheel had been rotated relative to the other. But if the inertia of the axle itself be taken into account, and this might be necessary with an axle of 5 inch diameter, then any impulsive couple applied simultaneously to both wheels would produce a torsion in the axle relative to the central cross-section, which would not be measured by Wöhler's apparatus[1]. It ought to have been shown that this torsion was in itself negligible. As first one wheel and then the other may lag, Wöhler's apparatus records twice the maximum integral torsion.

Der Apparat an der Achse von 3¾ Zoll Durchmesser ist so construirt dass 1 Zoll Zeiger-Ausschlag einer Bewegung von ·321 Zoll am Umfange des Rades von 36¼ Zoll Durchmesser entspricht; gegen die normale Lage des Rades beträgt also die Grösse der Bogen-Abweichung ·160 Zoll, oder der Torsionswinkel 30 Minuten.

Zu einer solchen Verdrehung ist eine am Umfange des Rades wirkende Kraft von 18¾ Centner erforderlich.

Bei dem Apparat der Achsen von 5 Zoll Durchmesser in der Nabe, deren Räder 36¾ Zoll Durchmesser haben, ist auf 1 Zoll Zeiger-Ausschlag die Bewegung am Umfange des Rades = ·228 Zoll, die Abweichung gegen die normale Lage also ·114 Zoll, und der Torsionswinkel = 21 Minuten. Um eine solche Verdrehung hervorzubringen, ist eine am Umfange des Rades wirkende Kraft von 44 Centner erforderlich (S. 642).

[1000.] Wöhler, having now measured the motion of his recording needles in terms of the force applied to the rim of the wheels, is able at once to reduce their *half*-maximum records to equivalent loads, and then to calculate from these loads the stresses in the axles. The journeys were made from Breslau, chiefly to Berlin and back, but also to Liegnitz, Lissa and Frankfurt. The wagons were four- and six-wheeled coal and covered goods wagons, and with two exceptions ran with goods trains. The axles were 3¾″ steel and 5″ iron axles. The wagons were not reversed before the return journey so that the axle experimented on was sometimes a fore- and sometimes a hind-axle. The weight of the wagons, their loads and dimensions are all recorded.

Wöhler gives no details of the calculations by which he deduces the flexural stress ('Faserspannung') from the equivalent load. Thus he says that a rim-load of 72 centners on the wheel of 36¼″ diameter for an axle of 3¾″ diameter produces a maximum stress of 252 centners per square inch. This can be easily verified if it be noted that T the tractive stress is connected with C the rim-load by the formula

$$T = \frac{16}{\pi}\frac{d}{\Delta^3} \times C,$$

[1] I have calculated the numerical value of this torsion in the paper referred to in the footnote of our p. 669. It is there shown to be practically small, but the torsional differences noted by Wöhler were also small.

where Δ is the diameter of the axle and d of the wheel, the longitudinal stress due to C being neglected. For the torsion of the same wheel, Wöhler says a rim-load of $29\frac{11}{16}$ centners produces a '*Spannung der äussersten Fasern*' of 52 centners.

Now if S be the *shear* and not the traction at the 'outer fibre' I find

$$S = \frac{8}{\pi} \frac{d}{\Delta^3} \times C,$$

and S has the value 52 centners.

So that there is a radical difference between the two stresses of 252 and 52 centners both of which Wöhler speaks of as *Faserspannung.* He proceeds as follows:

Die Möglichkeit des Falles vorausgesetzt, dass die grössten Kräfte auf Biegung und auf Verdrehung gleichzeitig wirkten, ist dann nach den vorstehend ermittelten Zahlen die grösste aus diesem Zusammenwirken resultirende Faserspannung der Achse $=\sqrt{252^2+52^2}=257$ Centner pro □ Zoll (S. 644).

I do not understand why the maximum stress at any point should be the square root of the sum of the squares of the maximum traction and shear.

The maximum stretch s_0 is, I think, given by

$$s_0 = \sqrt{\frac{T^2}{E^2} + \frac{1}{4}\frac{S^2}{\mu^2}},$$

and therefore the maximum traction, which is not of course in the direction of the 'fibre', by

$$T_0 = Es_0 = \sqrt{T^2 + \frac{1}{4}\frac{E^2}{\mu^2} S^2};$$

or, for uniconstant isotropy,

$$T_0 = \sqrt{T^2 + \tfrac{25}{16} S^2}.$$

In Wöhler's results the smallness of S as compared with T enables us practically to neglect its effect on the maximum stress.

Wöhler remarks that the maximum stress of 257 centners in these cast-steel axles of $3\frac{3}{4}''$ diameter would have produced set in iron axles, the elastic limit of which he takes at 180 centners per □″.

The above results were for four-wheeled wagons. For six-wheeled wagons he found that these stresses were increased in the ratio of 6 : 5 ; while for covered four-wheeled as compared with open four-wheeled wagons they were increased as 10 : 9.

For the 5″ diameter axle the maximum traction was 156 centners and the maximum shear 35 centners, so that the result appears rather close to the elastic limit of iron as stated by Wöhler above. Further, with these axles the maximum stress seems to have been often repeated. Wöhler reduces these stresses to percentages of the total load of wagon and cargo (S. 646).

[1001.] The number of repetitions of the maximum stress does not, Wöhler considers, exceed one per German mile[1], and he reckons the life of an axle at 200,000 miles. Hence he reduces his problem to the following one: To how great recurring positive and negative stresses can an axle of given dimensions be subjected 200,000 times without its breaking? He describes (S. 647–8) the method by which he proposes to answer experimentally this problem. Thus we see the origin of his later experiments on repeated loading. Two points are to be noticed in the problem as Wöhler states it. First, he supposes that all loads less than the maximum (and therefore unrecorded by his apparatus) do not contribute to the destruction of the axle. Secondly, he takes no account of the rapidity with which the loads are repeated. Would 200,000 loadings and reloadings of an axle in a day represent the same wear as 200,000 like maximum loadings occurring while the axle was running 200,000 miles, that is, spread out over its whole life? —I believe not. A load repeated many times a minute may lead to a vibratory accumulation of stress; this accumulation is impossible if the same load recurs only at long intervals. For both these reasons Wöhler's latter experiments do not seem to me so useful as they might otherwise have been.

The memoir concludes with tables of the numerical details of the experiments (S. 647–52).

[1002.] Wöhler: *Versuche zur Ermittelung der auf die Eisenbahnwagen-Achsen einwirkenden Kräfte und der Widerstandsfähigkeit der Wagen-Achsen. Erbkams Zeitschrift für Bauwesen*, Jahrgang X., S. 583–616. Berlin, 1860. The first portion of this memoir (S. 583–6) is entitled: *Versuche zur Ermittelung der auf die Wagen-Achsen einwirkenden Kräfte*, and it details experiments made with a slightly modified form of the apparatus referred to in our Arts. 998–1000, necessitated by its application to the carriages of passenger trains moving at a greater speed. Only flexure experiments were made, as it was considered that the first set of experiments had demonstrated that the torsional stress was of small account. In the case of the axles of passenger carriages the maximum stress in a 4·5″ iron axle was found to be 173 centners per □″, and it was reckoned that the mean stress was from 80 to 110 centners.

[1] About five English miles.

Ein Durchmesser der Achse von 4·5 Zoll in der Nabe darf daher bei Wagen der Art, wie der benutzte, für vollkommen ausreichend erachtet werden (S. 586).

The details of the experiments on the axles of passenger carriages are given on S. 603–4. Further experiments, confirmatory of the previous ones, on the $3\frac{3}{4}''$ steel and 5″ iron axles of goods wagons are given on S. 601–2.

[1003.] The second part of the paper is entitled: *Versuche zur Ermittelung der Widerstandsfähigkeit der Wagen-Achsen* (S. 586–600 with tables of results S. 605–16).

These are Wöhler's first series of experiments on repeated loading, and we postpone their consideration for the present, in order to deal with his complete researches in the decade 1860–70. The present experiments are on repeated flexure and the stress changed from zero to its maximum positive value through zero to its maximum negative value, and then back to zero again, once in about every four seconds (S. 586). Other matters in the memoir, not exactly bearing on repeated loading, are the erroneous treatment of the stretch-squeeze ratio on S. 592 where it is assumed that the volume of a bar under flexure does not change, the deduction of the form of the strained cross-section, and a not very lucid discussion of the relation of set to elastic strain. Wöhler holds that elastic strain is always in linear proportion to stress, and is in itself quite independent of the amount of set (S. 600).

Group D.

Memoirs relating more especially to Bridge Structure.

[1004.] F. W. Schwedler: *Theorie der Brückenbalkensysteme. Erbkams Zeitschrift für Bauwesen*, Jahrgang I., S. 114–123, 162–173, 265–278. Berlin, 1851.

The first part of this paper is purely theoretical, but does not present any novelty; the second deals with braced girders in general; and the third applies the rather complex formulae obtained to special systems (Neville's and Town's girders, and to Stephenson's and Fairbairn's tubes). There are no numerical examples, and I doubt some of the statements made (e.g. S. 277) and see little advantage in others.

[1005.] A series of papers also by Schwedler on the analytical calculation of stress in the members of latticed girders (*gitterförmige Träger*) will be found in *Zeitschrift des Vereins deutscher Ingenieure*, Jahrgang III., S. 37, 96, 135, 233, 297 (Berlin, 1859), but they have little real bearing on the theories of elasticity or of the strength of materials.

[1006.] Baensch: *Zur Theorie der Brückenbalkensysteme. Erbkams Zeitschrift fur Bauwesen*, Jahrgang VII., S. 35–50, 145–156, 289–308. Berlin, 1857. This memoir professes to be a continuation of F. W. Schwedler's paper in the volume for 1851 of the same *Zeitschrift*: see our Art. 1004. It applies the Bernoulli-Eulerian theory of beams to various cases of simple beams, to a few continuous beams, and to some cases of braced girders. There is a great deal of analysis in the paper, but from the theoretical standpoint no novelty. It may well be questioned whether practically it would not be easier to work out *ab initio* the theory of any given girder than to endeavour to apply formulae obtained in a memoir of this type for a great variety of special cases, none of which may really fit the exact type of girder it is required to construct.

[1007.] *Institution of Civil Engineers. Minutes of Proceedings.* Vol. XI., pp. 227–232, 1851–52. Appendix to a paper by A. S. Jee entitled: *Description of an Iron Viaduct erected at Manchester....*

This Appendix gives the details of experiments on cast-iron girders of varying cross-section so far as deflection and set are concerned—these were of I section; also the details of experiments on wrought-iron tubular girders—these latter might be described as I girders in which the upper flange was replaced by a tube of circular section to prevent buckling.

Further experiments by J. Hawkshaw on the absolute strength, deflection, etc. of other I cast-iron girders will be found on pp. 242–3.

[1008.] *British Association*, 1852, *Belfast Meeting, Transactions*, pp. 126–7. T. M. Gladstone gave some notes on the superior safety of malleable to cast-iron girders. He considered the reduction in weight compensated for the difference in price. He also treated of the proper relation between depth and span for a particular I-section, apparently however for the case of a special load.

[1009.] Poncelet: *Examen critique et historique des principales théories ou solutions concernant l'équilibre des voûtes. Comptes rendus*, T. XXXV., pp. 494–502, pp. 531–540, and pp. 577–587. Paris, 1852.

This is a very valuable criticism of the various theories of the arch propounded up to 1852, and is of peculiar interest as noting the extent to which these theories had applied the principles of elasticity to this very important but very difficult practical problem. The paper forms a most interesting historical *résumé* of the subject.

We may note that Navier seems to have been the first to apply the mathematical theory of elasticity to the calculation of arches (p. 532). He seems to have been the first also to state the rule of the 'middle third'—a result which follows at once from the core of any rectangular block subjected to loading on two opposite faces (p. 533). Méry in a memoir in the *Annales des ponts et chaussées* (1er Semestre, 1840, p. 50. Paris, 1840) follows Navier in applying the theory of elasticity to arches, but his memoir would hardly satisfy the more rigorous theorist whatever practical value it may have (p. 539). He makes use also of a graphical construction for the *line of pressure*, first introduced by Moseley, whose researches on this point were continued by Scheffler (see pp. 583–4). Most of the papers to which Poncelet refers do not appeal to the theory of elastic solids, but he insists that the theory of arches is really inseparable from this:

On comprend, en effet, d'après tout ce qui précède, que les deux questions de l'équilibre des voûtes et de la résistance élastique des solides, sont liées entre elles de la manière la plus intime, toutes les fois que l'on prétend sortir de l'hypothèse abstraite où l'on suppose aux voussoirs une continuité, une invariabilité de forme absolue. L'analogie même est telle, que l'on peut dire, sans trop s'avancer, que la théorie des voûtes et celle des solides élastiques courbés naturellement n'en constituent, en réalité, qu'une seule, considérée dans des conditions et sous des aspects différents (p. 586).

That the whole theory of arches would be revolutionised if we could solve the problem of the strains in a right six-face subjected to equal and opposite load systems on two parallel faces, seems to be Poncelet's view of the subject.

[1010.] H. Bertot: *Construction des ponts en arc. Mémoires et compte-rendu des travaux de la Société des Ingénieurs Civils,* Année 1858, pp. 298–303. Paris, 1858. This contains only an elementary statical method of finding the total stress across any cross-section of a doubly pivoted arched rib. The method is analytical but is of no importance.

[1011.] T. F. Chappé: *Account of Experiments upon Elliptical Cast-iron Arches. Institution of Civil Engineers, Minutes of Proceedings,* Vol. XVIII., pp. 349–362 (with discussion). London, 1858–9. The experiments were made on model arches of considerable size, but the method of experimenting and the castings appear to have been very inferior. The results obtained cannot therefore be taken as a true measure of the strength of cast-iron elliptical arches.

[1012.] J. Cubitt: *A Description of the Newark Dyke Bridge. Institution of Civil Engineers, Minutes of Proceedings*, Vol. XII., pp. 601–611. London, 1852–3. This contains some experiments on the deflection of Warren girders either partially or totally loaded.

[1013.] J. Poirée: *Observations sur la répartition de la pression dans la section transversale des arcs des ponts en fonte. Annales des ponts et chaussées*, 1er Semestre, pp. 374–95. Paris, 1854.

This article is of value for the details it gives of a number of experiments on the deflections of arched ribs due to temperature, to rapidly moving live-loads (railway engines and carriages, vehicles drawn at a trot, cavalry, etc.) and to sudden impacts. For a rapidly moving load the author considers his results confirm those of Willis and Stokes (p. 390): see our Arts. 1418*–22* and 1276*–91*.

[1014.] Further experiments on the deflection of arches due to slow and rapid live-loads will be found on pp. 1–7 of the volume for 1854, 2e Semestre, and on pp. 192–8 of that for 1855, 1er Semestre. There are also numerous papers in the volumes of this Journal for 1850–60 describing individual metal bridges, or dealing with the theory of the stability of arches, of which our space will not even permit us to cite the titles. They are of more interest from the standpoint of the historian of engineering, than from that of the historian of elasticity.

[1015.] H. Haupt: *Resistance of the Vertical Plates of Tubular Bridges. The Civil Engineer and Architect's Journal*, Vol. XVII., pp. 26–7. London, 1854. This is of no value for our purposes.

H. Cox: *On the Strength of Compound Girders*, *Ibid.*, pp. 122–125. This contains some interesting remarks on the theory of *Trussed Cast-Iron Girders*, with reference to Fairbairn's experimental results.

[1016.] J. Barton: *On the Economic Distribution of Material in the Sides, or Vertical Portion, of Wrought-Iron Beams. Institution of Civil Engineers, Minutes of Proceedings*, Vol. XIV., pp. 443–490 (with discussion). London, 1854–5. This paper propounds very obscure notions as to the stress in beams (e.g. p. 445 !), which can only be paralleled by certain observations put forward in the discussion. Thus one speaker 'altogether denied' that different forms of beams under flexure require 'different mathematical reasonings.' The vague use of the expression 'strains travelling in this or that direction' will produce despair in the mind of the scientific elastician. Indeed the whole problem, which engaged the minds of the practical men present, as to whether the strains in the web of a girder are horizontal or inclined at 45°, seems to point to a painful want of theoretical knowledge in the English engineers of that day. As a sample of the sort of obscurity to be found in the discussion I may cite the following:

Mr W. H. Barlow after referring to his untenable theory of a new element of strength in beams (see our Arts. 930–1) remarks:

Mathematicians had applied the axiom, "ut tensio, sic vis," in the case of transverse strains, in which continuous fibres were unequally strained, without considering the lateral action arising from the cohesion of the particles; this axiom, therefore, required modification (p. 480).

Perhaps pp. 485–6 containing remarks by Messrs R. Stephenson, W. H. and P. W. Barlow on the 'neutral axis' and absence or non-absence of strain there, showing as they do a want of any due appreciation of the difference between shearing and tractive stress, are the most remarkable picture that I have come across of the state of technical elasticity in 1854.

I may note that doubt was thrown by Mr Doyne on the accuracy of Hodgkinson's results for the beam of strongest section: see our Arts. 244*, 1431*, 875 and 1023.

[1017.] Wöhler: *Theorie rechteckiger eiserner Brückenbalken mit Gitterwänden und mit Blechwänden. Erbkams Zeitschrift für Bauwesen,* Jahrgang v., S. 122–161. Berlin, 1855.

This memoir may be considered as consisting of three parts. In the first (S. 122–141) after some not very lucid remarks on the method by which vertical load is transmitted by the bracing bars from any point of a girder to the supports, Wöhler deduces the stresses in the bracing bars from purely statical considerations and from hypotheses as to the reduction of systems of multiple bracing to single bracing. In the next part of his paper he deals with the flexure of the booms and the stresses in the bracing bars when the latter are supposed to be riveted to the boom and not merely pin-jointed. He concludes (S. 150) that the riveting has practically no influence on the strength of the girder. In a girder of 100 feet span with bracing bars 4 inches broad the ratio of increase of strength would be only 1/230. His conclusions as to the radii of the bent bracing bars in terms of the radius of the bent girder are similar to those of De Clercq and C. Winkler: see our Arts. 1026 and 1028. The memoir concludes with a lengthy discussion on girders with plate-iron webs (S. 154–61). This takes into account the shear in the web. The investigation is not particularly clear, and the simplicity of the problem (when once the hypotheses are accepted) by no means seems to warrant the great display of analysis. In the comparison of plate and lattice girders, with which Wöhler concludes his memoir, he attempts to show that there are within the limits of practical construction certain great spans possible in which the plate web girder would be superior to the lattice. In coming to this conclusion Wöhler takes into account the relative deflections of the two kinds of girders.

[1018.] J. M. Heppel: *On the Relative Proportions of Top, Bottom and Middle Webs of Iron Girders and Tubes. Institution of Civil Engineers, Minutes of Proceedings*, Vol. xv., pp. 155–194. London, 1855–6.

This paper is a fitting sequel to that of our Art. 1016, which indeed appears to have called it into being. The author remarks that in order to deal with the effect produced by forces on an elastic plate we must settle between two hypotheses which present themselves, viz.:

(i) That a force applied in a given direction causes no change in the dimensions of the material perpendicular to that direction.

(ii) That the application of force in any direction causes no change in the volume of the material.

The author remarks that "the simplicity alone of the former of these suppositions entitles it to preference." It is perhaps needless to remark that both are absolutely wrong. The paper itself leads to results, which if true, would be more easily obtained by the ordinary theory of elasticity, but the final assumption on p. 164 seems to me quite untenable, and indeed the results do not agree with Saint-Venant's theory for the case of a web without flanges or of a beam of rectangular cross-section.

[1019.] H. Lohse: *Versuche über das Zerknicken der Eisenstäbe in Gitterträgern. Erbkams Zeitschrift für Bauwesen*, Jahrgang VII., S. 573–580. Berlin, 1857.

This paper contains, I think, the details of the first experiments on a point which innumerable writers had been theorising about (see our Arts. 1017 and 1026–30). They were made in view of the construction with lattice girders of the *Rheinbrücke* at Cöln. The experiments were made on lattice girders treated as cantilevers, the bracing was single, double, triple and fourfold, the bars being riveted to each other. The load at which the bracing bars received a permanent set was in some cases noted, as well as the load at which they buckled (*zerknickten*), by which I think Lohse means total collapse. In one case three of the bars in tension were ruptured. It is noteworthy that in several cases the bracing bars bent elastically into an approximate S-form, a result which neither

De Clercq, Winkler, nor Wöhler take into account in their analysis (see our Arts. 1017, 1026 and 1028). In all cases the theoretical stresses in the bracing bars are calculated and tabulated. The experiments show the great increase of strength due to multiple-bracing and to the riveting together of the bracing bars. Lohse does not consider his experiments numerous enough for him to propound any general formula, but the numerical details and the general results are perhaps more likely to be of practical service than the lengthy analytical investigations to which we have previously referred.

[1020]. G. Wolters: *Résumé des résultats obtenus dans les épreuves de quelques ponts en fer. Annales des travaux publics de Belgique*, Tome xv., pp. 145–75. Bruxelles, 1856–7. This is translated into German under the title *Bericht über die Ergebnisse der Belastungsproben einiger eiserner Brücken* in the *Zeitschrift des österreichischen Ingenieur-Vereins*, Jahrgang x., S. 185–195. Wien, 1858. It gives details of experimental determinations of the deflection of various railway bridges in Flanders. The girders of the bridges were chiefly of cast-iron with openings in the web; there was one of plate-iron. The results do not seem to have any permanent theoretical importance.

[1021.] In the same volume of the *Annales* is an article by Houbotte entitled: *Expériences sur la résistance des longerons en tôle* (pp. 403–26). It is translated into German in the same volume of the *Zeitschrift*, S. 195–201. The girders were of plate-iron, and the object of the experiments described was to test the best relative dimensions for the flanges and web in the case of girders of I section. The span of the model girders was from 1·5 m. to 3 m. and their heights varied from ·2 to ·49 m. The load upon them was gradually increased up to rupture during a period amounting in some cases to several weeks; the duration of load, the elastic and set deflections were all noted. There is also one set of experiments on a girder in which the web was replaced by bracing. Houbotte concludes from this experiment that the bracing is more efficient than the plate-web. The rupture occurred in the plate girders by buckling of the web. Houbotte endeavours to construct a formula which will give the proper strength for the web of such a girder, but neither the range of experiments nor his method of obtaining his formula seems very satisfactory.

[1022.] A number of articles by J. Langer on wooden and iron lattice girders, the latter in the form of arches with braced ribs, etc. will be found in the *Zeitschrift des österreichischen Ingenieur-Vereins*, Jahrgang x., S. 113, 135, 152. Wien, 1858–9. Jahrgang xi., S. 69, 127, 153, 186, 206. Wien, 1859. These give a lengthy theory of the braced arch. Further projects for braced arches will be found in Jahrgang xii., S. 29, 91, 125 and 193. In several of these projects, graphical methods might now be usefully employed. The designs would form very good exercises in the calculation of stresses for advanced engineering students even at the present day.

[1023.] Thomas Davies: *Wrought and Cast Iron Beams. The Civil Engineer and Architect's Journal*, Vol. xx., pp. 20–23, and pp. 41–44. London, 1857. This paper was read at a meeting of the Architectural Institute in Edinburgh, February, 1856.

It commences with some account of the want of confidence felt in cast-iron beams, and of the superiority of malleable-iron beams owing to their lightness and sensible yielding before rupture. Fairbairn having given in his work on cast- and wrought-iron only one experiment on a "plate beam" (one of I section?) Davies proposes to supply this want of information with regard to the strength and elastic properties of wrought-iron beams, in order that they may be more generally understood and adopted.

The experiments given in the first part of the paper may have technical, but they hardly have theoretical or physical value; the load was applied over as much as $\frac{1}{6}$ of the length of the beam, and was brought into play by putting on the top flange iron railway bars "requiring two men at each end to lift them." The author agrees with Tate that the upper and lower flanges of a wrought-iron beam should have practically equal areas (p. 23).

The second portion of the paper criticises the results of Hodgkinson's experiments on the beam of greatest strength: see our Arts. 244*, 875 and 1016. The writer contends that the ratio of the sectional area of the flanges ought to be as 3·5 or 3 to 1 and not 6 to 1 as suggested by Hodgkinson. He enters into no theoretical investigation of the strength of such beams, nor does he adduce any experimental evidence beyond Hodgkinson's. He considers Hodgkinson's results erroneous because the latter left out of account the difference in the thicknesses of the webs of his individual beams when deducing conclusions from his experiments. It seems to me that Hodgkinson was right and quite justified in doing this, as the web added little to the flexural strength of the beam. Thus the ratio of the areas of the flanges ought to be nearly that of the compressive to the tensile strength of cast-iron, i.e. according to Hodgkinson about 6 : 1.

[1024.] Decomble: *Sur les meilleures formes à donner aux poutres droites en fonte. Annales des ponts et chaussées*, 2e Semestre, pp. 257–319. Paris, 1857.

This is a long memoir investigating a theory of the solid of greatest resistance, when the tensile and compressive strengths of the material are supposed different, and the solid is designed so that the ruptures of the stretched and squeezed 'fibres' occur at the same load. Apart from the assumption involved in applying the theory of elasticity to the phenomena of rupture, the discussion seems in several points very doubtful, and all that can be reached of value by a theory of this kind has been better given by Saint-Venant in his *Leçons de Navier*, pp. 102, 142–56, and our Arts. 176, 177, (*b*). There are, however, a number of interesting experiments on the rupture of cast-iron beams of various shapes and cross-sections, which may possibly have practical value still. The editors of the *Annales* remark in a note appended to the memoir:

Quoique la partie théorique du mémoire précédent soit en opposition, sur plusieurs points, avec les principes généralement admis, la commission des *Annales* a cru devoir le publier tel qu'il a été présenté par l'auteur, à raison des détails intéressants qu'il renferme sur les poutres en fonte et sur le moulage de la fonte en général (p. 319).

[1025.] *British Association. Report of Twenty-Seventh* (Dublin) *Meeting*, 1857.

The titles of two articles in this *Report* may be noticed: C. Vignoles: *On the Adaption of Suspension Bridges to sustain the passage of Railway Trains*, pp. 154–158. P. W. Barlow: *On the Mechanical Effect of combining Girders and Suspension Chains, and a comparison of the weight of Metal in Ordinary and Suspension Girders, to produce equal deflections with a given load*, pp. 238–48. Both these papers discuss the adaption or modification of suspension bridges when built for the transit of railway trains. They turn principally on stiffening the platform till it becomes a girder, or on special arrangements of the suspending bars. The bridges at Niagara and elsewhere built as girder suspension-bridges, had gone far to destroy the old mistrust in suspension-bridges for railway traffic; and the authors of the above papers endeavour to show that equal strength may be obtained with far less material from a suspension-girder than from a pure girder bridge.

[1026.] G. A. De Clercq: *Note sur les phénomènes de la flexion des poutres en treillis. Annales des travaux publics de Belgique*, T. xv., pp. 198–214. Bruxelles, 1856–7.

This is another of those memoirs which deal with the lattice-girders, which were rapidly taking the place of the older double-T girders with a solid web. The writer of the memoir supposes the bracing bars rigidly attached to the booms, and deduces by what does not seem to me very conclusive reasoning, that a bracing bar after flexure will take the form of a spiral of Archimedes (p. 201). C. Winkler (see our Art. 1028) had, I think, read De Clercq before writing the second part of his paper; he extends, however, the latter's results. The present paper is clearly written as compared with Winkler's, but it deals with a simpler case. At the same time to consider the special conclusions deduced by both these writers from their somewhat doubtful hypotheses would carry us beyond our limits.

[1027.] C. Knoll: *Zur Theorie der Gitterbalken. Eisenbahn-Zeitung*, Jahrgang XVI., S. 13–5. Stuttgart, 1859. This is an analytical calculation of the stresses in the bracing of lattice-girders with straight parallel booms.

[1028.] C. Winkler: *Theorie der eisernen Gitterträger. Försters Allgemeine Bauzeitung.* Jahrgang XXIV., S. 191–222. Wien, 1859.

This memoir on lattice-girders is divided into two parts. The first deals with the stresses in the booms and bracing bars when the bracing bars are not riveted to each other. In this case we have only to consider the flexure of the booms, for the bracing bars are, if buckling be excluded, in pure tensile or compressive stress. Winkler proceeds analytically to the discussion of the stresses, and points out an error of Scheffler's (see our Art. 651). The treatment appears sound, and the results, although having only special technical interest and application, may still be of service (S. 191–9).

In the second part of the memoir the bracing bars are supposed riveted or pinned where they cross each other, and the result is that these bars are now subjected to flexure. The calculations, here of course necessarily analytical, become more complex, and I confine myself to referring to Winkler's analysis which I have not verified (S. 199–206). How far his fundamental hypotheses—similar to those of De Clercq (see our Art. 1026),—approach the truth, especially for the second case stated on S. 199–200, I have no means of judging, they seem to me somewhat bold, not to say dubious. The memoir concludes with the application of the results obtained to a number of numerical cases of lattice-girders (pp. 206–22). The *exact* treatment of these lattice-girders, in which the frames have a great number of supernumerary bars, would be an extremely difficult analytical problem.

[1029.] B. B. Stoney: *On the Application of some new Formulae to the Calculation of Strains in braced Girders. Proceedings of the Royal Irish Academy,* Vol. VII., pp. 165–172. Dublin, 1862. This paper was read in 1859.

Pp. 165–9 deal by the simplest statical methods with the stresses in the diagonal bracing of a Warren girder when some or all of the nodes at the upper boom are loaded. There is nothing that calls for special comment.

Pp. 169–72 treat of *Lattice Girders* and use only ordinary statical processes. The discussion, however, seems to me obscure, especially the final paragraph. It is in many cases impossible to find the *exact* stresses in lattice-girders without appeal to the theory of elasticity, and this point does not seem to have been recognised by Stoney.

[1030.] Another paper by B. B. Stoney may be just referred to here: it is entitled: *On the Relative Deflection of Lattice and Plate Girders* and is published in the *Transactions of the Royal Academy,* Vol. XXIV., pp. 189–93 of Part I., *Science.* Dublin, 1871. The paper was read June 23, 1862. It does not seem to contain anything of sufficient theoretical interest or experimental value to require special notice.

[1031.] J. G. Lynde: *Experiments on the Strength of Cast-Iron Girders.* This paper was read before the Manchester Literary and Philosophical Society. An abstract of it will be found in *The Civil Engineer and Architect's Journal*, Vol. XXII., pp. 386–7. London, 1859.

Lynde made experiments on 89 girders, cast on their sides, and of the form recommended by Hodgkinson as that of strongest section (see our Art. 244*). One girder only was tested up to rupture, and Lynde remarks that no permanent set was visible up to that point (!). The girders were of large size (30 ft. 9 in. in span).

Hodgkinson (*Experimental Researches on Cast-Iron*, Art. 146) had given the following formula for W, the breaking central load in tons:

$$W = \frac{2}{3dl}\{bd^3 - (b - b')\, d'^3\},$$

where: l = span in feet,

b = breadth of bottom flange in inches,

b' = thickness of web in inches,

d = whole depth in inches,

d' = depth from the top of the beam to the upper side of the bottom flange in inches.

Lynde's single experiment on rupture would go to show that the coefficient 2/3 is too large for a large beam and should be replaced by ·625, as suggested by Hodgkinson himself in his Art. 147 for *large* beams.

[1032.] Marqfoy: *Mémoire sur les essais des ponts en tôle par l'électricité. Annales des ponts et chaussées*, 1859, 2e Semestre, pp. 74–89. Paris, 1859. This paper describes an apparatus for recording the deflection at various points of bridges under a rapidly moving load.

[1033.] Noyon: *Notice sur la restauration et la consolidation de la suspension du pont de la Roche-Bernard.* This paper is in the same volume of the *Annales*, pp. 249–329. It gives an account of the accident to this suspension bridge (see our Art. 936*) and also details of numerous experiments on the absolute strength of iron wire and cables.

[1034.] In the same periodical in the volume for 1860, 2e Semestre, are two articles on bridges which touch the limits of our subject. The first by Jouravski, entitled, *Remarques sur les poutres en treillis et les poutres pleines en tôle*, pp. 113–34, discusses in general terms the vibrations which occur in such bridges and their strength; the second by Mantion, *Étude de la partie métallique du pont construit sur le canal Saint-Denis...* pp. 161–251, gives a very full theoretical determination of the stresses etc. in all the different parts of a particular bridge.

[1035.] W. Fairbairn: *Experiments to determine the Effect of Vibratory Action and long-continued Changes of Load upon*

Wrought-iron Girders, pp. 45–8. *British Association, Report of Thirtieth* (Oxford) *Meeting*, 1860.

The experiments were made on a double-T plate girder of 20 feet span, the flanges being built up of plates and angle-irons, the total depth of section was 16″ and the calculated breaking weight 12 tons. The load was applied at the mid-section of the girder in a gradual manner at the rate of about eight changes per minute with the following results:

Load	Number of loadings	Total number of loadings	Mean Deflection
About 3 tons	596,790	596,790	·17″
About 3·5 tons	403,210	1,000,000	·22″
About 4·8 tons	5,175	1,005,175	·35″ (broke)

The beam was repaired after the last load and 1,500,000 additional loadings were given to it with a load of about 3 tons without its giving way.

It would appear, therefore, that with a load of this magnitude the structure undergoes no deterioration in its molecular structure; and provided a sufficient margin of strength is given, say from five to six times the working load, there is every reason to believe, from the results of the above experiments, that girders composed of good material and of sound workmanship are indestructible so far as regards mere vibratory action (p. 48).

It will be noted that Fairbairn's experiments differ from Wöhler's in their *slowness of repetition*, there was very little opportunity for accumulation of stress.

[1036.] P. Fink: *Allgemeine Betrachtungen über Biegungsfestigkeit und Biegungswiderstand zur Erzielung eines einheitlichen Standpunktes für die Beurtheilung verschiedener Brücken-Systeme. Zeitschrift des österreichischen Ingenieur-Vereins*, Jahrgang XII., S. 40, 69–77, 204–211. Wien, 1860. This paper does not seem to convey any information beyond what may be deduced from the ordinary Bernoulli-Eulerian theory of beams when the load is not perpendicular to the axis. I have made no attempt to investigate the accuracy of the lengthy formulae with which the memoir abounds.

GROUP E.

Researches on the Strength of Cannon and of Materials for Ordnance.

[1037.] *Reports of Experiments on the Strength and other Properties of Metals for Cannon, with a Description of the Machines for testing Metals, and of the Classification of Cannon in Service, by Officers of the Ordnance Department, U. S. Army.* Philadelphia, 1856. This folio volume of 428 pages is the first batch of a series of valuable technical researches in elasticity due to the United States Government. The more important portion of the present work is due to Major W. Wade, and it is sometimes cited as Wade: *On the Strength of Metals for Cannon.* A further group of reports by Rodman will be dealt with in the period 1860–70. The value for us of these reports lies not in their details as to cast-iron and bronze ordnance, which probably have little more than historical interest at the present day, but in the numerous experimental investigations on the strength of materials which are embraced in their pages. We can only afford space to note briefly some few of the facts recorded.

[1038.] The first report deals with cast-iron, and particularly, with the influence of the time of fusion and the number of meltings upon the strength. We mark the following conclusions:

(*a*) A prolonged exposure of liquid iron to intense heat augments its absolute strength. The strength increases as the time of exposure up to some not well ascertained limit between 3 and 4 hours (?). This result does not seem to be based on a sufficiently large range of experiments. The experiments made were on 8 cast-iron guns tested up to bursting (pp. 11–17).

(*b*) In experiments on the transverse strength of cast-iron bars it was found that the absolute strength as deduced from bars of circular cross-section was uniformly much higher than that from those of square cross-section cast from the same kind of iron. This is part of the old 'paradox in the theory of beams.' Casting at a high temperature gave greater strength than casting at a low one; a gradual increase of strength even up to 60 p.c. was found to result from increasing the time of fusion; this increase of strength was accompanied by a decrease of set, the set being measured for a given load somewhat less than the minimum breaking load (pp. 21–8). On p. 44 further evidence is given of the increase in both tensile and transverse strength by increasing the period of fusion. It is also shewn that rapid cooling increases transverse strength in small castings, and slow cooling increases tensile strength in large ones (p. 45).

(*c*) Some attempts will be found on pp. 77–88 to connect the tensile strength of a bar, the hydrostatic rupture pressure in a cylinder and the impulsive rupture pressure due to the discharge of powder in the same cylinder with one another. As no theory is proposed for this comparison, the experiments are rather vaguely directed and lead to no very definite conclusions. Wade takes for the transverse strength of a beam of length l and rectangular cross-section $b \times d$, when centrally loaded with w, the expression $wl/4bd^2$ or $1/6$ of the greatest traction in the extreme fibre. He has for mean results on p. 80: tensile strength of cast-iron = 22,133, transverse strength of cast-iron = 7370. Thus we should have for the rupture stress in the extreme fibre 44,220 or almost *double* the tensile strength. This is a good example of the so called 'paradox in the theory of beams.' The absolute strength calculated from flexure experiments upon a rectangular beam is by this misapplied theory double the tensile strength of the material: see our Arts. 173, 178, 930, 1043, 1049–53, etc.

Wade's process of calculating the resistance of a circular cylinder to hydrostatic pressure—i.e. by multiplying the pressure per square inch by the radius and dividing by the thickness of the cylinder—can hardly be considered satisfactory, when radius and thickness are commensurable. This is well brought out by the Table on p. 87, where the ratio of this resistance to the tensile strength varies greatly with the ratio of the thickness to the radius of the bore. But I doubt the accuracy even of Wade's experimental numbers, for when the ratio of the thickness to the radius of the bore remains constant, the internal bursting pressure does not bear the same ratio to the tensile strength, but varies from ·329 to ·602!

Further details of experiments on the bursting of musket barrels by hydrostatic pressure are given on pp. 92–107, but I have not been able to apply any theory (e.g. the formula in the footnote of our Vol. I. p. 550) to the numbers given because the proper details are not recorded.

(*d*) Some experiments on the effect which slow cooling and casting under atmospheric pressure have on the bursting strength of guns are given on pp. 129–34. They are neither numerous nor scientific enough to yield results of much value.

(*e*) A further report on the manufacture of 24-pounder iron cannon does not throw more light on the influence of the times of melting and fusion, pp. 145–8. For the proof bars of these castings the mean ratio of rupture stress in 'extreme fibre' to tensile strength = 1·65, so that it is considerably less than in the experiments considered under (*c*).

(*f*) The influence of height and bulk of the sinking head in bronze gun-metal castings on both density and tenacity is referred to on pp. 152–5 and may be compared with the more definite results of a somewhat later British Report: see our Art. 1050.

(*g*) On pp. 183–221 we have some interesting details of the relative durability of guns when cast solid and when cast hollow, in the latter case the interior was cooled by a flow of cold water. The hollow cast guns appear to have stood longer service than the solid cast guns, but the tenacity of specimens taken from the body of the former after bursting was not sensibly greater than that of specimens from the latter. This fact led Wade to suppose the difference in endurance to be due to differences in initial stress resulting from the different modes of cooling. Rodman (pp. 209–13) gives a not very lucid theory of initial stress deduced from an erroneous hypothesis of Barlow. The most interesting part, however, of this report is perhaps embraced by pp. 217–221, where Wade shows how experience and probability tend to demonstrate that initial stress due to cooling ultimately subsides. He cites a number of cases in which bodies held in a state of constraint obtain a gradually increasing set, apparently relieving them from this state, and he tries to show that guns retained after manufacture for long periods before proof, sustain a far greater proof than those tested directly. He accounts in this way for the hollow cast guns having greater endurance than the solid cast guns. The process of internal cooling he supposes produced less initial stress although no greater tenacity. Some of the details he gives are of interest in their bearing on elastic after-strain.

[1039.] The next portion of the volume (pp. 223–322) is entitled: *Report on the Strength and other Properties of Metals and on the Manufacture of Bronze and Iron Cannon*, 1854. This is the final report due to Major Wade. We proceed to note some points in it.

(*a*) The effect produced by remelting iron and by retaining it in fusion exposed to an intense heat for a long time is very fully considered on pp. 223–46. The quality of iron is as a rule very much improved by remelting and long continued fusion, but the effects vary from one kind of iron to another. The order of densities is almost invariably the order of tenacities or at any rate up to a certain limit of density[1]. As a sample of the sort of results reached, I cite the following (p. 234):

[1] This limit of increase of tenacity with increase of density does not seem to have been clearly proven. Thus on p. 244 it is supposed that Greenwood iron attains its maximum tenacity with a density of 7·27, but it was found later (pp. 246–7) that a density of 7·307 gave even higher tenacities: see our Art. 1086.

	Density.	Tenacity in lbs. per sq. in.
Amenia pig iron, 1st fusion	6·948	11420
Same iron, remelted, 6 hours in 2nd fusion	7·172	26310
Another parcel of Amenia iron, 2nd fusion	7·184	26237
Same iron remelted, 3rd fusion	7·322	34728

Thus it was found that the mass per cubic foot could be increased as much as 20lbs. and the tenacity in the ratio of 2·8 or even 3 to 1.

As a relation between small and large castings, Wade states that at least for one kind of iron (Greenwood) the strength of proof bars at any fusion may without material error be taken as an approximate measure of the strength of gun heads made of the same iron at the next fusion (p. 243).

(*b*) On pp. 248–9 we have a number of experimental details on transverse strength. It is not easy to identify the bars which correspond to those treated for tenacity. But it would seem as if the ratio of rupture stress in 'the extreme fibre' to tenacity was as low as 1·6 or even less: see our Arts. 936, 1038 (*c*), 1043 and 1052–3.

(*c*) We have next a series of experiments on torsion (pp. 250–6). So far as rupture is concerned what Wade records is really the value of $\frac{1}{16}\pi T_3$ for bars of circular cross-section or the S_3 of our Art. 1051 (*c*), T_3 being the absolute shearing strength. Or, if T_3 be taken $= \frac{4}{5}$ the tensile strength T_2, he records what ought to equal $\cdot 15708 T_2$. If T'_2 be the value of T_2 calculated from this, I find from Wade's summary of results on pp. 241 and 251 by recalculating his numbers, that for various kinds of cast-iron the ratio of T'_2/T_2 varies from 1·6 to 1·8, the mean value being very nearly 1·7. Or, with the notation of our Art. 1051, $S_3/S_2 = \cdot 267$; this differs but slightly from the mean value as found from the British cast-iron torsion experiments: see our Art. 1053.

Besides the absolute torsional strength, the torsional elastic strain and set were noted for a variety of loads as well as the load which produced an angular set of $\frac{1}{2}°$ in a length of bar equal to about 8 times the diameter. This appears to have been about $\frac{7}{10}$ of the rupture load.

Wade also made experiments on the torsional strength of wrought-iron and bronze. His mean value for $\frac{1}{16}\pi T_3$ for wrought-iron is 5465 and for bronze 5511 lbs. per sq. in.

(*d*) Then follow experiments on the torsional strain and rupture of prisms of square, circular and circular-annulus cross-sections. The mean results are given on p. 256. The mean strength of prisms of square cross-section is about ·811 times the mean strength of those of circular cross-section of equal areas. If Saint-Venant's theory of the fail-limit (see our Arts. 18 and 30) held up to rupture the ratio ought to be ·738. For the strength of a hollow circular cylinder, the ratio of the internal diameter of which to the external diameter is ξ, I find on

Coulomb's theory $\frac{1+\xi^2}{\sqrt{1-\xi^2}}$ times the strength of the solid cylinder of equal area. This gives the ratios of the strengths for $\xi = \frac{1}{2}$ and $\frac{3}{5}$ as 1·44 and 1·7. Wade finds for the corresponding ratios in these cases 1·22 and 1·45, thus considerably less than the theory of the fail-limit would give if extended to the rupture of cast-iron.

(*e*) On pp. 257–9 we have details of experiments on the crushing strength of various cast-irons and steels. For cast-iron the ratio of the mean crushing strength to the mean tensile strength is about 4·56. If the theory of uni-constant elasticity be extended up to rupture then the ratio should be 4. The cast-iron was in small cylinders the lengths of which were generally two and a half times their diameters and the fracture-surfaces made angles of 52° to 59°·6 with the bases. Probably the ends were held in by the friction of the bed-plates and the strength would thus appear to be increased. I expect the ratio of crushing to tensile strength, if both could be ascertained accurately, is not very far from 4 for cast-iron.

For cast-steel Wade gives (p. 258) the following values of the crushing strength in lbs. per sq. inch:

Not hardened	198,944
Hardened; low temper; chipping chisels	354,544
Hardened; mean temper; turning tools	391,985
Hardened; high temper; tools for turning hard steel	372,598

He does not give the tensile strength of these steels which were all samples cut from the same bar.

[1040.] (*f*) Pp. 259–67 deal with the *Hardness of Metals*. These pages were translated into French and published as a tract entitled: *Expériences sur la dureté des métaux* (Paris, Corréard, 1861), but without the name of author or editor. Wade commences with the following statement:

The comparative softness, or hardness of metals, is determined by the bulk of the cavities or indentations, made by equal pressures; the softness being as the bulk directly, and the hardness, as the bulk inversely (p. 259).

The form of the indenting instrument was a pyramid on a rhomboidal base. The longer diagonal of the base measured 1″, the shorter ·2″, and the height of the pyramid ·1″. The planes of the sides intersected at the penetrating edge (point ?) at an angle of 90°. Such is Wade's description of the instrument, but it seems to me that the real height is about ·098″, the difference is perhaps in the angle and probably within the limits of experimental error. According to the author the apparatus would have been improved by making the longest diagonal 1·25″ instead of 1″, and causing the faces to meet at 60° instead of 90°. Such a pyramid would make a longer indentation and mark minute differences more accurately (p. 266). A cone with a vertical angle of 90° made a cavity about equal in bulk to that produced by the pyramid under an equal pressure[1]. Wade found that for the same material the cavities made by his instrument under different pressures were nearly as the pressures raised to the power of 3/2. The *nearly* however neglects a divergence of about 17 per cent. for large pressures, although the accuracy for small pressures is remarkable. This suggests that the empirical law of Wade may be near the truth for indentations only producing *set*, but becomes increasingly inaccurate as the loads produce *separation* of the material. This want of distinction between set and separation of the particles of the material—Wade measures in each case the indentation due to a pressure of 10,000 lbs.—seems to me the most serious objection to the process. It has obvious advantages, however, over the scratching methods (see our Arts. 836–44), and if Wade's law of relation between the volume of the indentation and the pressure were a correct one for set, we could obviously avoid such pressures as produce separation and get a scientific measure of hardness. The method does not, however, seem applicable to the variation of hardness with direction in crystals, or again to what Hugueny has termed tangential hardness.

Hertz's theory of hardness makes, I think, the depth of the indentation which a sphere would make on a plane vary as the $(\text{pressure})^{\frac{2}{3}}$. Hence for small indentations the volume would vary approximately as the $(\text{pressure})^{\frac{4}{3}}$. This applied to Wade's numbers gives results more discordant than his $\frac{3}{2}$, but this is natural as a pyramid obviously has greater penetrating power than a sphere. Thus the general bearing of Hertz's investigation seems to confirm Wade's mode of experimenting.

[1041.] Suppose l to be the length of indentation when the whole volume V of the pyramid ($V = 3\cdot\dot{3}$ cubic tenths of an inch, $l = 10$ tenths of an inch with Wade's instrument) is sunk in a material under the given pressure p_0, then if l' be the length of the indentation for any other substance under p_0, Wade takes as a measure of the hardness of that substance Vl^3/l'^3 (p. 260). But this does not seem to me what he really intended, although he actually calculates his hardnesses from it. For he prints, l' being measured in tenths of an inch,

$$10^3 : \text{bulk } 3\cdot333 :: l'^3 : \text{bulk}$$

and he defines, as we have seen, hardness to vary inversely as bulk; we should thus have if H, H' be the two hardnesses:

$$10^3 : \frac{1}{H} :: l'^3 : \frac{1}{H'},$$

or

$$H' = H \cdot \frac{10^3}{l'^3}.$$

But Wade taking hardness to be *equal* to the inverse of bulk, makes a slip in inverting his ratio and really puts $H = 3\cdot3$, when it would seem more natural to put it $1/(3\cdot\dot{3})$. He has thus chosen to term the hardness

[1] This is Wade's statement, but it is I think hardly justified by the numbers in his table on p. 266.

of the material into which p_0 drives the whole volume of the pyramid $3\cdot\dot{3}$, or, hardness varying inversely as bulk to take an arbitrary coefficient of variation

$$=(3\cdot\dot{3})^2=11\cdot\dot{1}.$$

To add confusion to his numbers he remarks (p. 259):

The maximum indentation of the instrument $3\cdot\dot{3}$ cubic tenths, is therefore assumed as the type of extreme softness; and as the 0 of hardness (!!).

Wade's numbers would, perhaps, be more intelligible if divided by $11\cdot\dot{1}$. This, however, would still leave them dependent on Wade's particular pyramid. He suggests that a good standard of comparison might be obtained by finding the hardness of the silver coin of some given country and reducing all other hardnesses to this easily obtainable standard.

[1042.] As samples of his numbers we quote from p. 265 the following mean results:

		Density.	Hardness.
Cast-iron, proof-bars,	1st fusion	7·032	8·48
	2nd ,,	7·086	12·16
	3rd ,,	7·198	19·66
	4th ,,	7·301	29·52
Bronze	Seville		5·18
	Boston		4·73
Wrought-iron			11·03

[1043.] For any one investigating the relations which hold theoretically between density, tenacity, transverse strength, torsional strength, compressive strength and hardness, the table on p. 267 for upwards of 20 specimens of cast-iron would be invaluable. Want of space, however, compels me to cite here only the results for groups of 4 specimens arranged according to their densities (p. 268); but the inaccessibility[1] of these American Reports justifies at least this table in which I have corrected some of the numbers—

Group	Density	Strengths in lbs. per sq. inch				Hardness	Ratio of crushing to tensile strength
		Tensile	Transverse	Torsional	Crushing		
1	7·087	20877	6084	6176	99770	12·16	4·78
2	7·182	30670	7587	8341	139834	18·03	4·56
3	7·246	35633	8806	9659	158018	25·42	4·43
4	7·270	39508	9158	9827	159930	25·59	4·05
5	7·340	32458	9274	9065	167030	30·51	5·15
Mean	7·225	31829	8182	8614	144916	22·34	4·55

[1] I sought in vain for copies of these and other American Reports in England. I owe the copies I have used to the kindness of General S. V. Ben'et, Chief of Ordnance, Washington,—a kindness which I very fully appreciate.

Here except for the tensile and torsional strengths of Group 5, all the strengths and hardnesses increase with the density, although the laws of increase are not obvious. The ratio of compressive to tensile strength appears to decrease with the density till we come to the last group, where it suddenly increases. It must be remembered that Wade understands by the transverse strength $\frac{1}{6}$ of the 'tension in the extreme fibre' of a rectangular bar at rupture, and by the torsional strength $\pi/16$ of the shearing stress at the circumference of a bar of circular cross-section at rupture: see our Arts. 1049–53.

Thus with the notation of our Art. 1051, we have:

$$S_1/S_2 = \cdot 257, \qquad S_3/S_2 = \cdot 271, \qquad S_4/S_2 = 4\cdot 55$$

numbers greater in the last two cases, but in the first case considerably smaller than those of the British experiments.

[1044.] (*g*) Wade next records some experiments on the rupture of hollow cylindrical rings. These rings were burst by applying force to a conical frustum made of hardened cast steel inserted in them. By means of a shield of cast steel cut into segments and internally tapered to fit the frustum, the friction between the ring and the frustum was reduced to a minimum (p. 269–70). Wade found that when the external diameter was about double of the internal diameter the ratio of the tenacity computed from what he terms the 'central force' on the frustum to the tenacity obtained by a pure tensile tèst was for both cast-iron and bronze about as 4 : 1; when the ratio of the diameters was as 21 to 16 then the ratio of these tenacities was about as 2·6 : 1. Wade does not explain how he calculates the tenacity from the 'central force,' and he remarks that the divergence in the values of the tenacities is probably due to the friction. His theory is in general so weak, that it very possibly has failed him in the reduction of his numbers. Lamé's formula (see our Art. 1013*, ftn.) cannot be applied, when as in this case there is no longitudinal load on the cylinder. I find, however, that for a cylinder of isotropic material of radii a and b subjected to an internal pressure p, the maximum stretch s_0 would occur at the inner surface, $r = a$, and be given by

$$s_0 = \frac{p}{2\mu (b^2 - a^2)} \left\{ \frac{\lambda + 2\mu}{3\lambda + 2\mu} a^2 + b^2 \right\},$$

whence, if we put $s_0 = T_2/E$, T_2 being the tensile strength, and assume uniconstant isotropy, we have:

$$\begin{aligned} T_2 &= \frac{p}{4} \frac{3a^2 + 5b^2}{b^2 - a^2} \\ &= 1\cdot 92p, \text{ if } b/a = 2, \\ &= 4\cdot 02p, \text{ if } b/a = 21/16. \end{aligned}$$

It is, however, unlikely that Wade calculated the tenacity from the internal pressure and then from the 'central force' by any such formula as this. His method of calculation being unknown, the numbers he gives cannot be modified or used to test any theory.

(*h*) On pp. 272–4 will be found the details of experiments made with this conical frustum to test Barlow's hypothesis that the area of the cross-section of cylinders subjected to internal pressure does not change. The experiments were far too crude to efficiently demonstrate the erroneous nature of Barlow's assumption: see our Arts. 901, 1069 and 1076–7.

(*i*) We may note how Wade on pp. 274–5 draws attention to the very considerable ranges of density, hardness, tensile and compressive strengths to be found for different kinds of the same metal, and therefore to the importance of testing in every case samples of the metals which it is proposed to use for any given purpose.

[1045.] Wade after suggesting on pp. 278–80 chemical tests of the various types of iron which possess owing to repeated meltings such different elastic and cohesive properties, turns to the subject of bronze guns to which he devotes pp. 281–304. In these pages a great deal of information will be found as to the effect of position in the casting or of the size of the casting on the tenacity and density; thus gun-head samples have hardly half the strength of small bars cast with the guns, and as a rule less strength than small bars cast in quite different moulds. There is a good deal of interesting detail as to the exact effect of various methods of casting, but we cannot afford the space needful to discuss Wade's conclusions here.

[1046.] Pp. 305–22 contain a full account of Wade's testing machine for tensile, compressive, transverse and torsional strains. The description is of considerable historical interest as the machine has been the model of a good many others, even in this country. Following our usual rule we refrain, however, from discussing apparatus and refer the reader to the original paper, or to W. C. Unwin: *The Testing of Materials of Construction*, p. 127. London, 1888.

[1047.] The remaining Reports of the volume may be briefly noticed.

(*a*) On pp. 323–46 we have a report by Lieutenant Walbach on the tensile strength and density of specimens taken from the muzzles of nearly 3000 iron guns. He found that for metal of a high class with a tenacity of nearly 30,000 lbs. per sq. in. and a density of 7·21 there was a colour, structure and fracture quite different from those of a metal of a low class with tenacity of between 19,000 and 20,000 and a density of about 7·05. As a sample of the type of difference we may take the fracture described in the first case as "close and even, not hackly" and in the second as "rough, uneven and hackly" (p. 339). Some remarks on p. 344 on the general relation between increased density and increased strength are not without interest, but the whole Report has not much *physical* importance.

(*b*) This report is followed by one on the extreme proof by continuous firing of test guns (pp. 350–68); its contents appear only of interest for the art of gunnery.

(*c*) The volume concludes with three reports on the chemical analysis of specimens of cast-iron gun metal (pp. 370–428). In the first two reports by taking averages and classifying cast-iron into three classes it is shown that with decreasing density and tensile strength there is a decrease of combined carbon and an increase of silicium, and various suggestions are made as to the relation between the physical properties and chemical constitution of the metal (pp. 377—387). The effect of hot and cold blast on these properties is also noted (pp. 388–9)[1]. But in the Third Report (p. 394) the writers remark on the discordance which exists between the laws suggested connecting physical properties with chemical constitution and the results of their more elaborate investigations. They go so far as to throw doubt on the exactness of the physical investigations of density and tenacity and sum up with the words:

> the limited extent of our investigations prevents, at present, the establishment of any laws as to the relation of chemical composition and physical structure, in gun-metal (p. 394).

On p. 396 they give the chemical analysis of 32 specimens of cast-iron, but as they now suppress all data of tenacity and density, the results are not suggestive for further research on the relations between chemical composition and physical structure.

[1048.] Cast-Iron Experiments. *Report relative to a Series of Mechanical Experiments made under the direction of the Superintendent, Royal Gun Factories, and of Chemical Analyses under the Chemist to the War Department, upon various British Irons, Ores etc., with a view to an Acquaintance, as far as possible, with the most suitable Varieties for the Manufacture of Cast-Iron Ordnance; with an Appendix, containing similar Examinations of several Foreign and other Irons, carried on by order of the Secretary of State for War, dated* 9 *June,* 1856. London, 1858.

This Report appears to have been returned in June, 1858, although some of the experiments in the Appendix are dated as late as February 3, 1859, and so perhaps were added while the Report was being printed. The experiments were carried out under the superintendence of Colonel F. Eardley Wilmot, R.A. by the proofmaster Mr M'Kinlay. We need here only consider the

[1] In the Table of *Averages*, p. 388 in the 4th column for the *Total Carbon* of the *Cold Blast* read ·0417 for ·0407.

mechanical experiments and not the chemical analyses of the various irons and ores, which form the second part of the *Report.*

The *Report* is a folio volume consisting almost entirely of the numerical results of experiments on a great variety of cast-irons from all parts of the United Kingdom and some few foreign irons. The experiments are fairly comprehensive, but appear to have been made without any special regard to theory. They are thus very inferior in value to those of the *Iron Commissioners' Report* or of Kirkaldy on wrought-iron and steel.

A brief *résumé* of the results to be drawn from these experiments will be found in the *Mechanic's Magazine, New Series,* Vol. II., pp. 162–3, and another in the *Civil Engineer and Architect's Journal,* Vol. 22, pp. 397–8. Both, London, 1859.

[1049.] Experiments were made on the tensile, flexural, torsional and crushing strengths of a great number of specimens and with a view of testing the bearing of the results I cite the following table from p. 2 :

Strengths of Cast-Iron in lbs. per sq. inch.

	Specific Gravity of 850 specimens	Tensile, of 850 specimens: S_2	Transverse, of 564 specimens: S_1	Torsional, of 276 specimens: S_3	Crushing, of 273 specimens: S_4
Maximum	7·340	34,279	11,321	9,773	140,056
Minimum	6·822	9,417	2,586	3,705	44,563
General Mean[1]	7·140	23,257	7,102	6,056	91,061
Ratio of Strengths (from General Mean)	—	1	·305 (S_1/S_2)	·260 (S_3/S_2)	3·915 (S_4/S_2)

The difference between the maximum and minimum values fully justifies the remarks of the *Report* that :

The term "cast-iron" as describing any specific material does not convey to the mind of those connected with such experiments any more positive quality than what may be gathered from the use of the term "wood" in speaking of that material. The remarkable range of the various qualities of different samples is scarcely more marked in the latter than in the former; and in addition, the same iron treated in a different manner, as regards the apparently simple process of melting or cooling assumes a different character.

[1] In all cases of 51 samples or parcels.

No attempt was made to ascertain the result of mixing various brands of iron nor the special treatment which would improve the quality of any particular iron. Eighteen bars were cast of each iron, 22″ long and of cross-section 2″ square, nine were cast vertically and nine horizontally, three of each of these sets of bars were covered with sand to delay cooling as much as possible and kept in the mould till thoroughly cooled, three were cast 'in the usual way,' three were turned out of the mould as soon as set and exposed to currents of air. These processes are described in the tables as 'slow,' 'gradual' and 'quick' casting. The results of these various modes of casting show a distinct superiority of the bars cast horizontally over those cast vertically, and in a less marked degree of those cooled quickly over those cooled gradually or slowly.

It is to this rapid cooling and condensation that the superior strength of a two-inch bar, cast from a portion of the metal of which a gun is made is due (p. 4).

[1050.] Experiments were further made to show that the length of 'dead-head' does not add to the resisting power of metal. These experiments were made on cast-iron and 'on bronze or brass gun metal'. In the former case a cylinder 26′ long and 7″ diameter was cast vertically, and discs were cut from the top, centre and bottom, or at intervals of about 12′; out of these discs tensile specimens were taken. In the case of bronze there was 30″ distance between the specimens as they stood in the casting. The following results were obtained (p. 3):

	Cast-Iron (mean results)		Bronze	
	Tensile Strength	Specific Gravity	Tensile Strength	Specific Gravity
Top	29,778	7·217	35,600	8·539
Centre	27,650	7·263	38,704	8·545[1]
Bottom	28,648	7·324	49,401	8·814

Thus although the tensile strength of the cast-iron varied with the pressure at casting, it did not, like the bronze, shew a uniform increase with increase of pressure.

[1051.] Of the 18 bars referred to above, 12 were submitted to transverse test and 6 to torsional test, a tensile specimen and a small cylinder for specific gravity being taken from the end of the transverse specimen after the test, and a crushing specimen from the end of the torsional specimen after test.

[1] So in *Report*, but possibly a misprint for 8·645.

We will briefly note how the experiments were made in order to render the table in our Art. 1049 intelligible.

(*a*) *Transverse Strength.* The rectangular bars "were ground in the centre, so as to present a regular surface; this being necessary for obtaining a correct measure of fracture". The area of fracture was measured by taking the mean of three breadths, centre, top and bottom of the section, and multiplying by the mean height found in the same manner. When a load of 5000 lbs. had been applied, it was removed and the permanent set measured, and this repeated for each additional 5000 lbs. up to fracture; the deflections were also noted for the same increments of load. If L be the length, B the breadth, D the depth of the bar and W the central breaking weight, the report tabulates

$$S_1 = \frac{LW}{4BD^2}$$

as a measure of transverse strength. If T_1 be the *apparent* tensile strength in the 'extreme fibre' supposing the Euler-Bernoulli theory applied up to rupture:

$$T_1 = \frac{3LW}{2BD^2} = 6S_1,$$

i.e. *six* times the quantity recorded in the table in our Art. 1049.

(*b*) *Tensile Strength.* The specimens here were unfortunately made of varying diameter in order apparently to ensure breaking at a given central section: see our Art. 1146. Thus although the extensions were measured after a stress of 15000 lbs. at every additional 5000, these are of no real value owing to the irregular form of the specimen, and the results are only of value for the breaking load. If the rupture stress be T_2, the tables of the Report record:

$$S_2 = T_2.$$

(*c*) *Torsional Strength.* The test pieces were cylindrical in the centre and square at the ends for the purpose of fastening them, one end being "keyed to the standing part of the machine, the other to the moveable levers" (p. 9). From this description it appears to me not improbable that the pieces were subjected to both flexure and torsion, in which case the measure of strength adopted would not give a sound result. The tables tabulate: $S_3 = RW/d^3$, where R is the arm at which the weight W is applied and d the diameter. If we apply the theory of elastic torsion up to rupture, let T_3 be the absolute shearing strength, then by our Art. 18,

$$T_3 = \frac{16}{\pi}\frac{RW}{d^3} = \frac{16}{\pi}S_3.$$

(*d*) *Crushing Strength.* The specimens were ·6″ in diameter and 1·3″ in length and were taken from bars which had been subjected to the torsional test. It does not appear to have been noticed that this previous strain nearly up to torsional rupture may probably have had a sensible influence on the crushing strength. The squeeze of the material at 15,000 lbs. and the set at this load were noted and these quantities measured again with every addition of 5000 lbs. The rupture surfaces correspond fairly closely to fig. 5 of the frontispiece to our first volume. They show, however, that the bedded terminals were hindered by the friction from expanding fully. The tables of the Report record S_4, where $S_4 = T_4$ the ultimate crushing strength.

There are pictures of five samples of fracture-surfaces under the above different kinds of stress on pp. 8–10, and the work concludes with a number of diagrams representing, but not very clearly, the mean results of the tables of experiments. Tables *A* and *B* (pp. 154–6) give a *résumé* of the chief results for all the different kinds of cast-iron employed.

[1052.] We may note a few theoretical considerations, which flow from the formulae in the preceding article and from the table in our Art. 1049.

If uniconstant isotropy could be supposed to hold for cast-iron up to rupture, we should have the absolute shearing strength to the absolute tensile strength as 4 : 5, or

$$T_3 = \tfrac{4}{5} T_2.$$

Further T_4 would be given by

$$T_4 = 4T_2,$$

and

$$T_1 = T_2,$$

whence we ought to find :

$$S_1 = \cdot 1\dot{6} S_2,$$

$$S_3 = \cdot 157 S_2,$$

$$S_4 = 4S_2.$$

Of these results only the last is at all in accordance with the mean results of the table, which gives

$$S_4 = 3 \cdot 915 S_2.$$

This confirms the statement often made in the course of our work that for *practical* purposes the relation between the tensile and crushing strengths of cast-iron may be taken to be that deduced from supposing uniconstant isotropic elasticity to hold up to rupture.

Instead of the first result the table gives

$$S_1 = \cdot 305 S_2,$$

or

$$T_1 = 1 \cdot 83 T_2,$$

instead of $T_1 = T_2$. This is the so-called paradox in the theory of beams: see our Arts. 930–1. Recent experiments have given the ratio of T_1/T_2

for rectangular sections = 1·74, for circular sections = 2·03, showing that it varies with the form of the section.

If the result $S_3 = \cdot 260 S_2$ as given by the tables be correct, it shows that the ordinary theory of torsion certainly cannot be applied to cast-iron up to rupture, for there is little doubt that the shearing strength of cast-iron does not differ largely from the tensile strength.

If we express the above ratios in terms of the crushing strength as unity we have from data supplied in the Report:

	Crushing Strength	Tensile Strength	Transverse Strength	Torsional Strength
Elastic Theory	1	·25	·042	·039
General Means	1	·255	·078	·067
Butterley Iron (p. 181)	1	·201	·063	·038
New York Iron (p. 175)	1	·379	·097	·067
Charcoal Iron (p. 171)	1	·266	·093	·088
Blaenavon Iron (p. 147)	1	·183	·052	·022

The last four examples have been taken at random from the tables in the Report to show the great variations in the ratios of the different types of strength, and to demonstrate how rash it is to apply the ordinary theory of elasticity to determine the rupture stresses of a material like cast-iron.

[1053.] Let us apply the stress-strain relation which Saint-Venant has based on Hodgkinson's results and which does not assume Hooke's Law. For a rectangular beam under flexure, the method is discussed in our Art. 178. Let m_1 be not put equal to m_2, but their ratio taken as that of the ultimate tensile and crushing strengths. We shall following Saint-Venant then put for cast-iron $m_1 = 1$ and $m_2 = 4$ whence we find (*Leçons de Navier*, Table, p. 182):

$$T_1 = 1{\cdot}81\, T_2.$$

Further taking $m = 4$ in the corresponding torsion-formula in our Art. 184, we find:

$$T_3 = 1{\cdot}037\, T_2.$$

Whence reducing to the S-notation of the present discussion we have:

	S_1/S_2	S_3/S_2	S_4/S_2
Saint-Venant's theory	·302	·204	4
Mean results of experiments	·305	·260	3·915

Thus we see: that for practical purposes, Saint-Venant's formulae with the above values of the constants may be used, failing direct experiments, to find fairly good *mean* results for the transverse strength of cast-iron.

[1054.] Robert Mallet: *On the Physical Conditions involved in the Construction of Artillery, and on some hitherto unexplained Causes of the Destruction of Cannon in Service. Transactions of the Royal Irish Academy*, Vol. XXIII., Part I., *Science*, pp. 141–436. Dublin, 1856. This paper was read on June 25, 1855. A review and at the same time a criticism of the portions of Mallet's memoir bearing on the strength of materials will be found in Vol. XIX., pp. 325, 366, 389, 401 and Vol. XX., pp. 29–31 of the *Civil Engineer and Architect's Journal*. London, 1856–7.

This long memoir contains a great deal of interesting information with regard to the physical properties of the metals, notably iron. Some of the statements made, seem to me, wanting in scientific precision, but it is quite possible that they may be much more intelligible to one having a more intimate acquaintance with the appearance and the rupture surfaces of large masses of material. We shall note only one or two points referred to by the writer in his earlier chapters.

[1055.] *On the Bursting of Guns from internal Pressure.* The memoir notices that rupture invariably appears to have begun at some point on the inside; the gun opening out along one half a longitudinal section through this point, the opposite half being subjected in part to traction and in part to contraction,—this produces a characteristic point of inflexion in this half of the surface of rupture (p. 146).

[1056.] *Molecular Constitution of Crystalline Bodies.*

It is a law (though one which I do not find noticed by writers on physics) of the molecular aggregation of crystalline solids, that when their particles consolidate under the influence of heat in motion, their crystals arrange and group themselves with their principal axes, in lines perpendicular to the cooling or heating surfaces of the solid; that is, in the lines of direction of the heat wave in motion, which is the direction of least pressure within the mass (p. 147).

Mallet lays considerable stress upon this law and discusses it at some length in pp. 147–9 and Note E, pp. 353–57. If the law be true, it obviously has a very great bearing on the influences of the various processes of working on the strength of materials. It is not always quite obvious what is meant by *crystalline structure* and its opposite *fibrous condition* in the writings of technical elasticians, or whether they are distinctly related to molecular crystallisation. There is obviously a distinction between an *amorphic body or solid of confused crystallisation* (as expressed by Saint-Venant: see our Arts. 115 and 117 (*c*)) and a body in a 'longitudinally fibrous' condition as produced by drawing or rolling, but Mallet describes both as presenting no crystallisation (pp. 148–9). Thus he notes an experiment with a plate of rolled zinc which is nearly "homogeneous in structure [isotropic in structure?], or, if not so, presents fibres and laminae in the plane of the plate." This plate is laid upon a cast-iron plate which is then heated nearly up to the melting point of the zinc. The zinc is then said to assume a "crystalline structure,—the crystals now having their principal axes all cutting perpendicularly through the plate from side to side; in other words, *the planes of internal structure being in this......case absolutely turned round* 180° *of angular direction*." I suppose the 180° to be a slip for 90°.

The words "internal structure" here seem to point as much to *crystalline* as to elastic structure, and Mallet would seem to associate a 'fibrous condition' with crystalline axes in the direction of the fibre, and a 'crystalline condition' with crystalline axes perpendicular to the greatest dimension of a wire or plate. Now 'initial stress' due to working may produce aeolotropy, but it does not seem necessary to assume, that such stress really connotes an arrangement of crystalline axes in or perpendicular to the lines of initial stress. Indeed I think the identification of elastic aeolotropy having one or more planes of symmetry with crystalline structure, which is assumed by some English writers, is not without danger. That crystalline structure connotes a certain elastic structure may be perfectly true, but I do not see why the converse must necessarily hold. The passage of heat through a material, perhaps, changes its tensile strength, *when the temperature is thereby raised nearly to fusing point* (words omitted in Mallet's statement of his law, but which was apparently a condition of the experiments he quotes): see however our Arts. 692* (8), 876*, 953*, 968*, 1301*, and 1524*. What Mallet adds to this statement is, that the direction in which the heat is propagated through the metal affects the directions of greatest and least tensile strength and may interchange the two. At the same time it is not improbable that a much smaller change of temperature will produce a change in elastic structure, and alter the magnitude of the elastic constants and the directions of the planes of elastic symmetry.

If Mallet's law be true it would follow that many processes of annealing so far from producing isotropy may merely change the nature of the aeolotropy, and that further without very great precautions in the process of annealing, the question of rari-constant isotropy cannot be tested by experiments on annealed bodies, originally of fibrous structure. The process of annealing so far from producing Saint-Venant's 'amorphic' condition in place of the 'fibrous,' may produce Mallet's 'crystalline structure.' Mallet asserts (pp. 147–8) that a heat far below that of fusion will change an amorphic into a crystalline body, and that when a body cools "the principal axes of the crystals will always be found arranged in lines perpendicular to the bounding planes of the mass, that is to say, in the lines of direction in which the wave of heat has passed outwards from the mass in the act of consolidation (p. 147, § 10)." He adds nothing as to the *rate* at which the cooling is supposed to take place. The bearing of this remark, if true, on the labours of those experimenters who discard rari-constant isotropy on account of the evidence of multi-constancy found in *annealed wires* will be obvious to the reader.

One word more as to certain expressions used by Mallet in the statement of his law. He identifies the direction of the heat wave, I presume he means heat *flow*, with that of "least pressure within the mass" (pp. 147 and 353). I do not understand exactly what this pressure denotes. In the second page cited Mallet speaks of it as the "pressure...due to distortion or change of form by contraction or expansion." But this does not make it much clearer. Does he mean the direction of least initial traction? Even then I do not understand why the heat-flow always passes in this direction. According to the mode in which we apply heat to the body, it seems to me we can alter the direction of the heat-flow. If we could not, it is difficult to understand how the heat-flow could change the direction (in Mallet's phraseology) of the crystals, whose 'principal,' 'symmetric' or 'longest' axes are always in the direction in which the heat-flow has passed (p. 353). By "consolidation of particles" Mallet refers not only to a previously fused solid solidifying by cooling, but to the action of heat applied to the external surfaces of a body raised to a temperature even less than that of fusion (pp. 147–8).

[1057.] Chapter IV. of the memoir is entitled: *Molecular Constitution of Cast-Iron* (pp. 149–152). Mallet, after remarking that according to his previous law "the planes of crystallisation group themselves perpendicularly to the surfaces of the external contour", goes on to infer that when the contour presents either a re-entering angle, or a sharp change in direction, then a plane exists in the neighbourhood of the angle, in which there is *confused crystallisation*; this confused crystallisation he considers a source of weakness, and he terms the plane a *plane of weakness*.

Experiments seem to prove that such *planes of weakness*, ultimately of rupture, do really exist where Mallet has placed them, but I much doubt if they are due to "confused crystallisation." More probably they connote an *initial stress* due to the peculiarity of the cooling in these parts. Indeed if we followed Mallet's idea, as it appears exemplified in an experiment on lead on p. 148 (§ 12), it would seem that parallelism and not confusion of the directions of the crystalline axes would be a source of decreased tensile strength in directions perpendicular to the axes and so parallel to the surface of the casting.

[1058.] Chapter V. is termed: *Physical conditions induced in Moulding and Casting* (pp. 152–162). In this chapter Mallet points out that the size of the 'crystal' in the casting (and therefore its weakness) depends on the length of time the casting takes to cool. Hence the temperature of the molten metal ought to be only just above that requisite for fusion. He remarks also on the state of internal (initial) stress produced in large castings due to the different rates of cooling of adjacent parts. This points again rather to initial stress than to 'confused crystallisation' as a source of weakness. He cites Savart's memoir of 1819 (see our Art. 332*) and a memoir by Bolley upon the molecular properties of zinc (*Annalen der Chemie und Pharmacie*, Bd. xcv. S. 294) in support of his views. As an example of the evil of a long period of solidifying Mallet points out that a small bar which is part of a large casting and thus cools slowly is found not to be so strong as a bar of the same size cast alone under the same 'head' of metal (p. 162).

[1059.] We may note that on pp. 154–5 Mallet rejects Fairbairn's theory that a certain number of repeated meltings increases the strength of cast-iron (see our Art. 1098):

Indeed, these experiments (Fairbairn's), rightly considered, only prove what was well known before—that by continually remelting and casting into *small* pieces (i.e. imperfectly chilling) any cast-iron, we may gradually cause all its suspended carbon (in the state of graphite) to exude, as Karsten long ago proved, and so gradually convert the metal into an imperfect steel, with increased hardness and cohesion, and diminished fusibility, but with properties altogether unworkable and useless. No such result can occur when the metal is cast into large masses, nor any such improvement by repeated meltings, but very much the contrary (p. 154).

[1060.] Chapter VI. on the *Effects of Bulk and Fluid Pressure* and Chapter VII. on the *Quality of Metal* in reference to strength refer to practical points of casting and need not detain us. We merely remark that increase of bulk produces decrease, increase of 'head' or fluid pressure produces increase of both density and strength, while British irons show a tensile strength comparing favourably with foreign makes (pp. 162–172).

Chapter X. on the effect which heating the inside of a cylinder has in producing strain and ultimately rupture of the material is not very satisfactory from the theoretical point of view. With the aid of a somewhat more extended analysis more approximate results might I think have been obtained.

[1061.] Chapter XVIII. is entitled: *The General Relations of Elasticity to the Construction of Guns* (pp. 194–220). So far as the theory of elasticity is concerned this is not a very satisfactory chapter. Thus on p. 194 (§ 114) it is pointed out that 'linear' and 'cubic elasticity' have not a constant ratio, while in the following section (§ 115) the relation between them, and on p. 216 (§ 144) the relation between the slide- and stretch-moduli are given on the rari-constant hypothesis without a word of qualification. Similarly the thermal statements at the conclusion of § 114 and in § 116 strike me as very obscure. The following pages (pp. 198–207) are occupied with a reproduction of Poncelet's results on the cohesive and elastic resilience of bars, taken from the *Mécanique industrielle* (see our Arts. 981*–2*, 988*–991*). Mallet only reproduces those results which neglect the influence of the inertia of the metal, and makes no statement that he has done so. His application of these results for the longitudinal resilience of bars to the case of the cylinder of a gun in § 130, p. 208, seems to me quite unjustifiable. The elastic resilience of a massive hollow cylinder subject to internal impulsive pressure presents no great difficulties of analysis, but it certainly cannot be deduced from that of a bar without inertia, by supposing the latter bent into a ring!

[1062.] On pp. 211–219 are tables of the elastic strength and the coefficients of elastic and cohesive resilience of metals, chiefly extracted from Poncelet's *Mécanique industrielle*. Mallet draws attention, as Poncelet had already done to the importance of considering these coefficients of resilience rather than the cohesive strength of a material when we are judging its suitability for ordnance. At the same time I think he should have brought out more clearly that it is rather the elastic than the cohesive resilience which must be taken as a measure of suitability, otherwise the gun would rapidly lose its form and efficiency. Had he done so the disproportion in the efficiencies of cast-steel and wrought-iron of extreme ductility would not have appeared anything like so great as exhibited in the areas of the curves on p. 213. Thus in Table X., p. 219 'strong and rigid' wrought-iron bar has a greater elastic resilience than wrought-iron of 'mean strength and ductility,' while the cohesive resilience of the latter is much greater than that of the former. Similarly gun-metal has a less elastic resilience than either cast-iron or wrought-iron bar, but an immensely greater cohesive resilience. At the same time we must remark that Mallet's tables are not quite in accord (e.g. the results in Tables VII. and X.); this is perhaps due to the assumption of uni-constant isotropy in the calculation of some of the results.

[1063.] Chapters XIX. and XX. of Mallet's memoir are devoted to the physical properties of gun-metal or bronze (pp. 220–241). A table on p. 222 giving the physical properties and in particular the tensile strengths of various alloys of copper with zinc or tin is extracted from the author's *Second Report upon the action of Air and Water...upon Cast-Iron, Wrought-Iron and Steel*[1], *Transactions of the British Association*, Tenth (Glasgow) Meeting, 1840, pp. 221–308. London, 1841.

[1] These reports (1838–43) escaped my notice in working up the material for Vol. I., but only pp. 302–8 of the Second Report really concern us. On pp. 306–7 are the tables referred to (see also *Proceedings of the Royal Irish Academy*, Vol. II., pp. 95–6. Dublin, 1844). They give the specific gravity, tensile strength, hardness, order of ductility, order of malleability at 60° F., order of fusibility (the author does not state how these 'orders' were determined), nature of the fracture and commercial name, where known, of 21 alloys of copper and zinc and 14 of copper and tin, together with those of copper, zinc and tin themselves. On pp. 302–4 are details of the fracture and specific gravity of various kinds of cast-iron; on p. 304, of increase of density in cast-iron due to solidification under a considerable head of metal (4 to 14 feet); on p. 305, of decrease of density with the increase in bulk of a

[1064.] Chapter XXI. (pp. 242–3) deals with cast-steel. It contains a reference to Ignaz von Mitis' experiments, but nothing of importance for our present purposes: see our Art. 693*.

[1065.] Chapter XXII. is entitled: *Molecular Constitution of Wrought-Iron, and the Law of Direction of its Crystals or Fibre* (pp. 244–248). Here we have the same general statements as to crystalline axes to which I have objected in Art. 1056. On p. 245 the general law is stated:

In wrought, as in cast iron, the principal axes of the crystals, tend to assume the directions of least pressure throughout the mass while exposed to pressure and heat in progress of manufacture.

It appears by the remarks upon this law, that Mallet understands by the 'direction of least pressure' that in which the stress applied in the process of working is least, i.e. the direction of the 'fibres'[1] in a bar, plate or wire. Here again it seems to me that it would be safer to talk of an aeolotropy symmetrical with regard to certain planes rather than of the direction of the crystalline axes. Mallet notes (pp. 246–7) that in the case of a bar of wrought-iron of large cross-section, heat as well as working stress plays a part in determining the direction of the crystalline (elastic?) axes, and that the process of cooling tends to place these in directions perpendicular to the surface of the bar.

[1066.] Chapters XXIII.–XXV. (pp. 248–256) deal principally with the characteristics presented by large masses of forged iron. The author speaks of these masses as possessing confused crystallisation, or in other words being *amorphic*. He disputes the accuracy of Fairbairn's results cited in our Art. 1497* (ii), and refers to some experiments of Clarke's (*The Britannia and Conway Tubular Bridges*, Vol. I. p. 377) which gave for the mean tensile strength per sq. inch: with the fibres 20 tons, across the fibres 17 tons. Mallet holds that the tensile strength of bars cut out of a large mass of forged iron in *any* direction would also give a tensile strength of about 17 tons (pp. 249 and 253).

[1067.] Chapter XXVI. (pp. 256–260) deals with the point referred to in our Arts. 1463*–4*, 881 (*b*) and 970, namely the possibility of a change in wrought-iron from a 'fibrous to a crystalline state' by repeated loading or impacts. Mallet's general conclusion on this point is given on p. 257. He holds that no strain or impact which does not produce permanent change of form is capable of affecting any molecular alteration however often repeated, but:

It does appear certain from many well-observed phenomena, that instantaneous changes of molecular structure and reversals or transposition of

casting; on p. 308, of the specific gravity and fracture of a number of wrought-irons and steels.

[1] Mallet, p. 248, says: "I have used the term 'fibre' as being already long in use, and conveying well the character of this particular form of crystallisation to the eye; but it should be clearly understood that the 'fibre' of the toughest and best iron is nothing more than the *crystalline arrangement* of inorganic matter."

the crystalline axes can be produced in wrought-iron at ordinary temperature, by the violent application of mechanical force, producing suddenly change of form at one or more points of the surface of the mass...

He instances the effect of the blacksmith's 'nicking' with a blunt chisel the side of a bar of the toughest iron, which can then be easily broken, although without the 'nick,' it might have been sharply bent double without fracture. There is an attempt to explain this on the 'theory of direction of crystalline axes': see our Art. 1056.

[1068.] Chapter XXVII. (pp. 260–266) is concerned with the rupture of wrought-iron plates by impulses, such as the blow of a shot. In § 229 Mallet obtains a formula for the velocity V of the body which will certainly produce fracture. If u be the 'velocity of force transmission,' by which we are to understand the velocity of sound waves, and s_0 be the limit of safe stretch or squeeze, then if

$$V = u \times s_0$$

there will certainly be rupture. This is a result of Young's for *longitudinal impact* of beams (see his *Lectures on Natural Philosophy*, Vol. I. p. 144), but I do not understand how it can be straightway applied to the transverse impact of plates. Mallet applies it, however, taking $u = 13,000$ ft. per second, $s_0 = \frac{1}{25}$, and deducing that V is only one-third to one-fourth that of cannon-shot, so that the inevitable destruction of the iron plate follows. It is needless to add that in explaining the nature of the fracture of plates by shot he appeals to his crystalline law (pp. 265–6): see our Art. 1056. The subject of the rupture velocity for transverse impact on plates has been treated by Boussinesq in a memoir of 1882 (*Comptes rendus*, Vol. xcv. 1882, p. 123: see also his *Application des Potentiels*... pp. 487–90), which we shall consider in its proper place.

[1069.] After some chapters relating more closely to the construction of artillery, Mallet in Chapter XXXIII. (pp. 280–296) returns to our subject, dealing with the problem of constructing a gun by placing cylindrical rings of wrought-iron over each other, each new ring being shrunk on to the series of rings which form its core. It is well-known that a hollow cylinder subject to internal pressure, if homogeneous and without initial stress, will only sustain a certain definite pressure, however its thickness may be increased: see our Arts. 1013* (with footnote) and 1474*. Mallet proposes to raise this limiting pressure by putting the material into an initial state of stress. The theory of this initial state of stress is given in a *Note* by Dr Hart appended to the memoir to which note we shall return.

On pp. 284–5 Mallet cites five different formulae for the relation between thickness, safe tractive load and internal pressure. None of these agree with that I have given on p. 550 of Vol. I.; still less do they agree among themselves. Mallet makes no attempt to select any one of them as the correct one. He states with Barlow (*Transactions of Institution of Civil Engineers*, Vol. I. p. 136) that there is no thickness which will withstand an internal pressure equal to the safe tensile load. Dr Hart's formula gives the same result. As a matter of fact for uni-constant isotropy the thickness τ for internal pressure p and internal diameter d is given by:

$$\tau = \frac{d}{2}\left\{\sqrt{\frac{4T_0 + 2p}{4T_0 - 5p}} - 1\right\},$$

whence we find $p = \frac{4}{5}T_0$ as the limiting possible pressure.

[1070.] Chapter XXXIV. (pp. 296–9) is entitled: *On the Relations between Annealing and Tenacity.* It is based principally on Baudrimont's results: see our Arts. 830*–1* and 1524*. But there is very little evidence accessible on these points:

A rich reward awaits the physicist who, in a comprehensive manner, shall first, experimentally, attack the question of the molecular changes produced by hardening and annealing; it has been as yet almost unattempted (p. 297).

[1071.] The memoir concludes with a long series of notes, partly historical and partly statistical, of considerable general interest. I may draw attention to the following:

(*a*) Note S. (pp. 392–396). *Physical Constants of the Materials for Gun-founding.* This note gives some tables of information with regard to the ultimate strength of cast- and wrought-iron, cast-steel and bronze extracted from the *Ordnance Reports, United States Army*, 1856, and on the compression of bronze gun-metal from some experiments of Colonel F. E. Wilmot at Woolwich Arsenal made at Mallet's request (April, 1856). See our Arts. 1037–47 and 1050.

(*b*) Note W. (pp. 399–406). This note by Dr Hart professes to give the theory of the stress in a number of superposed metal cylinders (see our Art. 1069), but I have been unable to follow the analysis. If it be correct, which I very much doubt, at least the author should have clearly stated the meanings of the symbols he employs. After saying that the cylinder may be conceived as split up into 'cylindrical laminae,' he continues:

Let r be the radius of any of these cylinders, and $2P$ the corresponding force, the length of the cylinder being unity. Also let $r+u$ be the radius of the same cylinder when extended, then (according to the common theory):

$$\frac{dP}{dr} = -k\frac{u}{r}.$$

It would appear from what follows that the author means by the 'corresponding force $2P$' the expression which we should denote by $-2r \cdot \widehat{rr}$ and his equation then becomes

$$\frac{d\widehat{rr}}{dr} + \frac{\widehat{rr} - ku/r}{r} = 0.$$

This obviously assumes that the meridional traction $\widehat{\phi\phi}$ is equal to ku/r: see our Art. 120.

Similarly the second equation on p. 400 is

$$P = -k'r\frac{du}{dr}$$

or

$$\widehat{rr} = k'\frac{du}{dr}.$$

Thus it would seem that the author has either supposed the material to have no dilatation, or else assumed that the meridional and radial tractions are each proportional solely to the stretches in the same directions! The error is exactly that of Scheffler: see our Art. 655.

[1072.] On the whole Mallet's memoir presents much of interest and importance, but is painfully weak in analysis and even in elementary dynamical notions (e.g. equation (58), p. 269).

[1073.] British Association. *Report of Twenty-fifth* (Glasgow) *Meeting.* London, 1856. *Provisional Report of the Committee... appointed to institute an inquiry into the best means of ascertaining those properties of metals and effects of various modes of treating them which are of importance to the durability and efficiency of Artillery*, pp. 100–8. This does not appear to contribute anything of theoretical or permanent importance to the subject of our history, or to the science of gunnery.

[1074.] *Expériences faites en* 1856 *avec deux canons à bombes...en fonte de fer. Extrait du rapport fait sur ces expériences par M. von Borries. Annales des travaux publics de Belgique*, T. xv. pp. 427–56. Bruxelles, 1856–7. This is a translation of a portion of a report to the Prussian Government on the strength of two Belgian cast-iron cannon made at Liège. The cannon were tested to bursting. There is nothing that calls for special notice in the report.

[1075.] D. Treadwell: *On the Practicability of Constructing Cannon of Great Caliber, capable of enduring long-continued Use under full Charges. Memoirs of the American Academy*, Vol. VI. Part I. pp. 1–19. Cambridge and Boston, U.S., 1857. This memoir, after criticising the current methods of constructing guns of large size, proposes to form the caliber and breech of cast-iron, but to place outside these parts rings or hoops in one, two or more layers of wrought-iron; "every hoop is formed with a screw or thread upon its inside, to fit to a corresponding screw or thread formed upon the body of the gun first, and afterwards upou each layer that is embraced by another layer. These hoops are made a little, say $\frac{1}{1000}$th part of their diameters less upon their insides than the parts they enclose", and are placed on hot, being then allowed to shrink and compress. This method of constructing cannon appears to have been first suggested by Treadwell, and a process of building up guns by wrought-iron hoops has been largely used: see our Arts. 1069, and 1076–82. The memoir gives a few details of the relative strength of such cannon and of cast-iron cannon (pp. 13–16), and concludes by describing a process of avoiding 'lodgment' (pp. 16–18), and with a condemnation of the European process of 'piling or fagoting' for building up wrought-iron cannon.

[1076.] James Atkinson Longridge: *On the construction of Artillery, and other Vessels to resist great Internal Pressure. Institution of Civil Engineers. Minutes of Proceedings.* Vol. XIX. pp. 283–460 (with discussion). London, 1860. This is one of the numerous practical papers on artillery which contain statements with a good deal of bearing on physical and theoretical elasticity. There are frequent references in the course of the paper to Mallet's researches: see our Arts. 1054–72.

The author commences by saying that he intends to limit his remarks to methods of making a gun 'which gunpowder cannot burst.' He refers then to the difficulty of making the cylinders of large hydraulic presses sufficiently strong to resist a pressure of 3 or 4 tons per square inch, and refers to what he terms the explanation of this difficulty given by Professor Barlow, "with the clearness which distinguishes all the works of that accomplished mathematician." We have had occasion to mention this matter once or twice: see our Arts. 655, 901 and 1069.

[1077.] Barlow's formula for the strength of hydraulic presses, which at one time had worked its way into all hydraulic text-books for practical engineers, depends on the assumption that the volume of the cylinder does not change owing to pressure[1]. It was superseded in Germany ultimately by a formula due to Brix, based on the assumption that the thickness of the wall of the cylinder is not changed by the pressure. These two formulae, equally absurd in theory, maintained their places in the text-books long after Lamé had given more correct results[2]: see our Arts. 1012*–13* and footnote p. 550.

Our author proposes to make guns to withstand a very great internal pressure by placing coils of metal round the inner cylinder of the gun having initial stresses. Blakely, Sir William Armstrong and Mallet had, unknown to the author, been working on the same lines.

[1078.] The memoir commences by pointing out the extreme difficulty of making heavy guns of cast-iron, wrought-iron or steel. It notices how initial stresses are produced by cooling when metal is cast in large masses: see our Arts. 879 (*f*), 1039, 1056–8 and 1060. Further the difficulties inherent in the construction of wrought-iron and

[1] See Barlow's erroneous theory in a paper entitled: *On the force excited by Hydraulic Pressure in a Bramah Press. Institution of Civil Engineers, Transactions*, Vol. I., pp. 133–9. London, 1836.

[2] Rühlmann in his *Vorträge über Geschichte der technischen Mechanik*, Bd. I. S. 320, after remarking on the doubtful character of Barlow's formula, states that Brix's 'deserves much more confidence,'—apparently because it does not give a limit to the pressure possible for an infinite thickness. This approval was given so late as 1885!

steel guns are noticed, especially difficulties of good welding and hammering are referred to (pp. 287–96). The author then turns to the processes of construction suggested by Mallet and Blakely, consisting in putting on hoops of wrought-iron round the gun tube, which being put on hot, give, when cool, an initial tension. He considers that these processes of building up a gun are not satisfactory, because (i) they really would require an infinite number of infinitely thin hoops, and (ii) there is great practical difficulty in constructing the hoops with just the theoretically right radii. Longridge shows (pp. 301–3) that an error in workmanship of only $\frac{1}{500}$ of an inch in the radius of a hoop may make a very serious difference in the stress in the material when the internal pressure is applied. There is a mathematical theory of the proper values of the radii of the successive hoops given in the Appendix (pp. 329–335) by C. H. Brooks to which we shall return later. In a diagram on p. 297 curves of the stress across an axial section of a hollow cylinder are given. These curves are plotted out for the formulae of both Barlow and Hart (see our Arts. 1071 and 1077), so that in both cases they must be considered erroneous. The real curve would be obtained by plotting out, for values of r, the values of $\widehat{\phi\phi}$, the meridional traction, which can be deduced from the results of our Art. 120, or for isotropy from those of our Art. 1012*. Subtract the ordinates of this curve from a constant traction equal to the maximum to which we propose to subject the gun, and we have the initial tractions, which each point of the cylinder ought to be subjected to on the theory of Mallet and Longridge in order that we may have the strongest gun. There are I think obvious objections to this theory, of which I need only mention one, namely that it is not an equality of *stress*, but of *strain* (i.e. u/r: see our Art. 1080) that we ought to strive for, and that the former does not connote the latter: see our Arts. 1567*, 5 (c) and 321.

In order to obtain the exact traction initially required Longridge discards a finite and limited number of hoops, and proposes to use coils of wire, which he holds can be put on with the exact stress indicated by theory (p. 301). In the case of his experimental cylinders he put on his coils of wire with an initial tension deduced from Barlow's theory (p. 306). It is, therefore, difficult to believe that he constructed the strongest possible cylinder, even if we assume that the results for solid cylinders could be legitimately applied to wire coils, and that the test for maximum strength is equality of stress, not of strain, across an axial section.

Pp. 307–19 give details of the author's experiments on cylinders and guns bound with coils of steel or iron wire. Pp. 319–21 give the details of the construction of a small hydraulic press cylinder built up in this manner and of experiments upon it.

[1079.] An Appendix to the paper (pp. 322—337) contains various mathematical investigations. Thus on the "force of gunpowder," wherein it is shown that the pressure exerted can be 17 to 25 tons per square inch. Remembering that this is more or less of an impulsive pressure applied to the inside of the cylinder, and therefore theoretically might correspond in straining effect to a steady pressure of 34 to 50 tons, it would be little wonder if most guns ultimately burst by being thus continually strained beyond their elastic limit. It would not indeed much alter matters if the real impulsive pressure only reached a moiety of the above large value.

[1080.] The next portion of the *Appendix*, which is of interest for our present purpose is entitled: *Conditions of Stress of a Cylinder built up of Concentric Rings* (pp. 329–35) by Mr C. H. Brooks.

This investigation starts from expressions for the inner and outer meridian tractions in a hollow cylinder which agree with the values obtained from Lamé's formula (see our Art. 1012*). So far the theory seems likely to be more complete than Hart's, but, alas! the next stage is entirely erroneous. Brooks makes the following statement, which I cite with our notation:

Now if $\widehat{\phi\phi}$ be the tension at any radius r, and E the modulus of extension, then the extension of that radius is $\widehat{\phi\phi}\,.\,r/E$ (p. 330).

This is the error into which Scheffler, Hart, and Virgile (see our Arts. 122, 655 with ftn. and 1071 (*b*)) have all fallen, and which it still seems impossible to root out of the mind of the technical elastician.

Lamé's formula quoted by Brooks from Rankine involves a longitudinal traction in the cylinder, and thus if u be the radial shift, and the external and internal radii of the cylinder be r_1 and r_0, we easily find (see our Art. 1012*):

$$E\frac{u}{r} = \frac{E}{3\lambda + 2\mu}\,\frac{r_0^2 P_0 - r_1^2 P_1}{r_1^2 - r_0^2} + \frac{E}{2\mu}\,\frac{r_0^2 r_1^2 (P_0 - P_1)}{r^2 (r_1^2 - r_0^2)},$$

while $$\widehat{\phi\phi} = \frac{r_0^2 P_0 - r_1^2 P_1}{r_1^2 - r_0^2} + \frac{r_0^2 r_1^2 (P_0 - P_1)}{r^2 (r_1^2 - r_0^2)}$$

Whence in order that $u = \widehat{\phi\phi}\,.\,r/E$ we must have $E = 3\lambda + 2\mu = 2\mu$, an absurdity. Thus we need not inquire into the accuracy of the remainder of Brooks' investigation.

In the discussion which followed the author refers to Lamé's formula as the basis of Dr Hart's and Mr Brooks' investigations but he does not see how hopelessly the latter have misapplied it (p. 341).

[1081.] Pp. 338–460 are occupied by the discussion which was extremely long and somewhat discursive. I may draw attention to the remarks: p. 345, on the want of longitudinal strength in wire-bound cylinders—another obvious reason why Lamé's formula should not be applied to them; p. 358, on the difficulty of forging large masses without flaw; p. 360, that the pressure of gun-powder could reach 30 tons per sq. inch; p. 364, that there is less internal stress in large castings after they have been kept a long time, showing a very slow after-strain effect; pp. 385–7, on cooling hollow cast-iron cylinders from the inside and so obtaining an initial negative traction in the inner shells; p. 388, on a method of testing the pressure produced at various distances along the bore of a gun on discharge and so calculating the strength of material required at the corresponding sections; p. 443 and footnote, on the absolute tensile strength of cast-iron before and after remelting, also on its general average (p. 444); pp. 444–5, on the apparently slight influence of chemical identity on the identity of mechanical properties in iron. The impression made on my mind after reading the paper is the general want even so late as 1860 of theoretical training among practical engineers. The apparently universal acceptance in the discussion without the least enquiry of an erroneous theory is remarkable, and the need that experiments on the strength of materials should be conducted by those who have a real knowledge of the theory of the elasticity becomes very obvious. For example, throughout no distinction seems to have been drawn between impulsive external load and the resulting maximum internal stress; thus the absolute tensile strength of the material is spoken of as if it were the limit to be given to the internal pressure, which is quite false, were we even to suppose the gun to be elastic up to rupture, and its efficiency not destroyed by set.

[1082.] T. A. Blakely: *A mode of constructing Cannon, whereby the Strain produced by firing is distributed throughout the Mass of Metal.* This paper was printed in the *Journal of the United Service Institution*, whence it was reprinted in the *Civil Engineer and Architect's Journal*, Vol. 22, pp. 45–50, 81–3. London, 1859. *Idem. Strength of Guns and other Cylinders.* Extract of a paper read at the United Service Institution. *Civil Engineer and Architect's Journal*, Vol. 22, pp. 245–7. London, 1859.

The first of these papers contributes but little to our knowledge of stress in cylindrical bodies. The author quotes erroneous results of Barlow's and notes that a press or gun will only stand a certain limit of internal pressure, whatever its thickness. The whole theory of pressure in cylindrical bodies had been some time previously correctly worked out by Lamé and it is not to the credit of our Ordnance Department at that date, that its scientific knowledge should have extended no further than the range exhibited in this paper. Blakely notes experiments showing that cylinders subjected to internal pressure first rupture on the inside. His object in the paper is to advocate that system of building up guns which consists in putting on rings of metal of a diameter slightly smaller than that of the inner cylinder or tube over which they are placed. He suggests wrought-iron hoops over a cast-iron tube. There is considerable reference to the investigations of Mallet and Longridge: see our Arts. 1054 and 1076.

In the second paper Blakely cites results from the American *Reports of Experiments on Metals for Cannon* in order to show that a gun built-up of hoops shrunk over each other must be much stronger than a solid cylinder. The experiments cited are those of the work referred to in our Art. 1037.

[1083.] J. Cavalli: *Mémoire sur la théorie de la résistance statique et dynamique des solides surtout aux impulsions comme celles du tir des canons. Mémorie della R. Accad. delle Scienze di Torino.* Serie II. T. XXII. pp. 157–233 with three plates. Torino, 1865. The memoir was read on January 22, 1860.

This is one of several memoirs which were called forth by the publicity given to the results of Hodgkinson's experiments in the treatises of Love and others: see our Arts. 894, etc. The memoir is not without value although it contains some rather doubtful theoretical investigations. Considering the title of the paper and the fact that the major portion of it is devoted to the discussion of implements designed for the destruction of human life, it is a curious sign of the perverseness of even the scientific mind in 1860 to find the preface closing with the following words:

> La connaissance du calcul de ces vitesses, avec les principes les plus élémentaires de la mécanique rationnelle et des sciences en général, fourniront aux constructeurs le seul guide infaillible pour réussir dans les grandes et nouvelles constructions, que le Tout-Puissant ait donné à l'intelligence des hommes pour qu'ils sachent bien s'en servir dans les études et les travaux auxquels tout mortel doit se livrer à l'avantage de son espèce, fuyant l'oisiveté pour justifier son passage sur la terre (p. 168).

[1084.] Cavalli's memoir opens with a *Préface* which occupies pp. 157–168. It commences by quoting with approval certain principles stated by Love. These principles are chiefly deduced from Hodgkinson's experiments and may be summed up as follows:

(i) There is no exact proportionality between stress and strain for cast-iron.

(ii) Set begins for cast-iron with even the smallest loads, and the term elastic limit has thus no real meaning.

(iii) Both cast- and wrought-iron subjected to impact or vibration can support indefinitely loads very near to those capable of producing immediate rupture (p. 159).

The third conclusion seems to me founded on very doubtful evidence, the second is true only if the body has not been reduced to a state of ease, while the first will probably now be generally admitted.

Cavalli next proposes to replace the *elastic limit* by what he terms *la limite de stabilité.* This limit is, I think, what I have termed the yield-point (see our Vol. I. p. 889), as the following words indicate:

> Dans mes expériences à la flexion des barreaux on reconnaît nettement les flexions partagées en deux parties, retournantes les unes, restantes les autres dès leur commencement jusqu'à la rupture, et que chaque partie suit une loi différente mais régulière, dès la plus petite charge jusqu'à celle momentanée produisant la rupture. On découvre encore qu'il y a un terme

intermédiaire de la série de ces charges que les barreaux cessent de soutenir d'une manière stable, et où un mouvement de lassitude très-insensible d'abord commence et s'accroît ensuite rapidement au fur et à mesure qu'on se rapproche à la charge de la rupture, quoique le temps de l'essai soit très-court (p. 160).

Cavalli's experiments were made partly on flexure, partly on compression, and stress-strain curves were traced automatically. After each small increase of load the load was removed, and we thus have a very accurate representation of the relations of elasticity and set to increasing load. Cavalli's curves figured on Plates II. and III. are extremely instructive and are I think the earliest of their kind. Plate II. contains load-flexure diagrams for bronze, cast-iron and cast-steel; Plate III. contains compression diagrams for the same three materials. Roughly speaking these diagrams bring out the following points: (*a*) that both elastic strain and set follow laws the graphical representations of which give extremely regular curves; (*b*) that the state of ease can be extended almost up to absolute strength; (*c*) that the elasticity remains practically the same throughout this extension; (*d*) that there is a point at which set begins to increase with great rapidity: see our Vol. I., pp. 887–9, (5)–(8). With regard to (*d*) we note that for a considerable range of stresses the set curve is almost a straight line close to and parallel to the stress-axis, then it begins to slope more and more to this axis. The point at which this change takes place Cavalli calls the 'limit of stability' and he considers it ought to replace the 'elastic limit.' It seems to me that it is an important limit the knowledge of which is essential, but that it does not replace the 'elastic limit,' which notwithstanding Cavalli's statements (e.g. p. 162) has a real existence, only every stress exceeding the limit to the state of ease alters its value. In order to ascertain the exact point at which the bar ceases to sustain its load stably, Cavalli takes the limit of stability to be the point which is midway[1] between the point at which it is doubtful whether the curve of set has ceased to be parallel to the stress-axis and the point at which there is no doubt such parallelism has ceased (p. 181). He terms this point the limit of stability, because he holds apparently that for any load beyond this limit, the bar will continue to yield till after a longer or shorter time it ruptures (p. 175). Thus he considers the limit of stability to be the proper measure of strength for permanent loading, while for impulsive loading, lasting only during a very brief interval, it is allowable to pass this limit of stability, provided the stress still remains sufficiently below the absolute strength (pp. 176–7). This of course is the legitimate result of principle (iii) stated above, but that principle itself seems to me doubtful. Owing to the above statements we have associated Cavalli's 'limit of stability' with our yield-point although in some respects it seems to be closer to the point half-way between *B* and *C* on the diagrammatic stress-strain curve of our Vol. I., p. 890.

[1] Cavalli has 'le point intermédiare le plus près du second des dits points,' but this is very indefinite.

[1085.] Cavalli now notes that while in most cases the curve giving the relation between the elastic stress and strain is practically linear, that between the set strain and stress is represented by a curve which although perfectly regular has yet to be determined analytically. The sum of the areas of these stress-strain curves, however, gives the work done on the bar up to any given load, and Cavalli accordingly divides this work into two parts which we may term "elastic strain energy" and "ductile strain energy" (*travail élastique et travail ductile*). The energy which the body can absorb of the former kind, increases as the state of ease is extended; the energy of the latter kind is a definite quantity and can only be used once, although it may be consumed in parts on different occasions. Cavalli holds that the elastic and ductile strain energies are the true measures of the practical strength of materials; he expresses them in terms of the kinetic energy of a particle, of mass equal to that of the material, moving with velocities V (for the elastic strain energy) and W (for the united elastic and ductile strain energies). The values of V and W (*vitesses d'impulsion*) thus measure the *resilience* of the material, and their values at the limits of stability and rupture are tabulated on pp. 230–3 of the memoir for a considerable number of bars of bronze, cast-iron and cast-steel (wrought and unwrought), as ascertained by flexural and compressional experiments.

The above sufficiently indicates the general lines of Cavalli's investigations so far as they appear of real novelty or service, but a detailed criticism of his rather lengthy theoretical statements may be of service to other investigators, and I devote the next few articles to it.

[1086.] § I. of the memoir (pp. 168–75) is entitled: *De l'existence de la limite de stabilité au lieu de la limite d'élasticité.* This opens with a statement of the old 'paradox in the theory of beams': see our Arts. 173, 507, 542, 930–8, 1043 and 1051–3. Given a beam of rectangular cross-section of height h and breadth b, then if M be the breaking bending-moment, the absolute strength T (as deduced from an extension of the Bernoulli-Eulerian theory to rupture) is given by

$$T = 6\,\frac{M}{bh^2}.$$

Now Hodgkinson found that for cast-iron bars the factor 6 must be replaced by 2·63, or if T_2 and T_1 be the absolute strengths as calculated from traction and flexure respectively we have:

$$T_2/T_1 = \cdot 438, \text{ or (in the notation of Arts. 1051–3) } S_1/S_2 = \cdot 380.$$

But the American experiments on the metals for cannon (see our Art. 1043) show that the ratio of T_2 to T_1 varies with the density of the cast-iron, increasing up to a certain density and then rather strangely decreasing. Cavalli holds this decrease to be a result of defective experimental method (possible failure of exactly axial application of load which often occurs in *pure traction* experiments: see our Art. 1249* and Cavalli's memoir, pp. 169–70 and 160), and after rectifying the results, he obtains values of the ratio increasing from ·57 to ·72 with the density[1]. He uses this variation as a general argument against the ordinary theory of beams and as in some way suggesting the importance of his own investigations into the 'limit of stability,' because he supposes it to show the inapplicability of the theory based on the 'limit of elasticity'. He passes rather abruptly from this discussion to a description of his testing machine and automatic apparatus for drawing stress-strain diagrams (pp. 172–5).

[1087.] § II. of the memoir (pp. 175–87) is entitled: *Discussion des nouveaux principes à admettre, et déduction de la mesure du travail élastique et ductile, et de la vitesse d'impulsion que les solides peuvent supporter.* In this section the author first states and criticises the conclusions of Love, Hodgkinson, Belanger and Morin, and then states his own theory of resilience as the true test of resistance especially for the case of impulsive loading. He qualifies his previous statements as to the limit of rupture being the superior limit for impulsive stress (see the principle (iii) of our Art. 1084) by the rather vague reservation that the impulses must not succeed each other too rapidly, nor last for too long a time without interval of repose (p. 178). The following remarks indicate Cavalli's standpoint and deserve quotation:

Lorsqu'une seule portion du travail ductile l'épuiserait à chaque impulsion, le nombre ou la somme de ces impulsions ne devra pas dépasser la limite du travail ductile total, de sorte que ce nombre d'impulsions que le prisme pourra supporter à la limite prescrite se trouvera restreint.

Le choix entre les différents matériaux à employer dans les constructions se trouva par ces conditions soumis à un calcul qu'il faut savoir faire. L'on ne pourra pas dire d'avance qu'on doit dans telle sorte de construction employer les matériaux plus ductiles qu'élastiques et *vice versa* dans telle autre sorte de construction; on s'exposerait par un tel procédé à bien des méprises, comme l'abus des constructions toutes en fonte a fait ressortir, et comme il arriverait par l'abus de tout faire en fer forgé (p. 179).

[1088.] To apply his theory Cavalli proceeds thus: Let F be the load and x the elastic, y the 'ductile' deflection immediately under the load, then the elastic strain energy $=\frac{1}{2}Fx$, while the ductile strain energy $=\frac{1}{2}F\tau y$, where $\frac{1}{2}\tau y$ is the mean ordinate of the ductile stress-strain (or really of the load-deflection) curve. Now Cavalli's experiments were made on the flexure of a cantilever of length L and rectangular cross-section $b \times h$; hence, if T_0 be the maximum elastic stress in the beam:

$$T_0 = 6\,\frac{FL}{bh^2} \text{ and } x = \frac{4FL^3}{Ebh^3},$$

[1] The mean of the American results as rectified by Cavalli gives $T_2/T_1 = {\cdot}662$, the ratio as deduced from the hypothesis proposed by the Editor in a paper on the Flexure of Beams, *Quarterly Journal of Mathematics*, Vol. XXIV. p. 108, 1890, is for the case of a rectangular section ·667.

or, $$\tfrac{1}{2}Fx = \tfrac{1}{18}\frac{T_0^2}{E}bhL.$$

If D be the density of the material, Cavalli equates this to $\tfrac{1}{18}bhLD \times V^2$ and so finds:

$$V^2 = \frac{T_0^2}{ED}.$$

Thus Cavalli's V is at once determined by the density of the material and the *modulus of resilience*: see our Art. 363.

If W be the velocity corresponding to both elastic and ductile strain energies

$$\tfrac{1}{18}bhLD \times W^2 = \tfrac{1}{2}Fx + \tfrac{1}{2}F\tau y,$$

whence we have

$$W = V\sqrt{1 + \tau\frac{y}{x}},$$

where τ and y/x have to be determined by experiment for each material. According to Cavalli's results τ decreases by about a half between the limits of stability and rupture, so that Hodgkinson's experiments on cast-iron which made $y \propto F^2$ and give $\tau = 2/3$ cannot be accepted as generally true: see pp. 185–6 of the memoir and our Arts. 969* and 1411*.

[1089.] The reason apparently why Cavalli takes $\tfrac{1}{18}MV^2$ instead of $\tfrac{1}{2}MV^2$ as suggested by his definition of V (see our Art. 1085), is that he supposes $\tfrac{1}{2}MV^2$ to be the resilience of longitudinal elasticity. $\tfrac{1}{2}Fx$ in this case is equal to $\tfrac{1}{2}\frac{T_0^2}{E}bhL$ and this is nine times the above value. The following are Cavalli's *mean* results in metres per second[1] (p. 184):

Material		At Limit of Stability		At Rupture	
		V	W	V	W
Bronze	Extension	5·44	6·16	18·00	29·50
	Compression	14·03	15·43	34·74	48·80
Cast-Iron	Extension	8·16	8·79	16·4	18·3
	Compression	27·45	27·78	80·54	87·43

This table may be used to obtain the moduli of resilience, which are equal to DV^2 or DW^2 as the case may be.

[1] The second number in the first column of the table in the footnote, p. 184, should be 4·93 and not 5·60, I think.

Cavalli in some rather obscure reasoning on p. 186 appears to state that any portion of a body may receive a blow which gives it a velocity V (or W as the case may be) without ultimate (or immediate) danger. For example, the velocity given to the parts of the inner surface of a cannon ought not to exceed the value W. But I am unable to follow the argument, nor do I understand how the velocity, which if attributed to the entire mass would give an amount of energy equivalent to the strain-energy, is necessarily the velocity with which any part will commence vibrating. It must be noted that throughout Cavalli neglects the inertia of the vibrating parts, i.e. proceeds statically, and although this may give the maximum total flexure or compression fairly correctly, Saint-Venant has shown, that for the case of transverse impact at least it is very far from giving the correct value of the maximum strain, which depends on relative flexure or relative compression: see our Arts. 371, 406 and 412.

[1090.] § III. (pp. 187–96) is entitled: *De la position des fibres invariables dans les prismes soumis à la flexion.* This section rejects Hodgkinson's stress-strain relation for cast-iron, and asserts that the neutral axis does not pass through the centroid of the section because the stretch- and squeeze-moduli are unequal. This had in fact been previously discussed by Hodgkinson (see our Art. 234*), and there is nothing new or of real value in Cavalli's results. By taking P and Q as the resistances per unit area to extension and compression respectively and supposing the material perfectly elastic, Cavalli finds that the ratio Q/P must be in some cases as much as 6, if the absolute strengths as given by tractive and flexural experiments are to agree. He does not seem to have noticed that with his definitions and on his hypotheses, this would have made the squeeze-modulus six times the stretch-modulus (pp. 189–93)! Morin's hypothesis, which our author condemns,—i.e. that the resistances to compression and extension only begin to vary after the elastic limit is passed,—is certainly more reasonable than this!

Cavalli quotes a formula due to Roffiaen for the strength of a prism under flexure (see our Arts. 892 and 925), and applies his own results to a prism of circular cross-section. The treatment in both cases is obscure, not to say inadmissible.

[1091.] § IV. of the memoir is entitled: *Essai théorique de la résistance vive élastique et ductile des prismes par la vitesse d'impulsion des solides, suivi d'exemples pratiques* (pp. 196–229).

This introduces Tredgold's modulus of resilience $\frac{1}{2}T_0^2/E$ but attributes it to Poncelet. The investigation of the longitudinal resilience on p. 198 is obscure, because it is not obvious why Cavalli concentrates the mass of the rod at the free end. The results for a frustum of a cone on p. 199 seem to me still more doubtful. In treating of the flexure of a cantilever Cavalli concentrates *half* its mass at the free end (pp. 199–202) and applies his theory of the shifted neutral axis. In all these cases the inertia of the bar is neglected, but it has, as I have pointed out, the greatest influence on the maximum strain. On pp. 204–5, there is an unsatisfactory attempt to determine the time over which an impulse must be spread, in order that the whole and not a part of the bar may sustain the work due to the impulse. Cavalli finds that when the time of the maximum safe impulse is less than

$$t_1 = \frac{\pi}{2} \frac{x_1}{V},$$

where x_1 is the maximum shift of the free end and V the velocity discussed in our Art. 1088, then even with this impulse the bar will be injured at the part to which the blow is applied.

[1092.] Cavalli next passes to practical examples chiefly dealing with problems in gunnery and with the penetration of shot into iron-plates (p. 205 to the end).

As a sample of the somewhat loose style of reasoning as well as of grammar adopted in these pages, I cite the following example, which does not belong to the theory of gunnery:

Prenons à calculer un pont en poutres simples de fer sur un chemin de fer pendant le passage des trains : ces poutres fléchiront pour se redresser après le passage. De même que dans le calcul statique on ne considère que la moitié de la charge concentrée au milieu, l'autre moitié de la charge étant portée par les culées, l'on pourra considérer aussi ici que la moitié de la masse totale du pont et de la charge est concentrée au milieu, et tombant de la hauteur de la flexion entière ; soit pour plus de simplicité dans le calcul, que pour avoir égard aux secousses que l'irrégularité du mouvement du train causera au pont (pp. 215—6).

Considering the attention this problem had already received from Willis, Stokes and Phillips (see our Arts. 1276*–91*, 1418*–22*, 378–82 and 552–60), Cavalli's treatment is somewhat antiquated.

Without entering into an analysis of these individual problems, we may conclude our notice of Cavalli's memoir with citing a remark he makes on the testing of cannon. After noting that every impulse which exceeds the existing elastic limit uses up some of the reserve of ductile strain-energy in the material, and that every successive impulse uses up more of this surplus energy until either by raising the elastic limit the elastic strain-energy alone suffices, or the gun at last bursts, he continues :

L'épreuve des canons par des tirs surtout plus forts que ceux ordinaires, outre d'être embarrassante et très-coûteuse, prouve seulement qu'après ces tirs les canons qui l'ont subie sont moins bons qu'auparavant, sans pouvoir, pour plusieurs causes confirmées par l'expérience, nous rassurer d'après leur résistance sur celle des autres canons (p. 227).

To the memoir are affixed the tables of experimental results referred to in our Art. 1085.

Group F.

Strength of Iron and Steel.

[1093.] *A Series of Experiments on the Comparative Strength of Different Kinds of Cast-Iron, in their simple state as cast from the Pig, and also in their compounded state as Mixtures; made under the directions of Robert Stephenson, Esq., with a view to the selection of the most suitable for the various purposes required in the construction of the High Level Bridge.* The experiments were made at Gateshead, September, 1846, to February, 1847, and the results are published in the *Civil Engineer and Architect's Journal*, Vol. XIII., pp. 194–199. London, 1850. Only the numerical results—consisting of the loads and deflections through a certain range up to the breaking load, together with the initial series of sets—are given. No general conclusions are drawn, nor is there any graphical representation of results.

[1094.] *Rapport d'une Commission nommée par le gouvernement anglais, pour faire une enquête sur l'emploi du fer et de la fonte dans les constructions dépendant des chemins de fer: Annales des ponts et chaussées. Mémoires*, 1851, 1er Semestre, pp. 193–220. Paris, 1851. This is a translation by Busche of the report attached to the evidence of the Iron-Commissioners: see our Art. 1406*.

[1095.] In the volume of the *Annales des ponts et chaussées* for 1855, *Mémoires*, 1er Semestre, pp. 1–127 will be found a French translation of E. Hodgkinson's *Experimental Researches* (see our Arts. 966*–73*) by E. Pirel. Even at the present day the results of Hodgkinson's experiments reduced to French measure are not without special value.

[1096.] Dehargne: *Galvanisation du fer; avantages de l'emploi des fils galvanisés dans les ponts suspendus. Annales des ponts et chaussées. Mémoires*, 1851, 1er Semestre, pp. 255–88. Paris, 1851. On pp. 280–8 will be found details of experiments on the absolute strength of iron wire before and after galvanisation, and it is shown that the iron loses nothing of its strength or ductility by the process; some of the experiments show indeed a great increase of strength owing to galvanisation.

[1097.] *British Association*, 1852, *Belfast Meeting*, *Transactions*, p. 125. Notice of some experiments by Fairbairn then in progress to test the effect of repeated meltings on the strength of metals—and further to test the effect of temperature on unwrought iron plates.

[1098.] The experiments on the effect of repeated meltings referred to in the previous article form the subject of a paper communicated to the *British Association* in 1853, and printed on pp. 87–116 of the Report of the Hull Meeting for that year. The paper is entitled: *On the Mechanical Properties of Metals as derived from repeated Meltings, exhibiting the Maximum Point of Strength and the Causes of Deterioration.*

Fairbairn commences by thus stating the object of his investigation, undertaken at the request of the Association:

It is a generally acknowledged opinion, that iron is improved up to the second, third and probably the fourth meltings; but that opinion, as far as I know, has not been founded upon any well-grounded fact, but rather deduced from observation, or from those appearances which indicate greater purity and increased strength in the metal.

Those appearances have, in almost every instance, been satisfactory as regards the strength; and the questions we have been called upon to solve in this investigation, are, to what extent can these improvements be carried without injury to the material; and what are the conditions which bear more directly upon the crystalline structure, and the forces of cohesion by which they (*sic*) are united (p. 87).

[1099.] The first set of experiments were on the resistance of rectangular bars (in all cases of nearly 1 inch square cross-section and of 4 ft. 6 inches span) to a central transverse load. 18 successive meltings were undertaken of which the 17th melting was a failure, "the iron being too stiff to run into bars." Fairbairn reduces his results to a standard beam of 1 inch square cross-section and 4 ft. 6 in. span. He terms the product of the breaking load into the ultimate deflection, the *power of resisting impact.* He considers it *proportional* to the resilience, and it is entered in the table below as Proportional Resilience. The experiments were made on "Eglinton Iron, No. 3, Hot-blast." After each melting the rupture-surfaces were microscopically examined and presented interesting changes, in some cases figured in the memoir. Their general appearance is described in rather vague language, as: 'finely grained texture,' 'crystals of greatly increased density,' 'fine frosty appearance,' etc. One noteworthy change is the appearance of an internal core in the last meltings differing much in structure from the rest of the metal.

I reproduce the following summary of results (pp. 107–8):

No. of Meltings.	Specific Gravity.	Mean rupture load in lbs.	Mean ultimate deflection in inches.	Proportional resilience.
1	6·949	490·0	1·440	705·6
2	6·970	441·9	1·446	639·0
3	6·886	401·6	1·486	596·7
4	6·938	413·4	1·260	520·8
5	6·842	431·6	1·503	648·6
6	6·771	438·7	1·320	579·0
7	6·879	449·1	1·440	646·7
8	7·025	491·3	1·753	861·2
9	7·102	546·5	1·620	885·3
10	7·108	566·9	1·626	921·8
11	7·113	651·9	1·636	1066·5
12	7·160	692·1	1·666	1153·0
13	7·134	634·8	1·646	1044·9
14	7·530	603·4	1·513	912·9
15	7·248	371·1	·643	238·6
16	7·330	351·3	·566	198·8
17	lost	——	——	——
18	7·385	312·7	·476	148·8

It will be noted that the transverse strength decreases from the 1st to the 3rd melting and then increases to the 12th melting after which it rapidly decreases. The resilience also reaches its maximum at the 12th melting, but I should not feel inclined to lay much stress on any results obtained by a measurement of ultimate deflections.

[1100.] A second series of experiments was made on the compressive strength of the same iron after 18 meltings (pp. 109–113). The following results were obtained:

No. of Meltings.	Compressive strength in tons per sq. inch.	No. of Meltings.	Compressive strength in tons per sq. inch.
1	44·0	10	57·7
2	43·6	11	69·8
3	41·1	12	73·1
4	40·7	13	*66·0
5	41·1	14	95·9
6	41·1	15	76·7
7	40·9	16	70·5
8	41·1	17	lost
9	55·1	18	88·0

* In Experiment 13 the cube was not properly bedded, and so the result is erroneous. It would probably, Fairbairn says, have given 80 to 85 tons per sq. inch.

Up to the eighth melting it will be observed that the ordinary power of resistance to a crushing force, namely, about 40 tons to the square inch, is indicated. Afterwards, as the metal increases in strength, from the eighth to the thirteenth melting, a very considerable change has taken place, and we have 60 instead of 40 tons as the crushing force. Subsequently, as the hardness increases, but not the [transverse] strength, double the power is required to produce [crushing] fracture (p. 115).

Plate 3 at the end of the *B. A. Report* figures the rupture surfaces of the blocks crushed in the experiments.

The results are compared with those of Rennie, Rondelet and Hodgkinson: see our Arts. 185*–7*, 696*, and 948*–51*. The memoir concludes with a chemical analysis of the iron after different meltings by F. C. Calvert (pp. 115–116). From this analysis it appears that silicium increases, while sulphur and carbon fluctuate in percentage with the number of meltings.

[1101.] *British Association*, Report of Liverpool Meeting, 1854, *Transactions*, pp. 151–152. Letter of William Hawkes: *On the Strength of Iron after repeated Meltings.* The writer had made experiments on "Corbyns Hall Iron, No. 1, Hot-blast" with 29 successive meltings. His results do not present the regularity of change which marks Fairbairn's experiments: see our Art. 1099. They do indeed give a minimum and maximum of strength after the 6th and 12th meltings respectively, but these are followed again by a minimum at the 14th, a maximum at the 18th, a minimum at the 21st, and a maximum at the 24th, while the strength at the 29th is greater than after the first melting. There is thus no sign of deterioration following on any number of meltings, such as was manifested in Fairbairn's results: see our Arts. 1059 and 1099.

[1102.] F. C. Calvert: *On the Increased Strength of Cast-Iron produced by the use of improved Coke, with a Series of Experiments by* W. Fairbairn, *Institution of Civil Engineers, Minutes of Proceedings*, Vol. XII., pp. 352–381. London, 1853. Evidence is given in this paper as to the amount of influence which the method of preparation has on the elasticity, set and absolute strength of cast-iron.

[1103.] J. Jones: *Table of Pressures necessary for Punching Plate-Iron of various Thicknesses. The Practical Mechanic's Journal*, Vol. VI., p. 183. London and Glasgow, 1853–4. This table contains numerical details of apparently very careful experiments on punching plate-iron. It would still be of considerable service to any investigator wishing to test a theory of absolute shearing strength: see our Art. 184 (*b*). No theory is attempted in the paper itself.

[1104.] C. R. Bornemann: *Notiz über John Jones' Versuche über den Kraftbedarf zum Lochen von Kesselblechen. Dinglers*

Polytechnisches Journal, Bd. 140, S. 327–32. Stuttgart, 1856. The details of Jones' experiments on punching holes of various diameters in iron boiler-plates had been cited in the *Polytechnisches Centralblatt* for 1854 (see our Art. 1103). Bornemann gives a *résumé* of them, calculating the mean values, and he suggests the following empirical formula:

$$P = 62725 - 2822{\cdot}34\alpha,$$

where P is the punching stress per unit-area of sheared surface and α is the area of the sheared surface, P being measured in pounds per sq. inch and α in sq. inches. For a circular hole of diameter b in a plate of thickness τ, the total load $L = \pi b \times \tau \times P$ and $\alpha = \pi b \times \tau$ or

$$L = (62725 - 2822{\cdot}34\pi b\tau)\,\pi b\tau \text{ lbs.},$$
$$= (197056 - 27856 b\tau)\, b\tau \text{ lbs.}$$

Bornemann obtains the numerical coefficients by means of the method of least squares and he then compares the result with earlier investigations on punching strength,—e.g. those of E. Cresy (*Encyclopædia of Civil Engineering*, New Impression, Vol. II., pp. 1035 and 1708. London, 1861), which give considerably smaller values for L, of Fairbairn (*locus* ?), of Gouin et Cie. (see our Art. 1108). In round numbers we have for the punching strength in kilogrammes per square millimetre: Jones, 42; Cresy, 31; Fairbairn, 37; Gouin et Cie. 32. Bornemann concludes with the following table for English plate-iron

Resistance	to	punching	42	kilogrammes	per sq.	millimetre,
,,	,,	traction	40	,,	,,	,,
,,	,,	shearing	32	,,	,,	,,
,,	,,	crushing	25	,,	,,	,, .

The last number 25 I do not understand, as I should have expected the crushing strength to be greater than this.

[1105.] J. D. Morries Stirling: *On Iron, and some Improvements in its Manufacture. Institution of Mechanical Engineers, Proceedings*, 1853, pp. 19–33. London, 1853. This paper contains experiments on the transverse and tensile strengths of cast- and wrought-iron with the details of some experiments by Owen on the comparative strength of ordinary and 'toughened' cast-iron girders (p. 23 and Plate 4).

[1106.] Brame: *Note sur l'application de la tôle à la construction de quelques ponts du chemin de fer de ceinture. Annales des ponts et chaussées. Mémoires*, 1853, 1er Semestre, pp. 78–111. Paris, 1853. This contains some account of experiments by Brame on iron-plate with references to those of Hodgkinson, Fairbairn, Gouin et Cie., etc. see our Arts. 1477*, 1497* and 1108.

[1107.] Kirchweger: *Ueber die Prüfung des Stabeisens. Polytechnisches Centralblatt*, 1854, Cols. 1110–6. Leipzig, 1854. (Extracted from *Mittheilungen des Gewerbevereins für das Königreich Hannover*, 1853, S. 240.) This paper gives details of German experiments on the strength of English iron plates used for the girders of railway bridges. The plates were tested by boring rivet holes in them, which were then driven asunder by a conical steel wedge upon which a given weight was allowed to fall repeatedly from a definite height. The number of blows required for rupture was taken as a measure of the strength.

[1108.] Gouin et Cie. (*Expériences sur la résistance à la traction de tôles de diverses provenances et sur celle des rivets*). These are described in an article by Mathieu and Lavalley on the *Pont de Clichy* in the *Mémoires...de la Société des Ingénieurs civils*, Année 1852, pp. 153–7. Paris, 1852. A German translation appeared in the *Polytechnisches Centralblatt*, Jahrgang 1854, Cols. 525–6. The first part of the experiments deals with the absolute tensile strength of iron-plate parallel and perpendicular to the direction of the rolling. For charcoal raw iron there was on the average a fall from 3313 to 3240 kilog. per sq. centimetre; for coke raw iron a fall from 3657 to 2906. Hence the rolling has far less influence when the iron is prepared in the former fashion: see our Arts. 1497*, 879 (*d*) and 902.

The second part of the experiments deals with the absolute shearing strength of iron rivets of 8 to 16 millimetres diameter. The shearing strength averaged about 3200 kilogs. per sq. centimetre as compared with about 4000 kilogs. tensile strength, or very nearly in the $\frac{4}{5}$ ratio obtained by extending uni-constant isotropy to the rupture of wrought-iron.

[1109.] Collet-Meygret et Desplaces: *Rapport sur les épreuves faites à l'occasion de la réception du viaduc en fonte construit sur le Rhône, entre Tarascon et Beaucaire, pour le passage du chemin de fer, et sur les observations qui ont servi à constater les mouvements des arches sous l'influence de la température et des charges, soit permanentes, soit accidentelles; suivi de considérations sur le mode de résistance et sur l'emploi de la fonte dans les grands travaux publics. Annales des ponts et chaussées. Mémoires*, 1854, 1er Semestre, pp. 257–367. Paris, 1854.

This memoir contains an account of the viaduct over the Rhone at Tarascon, the arches of which were made of cast-iron.

The description is of interest, as these arches have been dealt with theoretically by Bresse: see our Arts. 520 (a) and 527. It is also, I think, the first bridge in which the strains due to changes of temperature were carefully measured (pp. 274–291). The exact deflections due to dead and live load were also very accurately ascertained (p. 280 and pp. 292–307). The diminution of the compressibility of the iron with the increase of the load, i.e. the non-proportionality of stress and strain within the elastic limit, seems to have been noted on the large scale of this bridge: see our Arts. 1411* and 935.

[1110.] On pp. 307–19 we have a comparison of theory and experiment. The authors give the following formula for the deflection f (deduced in *Note A*, pp. 360–4):

$$f = \frac{pr^4}{E\omega\kappa^2} \times \cdot 000{,}004{,}855,$$

where p = the weight of the arch per unit run of the horizontal, r the radius of its central axis, E its stretch-modulus and $\omega\kappa^2$ the usual moment of inertia of the cross-section about the 'central axis.' The values of f obtained from this formula were far from agreeing with those found by direct experiment. The authors accordingly argue that E ought only to be given one-half the value previously adopted for it from traction-experiments (p. 320). It must be remarked, however, that their theory of arched ribs is very far from satisfactory and that it ought to be replaced by Bresse's investigation: see our Arts. 514–31.

This discrepancy in their theory leads the authors to consider the details of a number of French and English experiments on cast iron. They show that its *tensile strength* varies with its quality and the dimensions of the test-piece to a very wide extent, and hence they appear to argue (p. 329) that its stretch-modulus can also have values varying from 6,000,000,000 to 12,000,000,000 kilogrammes per sq. metre. This does not seem very convincing, especially as the table (p. 327) of tensile strengths has been deduced from flexure experiments: see our Art. 1052. A more satisfactory investigation by direct experiment of the values of E follows on pp. 330–46. These values were found to vary from less than 3,000,000,000 to more than 12,000,000,000 kilogrammes per sq. metre, according to the material of the bar. The authors conclude that:

1° les barreaux de fonte des diverses usines essayés dans les mêmes circonstances donnent des valeurs de E peu différentes.

2° un barreau donne pour E des valeurs sensiblement différentes suivant qu'il est posé à plat ou de champ.

3° les barreaux d'une même usine donnent des valeurs de E très-différentes suivant les conditions des assemblages; posés sur deux appuis et chargés au milieu, ils donnent des valeurs de E plus grandes que lorsque étant posés sur deux appuis ils sont chargés à leurs extrémités, ou lorsqu'étant encastrés par un bout ils sont chargés à l'autre bout, et, dans ce cas, ils donnent des valeurs de E plus grandes que lorsqu'ils sont chargés debout, c'est-à-dire comprimés dans le sens de leur longueur (p. 337).

As a result of these conclusions Collet-Meygret and Desplaces consider that the best value of the stretch-modulus for the iron of the Tarascon viaduct ought to be obtained by comparing direct experiment on the bridge itself with the formula referred to above. They consider that it is the manner in which the iron is employed in the structure rather than its particular 'manufacture' which determines the value of its stretch-modulus.

[1111.] They especially note the difference between the elasticity of the core and periphery in the case of cast iron bars and conclude :

1° que de deux pièces semblables de la même fonte, la plus grosse donnera la plus faible valeur de E.

2° qu'une même pièce chargée de la même manière et sous les mêmes assemblages donnera, lorsqu'elle sera présentée sous différentes faces, des valeurs de E différentes, dépendantes du moment d'inertie de sa section, comparé au moment d'inertie de son périmètre.

3° que dans une même pièce de fonte on trouvera pour E une valeur d'autant moindre que dans les joints d'assemblage et par le mode de chargement, on laissera libre une plus grande portion du périmètre, de manière qu'une plus grande partie du métal extérieur, le moins élastique, soit entraînée par le métal intérieur, le plus élastique, au lieu de le retenir (pp. 340 1).

The authors suppose the periphery to have a thickness of ·005 metres, a stretch-modulus ϵ and an absolute tractive strength τ. Then if E and T be the like quantities for the core, they find from experiments on cast-iron bars such as were used in the Rhone viaduct in kilogs. per sq. metre,

$$\tau > 40,000,000, \qquad \epsilon > 12,000,000,000,$$
$$T < 20,000,000, \qquad E < 3,000,000,000.$$

Their remarks on the experiments leading to these results and the conclusions to be drawn from them are of considerable interest: see their pp. 341-6 and our Arts. 169 (*e*)-(*f*) and 974 (*c*). Similar differences probably hold for the temperature effect on the core and on the periphery, but the authors remark that as various physicists give values for the stretch per degree centigrade of iron, whether it be cast or wrought, varying only between ·000,011 and ·000,013, it is safe to neglect these differences and adopt the number ·000,012,2 to represent this stretch.

[1112.] With this value of the stretch or thermal coefficient and with the modified value of the stretch-modulus the authors (pp. 346–58) analyse the various elements of flexure due to temperature, to live and to dead load. They sum up their conclusions on pp. 358–60. Their remark, that the values of the elastic constants found by physicists experimenting on small bars of metal cannot be safely adopted for large masses of the same metal such as occur in great engineering structures, deserves from its obvious truth more attention than it has sometimes received (p. 359). To the memoir are appended (pp. 360–7) various notes which do not call for special mention here.

[1113.] G. Weber: *Versuche über die Cohäsions- und Torsionskraft des für Geschütze bestimmten Krupp'schen Gussstahls. Dinglers Polytechnisches Journal*, Bd. 135, S. 401–17. Stuttgart, 1855.

This memoir opens with some interesting details of the chemical constitution of gun-metal used in 1663 and later, and shows how the earlier metal would certainly not have stood the strength of modern (?) powder. Weber then proceeds to details of the tensile and the torsional strengths of steel (manufactured by Krupp, and in England, Salzburg and the Tyrol) of wrought-iron and of bronze or gun-metal. There is an interesting figure (Tab. VI., Fig. 2) giving a good picture of the stricture of a bar of Krupp's cast-steel for guns. It shows exceedingly well the relative amount of stricture at each cross-section and the total change at rupture of each dimension of the bar. The whole paper is an advertisement for Krupp, but probably a well-deserved advertisement.

[1114.] Details of various experiments on the strength and elasticity of steel with reference to the peculiar difficulties of casting it so that its quality is uniform throughout the piece, and with comparison of results obtained for wrought-iron will be found in the *Polytechnisches Centralblatt*, Jahrgang 1856, Cols. 1275–6, Jahrgang 1857, Cols. 35–44 (*Annales des Mines*, T. VIII., pp. 373–88, 1855) and Jahrgang 1857, Cols. 1128–38. All these have special reference to steel prepared by Uchatius' process. With regard to the strength and stricture of wrought-iron prepared by the Bessemer process an account of some experiments made at Woolwich will be found in *The Mechanic's Magazine*, 1856, p. 270.

[1115.] William Fairbairn: *On the Tensile Strength of Wrought-Iron at various Temperatures. British Association*, Cheltenham Meeting, 1856, *Report*, pp. 405–422.

These experiments are of very considerable interest, as in many structures of wrought-iron the material is subjected to very high temperatures or to a considerable range of temperatures. Fairbairn's first series are on the tensile strength of boiler plates with and against the fibres. From 0° to 395° Fahr. there seem to be only very slight fluctuations in the strength in the direction of the fibre, and these are not improbably due to experimental errors, or to weaknesses in the individual pieces. Roughly the strength fluctuates from 18 to 22 tons per sq. inch without any apparently regular variation with the temperature. I have little doubt that some of the fluctuation is due to want of exactly central pull in the arrangement adopted by Fairbairn. For tensile strength across the fibre there is a rise from about 18·7 to 20·4 tons per sq. inch for a change of temperature from 0° to 212°, a fall to 18·8 at 340° and to 15·3 at visible red heat, the latter result being considered by Fairbairn as too high. Here again the only safe conclusion seems to be that at "a dull red heat just perceptible in day-light" the tensile strength is much reduced. At what heat the maximum is reached is not rendered clear by the experiments (pp. 413–4).

[1116.] The second series of experiments relate to the tensile strength of rivet-iron. Here there was a more marked relation between strength and temperature. The experiments were on temperatures from – 30° to 435° and at 'red heat.' There was an increase here from 28·2 tons at – 30° to 37·5 at 325°, and at least a steady increase from 28·1 tons at 60° to 37·5 at 325°. After this there was a slight diminution at 435°, and a great drop to 16·1 tons (marked "too high") at red heat (p. 420).

The memoir concludes with a comparison of the increase in strength due to rise of temperature with that due to repeated fracture, and with some remarks on the stretch in bars of different lengths, which do not seem to me of much scientific value: see pp. 421–2 and our Art. 1503*.

[1117.] William Bell: *On the Laws of the Strength of Wrought- and Cast-Iron, Institution of Civil Engineers, Minutes of Proceedings*, Vol. XVI., pp. 65–81. London, 1857. A *résumé* of this paper will be found in the *Mechanic's Magazine*, Vol. 65, pp. 579–81. London, 1856. It is an endeavour to demonstrate from the many experiments which have been made on cast- and wrought-iron beams under flexure that theory and experiment are after all not so discordant as some have supposed. We may sum up the author's conclusions as follows:

(i) For slight strains theory and experiment coincide.

(ii) The ordinary theory of rupture practically coincides with experiment for wrought-iron beams, especially those of large size. [It should only do this if Hooke's Law practically holds for wrought-iron up to rupture.]

(iii) There is no reason for supposing the neutral axis shifts its position to any extent worth noticing before rupture.

(iv) There is a divergence between theory and experiment in the case of small cast-iron bars whose transverse strength is compared with their direct tensile strength, but the coincidence between these strengths for large girders is nearly exact.

I think the latter statement requires further demonstration. We should not expect such equality because Hooke's Law does not hold for cast-iron, even in the case of small strains, and certainly not up to rupture.

(v) That one of the chief failures of the ordinary theory occurs for cast-iron struts (rounded ends for length < 20 diameters, and flat ends for length < 50 diameters, p. 67).

[1118.] The author remarks that according to Hodgkinson the ratio of the tensile and compressive strengths of wrought-iron may for practical purposes be taken as unity In the case of cast-iron he apparently prefers a traction-stretch relation of the form (!)

$$\widehat{xx} = q + Es_x,$$

where q is a constant, to Hodgkinson's

$$\widehat{xx} = as_x - bs_x^2.$$

[1119.] In the discussion P. W Barlow laid stress on W. H. Barlow's 'explanation' of the paradox in the resistance of beams under flexure (see our Art. 931), and W. T. Doyne on his modification of Hodgkinson's rule for the section of the beam of maximum strength; see our Arts. 1016 and 1023. R. Sheppard communicated a method of noting the permanent set due to flexure by drawing lines on the faces of a beam of lead.

[1120.] On p. 83 ftn. of the same volume will be found some details of the tensile, transverse and crushing strengths of some iron manufactured in India.

[1121]. H. Wiebe: *Ueber die Festigkeit der Bleche und der Vernietungen. Zeitschrift des Vereins deutscher Ingenieure*, Jahrgang I., S. 255–268. Berlin, 1857. This paper gives details of the experiments of Fairbairn, Clark, and Gouin et Cie. on riveted iron-plates: see our Arts 1497*, 902, 1066 and 1108.

[1122.] B. Dahlmann: *Die absolute Festigkeit verschiedener Eisen- und Stahlsorten des königl. württemb. Hüttenwerks Friedrichsthal. Dinglers Polytechnisches Journal*, Bd. 143, S. 94–7. Stuttgart, 1857. Details are given of the absolute strength of cast-iron and steel made at a particular foundry, and as individual results they form only an advertisement of the same foundry.

[1123.] M. Meissner: *Mittheilung von Versuchen, welche zur Ermittelung der absoluten Festigkeit von Eisen- u. Stahlsorten in April 1858 ausgeführt worden sind. Polytechnisches Centralblatt*, Jahrgang 1858, Cols. 1195–9. Leipzig, 1858. (Extracted from the *Zeitschrift d. österr. Ingenieur-Vereins*, 1858, S. 88.) This paper merely contains details of the absolute tensile strength of various kinds of iron and steel, which might be useful, as others of the same type, to anyone writing a history of the gradual improvements in the preparation of iron and steel, but the results are of no permanent practical value and have no bearing on theory.

[1124.] *Vergleichende Zerreissversuche mit den Pöhlmann'schen, Webster-Horsfall'schen und Miller'schen Clavier-Stahlseiten. Dinglers Polytechnisches Journal*, Bd. 147, S. 460–1. Stuttgart, 1858. (Extracted from *Verhandlungen des nieder-österreichischen Gewerbevereins*, Jahrgang 1858, S. 54.) This contains details of the comparative strength of the steel pianoforte wires of different manufacturers, and is of no general interest at the present time.

[1125.] C. E. Browning: *On the Extension and Permanent Set of Wrought Iron when strained tensibly. The Engineer*, Vol. v. pp. 317 and 352. London, 1858. These two letters propound the thesis that the strength of wrought iron is increased by straining it to rupture. The writer apparently considers that the density and strength are alike increased by the drawing in of the cross-section in set, but the experiments he cites are certainly not conclusive, as it might well be argued that a bar would give way first at its *weakest* cross-section, and thus we might expect a greater load at successive ruptures: see our Art. 1503*. The proposal in the second letter to subject all the bars of braced girders and the cables of suspension bridges to a stress of 20 tons per square inch before using them, as a means of increasing their strength and reducing the weight of structures would hardly meet with favour, we think, from practical engineers.

[1126.] Völckers: *Ueber Festigkeit der Bleche, Zeitschrift des Vereins deutscher Ingenieure.* Jahrgang II., S. 17–20. Berlin, 1858. This paper contains the details of some experiments on the absolute tensile strength of iron plate with and across the fibre,—Völckers found with the fibre 100, across 91·3, in the diagonal 93·2 to represent the relative strengths of one kind of iron plate,—and on the loss of strength due to riveting. Völckers found the loss of strength due to punching rivet holes to be as 59·4 to 100, while Fairbairn had given it as 56 : 100 see our Art. 1500*. The author also gives some account of experiments on the loss of strength by heating. He concludes that there is little reduction of absolute strength up to about 300° C., but that temperatures from 500° to 700° C. enormously reduce the strength, the reduction amounting to one-half and even more. The memoir concludes with a comparison of the formulae of the Prussian, French and Austrian Governments for the thickness of cylindrical boilers. If n be the number of atmospheres internal pressures, τ the thickness in Prussian inches, d the diameter in Prussian feet, these formulae were:

$$\text{Prussian, } \tau = \cdot 018d(n-1) + \cdot 1$$
$$\text{French, } \tau = \cdot 0214d(n-1) + \cdot 113, \quad \text{(see our Art. 879 (c))}$$
$$\text{Austrian, } \tau = \cdot 0216d(n-1) + \cdot 114.$$

Völckers compares these formulae with the results obtained by him for the strength of riveted plates and concludes that even the Prussian formula gives theoretically sevenfold safety (p. 20).

[1127.] Another series of experiments on plate-iron by C. Schönemann will be found on S. 304–6 of the same Jahrgang of the *Zeitschrift* (*Resultate von Blech Versuchen*). The effects of temperature and of rivet holes in reducing strength were considered. Numerical results are given, but no general conclusions are drawn. Further experiments by Krame of a like kind will be found on S. 173 of the *Zeitschrift*, Jahrgang III., 1859.

[1128.] Robert Mallet: *On the Coefficients T_e and T_r of Elasticity and of Rupture in Wrought-Iron, in relation to the Volume of the Metallic Mass, its Metallurgic Treatment, and the Axial Direction of its constituent Crystals. Institution of Civil Engineers, Minutes of Proceedings*, Vol. XVIII., pp. 296–348 (with discussion). London, 1859. This is an interesting paper dealing principally with the influence of the bulk of a forging on its elastic and cohesive properties. The author on p. 298 states the three principal points of his inquiry as follows:

(i) What difference does the same wrought-iron afford to forces of tension and of compression, when prepared by rolling, or by hammering under the steam-hammer, the bars being in both cases large?

(ii) How much weaker, per unit of section, is the iron of very massive hammer forgings than the original, or integrant iron, of which the mass was made up

(iii) What is the average, or safe measure of strength, per unit of section of the iron composing such very massive forgings, as compared with the acknowledged mean strength of good British bar-iron in moderate market sizes?

Mallet holds the proper measure of strength in a bar of iron to be the "work done, whether by extension, compression, rupture, or crushing, by any force applied to it." Thus his T_e = the elastic resilience of the body $= \frac{1}{2}Es_0^2$, where s_0 is the limiting elastic stretch or squeeze, $= \frac{1}{2}s_0P_0$, where $P_0 = Es_0$.

"The value of the coefficient T_r," he continues, "is arrived at in the same way by substituting the corresponding values for P_0 and s_0, due to the moment of crushing or of rupture" (p. 299). This seems to me to suppose that the proportionality of stress and strain lasts up to rupture, which is indeed far from true for many materials. It would seem better to define T_r as the work done in rupturing a body, without expressing it in terms of the final stress and strain.

I find that Mallet calculates his value of T_r from the assumption that its value is half the product of the final strain into the final stress (see his Appendix, Tables I. and II., together with the remarks, p. 332). Thus it cannot be taken as a basis for some of the conclusions he draws from it. I think his own diagrams, pp. 318–19, should have shown him his error in this respect.

Mallet attributes the introduction of these coefficients T_e and T_r to Poncelet (see our Art. 982* and compare Art. 999* however) and re marks that for all crystalline substances, notably wrought iron (see our Art. 1065), their values will depend on the direction of the stress which produces rupture.

[1129.] He compares the strength of bars cut in different directions from massive forgings, and concludes generally that the latter are weaker than the rolled bars of moderate size of which the heavy forgings were built up. He likewise shows how the molecular arrangement far more than the metallurgical constitution affects the elasticity and strength of different kinds of iron.

Probably his *résumé* of the elasticities and strengths of cast- and wrought-irons, especially in regard to the longitudinal and transverse elasticities and strengths of large forgings would be useful even to-day: see Table V. of the Appendix. It is certainly of great interest to the theoretical elastician to see how far the distribution of elasticity depends on 'working,' and how widely the ordinary materials of construction diverge from isotropy.

[1130.] A. R. von Burg: *Untersuchungen über die Festigkeit von Stahlblechen, welche in dem Eisenwerke des Herrn Franz Mayr in Leoben für Dampfkessel erzeugt werden. Sitzungsberichte der mathematisch-naturwissenschaftlichen Classe d. k. Akademie der Wissenschaften*, Bd. 35, S. 452–74. Wien, 1859.

This memoir busies itself[1] with the safe use of cast-steel as a material then being adopted for boilers. Howell in England had introduced a 'homogeneous patent iron' especially intended for boilers, and the experiments detailed in this paper are on cast-steel plates (*Gussstahlbleche*) prepared by F. Mayr for a similar purpose and tested by the Vienna *Polytechnic Institute* at the request of the *Handelsministerium.* The experiments go to show that the tensile strength of cast steel plates is roundly double that of iron plates, both being of Austrian manufacture, and these experiments are shown to be well in accord with those of Fairbairn, Clark and Gouin et Cie.: see our Arts. 1497*, 902, 1066, and 1108. They give a tensile strength in the direction of the rolling (*Längenrichtung, Richtung des Walzens*)

[1] v. Burg gives an interesting foot-note on S. 454 on the difficulty of determining the exact factors in iron and steel which cause their very different elastic properties. All this difference is not due to the quantity of carbon (varying from ·625 to 1·9 p.c.), but has probably much to do with the state of crystallisation (Dalton attributed the difference almost entirely to the latter). Fuchs supposes iron dimorphic, consisting of a mixture of tesseral and rhombohedral crystals: wrought iron is chiefly tesseral, raw-iron rhombohedral. He attributes the difference between tempered and untempered steel to a transition from one form of crystallisation to the other. By annealing with increasing heat the tesseral replaces the rhombohedral crystallisation.

slightly greater than in the direction transverse to it (*Querrichtung*); this agrees with the results for iron-plates, of Clark, Gouin et Cie and von Burg himself, but not with the rather doubtful conclusions of Fairbairn as to Yorkshire and Staffordshire plates. The stricture was also greater and rupture more gradual in the case of bars cut in the direction of the rolling. Von Burg concludes his paper with some remarks on the elastic limit for steel plates, which are not, however, based on his own experiments. *Ausglühen* for two hours over a charcoal fire did not reduce on cooling the strength of steel more than 2 p.c. (S. 463–6).

[1131.] K. Karmarsch: *Ueber die absolute Festigkeit der Metalldrähte. Polytechnisches Centralblatt*, Jahrgang 1859, Cols. 1272–76. Leipzig, 1859. (Extracted from the *Mittheilungen d. Gewerbe-Vereins f. d. Königreich Hannover*, 1859, S. 137.)

The writer begins by referring to his experiments of 1824 (see our Art. 748*) in which he had shown that the process of drawing alters in a remarkable manner the absolute strength of metal in the form of wires:

Die Ursache der berührten Erscheinung liegt unstreitig in Folgendem: Wenn ein Draht feiner und feiner gezogen wird, vermindert sich seine Festigkeit—d. h. die zum Abreissen desselben erforderliche Zugkraft—nach Verhältniss seiner Querschnittsfläche oder des Quadrats seines Durchmessers. Zugleich aber findet ein Zuwachs an Festigkeit dadurch statt, dass das Metall zunächst an der Oberfläche, vermöge des Drucks in den Ziehlöchern verdichtet, wohl in der Textur vortheilhaft verändert wird. Da diese Wirkung unmittelbar am Umkreise des Querschnitts vor sich geht, so steht ihre Grösse im Verhältniss dieses Umkreises oder, was eben so viel sagen will, des Durchmessers.

Man darf sich daher die Festigkeit F eines Drahtes vom Durchmesser D als aus zwei Theilen zusammengesetzt vorstellen, von welchen der eine von dem Durchmesser, der andere von der zweiten Potenz des Durchmessers abhängig ist; d. h. man kann

$$F = aD^2 + bD$$

setzen, worin a und b aus der Erfahrung abgeleitete Coefficienten sind (Col. 1273).

Karmarsch then determines the constants a and b for a great variety of metal wires, but his method of selecting the results from which a and b are to be determined seems to me very unsatisfactory. He ought to have proceeded by the method of least squares, but he calculates a and b from a number of selected experiments by taking the arithmetical means.

The process of annealing reduces the values of both a and b. The coefficient b can amount in the case of ordinary iron or platinum wire to as much as one half of a and sinks in the case of lead to zero, or to an insensible quantity. The relation of the absolute strengths of annealed and unannealed wires is not the same for the same metal, but varies with the diameter of the wire. Further for wires of the same diameter but of different metals it varies from a maximum with platinum to a minimum with iron.

[1132.] *Admiralty Experiments on the Various Makes of Iron and Steel. Transactions of the Institution of Naval Architects.* Vol. I., pp. 169–70. London, 1860. Also *The Mechanic's Magazine, New Series*, Vol. III., pp. 156–7. London, 1860.

This paper records the results of a number of experiments on cast and puddled steel and on cable iron in the form of bolts and links as used for cables made at Woolwich Dockyard in 1859. With a side weld and $1\frac{1}{8}$ in. chain the best puddled steel bore 39 to 41 tons, while the best iron bore $41\frac{1}{4}$, to $43\frac{3}{8}$ tons. The difficulty of properly welding the steel seems to have told against the strength of steel as compared with iron cables in the experiments on links: see our Art. 1147. The numerical details might still be of service to any one working out a theory of the absolute strength of chain-cables: see our Art. 641.

[1133.] F. Schnirch: *Resultate einiger Versuche über die Festigkeit des Schmiedeisens und einiger Steingattungen. Zeitschrift des österreichischen Ingenieur-Vereins*, Jahrgang XII., S. 2–3. Wien, 1860 This paper contains nothing of permanent value.

[1134.] H. Tresca: *Procès-verbal des expériences faites sur la résistance des tôles en acier fondu pour chaudières. Annales des Mines, Mémoires*, T. XIX. pp. 345–65. Paris, 1861. This is attached to a *Rapport* by a Commission appointed to consider *les conditions spéciales d'épaisseur pour les tôles d'acier fondu employées dans la construction des chaudières à vapeur*, which occupies pp. 311–44 of the same volume. The Commission consisted of the engineers Combes, Lorieux and Couche, and they experimented on a boiler of cast steel plate presented by MM. Pétin et Gaudet to the Exhibition of 1855. The experiments on the material of this boiler and on plates of like material showed that the ductility and absolute strength of a plate were in inverse ratio; they also exhibited the now well recognised phenomenon of stricture. There are details (pp. 324–6) of further experiments on the strength and ductility of various kinds of steel plates and the evidence of various engineers with regard to their practical efficiency. At the request of the Commission further experiments were made by Tresca on bars cut from plates prepared by Pétin et Gaudet. These bars were tested for extension and absolute strength, and in various conditions as regards tempering and annealing. Tresca gives (Plate VI.) stress-strain diagrams, which show the rapid increase of stretch after the elastic limit is passed. For an untempered bar this limit was reached at a tensile stress T_1 of about 2477 kilogs. per sq. centimetre, the stretch-modulus being $E = 19{,}674{,}000{,}000$ kilogs. per sq. metre and the elastic limit a stretch of $s_1 = \cdot 001259$. Corresponding to rupture we have a traction T_2 of about 4873, with a stretch of $s_2 = \cdot 05493$. On the other hand after tempering and annealing we find for another bar with the same units,

$$T_1 = 6528, \quad s_1 = \cdot 00331, \quad E = 19{,}722{,}000{,}000,$$
$$T_2 = 8820, \quad s_2 = \cdot 00473.$$

Thus the elastic limit and the absolute strength are much raised by the process, but the stretch-modulus remains practically constant. Tresca was among the first to notice these facts and also to give well-drawn traction-stretch diagrams showing the life-history of individual material: see our Art. 1084 and Vol. I. p. 889.

Without entering more fully into the details of his individual experiments we may briefly indicate his conclusions:

1° Maximum stretch before rupture:

	Aciers doux	Aciers vifs
Before tempering	about ·04	about ·06
Tempered and Annealed	about ·005	about ·004

2° Stretch-modulus in kilogs. per sq. metre:

	Aciers doux	Aciers vifs
Before tempering	$17{,}273 \times 10^6$	$20{,}764 \times 10^6$
After tempering	$19{,}906 \times 10^6$	$19{,}199 \times 10^6$

Thus the stretch-modulus varies far less than the maximum stretch.

3° Tractions at elastic limit in kilogs. per sq. centimetre:

	Aciers doux	Aciers vifs
Before tempering	2400	2532
After tempering	7440	5056

4° Stretches at elastic limit:

	Aciers doux	Aciers vifs
Before tempering	·001,368	·001,236
After tempering	·003,767	·002,618

5° Elastic resilience (= area of elastic stress-strain diagram):

	Aciers doux	Aciers vifs
Before tempering	1·641 kilogs. per sq. cm.	1·565 kilogs. per sq. cm.
After tempering	about nine times the above.	about four times the above.

6° Absolute strength, means in kilogs. per sq. cm.:

	Aciers doux	Aciers vifs
Before tempering	5748	5318
After tempering	8562	7234

See Tresca's memoir pp. 361–5, and compare with the results of Brix and Wertheim cited in our Arts. 848*–58* and 1292*–1301*.

[1135.] 'Lloyd's' *Experiments upon Iron Plates and Modes of Riveting applicable to the Construction of Ships: Transactions of the Institution of Naval Architects*, Vol. I., pp. 99–104. London,

1860. This paper contains the details of a series of experiments on riveted joints made for 'Lloyd's' under the superintendence of W. T. Mumford, surveyor, in 1857. The experiments were arranged solely with a view to acquiring special information for iron shipbuilders, and seem both from the mode of experimenting and the comparative paucity of experiments, 22 in all, to be of little theoretical or even permanent practical interest. With two exceptions the joints were all butt-joints, the rivets were placed in single and double rows, being spaced at three, four, and four and a half diameters apart. In both lap- and butt joints spacing the back row exactly behind the front row gave better results than spacing midway, and four diameters apart seems to have been the best spacing. Lap-joint, double riveting, four diameters apart, reduced the strength of the plate in the ratio of 69·5 to 100, while butt-joint, double riveting four diameters apart, reduced the strength in the ratio of 76·5 to 100, in both cases the back rows were exactly behind the front-rows.

[1136.] J. Daglish: *On the Strength of Wire-Ropes and Chains. The Engineer*, Vol. XI., pp. 51 and 67. London, 1861. This contains the details of a paper read before the *Northern Institute of Mining Engineers* with the discussion upon it A further paper entitled: *On the cause of the Loss of Strength in Iron Wire when heated* will also be found on p. 67. These papers give some account of experiments on the absolute strength of wire ropes and iron chains. They show the reduction in strength produced by heating, by splicing and by ordinary socket joints in the case of wires. The experiments on chains do not give details of the links ($\frac{3}{4}$ inch wrought iron chains bore 15 to 24 tons). They show, however, the remarkable result that chains after being once tested and having borne a load of 18 to 22 tons may afterwards *break* with a less load of 16 to 20 tons. This does not tend to confirm the contention of Browning: see our Art. 1125. Daglish supposes the considerable weakening effect of heating wire ropes to a red heat as compared with the slight effect of the same treatment on chains to be due to the fact that the former are cold and the latter hot rolled. He does not believe, however, that the increased density due to drawing is the real cause of this difference in strength, for this difference in density he tries to show does not disappear on heating either wire or cold rolled iron to red heat. Such a process changes the density but sometimes increases, sometimes decreases it.

[1137.] David Kirkaldy: *Experiments on the Comparative Tensile Strength of Steel and Wrought-Iron* (made for Messrs

Napier and Sons). These are published in the *Transactions of the Institution of Engineers in Scotland*, Vol. II., 1859. A short *résumé* of Kirkaldy's results was also communicated to the British Association and will be found in the *Transactions of the Twenty-Ninth (Aberdeen) Meeting*, 1859, pp. 242–3. They were reprinted *in extenso* in *The Artizan*, Vol. XVIII., pp. 8–20 (pp. 9–20 are erroneously paged 321–332). London, 1860. The results are given in the above publications without comment and consist almost entirely of tables of numerical data. The experiments are some of the most comprehensive and thorough ever made on steel and wrought-iron.

The object sought in instituting the series of experiments about to be described was to ascertain the comparative strength of various kinds of steel and wrought-iron when subjected to a tensile strain, with the view of substituting homogeneous metal or steel for wrought iron in the construction of machinery, boilers, steam ships, etc.

Upwards of 540 specimens were tested and these were "indiscriminately collected from engineers' or merchants' stores except those marked *samples* which were obtained from the makers[1]." The object of this precaution was to avoid especially prepared test pieces. The experiments themselves made by perfectly reliable and disinterested engineers for their own practical information were among the first to give full details of stricture and fracture, and are of as great theoretical as practical interest. We shall not, however, attempt to analyse them in the above form, but note that all these results as well as others were embodied by Kirkaldy shortly afterwards in a work, the title of which is given in the following article.

[1138.] David Kirkaldy : *Results of an Experimental Inquiry into the Tensile Strength and other Properties of various kinds of Wrought Iron and Steel.* 1st edition, 1862, 2nd edition, 1864, Glasgow. This work, although falling a little outside our present period, to a great extent embraces experiments conducted several years previously and referred to in the preceding article Our references will be to the pages (1–227 and xvi plates) of the

[1] An attempt was made to suppress the publication of the results on the ground that the specimens had not been procured directly, and legal proceedings were threatened. That the attempt failed does not really affect the arguments in favour of an independent Government testing house.

second edition. The treatise may be said to do, and do in a more thorough manner, for wrought iron and steel what the English and American experiments had already done for cast-iron: see our Arts. 1406*, 1037 and 1048.

[1139.] Pp. 9–17 deal with the mode of collecting specimens (see our Art. 1137), with the form of the testing-apparatus (see Plate I., it was a somewhat primitive directly-loaded lever machine), with the preparation of the specimens and the measurement of their extension under stress (by a large pair of compasses with their points inserted in marks made by a centre-punch in the bar) and with the method in which the results are tabulated. The method of experimenting was throughout based on the desire to reach broad technical conclusions, rather than to make delicate physical measurements, and the methods adopted appear occasionally to have been rather rough and ready when judged from the physical standpoint. The experiments were directed to ascertain breaking stress, stricture, nature of rupture, rate of elongation under increasing stress, influence of treatment and of shape, strength of welded joints, effect of gradual and sudden stress on steel and iron in both bar and plate. Tables F–K give a summary of the numerical results, Plates II. to V. give reproductions of the rupture surfaces, and Plates VI. to XIII. represent the results graphically. We proceed in the following articles to give a brief *resumé* of some of the results of Kirkaldy's experiments together with the inferences which may be drawn from them.

[1140.] Section VIII. (§§ 29–70) deals with the tensile strength and stricture of wrought-iron. It opens with an historical account of the experiments on iron of Muschenbroeck, Lamé, Telford, Brunel, Fairbairn, Wade, Lloyd, etc. (see our Arts. 28* (δ), 1001*–4*, 1494*–1503*, 1037, and 1135) and points out the great divergence in the recorded results. This is attributed partly to difference in quality and partly to difference in methods of experimenting and stating experimental results. Kirkaldy remarks of these earlier researches:

In all former experiments the ultimate strength or breaking weight per square inch of the specimen's original area alone is given, and the various pieces are rated accordingly, the one that stands highest being considered the best.

It seems most remarkable that an element of the highest importance should have been so long overlooked, namely, the *Contraction* of the specimen's area [i.e. *Stricture*] when subjected to considerable strain [? stress], and the still greater contraction, at the point of rupture, which takes place in a greater or lesser degree as the material is soft or hard, and the consequent influence this reduction must have on the amount of weight sustained by the specimen before breaking. The apparent mystery of a very inferior description of iron suspending, under a steady load, fully a third more than a very superior kind, vanishes at once when we find that the former had the benefit of retaining to the last its original area only slightly decreased; whilst the latter on breaking was reduced to very nearly a fourth of its original area—the one a hard and brittle iron, liable to snap suddenly under a jerk or blow, the other very soft and tough, impossible to break otherwise than by tearing slowly asunder (pp. 23-4).

Kirkaldy is of course quite right in drawing attention to the importance of taking into account not only the absolute strength per unit of original area, but also the elongation and stricture in measuring the value of a certain class of metal, but the introduction of the words 'superior' and 'inferior' above would appear to suggest some test of the superiority or inferiority of the metal *not relative to the purpose to which it is to be applied.* In most cases the element of ultimate resilience will of course be of considerable importance: see our Arts. 1085 and 1128.

Kirkaldy following up the ideas suggested in the above quotation gives for bars full details of their total elongation, their general reduction of section other than at the section of fracture, their reduction at the section of rupture (or the stricture), and also of their absolute strength calculated to original, reduced and rupture sections. Further, the nature of the rupture is recorded. Similar details are then given for plates.

[1141.] Kirkaldy takes as his test of relative merit the absolute strength conjointly with the stricture, and it becomes important to ascertain what influence different methods of working have on one or both of these properties. He notes the following points:

(i) The size of the bar in rolled iron has far more influence on the absolute strength of 'inferior' iron than of iron of 'superior' quality.

(ii) Removing the skin does not alter the strength, or rough rolled bars are not stronger than turned ones: see our Art. 858*.

(iii) Reducing rolled bars by forging slightly increases the absolute strength, but decreases the stricture.

(iv) The absolute strength and stricture of iron-plates is greater in the direction in which they are rolled than across it (pp. 26-30): see our Arts. 1497*, 902 and 1108.

After dealing with *rolled* iron Kirkaldy turns to *hammered* iron and criticises the loose use of the term *scrap-iron.* He shows among other things, that the absolute strength and stricture are greater in specimens cut lengthwise than in those cut crosswise from crankshafts.

[1142.] Section IX. (pp. 33–5) deals with steel, and Kirkaldy insists on the same conjoint test (Art. 1141) of the value of different parcels. He states that the absolute strength and stricture of *puddled* steel plates are greater in the direction in which they are rolled, as in the case of iron plates, while the converse holds for *cast* steel plates (p. 92).

[1143.] Section X. (pp. 35–60) discusses with a criticism of the results of other writers the appearance of the rupture surfaces in iron. Kirkaldy enters into the history of the controversy as to the change from fibrous to crystalline structure by vibration: see our Arts. 1463*–4*, 881 (*b*) and 992. He cites the opinions of McConnell, Thorneycroft and Stephenson, as well as that of Roebling, the engineer of the Brooklyn bridge, who does not appear to have believed in the change. Kirkaldy himself considers that the nature of the rupture surface depends largely on the mode of rupture, and not on previous vibration. He holds that:

> the appearance of the same bar may be completely changed from wholly fibrous to wholly crystalline, without calling in the assistance of any of those agents already referred to—viz., vibration, percussion, heat, magnetism, etc., and that may be done in three different ways:—1st, by altering the shape of the specimen so as to render it more liable to snap; 2nd, by treatment making it harder; and 3rd, by applying the strain [stress] so suddenly as to render it more liable to snap from having less time to stretch (p. 53).

The *act of breaking* is really the determining cause and Kirkaldy's best demonstration of this was the actual breaking of the same bar with crystalline and fibrous fractures within a few inches of each other (pp. 53–4). Kirkaldy considers however, that any process of working that decreases the stricture of a specimen renders it more liable to *snap* or to take a crystalline fracture[1].

After considering the evidence brought forward by Kirkaldy and others with regard to crystalline and fibrous fractures, I am inclined to think that the difference really lies in the *extent* of the material which is subjected to a stress equal or nearly equal to the rupture stress. When only the material between two very close cross-sections is subjected to such stress then we get a crystalline fracture such as occurs in snapping; when a considerable extent of the material as in pure tensile strain is subjected to this limiting stress then the rupture is fibrous. This view would account for the crystalline fracture occurring in cases of vibration, for in such cases there is generally an 'accumulation of stress' due to stress waves at some particular cross-sections only. Almost the same result arises from the sudden blow of a hammer which also leads to a crystalline fracture.

[1] In this section Kirkaldy refers to the action of dilute hydrochloric acid in removing the impurities from the surface of a specimen and exposing more clearly to view the metallic portion and its texture. A like application to any planed section of a specimen which has been subjected to large stresses producing set will often bring to view the directions of maximum and minimum strain.

Section XI. (pp. 61–2) is devoted to the appearance of the rupture surfaces in steel. These are classified as *granular, granular and crystalline, crystalline and fibrous, granular and fibrous, fibrous.* The distinction between a fibrous and granular fracture is always due according to Kirkaldy to the slowness or suddenness of the act of breaking.

[1144.] Section XII. (pp. 62–9) is entitled: *Rate of Elongation under Increasing Strains* [? Stresses]. Kirkaldy holds that a slow application of load (i.e. one that leaves time to measure the stretches) does not lessen the absolute strength. I think Kirkaldy cannot be instituting a comparison with (p. 63) the sudden application of load, or else its slow application would certainly be remarkable not for lessening but *increasing* the apparent absolute strength: see our Arts. 988*, 970, etc.

The set and the ultimate stretches were measured, and Kirkaldy remarks that most of the specimens extended uniformly along their lengths nearly up to rupture just before which stricture began usually at one, sometimes at two, and in a few exceptional cases at three different places. The lateral dimensions of the specimens formed an important element in determining the value of the ultimate stretches (p. 69), i.e. in modifying the amount of stricture.

[1145.] Section XIII. (pp. 69–74) deals with the *Influence of various Kinds of Treatment.* Here Kirkaldy considers a number of interesting and practically valuable methods of altering the absolute strength and stricture of iron and steel. I remark that:

(i) The strength of steel is reduced by hardening in water, but is greatly increased by hardening in oil. This increase varies from 11·8 to 79 per cent. as we pass from soft steels slightly heated to hard steels highly heated (p. 70). The higher the temperature at which the 'hardening' takes place, the greater the increase provided the steel is not 'burnt.' Kirkaldy argues that the steel was also 'toughened' because when under great stresses it might be "repeatedly struck [? without breaking] with a rivet-hammer" (p. 70). I do not understand exactly what Kirkaldy means by 'toughened' here.

Further, steel plates hardened in oil and riveted are fully equal in strength to unriveted soft plates, or the hardening in oil more than counterbalances the loss of strength by riveting (p. 71).

(ii) In the course of the investigations on riveted steel plates, it is pointed out that the absolute shearing strength of steel rivets is about of the tensile strength. According to the uni-constant theory extended to rupture the former should be $\frac{4}{5}$ of the latter. As a mean from 17 rivets we find that the shearing is to the tensile strength as 63,796 is to 86,450 lbs. per sq. in., $\frac{4}{5}$ of the latter would have been 69,160 lbs. (p. 71). Kirkaldy questions whether the usual rule for iron rivets—that the diameter of the rivet should equal the combined thicknesses of the two plates to be joined—is a correct one.

(iii) 'Case hardened' iron bolts have less absolute strength than whole iron bolts. Iron highly heated and *suddenly* cooled in water has a greater absolute strength than before, but it is more liable to snap, i.e. exhibits less stricture. Iron like steel, when heated and *slowly* cooled loses in absolute strength: see our Arts. 692*, 1301* and 1353*. Cold-rolling increases the absolute strength but diminishes the stricture. Kirkaldy holds (cf. our Arts. 1136 and 1149) that the specific gravity is not raised by cold-rolling. Galvanising or tinning iron plates did not increase the strength of plates of the thickness (·375″ to ·186″) experimented on (pp. 73–4)

[1146.] Section XIV (pp. 74–7) discusses the effect of altering the shape of specimens. Kirkaldy states that we cannot compare the strengths of metals as given by different experimenters as we do not know the shape of their specimens, and instances Wilmot's Woolwich experiments, which are cited in Fairbairn's treatise on *Iron* (see our Art. 911), as diverging essentially from his own for Bessemer Steel (Wilmot's mean value for the absolute strength is 153,677 lbs. per sq. inch., Kirkaldy's, 111,460 lbs.). This divergence Kirkaldy shows to have arisen from the fact that the minimum cross-section in the test-pieces for the Woolwich machine occurred *only at one point.* He cites experiments to prove that grooving increases the absolute strength and decreases the stricture (see our Art. 1503*). This seems to me probable as it is very unlikely that the groove would be formed at the weakest cross-section, and the maximum stresses being confined to the neighbourhood of the groove there will arise according to the view expressed above (see our Art. 1143) a crystalline fracture, or at any rate one with less stricture.

[1147.] Section XV. deals with the comparative strength of screwed and chased bolts (pp. 77–80), and Section XVI. with the strength of welded joints (pp. 80–2). In the first case the strength of screwed bolts is found to be nearly proportional to their areas, with a slight difference in favour of the smaller area. The strength of the bolt is greater for a screw made with old than for one made with new dies, a result attributed by Kirkaldy to the hardening effect of an old and blunt die (p. 78). The loss in strength due to screwing is given in Table Q (pp. 174–9), or varies from about 7·5 p.c. for Govan bolts with old dies (about 17·8 p.c. with new dies) to about 23 p.c. for 'Glasgow B. Best' with old dies (about 33 p.c. with new dies). The results for the welding of iron are very inconclusive. In some cases the welded joint bore nearly as much as the uncut bar, and in other cases the strength was reduced fully one third (p. 80). Heating to the welding point and then cooling slowly without hammering was found to reduce the stricture very largely, but not the absolute strength. The welding of steel bars owing to their liability of being burnt is difficult and uncertain: see our Art. 1132.

[1148.] Section XVII. is concerned with *Suddenly Applied Strains* (pp. 82–6). Kirkaldy arrives at the conclusion that the breaking stress is considerably less when the load is suddenly applied, "though some have imagined that the reverse is the case" (p. 95). The decrease is, however, only about 18·5 p.c. instead of 50 p.c., so I imagine Kirkaldy's trigger apparatus did not really apply the load instantaneously. Like other experimentalists he does not distinguish in the case of sudden loading between the real breaking stress and the apparent load per unit area of cross-section.

He notes that the stricture is in the case of sudden application of load much reduced (p. 84). Turning to the effect of frost we find that the absolute strength is reduced when iron is frozen. *Provided the stress be suddenly applied*, there is a reduction of about 3·6 p.c. When the stress is gradually applied there is little difference. Kirkaldy attributes this to the warming of the iron by the drawing out of the specimen (p. 86). The experiments were not, however, nearly sufficient in number to be very conclusive. The difference between sudden and gradual loadings may, perhaps, explain the divergence between the conclusions reached by Joule and Kirkaldy: see our Art. 697 (*c*).

[1149.] Section XVIII. deals with the specific gravities of iron and steel (pp. 87–91). Kirkaldy found that the specific gravity of iron indicates generally its 'quality,' that it is *decreased* by wire drawing, cold rolling, and for some kinds by hot rolling in the ordinary way: see our Arts. 732 and 1136. It is also decreased by being drawn out by a severe tensile stress. In the case of steel, 'highly converted' steel has not the greatest density. Cast steel is denser than puddled steel, which is even less dense than some of the superior descriptions of wrought-iron (pp. 91 and 95).

[1150.] Section XIX. (pp. 91–100) gives a summary of the conclusions contained in the volume, and some general remarks on their practical application. Kirkaldy asserts that the truest measure of the quality of iron or steel is the breaking stress per unit area of the fractured surface (i.e. of the stricture) and appears to lay the greatest importance on this mode of comparison.

On pp. 106–187 we have the tables of numerical results which it is impossible to condense or analyse here; their importance has been long recognised by technical elasticians. For the absolute strength of wrought-iron we may, however, reproduce the mean results as given on p. 96 since they may be of service for later reference in our own work:

Strength in lbs. per square inch of original area.

	Highest	Lowest	Mean	
188 Bars, rolled	68,848	44,584	57,555	$=25\frac{3}{4}$ tons
72 Angle iron, etc.	63,715	37,909	54,729	$=24\frac{1}{2}$,,
167 Plates (length-ways)	62,544	37,474	50,737	} $=21\frac{3}{4}$,,
160 Plates (cross-ways)	60,756	32,450	46,171	

[1151.] In the *Appendix* will be found a series of extracts from articles and letters in *The Engineer* and other papers, which are not without interest as showing the general state of knowledge with regard to the elasticity of iron and steel about 1860. The volume concludes with plates of Kirkaldy's simple apparatus, and with more or less suggestive diagrams of the surfaces of rupture, showing the reduction in transverse diameter not only at the stricture, but also at other cross-sections of the specimens. There are graphical representations of some of the numerical results, and on Plate XIV. are given the distorted forms (approximately elliptical) after strain of circles drawn on the unstrained faces of a bar. These were obtained with a view of showing the relative longitudinal stretch and lateral squeeze. I find just about the strictured portion of a bar of cast steel from measurement of the semi-diameter that the circle of 1″ diameter has been converted into an oval of 1·125″ longitudinal and of 889″ lateral diameter. Hence the longitudinal stretch is ·125 and the lateral squeeze ·111, or the stretch-squeeze ratio η for the *set* of this bar equals ·89, which is far from agreeing with the ·25 which the uni-constant theory gives for η in the case of *elastic* strain.

GROUP G.

Strength of Materials, other than Iron and Steel.

[1152.] L. G. Perreaux: *Apparatus for testing and ascertaining the strength of yarn, thread, wire strings, or fabrics. London Journal of Arts (Conjoined Series).* Vol. 43, pp. 325–8. London, 1853. This contains a description of a patent for a testing machine. In order to prevent too great a shock upon the rupture of the material tested, one of the clamps holding the material sets in motion a fly-wheel on the release of the load and thus the shock is deadened.

[1153.] Houbotte. A testing machine invented by this engineer will be found described on p. 432 of the *Annales des travaux publics de Belgique*, T. XIII., 1854–5, or *Polytechnisches Centralblatt*, Jahrgang 1855, Cols. 1237–40. Leipzig, 1855. The machine was designed to ascertain the crushing strength of stone. Its peculiar novelty seems to be the gradual application of load by filling slowly a reservoir of water supported by the loading lever of the machine. Houbotte gives the details of various experiments on the crushing of stone blocks. He further made some few not very conclusive experiments on the increase of strength due to lateral support.

[1154.] T. Dunn. *On chain Cable and Timber Testing Machines. Institution of Civil Engineers. Minutes of Proceedings*, Vol. XVI., pp. 301 308. London, 1857. This gives an account of a testing machine made by Messrs Dunn and of some experiments made with it on bars and chains. This apparently was the first introduction into general use of fairly cheap testing machines. We refer in our Art. 1158 to the use made of one of these machines by Captain Fowke in the Paris Exhibition of 1855.

[1155.] W. Fairbairn: *On the Density of various Bodies when subjected to enormous compressing Forces. British Association*, Report of Liverpool Meeting, 1854, *Transactions*, p. 56. The author states that he has applied pressures of 90,000 lbs. per sq. inch to various substances. "Under this enormous pressure, clay and some other substances had acquired all the density, consistency and hardness of some of our hardest and densest rocks."

[1156.] W. Fairbairn: *Solidification of Bodies under Pressure: The Civil Engineer and Architect's Journal*, Vol. XVII., p. 394. London, 1854. The author gives further particulars of the experiments referred to in our Art. 1155. The tensile and compressive strengths of spermaceti and tin were found to be much increased by solidification under pressure.

Thus a bar of the former substance solidified under a pressure of 40,793 lbs. per sq. inch carried 7·52 lbs. per sq. inch more compressive stress than when solidified under a pressure of 6421 lbs. The tensile strength was again as 1 to 0·876 in favour of the more compressed bar.

Further three bars of tin were allowed to solidify, the first at the pressure of the atmosphere, the second at 908 lbs., and the third at 5698 lbs. per sq. inch. Between the two last there was an increase of tensile strength in the ratio of ·706 to 1, or an increase of about $\frac{1}{3}$ when solidified under six times the pressure.

Since the specific gravity of the metals increases at a less rate than the strength Fairbairn hopes from compression to insure not only greater strength but greater economy.

[1157.] Marcq: *Expériences faites sur différentes pièces de bois, à l'effet d'en déterminer le coefficient d'élasticité. Annales des travaux publics de Belgique*, Tome XIV., pp. 279–301. Bruxelles, 1855–6. This paper gives details of experiments on the flexure of various kinds of wood, such as may be bought in the market and not specially prepared for the purpose of experiment in small blocks as in the researches of Wertheim and Chevandier (see our Art. 1312*). Nothing is said about the state of moisture of the wood, or the position of rings and fibres relative to the plane of flexure; presumably the latter were always parallel to that plane, as the pieces were long. The author believed that he had found a real limit of elasticity up to which the flexures were proportional to the loads. This limit of elasticity, as measured by the load, bore to the rupture load the ratio ·43 for oak to ·33 for beech. After the limit of elasticity was passed the flexures increased more rapidly than the loads till the rupture load was approached, when this law was no longer true. For beams of large cross-section the stretch modulus was less than for pieces of small section, where the fibres are more continuous (p. 281). A load if removed and then after a short interval re-applied produced a greater flexure than on the first application. This seems a rather doubtful statement especially as it is not stated whether the load was above or below the elastic limit, when originally applied.

Either a stretch of ·0006 or a traction of 600,000 kilogs. per sq. metre may be taken as a safe limit of loading for all kinds of wood, even the poorest.

I do not cite here the values of the stretch moduli for the various kinds of wood (pp. 298–9) as the stretch-modulus of wood varies from tree to tree and with the state of dryness of the wood.

[1158.] Francis Fowke: *Results of a series of Experiments on the Strength and Resistance of Various Woods. Reports on the Paris Universal Exhibition, Presented to both Houses of Parliament by Command of Her Majesty:* Part I., pp. 402–525. London, 1856. This report contains the details of a long series of experiments on the specimens of various woods from Australia, British Guiana and Jamaica exhibited at the exhibition. The experiments were made with the aid of a hydraulic testing machine made by Dunn of Manchester: see our Art. 1154. The experiments were directed to ascertaining the following data: (i) the specific gravity of wood, (ii) the rupture strength under flexure, (iii) the crushing load in the direction of the fibre, (iv) the crushing load transverse to the direction of the fibre. The deflections for various loads are given, but as there is no reference to set, it is not certain that they give the true values of the stretch-moduli. In many cases the deflections are not proportional to the loads. Tables giving the final results for upwards of 80 specimens of wood will be found on pp. 514–25, and these might even now be useful for commercial purposes.

[1159.] Captain Fowke made further experiments on a much greater variety of woods exhibited at the International Exhibition of 1862. His results were published in 1867 by the Science and Art Department in a work entitled: *Tables of the Results of a Series of Experiments on the Strength of British, Colonial and other Woods.* The Report of 1855 was reprinted at the conclusion of this work.

Upwards of 3000 specimens were tested with a hydraulic machine due to Messrs Hayward, Tyler and Co. Each specimen was as nearly as possible 16 inches long and of square cross-section, 2″ side. The bearings for flexure seem to have been 12″ apart. The same series of experiments were made as at Paris, and an additional series was undertaken to ascertain the elasticity. The woods came from a wider range of colonies and from some European countries.

Table VIII. gives the deflection, divided into set and elastic strain, at every 1,120 lbs. (pp. 213–242) The stretch-moduli are not actually reckoned out. The elastic strains were not proportional to the loads, and it seems probable that elastic after-strain was not recognised and allowed for.

The work contains the most extensive series of experiments on wood hitherto made, and may for many purposes still be useful, but the experiments were conducted in a manner rather calculated to further commercial purposes than to put to the test any theories of the distributions of elasticity in wood: see out Arts. 1229*, and 308–15.

[1160.] *H. R. Storer: On Gutta Percha Tubes. Silliman's American Journal of Science and Arts. Second Series*, Vol. 21, pp. 445–6. New Haven, 1856. (Extracted from the *Proceedings Boston Society Nat. Hist.*, Vol. v., p. 268.) This paper gives details of the bursting strength of gutta percha tubes under water pressure. The tubes varied in diameter from 1″ internal, $1\frac{3}{16}$″ external, to $\frac{1}{4}$″ internal, $\frac{5}{8}$″ external diameter, and the bursting pressures varied from 266 lbs. to 760 lbs. per square inch. The smaller tubes had the greater strength.

[1161.] C. F. Dietzel: *Ueber die Elasticität des vulkanisirten Kautschuks und Bemerkungen über die Elasticität fester Körper überhaupt. Polytechnisches Centralblatt*, Jahrgang 1857, Cols. 689–94. Leipzig, 1857. This paper commences by general remarks on the influence of temperature, elastic after-strain etc. on elastic phenomena. It then criticises Boileau's experiments (see our Art. 851) on the ground that they left out of account the influence of after-strain. Dietzel gives an account of two series of experiments of his own on a vulcanised caoutchouc thread, in which he carefully distinguished fore-strain, after-strain and set. The loads were gradually increased from 1 to 29 grammes, and 24 hours were allowed for the action of the elastic after-strain. He found roughly speaking that the elastic after-strains were proportional to the loads, but that the elastic fore-strains were far from being so, increasing in a much more rapid ratio than the loads This was not due to the decrease in cross-section due to the increasing sets. The elastic after-strain developed in 24 hours decreased from about $\frac{1}{4}$ of the fore-strain down to about $\frac{1}{12}$ as the loads increased from 1 to 29 grammes. Dietzel remarks that Gerstner's Law (see our Art. 806*) does not appear to hold for vulcanised caoutchouc.

[1162.] A. von Burg: *Versuche über die Festigkeit des Aluminiums und der Aluminiumbronze* (*Legirung von* 90 *proc. Kupfer und* 10 *proc. Aluminium*) *Polytechnisches Centralblatt*, Jahrgang 1859, cols. 619–20. Leipzig, 1859. (Extracted from *Mittheilungen d. nieder-österr. Gewerbe-Vereins*, 1858, S. 530.)

Cast bars of aluminium gave an absolute strength of 10·96 kilogrammes per square-millimetre. Cold-hammered bars gave an absolute strength of 20·26, or reduced to the section of stricture, 28·67 kilogrammes per square-millimetre. The aluminium bronze had an absolute strength of 64·59 in one specimen and 49·62 in a second, the first being hot-hammered and the second only cast. Thus the absolute strength lies between those of iron and steel[1]

[1163.] C. Fabian: *Ueber die Dehnbarkeit des Aluminiums. Dinglers Polytechnisches Journal*, Bd. 154, S. 437–8. Stuttgart, 1860. Demonstration of the extensibility of aluminium by beating it into extremely fine leaves. See also the *Répertoire de chimie appliquée*, 1859, p. 435, where the discovery is attributed to the Parisian goldsmith Degousse, who had beaten aluminium to leaves as thin as those of gold or silver.

[1164.] Morin and Tresca: *Détermination du coefficient d'élasticité de l'aluminium. Annales des mines*, T. XVIII., pp. 63–6. Paris, 1860. This is an extract from the *Annales du Conservatoire des arts et métiers*, No. 2, presumably of the same year.

The authors after referring to the experiments of von Burg on the absolute strength of aluminium and aluminium bronze consisting of 90 p.c. copper and 10 p.c. aluminium (see our Art. 1162) remark that his results are not sufficient for the purposes of construction in the former material. They have accordingly determined its stretch-modulus. They find from flexure experiments that the stretch-modulus may be taken as equal to 6,757,000,000 kilogs. per sq. mm. and that the elastic limit was reached at about 8·16 kilogs. per sq. mm. For good iron we have $E = 20{,}000{,}000{,}000$ and the traction at the elastic limit 20 kilogs., while the density is about 7·7 as compared with the 2·5 of aluminium. Thus the comparative serviceability of the two metals is indicated.

[1165.] William Fairbairn: *Experiments to determine the Properties of some mixtures of Cast-Iron and Nickel. Memoirs of the Literary and Philosophical Society of Manchester*[2]. Vol. XV pp. 104–112. Manchester, 1860. This memoir read March 2, 1858, gives the results of experiments on the resistance to flexure of a mixture of cast-iron and 2·5 p.c. of nickel. This investigation was undertaken owing to

[1] Wertheim found for the absolute strength of steel wire 96 to 100 kilogs. per sq. mm., and for iron wire 62 to 55; thus the value for aluminium bronze, 64·59, is a little less than that of very strong iron.

[2] See also *Repertory of Patent Inventions*, London, Vol. 32, p. 156.

the fact that this percentage of nickel had been found in meteoric iron, which is "above all other, the most ductile." The ingots prepared, however, for these experiments were found to be widely different; for their "power to resist impact" was nearly one half less than those composed of pure iron. In this memoir the "power to resist impact" is, as in that discussed in our Arts. 1098–9, measured by the product of ultimate deflection and rupture load. The general result of two series of experiments is that the admixture of nickel reduces the strength of the cast-iron: see our Art. 1189.

[1166.] W. M. Ellis: *Results of Experiments on the Tensile Strength of Copper, Iron, Gun Metal, Yellow Metal and Bolts. The Artizan*, Vol. XVIII., p. 124. London, 1860. This is merely a table of the numerical results of experiments made by Ellis for the United States Government. The exact nature of the metals is not stated. As mean results for tensile strength we find:

Copper..36,000 lbs. per sq. inch,
Iron..52,250 ,, ,,
Gun metal (9 copper, 1 tin) ....17,400 ,, ,,
Yellow metal (19 copper, 6 spelter)...48,700 ,, ,,

[1167.] *Einige Bemerkungen zur Tragfähigkeit hölzerner Balken. Zeitschrift für Bauhandwerker*, Jahrgang 1860, S. 161–5. This paper gives some details of the best methods of cutting beams out of the tree, having regard to the variation of strength with the direction of the axis of the beam. It considers further the most advantageous forms of simple wooden trusses, etc.

[1168.] Vicat: *Mémoire sur l'emploi des ciments éventés comparés aux ciments vifs suivi de quelques observations sur les ciments brûlés ou cuits jusqu'à ramollissement. Annales des ponts et chaussées, Mémoires* 1851, 1er *semestre*, pp. 236–254. Paris, 1851. This paper gives some interesting practical details of the cohesion, absolute strength, etc., of various kinds of cements before and after immersion in water for various periods of time.

[1169.] J. M. Rendel: *Experiments on the relative Resistance to 'compression' of Portland and Roman Cement, etc. Institution of Civil Engineers. Minutes of Proceedings*, Vol. XI., pp. 497–502. London, 1851–2. This contains further experiments on the 'adhesive', 'cohesive', and 'cross-strain' strengths of cement. The paper is printed as an appendix to one by G. F. White on the subject of Portland cements. It deals solely with the strength of these cements under various kinds of stress, and has only practical value.

[1170.] J. Manger: *Untersuchungen über die Festigkeit von reinen und gemischten Cementen. Erbkams Zeitschrift für Bauwesen.* Jahrgang IX., S. 523–34. Berlin, 1859.

This paper gives details of the strengths of Medina and Portland cement. Specimens in the shape of small bars only 4″ long between the supports were tested by flexure. I doubt the possibility of calculating by use of the Bernoulli Eulerian formula the absolute strength from flexure experiments in which the length of the bar was not even four times the diameter, even if we suppose stress proportional to strain up to rupture. For the rest Manger's numerical results have no permanent interest. He concludes his memoir with a number of results as to the time various kinds of cement take to become hard.

[1171.] Tacke: *Versuche über die Festigkeit thönerner Röhren gegen inneren Wasserdruck. Hannöverische Bauzeitung* (*Architecten u. Ingenieur-Verein*) S. 308. Hannover, 1854.

This is an interesting experimental paper on the internal pressure at which earthenware pipes burst. The pipes were tested by water pressure, either 'dry', that is without previous soaking, or 'wet', that is after soaking for four days in warm water; they were from 2 to 3 feet long, 2 to 9 inches diameter and $1\frac{1}{8}$ down to $\frac{1}{2}$ inch thickness. In the case of some materials the strength of the pipes was enormously reduced by the process of soaking, in others it did not appear to have much influence. The following are some of the results:

	Substance of Pipe.	Length in feet.	Diameter in inches.	Thickness in inch.	Condition.	Bursting Pressure in Cologne pounds* per sq. inch.
1	Ordinary glazed earthenware	2	4	$\frac{5}{8}$	dry	94
2	,, ,,	2	6	$\frac{5}{8}$	,,	46
3	,, ,,	2	4	$\frac{11}{16}$	,,	110
4	'Porphyrmasse' burnt without glaze	3†	2	$\frac{7}{16}$	,,	142
5	,, ,,	3	3	$\frac{1}{2}$	,,	230
6	,, ,, (had two supporting rings)	3	4	$\frac{1}{2}$	,,	122
7	Same as (5)	3	3	$\frac{1}{2}$	wet	242
8	'Chamottthon (rather porous and yellow glaze)		6	$\frac{3}{4}$	,,	48
9	,, ,,		4	$\frac{11}{16}$	dry	166
10	,, ,,		3	$\frac{5}{8}$	,,	90
11	Same material, black glaze		9	$1\frac{1}{16}$	wet	42 (This had stood 118, dry)
12	,, ,,		6	$\frac{7}{8}$	dry	162
13	,, ,,		4	$\frac{3}{4}$	,,	102
14	,, ,,		6	$\frac{7}{8}$	wet	10
15	,, ,,		4	$\frac{3}{4}$	,,	34

* A Cologne lb. = 500 grammes. † Saxon feet, the others appear to be in English feet and inches.

These results might be useful in testing how far the theory of our Arts. 1012*–1013*, footnote, may be applied to rupture. The writer concludes that pipes of more than 6″ diameter ought to be made of cast-iron.

[1172.] Kraft: *Ueber die nôthwendige Stärke thönerner Wasserleitungsröhren. Polytechnisches Centralblatt*, Jahrgang 1859, Cols. 1445–6. Leipzig, 1859. (Extracted from the *Gewerbeblatt aus Württemberg*, 1859, Nr. 30.) This paper gives an empirical formula for the thickness of the sides of pipes, which is said to be based on experiments made in Ravensburg on pipes for the water supply. The formula, which is accompanied by a numerical table, is the following:

$$d = \tfrac{1}{2} w (a + 1),$$

where d is the thickness of the side of the pipe in *lines* (*Linien*), w is the internal diameter (*Lichtweite*) in inches (*Zoll*) and a is the internal water pressure in inches.

[1173.] *Institution of Civil Engineers, Minutes of Proceedings*, Vol. XIX., p. 276. London, 1859–60. Some details of experiments on the power of bricks to resist a crushing force will be found in an Appendix to a paper on the Netherton Tunnel.

[1174.] *Lateral Strength of Stone. The Civil Engineer and Architect's Journal*, Vol. XIII., pp. 269–270. London, 1850. Some account of experiments made in 1848 for Chester Railway Station on the flexural strength and ultimate deflection of slate and stone are here recorded. Only unreduced numerical results are given.

[1175.] W. R. Johnson: *Comparison of Experiments on American and Foreign Building Stones to determine their relative Strength and Durability. Silliman's American Journal of Science and Arts. Second Series*, Vol. XI., pp. 1–17. New Haven, 1851.

This memoir contains a general *résumé* of European investigations on the crushing strength of various kinds of stone together with accounts of experiments by C. G. Page, Dougherty and R. Mills on American stones. The American experiments were made on 2″ cubes, and the absolute crushing strengths as well as those relative to alum sandstone taken as 100 are recorded. Good tables in English measure are given of the results of Rennie (see our Arts. 185*–6*), Daniel and Wheatstone, W. Wyatt in England; Rondelet (see our Art. 696*), Gauthey, Soufflot and Perronet (see our Art. 28* (ζ)) in France. See especially pp. 14–15 of the memoir. Noting the discordance of the results obtained, Johnson concludes that the resistance to crushing must be some function of the number of units in the base of the column crushed, increasing with that number. He suggests the following law:

That the crushing strength of a cube varies as the product of the area of the base into the cube root of that area.

He tests this law by some ten cases, which he remarks do not conclusively prove it, but afford "a pretty strong presumption in its favor" (p. 17).

[1176.] Mohn: *Versuche über das Gleichgewicht und die rückwirkende Festigkeit von naturlichen Bausteinen. Notizblatt des Architecten und Ingenieur-Vereins*, Bd. I., S. 360. Hannover, 1853.

This paper contains details of experiments on the crushing strength of stone cubes, the sides of which were four Hannoverian inches long. In order to test the stones in a frozen condition some were placed in warm water and after becoming saturated submitted to a frost of 10° R. for several days. In some cases the strength appears to have been somewhat decreased by this freezing process, but no general law is obvious. The numerical results are somewhat irregular and have little more than local and temporal interest.

[1177.] Hodgkinson: *On the Elasticity of Stone and Crystalline Bodies. British Association*, Report of Hull Meeting 1853, *Transactions*, pp. 36–37. Hodgkinson refers again to the "defect of elasticity" in stone and cast-iron. It is not quite obvious what he means by "defect," but it is I imagine 'set' and not perfect elasticity with "defect of Hooke's Law." See our Arts. 969*, 1411* and Vol. I. p. 891 He refers to Lamé's "profound work" and remarks that its results do not apply to the bodies of which he is speaking—some of which are of primary technical importance.

[1178.] *Strength and Density of Building Stone. The Edinburgh New Philosophical Journal*, Vol. LVII., p. 371. Edinburgh, 1854. A few numerical details of the crushing strengths of sandstones, marbles and granites, extracted from a report on experiments made at Washington, U.S., are here published.

[1179.] Michelot: *Recherches statistiques sur les matériaux de construction employés dans le département de la Seine. Annales des ponts et chaussées, Mémoires* 1855, 2e *semestre*, pp. 189–212. Paris, 1855. This is only a report by Belgrand on a long memoir by Michelot, which as far as I am aware was never published. It refers on p. 209 to some experiments on the crushing of stone.

[1180.] Henry: *On the Mode of testing Building Materials and an account of the Marble used in the Extension of the United States Capitol. Silliman's American Journal of Science and Arts*, Vol. 22, pp. 30–38. New Haven, 1856. (Extracted from the *Proceedings of the American Association for the Advancement of Science*, August, 1855, *Providence Meeting.*) This paper describes the apparatus used by an American Commission to test the marble used in extending the Capitol. There is little that calls for general note here, except perhaps the statement on p. 33 that in crushing cubes the manner in which the ends are bedded is a fundamental factor in the apparent crushing strength. If the bases of the cube are free to expand, as is approximately the case if they be bedded on thin plates of lead, the crushing strength appears far less than if they be bedded on steel.

For example, one of the cubes, precisely similar to another which withstood a pressure of upwards of 60,000 lbs. when placed in immediate contact with the steel bed-plates, gave way at about 30,000 lbs. with lead interposed. This remarkable fact was verified in a series of experiments, embracing samples of nearly all the marbles under trial, and in no case did a single exception occur to vary the result.

Some remarks on cohesion and molecular attraction with which the memoir closes do not seem very lucid (pp. 36–38).

[1181.] A. Brix: *Zerdrückungs-Versuche zur Ermittelung der rückwirkenden Festigkeit verschiedener Bausteine. Verhandlungen des Vereins zur Beförderung des Gewerbfleisses in Preussen*, 1855, *Lief.* 2. *Dinglers Polytechnisches Journal*, Bd. 137, pp. 393–4. Stuttgart, 1855. This paper gives details of the crushing strength of various kinds of German stone. The loads at cracking and at crushing are given in each case. Details of earlier experiments by Brix will be found in the same *Verhandlungen* 1853, S. 1, 137, 203, and in the *Polytechnisches Centralblatt* 1853, Cols. 1308–9.

[1182.] W. Fairbairn: *On the Comparative Value of various kinds of Stone, as exhibited by their Powers of Resisting Compression. Memoirs of the Manchester Literary and Philosophical Society*, Vol. 14, pp. 31–47. Manchester, 1857. This memoir was read April 1, 1856.

It contains numerical values for the crushing loads of various kinds of granite, limestone and sandstone, and a comparison of these results with those of Rennie for stone, Hodgkinson for stone and wood, Latimer Clarke for brickwork, and Fairbairn himself for cast-iron see our Arts. 185*, 1445* and 953*.

There are three plates of rupture-surfaces. While the sandstones ruptured in wedges, the limestones formed longitudinal cracks or splinters. The strength of stone was about as "10 to 8 in favour of the stone being crushed upon its bed to the same when crushed in the line of cleavage." This applied to both sandstone and limestone (p. 39).

[1183.] Knight: *Strength of Building Stone. The Builder*, Vol. XVIII., p. 579. London, 1860. Details are given in this paper of the crushing strengths of various colonial building stones; they are taken from a treatise by Knight, presumably published in Victoria. Some account of experiments on the transverse strength of stone are also given.

[1184.] G. Cavalli: *Memoria sul delineamento equilibrato degli Archi in muratura e in armatura. Memorie dell' Accademia di Torino, Serie Seconda*, Tomo XIX., pp. 143–200. Turin, 1861. This memoir was read in 1858. It contains on pp. 183–7 values of the crushing strengths of a very great variety of Italian stones.

GROUP H.

Miscellaneous Minor Memoirs on topics related to the Strength of Materials.

[1185.] Bolley: *Ueber das Krystallinisch- und Sprödewerden des Schmiedeisens durch fortgesetzte Erschütterungen. Dinglers Polytechnisches Journal*, Bd. 120, S. 75–7. Stuttgart, 1851. Extracted from *Schweizerisches Gewerbeblatt* 1850, No. 5. This contains evidence in favour of the change of wrought-iron from the fibrous to the crystalline (*körnig*) condition by repeated impacts.

[1186.] P. W. Brix: *Ausdehnung des Gusseisens bei wiederholtem Erhitzen. Mittheilungen des Gewerbe-Vereins für das Königreich Hannover, Neue Folge*, Jahrgang 1853, Cols. 214–5 Hannover, 1853.

This short extract from a work on fuel by Brix contains some interesting statements with regard to the set produced in cast-iron bars by heating them. The fact that cast-iron after heating does not return to its old volume was first noted by Prinsep in the *Edinburgh Journal of Science*, Vol. x., pp. 356–7, 1829. Brix found that by continually heating a cast-iron bar there was after each heating more set, but in decreasing increments. The thermal set appears in this to resemble after-strain. Set produced by heating in a moderate fire 17 and more days gave an extension of 2 to 3 p.c. This fact deserves further investigation as its physical and practical consequences seem of much interest.

[1187.] L. Dufour: *Ténacité des fils métalliques qui ont été parcourus par des courants voltaïques. Bibliothèque universelle de Genève Archives des sciences physiques et naturelles*, T. 27, pp. 156–8. Genève, 1854.

Wertheim had noted the change in the stretch-modulus produced by sending an electric current through a loaded wire see our Art. 1306*. Dufour proposes to investigate the changes in absolute strength, if any, produced by passing a current for a long time through a wire. His results are not very conclusive. He finds a loss of strength in copper and a slight gain of strength in iron wire, the currents having run as long as 19 days. A much greater number of experiments would have to be undertaken to reach results of real value.

[1188.] Florimond: *Note sur les aimants de fer de fonte trempée et sur la fragilité des fils de laiton exposés à l'air sous l'influence de certaines variations de température. Bulletin de l'Académie Royale de Belgique*, 2me Série, T. VII., pp. 368–71. Bruxelles, 1859.

This paper merely puts on record that in 1848 "*après quelques jours de gelée suivi d'un brouillard*" the brass wires which bind the telegraph wires ruptured and the pieces falling to the earth broke into small bits of excessive fragility. In 1858 a similar phenomenon occurred with the brass ropes which worked the bells at the church of St Pierre in Louvain. Attempts to reproduce the phenomenon artificially failed. Florimond inquires what may be the peculiar crystallisation or disaggregation produced by these atmospherical changes in brass.

[1189.] Bri-Brachion (? Sir W. Armstrong): *The Cause and Prevention of the Deterioration of Wrought-Iron. The Chemical News*, Vol. II., pp. 183–4. London, 1860.

The author cites the fact that iron crystallises in cubes or octahedrons, and states his belief that such crystallisation takes place without melting and slow cooling, namely by the influence of frequently repeated vibrations. He quotes two French chemical writers to this effect (Pelouze and Frémy) and refers to the stock examples of railway axles and steam boilers. He then remarks that any impurity tends to hinder crystallisation. Hence he considers *pure* iron should not be used for structures subjected to frequent vibrations. To test whether iron is pure or not, he suggests magnetisation, pure iron losing immediately its magnetisation, but impure iron retaining it. He has himself tried as 'impurities' carbon, manganese, cobalt, zinc, chromium, tin and nickel, but his experiments lead him to believe that nickel is the most efficient, as it is not removed in the puddling furnace. As an example of the unsatisfactory nature of pure iron, he cites an experiment with a pure iron bar which was successfully tested with 80 lbs. before being submitted to vibration, but after the vibratory experiment it broke with a 'highly crystalline fracture' in three pieces on simply falling to the ground. Compare our Art. 1165.

[1190.] W. Lüders: *Ueber die Aeusserung der Elasticität an stahlartigen Eisenstäben und Stahlstäben und über eine beim Biegen solcher Stäbe beobachtete Molecularbewegung: Dinglers Polytechnisches Journal*, Bd. 155, S. 18. Stuttgart, 1860. *Polytechnisches Centralblatt*, Jahrgang 1860, Cols. 950–4. Leipzig, 1860. Lüders had noted that on Mägdesprunger bar-iron and on various soft kinds of cast-steel the surface after flexure is covered by a network of orthogonal systems of curves. These curves by means of a weak solution of nitric acid could be etched, and this process even repeated after several filings of the surface. At the same time these lines only exhibit themselves at places where the material has received great strain. Similar *Lüders' Curves* were shown to the Editor on I-bars of wrought iron a few years ago[1], and apparently corresponded very nearly with the lines of strain in beams under flexure as figured in the text-books. It would appear then that if a bar be bent beyond the elastic limit mechanical changes take place along the lines of strain and exhibit themselves in a system of orthogonal curves on the scale at the surface of the beam, or even further in, if the strain has been great, and acid be applied. Lüders attributes these curves to a *Molecularbewegung*, but does not associate them with the lines of strain. He had in *one* specimen found a third system of curves diagonal to the rectangular elements of the other two. He had observed these curves of strain after flexure in bars of pure tin (*tesserale Form*).

[1191.] *Summary.* The decade with which we have been dealing in this chapter is one of the most fruitful in the history of elastic theory and practice. Besides the large number of memoirs which have been dealt with in the last five hundred pages, it must be remembered that several of the most important publications of Lamé, of Saint Venant and of the older German elasticians, considered in previous chapters or in the following chapter of our *History*, really date from this period. Nor is the advance confined to any one branch of our subject. There is to be noted the beginnings of a real union between theory and technical practice in France and Germany, which has continued to bear fruit even to the present day, when its full value is also being realised in England by the establishment of numerous technical schools in which instruction in the strength of materials is given and research is scientifically carried on. In the department of physical elasticity we have to note that while great progress was made in the collection of facts, there was still too wide a divorce between theory and experiment. This is very obvious in the elaborate physical researches of Kupffer and Wertheim. Yet while these and other investigators to some extent failed to conduct their experimental inquiries in the

[1] Still more recently Mr J. B. Hunter, M.I.C.E., has sent me some splendid photographs and specimens of Lüders' curves produced by rust round holes punched in the steel plates of dredger buckets: see frontispiece to Part II. I look forward to these curves being used as a powerful mode of graphically analysing strain.

manner best calculated to advance scientific theory, they undoubtedly by their researches in such branches as thermo-elasticity, after-strain, and magneto-elasticity gave a great impulse to further theoretical and physical work. The development in this period of our knowledge of the physical properties of elastic materials showed the insufficiency of much of the accepted elastic theory, but for the establishment of a truer and more comprehensive theory we shall probably have to wait until we gain a wider acquaintance with the nature of intermolecular action and the part played by the ether in varying and adjusting that action.

In the technical researches of the period we find that the special problems of bridge structure and gun-making, notably the introduction of lattice-girders and composite cannon, largely influenced the direction of investigation, and incidentally led to the discovery of many important physical properties of iron and steel. On the technical side the researches of Bresse, Phillips and Kirkaldy form each in their peculiar fields models of what investigation in technical elasticity should be, and emphasise the special merits of the French and English systems of engineering training. In the sphere of terminology a great service was rendered by Rankine owing to his introduction or precise definition of a number of useful names for important elastic coefficients or conceptions. On the whole while the number of memoirs published was alarmingly great, the proportion which may be classified as absolutely worthless is extremely small. In many cases they contain important facts which have been forgotten in after decades only in order to be rediscovered in recent times. This is largely owing to the want of any easily accessible record.

Cambridge:
PRINTED BY C. J. CLAY, M.A. AND SONS,
AT THE UNIVERSITY PRESS.

Printed in the United States
By Bookmasters